WHITE CLOVER

This and other CABI Publications
can be obtained through any major bookseller or
direct from
C·A·B International
Wallingford, Oxon. OX10 8DE, UK.

WHITE CLOVER

Edited by
M. J. BAKER and W. M. WILLIAMS

CABI Publishing is a trading name of CAB International

CABI Head Office
Nosworthy Way
Wallingford
Oxfordshire OX10 8DE
UK

Tel: +44 (0)1491 832111
Fax: +44 (0)1491 833508
Email: cabi@cabi.org
Web site: www.cabi.org

CABI Publishing
875 Massachusetts Avenue
7th Floor
Cambridge, MA 02139
USA

Tel: +1 617 395 4056
Fax: +1 617 354 6875
Email: cabi-nao@cabi.org

Edited by
M. J. Baker and W. M. Williams
Grasslands Division
Department of Scientific and Industrial Research
Palmerston North
New Zealand

Front cover: *Trifolium repens*

ISBN-13: 978-0-85198-529-9
ISBN-10: 0-85198-529-7

First published 1987
Transferred to print on demand 2009

Printed and bound in Great Britain by
CPI Antony Rowe, Chippenham and Eastbourne

Contents

Preface

In many temperate regions of the world, white clover is the most important pasture legume. It shows wide adaptation to pasture situations in environments ranging from cool-temperate uplands in northwestern Europe and New Zealand to warm-temperate regions of the USA and even to sub-tropical parts of Australia. In these diverse environments the plant can fix large amounts of atmospheric nitrogen and provides outstandingly good feed for grazing livestock.

About ten years ago staff at Grasslands Division of the New Zealand Department of Scientific and Industrial Research, when assessing knowledge of the white clover plant, concluded that an exhaustive, systematic treatise was needed. Dr. R. H. Jackman of the division took on the job of editing this book and commissioned chapters from local authors. Unfortunately, ill-health forced him to give up the work, and the momentum of the book was lost for a while. But the task was then taken up by Dr. W. M. Williams and, later by myself when Dr Williams left the Division in 1982.

Thoughout the book the literature is reviewed comprehensively up until about 1982-83. References published after this time are incuded only selectively.

This book is the first specifically on white clover since Erith published in 1924. Its aim is to collate, review and evaluate existing knowledge of the biology of the white clover plant and so provide a reference text and a basis for future research. Much of the book is of direct relevance to pasture situations, as it considers the stresses imposed on the plant.

The appearance of this treatise is timely, coinciding as it does with widespread concern about consumption of fossil fuels in the agriculture of more developed countries. Concern is also mounting about the effect on freshwater quality of a burgeoning use of fertiliser nitrogen for intensification of grassland farming. In developing countries, where the legume-based approach to grassland development is probably the better alternative, scientists and students should find this text invaluable.

M.J. BAKER

The structure of the mature plant

R. G. Thomas
Department of Botany and Zoology, Massey University, Palmerston North, New Zealand.

Contents

1. Introduction

The emphasis in this chapter is on the basic structure of mature plants of white clover (*Trifolium repens* L.). Changes that occur during growth and development are described in Chapters 2 and 3; variations in size and in details of morphology and anatomy are discussed in Chapter 10. The account given here is based upon the detailed description of the structure of wild white clover plants from the south of England by Erith (1924). Data specifically from that source are indicated by (E). Additional observations by the present writer, and those made by him in confirmation of Erith's account, are indicated by (T). Photomicrographs used as illustrations for this chapter were prepared specially for it by D.H. Hopcroft (DHH) from material processed at the Electron Microscope Laboratory of the Applied Biochemistry Division, Department of Scientific and Industrial Research, Palmerston North, New Zealand.

2. Morphology of the vegetative plant

A 10-week-old plant grown under favourable conditions consists of a central primary axis of stem and root from which grow several secondary stolons. For about the first four weeks of growth, a seedling remains in a rosette form with an erect stem and very short internodes (see Chapter 2). Internodes which emerge from the apical bud of the orthotropic primary stem after that time usually become elongated and the stem grows horizontally. From that stage onwards the morphology, anatomy and growth habit of the primary stem are similar to those of the later-formed secondary stolons.

The primary root of a seedling develops into a tap root in a mature plant, often growing to a length of 30 cm or more with a maximum diameter of 5-8 mm after one year's growth (E). Tap roots bear several lateral roots up to 2-4.5 mm in diameter which in turn produce numerous thinner lateral roots. The primary root eventually dies, leaving an independent system of secondary stolons nutritionally supported by adventitious roots growing from their nodes.

(A) Stolon morphology

The procumbent stolon (Figure 1.1) is the basic structural unit of the plant. It consists of a series of internodes separated by nodes which form as a result of growth at the apical bud (Chapter 2, Section 2(A)). Each node bears a trifoliolate leaf with an erect petiole, two root primordia, and, during the purely vegetative stage of growth, an axillary bud (Figure 1.2) which is capable of growing into a lateral stolon. Production of lateral stolons leads to the vegetative spread of a plant. Adventitious roots forming at the nodes in contact with moist ground provide a degree of nutritional independence to each lateral stolon which increases its chances of survival when the older tissues die off.

The number of stolons on a plant increases with its age. In theory, these can all grow indefinitely at their tips, rooting at their nodes and gradually dying off at their bases, and developing lateral branches from their axillary buds. A mature plant thereby comes to consist of a hierarchy of stolons of differing origins and varying chronological and developmental ages which creates a problem of terminology. In describing the morphology of a plant, the age of its component stolons is frequently not known; and, in addition, in a kinetic study of growth, relationships between stolons change with time and the development of adventitious roots. Terms such as 'primary', 'secondary', 'tertiary', etc., are of little value because the hierarchical ranking of the oldest stolons on an old plant often cannot be determined. In any case, such terms are used for other purposes. To overcome these problems, it is suggested (T) that stolons be described simply as either 'main stolons' or 'lateral stolons', and that the four definitions below be adopted. These depend largely on the number of leaves produced on a branch rather than on stolon length and/or root growth, because internode elongation and root growth are very strongly influenced by environment.

Main stolon:

A stolon which bears (or has produced) at least 8 leaves with unfolded leaflets along its main axis *and* possesses adventitious roots at one or more nodes or at its base.

Axillary bud:

A vegetative shoot in a leaf axil which has produced fewer than two leaves with unfolded leaflets.

Lateral branch:

Any vegetative axillary outgrowth which has produced two or more leaves with unfolded leaflets. The apical bud of such a branch is often referred to by agronomists as containing an 'active meristem' (even when it is not growing).

Lateral stolon:

A lateral branch growing from a main stolon, or from another lateral stolon, which has produced 3 or 4 leaves with unfolded leaflets *and* has a stem which, from its base to the node of attachment of its youngest leaf with unfolded leaflets, is at least twice as long as the stipules of its subtending leaf, *or* has produced 5-7 leaves with unfolded leaflets and developed its own adventitious root system *or* has produced more than 7 leaves with unfolded leaflets but possesses no adventitious root system.

These definitions have the advantage that they can be applied equally to pot-grown plants without nodal roots and to field-grown plants. Applying them to the shoot depicted in Figure 1.1, the stolon axis would clearly be defined as a main stolon (MS); the axillary growth at node 3 would be an axillary bud (AB), and that at nodes 5 or 7 would be a lateral branch (LB); and the branch at the basal node would be a lateral stolon (LS). The axillary tissue at node 6 would be classed as a lateral stolon if its stem were at least twice as long as the stipule (S) of the subtending leaf, but otherwise would be referred to as a lateral branch. By implication, a main stolon always has, by virtue of its root system, the capacity to be self supporting for mineral nutrient and water immediately after being severed from its parent stolon, whereas a lateral stolon does not necessarily have this capacity.

(B) Internode morphology

Stolon internodes vary greatly in length, thickness and colour depending on their age, genotype and growing conditions. Internodes in thicker stemmed plants are commonly, but not invariably, longer than in those with thinner stems, often reaching a length of 40 mm or more. The presence or absence of anthocyanin is primarily determined genetically, some individuals having very deeply pigmented internodes while others have none. There is always more anthocyanin and chlorophyll on the better illuminated side.

(C) Leaf morphology

Leaves on stolons are arranged alternately (Erith 1924; Denne 1966), with their points of attachment most commonly oriented such that they grow from the sides of the stem, relative to the surface of the ground, rather than from the upper or lower surfaces. Their size varies greatly with genotype and environment. They are petiolate, stipulate, and normally trifoliolate (Figures 1.1 and 1.2) although leaves with more than three leaflets are occasionally produced, up to as many as 10 leaflets per leaf being recorded in some instances (E). Leaflets vary in shape from oval to obovate and obcordate and their margins vary from entire to quite heavily serrated. They are pinnately veined, with prominent midribs on their lower surfaces, and are either uniformly green or patterned with a whitish V-shaped mark (V-mark, Figure 2.13). The amount and distribution of anthocyanin in the leaves ranges from a total absence, through a partial covering of the surface by a relatively large area of red pigment, to an almost complete coverage of the leaf surface by anthocyanin in plants such as *Trifolium repens* var. *purpureum* (see Chapter 10).

Each leaflet is joined to the distal end of the petiole by a short petiolule (Figures 1.1 and 1.3) which acts as a pulvinule, causing movement of the leaflet on its petiole. Externally the petiolule is banded with transverse furrows which are associated with the mechanism of movement (Chapter 2, Section 5.(B)).

Petioles are relatively long, growing from an intercalary meristem at the end adjacent to the petiolules. In cross section they are somewhat concave on their adaxial and convex on their abaxial surfaces. The groove formed by the concavity on the adaxial surface is more pronounced in larger leaves and at the upper end of the petiole, and is shallower at the basal end. Each petiole is joined to the stolon by a sheathing leaf base from which grows a pair of membranous stipules, with acute apices, which are fused at their bases to form a stipular tube around the stem (Figure 1.2). Stipules frequently contain anthocyanin, particularly close to their veins.

(D) Axillary bud morphology

All vegetative stolons form an axillary bud at each node. Near the apical bud of a primary stolon (e.g. in the axils of leaves 1 and 2 in Figure 1.1) these axillary buds are small and completely enclosed within the sheathing leaf bases and stipular tubes at the bases of the petioles (Figure 1.2b). Further from the stolon tip the buds become visible without dissection (e.g. in the axil of leaf 3 in Figure 1.1). All these buds have the potential to grow into lateral stolons, more of them actually doing so under favourable than under unfavourable growing conditions (see Chapter 2, Section 3. (A)). Lateral stolons in turn produce axillary buds which can grow into further lateral stolons, and so on.

(E) Root morphology

Each young node on a stolon bears two adventitious root primordia, one situated on each side of the axillary bud (Figure 1.2a). In this way one root primordium is closer to the ground than the other. The upper primordium usually remains inactive, but under moist conditions the lower of the two grows out to form an adventitious nodal root.

The nodal roots often grow up to 15 cm long before lateral roots begin to grow from them to form a vigorously branching root system. In some plants these nodal root systems remain fibrous (e.g. in cv. Kentish Wild White) whereas in others (e.g. Ladino) the nodal roots undergo considerable secondary development to become taproots (Caradus 1977).

The structure and development of root nodules which form in the cortex in response to infection by nitrogen-fixing bacteria is described in Chapter 6. These can form on all roots, but are most abundant on lateral roots.

3. Anatomy of the vegetative plant.

The bulk of the plant body consists of ground tissue in which is embedded a vascular system and which is surrounded by a single-layered epidermis.

(A) Surface anatomy of the shoot.

In surface view the outer walls of many epidermal cells are seen to bear ridges which run predominantly parallel to their long axes. Ridging is most strongly developed on abaxial leaflet surfaces (Figure 1.7b) and is absent from adaxial surfaces (Figure 1.7a). Epidermal cells on stolons and petioles are sometimes quite strongly ridged (Hay et al. 1982) but occasionally they are almost smooth (Figure 1.8b). Transmission electron micrographs published by Hay et al. (1982) show the cuticle thickness over the surface of stolons to be relatively uniform, indicating that the ridging is the result of variations in thickness of the outer cell wall. These same micrographs reveal that 'very fine fibrillar elements....[extend] from the cellulose-cuticular interface through the cuticle almost to its outer edge'. Cuticle thickness varies over the shoot surface. Shepherd's (1964) statement that the cuticle is 'well developed on petioles but thin or absent on parts of the [leaf] laminae' agrees with Erith's (1924) description but no histochemical data exist in support of their contentions.

The upper surfaces of leaflets are covered with platelets of wax (Hall & Donaldson 1963 and Figure 1.7a). Similar platelets are absent from other parts of the plant (Figures 1.7b, 1.8b (T)), but wax is probably present in a less obvious form (see Section 2.(E)). The platelet-covered upper leaflet epidermis is far less wettable than the rest of the plant (Rentschler 1971).

Over most of the surface the epidermal cells are elongated in the direction of the axis they cover. Their shape is similar in petioles and stolons (Figure 1.8b) and over the midrib and major veins on the lower surface of leaflets, but differs markedly on petiolules (Section 3.(E)) and upper leaf surfaces (Figure 1.7a). They often contain anthocyanin (Carnahan et al. 1955), particularly on stipules and on regions of internode exposed to full sunlight.

All regions of the epidermis bear hairs and stomata. Hairs are of two types: relatively short club-shaped multicellular hairs and longer unicellular hairs (Erith 1924, Rohlfs 1950). The multicellular hairs are the most numerous, covering the abaxial leaf surfaces particularly densely in the apical buds (Figure 2.11), but becoming more sparsely distributed and shrivelled as surfaces expand during growth (e.g. Figure 1.3a). They possibly have the effect of decreasing desiccation in young organs. Under some environmental conditions the latter are covered by a mucilaginous layer (Figure 2.11) which can often be seen at low magnifications as a thin membrane overlying the surface of the upper ends of newly emerged petioles and is probably secreted by the multicellular hairs (see Chapter 2, Section 2. (C)i.). The nature of this layer is not known; it does not stain for lipids and seems to wash off in water (T). Long unicellular hairs are conspicuously abundant on the upper (adaxial) surfaces of the petiolules (Figure 1.3a).

Stomata are most abundant on adaxial leaflet surfaces (Section 2.(E)) and less so on petioles and stolons. In most regions they lie in the same plane as the epidermal cells (Figures 1.7b, 1.8b), but on the adaxial leaflet surface they are slightly sunken (Figures 1.7a, 1.9a). Guard cells bear well-developed cuticular ledges on their outer surfaces. The eisodal pores between the cuticular ledges open widest on the abaxial leaflet epidermis. Elsewhere they appear to open little and are often completely sealed over (e.g. Figure 1.8b). The resistance to gaseous diffusion through such stomata has not been determined.

(B) Internal anatomy of the shoot

In all organs of the shoot the ground tissue consists of parenchyma cells only; there is no clearly defined collenchyma. The cells of the outermost layer of the cortex adjacent to the epidermis are distinguishable from those within the body of the cortex, however, by slight thickenings at their outer corners and by a reduced chloroplast content (e.g. Figure 1.10c). They appear to represent a poorly developed hypodermis. Chloroplasts are found in all ground tissue cells in the shoot, and starch grains are usually present throughout, except in the youngest regions of the stolon tip where they tend to be confined to the cells of the innermost layer of the cortex adjacent to the bundle fibres (see below). The cells in the latter region of internodes, petioles and petiolules represent an incomplete starch sheath in which the starch grains move under the influence of gravity (T). As they age, these movable starch grains are replaced by calcium oxalate crystals. Similar crystals are abundant in the subepidermal cells of the stipules (E), and the bundle sheaths of leaflets (Figure 1.9a).

The vascular tissue is made up of vessels and tracheids in the xylem and sieve tubes and companion cells in the phloem; interspersed among these are parenchyma cells and gelatinous fibres. Similar fibres are also present in bundles outside the vascular tissues of the shoot (e.g. Figures 1.6, 1.9 and 1.10c). Erith (1924) described them as: long and narrow with pointed ends; they have very thick walls, the outer layers of which are lignified, the inner layers consisting of a gelatinous substance, which, in some cases, forms a comparatively narrow band inside the lignified wall, while in others it nearly completely fills the cell lumen. With chlor-zinc-iodine this inner thickening substance stains a beautiful wine-red colour, the outer lignified wall staining yellow.

(C) Vascular system of the shoot

Figure 1.4 is a diagram showing the vascular system of the stolon as if it were split along the lowermost side, and opened out. It is based upon a figure published by Devadas & Beck (1972) which has been modified slightly in the light of further study (T) to change its proportions, to show the vascular connexions of the nodal roots, and to indicate the position of the uppermost side.

Devadas & Beck have described the vascular system as follows: 'The vascular cylinder in the stem can be interpreted in terms of sympodial systems. Transverse sections..... taken at the middle of the internodes show about ten vascular bundles, which are well separated by interfascicular regions. Among these are four axial bundles [A1-4], a pair on either side of the longitudinal plane in which two rows of leaves occur. Sympodia A1 and A4 are on the same [lower] side of this plane and are identical (Figure 1.4). From the axial bundles of these [two] sympodia, a branch trace, a lateral leaf trace [R or L], and a median leaf trace

[M] diverge in acropetal sequence, repeated regularly at alternate nodes. The lateral traces diverge from the sides of these axial bundles that face each other, whereas the median and branch traces are on the sides facing axial bundles A2 and A3. Sympodia A2 and A3 are also similar to each other but differ from A1 and A4. Lateral leaf traces and branch traces diverge alternately from them. The lateral traces are produced on the sides of the axial bundles that face each other, and the branch traces come off the sides towards axial bundles A1 and A4. In addition, A2 and lateral traces R2 and R4.... are connected at regular intervals by vascular bundles. At nodes 3 and 5.... the connecting bundles diverge from the respective lateral traces and fuse to A2. There are no such connections between any of the other trace and axial bundles.

'The three traces that supply a particular leaf are parts of three different sympodia. For example, median traces [M2 and M4] come off axial bundle A1, and the lateral traces supplying the same leaves diverge from axial bundles A3 and A4. The lateral traces diverging from A3, and all median traces, traverse about two internodes, whereas laterals from A2 traverse a single internode from the points of their divergence up to the node at which they enter a leaf.'

'Each axillary bud is supplied by two branch traces each of which diverges from one of the two axial bundles on either side of the median trace which departs at the same node. Accessory bundles.... are also present, but in small numbers'.

Erith's earlier (1924) description of the vascular system is essentially in agreement with that by Devadas & Beck. It is not possible to determine whether the discrepancies result from errors of interpretation or differences between cultivars. Erith described the structure of an English wild type with thin stolons and relatively small leaves; Devadas & Beck did not indicate what cultivar they examined. A study of the vasculature of cv. Grasslands Huia clover (T) revealed a vascular system more closely similar to that described by Devadas & Beck than to that by Erith.

Regarding the vascular connexions of the nodal roots, Erith (1924) wrote that a nodal root 'appears to be in direct connection with the axillary branch and probably functions more particularly for this rather than for the main stolon'; experiments by Chow (1966) support the latter suggestion. Erith also noted that the vascular system of a nodal root joins the middle leaf-trace bundle at its node of entry to the stele (viz. two nodes basal to the node of attachment of the leaf to the stolon). The vascular connexions of nodal roots in cv. Grasslands Huia (T) are similar to those described by Erith.

The vascular system of the leaf of southern English wild white clover has been described by Erith as follows: 'Three bundles pass from the stolon into the leaf sheath [and thence into the petiole]..... Just below the point at which the petiole becomes free from the stipular tube a small branch arises from the right and left sides of the central strand: from the inner side of each of the two lateral bundles at about the same level a branch is also given off. These four small branches coalesce in pairs, the two resulting bundles travelling along the petiole parallel to the three large main bundles. In those petioles in which there are but four vascular strands, only one such small bundle is produced.

'The four or five bundles are connected by strong transverse anastomoses in the region to which the leaflet stalks [petiolules] are attached, a transverse section of the petiole at this point showing the bundles all united into one U-shaped band. From this bundle complex [typically] three bundles originate, each of which passes into the midrib of a leaflet.

...'The stipular bundles originate from the three bundles of the leaf sheath... From the outer sides of the two lateral bundles of the sheath two branches develop, one above the other; the higher of the two on each side extends into the two teeth of the stipular tube, branching sparingly below. The lower branches from each lateral bundle give rise to a series of fine bundles which traverse the side of the tube farthest from the petiole. One or two fine bundles are also produced from the central bundles of the leaf sheath and pass into the stipular tube.'

(D) Stem anatomy

A transverse section of a young fully elongated stolon internode typically shows a ring of eight, nine or (in thicker stolons) more, open collateral vascular bundles embedded in a parenchymatous ground tissue the cells of which contain chloroplasts and starch grains (Figures 1.5 & 1.6). Bundles of fibres, in which usually only the outer cells become thick walled and lignified, are present outside the phloem. Small cells between the protoxylem and the pith possibly represent incompletely differentiated fibres. Erith (1924) described cells in this position in primary stems and in petioles as being sclerenchymatous but they are usually neither thick-walled nor lignified (T). The cortex on the upper, more brightly illuminated, side of procumbent stolons is commonly wider, with larger cells, than the lower more shaded side (T). This is contrary to the incorrect report by Mange (1963) that it is wider on the lower side. Cells on the uppermost side of the internode contain more chloroplasts than on the lower side. Anthocyanin, which also forms predominantly on the upper side, is found in scattered epidermal cells but mainly in the single layer of subepidermal cells internal to them. The latter cells also contain a few chloroplasts. Stomata present in the epidermis are similar to those in the lower epidermis of leaves (Figures 1.7b and 1.8b). Interfascicular parenchyma cells are similar in size and shape to the cortical cells but only the 2 or 3 outermost of them contain chloroplasts, the innermost becoming lignified (see Figures 1.5, 1.6). Strong lignification also occurs in some of the parenchyma cells immediately internal to the vascular bundles.

The relationship between the vascular system in the stem and the appearance of vascular bundles as seen in a transverse section can be understood by comparing Figures 1.4 and 1.6a and imagining the section in the latter figure to have been cut midway between node 2 and node 3 in Figure 1.4. In any such section the three leaf-trace bundles from the node immediately above (viz. L_2, M_2 and R_2 in Figure 1.6a) show the characteristic structure which they possess in the petiole, and on this account can be readily distinguished.

In a study of wild white clover in southern England, Erith (1924) reported the number of radial rows of xylem tracheary elements in the vascular bundles (as seen in transverse sections) to be as follows. In a fully elongated internode near the stem apex, the xylem of the largest leaf trace (M) consists usually of 6 regular rows of tracheary elements and their accompanying parenchyma. The two other leaf-trace bundles (R and L) have a similar structure, but are rather smaller and generally possess only 5 rows of vessels. The other vascular bundles of the internodes have less xylem, but the groups of fibres outside the phloem are relatively much elongated tangentially. The vascular bundle on the uppermost side of the stolon lies between two leaf-trace bundles (M_2 and R_2 in Figure 1.6a). It is composite, being formed by the union of two bundles, and possesses only 8 tracheary elements, arranged in two isolated groups. Of the other bundles, one is rather larger than the rest, being also a composite bundle containing A3: in its xylem are about 9 tracheary elements in two isolated groups. The smallest bundles possess very few tracheary elements.

Erith's description gives an idea of the relative number of rows of tracheary elements in the various vascular bundles. The actual numbers vary with stolon thickness, however. Vascular bundles in larger stems contain more rows of tracheary elements than described by her, bundle M in a stolon of cv. Grasslands Huia having 8 rather than 6 rows (T) (Figure 1.5b), for instance.

In transverse sections of older internodes the general arrangement of the tissues is the same as in younger, but the formation of a cylinder of vascular cambium results in the production of rather weakly developed secondary tissues. Each bundle comes to have much more xylem and, in addition, two or three layers of lignified parenchyma are produced by the division of some of the cells of the interfascicular cambium, the xylem of all the bundles thus being connected by bands of secondary lignified parenchyma (Figures 1.5 & 1.6). The leaf-trace bundles from the node immediately above can still be distinguished by the smaller diameter of their tracheary elements and the regular arrangement of the latter in radial rows (E).

Early stages of secondary development are shown in Figures 1.5b & 1.6b. In both, a tangential layer of interfascicular parenchyma cells has de-differentiated to form interfascicular cambium. The latter subsequently divides to produce secondary interfascicular parenchyma towards the inside of the stem, and thick walled gelatinous fibres develop at the sides of, and to a lesser extent scattered through, the xylem (Figure 1.6b). At about the same time, subepidermal cells divide to produce cork cambium (Figure 1.6b) and up to about 4 layers of cork cells ((E), (T), and Hay et al. 1982).

Further aspects of the developmental anatomy of stolons are described in Chapter 2, Section 2. (E).

(E) Leaf anatomy

The anatomy of the leaflets is that of a typical mesomorphic leaf. Adult leaflets have stomata on both surfaces (see Section 2. (A)), those on the upper surface being more numerous but smaller than those on the lower. Erith quotes the frequencies as being 198 mm^{-2} on the upper surface and 136 mm^{-2} on the lower; Denne (1966) gives figures of 252 and 159 respectively. Shepherd's (1964) figures of 221 and 118 cm^{-2} are clearly incorrect by two orders of magnitude and the validity of his calculated stomatal and cuticular resistances is therefore uncertain. The surface appearance of upper and lower epidermes very different, the guard cells of the former being much smaller and slightly sunken so that they are partly hidden by adjacent cells, whereas those on the lower surface are larger and fully exposed (Figures 1.7 and 1.8). The latter figures also show the typical difference in size and shape between epidermal cells on the two surfaces.

The ultrastructure of wax deposits on leaflets is mentioned in Section 2.(A). Hall & Donaldson (1963) described the wax on the two surfaces as having a different structural appearance which does not seem to be the result of different chemical nature. Wax on the adaxial (upper) surface is in the form of platelets or spicules (Figure 1.8a) whereas that on the abaxial surface appears net-like. The former is very readily lost by wind abrasion (see Chapter 2, Section 2.(C)i).

The internal anatomy of leaflets (Figure 1.9a) has been described in detail by Erith (1924), and Denne (1966) has summarized it thus:

'The palisade tissue is from one to two layers thick, varying within the leaflet. The spongy mesophyll usually consists of four layers of cells which in transverse section appear to be compactly arranged with small intercellular spaces. In paradermal section they are seen to be irregularly shaped, having extended arms enclosing intercellular spaces which appear circular in this plane (Figure 1.9b).'

[When leaflets are V-marked, the palisade cells beneath the mark are shorter than in other parts of the leaflet (Carnahan et al. 1955), and the spongy mesophyll cells in the region of the V have no chloroplasts (E)].

'The vascular system is pinnate, having a parallel series of main lateral veins running out from the midrib toward the margin. Anastomoses between the main veins form a network of closed islets, and minor strands protrude blindly into the islets. The main lateral veins frequently end blindly in a tooth at the leaflet margin'.

Veins are all surrounded by parenchymatous bundle sheaths of a type referred to by Akita et al. (1969) as 'ordinary' as opposed to 'chlorophyllous.' In the midrib and larger lateral veins there are relatively large bundles of lignified fibres on both the adaxial and abaxial sides of the vascular tissue. As with starch sheath cells adjacent to fibres in mature stolon tissue, the bundle sheath cells in contact with fibres also often contain crystals of calcium oxalate. In some veins the fibres completely surround the vascular tissue; other smaller veins have no such associated fibres. Transfer cells with numerous filiform protuberances on the inner surfaces of their walls are associated with the xylem and phloem in the veins, as described for an unnamed *Trifolium* species by Gunning et al. (1968).

Petiole anatomy has been described by Erith (1924) and Rohlfs (1950). The ground tissue is similar to that in a stolon and typically contains 4 or 5 vascular strands (Section 2.(C).), three of which are larger than the others, arranged in a horseshoe near its periphery (E) (Figure 1.10b). In Ladino clover, with larger petioles, the mean number of vascular strands was found by Gausman & Fuelleman (1952) to range from 6.7 to 9.82. The vascular bundles resemble those in the stolons, possessing a relatively wide band of fibres on the outer side. Lignification of petiolar fibres is weak and decreases with distance from the base to almost nil at the upper (distal) end. In contrast, the thickness of the fibre walls increases with distance from the base, being similar at the distal end to the thickness of walls in the petiolules. (T). Immediately outside the band of fibres is a single layer of cells containing movable starch grains or (when older) crystals of calcium oxalate. In older and larger petioles the centre of the pith is often hollow.

The petiolules act as pulvinules, being causally involved in the 'sleep movements' of leaflets (Chapter 2, Section 5.(B)). They have the characteristic anatomy of pulvini, with a central vascular strand surrounded by a relatively wide zone of cortical motor tissue (Figure 1.3b). The vascular strand is similar to that in the leaflet midrib except that it is completely surrounded by a wide sheath of very thick-walled unlignified cells. In macerated material these are seen to be short fibres (often not more than 10x as long as broad) with relatively large, simple pits (T). Surrounding this sclerenchymatous sheath is a single layer of cells which contain movable starch grains when young (T) and calcium oxalate crystals when older (E). The cortical cells are larger towards the outside and possess numerous simple pits. Chloroplasts are more abundant in the innermost cells of the cortex (E). In longitudinal section the epidermal and cortical cells are unlike others on the plant, being of approximately the same transverse diameter as in the petiole but very short in the direction of the petiolule axis.

Details of the anatomy of leaf sheaths and stipules are given by Erith (1924).

(F) Root anatomy

Young roots (Figure 1.11a) consist of a central stele embedded in a parenchymatous cortex on the outside of which is a piliferous layer bearing root hairs. These frequently persist at a distance of more than 10 cm behind the root tip. Their functional effectiveness has not been investigated.

The stele, which is surrounded by a thin-walled endodermis, is triarch in the first-formed (primary) root and in lateral roots but is tetrarch or pentarch in the adventitious nodal roots (E). Sometimes its centre is occupied by a pith rather than by metaxylem. Phloem alternates with the protoxylem groups. Lateral root primordia arise in the pericycle opposite the protoxylem. Cells of the typical single-layered endodermis surrounding the stele have well developed Casparian strips but their walls remain thin and unlignified throughout the life of the root. Calcium oxalate crystals have not been reported to occur in roots.

Secondary vascular tissues arise early in the development of the primary root but usually later in nodal roots, which can grow up to 15 cm long without forming a vascular cambium. Secondary growth starts in the normal manner through the meristematic activity of the cells of the conjunctive parenchyma internal to the phloem groups. These cambial arcs are soon joined by meristematic cells of the pericycle outside the protoxylem, and a complete cambium cylinder is ultimately formed.

The cambium immediately outside the protoxylem at first produces only parenchymatous cells, these forming primary medullary rays. The rest of the cambium produces secondary xylem and phloem. Almost before the first cambium cylinder is complete, a phellogen arises in the outer layer of the pericycle (Figure 1.11b), forming periderm externally and a few layers of phelloderm internally.

Starch is abundant throughout the parenchyma cells of old roots. Chloroplasts form in the cortex and pericycle of roots exposed to light. Gelatinous fibres, mostly arranged in groups, are scattered throughout both xylem and phloem, being more abundant in nodal roots, in which they are particularly conspicuous outside each primary phloem bundle, than in the main primary tap root system. Transfer cells occur around the vascular tissue in root nodules (Pate et al. 1969).

4. The reproductive phase

During the reproductive phase, inflorescences are borne singly in the axils of some leaves, replacing the vegetative buds which otherwise form in those positions. The formation of such flower heads has little influence on the size and shape of subtending leaves or on the growth habit of the plant (see Chapter 3, Section 5).

(A) The inflorescence

Inflorescences consist of dense spheroidal racemose heads of small florets borne terminally on long peduncles (Figure 1.1, I). The peduncles are usually decumbent, and grow up almost vertically from the stolons to be longer than the petioles of the leaves in the axils of which they develop. They are green and longitudinally furrowed, the furrows being deep near the flower head and becoming gradually shallower towards the base, in which region the peduncle is clearly dorsiventral, with a flattened adaxial surface and a convex abaxial side. In the upper half the dorsiventrality is obscured by the development of ridges and deep furrows.

Anatomically the peduncle consists of a single layer of epidermal cells surrounding a ground tissue in which a number of vascular bundles are arranged in a single ring, each bundle being situated near the base of a ridge (Figure 1.10a); in most peduncles there is an additional small bundle on the adaxial side (E).

The epidermis bears stomata and both unicellular and multicellular hairs as described in Section 2.(A). The stomata and surface texture of the epidermal cells are similar to those in petioles and stolons. Anthocyanin forms in the single layer of subepidermal cells which are readily distinguishable from the main region of cortical cells by the relatively small number of chloroplasts within them.

Most peduncles are hollow in the lower half. In the upper half the ground tissue within the vascular ring and between the bundles contains very few or no chloroplasts. Starch is abundant in all the ground tissue cells near the base of the peduncle but decreases towards the top, being found in the upper region only in the cell layer adjoining the fibres on the outer side of each vascular bundle. The starch grains in the latter cells (which represent an incomplete starch sheath) move under the influence of gravity.

The closed vascular bundles resemble those of the petiole, possessing wide bands of lignified fibres towards the outside. They vary in diameter with the size of the ridges of the peduncle in which they occur. The interfascicular cells have thin walls in the upper part of the peduncle, but from the middle to the base their walls are thicker and lignified, the lignification being most pronounced near the base in which region the xylem of adjacent vascular bundles is connected by bands of lignified parenchyma cells 4 to 5 layers wide. A cork cambium develops in the subepidermal cells of the basal region. A more detailed study of the variation of peduncle structure along its length would be valuable as a basis for understanding the causes of peduncle lodging (see Chapter 3, Section 7.(E)).

The flower head has a comparatively short axis, about 5-10 mm long, upon which a number of pedicellate florets are arranged in a close spiral. Anatomically, the inflorescence axis is similar to the upper part of the peduncle, but fine vascular branches pass out from it into the pedicels and bracts of the flowers. Starch is absent (E).

Each floret arises in the axil of a minute, scarious, acute bract, about 1.5 mm long. Hairs are present on the axis among the florets, and are abundant at its base (E). The number of florets in an inflorescence varies considerably, frequently reaching as many as 140 (Thomas 1961) in some cultivars. The pedicels are shorter (about 1 mm long) at the base of the axis than those in the centre and at the apex, which are 1.5 to 3 mm long. Each bract consists of an upper and lower epidermis and a mesophyll tissue which is one cell thick except near the central vascular bundle, in which region it is composed of four layers of cells. A few stomata are present either side of its midrib.

(B) Floret morphology

Florets are hermaphrodite, medianly zygomorphic and perigynous (Figure 1.12). The whitish-green calyx consists of 5 sepals which are fused at the base to form a tube surmounted by 5 unequal teeth. Both the tube and the teeth are slightly longer on the posterior side. Prominent patches of anthocyanin are commonly present just below the angles between the teeth.

The calyx encloses a white or pinkish polypetalous corolla, the typical papilionaceous set of 5 petals comprising a larger posterior standard (vexillum) folded round two lateral wing petals (alae) which in turn enclose two anterior keel petals. The latter are fused along the distal part of their anterior edges to form a carina. Each of the wing and keel petals consists of a broader distal blade and a narrower proximal claw of a similar length. Near the bases of their blades, the wing and keel petals bear prominent swollen vesicular protuberances, those on the latter being the smaller. The extreme base of the standard and the lowermost third of the claws of the wing petals and keel are fused to the staminal tube. The significance of floret structure in relation to insect pollination is discussed in Chapter 3, Section 6.(A).

Within the perianth is a diadelphous androecium of 10 stamens, 9 of which are fused at their bases to form an incomplete staminal tube, leaving the tenth, posterior, one free. On the inner side of the staminal tube at its base is the nectary (Figure 1.13), which is accessible to bees by the openings on either side of the free stamen. The filaments are attached to the back of the anthers slightly less than half way up from the base (Figure 1.14). Dehiscence is introrse.

The centre of the floret is occupied by a typical leguminous carpel (Figure 1.16).

i. Calyx anatomy

The calyx consists of an epidermis enclosing a mesophyll that is one to two cells thick. The epidermal cells are chiefly oblong, about five times as long as broad, with anticlinal walls of wavy outline, and parallel undulating striations on their surfaces similar to, but more prominent than, the ridges on cells of the lower epidermis of leaflets. Those on the inner surface become irregularly thickened and lignified. The cells covering the vascular strands are almost square in surface view, with almost straight walls (E). Epidermal cells frequently contain anthocyanin, particularly over the veins and in the angles between the teeth where they give rise to the prominent red patches on the surface of the tube. A few short multicellular hairs occur on the tube and margins, particularly in the angles between teeth; a few long, almost unicellular, hairs are present near the base. Stomata occur on both inner and outer surfaces, more especially on the teeth. Chloroplasts are abundant only in the mesophyll cells of the teeth and around the vascular strands of the tube.

The calyx tube has 10 veins. Single unbranched vascular bundles run from its base to the apex of each tooth, and between each pair of these a similar unbranched bundle runs from the base of the calyx to near the angle between two teeth, at which point it divides, the two portions bending to the right and left, respectively, and joining the tooth bundles about halfway along the tooth. A few fine anastomosing strands connect the stronger bundles in the calyx tube. Crystals of calcium oxalate are found in the parenchyma immediately adjacent to the fibres of the vascular bundles. They are also abundant in the innermost layer of mesophyll cells of the calyx tube, beneath the thickened and lignified walls of the inner epidermal cells.

ii. Corolla anatomy

The epidermis of the inner face of most of the corolla consists of oblong cells, with long axes parallel to the long axis of the petal. The outer epidermis is similar, but the cells are somewhat larger, and the longitudinal walls more frequently wavy in surface view. The external walls bear fine longitudinal ridges similar to, but more strongly developed than, those on the surface of the stolons. Over the swollen parts of keel and wing petals the cells are somewhat broader, with thicker walls and even more prominent surface ridges. The epidermal cells commonly contain anthocyanin.

The margin of the proximal part of the petal is entire, that of the distal part being irregularly dentate, with slightly projecting ends of epidermal cells. The mesophyll varies in thickness from one to three layers of cells, and is absent from the margins. Its thin-walled parenchyma cells mostly contain colourless plastids but a few chloroplasts are present towards the bases of the petals. Within the vesicular regions of wing and keel petals, the mesophyll cells are longer and thinner than elsewhere and the intercellular spaces between them much larger. The arrangement of the vascular tissue in the mesophyll has been fully described by Erith (1924).

iii. Androecium anatomy

The epidermal cells of the staminal tube and filaments of the stamens are elongated in the direction of the axis. The very elongated cells in the middle regions of the filaments bear well developed longitudinal ridges. Nearer the anthers, where the cells are shorter, the ridges become transverse (Figure 1.14d). A transition from longitudinal to transverse ridging is seen on the cells of intermediate length. At the extreme distal ends of the filaments, where they become thinner adjacent to their attachment to the anthers, the epidermal cells are smaller and their surfaces smooth (Figure 1.14d). The ground tissue of the filaments is composed of long, narrow parenchymatous cells. Within it, large schizogenous cavities occur along the free margins of the staminal tube and in the flattened filament of the free stamen. Each filament has a delicate vascular bundle which extends from the base of the staminal tube into the connective of the anther.

The anthers are bilocular with thin-walled epidermal cells which tend to collapse at maturity (Figure 1.15a). Immediately internal to these is a conspicuous endothecium consisting of a single layer of thin-walled cells each of which has 5 to 7 rib-like thickenings extending outwards from its innermost wall into its anticlinal walls. Introrse dehiscence of each anther lobe takes place along a line of weakness which develops at the juncture between the connective and the endothecial cells (Figures 1.14a and b, 1.15).

Mature pollen grains are almost spherical. The exine is smooth and comparatively thin; in it are three small thinner circular areas, from one of which the pollen tube protrudes when the grain germinates. Each pollen grain in a dehiscing anther contains a tube nucleus and a generative cell which contains scanty cytoplasm and a narrow elongated nucleus, which in some stages is curved or spirally coiled, and is surrounded by a clearly defined thin membrane.

iv. Gynaecium anatomy

The gynaecium consists of a single carpel with its ventral side posterior (Figures 1.12 and 1.16). The two margins of the carpel adhere closely but are not organically united at the ventral suture. The ovary contains 3 to 7 ovules alternating with one another on the opposite edges of the ventral suture so as to stand in one longitudinal row. The lower florets on an inflorescence tend to contain approximately one more ovule per carpel than the upper (Dessureaux (1951)).

The outer epidermis of the ovary wall consists of polygonal cells with straight walls. A few stomata and scattered multicellular hairs are present. The inner epidermis is composed of narrow elongated cells with their long axes at right angles to the long axis of the ovary. The ground tissue between the outer and inner epidermes consists of two or three layers of cylindrical cells, arranged end to end in irregular rows, with their long axes running parallel to the long axis of the ovary. Wide irregular spaces exist between the rows, lateral contact between the cells occurring only at infrequent intervals. Small polygonal cells, each containing a crystal of calcium oxalate, are distributed here and there in the ground tissue (E).

Longitudinal rows of large, thin walled, bladder-like epidermal cells are present along the inner surface of the ventral suture with their free ends projecting into the loculus of the ovary (Figure 1.16 c-e). These extend for some distance upward along one side of a stylar canal which terminates just below the stigma. The cells of the outer epidermis of the style are oblong and narrow, 10-14 times as long as broad. On approximately the distal third of the style their outer walls bear very conspicuous longitudinal ridges. Their free ends, together with those of the cells of the ground tissue, form the papillose surface of the stigma (Figure 1.17). At maturity the interstices between the papillae are filled with a mucilaginous secretion (T) similar to that described for *Trifolium pratense* by Heslop-Harrison & Heslop-Harrison (1982).

Three vascular bundles enter the base of the gynaecium. Two run along opposite sides of the ventral suture, each giving off a single delicate branch into the funicle of each ovule and a small branch which extends for a short distance in the style. The third main bundle traverses the dorsal side of the ovary, and is continued along the same side of the style to a point just below the stigma (Figure 1.16b). Delicate anastomosing branches connect the larger bundles only in the upper part of the ovary.

The vascular traces of all the floral organs fuse to form a single vascular bundle in the pedicel (E).

The mature ovules are campylotropous with two integuments, and a short funicle attached near the middle of one side (Figures 1.16b and 3.15B). The micropyle is T-shaped (see Chapter 3, Section 6 (D)). The outer integument, the inner surface of which covers the distal end of the nucellus, varies in thickness. The thinnest part, which covers the middle region of the ovule opposite the point of attachment of the funicle, consists of 3 layers of parenchymatous cells (Figure 3.16), whereas near the micropyle and on the side to which the funicle is attached it has 8 to 10 layers of cells. The inner integument is shorter and completely hidden by the outer. It consists of 2-4 layers of isodiametric cells with thin walls.

The nucellus and embryo sac lie within a narrow, curved, sausage-shaped space 6-8 times as long as broad. Just before fertilisation the nucellus is more or less disorganised and shrivelled, except near the base, where it appears as a narrow column of tissue growing up from the chalazal end of the cavity within the inner integument, from which it is quite free. At the micropylar end the curved embryo sac is naked, and its thin wall stretches across the opening in the inner integument, the whole being completely covered by the outer integument except at the micropyle. The egg apparatus consists of a pair of synergidae and the egg cell, close to which are the two polar nuclei, or the fusion nucleus resulting from their union, which takes place in the neighbourhood of the egg apparatus. The three antipodal cells are very small.

A fine vascular strand with a single line of tracheary elements traverses each funicle and enters the ovule, passing to near the centre of the chalaza.

Seed and fruit structure are described in Chapter 3, Section 6.(D).

References

Akita, S.; Miyasaka, A.; Murata, Y. (1969) Studies on the differences of photosynthesis among species I. Differences in the response of photosynthesis among species in normal oxygen concentration as influenced by environmental factors. *Proceedings of the Crop Science Society of Japan* **38**, 507-523.

Caradus, J.R. (1977) Structural variation of white clover root systems. *New Zealand Journal of Agricultural Research* **20**, 213-219.

Carnahan, H.L.; Hill, H.D.; Hanson, A.A.; Brown, K.G. (1955) Inheritance and frequencies of leaf markings in white clover. *Journal of Heredity* **46**, 109-114.

Chow, H.-S. (1966) The structure and growth habit of white clover in relation to its persistence. *PhD Thesis, University of New Hampshire*, 53 pp.

Denne, M.P. (1966) Leaf development in *Trifolium repens. Botanical Gazette* **127**, 202-210.

Dessureaux, L. (1951) Ovule formation as a factor influencing seed-setting of Ladino white clover. *Scientific Agriculture* **31**, 373-382.

Devadas, C.; Beck, C.B. (1972) Comparative morphology of the primary vascular systems in some species of Rosaceae and Leguminosae. *American Journal of Botany* **59**, 557-567.

Erith, A.G. (1924) White clover (*Trifolium repens* L.) *Duckworth, London.*

Gausman, H.W.; Fuelleman, R.F. (1952) Distinguishing Ladino clover from white clover plants. *Agronomy Journal* **44**, 96-97.

Gunning, B.E.S.; Pate, J.S.; Briarty, L.G. (1968) Specialised 'transfer cells' in minor veins of leaves and their possible significance in phloem translocation. *Journal of Cell Biology* **37c**, 7-12.

Hall, D.M.; Donaldson, L.A. (1963) The ultrastructure of wax deposits on plant leaf surfaces. I. Growth of wax on leaves of *Trifolium repens. Journal of Ultrastructural Research* **9**, 259-267.

Hay, M.J.M.; Dunlop, J.; Hopcroft, D.H. (1982) Anatomy and development of the superficial layers in stolons of white clover (*Trifolium repens*). *New Zealand Journal of Botany* **20**, 315-324.

Heslop-Harrison, J.; Heslop-Harrison, Y. (1982) Pollen-stigma interaction in the Leguminosae: constituents of the stylar fluid and stigma secretion of *Trifolium pratense* L. *Annals of Botany* **49**, 729-735.

Mange, M. (1963) Action de l'acide gibbérellique sur *Trifolium repens* L.: observations sur quelques modifications morphologiques, physiologiques et anatomiques. *Annales Scientifiques de L'Université de Besançon, 2ᵉ Série, Botanique* **19**, 43-53.

Pate, J.S.; Gunning, B.E.S.; Briarty, L.G. (1969) Ultrastructure and function of the transport system of the leguminous root nodule. *Planta* **85**, 11-34.

Rentschler, I. (1971) Die Wasserbenetzbarkeit von Blattoberflächen und ihre submikroskopische Wachsstruktur. *Planta* **96**, 119-135.

Rohlfs, L. (1950) Anatomie des pétioles d'une variété de *Trifolium repens* L. á 3 ou 4 folioles. *Annales des Sciences Naturelles, Botanique, 11ᵉ Série*, **11**, 197-208.

Shepherd, W. (1964) Paths and mechanisms of moisture movement in detached leaves of white clover (*Trifolium repens* L.). I. Losses of petiole moisture direct from petioles and via laminae. *Annals of Botany* **28**, 207-220.

Thomas, R.G. (1961) The influence of environment on seed production capacity in white clover (*Trifolium repens* L.). I. Controlled environment studies. *Australian Journal of Agricultural Research* **12**, 227-238.

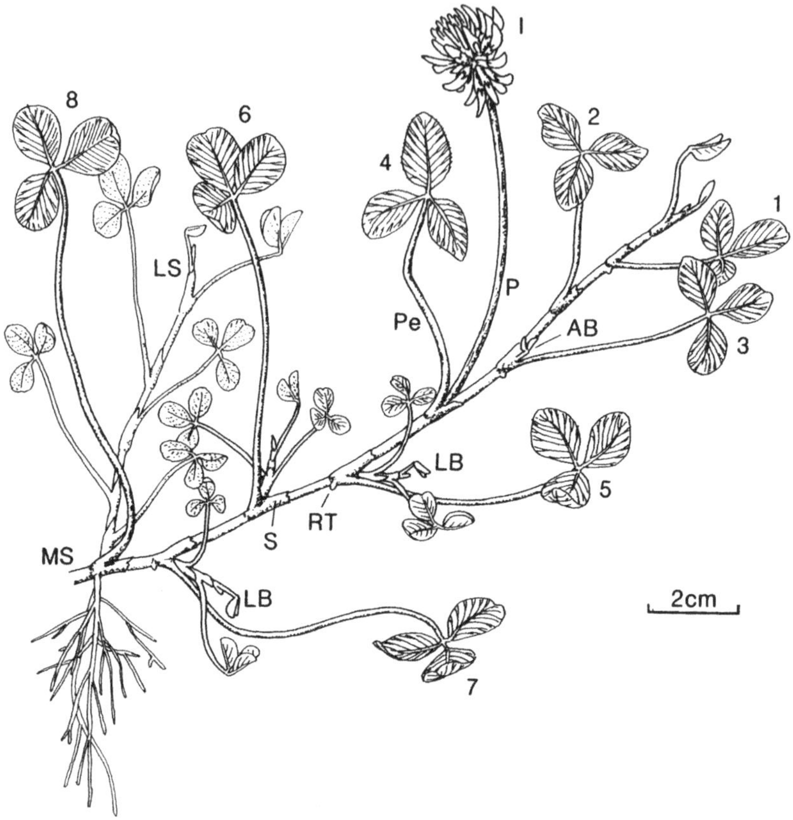

Fig. 1.1. Drawing of a main stolon (**MS**) of white clover, showing axillary buds (**AB**), lateral branches (**LB**), and a lateral stolon (**LS**). S = stipule; Pe = petiole; RT = nodal root primordium; I = inflorescence; P = peduncle. Emerged leaves on the main stolon, and the nodes bearing them, are numbered 1 to 8 [T].

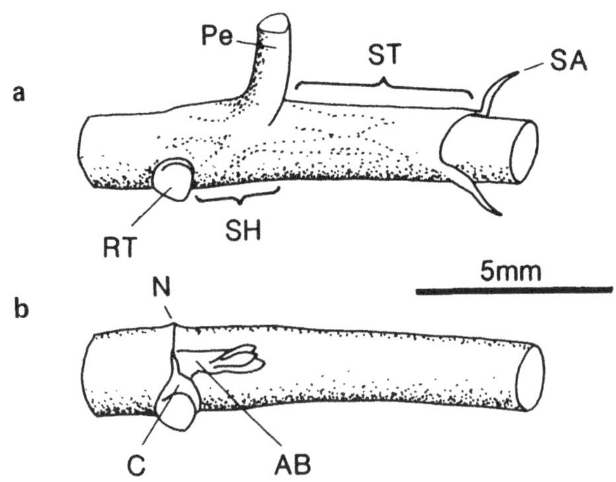

Fig. 1.2 Detail of node 2 on plant depicted in Fig. 1.1 (a) Intact, showing nodal root primordium (**RT**), petiole base of leaf 2 (**Pe**), leaf sheath (**SH**), stipular tube (**ST**), and stipule apex (**SA**). [T]. (b) Axillary bud (**AB**) revealed by removal of leaf base and stipules. N = the line of attachment of the leaf to the node; C = cortical tissue exposed by penetration of the root primordium through the stolon surface. [T].

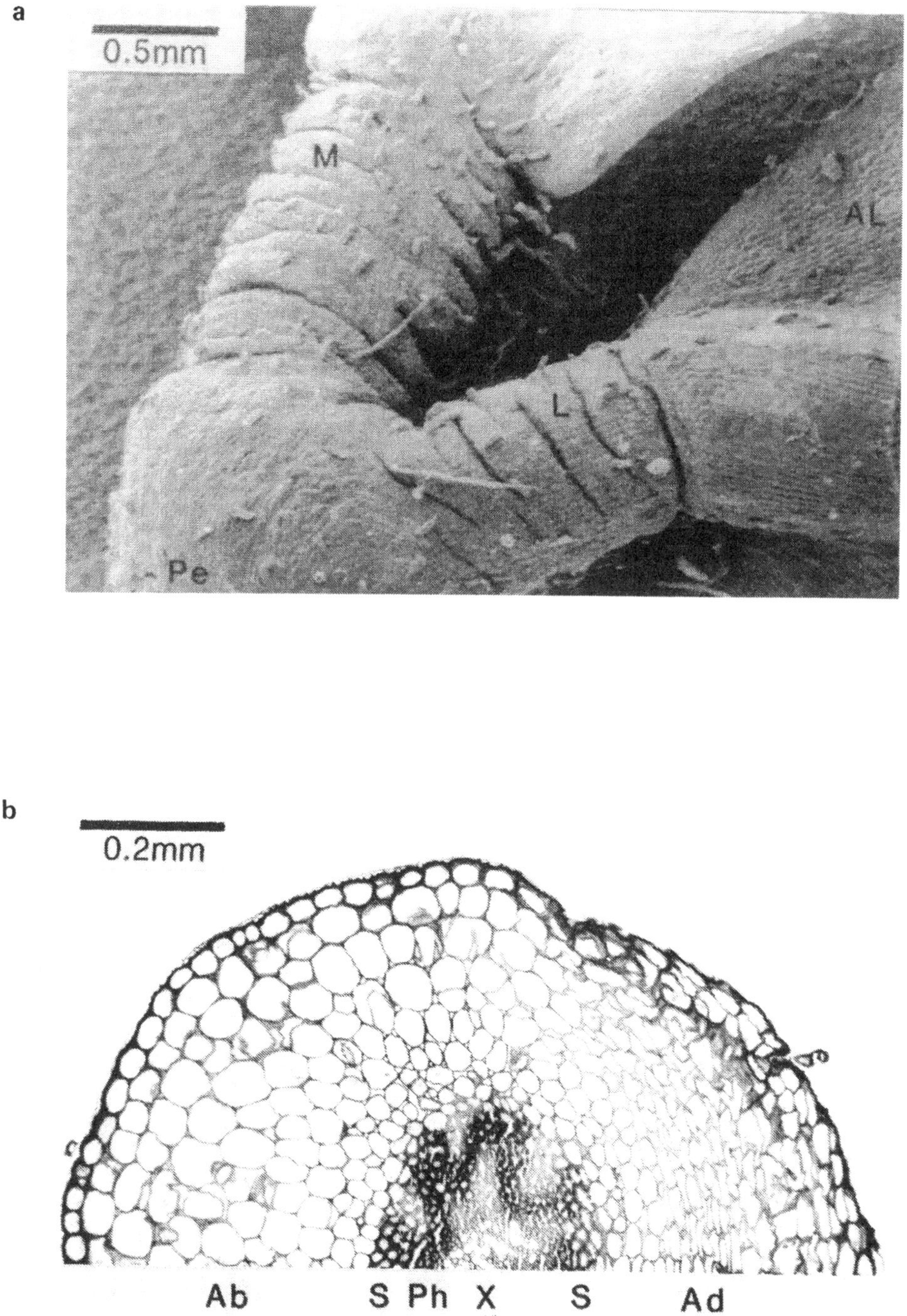

Fig. 1.3 (a) Scanning electron micrograph of petiolules placed in Karnovsky's fixative at night when the leaflets were in the closed position. M= petiolule of middle leaflet; L = petiolule of lateral leaflet; Pe = distal end of petiole; AL = abaxial surface of lamina at base of leaflet. [D.H.H.]. (b) Light photomicrograph of transverse section of a half petiolule fixed at night. Ab = abaxial cortical cells; Ad = adaxial cortical cells; S = sclerenchyma; Ph = phloem; X = xylem. [D.H.H.].

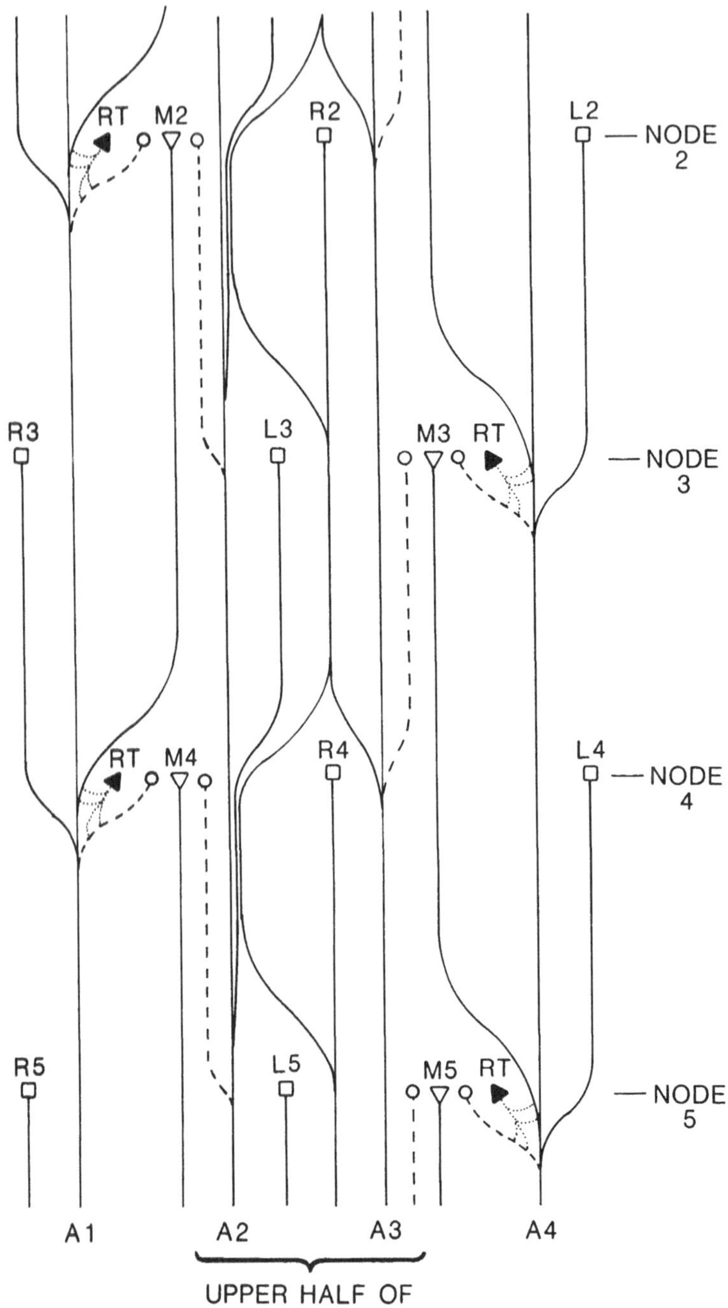

Fig. 1.4 Diagram of vascular system of stolon (see text, page 5, for explanation). Bud traces (O) are shown as broken lines, vascular connexions of nodal roots (**RT**) as dotted lines. M, L and R are leaf traces [T].

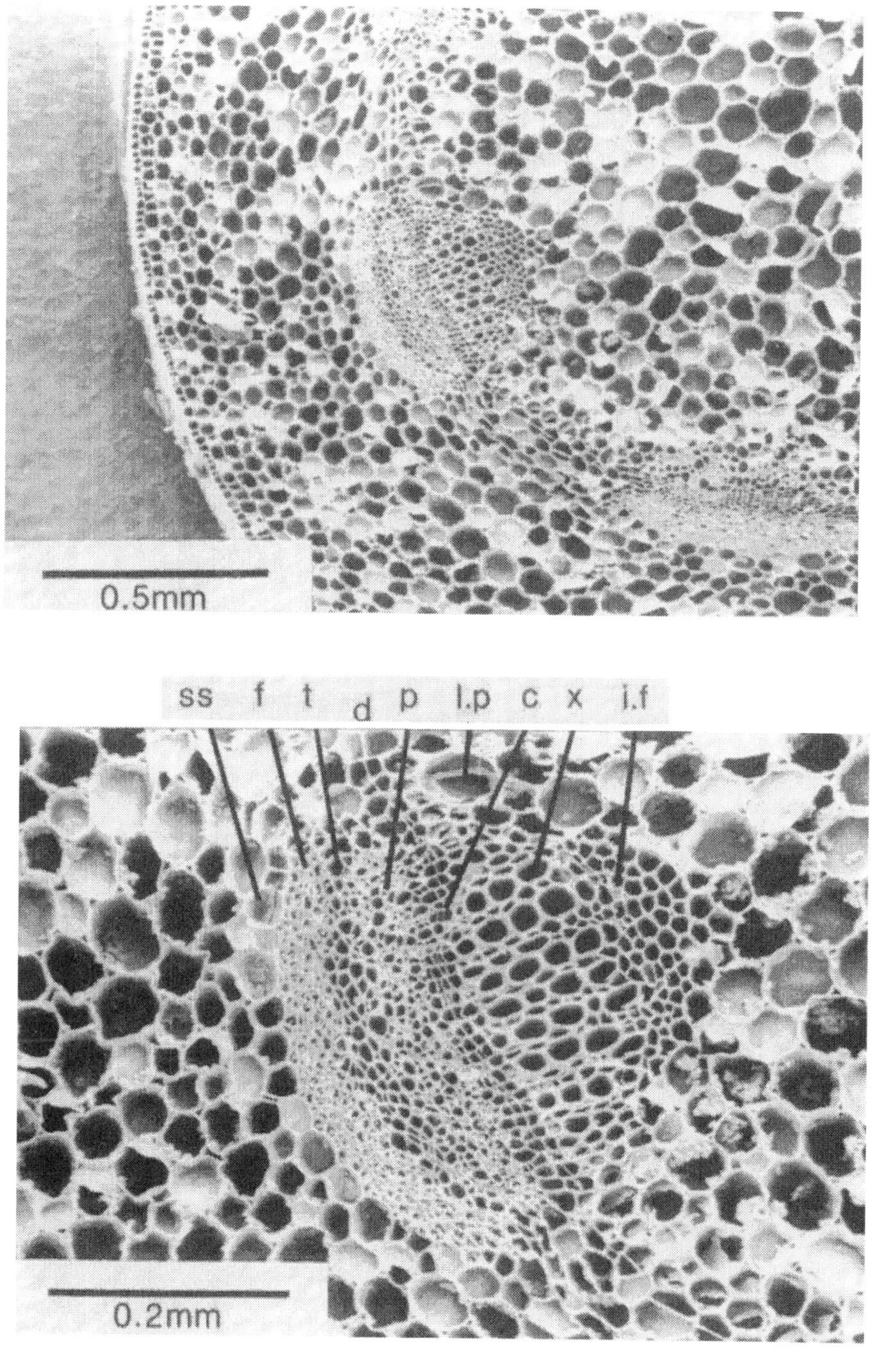

Fig. 1.5 Scanning electron micrographs of a transverse section of a stolon of cv.Grasslands Huia cut mid-way along the internode (1) proximal to leaf 1 (see Fig. 1.1). The single vascular bundle shown in (b) is the equivalent to M2 in Fig. 1.6(a). ss = incomplete starch sheath; f = gelatinous fibres; t = thin walled (incompletely differentiated) fibres; p = phloem; d = division of parenchyma cells to form interfascicular cambium; l.p = lignified parenchyma; x = xylem; c = fascicular cambium; i.f = incompletely diffferentiated fibres(?). [D.H.H.].

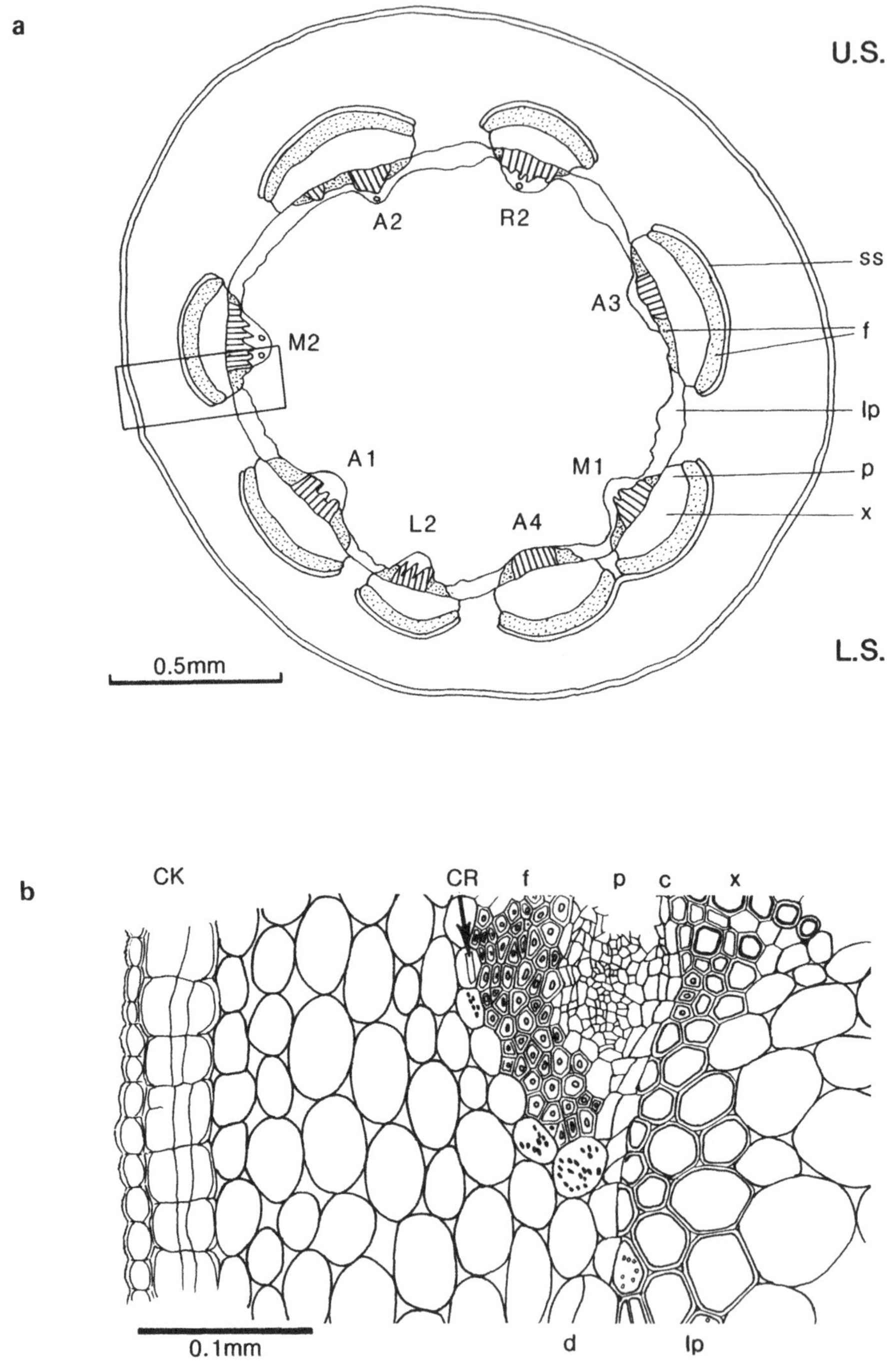

Fig. 1.6 Transverse sections of stolons of cv. Grasslands huia (a) cut midway along the internode (2) proximal to leaf 2 (see Fig. 1.1). A, M, R, and L identify vascular bundles with those shown in Fig. 1.4. U.S. and L.S. indicate the uppermost and lowermost sides respectively relative to the soil surface. (b) Cut mid way along the internode (8) proximal to leaf 8 (see Fig. 1.1) in the region indicated by the rectangle in (a). CK = cork cambium forming in subepidermal cells; CR = calcium oxalate crystal in incomplete starch sheath. For key to other labelling in both (a) and (b), see legend to Fig. 1.5 [T].

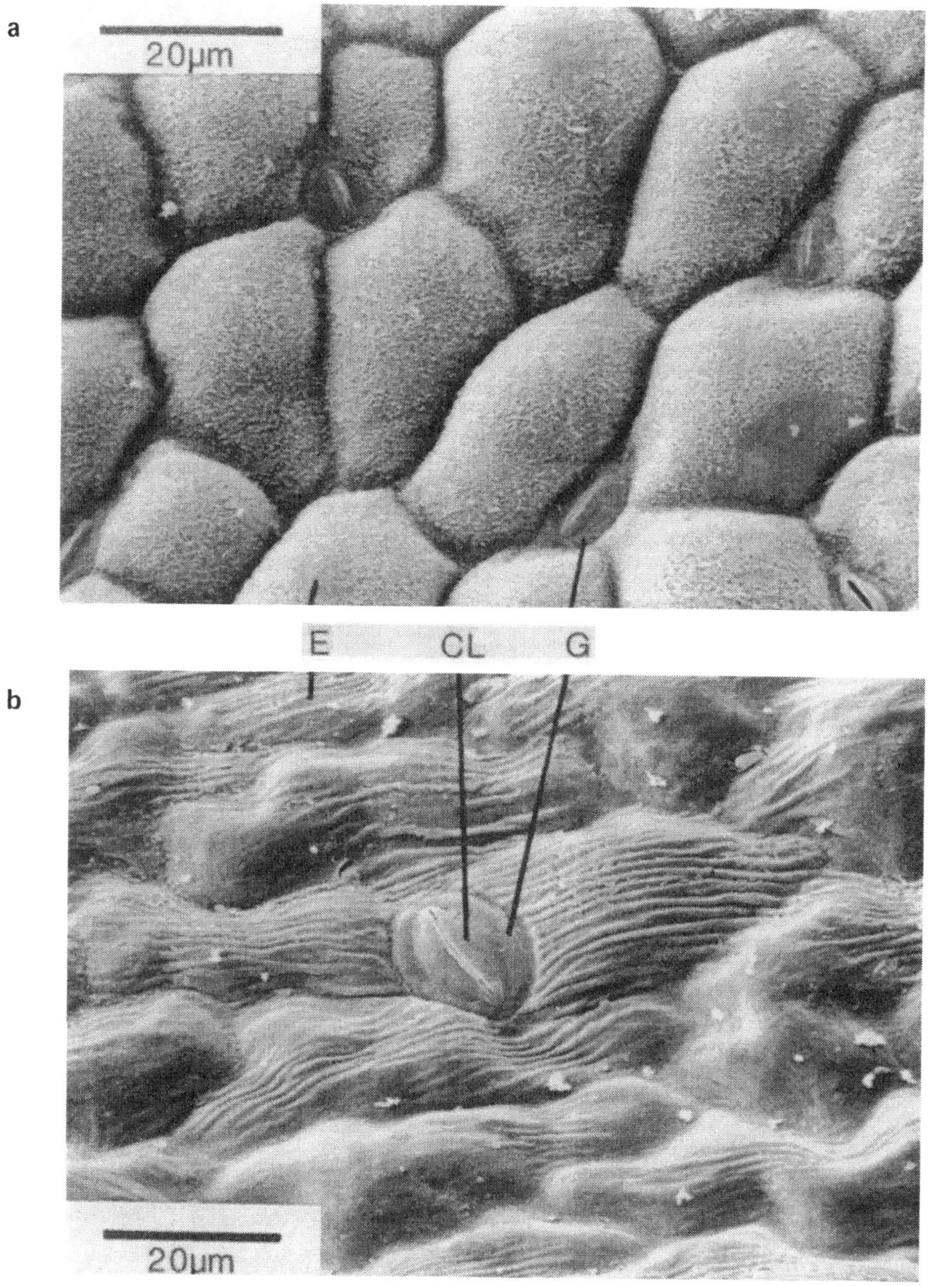

Fig. 1.7 Scanning electron micrograph of: (a) the upper epidermis and (b) the lower epidermis of a leaf lamina fixed in Karnovsky's fixative, showing epidermal cells (**E**); stomatal guard cells (**G**); and cuticular ledges (**CL**) on guard cells. Closure of eisodal pores might have been caused by slight shrinkage of tissues during critical point drying. [D.H.H.]

a

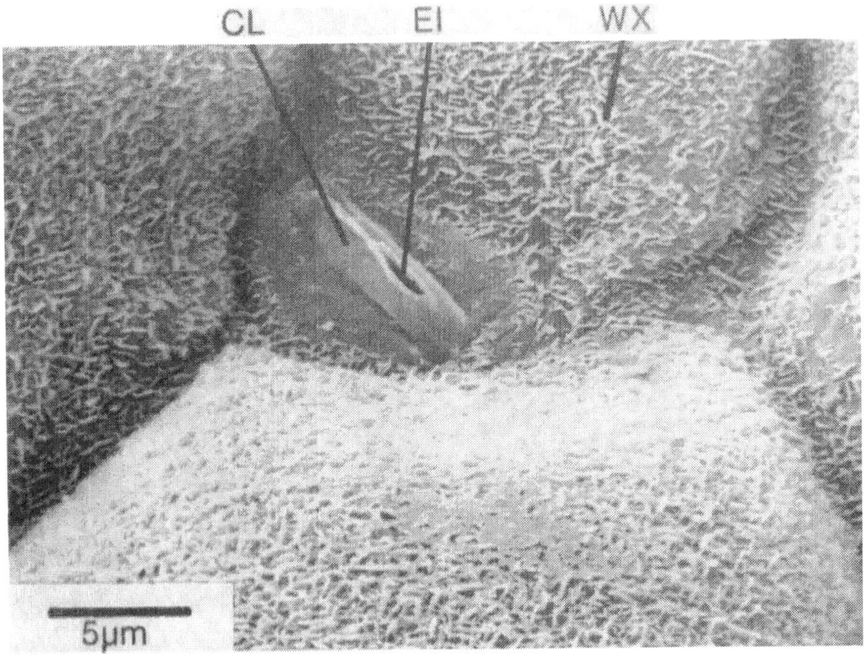

b

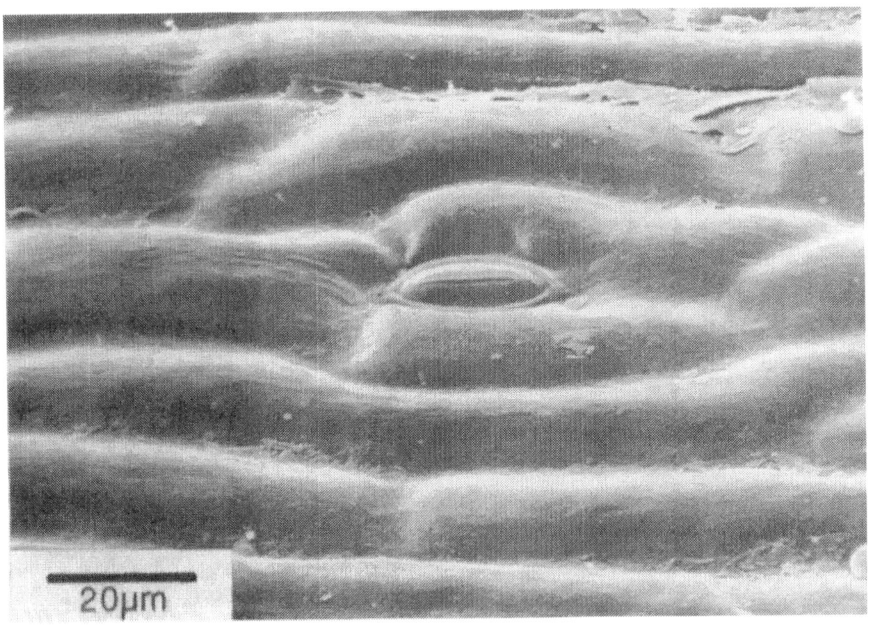

Fig. 1.8 Scanning electron micrographs of: (a) the upper epidermis of a leaf lamina fixed in Karnovsky's fixative showing wax platelets (**WX**) and an open eisodal pore (**EI**) between cuticular ledges (**CL**), and (b) the surface of the internode (1) proximal to leaf 1 on a main stolon (see Fig. 1.1) showing a stoma with closed eisodal pore. [D.H.H.]

a

b

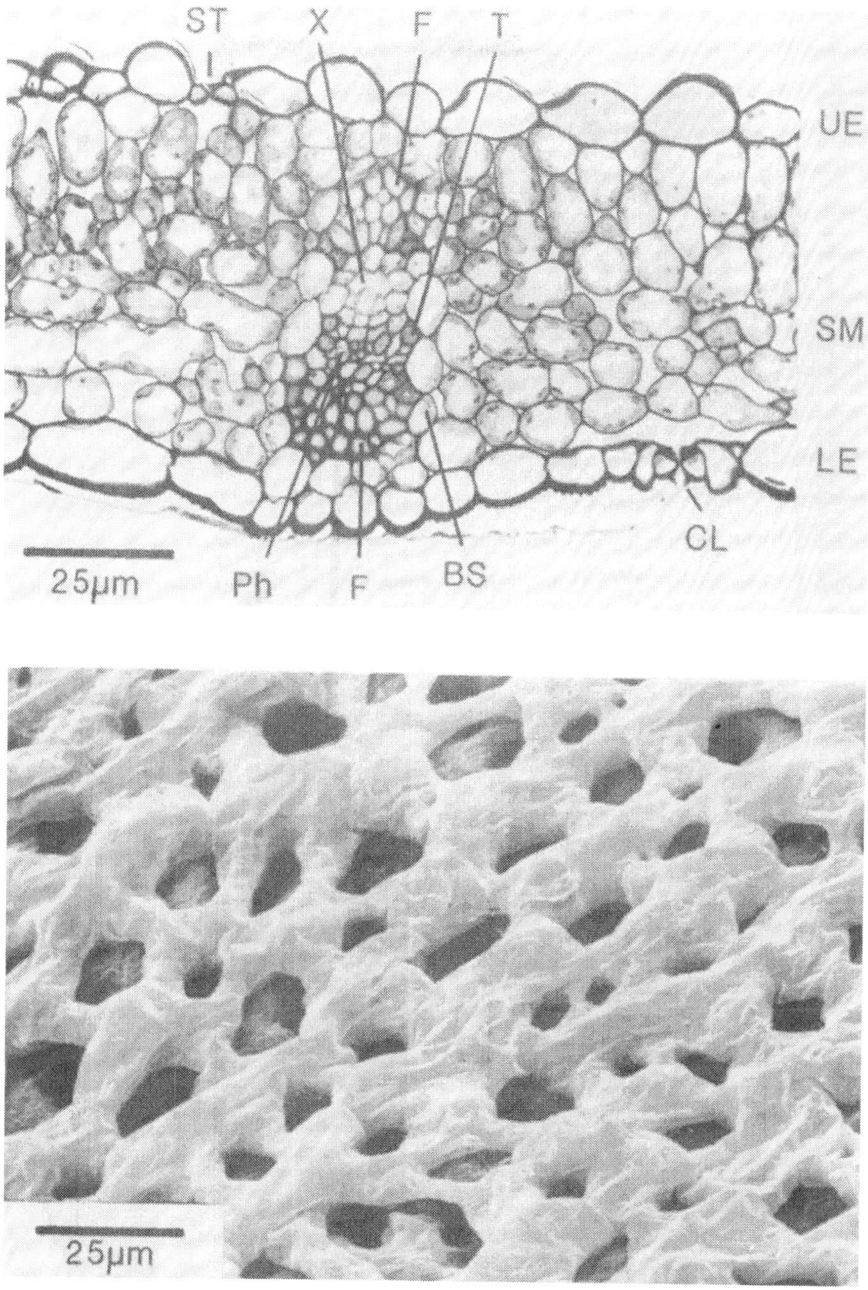

Fig. 1.9 (a) Light micrograph of a transverse section of the lamina of a mature leaf of cv. Grasslands Huia cut through a main lateral vein and stained with toluidine blue. UE = upper epidermis; ST = stoma; BS = bundle sheath with crystals of calcium oxalate in cells near the abaxial and adaxial lamina surfaces; F = gelatinous fibres; X = xylem; Ph = phloem; T = transfer cell; LE = lower epidermis; CL = cuticular ledge on guard cell; SM = spongy mesophyll. [D.H.H.] (b) Scanning electron micrograph of surface view of leaf lamina spongy mesophyll as revealed by peeling off the lower epidermis. IC = intercellular space. [D.H.H.]

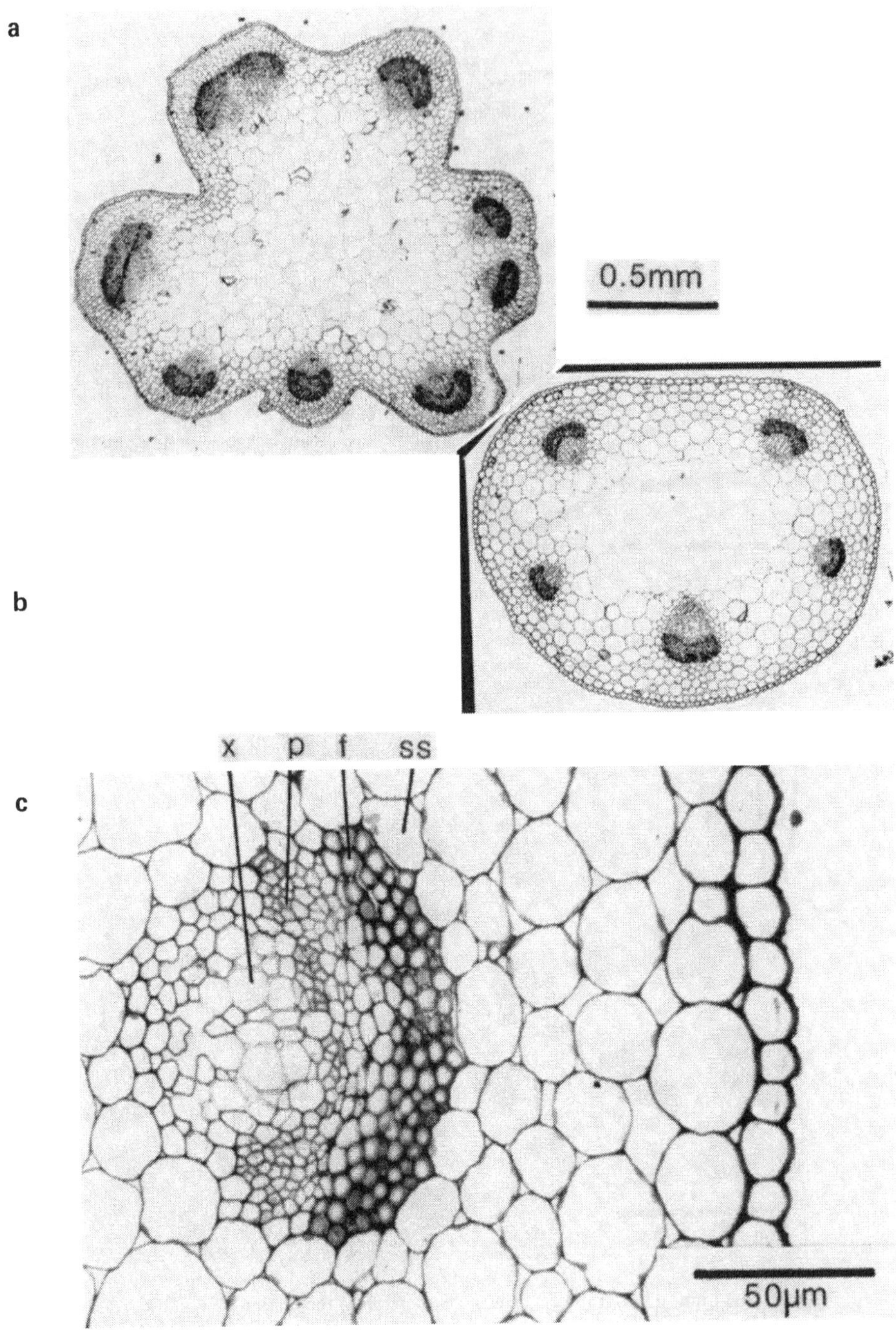

Fig. 1.10 Light micrographs of transverse sections of: (a) a peduncle cut at its midpoint; (b) and (c) a petiole cut at its midpoint. Labelling is as for Fig. 1.5. The vascular bundle shown at higher magnification in (c) is the small one on the right of (b). [D.H.H.]

a

b

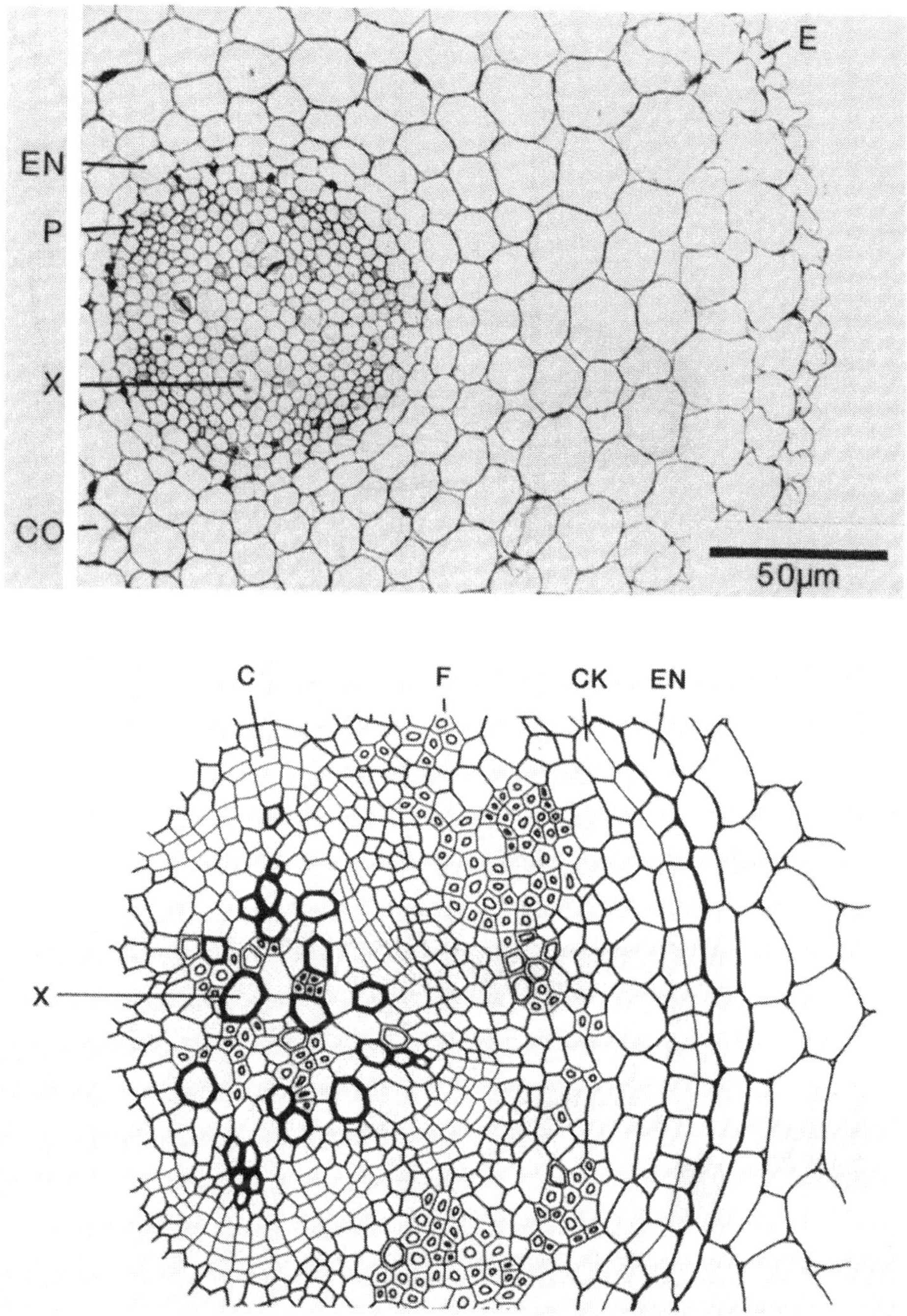

Fig. 1.11 (a) Light micrograph of part of a transverse section of a lateral root cut approximately 20 mm behind its apex and stained in toluidine blue. EN = endodermis (with Casparian strips discernible as more lightly stained areas in the radial walls); P = pericycle; X = differentiating metaxylem; CO = cortex; E = epidermis (piliferous layer). [D.H.H.] (b) Transverse section of part of an older root than in (a). F = gelatinous fibres; C = cambium; CK = pericycle cells dividing to form cork cambium. Other labels are as for (a). [Redrawn from Erith, 1924].

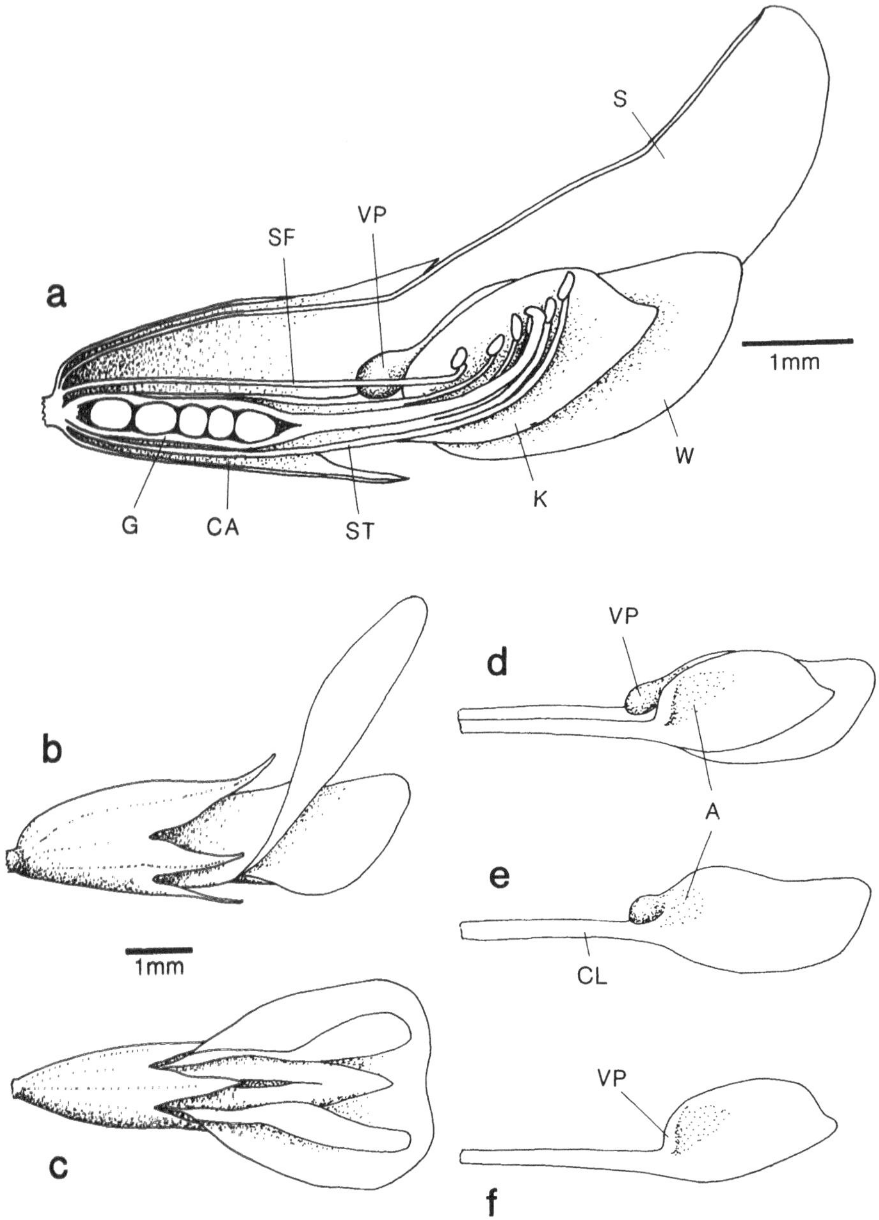

Fig. 1.12 (a) Half floret of white clover; (b) side view of intact floret; (c) abaxial (anterior) view of floret; (d) isolated wing and keel petals as viewed from the interior of the floret showing their relationship to each other in the intact floret; (e) isolated wing petal as in (d) with keel petal removed; (f) isolated keel petal as in (d) with wing petal removed. (b) to (f) are drawn to the same scale. CA = calyx; S = standard petal; W = wing petal; K = keel petal; VP = vesicular protuberance; A = area of adhesion of wing petal to keel petal; CL = claw of petal; SF = free stamen; ST = staminal tube; G = gynaecium. [T].

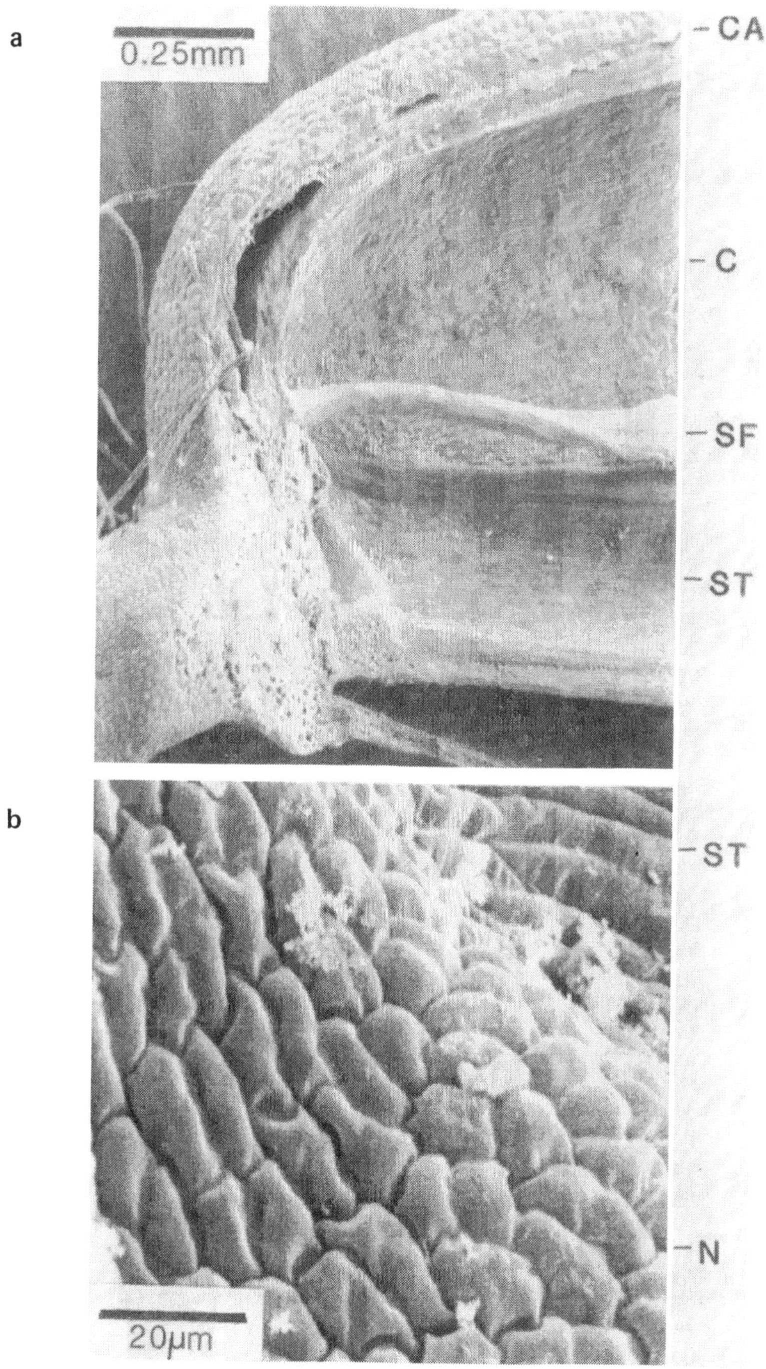

Fig. 1.13 Scanning electron micrographs of: (a) Interior of basal region of half a floret at a stage just before anther dehiscence. The ovary has been removed to expose the nectary (**N**) at the base of the staminal tube (**ST**). Other labelling is as in Fig. 1.12. (b) cells of the nectary shown in (a). [D.H.H.]

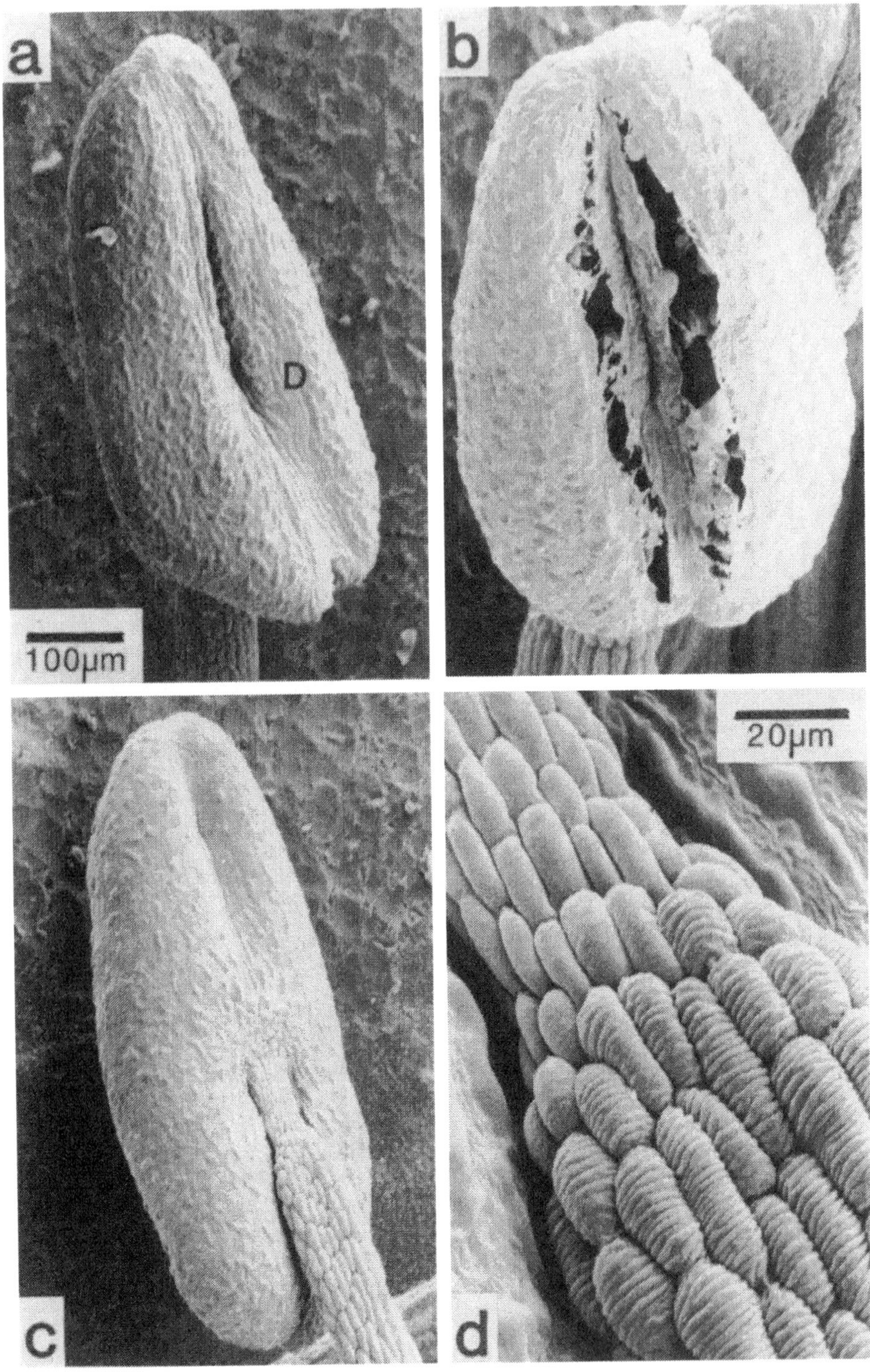

Fig. 1.14 Scanning electron micrographs of stamens: (a) shows the lines of dehiscence just before, and (b) soon after, dehiscence; (c) shows the attachment of the anther to the filament and (d) the distal end of the filament immediately below its point of attachment to the anther. D = line of dehiscence. [D.H.H.]

a

b

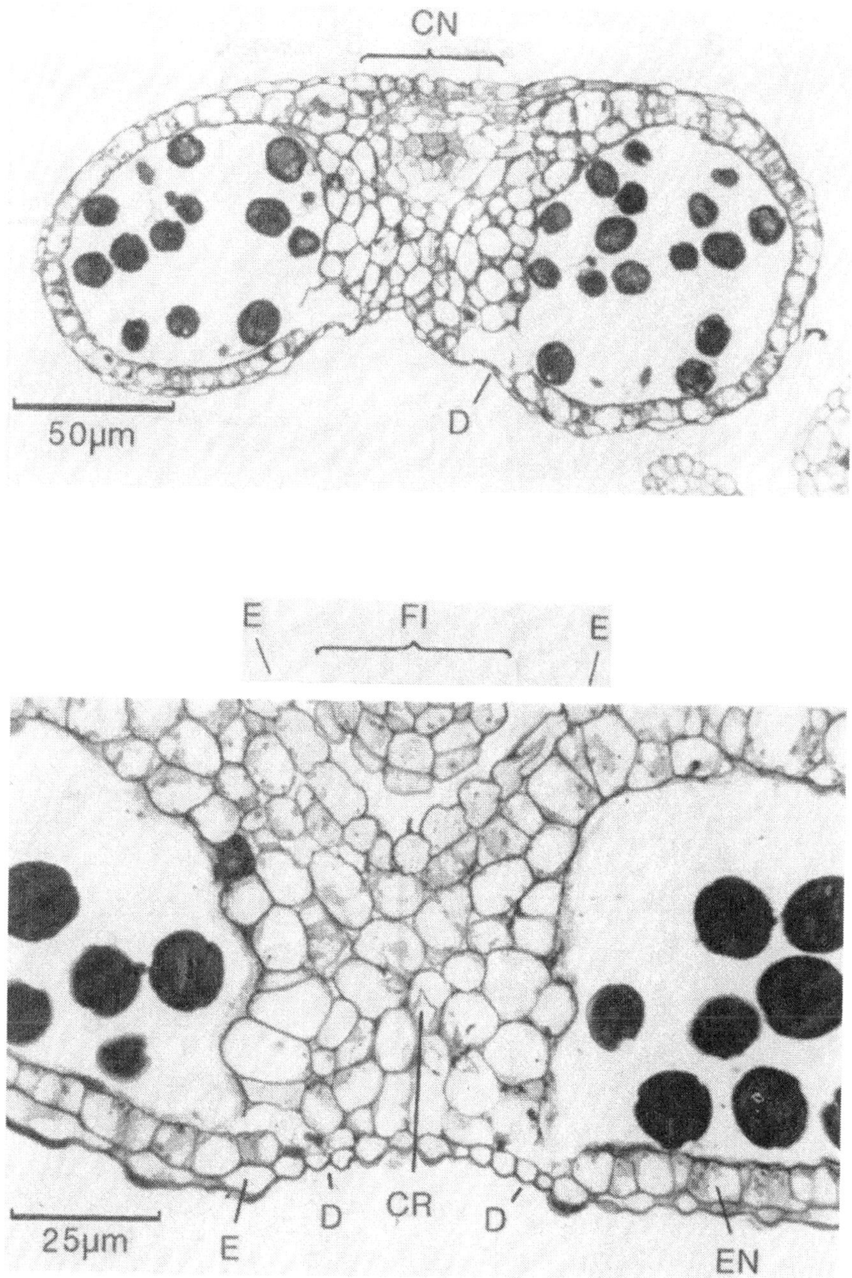

Fig. 1.15 Light micrographs of transverse sections of anthers. (a) An older anther at the stage shown in Fig. 1.14(a), sectioned above the point of attachment of the filament. (b) A younger anther, showing breakdown of the cells internal to the line of dehiscence, sectioned below the point of attachment of the filament. In both (a) and (b) pollen grains are fully formed. CN = connective; D = line of dehiscence; EN = endothecium; E = epidermis; CR = calcium oxalate crystal; Fl = filament. [D.H.H.].

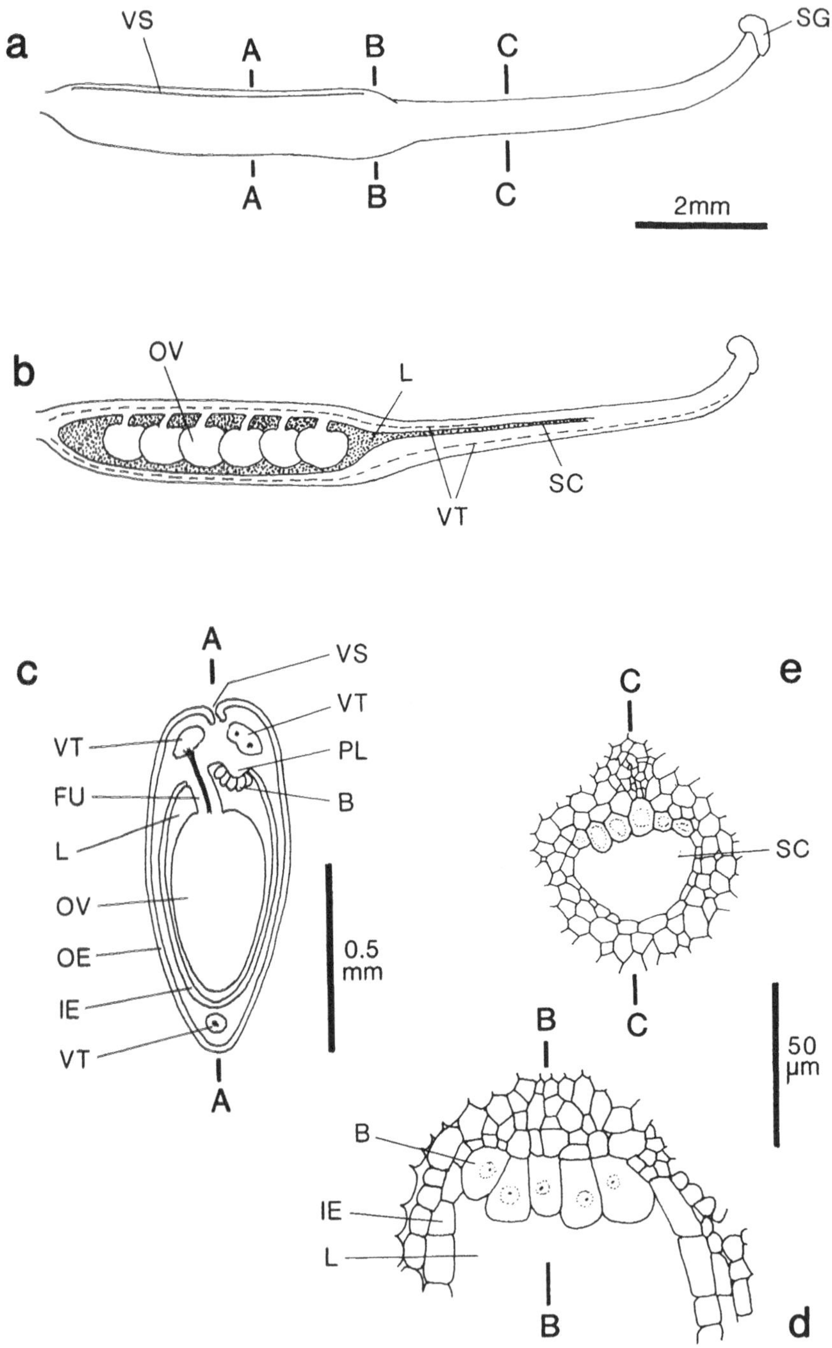

Fig. 1.16 Carpel structure. (a) Intact carpel; (b) diagrammatic representation of half a carpel; (c) transverse section midway along ovary at A-A in (a); (d) transverse section through distal end of ovary at B-B in (a); (e) transverse section through style at C-C in (a). VS = ventral suture; SG = stigma; L = loculus; OV = ovule; VT = vascular trace; SC = stylar canal; PL = placenta; FU = funicle; OE = outer epidermis of ovary wall; IE = inner epidermis of ovary wall; B = 'bladder-like' epidermal cells. [(a)-(c) original [T], (d) and (e) redrawn from Erith, 1924].

a

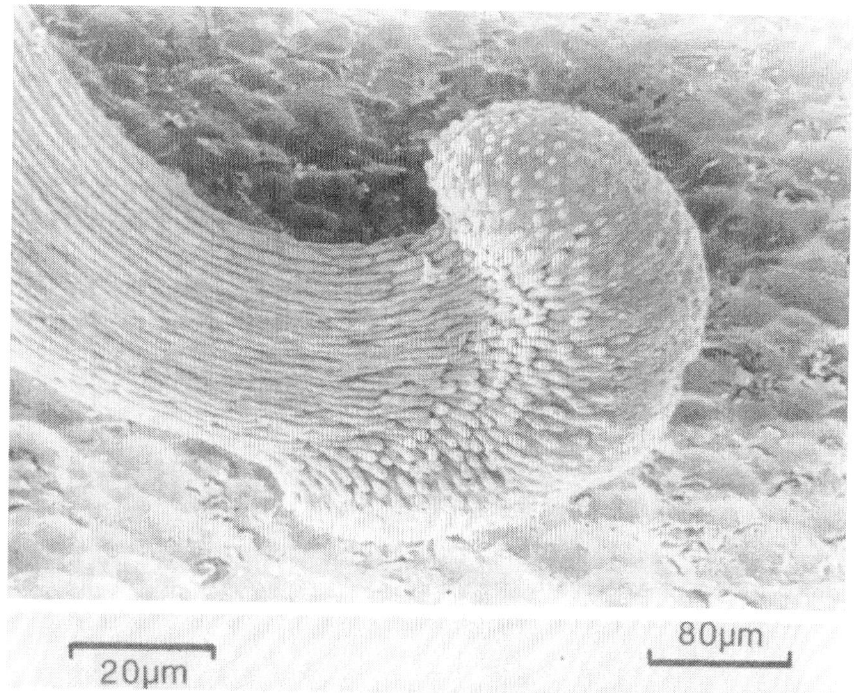

b

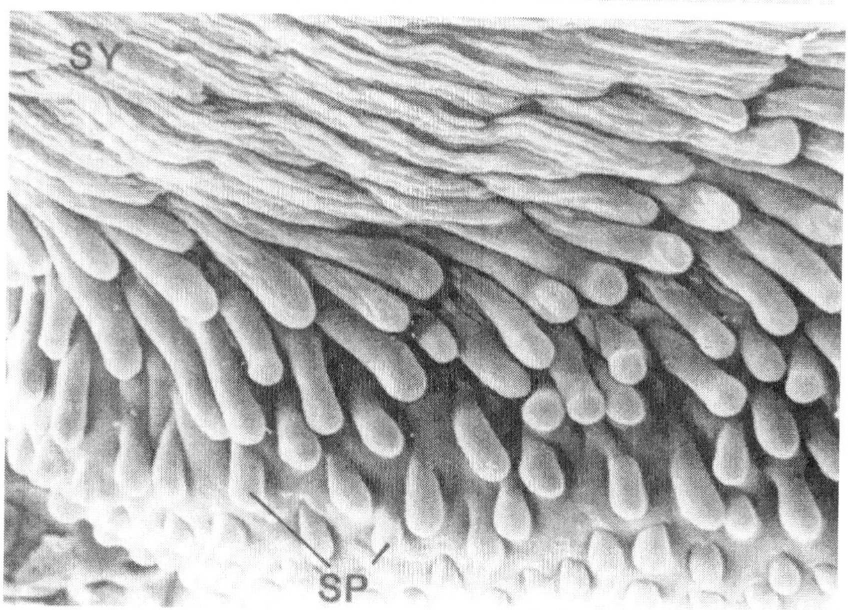

Fig. 1.17 Scanning electron micrographs of: (a) whole stigma and (b) portion of stigma, removed from a young floret at a stage just before anther dehiscence. SP = stigma papillae; SY = heavily ridged stylar epidermal cells. [D.H.H.]

Vegetative growth and development

R. G. Thomas
Department of Botany and Zoology, Massey University, Palmerston North, New Zealand.

Contents

The account that follows is mainly concerned with a description of the sequence of growth and development of vegetative organs of the plant and a discussion of the manner in which this is controlled internally. Consideration of the influence of environment is mostly restricted to its effects on plant movement. More detailed discussions of environmental effects on vegetative growth are given in Chapter 4. As in Chapter 1, unpublished data obtained by the present writer are indicated by (T). The scanning electron micrographs in Figures 2.4, 2.5 and 2.11 were prepared for this publication from material processed in the Electron Microscope Laboratory, Applied Biochemistry Division, Department of Scientific and Industrial Research, Palmerston North, New Zealand by D.H. Hopcroft (D.H.H.).

1. Seedling growth

The embryo in a mature seed consists of two cotyledons between which are hidden the stem apex and one leaf primordium. At this stage the cotyledons have no stomata and the vascular strands in the embryonic axis and cotyledons are blocked out but their cells are not differentiated into tracheary or sieve tube elements.

After seed germination, the embryo develops into a seedling with the following characteristics (Erith, 1924). The cotyledons are epigeal, with broad petioles united at their bases to form a cotyledonary sheath enclosing the plumule. The sheath is slightly higher on one side than the other. The root system develops faster than the plumule, three rows of lateral rootlets (one opposite each of the protoxylem groups in the triarch primary root - Chapter 1, Section 2.(F)) having appeared by the time of plumule emergence from the cotyledonary sheath. The first bacterial nodules form at the top of the primary root at about the time of lateral root emergence. Leaf emergence and unfolding take place after the establishment of the root system. The first leaf is unifoliolate, approximately 4 mm wide, and rather cordate. All subsequent leaves are trifoliolate.

During the early stages of development the seedling has a rosette habit with a single primary stem and no apparent internode elongation. The pattern of initiation of axillary bud and leaf primordia over this period is shown diagrammatically in Figure 2.1 (T). Major morphological changes that occur over the first few weeks of the seedling's life include an increase in size in successively produced leaves, a marked increase in the number of immature leaf primordia present in the apical bud, and the eventual development of stolons. These early stages of development can be conveniently divided into two phases. During the first of these the rate of leaf emergence is low (about one per week) and, as the rate of leaf initiation at the main stem apex exceeds the rate of leaf emergence, the number of unexpanded leaf primordia at that apex gradually increases from one in the newly emerged seedling to five or six by the end of the third week. During this phase also, buds are initiated in the axils of the older leaves on the main stem and these in turn initiate further leaf primordia. As a result, there is a rapid increase in the number of leaf primordia on seedlings in this third week of life (see Figure 2.2).

During the first phase, the external appearance of the plants alters little. A second phase is marked by an increase in the rate of leaf emergence per plant as the leaf primordia initiated on axillary buds start to appear, this growth activity being accompanied by elongation of the internodes on the axillary shoots with the resultant formation of secondary lateral stolons. At this stage, which begins at about the end of the fourth week under good growing conditions (Figure 2.1), the seedling can be considered to have reached its mature form. By about the seventh week, adventitious roots are growing from the nodes of lateral stolons, of which there can be up to five per plant, and these subsequently continue growth to become main stolons (Chapter 1, Section 1.(A)). The original primary stem then either begins to elongate beyond the levels of insertion of the lateral stolons, or, commonly, stops growing altogether.

The anatomical features of the young seedlings are described by Erith (1924).

2. Vegetative growth of stolons

The basic structural component of a mature white clover plant is the stolon (Chapter 1, Section 1.(A)). Its final form is the result of the initiation, growth and development of leaf, axillary bud and internode primordia in the apical bud. Only on the basis of a full knowledge of the growth processes within this region is it possible to understand the post-emergence growth of leaves and internodes.

(A) The apical bud

The apical bud can be conveniently defined in white clover as that region of a stolon distal to the youngest leaf with unfolded leaflets (T). It is readily seen by dissection and observation at a magnification of x 30 under a dissecting microscope to be comprised of 5-8 (very commonly 6) leaf primordia, all of which except the two youngest normally subtend axillary bud primordia (see Figure 2.3). The number of leaf primordia in the bud varies little under constant environmental conditions even when the rate of leaf initiation is at its highest or lowest. This lack of variation is a consequence of the rate of initiation of leaf primordia equalling the rate of leaf emergence. Only when plants are moved from one environment to another, e.g. from long to short days, or from cool to warm conditions, does leaf emergence rate come to differ from leaf initiation rate. In these circumstances the number of leaf primordia within the apical bud decreases when the rate of leaf emergence exceeds the rate of initiation of new primordia at the stem apex and vice versa. The number of leaf primordia in the apical bud then remains constant at a new level which varies to some extent with environment.

In natural environments the rate of initiation varies from 0 to approximately 2 leaves per week at latitude 40°S. in New Zealand (Thomas 1979) although under extremely favourable conditions of warm continuous light, plastochrons as short as two days have been recorded (T). In one study testing the effects of daylength and temperature conditions, the number of leaf primordia in the apical bud ranged only from 6.14 in cool short days to 7.10 in warm long days (Thomas 1962). The greatest reduction in number is brought about by growth at low light intensities.

To facilitate discussion, the convention of designating the youngest leaf primordium at the apex P_1, and successively older ones P_2, P_3, etc. will be used. For example, on a stolon on which there are 7 leaf primordia in the apical bud, P_7 will be the oldest primordium and P_8 the youngest leaf with unfolded leaflets.

All the leaf primordia except P_1 possess stipules. These constitute a high proportion of bud tissue and are relatively large in comparison with the leaf laminae and petioles at younger nodes. All apical tissue distal to P_6 is frequently completely enclosed by the stipules of P_6. Likewise P_1 to P_4 are enclosed by the stipules of P_5, and P_1 to P_3 by those of P_4. Within the bud the two halves of each leaflet of the leaf laminae are closely adpressed, as shown in Figure 2.4, and remain so as primordia grow from P_3 to P_7. Axillary buds subtended by leaf primordia P_2 to P_6 are numbered ax_2 to ax_6 respectively. P_1 does not subtend an axillary bud in vegetative plants (see Section 2.(D)).

(B) The apical meristem of the stolon

The morphology of the stem apex of white clover is shown in Figure 2.5. Its anatomy, as described by Denne (1966a) for plants of cv. Grasslands Huia, is very similar to that of other *Trifolium* species such as *T. subterraneum* (Williams and Bouma 1970). Denne considered the apex to be divisible into five regions (shown in Figure 2.6) as follows:

Region S. Includes the central group of cells (which divide in all planes) together with the overlying layers at the summit.

Region F_1. The flank meristem of the apex, from which is produced the next leaf primordium to appear. This region is four cell layers deep, the periclinal divisions that precede leaf initiation usually appearing first in the third or fourth layer below the surface of the apex.

Region F_2. The peripheral part of the apex producing the leaf primordium which is next to appear after that produced by region F_1. Since the phyllotaxis is approximately distichous, this region appears opposite F_1 in median longitudinal section. As in F_1, this region is four layers deep.

Region R. The rib meristem.

Region A. The axillary region, producing both the axillary bud and part of the sheath of the leaf primordium of the opposite side. The extent of this region is clearly defined by the shell zone, which appears as an arc of narrow cells in median longitudinal section.

In controlled environments Denne found the mitotic index at the apex to be highest in region F_1 and lowest in region A. Averages she obtained for all regions are given in Table 2.1, together with the duration of mitotic cycle estimated from the accumulation of metaphases following treatment with colchicine.

In the same series of investigations Denne (1966b) found a slight but definite diurnal fluctuation in mitotic indices in all regions except S, the summit of the meristematic dome. This fluctuation was greatest and the mitotic index highest in region F_1, where it ranged from a low of 3 towards the end of a 12-h dark period at 14 °C to a high of 5 towards the end of a 12-h light period at 28 °C. The same pattern of fluctuation occurred with 12-h and 8-h photoperiods at a constant temperature of 21 °C. These fluctuations in mitotic index were not correlated with the stage of the plastochron, which was completely independent of the diurnal changes in environmental conditions.

Table 2.1 Variation of mitotic index and duration of mitotic cycle in different regions of the stem apex (see Fig. 2.6 for explanation). [From Denne 1966a].

	Region of apex				
	S	F_1	F_2	R	A
Mitotic index	2.26	3.56	2.55	1.69	1.53
Duration of mitotic cycle (h)	108.3	69.4	86.9	136.5	211.6

(C) Leaf growth and development

Most studies of leaf growth and development have been made by making measurements of size and developmental stage at successive nodes. Under controlled environment conditions, when the rates of leaf initiation and emergence are constant, the changes from node to node represent changes in time.

i. *Leaf lamina growth*

The post-initiation growth and development of leaf primordia and leaflets has been studied in detail by Denne (1966c) in a clone of cv. Grasslands Huia white clover grown in 16-h photoperiods at day/night temperatures of 25/16°C. The following account is based on her paper.

After leaf initiation, the leaf primordium extends vertically from the meristematic dome by continuous division of cells throughout its length. Meanwhile it expands horizontally around the apex. By the time it has completely encircled the apex, the next primordium has been initiated on the opposite side of the dome (Figure 2.7). At this time (about day 4 after leaf initiation) three localized centres of active growth produce the three leaflets at the tip of the primordium. When a leaflet is about 0.1 mm long (about day 6 to day 7), growth becomes more active in two vertical strips arising close together on the side nearer the apex. These form the halves of the lamina.

No distinct marginal or submarginal initials are recognizable in these vertical strips. The outermost layer divides exclusively anticlinally to produce upper and lower epidermis. The inner layers of the margin produce a lamina with four layers by the time the leaflet is 0.2 mm long (day 9). From then on, divisions are almost all anticlinal, constituting the plate meristem. The layer nearest the upper epidermis forms palisade tissue and divides exclusively anticlinally until a late stage (about day 30), when some cells divide periclinally to produce a two-layered palisade. The other three layers form spongy mesophyll and vascular tissue. The fourth layer of spongy mesophyll is formed by periclinal divisions in one of the two innermost layers before about day 15.

Leaflet growth rate is shown in Figure 2.8. The relative growth rate is apparently highest immediately after initiation and declines as P_1 grows to P_4 (day 16 in Denne's investigation). It is then approximately constant during growth from P_4 to P_6 (Thomas 1962) and declines rapidly after about day 29 (Figure 2.8). Maximum leaflet length and breadth are reached by about the ninth or tenth nodes after initiation (day 36 to 40 when the plastochron is 4 days).

Rates of cell division in the plate meristem are shown by Figure 2.9. Under the conditions of Denne's study, they were high in all layers for the first 23 days (about 6 plastochrons after initiation). Cell division slowed down and ceased first in the lower epidermis, then in the upper epidermis and spongy mesophyll, and lastly in the palisade cells. In the upper epidermis, longer continuation of cell divisions leading to stomatal initiation resulted in the final numbers of stomata being higher, and the guard cells smaller, in the upper than in the lower epidermis. All cell division had virtually ceased by 30 days (7.5 plastochrons) after leaf initiation. Rapid cell expansion in all layers started on approximately day 25, just before cell division began to slow down (Figure 2.10).

The development of the leaf vascular system is described by Denne as follows: The midrib of the leaf 'appeared shortly after leaflet initiation, but lateral veins did not become distinguishable until about day 12. Main laterals appeared first at the tip of the leaflet, then progressively toward the base, larger laterals appearing first, then smaller ones filling in between them. But before these lateral strands became visible, the margin of the lamina became wavy, from tip toward base, the protruding humps becoming teeth into which the lateral veins projected. A second series, and in the midpart of the leaflet a third series, of marginal teeth were subsequently formed between the first ones.

'Main lateral strands were several cells wide by the time they were distinguishable from surrounding mesophyll. Minor strands first appeared as a single row of mesophyll cells, distinguished by their densely staining cytoplasm. Successive strands usually bisected the space between previous strands into two more or less equal areas. The network of closed vein islets was produced as minor strands arose connected at both ends to existing strands. The final strands to be initiated appeared to originate as free-ending strands protruding into the islets; although some of these may have arisen from breakage of complete islets, this did not seem to be general. The final blind ending strands were produced from about day 27 to day 32, subsequent increases in total vascular length resulting from elongation of existing strands as the leaflet expanded.'

In a mature leaf the midrib and larger lateral vascular strands have well developed fibres on their adaxial and abaxial sides (Figure 1.9a) and are connected to the epidermes by bundle sheath extensions. Smaller strands are without fibres and are surrounded by photosynthetic mesophyll cells. The extent to which this difference is causally related to the time of formation of vascular strands or to their size has not been studied.

No mention is made by Denne of the extremely conspicuous multicellular hairs that develop on leaf laminae and petioles at an early stage. These are absent in leaf primordia P_1 and P_2 (Figure 2.5) but are present at a high density in P_3 and P_4 (Figures 2.4 and 2.11), forming well ahead of the stomata. Initiation of hairs clearly also stops at an early stage of leaf development. As leaves expand, the hairs become widely spaced and eventually dry up (Figure 1.3a) giving the impression that the leaf surfaces are glabrous. Erith (1924) is incorrect in stating that as the hairs increase in size the 'cuticle is forced up and eventually stripped from the epidermis, carrying with it the majority of the hairs which are broken off at their bases [so that the leaf surface] becomes glabrous with the exception of a few isolated hairs'. As mentioned in Chapter 1, Section 2 (A), the material which covers the hairs in Figure 2.11b does not seem to be lipoidal: it fails to take up Sudan stains (T). Instead, it is readily observable that the surfaces of young tissues covered in hairs are highly wettable, and it is therefore most probable that the surface film is a layer of mucilage secreted by the multicellular hairs (T). These hairs and the associated surface film probably have the effect of decreasing desiccation in young tissues. Their low density on older organs suggests that they have little adaptive significance on mature organs.

The wax on leaf surfaces is described in Chapter 1, Section 2.(E). It is readily removed by brushing the surfaces, whereupon the rate of cuticular transpiration doubles (Hall & Jones 1961). After removal, platelets of wax reform on leaves within a few days (Hall & Donaldson 1963). The ability to deposit wax on the surface exists only as long as a leaf is still growing, however. The structure of the wax so produced has been reported to be little influenced by environment (Rentschler 1971).

ii. Petiole growth

The early stages of petiole growth in white clover have not been described in detail. Figure 2.7. shows the petiole to be absent in 9-day-old P_3 leaf primordia. It begins to become clearly differentiated from the lamina at P_4.

Starting from the stage at which the petiole is first measurable without dissection of the apical bud, Mange (1963) showed that plotting petiole length against time over a 15-day period gave a sigmoid curve. The kinetics of such post-emergence elongation are shown in Figure 2.15. Fawkner (1967) studied the pattern of elongation by marking young petioles with nine equally spaced Indian ink lines when they were 20 mm long and following the subsequent elongation of the regions between the lines for seven days. His results show that the upper regions of the petiole, near the point of attachment of the lamina, elongated more than the lower. Elongation of the basal segments resulted almost solely from cell elongation, whereas the apical regions grew both by cell elongation and cell division. Cell division stopped earliest at the petiole base but continued at its upper end almost up to the time when overall growth in petiole length stopped (Figure 2.12).

In a mature petiole it is only the uppermost, youngest, region which contains starch grains that are movable under the influence of gravity. As petiolar tissue matures, these starch grains, which are located in the incomplete starch sheath (Figure 1.10c), break down and tend to be replaced by crystals of calcium oxalate. Conversely, with increasing maturity, the other cells of the cortex accumulate non-movable starch grains as a result of photosynthesis.

The ultimate thickness attained by a petiole varies with genotype and environment. The fundamental number of vascular bundles per petiole is three, consisting of one from each leaflet, but petioles normally contain one or more supernumerary bundles branching from these. Gausman & Fuelleman (1952) found that the petioles of Ladino clover contained an average of 7.65 vascular bundles, whereas those of a wild clover from Illinois contained only 4.22. The possibility that the difference in number is mainly related to petiole size, which is greater in Ladino clover, is strongly suggested by a report by Gibson & Schultz (1953) that the number of vascular bundles in Ladino petioles grown under dry conditions (which decreased their size) was the same as the number in plants with genotypically smaller leaves grown under moist conditions (which increased their size). Further investigations are required to determine the extent to which bundle number is independent of petiole size.

iii. Leaf emergence

Leaf emergence can be defined as the appearance of a new leaf out of the stipular sheath surrounding it in the bud; it most frequently involves the lamina of P_6 growing out of the stipular sheath of P_7. As P_6 becomes the new P_7, its leaflet halves remain adpressed to each other. These leaflets normally open as P_7 grows to become P_8, in which case P_8 is the youngest leaf with unfolded leaflets. The rate at which leaflets open after emergence is usually slower under conditions of lower light intensity but no detailed study of the effect of light on leaflet opening has been made.

Emergence of the leaf and its subsequent development to maturity are accompanied by leaflet expansion and petiole elongation. In developmental studies, Carlson (1966a) and Denne (1966b) independently and simultaneously described the various stages through which a leaf passes after emergence. Carlson, working with Ladino clover, divided the development of the leaf into 10 stages, 0.1, 0.2, etc. up to 1.0, and showed that the times taken to pass from each one to the next were approximately equal. The stages are illustrated in Figure 2.13, in which at stage 0.1 approximately 25% of the leaf lamina is visible as it emerges from the stipules of the leaf one plastochron older. The leaf at stage 1.0 has 'leaflets 95% unfolded or slightly cupped', but petioles are only 70-75% elongated at this stage. A published graph shows the progression from stage 0.1 to 1.0 to take about 10 days in seedlings growing at 16°C, and about 7 days at 24°C.

Denne (1966b), studying cv. Grasslands Huia, described only three stages of development, which she termed categories a, b and c. In each category the youngest emerging leaf is designated L1, and the leaf one plastochron older L2. In category 'a' L1 is stage 0.1 in Carlson's scheme, and L2 is 0.4 to 0.5; in category 'b' L1 is 0.2, and L2 is 0.6; in category 'c' L1 is 0.3 and L2 is 0.7.

The data of both Denne and Carlson show that progression from stage 0.1 to stage 1.0 takes about two plastochrons.

iv. Leaf numbering systems

The numbering systems used to indicate the position of a leaf along a stolon have varied from one investigation to another but are basically of two types.

In the first type of system, which is probably the most widely used, the youngest leaf with unfolded leaflets is referred to as number 1 and leaves at successively older nodes are given successively higher numbers from 2 upwards (as in Figure 1.1). This system, and its variants, have been used mostly in short-term studies not involving growth, such as translocation experiments by Hoshino et al. (1965) and Harvey (1970), and in field studies in which past changes in stolon development have been described on the basis of observations made at only one point in time (e.g. Thomas, 1980). It is of very little value for kinetic studies because it does not provide a means of numbering leaves which emerge after the start of an experiment.

In the second type of system, which has been that most commonly used for studies of stolon growth and development over periods of several weeks or more, the youngest leaf with unfolded leaflets at the start of a period of growth is given the number 1 and successively emerging leaves are numbered 2, 3, 4 and so on. Such a system was used by Kawanabe et al. (1962) and also, in addition to the first system described, by Thomas (1980). It has the obvious disadvantage that leaves older than the youngest with unfolded leaflets remain unnumbered.

Partly to overcome the problems arising from these two frequently used systems, Thomas (1972) devised a third system which combined the two by using sequences of both numbers and letters. The oldest leaf with still folded leaflets was numbered 1 and successively younger leaves numbered 2, 3 and so on. The youngest leaf with unfolded leaflets was referred to as A and successively older leaves as B, C, etc. While this dual system is satisfactory, it has the disadvantage that it has no component in common with that most widely used (viz. the first described) in which successively higher numbers are allotted to successively older leaves.

To avoid confusion, it is recommended that a new dual system of letters and numbers be introduced and adopted for all future studies. In this system:

(a) the youngest leaf with unfolded leaflets would be designated number 1 and successively older leaves allotted successively higher numbers, and

(b) the oldest leaf with leaflets still folded would be termed A and successively younger leaves would be referred to as B, C, D and so on.

Nodes would be referred to by the numbers or letters used to identify the leaves attached to them, and each internode would be given the number or letter of the node immediately distal to it. This recommended system is used hereafter in this chapter.

(D) Axillary bud initiation and growth

The initiation of vegetative axillary bud primordia typically takes place by the formation of a shell zone (see Section 2.(B)) in the axil of leaf primordium P_2. Periclinal divisions of cells within this zone then lead to the formation of an axillary mound. Axillary bud primordia so formed can be seen clearly in dissected apical buds examined at a magnification of x 30, the youngest (ax_2 or ax_3) being discernible in the axil of either P_2 or P_3 (Figure 2.5) depending on the vigour of the stolon and on its environment. Commonly, the youngest buds are nearer the summit of the apical meristem in higher light intensities, longer photoperiods and lower temperatures, and further from the summit at lower light intensities, shorter photoperiods, higher temperatures (Thomas, 1962) and in poorer mineral nutrient conditions.

The effect of apical dominance on bud growth is not apparent within the apical buds, axillary buds increasing in length approximately exponentially from node to node between P_2 and the youngest leaf with unfolded leaflets (at node 1, see Figure 2.14). During the period of exponential growth, leaf primordia are initiated on the axillary bud primordia at approximately the same rate as at the stem apex, with the result that each axillary bud tends to bear one more leaf primordium than that at its adjacent younger node.

The growth rate of axillary buds then declines at nodes bearing leaves with unfolded leaflets (nodes 1, 2, 3, and so on) so that they are relatively inactive at the older nodes. Outgrowth of axillary buds into lateral stolons usually does not occur until several nodes basal to the apical bud, the exact point depending on stolon vigour and growing conditions (see Section 3.(A)).

(E) Internode growth and development

Internode elongation takes place in the region immediately basal to the apical bud, being most rapid in the region of leaf emergence from the bud (viz. as internode B elongates to become internode A) and is virtually complete by the time it becomes internode 2 (Thomas 1972) (see Figure 2.15).

Up to the length of approximately 2 mm, at which stage it is still enclosed by the stipular sheath of the proximal leaf internode, growth is largely by cell division. Over the next plastochron, as internode B elongates to become internode A, there is approximately one more division per cell (less in epidermal cells, which stop division ahead of those of cortex and pith), but most of the increase in length results from cell elongation (see Table 2.2). The final stage of elongation during the next plastochron is the result of cell elongation only.

Table 2.2 Average cell lengths and numbers of cells in internodes 1, A and B (see Fig. 2.15 for explanation) in cv. Grasslands Huia. [Modified from Thomas 1972].

		Cell number			Cell length (μm)		
	Internode length (mm)	Epidermis	Cortex	Pith	Epidermis	Cortex	Pith
Internode 1	19.4	306	191	211	65	106	97
Internode A	14.6	296	203	228	49	72	69
Internode B	2.1	198	109	112	11	19	19

Transverse sections of internode B show starch to be present in the incomplete starch sheath at that stage, but the starch grains do not appear to be freely mobile under the influence of gravity until internode A (T).

The time of onset and rate of secondary development vary greatly with environment and probably with genotype also, but detailed studies have not been made. In cv. Grasslands Huia, fascicular cambium is initiated in internode A, but initiation of interfascicular cambium, which begins in the lower side of the internode, typically does not occur until internode 1. Fibre thickening and lignification of fibres and parenchyma likewise take place while internode A grows into internode 1. Cork cambium initiation starts in the subepidermal cells on the lower side of internode 2 or in older tissue. Periderm development has been described by Hay et al. (1982).

(F) Nodal root initiation and growth

Root initiation and early growth have been little studied in white clover. Two root primordia are present at each node as it emerges from the apical bud, one each side of the axillary bud. These are absent from the third or fourth youngest node (bearing leaf primordia P_1 to P_3 or P_4), but are initiated at the node bearing P_4 or P_5.

The lowermost of each pair of root primordia grows out sequentially through the cortex, the epidermis, and the stipular sheath to form a mound visible externally at node 1 (see Figure 1.2.). The uppermost is delayed in its growth, usually to the extent that it never appears externally. Factors controlling growth of this upper primordium have been little studied, but results of preliminary experiments by Ueno & Tsuchiya (1968) suggest that auxins might be involved. They found slightly enhanced growth of the upper primordium in response to applied 0.5% indole butyric acid.

When the root primordium at a node is in contact with a moist substratum it grows out into a nodal root (see Section 3.(A)). Each nodal root branches to produce a root system which has a vascular connexion with the axillary bud at that node. Because only one of a pair is in constant contact with the substratum, there is normally only one nodal root per node.

The structure of the root apex is of the same type as described for *Pisum* and other members of the Leguminosae (Erith, 1924).

(G) Summary : Growth at the Stolon Tip.

Data on the growth of leaves and internodes are collated to show the relationships between them in Figure 2.15. The data from which the graphs were drawn are taken from an experiment in which plants were grown in natural winter short days in a temperature-controlled greenhouse. Those for lamina length have been published previously (Thomas 1961); those for petiole and internode lengths (previously unpublished) were obtained from the same plants. On most stolons there were seven unexpanded leaves and leaf primordia younger than the youngest unfolded leaf. This has therefore been labelled P_8 in the figure.

Within the apical bud, up to the time of leaf emergence (at leaf primordium P_6) growth is largely by cell division; but after leaf emergence, elongation growth of leaflets, petioles and internodes is mostly the result of cell expansion (except at the upper end of petioles). Internodes, which are the last organs to begin rapid elongation, are the first to stop; petioles continue to elongate for a longer time than leaflets. Final sizes attained by the various organs are strongly influenced by environment and genotype but their relative durations of growth differ little.

3. Internal coordination of vegetative growth and development.

Growth of individual organs is influenced by the activities of all other parts of the plant: stem apex, leaves, root system, etc. It is thus essential to understand organ growth in the context of the correlative influences within the whole plant.

(A) Apical dominance

The internal control of apical dominance in *Trifolium repens* has been little studied. Typically, stolons of actively growing plants appear as in Figure 1.1, outgrowth of axillary buds not occurring near the apical bud but lateral branches developing at the older stolon nodes.

Immediately after their initiation, the growth of axillary buds within the apical bud of a stolon proceeds in an uninhibited manner for five or six plastochrons (Thomas 1962; see Section 2.(D)). Their relative growth rate then begins to decline at approximately the time when the leaflets of their subtending leaves unfold and, presumably, become actively involved in photosynthesis. From this time onwards growth of the axillary buds is repressed by the presence of the apical bud. Complete removal of the apical bud allows rapid growth of the youngest one, two or three axillary buds remaining, as does the excision of leaf primordia P_6, P_5 and P_4 from the stolon apex (T). It is thus possible that the inhibitory role of the apical bud is the result of repressive influences associated with young leaf primordia rather than the apical meristem itself. Removal of older leaves from stolons has no marked stimulatory effect on axillary bud growth. The repressive effect of the apical bud is much weaker when the availability of mineral nutrients to the plant is higher (T).

Nodal roots do have an effect on the growth of otherwise inhibited axillary buds situated six or more nodes basal to the youngest leaf bearing unfolded leaflets. When plants were grown in pots for six months under poor mineral nutrient conditions and stolons prevented from making contact with a moist substratum, no nodal roots grew out and axillary buds remained completely inactive. When nodal roots were stimulated to grow from randomly selected nodes along these stolons by providing them with distilled water applied in wads of cotton wool, axillary buds at these nodes, and these nodes only, began to grow. Removal of the nodal roots as soon as they began to elongate prevented the stimulation of axillary bud growth. Thus the bud growth was directly correlated with nodal root growth and not a direct response to water itself. This was confirmed in a further experiment in which stolons were grown in translucent plastic containers in which the atmosphere was saturated with water vapour. This treatment stimulated the outgrowth of nodal roots but did not lead to the growth of axillary buds when nodal roots were excised as soon as they showed signs of growth (T).

From these experimental results it seems clear that nodal roots stimulate axillary bud growth by some means other than by providing a supply of mineral nutrients. Further experiments are required to determine whether they act by removing an inhibitory influence or by supplying a promoter such as a cytokinin (shown by Wickson & Thimann (1958) to have a stimulatory effect on bud growth in *Pisum*, for instance).

On the basis of the observations described above, together with the fact that axillary buds are typically released from apical dominance at the basal ends of stolons (Figure 1.1), the growth of axillary buds into lateral stolons can be explained as being a response to two opposing influences. These are an inhibitory effect of the stolon apex and a promoting effect of roots. As a basis for further experimental investigation, it is suggested that both the main basal root system and the nodal roots act in the same way to promote the outgrowth of axillary buds into lateral stolons. The promoting effect of mineral nutrients might then be simply the indirect result of increased root growth. The degree of outgrowth of axillary buds immediately behind the apical bud of a stolon would be determined by the balance between inhibitory influences of apical origin and promoting influences from root tissues.

(B) Stolon elongation and anatomy

Stolon elongation results predominantly from the extension of internodes immediately basal to the apical bud. Maximum elongation only occurs in the presence of both apical and basal tissues.

Kawanabe et al. (1962) showed that defoliation markedly decreases internode elongation and stolon thickness in Ladino clover. In a more detailed study, Thomas (1972) compared internode elongation in intact plants of cv. Grasslands Huia with that in excised apical stolon pieces and in stolons from which the apical bud had been removed. At the start of each experiment the length of the youngest visible internode (internode B, see Section 2.(E)) was approximately 2 mm.

Table 2.3 Percentage increase in length of internode B following various excision treatments. [From Thomas 1972].

Treatment	Time after treatment (days)		
	2	4	6
1 Intact	123	292	375
2 Decapitated	19	57	73
3 Excised (Short)	56	56	60
4 Excised (Long)	23	56	113

The results of treatments are shown in Table 2.3. Short apical pieces were excised at the node immediately basal to internode B; long apical pieces were excised at the node proximal to the first fully elongated internode (internode 2, Figure 2.15, Section 2.(E)). Stolons were decapitated by removal of all tissues apical to internode B. Leaves were not removed. From the table it is clear that both decapitation and excision had a strong depressing effect on internode elongation. Elongation of excised pieces stopped after two days when they were short but recovered slightly in the longer pieces, presumably as a result of some positive influence from the fully elongated internodes and/or leaves. Internode elongation on decapitated stolons was very stongly retarded during the first two days after treatment but recovered slightly in the next two days. Clearly internode elongation is under the stimulatory influence of tissues both in the apical bud and further down the stolon.

In an attempt to assess to what extent the reduced elongation in excised pieces was the result of a lack of unknown factors supplied via the xylem, growth in excised pieces was compared with that on stolons of intact plants which had been girdled by treatment with hot wax to kill living cells. The retarding effect of girdling on elongation of internode B was similar to that of excision, and therefore indicated that the retarding effect of the removal of basal tissues was not the result of deprivation of promoters supplied via the xylem from the basal region of the stolons. This result was interpreted to indicate that either basally-supplied promoters reach the young unexpanded internode via live cells or girdling prevents the removal via the phloem and/or parenchyma cells of apically produced inhibitors which accumulate when the base is removed.

The effect of leaves on internode elongation was found by Thomas (1972) to be less than that of basal stem and/or root tissues. Defoliation of whole plants to leave only leaf primordia P_1, P_2 and P_3 remaining on stolons, reduced elongation of internode B (see Section 2.(E)) to 17 mm in 7 days compared with 23.5 mm in controls. In the same experiment excision of fully foliated stolons reduced elongation to 6.5 mm. It thus appears that while photosynthate supplied by the leaves might be one limiting factor for internode elongation, an additional unknown factor supplied by basal tissue is also required.

As with internode elongation, there is some evidence that the development of the anatomical features of an internode is under the partial control of basal tissues. Excision of stolon pieces from the parent plant results in a very strong reduction in the amount of fibre wall thickening and lignification and in the formation and division of cambial cells over a 10-day period (Thomas, 1972). Unlike internode elongation, however, girdling of stolons remaining attached to plants had no inhibitory effects on cambial cell activity, and it had a much less pronounced retarding effect than excision on fibre thickening. Division of cambial cells and fibre development thus appear to be strongly influenced by factors supplied from the base of the plant via the xylem.

(C) Leaf initiation and growth

Initiation of leaf primordia is influenced both by older leaves and by basal tissues. Removal of all leaves and primordia from a stolon except for leaf primordia P_1, P_2 and P_3 resulted in a slight increase in leaf initiation rate (Thomas 1972). Removal of P_1 to P_6 from an apical bud approximately doubles the rate of initiation of the next two or three leaf primordia to form (T), as has been reported to occur in other genera. Young leaves thus retard leaf initiation. Whether this is the result of competition for promoters at the stem apex or the production of regulatory compounds by existing leaf primordia is not known.

Basal tissues, however, promote leaf initiation, the rate of initiation being reduced from 2.6 over a 7-day period in intact controls to 0.8 in excised stolon tips and 1.1 in girdled stolons (Thomas 1972). Initiation of leaf primordia thus appears to be stimulated by basal tissues in the same way as internode elongation, i.e. communication between base and apex is effected by means of live cells rather than by xylem.

The correlative influence of other organs on the growth of young leaves has been investigated in two different systems by Carlson (1966b) and Thomas (1972).

Carlson recorded the effect of defoliating seedlings of Ladino clover. Treatments involved removal of all leaves except those at his stage 0.2 (see Section 2.(C)iii.) and younger. The youngest leaves remaining were then removed as they individually reached stage 0.9. He found that leaf removal reduced the leaflet area, dry weight and petiole lengths attained by subsequently emerging leaves, leaflet area of individual leaves being reduced from c. 12 cm^2 in controls to c. 2 cm^2 in defoliated seedlings.

In contrast to the decreased size at maturity of leaves formed on defoliated plants, Carlson found the post-emergence rate of leaf development to be accelerated. The time taken for leaves to grow from stage 0.2 (just emerged) to 0.9 (leaflet expansion) dropped from a mean of 4.2 days in fully foliated plants to 3.7 days in defoliated plants. A similar effect of defoliation on petiole length was observed in stolons of cv. Grasslands Huia by Thomas (1972) when all leaves except the three youngest leaf primordia, P_1, P_2 and P_3, were removed.

In Carlson's experiment defoliation decreased the rate of leaf emergence, the interval between successive leaves reaching stage 0.2 being 2.3 days in control plants and 5.2 in defoliated plants. This decrease could be the result of either a decrease in leaf initiation rate or a decrease in emergence rate without a concomitant decrease in leaf initiation (which would inevitably lead to an accumulation of leaf primordia in the apical bud). The observation, mentioned earlier, that removal of P_1 to P_6 from an apical bud enhances subsequent leaf initiation seems at first to be either in conflict with Carlson's results or to point to the possibility of an accumulation of leaf primordia at the apex following defoliation. It is equally possible, however, that the rate of leaf initiation is controlled by the young leaves at the stage of emergence (P_5 to P_7), emerging leaves having a somewhat inhibitory influence on leaf initiation. Slowing the rate of emergence would then retard leaf initiation; removal of P_5 to P_7 would accelerate it. Carlson has suggested that older leaves might retard the rate of development of younger ones by competing with them, possibly for growth regulators and/or photosynthate or by producing an inhibitor of leaf expansion and unfolding. Older leaves might thus affect leaf initiation rates indirectly through their effect on younger leaves.

Growth of emerging leaves is also affected by the basal tissues of the plant. In experiments with stolons, Thomas (1972) found elongation of the petiole of leaf A (see Figure 2.15) over a 4-day period to be decreased from 39.6 to 16.1 mm by excision of the stolon tips from the base of the plant at node 2. Final leaflet length of leaf A was similarly reduced from 23.3 to 18.8 mm and of leaf B from 20.3 to 11.1 mm (T). Whether the influence of the base reached the tip via the xylem or via live cells was tested by hot-wax ringing of the stem to kill all live cells. Ringing the stem at internode 1 had a similar effect to excision at node 2 but ringing nearer the base of the plant, at internode 5, had only a slight effect. It was therefore concluded, firstly, that post-emergence growth of leaves is not under the direct influence of the root system because communication via the xylem is inadequate on its own to maintain leaf growth, and, secondly, that growth requires communication between the young leaves and the older regions of the stolon via live cells. The cells of leaves and young internodes of excised stolon tips in this experiment were found on sectioning to be packed with starch grains. It seems more likely therefore that the older leaves exert their promoting influence by means of growth regulators rather than photosynthate.

(D) Root Growth

The effect of the rest of the plant on the growth of its root system is demonstrated by experiments by Evans (1973) in which seedlings of white clover were grown in glass-fronted containers to facilitate observation of the root system. Plants were defoliated every second day to 25, 50 or 100 mm above ground level. Root elongation in these plants was compared with that of non-defoliated controls (Figure 2.16). Clearly, growth of roots is strongly influenced by the presence of foliage, but whether this acts by providing photosynthate or growth regulators, or both, has not been investigated.

(E) Summary: Correlative control of growth.

Our present understanding of correlative control of growth in white clover is summarized in Figure 2.17. This represents a model that could profitably be used as a basis for future investigations. The evidence accumulated suggests that leaves are a source of a stimulus for growth of roots, internodes and apical buds. Whether their stimulatory effects accrue solely from photosynthate or from a mixture of photosynthate and growth regulators has yet to be elucidated. Mature leaves also appear to have an inhibitory effect on post-emergence expansion of young leaves, either by producing an inhibitory compound or competing for promoter(s). In addition, the apical bud acts as a source of promoter (probably growth regulator) necessary for internode elongation.

The apical bud clearly exerts an inhibitory influence over axillary bud development at emerged nodes, which is counteracted by the stimulatory effects of growing nodal roots. And within the apical bud itself, the data available support the existence of a feedback control of the rate of initiation of leaf primordia by the growth of emerging leaf primordia. This might provide the mechanism by which the number of unexpanded leaf primordia within the apical bud remains constant, very young leaf primordia producing promoter(s) which stimulate the growth of emerging leaves which in turn exert an inhibitory control over leaf initiation. If this hypothesis is correct, removal of emerging leaves should increase the rate of leaf initiation and removal of the very young leaf primordia reduce the rate of leaf emergence. The former is supported by the evidence available; the latter possibility remains to be tested, although the stimulatory effect of the apical bud on internode elongation points to the likelihood of its correctness.

The role of plant growth regulators in the correlative control of growth is discussed in the next Section.

4. Influence of growth regulators on growth and development.

Information on the internal control of white clover growth by growth regulators has been based almost entirely on observations of responses to applied compounds. The work of Cohen and Dovrat (1976), in which gibberellin and inhibitor levels were estimated in stolon tips and leaves, is an exception. These workers extracted gibberellin-like compounds by diffusion from stolon apices (each with one expanded leaf) into agar and found GA_3 equivalents ranging from 0.4 μg/100 g fresh weight in 8-h photoperiods to 4.9 in 16-h photoperiods. There was little or no GA activity in mature leaves. Considerable inhibitor activity occurred in methanol extracts of mature leaves, however, abscisic acid occurring at concentrations of 3.05 to 7.66 μg/100g fresh weight.

Investigations into the levels of auxins and cytokinins have not been reported, although Cumming (1959) has published evidence for the presence of auxins in stem tips of *Trifolium pratense.*

(A) Apical dominance

Effects of auxin on apical dominance in white clover have not been reported. Studies with other growth regulators, however, have shown that GA_3 at 500 mg/l, decreases the number of axillary buds which grow out into lateral branches (Fletcher and Martin 1962, Mange 1963), the number in GA-treated plants compared with controls being in the ratio of approximately 1:5. Abscisic acid, on the other hand, increased the number of lateral branches per plant, when applied through the roots or to the leaves at 10 mg/l, from 1.1 to 3.6 in cv. Tamar and from 7.4 to 9.2 in cv. Tammisto (Cohen and Dovrat 1976). The same authors reported CCC at concentrations of up to 2×10^3 mg/l to have no effect on branching.

(B) Plant size

The ultimate size of a plant organ is the product of the number of cells of which it is comprised and the average volume of those cells. Plant growth regulators influence both these parameters.

i Leaf area and thickness

Studies of the effects of growth regulators on leaf area in white clover have been restricted to the influence of applied gibberellins and inhibitors. The effects of auxins and cytokinins are undocumented.

Most reports of the effects of gibberellins indicate that GA_3 increases leaf area. Fletcher and Martin (1962) found leaf area of cv. Kent Wild White to be increased from the control level of 1.8 cm^2 to 3.5-5.4 cm^2 after spraying with GA_3. Gibberellin concentration had no significant effect over the range 25-100 mg/l, although their data indicate higher concentrations to be somewhat less effective than lower. The surface area of epidermal cells was uninfluenced by GA_3, the difference between treated and control leaves being wholly the result of an approximately three-fold difference in the number of epidermal cells. Ladino clover responded similarly.

These results are similar to those obtained by Mange (1963) who found leaflet length in plants from a wild population in France to range from 20 to 40 mm in GA_3-treated plants compared with 12-15 mm in controls. The length-breadth ratio was also affected, ranging from 1.5-1.8 in GA_3-treated plants to 1.0-1.3 in controls. Results reported by Cohen and Dovrat (1976) for plants grown in 8-h photoperiods agree only partly with these, GA_3 increasing lamina length in cv. Tamar from 28 mm in controls to 36 and 39 mm with 5 and 10mg GA_3/l; GA_3 had no effect on lamina length in cv. Tammisto at concentrations of up to 50mg/l. GA_{4+7} had no effect on leaf size in either cultivar. It is not possible to assess the extent to which differences in daylength might account for the differences in results obtained, as neither Fletcher and Martin nor Mange stated the time of year at which their experiments were performed.

Growth inhibitors are reported to have little effect on leaf expansion. ABA and CCC applied as foliar sprays or via the root system had no effect on leaf size at 1-20 mg/l (Cohen and Dovrat 1976); Calder et al. (1973) reported that Alar-85 (daminozide) sprays at concentrations of 2500-15 000 mg/l decreased leaf size and increased leaf thickness by 20%, largely as the result of an increase in the number of layers of palisade cells.

ii. Internode and petiole elongation

Elongation of internodes and petioles is controlled in large measure by their adjacent tissues (Sections 3.(B) and 3.(C)). Internode elongation is decreased by removal of apical tissues, leaves and basal stem tissues and roots; petiole elongation is likewise decreased by removal of apical tissues, leaf laminae, and basal stem and root tissues. Some indications of the possible nature of the regulatory compounds involved is provided by experiments with applied growth substances.

Elongation growth of both internodes and petioles in intact plants is strongly stimulated by application of GA_3 (Fletcher et al. 1958). Both Fletcher and Martin (1962) and Mange (1963) found petiole length to be approximately doubled by application of GA_3, irrespective of concentration between 25 and 1000 mg/l, the increase in length being the result of a doubling in cell number rather than an increase in cell length. Internode length was more than doubled by 500 mg GA_3/l (Mange, 1963). The lowest concentration found to produce marked stimulatory effects was 1 mg/l (Thomas 1972). A similar effect of GA_3 was shown for *Trifolium fragiferum*, internode length being approximately doubled by application at 1000 mg/l (Bendixen and Peterson, 1962).

Application of IAA to intact plants of white clover has little effect on internode length at concentrations of 10^{-2} and 10^{-6}M. A similar lack of effect at 10 mg/l was reported for *T. fragiferum* (Bendixen and Peterson 1962) although in this case higher concentrations (100 and 1000mg/l) were somewhat inhibitory, as was 10 mg/l of the anti-auxin TIBA. As in several other much studied systems, such as pea stem segments and cereal coleoptiles, IAA has a stimulatory effect on elongation of excised segments of young internodes of white clover, the optimum concentration for elongation being 10^{-4} M (T).

The possible role of auxins and gibberellins in controlling internode elongation in intact plants was tested by comparing growth of intact stolons with that of stolons decapitated above a rapidly elongating internode approximately 2 mm long (Thomas 1972). Decapitated stolons were then treated with 10^{-4}M IAA and/or 1 mg GA_3/l incorporated into small agar blocks which were applied to the cut surfaces. In this system both IAA and GA_3 increased internode elongation at least twofold compared with decapitated controls. IAA was more effective than GA_3 during the first 24 hours, GA_3 becoming relatively more effective on the third day after decapitation.

These results suggest that internode elongation might be promoted in intact plants by auxins and gibberellins diffusing basipetally from apical buds. That gibberellins are present in white clover apices has been shown by Cohen and Dovrat (1976); the presence of auxins in apical buds has not been demonstrated but it is well documented for several other species, including *Trifolium pratense* (Cumming 1959).

The nature of the regulatory influence of leaves and basal tissues on the elongation of young petioles and internodes remains unknown. In the absence of basal tissues, neither benzyladenine at concentrations up to 1 mg/l nor GA_3 at 1 or 100 mg/l had a stimulatory effect on elongation (Thomas 1972).

Effects of other growth regulators on elongation have been little studied. Alar-85 applied at concentrations of 2500 to 15 000 mg/l is reported to decrease internode elongation (Calder et al. 1973); CCC has apparently no effect on the growth rate of stolons but at 2000 mg/l it is phytotoxic (Cohen and Dovrat 1976).

iii. Root:shoot ratios

Effects of growth regulators on root growth have been little studied in white clover. Fletcher and Martin (1962) compared dry weights of shoots and roots in seven cultivars treated and untreated with 500 mg GA_3/l. Both were about double in the controls what they were in the treated plants. Root-shoot ratios were reduced from 0.208, 0.190 and 0.138 in control plants of New Zealand White, Ladino and Kent Wild White to 0.186, 0.138 and 0.114 respectively in GA_3 treated plants.

(C) Plant anatomy

Studies of the effects of growth regulators on anatomy have been made in intact plants by Mange (1963) and Calder et al. (1973) and in excised plant pieces by Thomas (1972). The results of these broadly agree but are unexpectedly conflicting in some respects.

The effects of GA_3 on stem anatomy have been studied by Mange and Thomas. Mange found that the erect stems developing after treatment with GA_3 were radially symmetrical, in contrast to the dorsiventral symmetry observed in horizontal controls. 500 mg GA_3/l prevented the formation of a periderm and increased the degree of lignification, particularly in interfascicular parenchyma cells. Interfascicular cambium development was delayed in gibberellin-treated stems.

In a rather differently designed experiment in which excised stems were treated with GA_3 supplied through the cut basal end, Thomas (1972) similarly found that GA_3 increased lignification of interfascicular parenchyma cells but that the inhibitory effect on interfascicular cambium development described by Mange was not apparent. In these experiments excision of the apical region of the stem immediately basal to the first fully elongated internode led to a marked subsequent reduction in the development of gelatinous fibres and lignification of interfascicular parenchyma cells, and a slight reduction in the development of interfascicular cambium in the elongated internode. GA_3 brought the development of gelatinous fibres and lignification of interfascicular parenchyma cells close to the level found in intact control plants, but had no effect on the development of interfascicular cambium.

In the same experiment Thomas found benzyladenine supplied to excised stolon tips via their cut bases at concentrations below 0.1 mg/l had little effect, but at 0.1 and 1.0 mg/l benzyladenine inhibited fibre wall thickening and lignification and promoted the division of cambium cells. The stimulatory effect of 1 mg GA_3/l on fibre thickening and lignification was prevented by simultaneous treatment with 0.1 mg benzyladenine/l.

Thomas concluded from these experiments that the basal tissues necessary for both fibre development and maximum cambial activity might act by providing a combination of cytokinins and gibberellins to the growing region of the stolons. No clear explanation of the difference in response to gibberellins found by Thomas and Mange is evident. In the system used by Thomas, however, growth would have been limited by the lack of all basally supplied regulators, whereas in the intact plants used by Mange regulators supplied by the roots and stem bases would have been present in non-limiting quantities.

The effect of growth retardants might be expected to be opposite to that of gibberellins. Calder et al. (1973) found that the development of interfascicular cambium and the degree of lignification of cells in stolons of intact plants were both markedly reduced by Alar-85 sprays at 7500 mg/l. The effect on lignification in this case is the opposite to that of GA_3. The expectation that Alar and GA_3 have opposite effects is not supported by the reported observation by Calder et al. that Alar stimulated the development of phloem fibres, Thomas (1972) having reported that development of these fibres (in excised stolons) was enhanced by GA_3. It should be noted here, however, that Calder's paper contains errors (Calder, personal communication): Section 5 on his page 275 refers to samples collected from clover plants (not alfalfa) and the top and bottom photomicrographs in his Figure 6 have been transposed (the bottom section being untreated and the top Alar-treated). This figure does not provide strong support for the statement that Alar treatment 'appeared to stimulate the development of phloem fibres.'

5. Movement

White clover exhibits a wide range of movements in response to its environment, but these have been surprisingly little studied in any detail.

(A) Tropic movements

Except where otherwise indicated, the following brief account is based entirely upon observations and unpublished results of experiments by the present writer.

The basic growth form of a white clover plant is the result of tropic responses to gravity. When plants are placed in total darkness, petioles, peduncles and stolon tips all grow vertically upwards. The negatively geotropic response of petioles and peduncles is quite rapid. Excised petioles from young leaves with opened leaflets often bend upwards through 45° within 3 hours of being placed horizontally in a humid atmosphere in the dark and significant bending can be seen within 90 minutes. The response is not markedly different if the leaflets of the lamina are still attached. As with similar organs of other plants, bending growth occurs most strongly in the region of most rapid cell elongation in the zone immediately below the point of attachment of the leaflets.

The bending response of stolons similarly occurs in the region of most rapid cell expansion (i.e., the internodes either side of the youngest leaf with unfolded leaflets - see Figure 2.15), but it is very much slower than in petioles. No bending was seen in excised internodes kept for up to 3 days in a chamber in dark, humid conditions; attached internodes bent upwards only about 15° on average during that time. Young petioles, peduncles and stolons all have equally well developed incomplete starch sheaths, however, the cells of which contain movable starch grains which might act as statoliths (e.g. Chapter 1, Section 2.(B) and Chapter 2, Section 2.(C)ii.).

The degree of bending of stolon tips varies greatly. When stolon tips of plants otherwise maintained in the light are placed in darkness for several days, they bend upwards very much more and become erect or nearly so, whereas stolons of plants placed wholly in the dark bend upwards much less. Internode elongation at stolon tips is very much less in completely darkened plants than at the darkened stolon tips of plants otherwise kept in the light. The relationship between growth, bending and nutrient availability warrants further investigation.

Geotropic responses of petioles, peduncles and stolons are modified by light. In pot-grown plants of white clover, stolons growing out over the edges are typically approximately horizontal in the light. At high light intensities (such as full summer sun) the stolons grow downwards often at an angle of about 45°. That this downward bending is not caused by the weight of the stolon is seen when such plants are transferred to low light intensities, whereupon the stolon tips start to grow upwards. This modifying effect of light intensity on the response to gravity is demonstrated by the results of an experiment in which stolon angle above the horizontal was found to average -7.0, + 16.5, +45.5 and +41.9° after 14 days growth in controlled environment cabinets at light intensities of 20 000, 3800, 1800 and 450 lux, respectively.

In a preliminary study of the action spectrum of the light response, excised petioles were found to bend upwards equally (about 38° in 3 hours) from their original horizontal position when placed in darkness or exposed to approximately 50 $\mu Em^{-2}sec^{-1}$ of red or blue light. At an equal intensity of light at wavelengths greater than 700 nm (far-red light), however, petioles bent upwards 57° in 3 hours. That far red light stimulated geotropic bending suggests, but by no means proves, that phytochrome might be involved. The optimum wavelength for inhibition of the geotropic response was not determined.

The tropic responses of white clover petioles, peduncles and stolons are all similar, varying only in degree. All three are negatively geotropic and in all three the strength of the geotropic response decreases as light intensity increases. At high intensities in the natural environment stolons grow closely adpressed to the ground and petioles tend to grow horizontally, especially near their bases. At low light intensities (and in tall pasture where, additionally, far-red light predominates) stolons tend to grow upwards and petioles and peduncles to grow vertically.

The involvement of plant growth regulators in these responses is suggested by the stimulatory effect of gibberellic acid on negatively geotropic growth. Stolons of light-grown plants grow vertically upwards rapidly in response to treatment with GA_3 (Fletcher & Martin 1962). A clear growth response to GA_3 is noticeable within 2 days of treatment. On plants of several different cultivars growing at full natural light intensity in spring in Palmerston North, New Zealand, stolons sprayed with 10 mg GA_3/l grew vertically upwards to a height of 30 cm, though such stolons were weak and tended to fall over well before they attained this height. At a low artificial light intensity (2500 lux) stolons treated with 500 mg GA_3/l grew to a height of at least 50 cm (Mange 1963).

As would be predicted, growth retardants have the opposite effect to gibberellins. The negatively geotropic tendencies of Ladino clover petioles were decreased by application of Alar-85 (2500-15 000 mg/l) which caused the plants to grow extremely close to the ground and so give 'a very flattened appearance' (Calder et al. 1973).

The response of stolons to GA_3 is similar to, but far greater than, their response to darkness, both with regard to their vertical elongation and the repression of axillary buds (see Section 4.(A)). Leaves open normally in response to GA_3, however, whereas their leaflets remain closed in darkness.

The relationship between GA_3, light intensity and the geotropic response should be studied more fully.

Effects of other growth regulators on geotropic responses in white clover have not been published. Preliminary investigations indicate that, as reported for *Trifolium fragiferum* (Bendixen & Peterson 1962), neither IAA nor TIBA has any effect on the geotropic response of stolons. Naphthaleneacetic acid, similarly, appears to have no effect.

In addition to movements in response to gravity, it is commonly observed and reported that leaves of white clover plants 'follow the sun' (e.g. Brougham 1958) by orienting the plane of their laminae at right angles to the source of brightest light in the sky. For plants growing in unilateral shade (such as under a tree) leaf laminae are permanently oriented at right angles to the patch of sky from which radiation is received. Investigations into the mechanism of this response have not been reported.

(B) Circumnutation and nastic movements

Darwin (1880) described circumnutation of the pedicels at the base of florets and in the petiolules at the base of leaflets of white clover, but further studies of these movements have not been reported. It seems not unlikely that the 'sun-following' movements of leaf laminae are, at least in part, caused by circumnutation of the uppermost ends of the petioles.

Darwin (loc. cit.) also described his studies of nastic movements of pedicels and leaflets. As each floret within an inflorescence withers, its pedicel bends downwards so that it becomes parallel to the inflorescence axis. It was shown experimentally that this bending is an epinastic movement, occurring despite the reversed influence of gravity when flower heads were inverted in the dark. He also found a positively phototropic response of pedicels which prevents them from bending downwards when they are young.

The most widely cited of Darwin's observations relate to nyctinastic ('sleep') movements of leaflets. He described this movement as follows:

'If we select a leaf of *T. repens* having an upright petiole, and with the three leaflets expanded horizontally, the two lateral leaflets will be seen in the evening to twist and approach each other, until their upper surfaces come in contact. At the same time they bend downwards in a plane at right angles to that of their former position, until their midribs form an angle of about 45° with the upper part of the petiole. This peculiar change of position requires a considerable amount of torsion in the pulvinus. The terminal leaflet merely rises up without any twisting, and bends over until it rests on and forms a roof over the edges of the now vertical and united lateral leaflets. Thus the terminal leaflet always passes through an angle of at least 90°, generally of 130° or 140°.'

Of the first foliage leaf of a seedling of white clover, which is usually unifoliolate, Darwin wrote: 'It always rises, and either stands vertically at night or more commonly bends a little over so as to expose the lower surface obliquely to the zenith, in the same manner as does the terminal leaflet of the mature [later-formed] leaf.'

The effect of the environment on opening and closing of white clover leaves has been studied more recently by the present writer (T) and by Scott & Gulline. Leaves transferred from 24-h cycles consisting of 12 h light alternating with 12 h dark (12L:12D) to continuous darkness or to continuous light at intensities between 120 and 4000 lux continue to close and open at approximately the same times as control leaves still receiving 12L:12D cycles (T). In continuous darkness at constant temperature and humidity, opening and closing continues with a periodicity of about 24 h for 5 or more days (Scott & Gulline 1972). At a light intensity of 120 lux, the rhythm of opening and closing is maintained for slightly longer, but at higher light intensities movement stops after about 3 days or less (T). The existence of such a free-running circadian endogenous rhythm suggests that the movement is under the control of a basic oscillator or biological clock.

After rhythmic movement has stopped in continous light or dark, it can be restarted by exposure of the leaves to dark or light, respectively. In a plant in which rhythmic movements are still occurring, the phase of the rhythm can be reset both by changes from light to dark, and vice versa, and,if the plant is in continuous light, by changes in temperature (e.g. from 21° to 8.5°C, Scott & Gulline 1972). The time of leaf closure is brought forward by about 2 h when plants previously grown in 12L:12D cycles are placed in darkness after only 4 h light (T), and interruption of a 16-h dark period by a 90-minute light break after 6.5 h delays the opening of leaves at the end of the dark period by about 3 h (T).

The regulation of leaf opening and closing once a rhythm has been set in motion is predominantly under the control of internal rather than environmental factors. Over the temperature range 8.5 to 21.5°C, the period of oscillation increased only from 25.3 to 27.5 h in continuous light and from 23.2 to 26.7 h in continuous darkness (Scott & Gulline 1972), and substitution of darkness by light of intensities ranging from 4000 to 120 lux had no significant effect on the time at which leaves opened in plants growing in 12L:12D cycles (T).

Outside the range 8-22°C, however, leaf movement was described by Scott & Gulline (1972) as 'sluggish and aperiodic' and results of other experiments show that extreme diurnal fluctuations in temperature can also give rise to anomalous effects. For example, leaves were found to open fully (and 2 h earlier than at 21°C) in continous darkness at 7°C following a 12 h light period at 35°C, whereas after a similar 12 h light period at 28°C they remained closed during 24 h darkness at 7°C (T). The extent to which leaf movement is controlled by environmental rather than internal factors in naturally cold winter or hot summer conditions has not been investigated.

The nyctinastic movement of white clover leaflets is caused by changes in turgidity of the cells on opposite sides of the pulvinules (Chapter 1, Sections 1.(C) and 2.(E)) at the bases of the leaflets. Swelling of cells on the abaxial sides of the pulvinules relative to that on the adaxial sides causes leaves to close, the reverse process causing them to open. Changes in turgidity continue in excised leaves and excised pulvinules in the same way as in intact plants (Scott & Gulline 1972). Conspicuous transverse furrows on both the abaxial and adaxial sides (Figure 1.3a) are probably causally related to movement of pulvinules.

Within pulvinules, changes in turgidity are correlated with changes in potassium ion concentration and membrane potential (Scott & Gulline 1975; Scott et al. 1977). During the day the potassium ion concentration is higher in the adaxial (turgid) cells than in the abaxial cells. At night the situation is reversed: the change in potassium ion content takes place about 3 h before the change in turgidity, and membrane potential starts to change several hours before this.

More recently, a century after Darwin's work, Bollig et al. (1978) have used white clover pulvinules in fundamental investigations of the cause of circadian rhythms.

Whether the sleep movements of leaves have an adaptive significance has been, and still is, the cause of considerable speculation. Darwin (1880)'exposed on two occasions during the summer to a clear sky several pinned-open leaflets of *Trifolium pratense,* which naturally rise at night, and looked at them early on several successive mornings, after they had assumed their diurnal positions. The difference in the amount of dew on the pinned-open leaflets and on those which had gone to sleep was generally conspicuous; the latter being sometimes absolutely dry, whilst the leaflets which had been horizontal were coated with large beads of dew. This shows how much cooler the leaflets fully exposed to the zenith must have become, than those which stood almost vertically, either upwards or downwards, during the night.' The leaves of white clover are presumably affected similarly. The extent to which Darwin's observations might have been brought about by differences in wettability of upper and lower leaf surfaces (Chapter 1, Section 2.(A)) rather than differences in temperature remains undetermined.

References

Bendixen, L.E.; Peterson, M.L. (1962) Physiological nature of gene-controlled growth form in *Trifolium fragiferum* L. II. Auxin-gibberellin relationships to growth form. *Plant Physiology* **37**, 245-250.

Bollig, I.; Mayer, K.; Mayer, W.E.; Engelmann, W. (1978) Effects of cAMP, theophylline, imidazole and 4-(3,4-dimethoxybenzyl)-2-imidazolidone on the leaf movement rhythm of *Trifolium repens* - a test of the cAMP hypothesis of circadian rhythms. *Planta* **141**, 225-230.

Brougham, R.W. (1958) Interception of light by the foliage of pure and mixed stands of pasture plants. *Australian Journal of Agricultural Research* **9**, 39-52.

Calder, F.W.; Canham, W.D.; Fensom, D.S. (1973) Some effects of Alar-85 on the physiology of alfalfa and Ladino clover. *Canadian Journal of Plant Science,* **53**, 269-278.

Carlson, G.E. (1966a) Growth of clover leaves - developmental morphology and parameters at ten stages. *Crop Science* **6**, 293-294.

Carlson, G.E. (1966b) Growth of white clover leaves, after leaf removal. *Proceedings of Xth International Grassland Congress* 134-136.

Cohen, Y.; Dovrat, A. (1976) Gibberellins and inhibitors in relation to flowering of white clover (*Trifolium repens* L.) *Journal of Experimental Botany* **27**, 817-826.

Cumming, B.G. (1959) The control of growth and development in red clover (*Trifolium pratense* L.) III Endogenous diffusible auxin. *Canadian Journal of Botany* **37**, 1049-1062.

Darwin, C. assisted by F. Darwin (1880) The power of movement in plants.

Denne, M.P. (1966a) Morphological changes in the shoot apex of *Trifolium repens* L. I. Changes in the vegetative apex during the plastochron. *New Zealand Journal of Botany* **4**, 300-314.

Denne, M.P. (1966b) Morphological changes in the shoot apex of *Trifolium repens* L. 2. Diurnal changes in the vegetative apex. *New Zealand Journal of Botany* **4**, 434-443.

Denne, M.P. (1966c) Leaf development in *Trifolium repens. Botanical Gazette* **127**, 202-210.

Erith, A.G. (1924) White clover (*Trifolium repens* L.). London, UK; Duckworth.

Evans, P.S. (1973) The effect of repeated defoliation to three different levels on root growth of five pasture species. *New Zealand Journal of Agricultural Research* **16**, 31-34.

Fawkner, J.J. (1967) A study of the development of the petiole of *Trifolium repens* L. and its morphological response to light quality. *M. Agr. Sc. Thesis, Massey University, New Zealand.*

Fletcher, W.W.; Alcorn, J.W.S.; Raymond, J.C. (1958) Effect of gibberellic acid on the nodulation of white clover (*Trifolium repens* L.) *Nature, UK* **182**, 1319-1320.

Fletcher, W.W.; Martin, D.J. (1962) The effect of gibberellic acid on the germination, growth and development of *Trifolium repens* L. *Journal of Agricultural Science, UK* **58**, 235-241.

Gausman, H.W.; Fuelleman, R.F. (1952) Distinguishing Ladino clover from white clover. *Agronomy Journal* **44**, 96-97.

Gibson, P.B.; Schultz, E.F. (1953) The effect of environment on the number of vascular bundles per petiole in white clover. *Agronomy Journal* **45**, 123-124.

Hall, D.M.; Donaldson, L.A. (1963) The ultrastructure of wax deposits on plant leaf surfaces.I. Growth of wax on leaves of *Trifolium repens. Journal of Ultrastructural Research* **9**, 259-267.

Hall, D.M.; Jones, R.L. (1961) Physiological significance of surface wax on leaves. *Nature, UK* **191**, 95.

Harvey, H.J. (1970) Patterns of assimilate translocation in *Trifolium repens.* In: *White Clover Research.* Ed. J. Lowe. *Occasional Symposium No. 6. British Grassland Society* pp. 181-186.

Hay, M.J.M.; Dunlop, J.; Hopcroft, D.H. (1982) Anatomy and development of the superficial layers in stolons of white clover (*Trifolium repens* L.). *New Zealand Journal of Botany* **20**, 315-324.

Hoshino, M.; Nishimura, S.; Okubo, T. (1965) Studies on the assimilation and translocation of $^{14}CO_2$ in Ladino clover. *Proceedings of the Crop Science Society of Japan* **33**, 130-134.

Kawanabe, S.; Yoshihara, K.; Okada, T.; Ueno, M.; Hidaka, M. (1962) Studies of summer depression of pasture crops. 2. Diagnosis of summer depression of Ladino clover by means of the internode length and thickness of stolons. *Journal of the Japanese Society of Grassland Science* **8**, 211-220.

Mange, M. (1963) Action de l'acide gibbérellique sur *Trifolium repens* L.: observations sur quelques modifications morphologiques, physiologiques et anatomiques. *Annales Scientifiques de l'Université de Besançon, 2e Série, Botanique* **19**, 43-53.

Rentschler, I. (1971) Die Wasserbenetzbarkeit von Blattoberflächen und ihre submikroskopische Wachsstruktur. *Planta* **96**, 119-135.

Scott, B.I.H.; Gulline, H.F. (1972) Natural and forced circadian oscillations in the leaf of *Trifolium repens*. *Australian Journal of Biological Science* **25**, 61-76.

Scott, B.I.H.; Gulline, H.F. (1975) Membrane changes in a circadian system. *Nature, UK* **254**, 69-70.

Scott, B.I.H.,; Gulline, H.F.; Robinson, G.R. (1977) Circadian electrochemical changes in the pulvinules of *Trifolium repens* L. *Australian Journal of Plant Physiology* **4**, 193-206.

Thomas, R.G. (1961) The relationship between leaf growth and induction of flowering in long-day plants (LDP). *Die Naturwissenschaften* **48**, 108

Thomas, R.G. (1962) The initiation and growth of axillary bud primordia in relation to flowering in *Trifolium repens* L. *Annals of Botany,* **26**, 329-344.

Thomas, R.G. (1972) The role of basal and apical factors in the coordination of growth in the stems of white clover (*Trifolium repens* L.). In: *Plant Growth Substances* 1970. Ed. D.J.Carr. Berlin, W. Germany; Springer Verlag 624-632.

Thomas, R.G. (1979) Inflorescence initiation in *Trifolium repens* L.: influence of natural photoperiods and temperatures. *New Zealand Journal of Botany* **17**, 287-299.

Thomas, R.G. (1980) Flowering responses of white clover (*Trifolium repens* L.) growing in arctic Norway. *Acta Agriculturae Scandinavica* **30**, 51-58.

Ueno, M.; Tsuchiya, S. (1968) Rooting habit of adventitious roots in white clover. *Journal of the Japanese Society of Grassland Science* **14**, 116-118.

Wickson, M.; Thimann, K.V. (1958) The antagonism of auxins and kinetin in apical dominance. *Physiologia Plantarum* **11**, 62-74.

Williams, R.F.; Bouma, D. (1970) The physiology of growth in subterranean clover. I. Seedling growth and the pattern of growth at the shoot apex. *Australian Journal of Botany* **18**, 127-148.

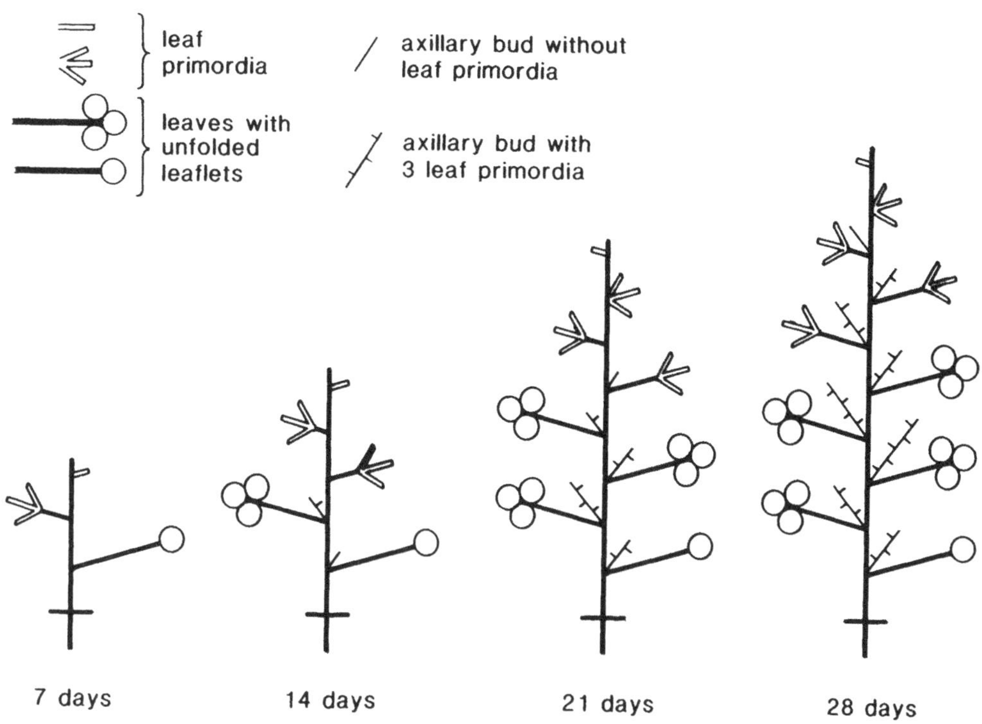

Fig. 2.1 Stylized diagrams showing the development of seedlings of cv. Grasslands Huia dissected 7, 14, 21 and 28 days after seeds were sown. Seedlings were grown in 16-h photoperiods in a warm glasshouse. During the period of growth, internodes on the main stem and lateral stolons remained short. [T].

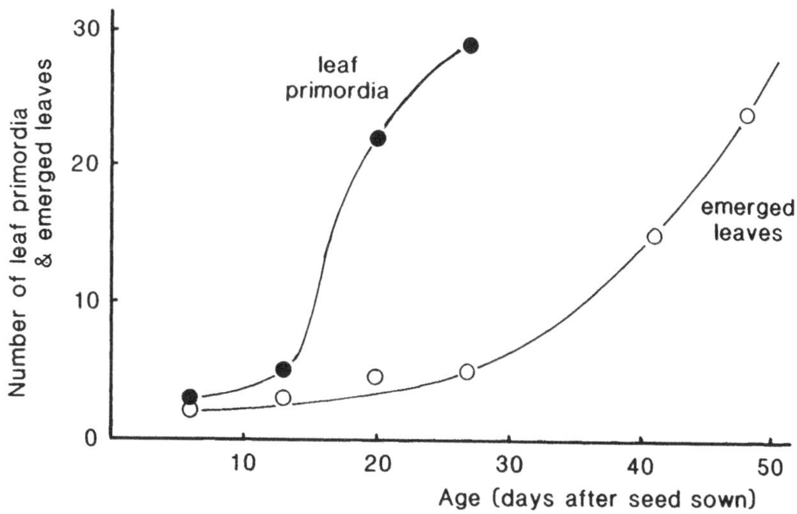

Fig. 2.2 Increase in number of leaf primordia and emerged leaves on seedlings of cv. Grasslands Huia grown in continuous light in a warm glasshouse [T].

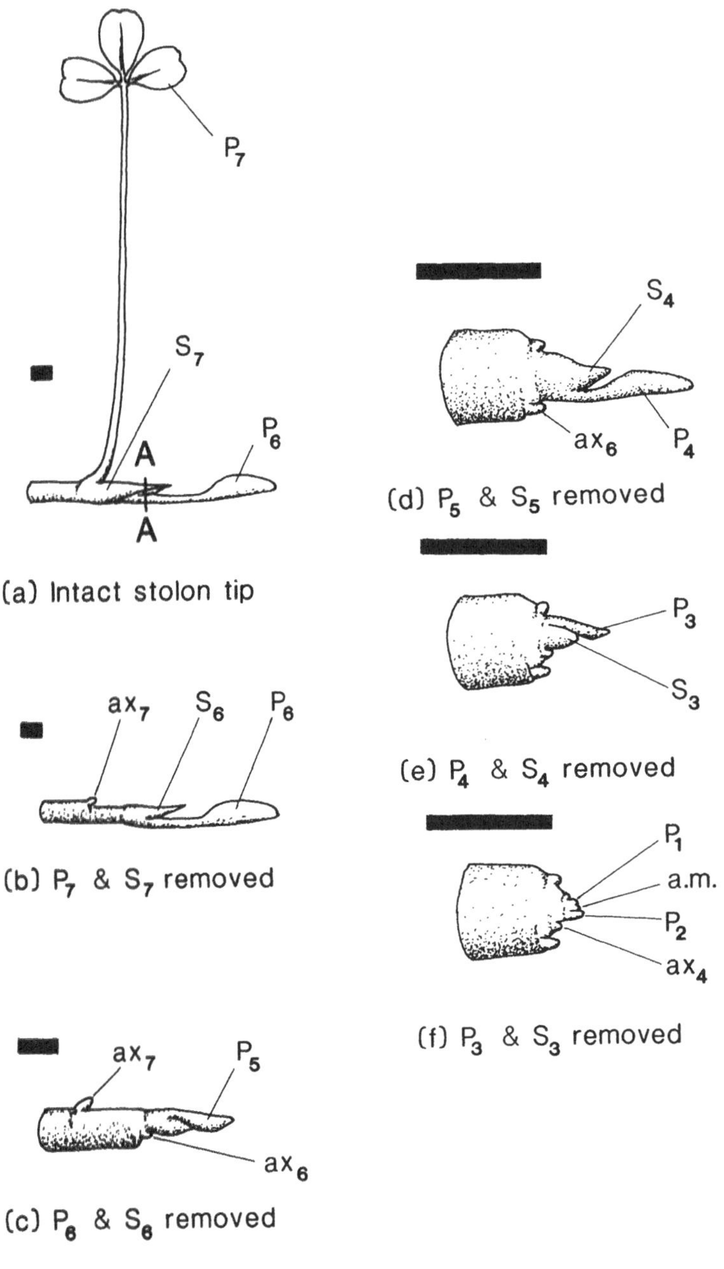

Fig. 2.3 Structure of the apical bud of a stolon as revealed by successive stages of dissection. Leaf primordia and leaves present in the apical bud of the stolon tip in (a) are labelled P_1-P_7 in sequence from youngest to oldest, their axillary buds are labelled ax_1-ax_7 respectively, their stipules S_1-S_7, and the stem apical meristem a.m. The scale is indicated by horizontal bars, each of which represents 4 mm. [Redrawn from Thomas 1980].

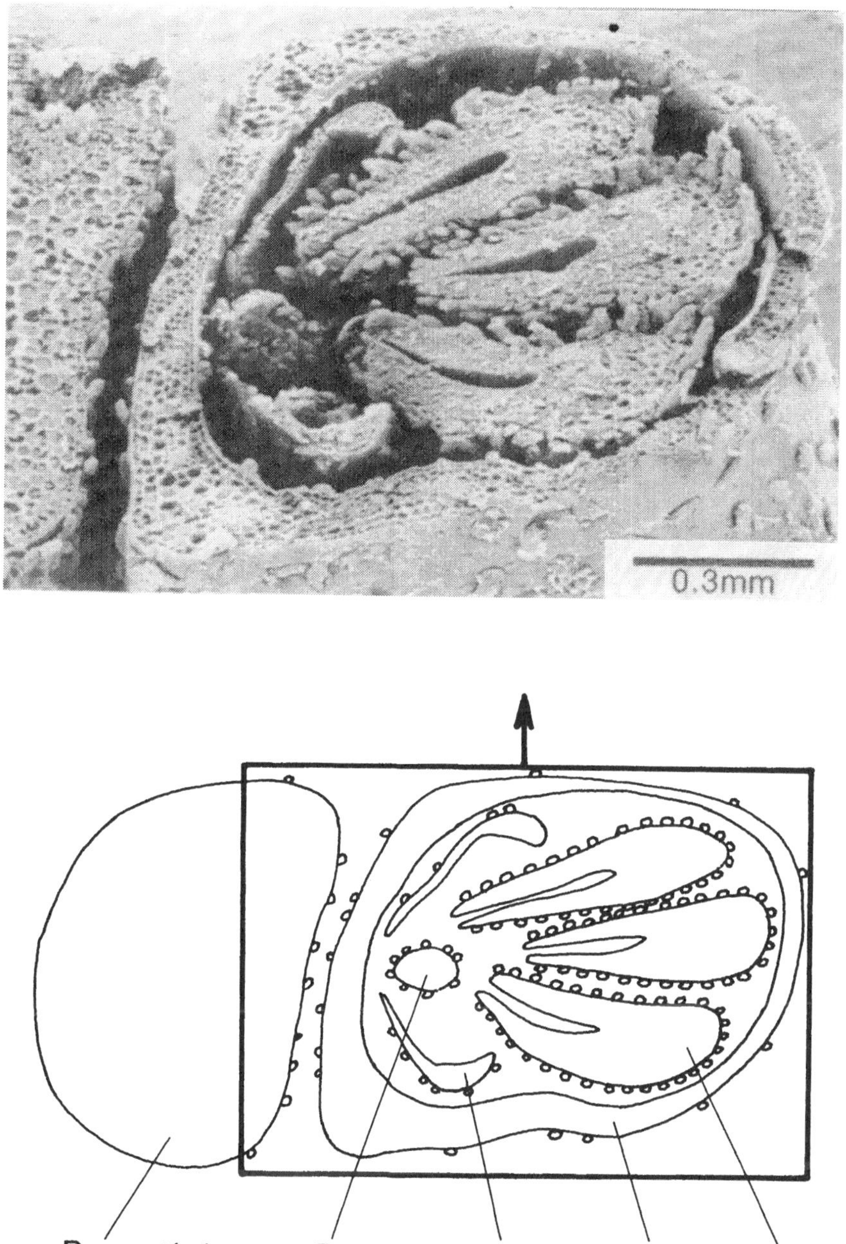

Fig. 2.4 Scanning electron micrograph to show leaf vernation in an apical bud. The photograph shows the internal morphology of the apical bud revealed by cutting along the line A-A in Fig. 2.3(a) and discarding the distal tissue. Labelling is as for Fig. 2.3 [D.H.H.]

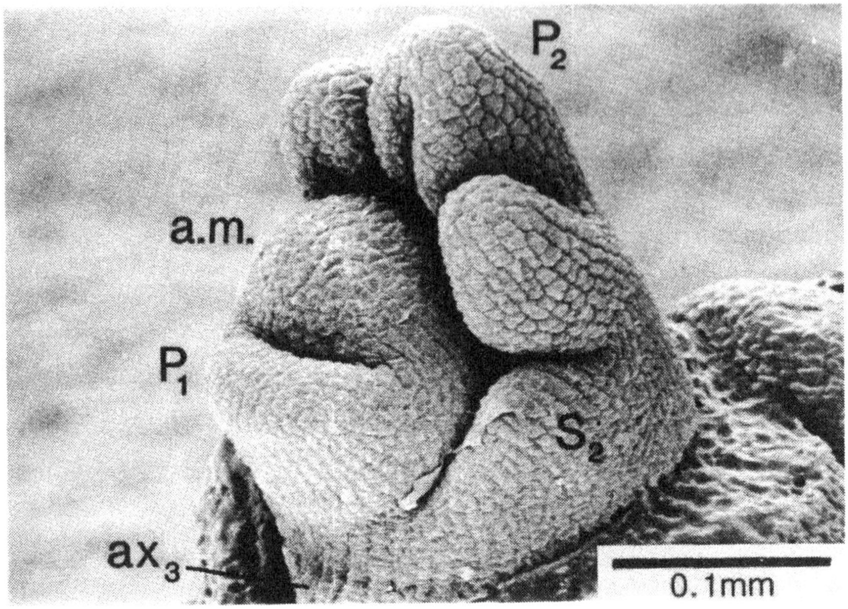

Fig. 2.5 Scanning electron micrograph of the apical meristem and adjacent tissues at a stolon tip. P_1 and P_2 are the two youngest leaf primordia, S_2 the stipule of P_2, ax_3 the bud in the axil of leaf primordium P_3 (removed), and a.m. the apical meristem [D.H.H.]

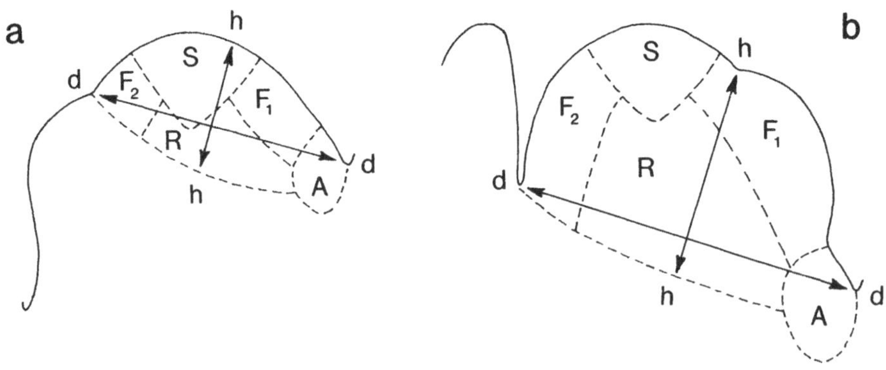

Fig. 2.6 Diagrams of median longitudinal sections of stem apices showing the apical regions at the beginning of a plastochron (a) and just after the end of a plastochron (b). d-d is the basal diameter and h-h the apex height. [Redrawn from Denne 1966a].

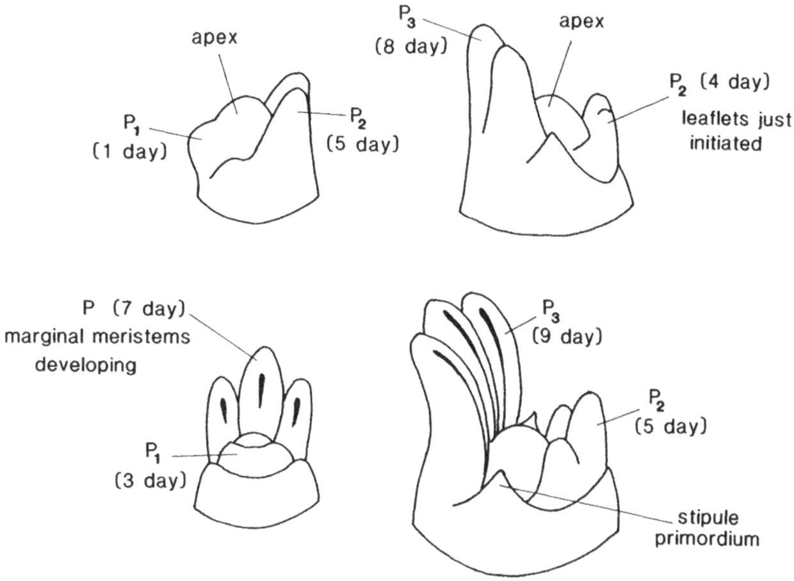

Fig. 2.7 External features of developing leaf primordia (P_1-P_3). [Redrawn from Denne 1966c].

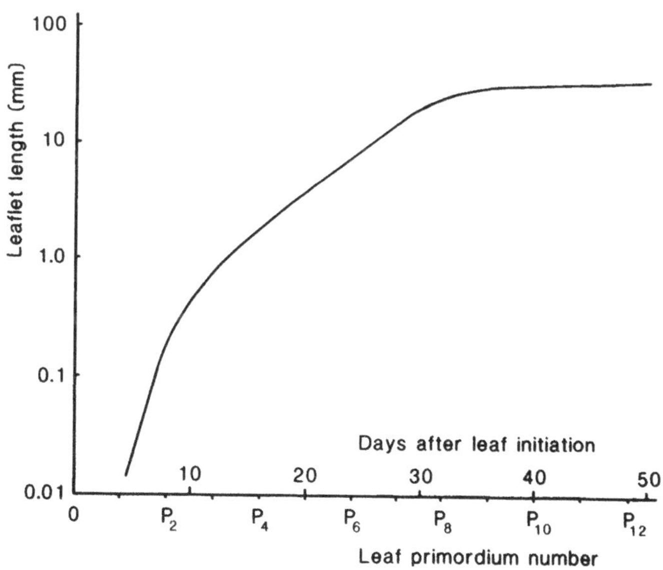

Fig. 2.8 Rate of leaflet elongation within the apical bud. See text for growing conditions. [Redrawn from Denne 1966c].

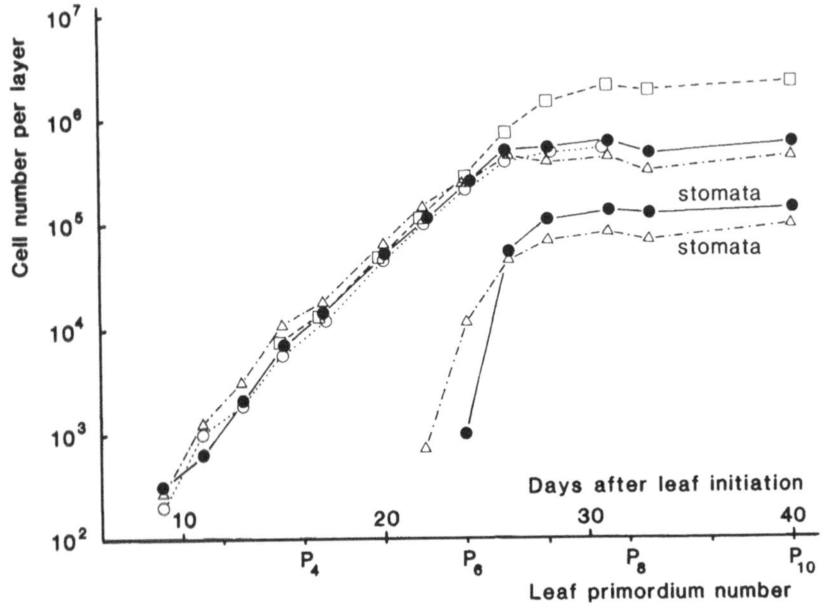

Fig. 2.9 Mean cell number per layer in developing leaflets. Numbers of stomata (including stomatal initials) are not included in epidermal cell number but shown separately. ●——●, upper epidermis; □-------□, palisade mesophyll; ○·······○, spongy mesophyll; △-·-·-·-·-△, lower epidermis. [Redrawn from Denne 1966c].

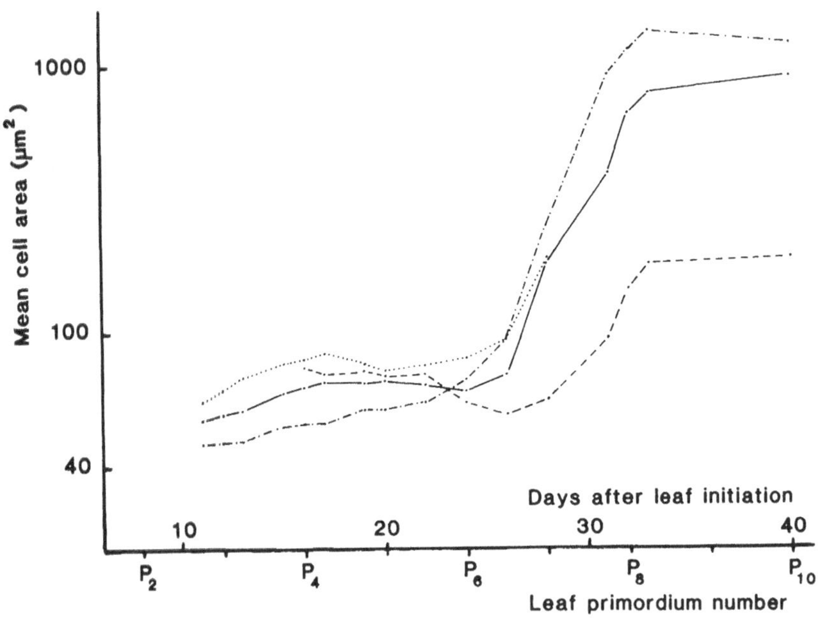

Fig. 2.10 Mean cell area in developing leaflets. ·——·, upper epidermis; ·------·, palisade mesophyll; ·······, spongy mesophyll; ·-·-·-·-·, lower epidermis. [Redrawn from Denne 1966c].

a

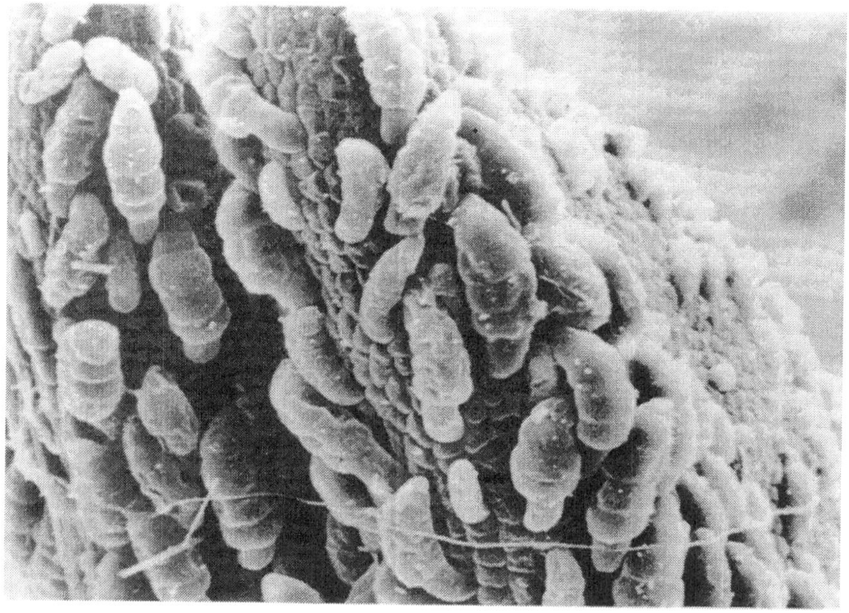

b

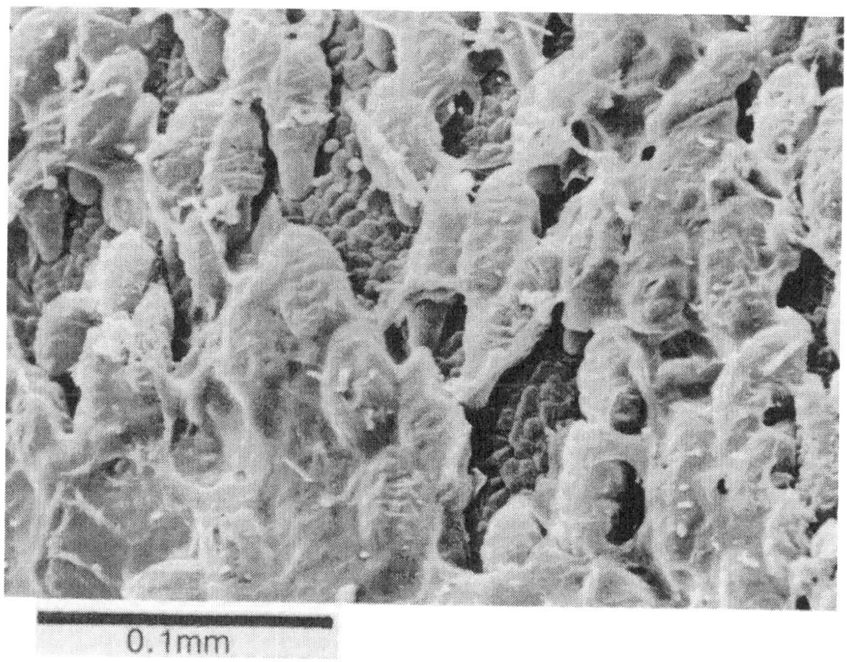

Fig. 2.11 Scanning electron micrographs of hairs on the surface of leaf primordium P_4. (a) Multicellular hairs at various stages of development in the midrib region near leaflet apices. (b) Multicellular hairs overlain by mucilage (?-see text) on the abaxial leaflet surface. No stomata are yet visible on the epidermis [D.H.H.]

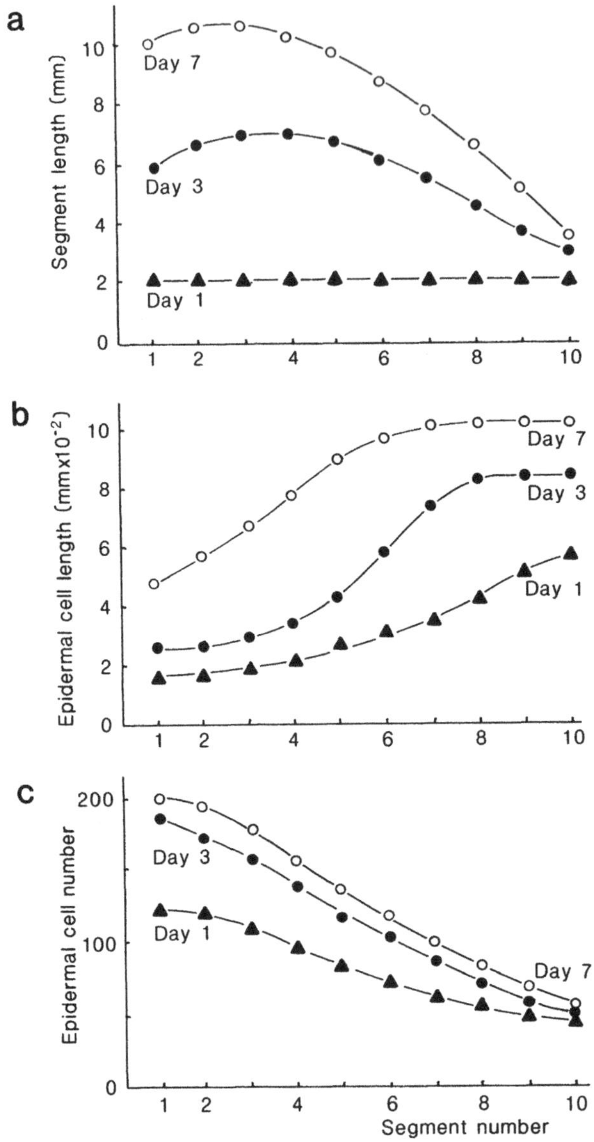

Fig. 2.12 Growth of petioles. Petioles 20 mm long were marked off into 2 mm segments by Indian ink and the segments numbered from 1 (apical, youngest) to 10 (basal, oldest). Length (a), mean epidermal cell length (b), and number of epidermal cells per segment (c) were recorded over a 7-day period for each segment. [Redrawn from Fawkner 1967].

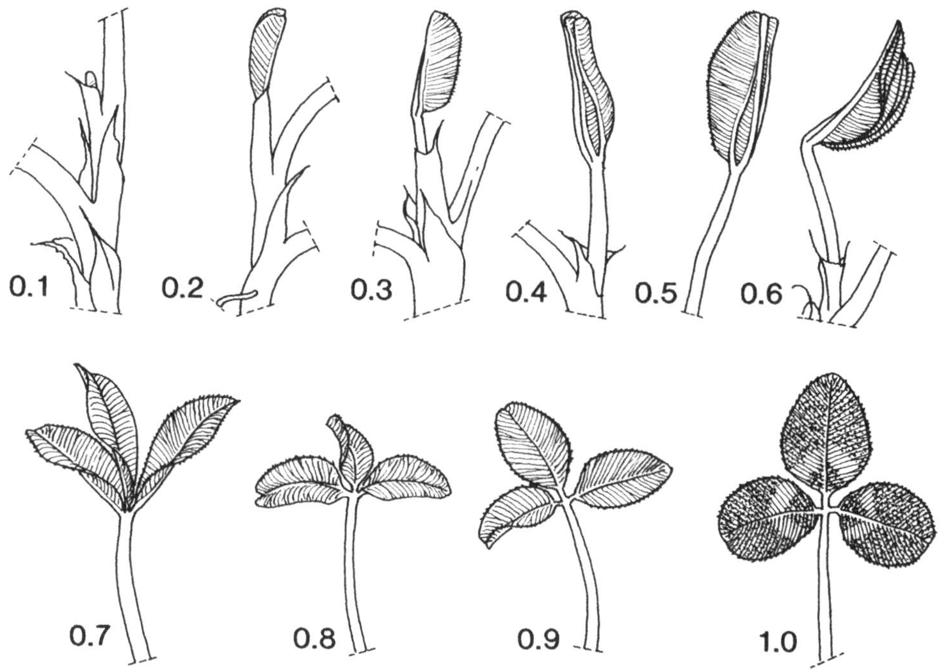

Fig. 2.13 Ten stages of morphological development of Ladino clover leaves. [Redrawn from Carlson 1966a].

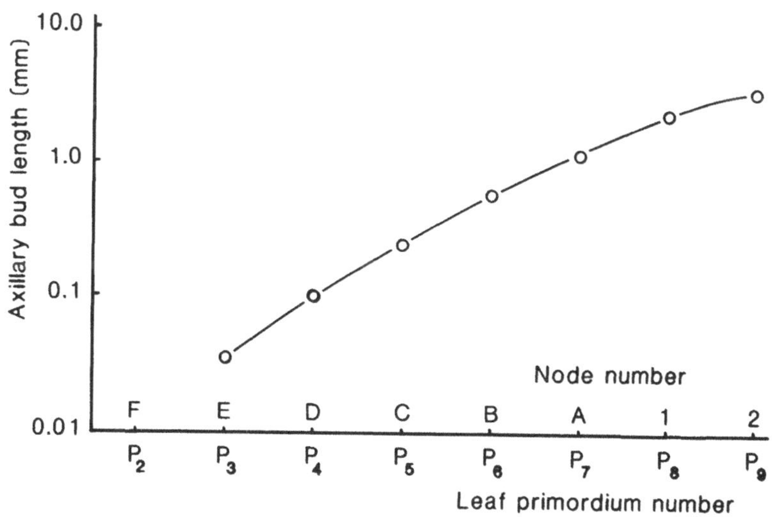

Fig. 2.14 The length of axillary bud primordia at consecutive nodes in plants of cv. Grasslands Huia growing in warm short days. [Redrawn from Thomas 1962].

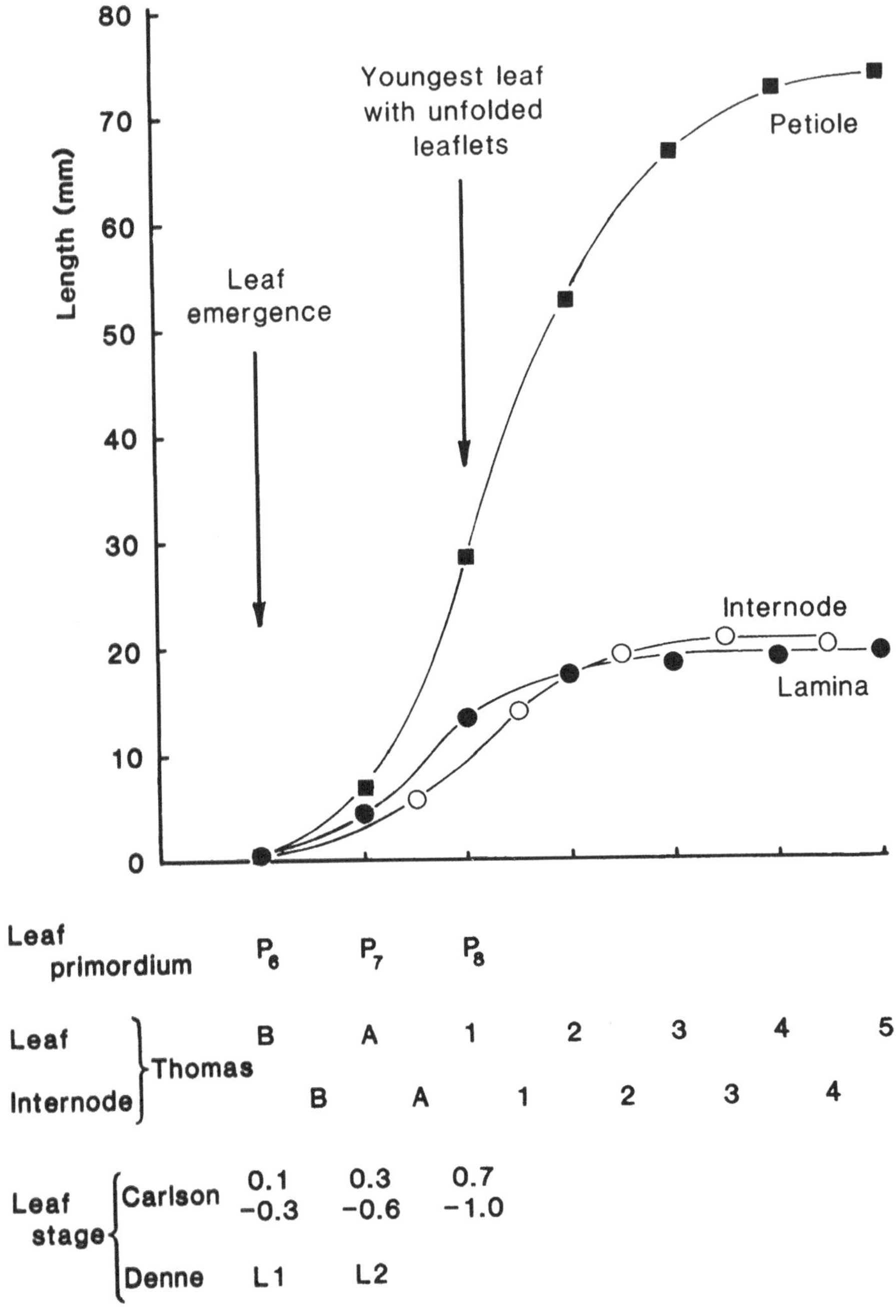

Fig. 2.15 Growth of leaves and internodes from the time of leaf emergence onwards, showing the relationships between the numbering systems used by Carlson (1966a), Denne (1966b), and Thomas (this publication)

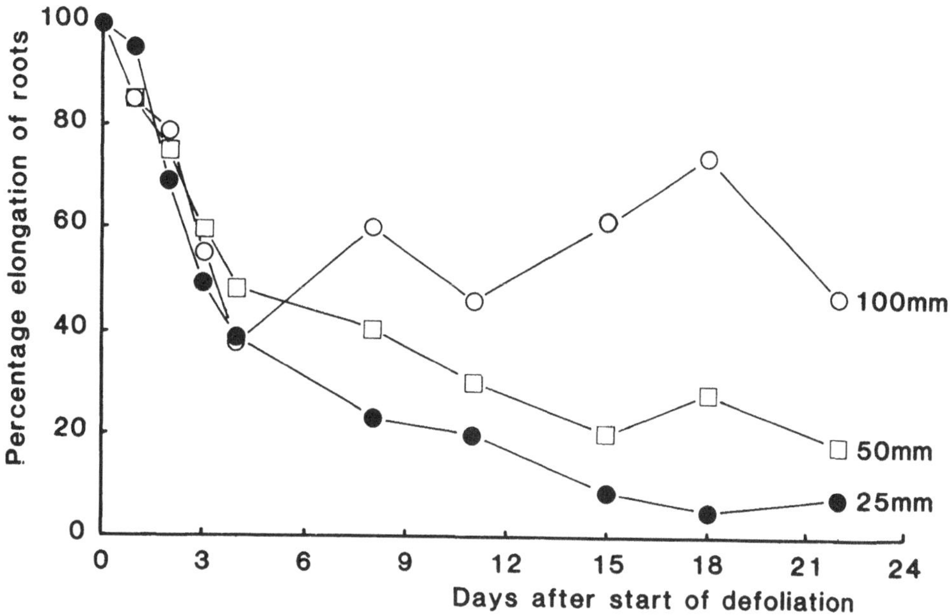

Fig. 2.16 Effect of defoliation to 25, 50 and 100 mm every second day on root elongation (expressed as percentage of that on undefoliated control plants). [Redrawn from Evans 1973].

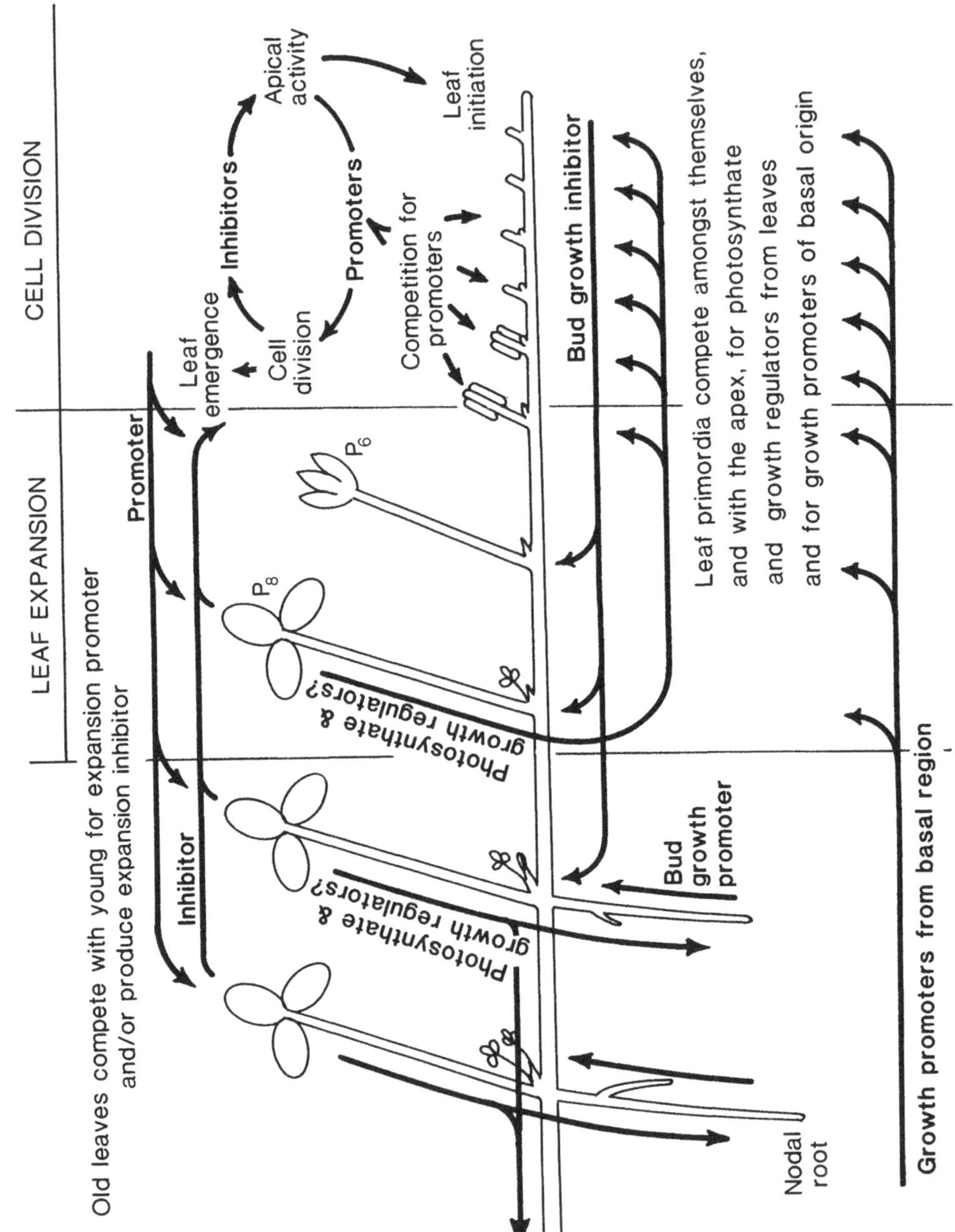

Fig. 2.17 Possible mechanisms of correlative control of stolon growth [T].

Reproductive development

R. G. Thomas
Department of Botany and Zoology, Massey University, Palmerston North, New Zealand.

Contents

A full understanding of reproductive development is fundamental to an appreciation of the factors influencing seed yield. As a basis for such an understanding, this Chapter includes a description of the developmental sequence from inflorescence initiation to seed maturation, and a discussion of the influences of both internal and environmental factors on all its phases. In a concluding section the developmental factors influencing seed yield are discussed in the light of present knowledge, and directions for further research are indicated.

Throughout this account, previously unpublished data and observations by the writer are indicated by the initial (T). The scanning electron micrographs in Figure 3.17 were prepared specially for this publication from material processed in the Electron Microscope Laboratory, Applied Biochemistry Division, Department of Scientific and Industrial Research, Palmerston North, New Zealand by D. H. Hopcroft (D.H.H.)

Frequently used abbreviations include: SD for short days, SDP for short-day plant, LD for long days, LDP for long-day plant, and CL for continuous light. Other abbreviations are explained at their point of introduction.

1. Inflorescence initiation

During vegetative growth of a stolon, the youngest axillary bud primordium visible at a magnification of x 40 under a dissecting microscope is usually situated in the axil of the third youngest leaf primordium (Thomas 1962). A bud is not apparent in the axil of the youngest leaf primordium, and usually not in the second youngest (Figure 2.6). The first morphological sign of a change from a vegetative to a flowering state is the precocious formation of a bud in the axil of the youngest leaf primordium. When a plant is transferred into conditions strongly favouring reproductive development, such a bud is initiated in the axil of the next leaf primordium to form at the apex of a stolon. The youngest leaf primordium and its subtended bud together form a so-called 'double ridge'. This pattern of development is similar to that described for other *Trifolium* species (Aitken 1955, 1960). The precocious axillary buds, which represent inflorescence primordia, develop rapidly into flower heads in the sequence shown in Figure 3.1

(A) Nodal pattern of inflorescence initiation

Entry into the reproductive stage of development does not directly involve the cells of the dome of the stolon apical meristem. These remain vegetative, initiating further foliage leaf primordia. Under favourable flowering conditions a cyclic sequence of development occurs at the apex, successively produced leaf primordia subtending either vegetative or reproductive axillary buds. Under weaker flowering conditions the stem apex commonly initiates a sequence of one or two nodes bearing inflorescence primordia followed by one, two or three vegetative nodes, this pattern being repeated several times. When the flowering stimulus is stronger, as in continuous light, the ratio of flowering to non-flowering nodes can be very much higher. This ratio is under genetic as well as environmental control, some genotypes producing an inflorescence at every node under optimal conditions for inflorescence initiation while in others the highest ratio of flowering to non-flowering nodes is 1:1 or less.

If an inflorescence-bearing node is indicated by the letter F and a vegetative node by the letter V, the nodal pattern of initiation can be depicted by a sequence of letters, with the oldest node on the left. Thus the patterns in weakly flowering plants, such as those growing in arctic Norway and described by Thomas (1980a), can be expressed as: Fvvvvvvvvv, FFvvvvvvvv and FFvvvFvvvvvv. The patterns given by Booysen and Laude (1964b) for Ladino clover plants transferred from short days to long were: FFvFv, FvFFv, and FvvFv. The pattern occurring in the most strongly flowering genotypes is: FFFFFFFFF..... The internal mechanism by which the pattern is controlled requires investigation.

The location of both vegetative axillary buds and inflorescences in the same morphological position on the stem leads inevitably to an inverse relationship between the numbers of each. Since each axil only ever forms one bud, the more inflorescences are produced, the fewer vegetative axillary buds are present. In this way reproductive growth reduces the potential for vegetative growth (see Section 5.(C)).

2. Inflorescence and floret growth and development

Immediately after an inflorescence is initiated as a precocious axillary bud, its cells divide rapidly to produce a roughly hemispherical mound(Figure 3.1a F_1). This soon grows to be similar in size and shape to the apical meristematic dome and then overtops it, becoming somewhat conical with a flattened adaxial side. Floret primordia are then initiated as small mounds subtended by bracts on the abaxial side of the inflorescence base (Figure 3.1c, F_1). As the young inflorescence continues to grow longer and broader, further floret primordia are added both acropetally and on the adaxial side until the whole somewhat conical structure becomes covered by them and their associated bracts. At first the bract primordia elongate faster than the floret primordia so that the young florets become hidden by the bracts.

The following account of floret development is condensed from Erith (1924), and, where indicated, incorporates data of Rembert (1977) and Chubirko (1963, 1965). Erith's data were obtained from plants of English wild white clover; Chubirko's plants were of Ukrainian origin (Zakarpatskoi), but Rembert did not specify the source of his material.

A floret arises as a hemispherical mass of undifferentiated meristem, and attains a size of about 90 by 90 μm before there is any sign of the development of the floral organs. These arise in the order: sepals, stamens, petals, carpel.

Calyx The calyx arises as a zone of tissue near the base of the receptacle and is initially gamosepalous. Later the free parts of the sepals arise from this zone. The sepal occupying the anterior (abaxial) position (Figure 1.12b) appears first and is closely followed by those on either side. The two posterior sepals form last (Figure 3.2b). The zygomorphy of the calyx is thus established very early in its development.

As development proceeds, growth takes place both in a basal meristematic zone, which grows up as a tube, and in the free parts of the sepals. For a time the anterior side remains longer than the posterior (Figure 3.2e). Later, however, growth on the posterior side is faster and when the flower is fully developed the latter is about 1 mm longer than the anterior side (Figures 1.12a, 3.2i).

Androecium At about the time of origin of the posterior sepals the first stamen appears as a small outgrowth from the receptacle immediately distal to the anterior sepal. This is followed by the appearance of two more stamen primordia, one on either side of it, and shortly afterwards 2 stamens arise on the posterior side of the receptacle, thereby completing a whorl of 5 stamens opposite the sepals. The 3 anterior stamen primordia are thus initially larger than the two posterior (Figure 3.2b,c).

About the time of origin of the 2 posterior stamens of this whorl, or sooner, a second whorl of 5 stamens arises alternating with the first. The members of this later-formed whorl arise in the same order as those of the first (viz. development proceeds from the anterior to the posterior side of the receptacle), and they arise from the same meristematic zone as those of the first, being interposed between them.

Stamen primordia soon become differentiated into anthers and filaments, the upper portions enlarging considerably so that by the time they are approximately 80 μm long, anthers and filaments are clearly distinguishable, the former being about 60 μm and the latter about 20 μm long. When the first-formed stamens have reached this stage of development, those in the later-formed whorl are still undifferentiated. Once the anther and filament have become equal in length, the anther, hitherto terminal, begins to assume a dorsiventral position through the downward and slightly outward growth of the internally developing pollen mothercells (Figure 3.3a).

At a later stage in floret development, the meristematic zone from which the stamens arise begins to grow up with the exception of that portion of it occupied by the posterior stamen of the shorter whorl. The result is that 9 stamens are carried up at the end of a tube which is open on its posterior side (Figure 3.3b). The odd posterior stamen, which is free, remains shorter. As growth proceeds, the anterior side of the tube grows at a greater rate, and the difference in length of alternating stamens disappears, so that in an open flower there are not two obvious whorls of stamens but one whorl consisting of stamens of irregular lengths, those on the anterior side being the longest. The apparent development of two whorls of stamens seen in transverse sections of flower buds is caused by the longer stamens bending inwards as a result of crowding.

The development of the anther and the differentiation of anther wall, tapetum, sporogenous tissue and pollen grains follow generally the course described by Esau (1965) and others: only a single microsporangium, however, is produced in each anther lobe (Figure 3.3). Seen in transverse sections, as development of the very young anther proceeds, two or three of its sub-epidermal cells (archesporial cells, ac) near two opposite sides enlarge radially and divide periclinally (Figure 3.3c,d). The inner components of each divided cell are the primary sporogenous cells (s); the outer are the primary parietal cells (p). The latter, together with neighbouring cells, divide again periclinally to give rise to two layers (x, y, Figure 3.3e), the inner (x) becoming the tapetum, the outer layer (y) dividing again to produce the so-called 'middle layer' (y_1) in contact with the tapetum, and the endothecium (y_2) adjoining the epidermis. Thus the sporogenous tissue becomes completely surrounded by three concentric layers of cells in addition to the epidermis (x, y_1, y_2, Figure 3.3f). Chubirko's description agrees with that of Erith.

The anther then grows rapidly, chiefly by expansion of the cells already formed. The middle layer breaks down and soon disappears; the tapetal cells become greatly inflated, and their protoplasmic contents begin to degenerate and collect in irregular masses within their cell cavities (Figure 3.3g). The epidermis also loses its cell contents, and the cell walls of the endothecial layer gradually develop their characteristic thickened bars (Figure 1.15).

About the time that the middle layer and the tapetum begin to degenerate, the pollen mother-cells become free and their nuclei undergo meiosis. The development of the pollen grains in the whorl of long stamens is always slightly in advance of that in the shorter whorl.

Corolla The petals appear later than most of the stamens. The two keel petals arise first, just before the posterior stamens of the second whorl, and are followed by the wing petals and then the standard. In the very youngest florets the standard, having formed later, is shorter than the other petals (Figure 3.2c), but it grows faster and by the time the floret is approximately 0.7 mm long it is the longest of the petals (Figure 3.2f, h).

At first the petals grow very slowly, and remain shorter than the sepals, stamens and carpel. Even when the calyx has almost reached its full development the petals are not visible externally (Figure 3.2g), but at about the time that the former reaches its maximum length the corolla grows rapidly and becomes longer than the calyx. The claws of the petals (Figure 1.12d) develop later than the blades.

Gynaecium The carpel arises at the apex of the floral axis, and at first appears as an almost hemispherical mass of meristem on which a shallow depression develops on the posterior side (Figure 3.4a,b). It gradually elongates and grows in circumference, its thick convex margins surrounding a deeper open cavity (Figures 3.2c, 3.4c). As it increases in size, the margins closely approach each other to enclose a hollow space - the cavity of the ovary. At this stage the edges of the carpel do not quite meet, and there is no style or stigma (Figures 3.2d, 3.4d).

Later, when the carpel is about 0.8-1 mm long, the apex elongates, and a short, blunt style is produced, the tip of which broadens into the stigma (Figure 3.4e). The margins of the carpel then come into contact, subsequently becoming closely pressed together, but there is no organic fusion of the edges. The placenta forms a little way from the margins (Figure 1.16c).

Ovule primordia arise by divisions of the epidermal and subepidermal cells of the placenta before conduplicate folding of the carpel is complete. Near the base of each, an annular outgrowth (Figures 3.4e, 3.5b) grows upwards into an inner integument (Figures 3.4f, 3.5c), and an outer integument then arises below this (Figure 3.5b). The latter grows more rapidly, enclosing the former and reaching the apex of the nucellus by the time the inner integument has surrounded only the basal half (Figures 3.4f, 3.5c). Scanning electron micrographs of ovules at this stage of development show what Rembert (1977) describes as a 'lobed structure' to the inner integument. This, it is stated, 'lends credence to the idea that integuments originated from a fusion of parts'. Examination of ovules at this developmental stage under the light microscope (T) shows clearly, however, that each so-called 'lobe' is an individual cell and there is no basis for considering the inner integument to be lobed.

Ultimately the inner integument extends to the tip of the nucellus and almost encloses it; at the same time the margin of the outer integument farthest from the funicle grows over and completely covers both inner integument and nucellus, the only opening left being the micropyle (Figure 3.4f) which Rembert (1977) describes as having the appearance of a 'T' in surface view, the top of the 'T' lying nearer to the funicle. This shape is retained in the mature seed (Figure 3.17).

Ovule primordia begin growth at right angles to the margin of the carpel, but very soon begin to curve towards the carpel apex. By the time the funicle is morphologically distinguishable at the proximal end of the ovule, the long axis of the nucellus lies almost at right angles to it (Figure 3.4e,f). Further change in the relative positions of funicle and ovule results from the rapid growth of that half of the integuments and chalazal region which is farthest from the funicle, so that the ovule ultimately becomes campylotropous (Figures 3.4f, 3.17).

Very young ovule primordia, up to about 30 μm long, consist of a clearly differentiated epidermis enclosing a mass of almost uniform meristem. Erith states that a single sub-epidermal archesporial cell is marked out in the apical region of the nucellus, and from it a tetrad of megaspores is produced by two successive periclinal divisions. These divisions begin when the ovule is about 50 μm long and 30 μm broad, shortly before the formation of the inner integument. No parietal cell is cut off, and the upper cell of the tetrad is in immediate contact with the epidermis (Figure 3.5a).

When the tetrad is completely formed, periclinal divisions occur in the two or three epidermal cells covering it, the epidermis thus becoming a double layer at that point. At about the same time the divisions leading to the formation of the inner integument begin (Figure 3.5b). The above account by Erith (1924) differs from those of Chubirko (1963) and Rembert (1977) both of whom describe the archesporium as multicellular. Rembert states 'in all species of *Trifolium* investigated, an archesporial cell divides to form a parietal cell and a primary sporogenous cell. In *T. repens* several of the archesporial cells may divide to produce parietal cells and primary sporogenous cells which may in turn enlarge to form several megasporocytes in the nucellus of a single ovule.' Chubirko considers that the 'reduction division frequently occurs in several archesporial cells simultaneously', whereas according to Rembert 'apparently only one megasporocyte undergoes meiosis to produce a linear tetrad of megaspores.'

Later, according to Erith, 'one of the cells of the tetrad - generally the third of the axial row - grows larger than the rest and functions as the embryo-sac, the two upper and the basal megaspores degenerating' (Figure 3.5b). Rembert (1977), on the other hand, stated that 'three of these megaspores degenerate leaving the chalazal spore to function in megagametogenesis. Evidence that the third megaspore functions in *T. repens* could not be demonstrated in my material.'

All three authors agree that the embryo-sac in its subsequent development and growth is similar to the 'normal' *Polygonum* type (Maheshwari 1950). While the embryo-sac is developing, the rest of the cells of the nucellus, especially those at the micropylar end, degenerate and are absorbed (Erith 1924, Chubirko 1963). The differences between the three authors' descriptions could have resulted from their use of genetically different plants and/or different environmental conditions. Clearly there is need for reinvestigation of embryo-sac formation.

The relationship between the stages of floret development and the nodal distances of flower heads from the stolon apical meristem is shown in Figure 3.6 for plants of cv. Grasslands Huia growing outside in late November in Palmerston North (T). Up to the time of their emergence, viz. while they were still within the apical buds of the stolons, the relative rates of growth in length of inflorescence primordia were approximately equal to those of leaves (Thomas 1962). Peduncle elongation then continued after extension of the subtending petioles had stopped, with the result that mature flower heads were raised above the leaf canopy.

The first floret primordia were initiated on inflorescence primordia when the latter reached a developmental age of approximately 3 plastochrons, and the inflorescences became completely covered by floret primordia at an age of about 5 plastochrons (viz. in the axil of leaf primordium P_5). Organogenesis in the first-formed floret primordia proceeded as shown in Figure 3.6, the first ovules being initiated on the inflorescence in the axil of leaf P_5. Meiosis in the first-formed anthers and ovules took place one to two plastochrons before inflorescence emergence. After inflorescence emergence, which occurred either at the node bearing the youngest leaf with unfolded leaflets or at one or other of the nodes immediately adjacent, very rapid peduncle elongation was closely followed in the oldest florets by cessation of sepal elongation and rapid elongation of petals, filaments and styles. As a result, petals protruded beyond the sepals to show their white colour and the stage of blossoming was reached. Anther dehiscence occurred soon after the petals extended beyond the tips of the sepals. Petal emergence then progressed acropetally on the flower head. As flowers aged, their corollas, which were persistent, became flushed with pink, and ultimately withered and turned brown. At the same time, irrespective of whether or not fertilization had taken place (but, as reported by Atwood (1940) and Darwin (1880), more rapidly after fertilization), the pedicels of the florets reflexed.

The extent to which the relationship between stage of development and nodal position is influenced by environment and genotype has not been studied. Erith's (1924) statement that anther dehiscence takes place 'while the corolla is closed and scarcely extended beyond the calyx tube' suggests that it sometimes takes place slightly earlier than shown in Figure 3.6, but without further study the degree and cause of any variation which occurs remain unknown. The effects of environment on overall rates of development are described in Section 4.(A)).

(A) Abnormal inflorescence development

Abnormal inflorescence development is frequently described in white clover (e.g. Becker, 1941; Kuester 1942; Penzig, 1921; Vasil & West 1971). The following account is based on that of Erith (1924).

One of the commonest abnormalities is the extension of internodes between whorls of florets in the flower heads. Often the terminal internode alone is considerably elongated, carrying up a single floret (Figure 3.7a), or a group of florets, some distance above the rest. More frequently there is extension of one or more intermediate internodes, resulting in the separation of the florets into 2 or 3 distinct groups (Figure 3.7b). Abnormal pedicel elongation also occurs (see Section 4.(B)). In some inflorescences the axis is branched, each branch terminating in a small inflorescence. Erith cited a branched inflorescence in which 'small secondary flowerheads containing minute, rudimentary, regular flowers were produced at the ends of very short peduncles arising in the axils of the foliaceous carpels of the flowers near the apex of the original inflorescence. In addition, in a few flowers, small, almost sessile, inflorescences were present in axils of the leaf-like carpels, and were still enclosed by the stipules of the latter. The calyces of these flowers were burst as a result of the developing inflorescences inside them.'

Teratological variations of florets take many forms. Sometimes the floret axis elongates to produce an entire inflorescence or a vegetative shoot. In other cases irregular or vexillar peloria of the florets is brought about by the wing and keel petals developing exactly like the standard. Meristic variation in the parts of the florets is quite common. Sometimes there is only a slight increase in number of some of the parts, but partly or fully double florets have occasionally been seen. Increase in number occurs more frequently in the carpels than in the rest of the floral organs, flowers containing two or three apocarpous or syncarpous carpels being quite common (Figure 3.7e). Reduction in number of the floral organs is not uncommon, florets having been described in which the wing or keel petals were absent, the number of stamens reduced, or, in some cases, the carpel was absent.

The calyx is frequently abnormal, exhibiting partial or complete phyllody. The calyx teeth are often larger and greener than in normal florets and sometimes dentate at the margin (Figure 3.7c); often each tooth is transformed into a leaflet with a relatively long petiole (Figure 3.7d). Sometimes the two large equal teeth form one ternate leaf and the three remaining teeth another. Occasionally sepals remain unfused during their development, the calyx thus becoming polysepalous. A floret has also been described in which the sepals were replaced by overlapping scales, each with a floret in its axil. In the same floret the petals were similar to the sepals and the carpel was foliaceous.

The corolla varies less than the calyx. In virescent florets possessing a leaf-like calyx and/or a foliaceous carpel, the petals are often reduced to membranous scales. Sometimes the standard is replaced by a simple or ternate leaf, this change being accompanied by a suppression of the other parts of the corolla.

Abnormality of the stamens is comparatively rare. Dialysis of the filaments is the most common variation, and is generally correlated with phyllody of the carpel. Sometimes the free filaments are dilated and somewhat leaf-like. Occasional phyllody of the stamens has been reported.

Phyllody of the carpel is very frequent, resulting in its transformation into a simple or compound leaf with 2, 3 or more leaflets (Figure 3.7c,d). The petioles of these carpellary leaves vary in length, and stipules similar to those of the foliage leaves are always present at their bases. Often a bud arises in the axil of a foliaceous carpel and develops into a secondary inflorescence or a second imperfect carpel. Sometimes a carpel, which appears normal, protrudes some distance beyond the calyx tube due to excessive elongation of the ovary or of the region of the receptacle immediately proximal to it. Quite often the margins of the ventral suture of an otherwise normal carpel are free for a part or the whole of their length, causing the carpel to become flat and more leaf-like. Abnormal ovules also arise, associated with virescent carpels. The least leaf-like carpels sometimes bear abortive or foliaceous ovules along their free margins, but no ovules are bourne on the most leaf-like carpels. The virescent ovules are usually shortly stalked bi-lobed green leaf-like structures (Figure 3.7f).

The causes of the above abnormalities are not known, but they are often considered to result from physiological imbalances brought about by pathogenic organisms. One form of abnormality, however, seems to be readily interpretable as a physiological reversion to the vegetative state, and this is considered separately in the next section.

i. Reversion of inflorescences to the vegetative state

Most inflorescence primordia initiate floret primordia over their whole surface and develop into rather globular flower heads comprised solely of normal florets (Figure 3.1). Occasionally, however, the tip of an inflorescence primordium reverts to the vegetative state with the result that the base of the inflorescence develops normally to produce fully functional florets while the apex, which is the last region to differentiate, initiates foliage leaf primordia instead of bract primordia and grows out into a complete normal vegetative stolon. When such reversion occurs early, the inflorescence base produces very few florets, whereas inflorescence primordia that revert late produce many florets before their apices begin to grow into vegetative stolons. This type of reversion has been observed in many cultivars of *Trifolium repens* including Tamar and Grasslands Huia as well as the closely related *T. occidentale* (T). Attempts to bring about reversion of young inflorescence primordia to vegetative axillary buds before they have initiated floret primordia have so far failed.

The cause of reversion is by no means clear, but it is most frequent under glasshouse and controlled environment conditions in the last few flower heads produced before inflorescence initiation stops in long days. Thus, when plants of Grasslands Huia are transferred from short photoperiods into long, inflorescence initiation is strongly stimulated and the first few inflorescences formed are usually perfectly normal. After three weeks or more in warm, long days, inflorescence initiation begins to stop and the proportion of flower heads reverting increases (T). A similar reversion has been observed by Beatty & Gardner (personal communication) in a clone of Arizona wild white clover growing in long days. It is thus possible that the internal factor which leads to the eventual cessation of inflorescence initiation in long days (see Section 3.(A)ii.(b)) also brings about reversion of the apices of inflorescence primordia to the vegetative phase.

3. Physiological control of inflorescence initiation

White clover has been grown as a pasture plant for several hundred years and the production of seed for resowing is thus of long-standing interest. Before the middle of this century, very little research effort had been directed towards a study of the physiology of flowering and seed set. A glance at the bibliography at the end of this chapter, however, will show that following a period of rather superficial work during the early 1950s, the late 1950s and early 1960s saw several workers, including Britten in Hawaii; Beatty and Gardner, Gibson and Laude and co-workers in the United States; Haggar, Jewiss and Zaleski in Great Britain; Kawanabe et al. in Japan; and Thomas in New Zealand, embark quite independently upon studies of the physiological control of flowering. This spate of activity declined as quickly as it began, with the result that much of the present account of reproductive physiology is based on work carried out in the 1960s.

Most of our understanding today of the physiology of flowering is the result of studies on the effects of the environment on flowering (Section 3.(A)), relatively little information being available on the mechanisms of internal correlative and hormonal control (Sections 3.(B) and 3.(C)) Considerable scope remains for further investigations into the latter aspects.

Measurement of flowering intensity. As a means of assessing responses to treatments, several different measures of flowering intensity, some of which are more valuable than others, have been used. For fundamental research into the mechanisms controlling inflorescence initiation, the most useful measures are those yielding information on the time and intensity of initiation. It is unfortunate in this respect that most investigators have only obtained data on the time or extent of blossoming, from which it is usually only possible to estimate the time of inflorescence initiation rather inaccurately.

Intensity of initiation is of most value if expressed as the average number of inflorescences initiated at nodes formed after the start of experimental treatment. Expressing the intensity as number of inflorescences per stolon without taking into account the number of sites (nodes) available for their formation is an unsatisfactory measure if the rate of leaf initiation varies between treatments (which it frequently does). This, in turn, however, is a better index of initiation intensity than 'number of inflorescences per plant', which does not take into account the number of stolons. Least useful of all in physiological terms is 'number of flower heads per unit area.' While such measures can be of agronomic value, in physiological terms 200 flower heads per m^2 per field plot in which the density of stolons is 100 per m^2 indicates a stronger inflorescence initiation response than 300 flower heads per m^2 on 200 stolons. Measures of intensity must be expressed relative to the number of sites (nodes) at which inflorescences can be initiated. In addition, it is important to bear in mind that different stolons at different positions on a plant have different capacities to produce inflorescences (see Section 3. (B)).

Measures of time (date) of inflorescence initiation are often less informative than measures of intensity. In controlled environment experiments, inflorescence primordia often develop at the first nodes formed at stolon apices after the start of conditions favourable to their initiation, regardless of the strength of the stimulus to initiation (see for example, Ridley & Laude (1968) and Thomas (1981d)). Only with photoperiods close to the threshold, or conditions in which progression towards the flowering phase is slow and gradual, as with the natural transition from winter to spring, is the time of inflorescence initiation a measure of the intensity of the flower-inducing stimulus. Data obtained on the time of initiation can be usefully expressed as either date of initiation, days to initiation from the start of treatment, or number of nodes formed per stolon from the start of inductive treatment to the onset of initiation.

Data on date of blossoming, such as time to first appearance of petal colour, which are commonly collected, are the least useful as an aid to physiological interpretation although of obvious value for agronomic purposes. The time of inflorescence emergence is directly dependent on the rate of leaf emergence (see Section 4. (A)). Only if full data are also obtained on rates of leaf initiation and emergence over the months preceding the emergence of inflorescences can the time of emergence of an inflorescence be used as a basis for the fairly accurate estimation of the time of its initiation (e.g. see Thomas 1979a).

(A) Environmental control of inflorescence initiation

To date, white clover has been generally regarded as a long-day plant (LDP). This gives a rather misleading conception of its physiological response to environment, however, firstly because inflorescence initiation is strongly influenced by low temperature (see Section 3.(A)iii.) as well as by long days (LDs), and secondly because many cultivars behave more as short-long-day plants (SLDPs) than as LDPs. The relative importance of LDs and low temperature seems to vary with the geographical origin of cultivars, plants of higher latitude origin tending to be more sensitive to LDs and those from lower latitudes tending to be more responsive to low temperatures (Section 3.(A)v.).

Investigations into the photoperiodic behaviour of white clover which have shown it to be a LDP have mostly involved the transfer of vegetative plants into various test conditions after pretreatment with warm short days (SDs). Observations so obtained have provided information on the plants' sensitivity to LDs (Section 3.(A)i.). Most of those experiments have not tested the effect of pretreatment conditions on the subsequent response to LDs, but from more detailed studies it is apparent that pretreatment with SDs is a prerequisite for inflorescence initiation and that some cultivars at least are better described as SLDPs (Thomas 1961a) than as LDPs. Such plants stop initiating inflorescences when they are maintained in warm LDs (Section 3.(A)ii.).

i. Promotion of inflorescence initiation by long days

Many cultivars have been reported to be LDPs. These include: Dutch white (Adachi & Suzuki 1968), Grasslands Huia (Thomas 1961c, 1979a), Kent Wild White (Haggar 1961, Thomas 1982), Ladino (Laude et al. 1958, Beatty & Gardner 1961, Booysen & Laude 1964a, Adachi & Suzuki 1968, Thomas 1982), Louisiana (McCloud & Cole 1951, Gibson 1957), S100 (Jewiss 1962, 1963), S184 (Zaleski 1964a) and Tamar (Cohen & Dovrat 1976).

(a) Critical daylength

The critical daylength for the LD reaction varies greatly with genotype and with temperature during LD treatment (see Section 3.(A)i.(e)). It is also decreased by pretreatments such as low temperatures (Section 3.(A)iii.(b)) and SDs (Section 3.(A)ii.(a))

Table 3.1. Effect of photoperiod on percentages of plants which initiated inflorescences after pretreatment with short photoperiods.

Cultivar	Photoperiod								Light intensity (lux)	Day/night temp. (°C)
	8	10	12	14	15	16	20	24		
[1]Ladino	—	—	—	0-31	88	100	100	—	13 000	24/18
[2]Kent	0	—	—	—	—	58	67	100	5000	18/18
[3]Tamar	20	—	90	—	—	98	100	—	18 000* +750	22/14*
[3]Tammisto	0	—	13	—	—	80	100	—	18 000* +750	22/14*
[4]Grasslands Huia	0	0	3	100	—	100	—	100	20 000	21/21

[1]Laude et al (1958) [2]Haggar (1961) [3]Cohen & Dovrat (1976) [4]Thomas (1981d)

— Signifies no data obtained
* Plants received 24-h cycles with an 8 h main light period at 18 000 lux and 22°C followed by supplementary light periods of 0-12 h duration at 750 lux and 14°C. Dark temperatures were 14°C.

The results of four studies in which the critical daylength was determined under warm conditions are summarized in Table 3.1. In only two of these (Haggar 1961 and Thomas 1981d) were temperatures during pretreatment conditions adequately described. From the table it can be seen that all the cultivars tested showed strong inflorescence initiation at daylengths of 16 h and above and weak initiation in SDs. There was no common critical daylength, however.

Differences in responses of cultivars can only be said to exist with certainty between cv. Tamar and Tammisto; only in these has the critical daylength been compared under identical test conditions and after identical pretreatments (Cohen & Dovrat 1976). Differences between genotypes were detected in cv. Ladino (Laude et al. 1958), some clones exhibiting slow progression towards inflorescence initiation in 14-h photoperiods and others not initiating at all. Beatty & Gardner (1961) also found that amongst four clones of Ladino, one had a critical daylength of less than 14 h, while the other three had critical daylengths of 14, 15.5 and 16 h. Likewise with Tamar the results indicate that 20% of genotypes had critical daylengths of 8 h or less, while in the other 80% they were >8 h; and in Tammisto 13% had critical daylengths of 12 h or less, while in 87% they were >12 h.

The critical daylength is affected by defoliation in plants growing at 21 °C. Plants of Grasslands Huia clone C, pretreated in 8-h short days, were exposed to photoperiods of 8, 12 or 16 h for 9 days, and then returned to 8-h days (T). Immediately before the start of the 9-day experimental treatments, half the plants were defoliated. The percentages of stolons that initiated flower heads in response to 8-, 12- or 16-h photoperiods were 1, 19 and 100 respectively in intact and 9, 73 and 100 in defoliated plants.

(b) Number of long-day cycles required

White clover plants grown in warm SDs and exposed to a brief period of LD induction will initiate inflorescence primordia immediately after transfer back into SDs. Stolon apices soon revert to the vegetative phase of growth, however, and inflorescence initiation only continues for a few days when photoperiods are short. The minimum number of LDs required to induce inflorescence initiation varies with genotype and the length of the LD photoperiods.

In one experiment with Grasslands Huia clone A, exposure to a single 32-h period of continuous light (CL) was sufficient to induce initiation in 94% of stolons (Thomas 1961a). On another occasion (T) the same CL treatment induced initiation in only 52% of stolons in clone A and 11% in clone C. All clones of Ladino clover studied by Booysen & Laude (1964a) initiated inflorescences in SDs following three 16-h LDs, but only one initiated inflorescences (and that very weakly) after two 16-h days.

Table 3.2. Effect of the number of long-day cycles and their photoperiods on the average number of inflorescences initiated per stolon in Grasslands Huia (Thomas 1981d).

Photoperiod length (h)	No. of long days		
	1	2	3
16	0	0	0.54
20	0	0.87	1.05
24	0	0.88	1.14

The minimum number of LDs required for induction increases as the length of photoperiod decreases (Thomas 1981d). Two were sufficient to induce initiation in Grasslands Huia clone C when the photoperiod was 20 or 24 h but three were required when plants were in 16-h photoperiods (Table 3.2). These data show that initiation of inflorescences stopped very soon after plants were transferred back to SDs from LDs. In the cases in which the average number of inflorescences produced per stolon was less than one, it is probable that they were initiated during the LD treatment and that none were initiated during SDs. In the plants receiving three days of CL, however, an average of 1.14 inflorescences formed per stolon, and by implication some stolons must have produced inflorescences at two or more nodes. It seems possible that the second of these inflorescences was initiated after return to SDs. Failure to initiate after exposure to CL for 32 h in this experiment indicates the need to investigate the effect of other factors such as pretreatment conditions and mineral nutrient supply on sensitivity to LDs (see Section 3.(A)iv.).

Photoperiodic sensitivity is also affected by leaf age (T), the youngest leaf with unfolded leaflets being more sensitive than successively older ones in Grasslands Huia clone C (Table 3.3). Relatively old leaves, such as leaf 8 (see Chapter 2, Section 2.(C)iv. and Figure 1.1), are also capable of perceiving the photoperiodic stimulus, however.

(c) Effects of light quantity and quality

Most of the investigations of responses to daylength in white clover have not distinguished clearly between photoperiodic effects and possible photosynthetic effects of light. Cohen & Dovrat (1976) showed that an 8-h supplementary light period (SLP) at the relatively low intensity of 750 lux promoted inflorescence initiation in cv. Tamar and Tammisto when it was preceded by an 8-h main light period (MLP) at 18 000 lux. For Grasslands Huia clone C an 8-h SLP at an even lower intensity (450 lux) from a mixture of tungsten filament and daylight fluorescent lamps was found to be adequate to induce initiation following MLPs of 3800 lux or more (Thomas 1981d). Sensitivity of the response to such a low intensity as this, which was almost certainly below the light compensation point for photosynthesis, supports the conclusion that the mechanism of the long-day reaction is photoperiodic rather than photosynthetic. A probable concurrent need for photosynthesis is suggested, however, by the requirement of higher light intensities during at least some part of the long photoperiods. The lowest light intensity that led to initiation when given throughout a 16-h photoperiod was 1800 lux (Table 3.4). No initiation occurred in plants receiving an intensity of 450 lux throughout the long photoperiods.

Table 3.3. Effect of leaf age on the percentage of stolons initiating inflorescences in response to a long-day stimulus.†

Node bearing remaining leaf	No. of 16-h days	
	4	6
Control	7.7	no data
1	87.5	no data
2	53.8	84.2
4	18.2	100
6	33.3	86.6
8	no data	41.6

†Plants pretreated with warm 8-h days were exposed to four or six 16-h days and then returned to short days. Immediately before the start of long-day treatments, they were partially defoliated to leave one leaf on each stolon. Controls were completely defoliated. Age of leaf is indicated by its nodal position relative to the stolon apex (see Figs 1.1 & 2.15). (Thomas, unpublished).

The photoperiodic nature of the LD reaction is further demonstrated by results of light break experiments. In Grasslands Huia clone C (Thomas 1981d), plants remained vegetative in 12-h photoperiods at 21 °C, but inflorescences were initiated on 33% of stolons growing in 10-h photoperiods with a 2-h light break in the middle of the dark period. Substitution of two 1-h light breaks for the single 2-h one increased initiation to 47%, and four 30-min light breaks increased it to 75%.

Effects of light quality have not been reported. Preliminary studies indicate that blue and green wavelengths are more effective than red during the SLP (T).

Table 3.4. Average number of inflorescences initiated per stolon in 16-h photoperiods in plants of Grasslands Huia transferred from 8-h days.†

SLP light intensity (lux)	MLP light intensity (lux)			
	20 000	3800	1800	450
20 000	1.6	1.6	1.0	1.0
3800	1.3	1.0	0.6	0.8
1800	1.1	0.7	0.6	0
450	1.1	0.2	0	0

†Each 16-h photoperiod consisted of a daily 8-h main light period (MLP) at one of four different light intensities followed by a daily 8-h supplementary light period (SLP) at one of four intensities (Thomas 1981d).

(d) Perception and transmission of the long-day stimulus

Experiments in which all other parts of plants of Grasslands Huia were in SDs while leaf laminae received 16-h days have shown that, as in other species, the leaves perceive the photoperiodic stimulus (T). It is not known whether the stolons are also perceptive. As shown in Table 3.3, perception occurs in live leaves of all ages, but is maximal in the youngest leaf with unfolded leaflets at node 1 (T). The inductive stimulus can be transmitted both acropetally from a leaf to the apical bud on the same stolon and basipetally from the leaves on a main stolon to the apical bud of a lateral stolon. It has also been shown to be transmitted across a graft union (T).

(e) The influence of temperature on the long-day response.

Data on the effects of temperature on inflorescence initiation in response to LDs seem to conflict. Roberts & Struckmeyer (1938, 1939) reported that the intensity of blossoming of three unspecified varieties of white clover in glasshouses in LDs increased as minimum dark temperatures decreased from 24° to 18° to 13°C. Ridley & Laude (1968) found, however, that inflorescence initiation in Ladino clover clones was stronger at higher than lower temperatures: in 16-h photoperiods, 76.4% of the nodes formed at 32°C initiated inflorescences, whereas only 37.5% did so at 15°C.

Table 3.5. Effect of temperature on the average number of inflorescences initiated per stolon in 12-h photoperiods in Grasslands Huia clones A and C (Thomas, unpublished).

Dark temperature (°C)	Light temperature (°C) 14	21	28	35
7	1.75	0.62	0.35	2.27
14	0.0	0.02	0.22	0.45
21	0.2	0.15	0.0	0.0
28	1.65	1.05	1.4	no data

One possible explanation of the apparent discrepancy between these two sets of results is suggested by the data in Table 3.5. These were obtained from plants of Grasslands Huia grown in 12-h photoperiods at a wide range of combinations of day and night temperatures (T). Photoperiods of 12 h are very close to, or even below, the critical daylength for inflorescence initiation at temperatures near 21°C and the data could thus be considered not to be comparable to those obtained from plants grown in 16-h photoperiods, particularly as inflorescence initiation is strong at 21°C in the latter conditions. The general pattern of response to temperature, however, which showed inflorescence initiation to have been much stronger at both low (7°C) and high (28°C) dark temperatures than at those between, should be noted. This had the consequence that at a day temperature of 28°C, inflorescence production increased as night temperatures decreased from 21 to 7°C and it decreased as night temperatures decreased from 28 to 14°C. The latter result agrees with the finding of Ridley & Laude that inflorescence production was higher at a dark temperature of 32°C than at 15°C, and the former concurs with that of Roberts & Struckmeyer. The apparent discrepancy can thus be explained as probably having been caused by the use of an inadequate number and range of temperature treatments. This hypothesis should be tested by determining initiation responses in 16-h photoperiods at the temperatures shown in Table 3.5.

A second possible explanation of the difference between the results obtained by Roberts & Struckmeyer and those of Ridley & Laude is that it might have arisen from their use of different lines of clover. In a comparison of the effects of temperature on cv. Grasslands Huia and a Spanish ecotype (Thomas 1980b) grown in 16-h days, the average percentages of stolons initiating inflorescences in the former were found to be 36, 61 and 92 at day/night temperatures of 17.5/10, 22.5/14, and 27.5/18°C respectively (Table 3.9). Under the same conditions, the percentages of initiation in seedlings of the Spanish ecotype (C1067) were 24 and 6 at temperatures of 17.5/10 and 22.5/14°C respectively. Higher temperatures were also found to favour inflorescence initiation in some lines, while having the opposite effect on others when nine lines of white clover were compared at high summer glasshouse temperatures and in cooler natural summer conditions (Thomas 1982). In this study the percentage of nodes initiating inflorescences in late November in New Zealand was higher under glass in 'Ladino' and a Russian line from Kalinin but very much lower under glass than outside in the Spanish ecotype (C1067) and in cv. Tamar. This suggests that high temperatures favour inflorescence initiation in lines showing a strong LD response (such as Ladino, the Kalinin line, and Grasslands Huia) but that LD initiation is stronger at lower temperatures in lines such as Tamar and C1067 which have a weaker LD response (see Section 3. (A) v.).

The site of action of temperature on the LD initiation response has been investigated by Ridley &

Laude (1968). They concluded from experiments in which leaves and stolon tips of Ladino were exposed to different temperatures 'that the initiation process of flowering at the stolon apex is more sensitive to temperature than is the induction process associated with the leaves.' Their results show, however, that the numbers of inflorescences produced after twelve 16-h photoperiods were 4.4, 2.4, 1.0 and 1.0 when temperatures of $\frac{\text{apical bud}}{\text{rest of plant}}$ were $\frac{32^\circ}{32^\circ}, \frac{32^\circ}{15^\circ}, \frac{15^\circ}{32^\circ}$, and $\frac{15^\circ}{15^\circ}$ respectively.

Thus, while it is true that increasing the temperature of the 'rest of the plant', which included the leaves, from 15 to 32°C had no effect on inflorescence initiation when the apical bud was held at 15°, the same increase nearly doubled the number of inflorescences (from 2.4 to 4.4) when the apical bud temperature was 32°. It seems from these results more likely that temperature can affect processes in both regions but that the rate of reactions in the apical bud is the limiting factor at 15°C.

(f) The nature of the light and dark reactions.

In a long-day plant there are always two possible types of mechanism of photoperiodic control, namely (a) that flower-promoting substances are only synthesised in the long light periods of LDs, and long dark periods are neutral, and (b) that flower-promoting substances are synthesised irrespective of the lengths of the light periods, but that long dark periods are antagonistic to their activity. Data available for white clover (T) support the latter possibility. When plants of Grasslands Huia clone C, previously grown in 8-h photoperiods, were exposed to 7 cycles of each of the treatments shown in Table 3.6, they remained vegetative when 16-h light periods were alternated with 16-h dark periods, but initiated inflorescences when 8-h light periods alternated with 8-h dark periods. Clearly inflorescence initiation only occurred with short dark periods, and long dark periods were apparently antagonistic. Unfortunately the design of this experiment does not allow a distinction to be made between the possibilities: (a) that long dark periods were antagonistic to initiation and (b) that the effect on initiation was caused by perturbation of the photoperiodic time-measuring mechanism by cycle lengths other than 24 h.

Table 3.6. Inflorescence initiation in plants of Grasslands Huia in response to seven cycles of light and dark of different durations (Thomas, unpublished).

| Photoperiod treatment | | Duration of | % stolons |
Hours light	Hours dark	light/dark cycles	initiating inflorescences
8	16	24 h	0
8	8	16 h	100
16	16	32 h	0
16	8	24 h	100

The hypothetical antagonistic role of long dark periods can be considered to be supported by the data on the effect of temperature on inflorescence initiation in 12-h photoperiods shown in Table 3.5. Plants of Grasslands Huia clones A and C grown at light temperatures of 28 and 35°C initiated inflorescences very poorly in response to night temperatures of 21°, slightly better at 14°C and strongly at 7°C (T). Such results can be interpreted, as in the case of *Trifolium subterraneum* (Evans 1959), to indicate that low dark temperatures act to decrease reactions antagonistic to inflorescence initiation. This conclusion is further supported by Table 3.7 which shows that low night temperatures in combination with high day temperatures lead to initiation in 8-h photoperiods (T).

The data in Table 3.5 show, however, that the effect of temperature on the dark reaction is much more complex than suggested above. Firstly, it clearly varied with the light temperature; secondly, there was no steady gradient of increasing inflorescence initiation as the dark temperatures decreased from 21 to 14 to 7°C; and, thirdly, inflorescence initiation was strongly promoted by a dark temperature of 28°C. The latter is in agreement with the findings of Ridley & Laude (1968) who reported that for Ladino plants growing in 16-h photoperiods, the percentages of nodes initiating inflorescences at day/night temperatures of 32/32, 32/15, 15/32 and 15/15°C were 76.4, 34.7, 65 and 37.5 respectively, and commented that 'the association of warm dark-period temperatures with increased flowering is in disagreement with the concept of an inhibitory process occurring during the dark periods in long-day plants.' This comment was based on the supposition that dark is either inhibitory or promoting, but not both. There is no theoretical reason, however, why inhibition and promotion could not act simultaneously. Considered together, the apparently conflicting results of Ridley & Laude and those of the other experiments described earlier are perhaps best explained by the hypothesis that at least two temperature-sensitive processes occur during the dark period, one of which is antagonistic to inflorescence initiation and the other of which promotes. It seems probable from preliminary investigations (T) that both low and high temperatures act in their different ways at the apical buds (Section 3.(A)i.(e)) rather than on the photoperiodic responses in the leaves.

The possibility that the flower-promoting influence of LDs requires long light periods rather than short dark ones is not supported by available data. The results in Table 3.6 show that inflorescence initiation can occur in plants exposed to short light periods provided that the dark periods are also short. The lack of response to temperature during the light period reported by Ridley & Laude (1968), and supported by data in Table 3.5 for plants grown at night temperatures of 14-28°C, suggests that either the light reactions are temperature-compensated over that range or that no reactions directly stimulatory to inflorescence initiation occur during the light.

Table 3.7. Interaction of temperature and photoperiod on the average number of inflorescences initiated per stolon in Grasslands Huia (Thomas, unpublished).

Photoperiod	Day/night temperatures 21/21	28/14	35/7°C
12 h	0.0	0.43	2.46
10 h	0.0	0.0	0.92
8 h	0.0	0.0	0.53

ii. The short-long-day reaction

Plants of cv. Grasslands Huia transferred from short days into 16-h photoperiods at 21°C cease inflorescence initiation after about 3 weeks if maintained in long days, both in controlled environments (T) and under glasshouse conditions (Thomas 1961a, 1979). These observations imply that at the time of transfer from SDs into LDs the plants are competent to initiate inflorescences but that competence is gradually lost in LDs. A similar loss of competence has been reported in Louisiana S1 by Gibson (1957; Section 5.(C)). He showed that plants grown in field plots with photoperiods artificially extended to 19 h in Alabama, not only flowered earlier than controls in natural photoperiods but also stopped flowering earlier, so that by the end of May the intensity of flowering had halved. These plants 'had at least partially reverted to vegetative growth prior to the occurrence of hot dry summer weather.' Cessation of initiation in the natural environment in summer at a time when photoperiods exceed the critical duration for the start of initiation, has also been reported for several lines growing in England and New Zealand (Thomas 1981a and 1982). In wild populations growing in arctic Norway (Thomas 1980a) inflorescence initiation stopped in summer while plants were still receiving natural continuous light.

A similar loss of competence seems not to occur in all lines of white clover, however. Within the Ladino type cv. Pilgrim, for instance, earlier flowering genotypes flower for a longer period than those that start to flower late, giving the impression that Ladino plants initiate inflorescences as long as the photoperiod exceeds the critical daylength (Laude et al. 1958). The data for Ladino in Figure 3.10, as well as results obtained by Mohamed (1981) for AC20, a line derived from Ladino, also support this conclusion.

(a) Restarting inflorescence initiation after cessation in long days.

The observation that competence to initiate inflorescences is lost in LDs in many lines (Section 3.(A)ii.) raises two questions, namely 'what causes the loss of competence?' and 'how can competence be regained?'. Answers to the second question provide a basis for answering the first.

Inflorescence initiation can be restarted in Grasslands Huia plants in which it has stopped in LDs in at least three ways:

1. By returning them to a period of SD treatment followed by re-exposure to LDs (Thomas 1961a, Jewiss 1962, 1963).
2. By defoliation (Thomas 1981a)
3. By transferring them to longer photoperiods, e.g. from 16 h to continuous light (Thomas 1981d).

The effectiveness of short days. The effectiveness of pretreatment with SDs on subsequent inflorescence initiation in LDs is influenced by the number of SDs given and their photoperiod. The critical daylength for the induction of competence by SDs lies between 14 and 16 h at 21 °C in Grasslands Huia clone C. Plants transferred into 16-h photoperiods after pretreatment for 4 weeks in photoperiods of 8, 10, 12, 14 and 16 h produced 1.22, 1.55, 1.25, 0.50 and 0.03 inflorescences per stolon respectively (Thomas 1981d).

Experiments in which the number of SD cycles given has been varied indicate that a 4-week SD treatment leads to maximal LD response. Under some circumstances a period as short as 3 SDs has led to a response (e.g. Thomas 1961a), while under others 4 weeks of SDs have led to only a relatively weak response. The reasons for this are uncertain. From preliminary studies it seems that light intensity during SDs is not a main causal factor in the variability (T). Temperature might be more important, however. In experiments with cv. Tamar (T), inflorescence initiation in warm LDs was more vigorous following natural cool autumn SDs (1.06 inflorescences per stolon) than after warm autumn SDs inside a glasshouse (0.34 inflorescences per stolon).

Experiments using light break treatments show that the reaction causing induction of competence in SDs is truly photoperiodic (Thomas 1981d). Vegetative plants of Grasslands Huia clone C, transferred into 8 h photoperiods for 4 weeks at 21 °C after growth in 16-h days, produced an average of 0.97 inflorescences per stolon on return to LDs. Similar plants transferred into 6.5-h photoperiods with a 1.5-h light break at the midpoint of the dark period failed to initiate inflorescences on return to LDs.

The effect of defoliation. The observation (T) that defoliation of Grasslands Huia plants causes inflorescence initiation in 12-h photoperiods (Section 3.(A)i.(a)) indicates that leaves might have some inhibitory effect on initiation. Two other types of experiment point to an inhibitory role of leaves. Firstly, defoliation at the time of transfer of plants into 16-h days from 8-h days at 21°C increased the percentage of flowering stolons (Table 3.8, (T)). Secondly, inflorescence initiation can be restarted in plants that have stopped initiation in warm LDs by removal of their leaves. In one experiment using Grasslands Huia clone C, a single defoliation of plants growing in 16-h days at 20 000 lux and 21 °C allowed initiation to restart on 41% of stolons, while undefoliated controls remained entirely vegetative (T). In the same experiment repeated removal of leaves led to a restart of initiation in 78% of stolons. A similar effect of defoliation has been shown for several cultivars in the field in summer in southern England (Thomas 1981a).

The effect of basal tissue. In cv. Grasslands Huia it has been frequently observed (T) that when cuttings are taken from stolon tips of wholly vegetative or weakly flowering plants at or after midsummer, there is a brief burst of inflorescence initiation once an active root system has formed and vigorous stolon growth and elongation begun. Such a response is shown in Figure 4(a) of Thomas' 1979(a) paper in which it can be seen that after newly rooted cuttings were potted up in January (during the southern hemisphere midsummer) there was a peak of inflorescence initiation in early February in clones A and C. In undisturbed intact plants there was a decline of initiation from December through to May.

The attachment of a stolon apex to the base of a mature plant thus prevents its initiation of inflorescences under environmental conditions in which initiation can occur when the basal tissues are removed.

Table 3.8. Effect of partial defoliation on inflorescence initiation in response to long days (mean of two experiments).†

Defoliation treatments*								Average % stolons with inflorescences
8	7	6	5	4	3	2	1	
—	—	—	—	—	—	—	—	4
—	—	—	—	—	—	—	L	57
—	—	—	—	—	—	L	L	53
—	—	—	—	L	L	L	L	43
—	—	L	L	L	L	L	L	21
L	L	L	L	L	L	L	L	29

*Presence (L) or absence (—) of leaves at nodes 1 to 8 (see Fig. 2.15) immediately before the start of long-day treatment.

†Plants pretreated with warm 8-h short days were exposed to either three (in one experiment) or four (in the other) 16-h photoperiods and then returned to short days (Thomas, unpublished).

Whether the failure to initiate inflorescences is the result of an inhibitory influence of the basal tissues on initiation at the apex or whether the supply of necessary precursors or cofactors by the old root system is limiting is quite unknown. Nor is it certain that the stimulatory effect of taking cuttings is not simply a result of removing inhibitory leaves (see earlier). It seems not unlikely, though, that old basal tissues influence inflorescence initiation by providing both an inadequate supply of promoters and an excess of inhibitors.

(b) The cause of cessation of inflorescence initiation in long days

Based on the observations described in the previous section, the following three hypotheses as to the cause of the cessation of inflorescence initiation in LDs can be proposed.

(a) It is possible that a flowering precursor substance that is produced in SDs is necessary for continued inflorescence initiation in LDs. When the level of this hypothetical precursor falls below a certain point, inflorescence initiation will stop. Renewed exposure to SDs might allow the build up of such a precursor.

(b) An inhibitory effect of leaves might build up in LDs and decline in SDs.

(c) An inhibitory influence might be exerted by inflorescences themselves. Formation of some inflorescences might thereby prevent the formation of others.

The evidence available does not support the first possibility (Thomas 1981d). The observations that inflorescence initiation is restarted by transfer from 16-h photoperiods to continuous light and by defoliation cannot be explained simply on the basis that the plant is short of precursor. Neither of these treatments could be envisaged to lead directly to an increase of a precursor which is synthesised in SDs. Instead, the data favour strongly the possibility that initiation stops as a result of inhibitor accumulation. The third possibility remains to be critically tested, but is supported by the results of one experiment reported by Kawanabe et al. (1963).

(c) The mechanism of the short-long-day response

Any model designed to explain the mechanism of the short-long-day response must take into account the following facts:

(i) transfer from short to long days causes inflorescence initiation,
(ii) prolonged exposure to 16-h photoperiods leads to a cessation of initiation,
(iii) transfer of vegetative plants from 16-h days to continuous light restarts initiation.

Such a model, depicted in Figure 3.8, has been suggested by Thomas (1981d). In this, initiation is envisaged to be controlled by the ratio of inhibitors (I) to promoters (P). In SDs there is a low level of both I and P, but I exceeds P. Transfer from SDs to 16-h days immediately (within 1 to 3 days) increases the level of P so that it exceeds I. Continued exposure to 16-h days increases the content of I until it once again exceeds P, and initiation stops. Inflorescence initiation can then be restarted both by transfer to continuous light (which increases P further to exceed I) and by defoliation (which decreases the level of I to below that of P).

Experiments in which single leaves are exposed to LDs on plants otherwise wholly in SDs show that a LD flower-promoter (P) is translocatable both acropetally and basipetally along stolons of Grasslands Huia clone C, but experiments designed to test for the presence of a translocatable inhibitor have so far failed to prove that one exists (T).

Data presented in Section 3.(A)i.(f) were interpreted therein as indicating that long dark periods in SDs are antagonistic to inflorescence initiation. It is clearly unlikely that this dark antagonism is the same as the inhibitory influence described above, as the latter must increase in LDs (with short dark periods) whereas the former increases in SDs. It is thus necessary to extend the model above by incorporating a 'dark antagonism'. The overall response to photoperiod then becomes explicable on the basis of three reactions, namely:

(i) production of promoter (P) in the light,

(ii) production of inhibitor (I) in the light,

(iii) activity of a dark antagonist (A) which either leads to the breakdown of P or to the production of a second inhibitor.

Effects of temperature on inflorescence initiation in response to long days are discussed in Sections 3.(A)i.(e) and 3.(A)i.(f). The evidence for the existence of a light inhibitor (I) as well as a dark antagonist (A), presented here, assists interpretation of the data in Table 3.5 discussed in those sections. Thus, if dark temperatures were held low (7°C), thereby decreasing the effect of dark antagonists (A), initiation of inflorescences would be expected to depend strongly on the light reactions. If, under these conditions, inflorescence initiation was limited by the level of light inhibitor (I), low light temperatures would be expected to enhance initiation relative to the amount occurring at higher light temperatures by decreasing inhibitor synthesis. The observed effects of day/night temperatures of 14/7, 21/7 and 28/7°C in Table 3.5 are interpretable in this way. The strongly stimulatory effect of the day/night temperature combination of 35/7°C remains unexplained, however, and further investigations into the site and mode of action of such temperatures are clearly necessary to increase our understanding of the process of inflorescence initiation.

In spite of some uncertainties regarding interpretation of responses to temperature, it can be said in summary of the information available on the short-long-day response that the following hypothesis, which also incorporates an explanation of the long-day reaction discussed in Section 3.(A)i.(f), is in general supported by the available evidence. During the light period a promoter (P) forms in leaves and this is later translocated to the stolon apices. Reactions stimulated by P at the stem apex are enhanced by high temperatures in the dark (Ridley & Laude 1968). Long light periods also act via the leaves to exert an inhibitory influence (I) which is greater at higher light temperatures. The strength of inflorescence initiation at the stem apex depends on the ratio of P to I (Thomas 1981d). The action of P at the stem apex is additionally antagonised by long dark periods, the effect of which is decreased by low dark temperatures (T).

iii. Promotion of inflorescence initiation by low temperatures.

Most plants of white clover tested have shown some inflorescence initiation in response to low temperatures, both 'direct' and 'indirect' responses (Lang 1965) having been reported. In plants showing a direct response, initiation occurs during natural cool conditions in winter. A direct response has been reported in cv. Grasslands Huia (Thomas 1979a), a Hawaiian line (Britten 1960, 1961), Spanish C1067, Kent Wild

White, Ladino, Grasslands Pitau and Louisiana S1 (Thomas 1980b, 1982). Indirect responses are classically referred to as vernalisation. In these cases low-temperature pretreatment enhances inflorescence initiation in subsequent warm conditions. Few clear instances of indirect responses have been reported for white clover, but they have been shown to occur in Kent Wild White (Haggar 1961), and the existence of indirect responses is indicated by observations on several other lines (Thomas 1982) (see Section 3,(A)iii.(b)).

(a) Direct effects of low temperatures

Genotypes of white clover originating from high latitudes tend to be LDPs and do not initiate inflorescences in warm SDs. In most genotypes of all lines tested, however, inflorescences are initiated earlier in the spring under natural cool conditions than in the same rather short photoperiods under artificially warm (glasshouse) conditions. These include genotypes of two Russian lines (Jygeva-4 and Kalinin, Thomas 1982), Ladino (Laude et al. 1958, Thomas 1982), S184 (Zaleski 1964), Kent Wild White (Thomas 1982) and Grasslands Huia (Thomas 1979a). The time of inflorescence initiation in a population of plants of the latter cultivar observed in the field over 12 months in New Zealand is shown in Figure 3.9. In the same study (T), a similar pattern of initiation over the year occurred in populations of cv. S100, Kent Wild White, Milkanova, Grasslands Pitau and Grasslands G.18. Compared with plants grown in artificial warm conditions throughout, plants of these cultivars grown in natural cool winter conditions at latitude 40°S. in New Zealand initiate inflorescences about 3 months earlier.

Genotypes originating from low latitudes also initiate inflorescences in cool short days but tend to do so in the decreasing daylengths of autumn rather than in early spring (see Figure 3.10). Such lines include Louisiana S1, Tamar, a Spanish ecotype (C1067) (Thomas 1982) and natural Hawaiian populations (Britten 1960, 1961).

The time taken to form inflorescence primordia which are detectable by dissection is much longer in cool SDs than in warm LDs. In one experiment, the results of which are shown in Table 3.5, the first primordia were not detectable until 5 weeks after the start of treatment with day/night temperatures of 14/7 °C in Grasslands Huia clones, whereas primordia become visible microscopically within 7 days of transfer from warm SDs to warm LDs.

Effective Temperatures. Temperature regimes effective in causing a direct initiation response in cv. Grasslands Huia clones grown in 12-h photoperiods are shown in Table 3.5 (T). Day/night temperatures of 14/7°C induced strong inflorescence initiation and 21/7°C had some promoting effect, but both 14/14 and 21/14°C were completely ineffective.

Other observations are in agreement with these findings. Using a combination of glasshouse and natural temperatures in winter, Thomas (1979a) showed that initiation was directly induced by natural low night temperatures (-2°) but not by warm nights (19°) in clones of Grasslands Huia irrespective of whether the day temperatures were warm (24° in a glasshouse) or cool (13° in natural conditions). In a comparison of flowering responses of populations of seedlings of Grasslands Huia, Grasslands Pitau and Spanish C1067, growing in 12-h photoperiods (Table 3.9), day/night temperatures of 12.5/6° led to strong initiation in all three lines, but 17.5/10° caused no initiation in the former. A regime of 17.5/10° led to weak initiation in Grasslands Pitau and Spanish C1067 (on means of 16% and 7.5% of stolons respectively). Britten (1960, 1961) similarly found that lowering the night temperatures to about 8 °C led to a direct initiation response in Hawaiian lines grown in 12-h photoperiods.

In the investigations described above, temperatures below 12.5°C have been shown to induce inflorescence initiation when given during the dark period, but their effect during the light period has not been tested. It is therefore invalid to conclude that such temperatures are only effective when given during the dark phase of a 24-h cycle. Until the effect of low temperatures has been ascertained during the photoperiod, any discussion of their possible mode of action is premature.

(b) Indirect effects of low temperatures.

No cases have been reported in which white clover plants have been found to have an obligate requirement for low temperature pretreatment (vernalisation) before they become capable of initiating inflorescences in LDs. Inflorescence initiation in LDs is sometimes enhanced by low temperature pretreatment, however. Haggar (1961), for instance, stated that in Kent Wild White in England 'seed vernalisation [at 3°C] for 15 days resulted in earlier flowering and increased the number of inflorescences per plant, when grown under natural long-day photoperiods in early summer'. In another series of experiments in which 'vegetative stolon cuttings were given successively longer periods of cold temperatures in an unlit refrigerator [at 3°C]' earliest flowering in LDs at 18°C occurred in plants which had received 1 to 2 weeks of low temperature.

In cv. Grasslands Huia, clones transferred into warm 12.5-h photoperiods from natural cool conditions in which they were initiating inflorescences continued to initiate for 2 to 4 weeks under conditions in which plants maintained in the warm throughout failed to initiate (Thomas 1979a). Initiation continued similarly in Spanish C1067, Louisiana S1 and Tamar clones transferred inside from the cold in mid-winter (Thomas 1982).

In Tamar, a cultivar which has a relatively short critical daylength under warm conditions, a period as short as two weeks at natural low winter temperatures greatly increased subsequent initiation when plants were transferred into warm SDs (T). Control plants maintained in warm SDs throughout were almost entirely vegetative. In this case low temperatures acted indirectly by lowering the critical daylength under subsequent warm conditions.

Investigations into the devernalising effects of high temperatures have not been reported. It is possible that the abrupt cessation of inflorescence initiation which occurs when initiating Grasslands Huia plants are transferred from cool natural conditions into a heated glasshouse in spring (Thomas 1979a) is the result of devernalisation, but alternative explanations cannot as yet be eliminated.

(c) Mechanism of action of low temperatures

A frequently suggested mode of action of low temperature is the production of a hypothetical compound, 'vernalin', which acts as a precursor for synthesis of a flower promoter in warm conditions. In white clover, many genotypes are capable of initiating inflorescences immediately on transfer to LDs after pretreatment with warm SDs. In these instances it must therefore be assumed that all necessary precursors are already present in warm SDs and it seems unlikely that low temperature acts by producing a precursor. Taking into account the evidence accumulated that inflorescence initiation in white clover is influenced by inhibitors as well as promoters (Section 3.(A)i.(f)), with our present state of knowledge it seems more likely that low temperature might act either by increasing the level of promoter (P) itself or by decreasing the level of initiation antagonists (A). But if low temperature acts directly by increasing the level of P, the continuation of initiation in warm SDs after cool pretreatment, which occurs in both Grasslands Huia and Tamar (as well as several other cultivars, at least), would not be expected because warm SDs lead to an immediate cessation of initiation in plants transferred from warm LDs in which production of P is actively occurring (see Section 3.(A)ii.). It is thus more likely that low temperatures act by decreasing the level of initiation antagonists hypothesised in Section 3.(A)i.(f). Such a hypothesis is probably far simpler than reality, bearing in mind the complexities of plant organisation and the diversity of responses to temperature (Section 3(A) i.(e)), but it provides a firm basis for further investigations into the mechanism of action of low temperature. Preliminary studies (T) indicate that apical buds rather than leaves are sensitive to low temperature and that the effects are not translocatable, but there is scope for considerably more exploration of the response mechanism.

iv. Effects of nutrient availability on inflorescence initiation.

While several studies have been made of the effects of mineral nutrient availability on seed production potential in white clover (e.g. Haggar et al. 1963) there is very little information on the effect of mineral nutrition on inflorescence initiation in response to daylength or temperature.

In one investigation, when plants of Grasslands Huia clone C were fed different daily quantities of a complete mineral nutrient solution, there was a clear positive correlation between the amount of nutrient given and the number of flower heads initiated per stolon when plants were transferred from SDs to LDs (T). It is not known in this case whether nutrient level affected the SD pretreatment or the LD initiation process, or both.

Table 3.9. Effect of temperature and nitrogen source (inorganic N or symbiotic N) on percentage of stolons producing inflorescences in plants grown throughout in 12-h photoperiods or transferred from 12-h to 16-h. (From Thomas 1980b).

		Percentage inflorescence initiation					
		Grasslands Huia		Grasslands Pitau		Spanish C1067	
Day length (h)	Temperature (day/night) °C	Inorg. N	Symb. N	Inorg. N	Symb. N	Inorg. N	Symb. N
12	27.5/18	10	0	10	0	0	0
	22.5/14	0	7	0	9	0	0
	17.5/10	0	0	25	8	15	0
	12.5/6	84	82	69	76	94	94
16	27.5/18	98	86	no data		no data	
	22.5/14	74	48	63	61	6	6
	17.5/10	70	3	71	34	29	20

In another experiment (Thomas 1980b) in which inflorescence initiation was compared in cv. Grasslands Huia, Grasslands Pitau and Spanish C1067, marked effects of nitrogen source were observed. Plants similarly provided with all other essential mineral nutrients were supplied with nitrogen either by inoculation with *Rhizobium* or by feeding inorganic nitrogen in the absence of nodulation. Responses to the different nitrogen sources are shown in Table 3.9. When plants were grown in 12-h photoperiods there was little effect of nitrogen source on inflorescence initiation at any temperature, particularly at that temperature regime which evoked the strongest initiation response, viz. 12.5/6°C. When plants were transferred from 12-h photoperiods into 16-h days, a pronounced effect of nitrogen source became noticeable, plants which were wholly dependent on N-fixation as a nitrogen source initiating many fewer inflorescences than those receiving inorganic nitrogen. This was particularly marked under cooler conditions and in Grasslands Huia. These results point strongly to a difference in nitrogen requirements between initiation in SDs as a direct response to low temperatures and initiation in response to LDs.

More detailed studies of the interactions between nutrient availability and inflorescence initiation would contribute greatly to our understanding of the internal reactions controlling initiation.

v. *Variations in responses to environment related to latitude of origin*

Significant variations exist in physiological responses between populations of different origins. In general it can be said that those of relatively high latitude origin behave as long-day plants in their natural environment, initiating inflorescences as the photoperiods increase in the spring, and that those of lower latitude origin tend not to behave as long-day plants but, instead, initiate inflorescences from autumn through winter and into spring and stop as daylength and temperature increase. The adaptive significance of these differences is clear. At higher latitudes where winters are cold and the growing season is relatively short, initiation of flower heads early in spring has the advantage that it allows maximum time for seed maturation. Likewise, cessation of inflorescence initiation relatively early in the growing season prevents the inefficient utilization of energy for production of flowers at a time too late for mature seeds to be developed before winter. Plants from higher latitudes tend to be short-long-day plants in which spring long days stimulate a brief period of inflorescence initiation followed by cessation, possibly as a result of an inhibitory effect of long days as summer progresses. Plants of lower latitude origin have often evolved, or been artificially selected, in environments with dry summers. In these, early inflorescence initiation near the start of the growing season in autumn allows maximum head production and seed development before the onset of drought. Plants from high altitudes, which naturally experience colder winters, tend to behave like high-latitude plants, irrespective of their latitude of origin.

The effects of different adaptive responses are clearly seen when the times of inflorescence initiation are compared at a common site in which daylength in summer is long enough to allow initiation in high latitude lines, and in which winter temperatures are relatively mild. In one such comparison (Chamblee 1954) plants of Ladino clover originating in Oregon (latitude 45°N) and California (35°N) were grown from seed in North Carolina (35°N). Flowering in the Californian plants was more vigorous than in those from Oregon, but it is not known whether the different responses were caused by the Californian plants having a shorter critical daylength (as suggested by Chamblee) or a lesser requirement for winter cold.

Results of another comparison made at Palmerston North, New Zealand (latitude 40°S), are shown in Figure 3.10 (Thomas 1980b, 1982). From these results it can be seen that time of initiation in cool SDs tended to be later in the year in plants with a higher latitude of origin. Cv. Louisiana S1 and Tamar began to initiate soon after the photoperiod shortened to less than 12 h in autumn. Some, but fewer, plants of Spanish C1067 also did so. The earliest initiation in populations of Kent Wild White and Ladino began in July, but in most genotypes in these populations initiation was delayed until August or later. A subsequent comparison of the 'high-latitude' cultivars Milkanova and S100 in Palmerston North showed their response in natural conditions to be similar to that of Kent Wild White and Grasslands Huia (T). Direct initiation in plants from higher latitudes seems to require more exposure to cold than in those from lower latitudes.

When plants from higher latitudes are grown in a heated glasshouse throughout the year, inflorescence initiation does not normally take place until the critical daylength of 14 h or more is reached in spring (Thomas 1982). If such plants are grown at low latitudes where the longest summer photoperiods only just reach this length, inflorescence initiation is weak. This showed clearly in a comparison of inflorescence formation in the Finnish cultivar Tammisto at latitudes 40° and 32°N in the western USA (Valle, 1963): inflorescence initiation was strong at the higher but very much reduced at the lower latitude.

Plants originating from lower latitudes tend to initiate inflorescences under warm conditions at shorter daylengths than those from higher latitudes. In some cultivars (e.g. Tamar) initiation has even been reported in warm 8-h photoperiods (Cohen & Dovrat 1976, and Table 3.1). The extent to which such initiation in warm SDs is dependent upon previous exposure to cool conditions is uncertain. Plants of Tamar, Louisiana S1 and Grasslands Huia, in which initiation was reported to take place in artificially warm, short natural photoperiods (Thomas 1982), had all received cool SDs immediately before, as had the plants of Tamar reported by Cohen & Dovrat (1976) to have a short critical daylength. And flower head initiation did not occur in clones of Tamar transferred from warm 12-h photoperiods (day/night temperatures of 25/16°C) to artificial SDs at 27/24°C (T). Experiments described by Beatty & Gardner (1961) on a clone of common white clover obtained from an established planting in Arizona suggest that cool pretreatment is not always necessary, however. Their observation that inflorescences formed on plants said to have been grown at 25°C in a heated glasshouse in winter, led them to term this clone a 'short-day type'. Unpublished data obtained by the same authors indicate that the optimal photoperiod for flower initiation in the same clone was 12 h, the total number of flower heads on 3 plants in each of 5 treatments being 11, 44, 7, 3 and 2 in photoperiods of 10, 12, 14, 16 and 18 h respectively. Based on these results, the term 'daylength-intermediate type' would have been more apt than 'short-day type'.

Within lines there is considerable genotypic variation. From Figure 3.10 it can be seen that a few plants of cv. Tamar, for instance, started initiation in April whereas most did not begin until May. Likewise most plants of Grasslands Huia stopped initiation in November, though some continued into January. Critical daylengths for individual genotypes of Kent Wild White under warm conditions have been found to range from about 14 h to more than 16 h (T). Similarly, data for the Pilgrim cv. of Ladino, reported by Laude et al. (1958), show that at Davis, California (c. 38°N), the date of first petal emergence in several different clones overwintered in a warm greenhouse ranged from June 13 to July 14, indicating a range of critical daylengths. More notable, in addition, was the variation in flowering vigour among the clones, the total number of heads produced per plant over the summer ranging from 102 in the earliest to flower to only one in the latest. In this case the number of heads produced seems to have been related directly to the time available for their initiation, i.e. the period over which the daylength exceeded the critical. Within any line, however, including Ladino, there is very great variation in the proportion of nodes producing flower heads under favourable conditions. Within a population of cv. Olwen growing at Hurley in southern England, for instance, the number of heads produced ranged from one at every node in some individuals to one at only every third node along a stolon in others (Thomas 1981a).

The cause of the differences between cultivars is not known. Available evidence suggests, as summarized in the model presented in Section 3. (A) ii.(c), that inflorescence initiation is controlled by a promoter (P), an inhibitor (I) synthesised in the light, and an antagonist (A) synthesised in the dark. On the basis of this model, the greater tendency of plants of low-latitude origin to stop initiating inflorescences in warm LDs compared with those from higher latitudes could be the result of lower levels of P or higher levels of I (or both) in the former. The earlier initiation of inflorescences in autumn in these same plants suggests that their level of dark antagonist is also lower. These hypotheses remain to be tested.

(B) Correlative control of inflorescence initiation

The study of mechanisms of the internal control of inflorescence initiation has been grossly neglected to date. It is frequently observed that heavy shade within a dense canopy reduces the intensity of flowering, but the physiological basis of this effect is largely unexplored. Defoliation has been shown to stimulate inflorescence initiation (e.g. Thomas 1981a, and Section 3. (A) ii.(a), even though leaves are the main sites of photoperiodic perception, and shading has a repressive effect on the growth of axillary buds into lateral stolons (Mitchell 1956; Zaleski 1970). But the relationship between foliage, light intensity, growth of lateral stolons, and inflorescence initiation is in urgent need of elucidation. The data available on the correlative control of inflorescence initiation on axillary buds and lateral stolons, presented below, indicate the potential gain in understanding which might arise from an investigation of this relationship.

Axillary buds within the apical bud of a stolon grow in an uninhibited fashion, initiating leaf primordia at approximately the same rate as does the apical meristem itself (Chapter 2, Section 2.(D) and Thomas 1962). When a vegetatively growing stolon is transferred into environmentally strongly inductive conditions, however, an inflorescence primordium is initiated in the axil of the next leaf primordium to form at the apical meristem, and others are initiated in the axils of the next leaf primordia to form at the apical meristems of each of the lateral branches which have rapidly elongating internodes; but none are formed in comparable positions on axillary buds within and immediately behind the apical bud.

The failure of inflorescence primordia to develop in the axils of newly formed leaf primordia at the apices of the 7 to 10 axillary buds closest to the apex of a stolon is apparently not the result of the slow growth of those buds. Instead there seems to be a correlative inhibition within the apical buds which prevents inflorescence formation on axillary buds, and this inhibition is associated with the presence of the organs at and immediately behind the apical meristem. Removal of the third youngest leaf primordium (P_3) (see Chapter 2, Section 2.(A)),and the tissues distal to it,from the stolon tip of a plant of Grasslands Huia allowed the initiation of an inflorescence primordium in the axil of the next-formed leaf primordium on the youngest remaining axillary bud (subtended by P_4) when the stolon was placed in long days (T). Inflorescence initiation did not occur in that position on intact plants. Apparently it is the apical tissues at the stolon tip which prevent the initiation of inflorescence primordia on the axillary buds within the apical bud.

(C) Effect of growth regulators on inflorescence initiation

Until 1976 no published reports indicating that inflorescence initiation in white clover could be stimulated by treatment with growth regulators had appeared. In the LDP *Trifolium pratense,* inflorescence initiation was reported to be hastened at photoperiods close to the critical daylength by treatment with indoleacetic acid (IAA) at 50 mg/l (Cumming 1959), but this only occurred when plants were pretreated at low temperatures; treatment with IAA in shorter photoperiods was ineffective. In the same species, evidence that gibberellic acid (GA_3) might also be involved in inflorescence initiation has been accumulated by Stoddart (1962, 1966). Applied GA_3 will not induce inflorescence initiation in SDs in *T. pratense* (Stoddart 1959), but in one mutant which fails to initiate inflorescences in LDs, the gibberellin content was found to be considerably lower than that of normal plants. In this clone, application of GA_3 causes initiation to occur in LDs.

In white clover no published reports are available of the influence of auxins on inflorescence initiation, but in a preliminary trial there was no clear indication of any effect of naphthaleneacetic acid or triiodobenzoic acid sprayed onto an Arizona clone at 10 mg/l (Beatty & Gardner, personal communication). Attempts to induce initiation in Ladino (Ridley 1967) and LD clones of Grasslands Huia (T) by application of GA_3 in warm SDs have proved unsuccessful. Unlike Stoddart's non-flowering red clover mutant, Grasslands Huia clone B (Thomas 1961a), does not initiate inflorescences when transferred from warm SDs to warm LDs and does not do so in response to applied GA_3 (T). Initiation was apparently reduced by spraying with 200 mg GA_3/l in the experiments with the Arizona clone mentioned above.

Not all experiments with GA_3 have yielded negative results, however. Recently it was found (T) that the start of inflorescence initiation was hastened by 2 to 3 weeks, and the number of inflorescences per stolon approximately doubled, when clones of several high-latitude cultivars growing in the open during winter in Palmerston North were sprayed thrice weekly with GA_3 at 10 or 100 mg/l. This only occurred in clones in which the critical daylength was lowered by exposure to low temperature and not in those in which it was unaffected. In a more important study, Cohen & Dovrat (1976) showed that in the low-latitude cultivar Tamar, inflorescence initiation could be induced in vegetative plants growing in 8-h SDs by treatment with 20 to 2000 mg GA_3/l. Under identical conditions, however, the high-latitude cultivar Tammisto, which has a longer critical daylength than Tamar (Section 3. (A) i.(a), failed to initiate inflorescences. Neither cultivar responded to GA_{4+7}.

That gibberellin content (which is low in SDs and higher in LDs - see Chapter 2, Section 4.) is not the sole regulatory factor in Tamar is indicated by the observation that the number of inflorescences produced in LDs without applied GA_3 was two or more times that formed in response to GA treatment in SDs. And in Tammisto some other factor is clearly involved besides gibberellins. Cohen & Dovrat suggested that one possible additional regulatory factor is the level of inhibitory compounds in the leaves. They found the inhibitor content in Tammisto at day/night temperatures of 22/14°C to be twice as high in SDs as in LDs, whereas in Tamar it was uninfluenced by photoperiod, remaining at Tammisto's LD level. This led them to the hypothesis that 'the lack of response to applied GA_3 in Tammisto in short days is therefore likely to be the result of interference by the relatively high level of inhibitors.' Neither Cycocel (chlormequat) nor abscisic acid, the latter of which was detected amongst the inhibitors and the level of which doubled in SDs in Tammisto, had any inhibitory effect on inflorescence initiation when applied in LDs, however. Attempts by Mohamed (1981) to influence inflorescence production by treatment with Cycocel and Alar 85 (daminozide) were similarly unsuccessful.

The results of these studies on growth regulators are of great interest in relation to the hypothetical model presented in Section 3. (A) ii.(c). Two important postulates of this hypothesis are that LDs greatly enhance synthesis of an inflorescence initiation promoter (P) which is present only at low concentrations in warm SDs, and that an antagonist of initiation (A) is synthesised in long dark periods. Additionally, in the discussion of the possible physiological causes of differences between plants originating from high and low latitudes in Section 3.(A) v., based on this hypothetical model, it was suggested that the level of A might be higher in plants from high latitudes than in those from lower latitudes.

Several predictions can be made from this model, namely that:

1. In warm SDs the level of A would be much higher in Tammisto than in Tamar.
2. In LDs there would be a high level of P in both cultivars, and if the level of A were artificially raised, it would not necessarily be expected to overcome the stimulatory activity of P.
3. In warm SDs the ratio of A to P would be much higher in Tammisto than Tamar so that any

applied compound favouring P synthesis or P action would tip the balance towards inflorescence initiation more readily in Tamar.
4. Any factor, such as low temperature, which lowers the critical daylength by decreasing A would increase the sensitivity to compounds enhancing P synthesis or action.

The observed effects of photoperiod on the concentrations of naturally occurring growth regulators and the responses to artificially applied substances are in agreement with these predictions. In relation to prediction 1, it is of interest that the level of abscisic acid (ABA) in warm SDs is higher in Tammisto than Tamar. This suggests that the antagonist might be ABA or a related compound. The unexpected finding that applied inhibitors including ABA had no inhibitory effect on inflorescence initiation is satisfactorily explained by prediction 2. It in no way weakens the possibility that the antagonist is ABA.

Lack of response to gibberellins in many cultivars strongly suggests, as proposed by Stoddart (1962) for *Trifolium pratense*, that their action in promoting initiation is secondary to that of another photoperiodically sensitive promoter (P). Their slight stimulatory effect on Tamar in warm SDs, which is not seen in Tammisto, is thus explained by prediction 3. Likewise, their promoting effect on inflorescence initiation at low temperatures in cultivars sensitive to such temperatures but not in those which are not (T) would be expected from prediction 4.

The closeness of fit between the results of experiments with growth regulators and predictions arising from the hypothetical model, which was independently developed from studies of environmental effects, is encouraging and suggests further ways in which the model could be tested.

4. Physiological control of post-initiation inflorescence development

As with most other plant species studied, far more attention has been paid to the internal and environmental causes of the transition of white clover plants from the vegetative to the reproductive phase than to the factors influencing the development of inflorescences once they have been initiated. Our knowledge of the latter is, as a result, limited.

(A) Environmental control of inflorescence and floret development

The emphasis in studies on the effect of environment on inflorescence development has been on the grosser aspects such as rates of growth and ultimate size attained rather than on details of responses in the more functionally important components of florets such as embryo sacs and nectaries. There is considerable scope for further investigations into the effects of environment on carpel fertility.

i. Inflorescence emergence and growth

The rate of inflorescence emergence is completely dependent on the rate of leaf emergence. Thus, if the rate of leaf emergence were as slow as one per fortnight, as in mid-winter in New Zealand (Thomas 1981b), and petal emergence did not occur until the 9th node after initiation (see Figure 3.6), it would take 18 weeks from initiation to 'first petal colour.' Assuming that the average rate of leaf initiation over the period from inflorescence initiation to emergence was as low as this, which it certainly would not be in New Zealand (see Thomas 1981b), flower heads initiated in April in New Zealand or October in England would not emerge until August or March respectively. A real example is shown in Figure 3.11 for cv. Grasslands Huia plants growing in New Zealand (Thomas 1979a): in this case flower heads which were initiated in June did not emerge from the apical buds of stolons until September. In contrast, when the rate of leaf emergence is as high as two per week, as it is in summer, the period from inflorescence initiation to first petal emergence can be as short as four weeks.

The rate of inflorescence emergence is strongly affected by both temperature and daylength. This is demonstrated by the results of an experiment (Thomas 1961c) in which plants of Grasslands Huia were transferred from the field in Palmerston North in mid-August to controlled environments where they received all combinations of 10, 20 and 30°C with photoperiods of 10, 14 and 18 h. The number of days from the time of transfer to the first appearance of petal colour is shown in Table 3.10. The rate of inflorescence growth was decreased by low temperature and to a lesser extent by short days.

Table 3.10. Effect of temperature and photoperiod on (a) mean number of florets per flower head, (b) mean corolla length (mm), (c) mean number of ovules per carpel, and (d) number of days from the start of treatments to the first appearance of white petals on flower heads.†

Temperature (°C)	30			20			10	
Photoperiod (h)	18	14	10	18	14	10	18	14
(a) Florets/head	69	57	36	75	75	57	75	69
(b) Corolla length	8.6	8.2	7.6	10.9	10.3	8.9	11.8	11.3
(c) Ovules/carpel	4.4	4.3	4.3	5.4	5.2	4.8	5.5	5.4
(d) Days to blossom	17	22	30	22	23	26	33	49

†Note: inflorescences were already initiated and were developed to the same extent in each experimental group before the start of treatments (Thomas 1961c).

The effect of temperature on floret opening has not been described in detail. In warm summer conditions in southern England, Erith (1924) observed that two or three horizontal zones of florets opened each day on inflorescences of wild white clover, a whole flower head requiring 7 or 8 days for completion of anthesis. Bearing in mind possible differences in head sizes, this seems comparable to the rate of 7-8 florets 'opening' per day recorded at Davis, California (Vansell 1951). In cooler conditions the rate is very much slower (T).

ii. Inflorescence and floret size

Environment has a strong effect on the size of flower heads and their component parts. When plants on which inflorescences had already been initiated in cool natural winter days were transferred into controlled environment conditions (Thomas 1961c), LDs and low temperatures both independently increased the number of florets per flower head and the size of the florets as indicated by corolla length and ovule number per carpel (Table 3.10). Data for the same genetic material growing in field plots in Palmerston North showed floret number per head to decrease from 61.9 in early November to 51.2 in January and corolla length to decrease from 10.86 to 9.35 mm over the same period (Thomas 1981c). Number of florets per inflorescence was similarly found to decrease from spring to summer in field plots of Grasslands Huia and Grasslands Pitau in New Zealand by Clifford (1979).

In controlled environments, lengths of peduncles and the petioles of their subtending leaves were likewise shown to be strongly affected by temperature (Thomas 1961c). In LDs peduncle and petiole length at 30, 20 and 10° averaged 123 and 76, 202 and 150, and 246 and 195 mm respectively, and the ratio of peduncle to petiole length was thus higher at 30° (1.63) than at 20 or 10° (1.34). In short days at 30 and 20°C the ratio was reduced to 1.21; no data were obtained for cool SDs. Decreases in petiole and peduncle lengths were also recorded in the field for Grasslands Huia as temperatures increased over the period from November to January in Palmerston North (Thomas 1981c), but in this case it was not possible to distinguish direct effects of temperature from a concomitant increase in water stress.

iii. Pollen Fertility

No studies have been made of the effects of environment on development of the female gametophyte or ovule as a whole and only two on pollen fertility. Thomas (1961c) showed that in clones of Grasslands Huia growing in controlled environments, the percentage of sterile pollen grains was strongly affected by temperature, averaging 10.8, 11.2 and 42.0 at 30, 20 and 10°C respectively. The effect of daylength was slight. This effect of temperature was reflected in a parallel study of pollen abortion in a population of wild white clover growing outside in Palmerston North. In these plants pollen sterility decreased from 41% in August (southern winter) to 7% in February (southern summer) (Thomas 1981c).

iv. Inflorescence and floret abortion

Inflorescences initiated in LDs but which undergo most of their subsequent development in SDs develop normally if the light intensities in the short photoperiods are relatively high. Under certain environmental conditions (Thomas 1961c), however, and in certain genotypes (Dunn et al. 1962), inflorescence primordia fail to develop fully into fertile inflorescences. Low light intensities and warm short photoperiods in particular seem to favour the abortion of inflorescences. In all cases of abortion recorded, observations indicate that inflorescence development proceeds normally for approximately the first six plastochrons. Attempts to induce an arrest of development before this stage have failed (T). Thus the initial stimulus to inflorescence formation is apparently strong enough to carry development through six plastochrons. After this stage, however, environmental conditions do have an effect.

When arrested development occurs early, the inflorescences remain contained within the stipular sheath of their subtending leaf and do not become visible without removal of the sheath. Without such removal the false impression is thus gained that no inflorescences have formed. Later cessation of development results in the death of floret buds while they are still immature when the peduncle is about 2 cm long. In these cases there is little or no further development after the leaflets of the subtending leaf have unfolded. It is common for the upper florets of an inflorescence to abort while the lower ones develop fully.

(B) Internal control of inflorescence and floret growth

The only aspect of the correlative control of inflorescence growth which has been studied in any detail is the influence of the flower head on the growth of its subjacent peduncle (Booysen & Laude 1964b). These workers found that once an inflorescence has developed to the stage of emergence from the stipules of the apical bud, i.e. about five or six plastochrons after its initiation, further elongation of its peduncle is dependent on the presence of the flower head. Excision of the young flower head at this stage reduced subsequent peduncle elongation in Ladino clover from 17.7 cm in the intact controls to 1.25 cm. Most of the effect of the flower head was brought about by the floret buds, removal of which alone was nearly as effective as removing the whole head. The stimulatory effect of flower heads on peduncle elongation appears not to be dependent on the development of fully fertile florets. In sterile plants of Ladino described by Dunn et al. (1962), peduncle elongation appeared normal in spite of the failure of florets to produce ovules or fertile pollen; it therefore seems probable that the elongation of the peduncle is under the control of florets at a relatively early stage of their development.

Young flower heads possibly control the elongation of their subjacent peduncles by supplying gibberellins and auxins. Booysen & Laude (1964b) found that substitution of excised flower heads by application of IAA at 1000 mg/l or GA_3 at 10 000 mg/l in lanolin resulted in elongation of decapitated peduncles to 4.83 and 6.43 cm in IAA- and GA_3-treated peduncles respectively. The combination of IAA and GA_3 in lanolin boosted elongation to 12.38 cm, indicating an additive effect.

Few other effects of growth substances on the post-initiation development of inflorescences and florets have been reported. Calder et al. (1973) noted that high concentrations (15 000mg/l) of Alar-85 led to the formation of malformed flower heads in which the growth of some pedicels continued after the drying out of all other pedicels and small florets. 'The last-formed florets on the abnormal [flower heads] were inclined to be larger than normal, but in most cases the 'petals' were green and remained so' (Calder, personal communication). Similar concentrations of Alar-85 and Cycocel were found by Mohamed (1981) to have little effect on flower head development, and a preliminary study revealed no marked response to GA_3 when it was applied to newly-emerged inflorescences (T).

5. Relationships between flowering and growth

The relationships between flowering and growth in white clover are complex. The earliest response associated with the initiation of inflorescences is a stimulation of growth, but this is followed by decreased vigour of vegetative growth in strongly flowering plants and decreased persistence of plants from one growing season to the next.

(A) Stimulatory effects of flowering on growth

In many plants the size attained by successively produced leaves formed after the onset of flower initiation decreases from node to node. In white clover this occurs to only a very slight degree (Section 5.(B)); the final size attained by leaves formed during the reproductive phase is not markedly different from that of those formed during the vegetative phase. Transfer from short days to long days, which causes inflorescence initiation, leads to a transitory increase in growth rate of leaf laminae, petioles and internodes, however (Thomas 1961b). When plants previously grown in short days were transferred to 0, 2, 4 or 6 days of continuous light (CL) before return to SDs, the number of inflorescences initiated at the first three nodes formed after the start of CL treatment, and the growth rate of leaves and internodes formed at the same time, were both positively correlated with the duration of CL. The growth rates of, and final sizes attained by, immature leaves and internodes already present in the apical buds of stolons at the start of CL treatment were unaffected by it. The data obtained in this investigation do not allow conclusions to be reached as to the nature of the causal relationships between growth rates, inflorescence initiation and light treatments. They do show, however, that the stimulatory effect on growth associated with the initiation of inflorescences is localised in the newly-formed tissues in their immediate vicinity.

In addition to the correlation between inflorescence initiation and growth rates, a strong correlation exists between the presence of inflorescences and the final size attained by associated internodes. It is commonly observed that the internode immediately proximal to an inflorescence, particularly the first one formed at the onset of the reproductive phase, is longer than the average on a given stolon and that the internode distal to the inflorescence is shorter than average (Kawanabe et al. 1963, Booysen & Laude 1964b, Thomas 1980a). In a detailed study of this correlation, similar to that by Thomas (1961b) referred to above, Booysen & Laude demonstrated a correlation between the lengths of mature internodes in Ladino clover and the presence of inflorescences. Figure 3.12 shows the effect of transfer from short days into 0, 2, 3 or continuous 16-h long days on lengths of mature internodes relative to inflorescence positions. Plants remained vegetative after 2 LDs, whereas up to 2 inflorescences formed per stolon after 3 LDs and 4 in continuous LDs. Internodes immediately proximal to the first-formed inflorescences were approximately 40% longer than average, whereas lengths of those on plants which remained vegetative after transfer to 2 LDs were the same as those remaining in SDs throughout. Because increased internode length only occurred in association with inflorescence initiation, and not in response to 2 LDs, Booysen & Laude concluded that it was causally linked to inflorescence formation rather than being a direct response to a sudden increase in daylength. This conclusion is supported, but not proven, by observations of plants growing in the field (e.g. Thomas 1980a) which show the same relationship between final internode length and inflorescence position to exist under conditions in which natural increases in photoperiod occur only slowly.

The final lengths reached by internodes are determined while they are still within the apical bud; excision of flower heads from inflorescences at the time of their emergence from between the stipules of the apical bud does not prevent the internode proximal to the inflorescence becoming longer than average (Kawanabe et al. 1963; Booysen & Laude, 1964b). Whether the greater length of these internodes is the result of increased cell number or cell length has not been ascertained. It seems likely, however, that the formation of an inflorescence is associated with an increase in the number of cells in the immediately proximal internode rather than with an increase in their average length, because the number of cells along the axis of an internode is largely determined (at least in vegetative plants) before its emergence from the apical bud, whereas the greatest increases in cell lengths take place after internode emergence (Thomas 1972, see Chapter 2, Section 2.(E)).

(B) Decreased growth vigour associated with flowering

Despite the stimulation of vegetative growth associated with the onset of the reproductive phase (Section 5.(A)), vegetative growth in the field usually declines after the start of blossoming. The extent to which this is a result of inflorescence formation rather than a direct response to adverse climatic conditions is uncertain.

Several observations and experiments point to an inhibitory effect of inflorescences. The internode on the distal side of an inflorescence is frequently shorter than average (Section 5.(A)) as also are the leaf lamina and petiole at the adjacent node on that side (Kawanabe et al. 1963). Experiments by the latter workers indicated strongly that, as with thc stimulatory effect described in the previous section, this inhibitory effect of inflorescences occurs while they are still at an early stage of development within the apical bud.

The same experiments by Kawanabe et al. suggested that there is also competition for nutrient between older inflorescences and vegetative organs. After a 3-week treatment period, the dry weights of lateral stolons and roots of Ladino clover were found to be higher in plants from which inflorescences had been removed immediately after their emergence from the apical buds than in intact control plants. The increase in dry weight of stolons and roots in the treated plants was approximately the same as the dry weights of flower heads and peduncles remaining in the untreated plants.

The extent to which these competitive effects are dependent on seed development is unclear. Dunn et al. (1962) observed, for instance, that stolon elongation in two mutants of Ladino clover that produced sterile flower heads was twice as great as in normal plants, supporting the hypothesis that seed production, rather than flower head production, per se, reduces vegetative growth. On the other hand, under glasshouse conditions in which seed set does not occur, prolifically flowering pot-grown plants frequently show a decline in internode elongation after the production of several flower heads on each stolon which does not occur in vegetative plants growing in the same conditions (T). At present it seems probable that different inhibitory influences are associated with each of the three stages of inflorescence development, namely pre-emergence, post-emergence and post-fertilisation, but a full understanding of the influence of inflorescences on vegetative growth is not possible with the limited data available.

In an attempt to distinguish between the possibilities that vegetative growth in summer is adversely affected by either climatic conditions or the presence of flower heads, Kawanabe et al. (1963) compared growth of Ladino clover with and without flower heads over the period from spring to autumn. In the group of plants without flower heads, all inflorescences that had emerged from the apical buds of stolons were removed at 5-day intervals throughout the period of the experiment. Removal of flower heads had little effect on the overall decline of growth during the summer, suggesting that under the conditions of this experiment adverse climatic factors might have been the primary cause of reduced growth. The alternative explanation, that reduced growth might have been caused by the presence of immature inflorescence buds in the apical buds of the stolons, cannot be ruled out, however. It will not be possible to distinguish between these alternatives until the growth of completely vegetative plants is compared with that of genetically similar plants in which no inflorescence initiation is taking place.

(C) Flowering and persistence

The chances of a white clover plant persisting from one growing season to the next are strongly influenced by its strength of flowering. Four possible reasons for this are that:

1. The initiation of inflorescence primordia has an inhibitory effect on growth (Section 5.(B)).
2. The growth of lateral stolons and roots is decreased as the result of competition for nutrients between unfertilised flower heads and vegetative organs (Section 5.(B)).
3. The growth of lateral stolons and roots is decreased as the result of competition for nutrients between developing seeds and vegetative organs (Section 5.(B)), and
4. The more nodes that are committed to inflorescence formation, the fewer there are that

possess vegetative axillary buds and therefore retain the capacity to produce new lateral stolons (Section 1)

In discussions of persistence, the last of the above possibilities is most commonly suggested as the reason for its inverse relationship with flowering and the first three possibilities are often overlooked. They are probably of far greater significance than is generally acknowledged, however, and should not be forgotten. The relationship between persistence and flowering vigour was shown clearly in an investigation by Gibson (1957) who grew lines of Ladino and Louisiana clover at latitude 32° 50'N. in Alabama. At this

Table 3.11. Relationship between flowering vigour and persistence in dry summer conditions, measured by the number of live stolons/m² on 29 July.†

Genotype flowering vigour	Average number of live stolons	
	Natural day length	Extended day length
Weak	28	8
Medium	25	23
Profuse	11	17

†Plants were grown outside in natural photoperiods or with day length artificially extended to 17-19 h. (From Gibson 1957).

latitude some lines of Ladino flower extremely weakly, probably because the maximum photoperiod attained in midsummer (14 h) is below the critical daylength for their LD reaction (see Section 3.(A)i.(a)). Gibson compared persistence in plants of this type (weakly-flowering) with that in profusely-flowering Louisiana and in more freely flowering Ladino lines (medium-flowering), which presumably had shorter critical daylengths. In natural photoperiods, profusely-flowering Louisiana plants were less persistent in late July, as measured by the number of live stolons present, than either group of Ladino (Table 3.11). Whether this difference was solely the result of differences in flowering vigour or brought about by other genetic differences is uncertain.

A relationship between flowering and persistence was more clearly indicated by comparing the persistence of plants grown in natural photoperiods with others in which the natural daylength during spring was artificially extended to 17-19 h. In Louisiana plants, such extended LDs brought the peak period of flowering forward from May to April (see Figure 3.13) so that by the end of May the number of flower heads emerging was greatly reduced. This earlier flowering was correlated with increased persistence. Inflorescence initiation presumably stopped in these plants during June (see Section 3.(A)ii.) and Figure 3.10), and Gibson suggested 'that the reversion from flowering to vegetative stage enabled the plants to tolerate the [hot dry summer weather] better than plants ....that were in full blossom.' An inverse relationship between flowering vigour and persistence was also found in Ladino. The weakly-flowering line persisted better in natural daylengths than with extended photoperiods in which it flowered more strongly. Medium-flowering Ladino persisted equally well in both natural and extended daylengths, however.

Unfortunately, it is not possible to distinguish from the results of Gibson's experiment which of the four aspects of inflorescence development mentioned earlier led to the strongest reduction in persistence. Dissection of apical buds during June and July to determine their reproductive state would have greatly assisted interpretation. Without further detailed investigation our understanding of the nature of the relationship between flowering vigour, vegetative growth and persistence must be considered very inadequate.

6. Pollination, fertilisation and seed formation

Once a fertile flower head has reached maturity, the formation of ripe seeds is dependent on the successful completion of the steps from pollination through to fertilisation, endosperm formation and embryogenesis. This sequence of events is of obvious significance for seed production, the problems of which are discussed in Section 7. The account which follows reveals the inadequacies of our present knowledge of this sequence and points to the need for further detailed studies to provide a sound basis for the development of measures to improve white clover seed production.

(A) Pollination

White clover has a well developed genetic gametophytic self-incompatibility mechanism (Williams 1931, Atwood 1940, 1941a) with only a small proportion of plants in a population being quite strongly self-compatible (see Chapter 12). Cross-pollination is thus essential for significant seed set within a population and the most important natural pollinating agents responsible for this are bees (Darwin 1876, Erith 1924).

Although the anthers of a floret dehisce before its petals are fully elongated (see Figure 3.6), autogamous self-pollination is infrequent. Seed set without the aid of an external pollinating agent does occur regularly in some genotypes, however (Atwood 1943). Within a population of plants in which the transfer of pollen from anthers to stigmas by an external agent was prevented, the average number of seeds set per flower head was found to be only 0.5 by Atwood (1941a) and 2.7 by both Scullen (1952) and Dunavan (1953). Other investigators have found the seed set under such circumstances to be lower, possibly as a result of differences in either temperature (see Section 6.(C)) or genotype, averaging only between 0 and 0.05 per flower head (Darwin 1876, Erith 1924, Ware 1925, Williams 1931, Green 1956).

When florets are artificially self-pollinated, either by hand pollination or by gently rolling the heads between thumb and fingers (rubbing), the amount of seed set is increased. Atwood (1941a) obtained an average of 5.7 seeds per head by hand self-pollinating and by rubbing, 90% of his plants then setting some seed. Harberd (1963) found 3% of selfed carpels set seed. Other workers have had less success, however, Ware (1925) obtaining virtually no seed by selfing and Williams (1931) finding seed set in only 25% of plants. Once again, the differences reported probably result largely from differences in temperature and genotype.

White clover florets are so constructed that, according to Knuth (1908), bees alone are able to work them effectively so as to bring about regular pollination. A visiting bee alights on the wing petals, each of which is adherent, just distal to its vesicular protuberance, to the corresponding keel petal, so that the wings and keel petals move simultaneously (Figure 1.12). As the bee thrusts its head beneath the standard towards the nectar at the base of the staminal tube, it depresses both wings and keel. During this process the vesicular protuberances on the wing petals slide around the bases of the staminal filaments and open the keel laterally away from the anthers with minimal disturbance to anthers and stigma (T). The latter then come into contact with the bee's underside (Figure 3.14). The insect gains access to the nectary by inserting its proboscis into the staminal tube through the gaps either side of the free posterior stamen.

After a visit, the parts of the floret return to their original position, this movement according to Müller (1873) being chiefly caused by the action of the standard and wing petals. He states that the broad claw of the former grasps the other petals, stamens and pistil, and, because of its elasticity, guides them back to their original position, the elasticity of the vesicular processes on the wings assisting in the movement.

Macleod's (1894) explanation of the floral mechanism differs somewhat from the above. He considered that the floret consists of two cooperating parts, the wings and keel. Although moving simultaneously, these are able to execute independent movements, but by cooperation achieve greater certainty of movement and so minimise exhaustion. In support of this he found that even if the wings and keel of an unvisited floret are pressed downwards thirty or forty times in succession the mechanism is not exhausted, and the reproductive organs always return into the keel. If, however, the pressure is exerted several times in succession on the wing petals alone or the keel alone, these parts resume their original positions slowly and the mechanism shows signs of exhaustion.

Personal observations (T) of petal movement suggest that the return of the wing and keel petals to their original positions after a bee's visit is largely caused by the adhesion of wings to keel and the presence of vesicles. Adhesion of these petals to each other prevents their movement relative to each other when they are pushed downwards, with the result that the claw of each wing petal is slightly stretched and that of the keel compressed. It seems probable that the tension and compression so induced provide the force causing the displaced petals to be drawn back up into their original positions. The rounded vesicles on the wing petals clearly act as spacers separating the wings from the standard (Figure 3.14). As a result, the area of contact between wings and standard is relatively small. Together, the curved surface of the vesicles and their small area of contact with the standard reduce the frictional resistance to the wing and keel petals sliding over the surface of the standard during and after a visit by a bee.

Observations of bee foraging behaviour on white clover show that the attractiveness of inflorescences and individual florets varies greatly. Heinrich (1979) noted that bumblebees often approach 'flowers' (flower-heads?) and then veer off at the last moment. He suggested that they can detect the presence or absence of nectar by scent, visiting preferentially those 'flowers' that contain nectar. Oertel's (1961) comment that blossoms only appeared attractive to honeybees at temperatures of 27°C or more is possibly explicable on the basis of a stimulatory effect of warm (or dry) conditions on nectar secretion.

White clover produces only rather small quantities of nectar compared with many other forage legumes. Nectar is reported to be absent at the time of floret opening, to accumulate slowly over the next 24 h, and then to continue to build up over the next few days (Vansell 1951). Weaver (1965) stated that in a comparison of bee foraging behaviour on white clover and hairy vetch (*Vicia villosa*) growing in the same conditions, florets of the former contained an average of only 0.1 μl of nectar compared with 0.8 μl in the latter. In the same paper he reported nectar levels per floret to range from as low as 0.02 μl in uncaged plants up to 0.34 μl in plants caged to exclude bees.

Variations in nectar content of florets are probably one important cause of the variation in number of florets per flower head foraged by bees. On each flower head visited, the number of florets entered by individual honeybees ranges from 3 to 8 (Oertel 1961) and commonly averages approximately 3.5 (Weaver 1965). The relationship between nectar content and bee foraging is indicated by the results of an experiment in which the behaviour of bumblebees on uncaged plants was compared with that on adjacent plants which had been caged to exclude bees for the preceding 24 hours (Morse 1980). On the day of the observations, the mean nectar content of florets on caged plants was 0.1 μl, but in florets on uncaged plants visited by bees the previous day it was too low to be detectable. Bumblebees visited an average of 5.4 florets per flower head on the previously caged nectar-rich plants compared with only 3.2 on the nectar-poor uncaged ones.

The pattern of bee foraging on white clover suggests that a high proportion of self-pollination probably occurs, at least in some circumstances. Harberd (1963), for instance, described individual plants up to 20 metres across growing on sand dunes, and plants growing on free-draining soils in arctic Norway similarly often cover areas of well over a metre in diameter (Thomas 1980a). In such plants the chance of self-pollination is very high indeed because the distance over which bees are recorded as transferring pollen from one flower head to another is rather short. When genetically distinctive plants of normal V-marked S100 and S100 Nomark white clover were grown in adjacent plots, Roberts (1967) found that only 2% of seed set resulted from fertilization by pollen transferred over a distance of more than 24 m and 5% from transfer over a distance of more than 5 m. Studies of the movement of bees from flower head to flower head support this conclusion. Weaver (1965) found that the distances between successive flower heads visited by honeybees were: 2.5 cm or less in 7.4% of visits, 2.5-5 cm in 11.5%, 5-7.5 cm in 21.8%, 7.5-10 cm in 20.2%, 10-15 cm in 13.2%, 15-20 cm in 7.3% and 20-25 cm in 5.0%. In other words 60.9% of flower heads visited were within 10 cm of the previous head pollinated. Only in 13.6% of cases was the distance between successively visited heads more than 25 cm and in these the frequency of visits decreased with distance. The proportion of successive flower heads visited which are of the same genotype must therefore be very high. In addition, as mentioned earlier, several florets are usually visited successively by bees on each flower head.

Investigations to date suggest the existence of a close relationship between bee foraging behaviour, nectar production and pollination, and these must all be influenced by environmental conditions. The results available do not provide us with a clear understanding of these relationships, however, and further detailed studies are required, as are data on the percentage of self- to cross-pollination which occurs under field conditions.

(B) Pollen germination and pollen tube growth

The interaction between the stigma and pollen germination in white clover has yet to be elucidated. It is not known how much, if any, of the pollen shed within a floret before anthesis comes into contact with the stigma. Preliminary observations (T) suggest that pollen on the stigma does not germinate until the cuticle covering the stigma papillae is ruptured and releases the mucilaginous secretion beneath. Presumably pollen germination in 'rubbed' flower heads results from the combined release of mucilaginous matter by mechanical damage and the transfer of pollen grains from dehisced anthers to stigma surfaces. Visits by bees would have a similar effect to manual 'rubbing'. The report by Rinderer et al. (1981) that the intensity of honeybee movement during visits to flower heads was positively correlated (albeit weakly) with seed set also suggests the importance of mechanical damage in stimulating pollen germination.

Compatible pollen grains germinate quickly on the stigma. The tube nucleus and generative cell then pass into the growing pollen tube where the latter divides to give rise to two male gametes (Chubirko 1965). Pollen grains also germinate within an hour at about 18 °C when placed on the surface of a 25% sucrose solution in an open watch glass but will not do so in enclosed hanging drops of solution (Erith 1924).

Data on comparative rates of germination of compatible and incompatible pollen on stigmas are conflicting. Atwood (1941b,c) and Cohen and Leffel (1963) reported germination rates of compatible pollen to be faster than incompatible, but Chen and Gibson (1973) found little difference in germination rates. The mean rate of pollen tube growth is faster following compatible cross-pollination, although pollen tubes of occasional individual incompatible grains grow as fast as the compatible tubes (Atwood 1941b, Chen & Gibson 1973). Under unspecified warm conditions, Atwood (1941c) observed that pollen tubes of compatible grains had traversed 70-80% of the distance from stigma to ovary within an hour of pollination, growing on the walls of the cells lining the stylar canal and becoming restricted to the ventral side of the canal as they neared the ovary (see Figure 1.16). Growth of the tubes of compatible pollen grains slowed down towards the base of the style but many then proceeded to grow along the placenta, reaching the second ovule from the distal (stylar) end of the ovary by about 4 hours and the fourth by about 6 hours after pollination. Incompatible pollen tubes usually failed to grow out from the base of the style into the ovary. Based on these observations, Atwood concluded that there are two 'interference zones' active in preventing the growth of incompatible pollen; namely one on the stigma and the other about three quarters of the distance down the style. Which, if either, of these is responsible for the apparent competition between cross-compatible and self-compatible grains on a common stigma (Atwood 1943) is not known.

Temperature has a strong influence on pollen germination and pollen tube growth. Comparing plants at constant temperatures of 15, 25 and 35 °C, Chen and Gibson (1973) found that germination of pollen grains on compatible and incompatible stigmas was faster at higher temperatures, but detected no differences between germination rates in compatible and incompatible crosses at any temperature. Rates of pollen tube growth were also strongly influenced by temperature: from the time of compatible cross-pollination, pollen tubes took 20 h to grow down the style to reach the ovary at 15 °C compared with only about 2 h at 35 °C. At all three temperatures the pollen tube growth following incompatible self-pollination was slower than after cross-pollination. Pollen tubes reached the ovules at each temperature in compatible crosses, but only did so at 35 °C in incompatible self pollinations. At 15 °C, tips of incompatible pollen tubes swelled (and sometimes burst) in the upper or middle part of the style; at 25 °C they reached the lower region of the style or the upper (distal) end of the ovary but mostly swelled and burst at that stage. No fertilisation occurred after incompatible self-pollination at 15 and 25 °C, but seed was set at 35 °C. These responses to temperature suggest possible causes of the discrepencies and variations in the results obtained by different workers, and by the same worker in different years.

(C) Fertilisation, endosperm growth and embryogenesis

Cross-pollination usually results in fertilisation, but the quantity of seed set after self-pollination depends on genotype and temperature. Chen and Gibson (1973) reported no seed set after incompatible self-pollination at 15 or 25 °C but the amount of fertilisation and early endosperm and embryo growth at 35 °C was almost as great after incompatible self-pollination as after compatible cross-pollination. High temperatures act to break down the incompatibility barrier in incompatible crosses. The amount of self-fertilization occurring in the field is generally reported to be low (Section 6.(A)), but the observation by Oertel (1961) that blossoms only appeared attractive to bees at temperatures of 27°C or more suggests that it might be more frequent than commonly imagined.

Before fertilization occurs, the polar nuclei fuse near the egg apparatus and the antipodal cells become almost disintegrated. In natural summer conditions in England, fertilization takes place 15-20 h after pollination (Erith 1924). Atwood (1941c) observed 2-nucleate proembryos in embryo sacs within 16 hours of pollination. The effect of temperature on the rate of pollen tube growth (Section 6.(B)) has a strong influence on the time taken from pollination to fertilization. At controlled temperatures of 15, 25 and 35°C, the percentages of ovules showing signs of embryo and endosperm development were 0, 48 and 65 respectively 16 h after compatible pollination and 30, 71 and 82 after a further 8 h (Chen & Gibson 1973).

Division of the primary endosperm nucleus formed as a result of double fertilization begins at about the same time as the first divisions of the zygote. At 25°C there were 5.2 nuclei in the endosperm 16 h after pollination (Chen & Gibson 1973). Rapid free nuclear divisions occur near the periphery of the embryo sac in the region of the growing embryo, but few near the antipodal end. Cell walls form between the nuclei at about the time the cotyledons become visible. The tissue thus produced never fills more than half the embryo sac (Figure 3.15B,c). Fully developed endosperm cells have thin walls, contain a nucleus with a relatively large nucleolus, and never contain starch grains. Fully developed ovules retain very little nucellar tissue and no traces remain in the ripe seeds (Erith 1924).

The sequence of embryo development is shown in Figure 3.15A. The first division of the zygote to produce the proembryo is transverse. The inner cell divides typically to produce a dicotyledonous embryo, whereas the outer divides in several planes to produce a massive multicellular suspensor which never consists of a single row of cells. The embryo and suspensor are not very clearly delimited, although the nuclei and nucleoli of the suspensor are smaller than those of the embryo and the suspensor cells soon show signs of degeneration. There is no hypophysis and the suspensor does not contribute to the formation of the root cap (Erith 1924, Guignard 1880).

The rate of embryo development has been described in unspecified warm conditions by Erith (1924) and Atwood (1941c) and at temperatures of 15, 25 and 35°C by Chen and Gibson (1973). Results obtained by the former two workers are similar to those found at 25°C by Chen and Gibson. Atwood (1941c), in the fullest account of the three, reported the proembryo to be 2-celled 18-20 h after pollination, 5- to 6-celled 10h later and 10- to 16-celled 36 h after pollination. 48 h after pollination the embryo and suspensor were distinct and 5-8 days later the embryo was U-shaped.

Embryos in both immature and mature seeds are green (see Section 6. (D) iii.) but this colour is lost during seed ripening. In the ripe seed the cells of the embryo show very little differentiation. Stomata are absent from the cotyledons and although the position of future vascular strands is indicated by the presence of elongated procambial cells, there is no development of the characteristic wall thickening of protoxylem tracheary elements. The cells throughout contain minute oval starch grains. A single leaf primordium is present at the stem apex of the plumule.

(D) Seed and fruit development

Development of an ovule into a seed proceeds as a response to fertilisation. In compatible crosses Atwood (1941c) described the micropyle as seeming shut 24 h after pollination 'thus preventing more pollen tubes from entering,' but he did not comment on its condition after incompatible pollination. Pollination or pollen tube growth alone, as occur with incompatible self-pollination, do not stimulate ovule growth. The structure of the mature seed is shown in Figures 3.15Bd and 3.17a-c. Externally it is approximately cordate, and in wild white clover is about 1 mm wide and 0.5 to 0.7 mm thick when ripe (Erith 1924). The testa is smooth and pale yellow or, less commonly, brown and of uniform thickness over most of the seed. A T-shaped micropyle and a small roundish hilum are in close proximity at one end. At this end the testa is thicker than elsewhere, a well-developed strophiole developing from the chalazal tissue beneath the hilum.

i. The testa

Extensive changes occur in the integuments of the ovule as they develop into the testa of the seed. Over much of the ovule surface the two fused integuments are each about three cells thick. After fertilisation, anticlinal divisions in the outer cells of the integuments continue after division in the inner cells below has stopped. This is apparent by comparing the anatomy of the testa in an unfertilised ovule and a developing seed. In the sequence shown in Figure 3.16a-d the relative numbers of cells in the outer epidermis (x) of the outer integument, the adjacent sub-epidermal layer (y) and the inner epidermis of the inner integuments (z) change from 14:10:11 to 18:10:4 in 3.16a and b respectively. Ultimately, in 3.16c and d the number of outer epidermal cells relative to subepidermal cells increases to more than 3:1. These changes accompany a 3- to 4-fold increase in average ovule diameter. While the outer epidermal cells continue to divide, the innermost cells become stretched and collapse, and the middle lamellae of the cells nearer the outside break down to various degrees leading to the formation of intercellular spaces. These are particularly apparent in the subepidermal layer (y), the cells of which expand more at their innermost than outermost ends so that large intercellular spaces (i) form immediately beneath the outer epidermis. The mature form of these cells, which are variously referred to as 'osteosclereids' (Watson 1948) or 'hour-glass' and 'pillar' cells (Hyde 1954), resembles the 'button' stage of mushroom development. Seen from the outside, the whole subepidermal layer can be envisaged as a sheet of cells made up of inverted button mushrooms in which the 'caps' are contiguous and the short, relatively thick, 'stalks' project outwards holding the epidermis away from the main body of the seed. At maturity these cells are dead, with thin walls at their cap ends and thick walls to their stalks so that in surface view the subepidermal layer appears as in Figure 3.16f.

In the mature seed two layers of intact cells remain beneath the subepidermal cells (Figure 3.16e). The outer of these, the innermost layer of the three-layered outer integument, consists of thin-walled parenchyma cells (n). The testa down to and including this layer can be readily stripped as a separate skin from unripe seeds. For doubtful reasons this inner layer has been termed the 'nutrient layer' (Watson 1948) in other *Trifolium* species. Beneath it, the outer epidermis of the inner integument (m) remains intact as a layer of parenchyma, the outer walls of which are slightly thickened and strongly cutinized. The equivalent layer in other *Trifolium* species is termed the 'aleurone layer' by Watson (1948). At maturity these cells retain cytoplasmic contents, but whether they contain stored protein is unclear. The remains of other cells crushed between the testa and the mature embryo originate mainly from the innermost cells of the inner integument (Erith 1924) but must also include fragments of degenerated nucellar and endosperm cells.

The outer epidermal cells of the outer integument develop to provide the tough, impermeable, exterior to the seed. In the unfertilised ovule they are approximately isodiametric (Figure 3.16a), but after fertilisation they divide anticlinally and elongate perpendicularly to the ovule surface to become 4 to 5 times as long as broad. The elongated cells, which are referred to as epidermal palisade cells (Erith 1924, Hyde 1954) or Malpighian cells (Watson 1948), become peculiarly thickened by the formation of 6 to 8 tapering ribs which extend along the cell walls from the outer (apical) end towards the inner (basal) end. In surface view these ribs give each cell a rather stellate appearance (Figure 3.16j). At the outer ends of the cells the ribs are so broad that the cell lumen is almost obliterated, and the lumens thus broaden towards the cell bases (Figure 3.16 e,g,h). The outer cell walls, which are very thick, forming a capping layer on the outside, are overlain by a thin cuticle (k). An abrupt change in the refractivity of the cell walls about 8μm below the external surface (Erith 1924) produces a very distinct 'light line' (L) similar to that seen in other leguminous seeds when examined as sectioned material under a light microscope.

ii. The hilum

Hilum development in white clover is undescribed. Hyde (1954) studied hilum structure in white clover and *Lupinus arboreus* but published a description only of the latter, commenting that 'the anatomy of the testa in the region of the hilum is essentially the same [in the two species]'. His description differs greatly from that of Erith (1924). Re-examination of the hilum of white clover (Figure 3.17c) shows Erith's description to be inaccurate and that its structure is indeed similar to that of *Lupinus*, although it differs in several minor respects (T).

The hilum is a round to oblong scar about 150 μm across, lying in a depression (Figure 3.17 a,b). Tissue derived from the funicle overlies the epidermis of the testa in this region. In contact with the epidermis is a counter-palisade layer of funicular origin (Hyde 1954) composed of cells elongated in the direction perpendicular to the seed surface and having walls heavily thickened except towards the outer ends. Overlying these counter-palisade cells is a layer of parenchyma cells. In the same plane as the micropyle there is a groove in the funicular tissue coinciding with a fissure in the testa epidermis. Immediately below and alongside the fissure lies a bar of tissue composed of short tracheids with large bordered pits (the 'tracheid bar'; Hyde 1954). This is surrounded by rather irregularly shaped aerenchyma cells referred to by Hyde as 'stellate cells.' The intercellular spaces in this tissue are continuous with those of the subepidermal layer which underlies the epidermis everywhere except in the region of the hilum.

iii. Physiological changes during seed development

Seed development in all plants involves a highly coordinated sequence of events. In white clover the ovule at anthesis is green and probably capable of photosynthesis. A vascular connexion between the carpel and the chalaza via the funicle presumably supplies water and nutrients from the parent plant. At this stage the cells of the outer integument are definitely still meristematic but the condition of the inner integument and chalazal cells is not known.

After fertilisation, growth of the embryo, endosperm and ovule proceed in a coordinated way. The youngest embryos appear to contain no chlorophyll, presumably being fully dependent on the endosperm for nutrient. By the time the embryo reaches the start of the heart stage of development (Figure 3.15Ac) it is beginning to synthesise chlorophyll and starch grains start to appear in the cells of the subepidermal layer of the testa. Starch grains become particularly abundant in the region below the hilum as the embryo continues to grow. Once it is fully grown and the ovule reaches its maximum size, this starch has mostly disappeared. Erith (1924) suggests that this starch is used as nutrient for embryo growth, but it seems just as likely that it is either simply redistributed into the cotyledons as stored material or utilised as raw material for continued thickening of the outer epidermal cells of the testa. At this point all the cells of the embryo except those at the extreme radicle tip, and all the cells of the outer integument except the outer epidermis, contain chlorophyll, but the strophiole and inner integument cells do not. Development of the outer epidermal cells of the testa is still incomplete when the ovule first reaches maximum size. Additional thickening is laid down after this time and only later does the light line become distinct.

Post-fertilisation ovule development is rapid (Hyde et al. 1959). At natural spring/summer temperatures in New Zealand, about 50% of seeds are viable within only 12 days of pollination although they are still green. At this stage the seeds are deficient in nutrient reserves, however, and are unable to withstand adverse storage conditions or sowing below a depth of about 2.5 cm (Hyde 1950). Maximum viability is reached about 20 days after pollination, at which time the seed colour begins to change from green to yellow. For about two weeks after viability is reached, seed dry weight continues to increase while moisture content remains fairly constant (Figure 3.18 and 3.19).

As the seeds develop, their testas become impermeable to water and gases. Such seeds with impermeable testas are unable to germinate and are termed 'hard'. From the data of Hyde et al. (1959) it is clear that: (1) hardness develops about 5 to 7 days after viability is reached, (2) dry weight continues to increase after hardness is attained, and (3) that most moisture loss from the seed during ripening on the plant occurs after the seeds become hard. The latter conclusion implies that during the development of hardness, water is either lost outwards through the testa without being able to enter the seed in the reverse direction, or lost mostly through some other part of the seed surface such as the hilum.

Experiments in which hilums of red clover seed were blocked after ripening (Hyde 1954) showed that water loss into the dry atmosphere of a desiccator was thereby prevented; in other words, in ripe seeds the testa was impermeable both outwards and inwards. Hyde also concluded from studies of lupin and red clover seeds removed from their pods before ripening that nearly 90% of the water loss occurs through the testa epidermis during ripening. Assuming the data in his 1959 paper regarding the relative times of water loss and development of hardness to be correct (and there is no reason to believe otherwise), we are presented with three possibilities, namely, that:

(1) During seed ripening the testa allows the outward escape of water vapour while at the same time preventing inward movement,

(2) Testas become fully impermeable in both directions only after hardness is attained,

(3) The ripening processes within the pod (as described by Hyde et al. 1959) differ from those occurring when seeds are removed from the plant before ripening (as described by Hyde 1954).

Clearly there is need for a detailed study of seed ripening in white clover in which parallel observations are made of the pathway of water loss and the development of hardness. The lack of data on the latter in Hyde's 1954 paper makes interpretation of his 1959 results very difficult.

(a) The role of the hilum

The following three observations on the relationship between the atmospheric moisture content during seed ripening, the degree of seed hardness, and germination in white clover were reported by Hyde (1954):

(1) Seeds ripened off the plant at high (70%) relative humidity (RH) subsequently germinated more rapidly (80% in 2 months) than those ripened at lower RH. The intensity of hardness was such after ripening at 30% RH that only 3% germination occurred within 18 months.

(2) Seeds ripened at low RH absorbed little water vapour when transferred to 70% RH for 71 days, and they germinated very poorly when tested subsequently. Similar seeds exposed to RH that was gradually increased from 45 to 70% over the 71-day period absorbed significant amounts of water vapour and showed about a six-fold increase in germination when tested subsequently.

(3) The moisture content of seeds decreased when they were transferred to a drier atmosphere than that in which they had been previously stored but did not increase when they were transferred to a moister atmosphere.

In the same paper, Hyde described the structure of the hilum in *Lupinus arboreus* and showed that it opens and closes in response to changes in the relative humidity of the external atmosphere. He suggested that the counter-palisade layer acts as motor tissue which dries and shrinks, drawing the margins of the hilar fissure up and apart when the external relative humidity falls, and swelling to close the fissure under moist external conditions. As the structure of the hilum in white clover is similar, Hyde suggested the same mechanism applies there also, and interpreted the three observations listed above on this basis. He wrote: 'the hilum continues to function as a hygroscopic valve opening or closing according to whether the moisture tension in the counter-palisade is greater or less than that in the epidermal cells, provided that the difference in moisture tension exceeds the minimum required to bring about effective movement. Because of this valve action of the hilum, progressive drying can take place intermittently, and hard seeds tend to acquire and to retain a moisture content in equilibrium with the lowest relative humidity to which they have been exposed.'

Hyde's hypothesis would certainly explain the unidirectional movement of water, but it should be pointed out that his experimental designs do not rule out the alternative (or additional) possibility that changes in relative humidity act by affecting the degree of testa impermeability in regions other than the hilum. A much fuller investigation is required to elucidate all the factors involved, and in particular to determine whether the higher germination ability of seeds ripened at 70% than 30% RH is a result of the hilar fissure remaining open wider in the former than the latter.

iv. The fruit

A detailed description of the post-fertilisation changes which occur as a carpel develops into a fruit is not available. The mature fruit is covered by the brown, shrivelled remains of the persistent calyx and corolla. It is a short legume with pale brown membranous walls comprised of rather disorganised ground tissue sandwiched between inner and outer epidermal cells with sinuous anticlinal walls. The dead style remains at the apex. The legume contains from 1 to 7 fully developed seeds. Where several are present, those at the apex and base are generally arranged with their flattened sides parallel to the plane passing through the dorsal and ventral sutures, the rest being at right angles to this. Dehiscence takes place along the ventral suture.

7. Seed production

The seed yield of white clover varies with environment, from year to year, and with genotype. Genetic variation provides a basis for selecting new cultivars with reliably higher seed yields. To be effective, such selection should be built on a sound knowledge of the developmental and environmental factors contributing to the variation, and emphasis in this section is therefore placed on the extent of this knowledge rather than on the genetic variation itself, which is described further in Chapter 12.

Full understanding of environmental effects on seed yield can only be attained as the result of an analysis of the effects of environment on each of the separate components of yield. These components are:

1. number of flower heads per unit area,
2. number of florets per flower head,
3. number of seeds per floret,
4. weight of ripe seeds,
5. proportion of ripe seeds harvested.

The factors influencing each of these are discussed below.

(A) Number of flower heads per unit area.

The number of flower heads formed per unit area at any given time is affected by the strength of the inflorescence-initiating stimulus, the vigour of the response to that stimulus, the number of sites at which flower heads can be initiated, and the rate of leaf emergence.

Two sets of environmental conditions favour flower head initiation, namely cool conditions (as occur generally in autumn/winter/early spring - Section 3.(A)iii.) and a combination of long days and warm temperatures (as occur in summer - Section 3.(A)i.). Lines of white clover from different geographic origins respond differently (Section 3.(A)v.), but there is a tendency in all lines for flower initiation to decline and in many cases stop altogether with the approach of midsummer.

The vigour of response to environment is affected by nutrient supply (Section 3.(A)iv.), fewer flower heads being initiated in response to a given level of long-day stimulus under poorer nutrient conditions than under richer. Inflorescence initiation in response to low temperatures has been reported to be enhanced by the addition of inorganic nitrogen under controlled environment conditions.

Flower heads are initiated at actively growing stolon apices. The potential number of heads produced per m^2 is determined by the number of growing stolon tips, including laterals, in that area. Factors increasing the number of sites for initiation will thus favour an increase in head density. The release of axillary buds from inhibition so that they grow independently and become sites for flower head initiation is favoured by high nutrient availability, high light intensity (Mitchell 1956, Zaleski 1970), and removal of the apical bud from the main stolon (Chapter 2, Section 3.(A))

Theoretically, the maximum rate of inflorescence emergence must occur when both leaf emergence and inflorescence initiation rates are at their peak. For leaf emergence this occurs in warm, moist, long days (Brougham 1962, Thomas 1979a), and in New Zealand the main period of flower head initiation is during October/November for almost all lines studied (Thomas 1980b, and Figure 3.10). Assuming a rate of leaf emergence of 1.5 per week over this period (Thomas 1979a), and that inflorescences emerge about 9 plastochrons after their initiation (Figure 3.6), these heads would be expected to emerge 6 weeks after initiation, i.e. from late November to December. These expectations are in complete agreement with observations on the time of peak inflorescence emergence in cv. Grasslands Huia (Figure 3.11b, and Clifford 1980).

(B) Number of florets per flower head

The number of florets per flower head is higher in long days than short, but is decreased by high temperature, particularly at photoperiods of 14 h or less (Section 4.(A)ii.). Under field conditions in New Zealand, where the maximum daylength does not exceed 16 h in midsummer, Clifford (1979) found the average number of florets per head in Grasslands Huia to decrease during spring from 82 in September to 56 in December (Table 3.12). Decreasing water availability over this period would probably have had a detrimental effect on head size, despite Clifford's statement that watering after cutting to a height of 2.5 cm in mid-December did not increase any components of seed yield, because many of the flower heads that emerged after that cutting would already have been well developed before the irrigation treatment began.

In warm, short days and in shade, complete or partial abortion of young inflorescences occurs before they reach maturity, reducing the number of functional florets per head.

(C) Number of seeds per floret

The number of seeds set per floret is dependent upon the number of ovules per carpel and the proportion of these which set seed. The latter is dependent upon:

efficiency of pollination.
fertility of pollen,
fertility of ovules,
effectiveness of fertilization,
endosperm growth, and
embryo growth.

Number of ovules per carpel. The carpel normally contains 3-7 ovules (Chapter 1, Section 3(B)iv.). but the number is strongly influenced by environment, being lower at 30°C than at 20 or 10°C (Section 4.(A)ii.). In the field in New Zealand little change in ovule number has been observed from spring to early summer, but under non-irrigated conditions there is a pronounced decrease after midsummer - from 5.9 to 4.5 ovules per carpel in one reported case (Thomas 1981c) - which possibly results from a combination of high temperatures and decreased water supply.

The proportion of ovules which sets seeds is normally rather low. Under a wide range of conditions and at apparently optimal conditions for pollination, Dessureaux (1951) found that on average less than half the ovules per floret developed into seeds in Ladino clover. Other investigators including Zaleski (1961), van Bogaert (1977), and Clifford (1979) have reported similar low percentages of seed set. The causes of such low seed set are not known. Reports on the effects of environment are conflicting. In each of three growing seasons in England, Zaleski (1961) found the average number of seeds per floret to be the same in uncut plants of S100, S184 and Kent Wild White as in plants that were cut in early May and in others cut in early June. Similarly, the mean number of seeds per floret in Grasslands Huia and Grasslands Pitau was uninfluenced by the time of closing for seed production between mid-September and mid-December in New Zealand (Clifford 1979; and Table 3.12). In contrast, van Bogaert (1977) found the average number of seeds per pod to range from 1.56 to 3.66 in a cooler, wetter, summer and from 2.58 to 4.92 in a warmer, drier, summer in Europe. In the warmer summer 97.6% of florets produced seed compared with 82.2% in the cooler. The mean percentage of ovules setting seed ranged from 43 to 82 (mean 65.1) in the warmer summer and from 25.9 to 61.1 (mean 42.1) in the cooler.

Van Bogaert suggested that the differences observed between years might have resulted from differences in pollinating activity of bees, which was probably greater in the warmer year. Results obtained from an experiment in which flower heads were allowed to develop for four weeks in controlled environments at either 13 or 28°C before being transferred into a natural summer environment for pollination by bees showed, however, that the number of seeds set per carpel was higher in the flower heads pretreated at the lower temperature (T). This indicates that the interactions between factors acting in the field are much more complex than suggested by van Bogaert. It is very likely that environmental factors such as light intensity, temperature, atmospheric vapour pressure deficit and soil water stress have direct effects on developmental events in the florets in addition to acting indirectly via bee activity. Clearly, an understanding of the causes of low seed set can only come with greater knowledge of the sequence of events from pollination to embryo growth and the ways in which this is influenced by environmental conditions.

Efficiency of pollination. Seed set in the field is strongly dependent on pollination by bees (Section 6.(A)). Warm, dry conditions which favour bee foraging are thus necessary for good pollination. The extent to which the numbers of bees per unit area limit the amount of seed set requires further study, however. In New Zealand, 50 honey bees (*Apis mellifera*) per hectare were found to give seed set in 5% of florets, while nearly all florets set seed at a bee density of 10 000 per hectare (one per m^2) (Palmer-Jones et al. 1962; Forster 1974). In North America and the United Kingdom, however, data obtained by Bohart (1960) and Haggar & Holmes (1963) respectively suggest that in these regions a bee density up to three times as high might be required for fully effective pollination. It is probable that the density required varies with climatic conditions.

Fertility of pollen and ovules. Pollen fertility is strongly influenced by environment (Section 4.(A)iii.). In one controlled environment study (Thomas 1961c) the percentage of empty pollen grains in cv. Grasslands Huia was found to increase from 18% at 30°C to 46% at 10°C. Responses in individual clones differed greatly, sterility in one increasing from 14% at 30° to 98% at 10°C. Parallel studies in the field have shown pollen sterility to increase from about 5% in summer to about 50% in winter in New Zealand (Thomas 1981c). Seed set must be strongly influenced by pollen sterility under cool conditions in the more temperature-sensitive genotypes.

No data are available on the influence of environment on ovule fertility.

Fertilisation, endosperm growth and embryo growth. Extremely little is known of the effects of environment on fertilisation. High temperatures are known to allow self-fertilisation in otherwise self-incompatible plants (Section 6.(C)), but effects of temperature on the percentage of ovules fertilised in compatible crosses have not been reported. In fact no information is available to indicate what percentage of ovules is normally fertilised at any temperature.

The percentage of ovules developing into seeds is dependent on both the percentage fertilised and the percentage of these which develop into mature seeds. Failure of fertilised ovules to develop into seeds could result from failure of endosperm to grow and/or arrested development of embryos. Atwood (1943) suggested that abortion of some developing seeds might arise as a result of competition for nutrient within the flower heads. This hypothesis has not been thoroughly tested. The extent to which the low proportion of ovules developing into ripe seeds is attributable to each of the factors mentioned awaits investigation.

(D) Weight of ripe seeds

An important component of the weight of ripe seeds is the quantity of stored material. Conditions leading to poor embryo growth, such as shortage of mineral nutrients or light, would be expected to result in the formation of relatively light seeds. So also would premature ripening caused by water shortage before the content of stored compounds reaches its maximum.

Experiments have not been performed in controlled environments to test these hypotheses but field observations certainly support them. The 1000-seed weight of the earliest flower heads to emerge in the spring is higher than that of late-emerging heads. Zaleski (1961) found a reduction from 0.60 g to 0.56 g from early spring to mid-summer in heads of cv. S100,S184 and Kent Wild White in England; and Clifford (1979) reported a reduction from 0.70 g to 0.60 g in Grasslands Huia from September to December in New Zealand (Table 3.12). The combination of warmer and drier conditions in summer would certainly have hastened ripening and seems likely to have contributed significantly to the lower seed weight. That soil water shortage is not the main cause of decreased seed weight in summer is indicated by Clifford's statement that irrigation did not increase seed weight. High temperatures might nonetheless act directly to reduce the water content of flower heads by evaporation, even though the root systems are amply supplied with water. In addition, high summer temperatures lead to the production of small leaves which in turn might result in decreased photosynthesis. The degree to which flower heads are photosynthetically self-supporting has not been explored.

No data are available that indicate a clear effect of nutrient level, but Zaleski (1961) reported that the 1000-seed weight in plants grown at higher density from seed sown at 6 kg/ha was slightly, but significantly, lower (about 0.57 g) than in plants grown from seed sown at 1.5 kg/ha (about 0.59 g). Higher plant density would have decreased both light intensity and mineral nutrient availability as a result of competition. Clifford (1977) found no significantly different seed weights on plants grown from seed sown at rates of 1.5 and 0.75 kg/ha.

Table 3.12. Effect of closing date on seed yield, and four components thereof, in Grasslands Huia in New Zealand. (Modified from Clifford 1979).

	Closing date			
	mid Sept.	mid Oct.	mid Nov.	mid Dec.
Seed yield (kg/ha)	440	379	504	227
Inflorescences/m^2	508	563	788	580
Florets/inflorescence	82	73	66	56
Seeds/floret	2.15	2.15	2.28	2.15
1000-seed weight (g)	0.70	0.69	0.65	0.60

(E) Proportion of ripe seeds harvested

Seed yields in white clover are invariably reduced significantly by losses at harvest. It is not uncommon for only about half the seeds set to be gathered at harvest (Zaleski 1964b), and in the UK losses of as much as 75% have been reported (Haggar et al. 1966). Factors contributing to such losses include wet weather at harvest time and poor recovery of seed heads both as a result of the method of harvesting and the lodging (or brackling: Mohamed 1981) of peduncles.

White clover peduncles have a high tendency to lodge during the process of seed maturation, with the result that seed heads sink down into the foliage canopy. The position of such heads close to the ground renders them difficult to harvest, and in wet weather the humid microclimate beneath the canopy leads to seeds germinating before they are shed. Factors leading to the development of erect peduncles which hold the flower and seed heads well above the canopy of leaves have the advantage of increasing accessibility of heads to pollinators, preventing premature seed germination and facilitating harvesting.

The ratio of peduncle to petiole length is an index of the accessibility of flower heads for pollination. In controlled environments, although peduncles are shorter in warm conditions than in cool, the ratio of peduncle to petiole length is greater in warm, long days than in cool, short days (Thomas 1961c). The same trend is seen in the field: Zaleski (1961) showed that apparent heights of peduncles averaged 24.7 cm over two years on the heads which formed earliest in the spring, but were reduced to 21.57 and 14.03 cm respectively on heads that emerged after defoliation on 8 May and 2 June in England. The ratios of peduncle heights to petiole heights in the three groups were 1.13, 1.23 and 1.21 respectively. Environmental effects on peduncle lodging have not been investigated.

(F) Optimum conditions for seed production

The optimum conditions for each component of seed yield differ, so that the overall optimum for seed production must be a compromise between them. These individual optima and the times of year at which they prevail in cool-temperate climates are summarised in Table 3.13. This shows clearly that the most favourable conditions for almost all seed yield components occur during late spring and early summer. Rates of flower head initiation and emergence and the activity of pollinators are lower earlier than this, and after mid-summer there is marked decline in head initiation, emergence, number of florets per head, number of ovules per floret and seed weight.

There is no single most important component of seed yield. In practice, the most significant component is commonly considered to be the number of flower heads produced per unit area per unit time, viz. flower head density (Zaleski 1961; Huxley & Brink 1979). However, given a high head density, seed yield can be strongly influenced by number of florets per head (Clifford 1979) and by the number of seeds per floret (Mohamed 1981). The suggestion from Mohamed's data that low inflorescence production might be compensated by high seed yield per floret points to the need to investigate nutritional limitations to seed production.

(G) Seed production practice

It is not within the purview of this chapter to consider in any detail management or harvesting practices relating to seed production. These have been discussed by Zaleski (1970). There is some merit, however, in considering the significance of the previously described environmental influences on inflorescence development in relation to such practices.

To obtain maximum seed yields within a given area of white clover three criteria must be met. Firstly, there must be rapid emergence of the highest number of flower heads possible per unit area. Secondly, conditions must be ideal for full flower head development to anthesis and seed ripening. And thirdly, the loss of ripe seeds must be minimal, both before and at harvest.

Head number is determined by the number of potential sites (actively growing stolon tips) and the ability of these to respond to the floral stimulus. The number of stolon tips per unit area increases under conditions of good light penetration of the canopy and a plentiful supply of mineral nutrient (Haggar et al. 1963). Defoliation by cutting, grazing and chemical defoliants thus increases the number of sites of flower head production. Use of heavily weighted discs to cut up plants in the field (e.g. Lay 1980) should also increase the number of stolon tips.

The provision of additional mineral nutrient as a short-term boost to stolon tip production carries with it the danger that it has the long-term effect of increasing leaf growth and, in a mixed pasture, grass growth. This shading effect of competitive foliage can lead to a decrease in the number of flower heads per unit area (Haggar et al. 1963). Similarly, excessive foliage growth resulting from irrigation in summer can reduce both the number of flower heads per unit area and the rate of seed ripening (Adachi & Suzuki 1968). Whether the decreased flower head density at higher nitrogen and soil moisture levels is the result of reduced inflorescence initiation per stolon or fewer nodes (potential flower head sites) per unit area is uncertain.

Attempts to increase the number of flower head sites by increasing the density of planting can also lead to reduced head initiation as a result of overcrowding rather than gains in flower head number per unit area. Over three seasons Zaleski (1961) found the average number of brown flower heads at harvest to be 700, 647 and 615 per m^2 when seeds were sown at rates of 1.5, 3 and 6 kg/ha respectively. Clifford (1977) found that excessively low seeding rates (of e.g. 0.75 kg/ha), however, led to fewer initiation sites per unit area and reduced head density thereby.

Sensitivity of stolon tips to the floral stimulus is enhanced by a good supply of mineral nutrient (Section 3.(A)iv.) and high light intensity. Again, application of mineral nutrient can in practice act to decrease initiation by causing production of excessive foliage. Defoliation, however, only acts positively, and has the added theoretical advantage of acting as a direct stimulant to flower head initiation after the peak initiation period is past (Section 3.(A)ii.(a)). The second flowering in the aftermath following the main spring flowering is probably strongly stimulated by this means. The stimulatory effect of defoliation has not been observed early in the growing season.

Table 3.13. Optimal conditions for components of seed yield

Seed yield components	Contributing factors	Environmental conditions favouring yield components	Most favourable time of year				
			early spring	mid spring	late spring	early summer	mid summer
Inflorescence no./m^2	No. of active stem apices	High light & nutrient					
	Strength of flowering stimulus	High light, warm (≥ 25°C)					
		long days, *or* cool (≤ 10°C)					
	Plant vigour	High nutrient					
Rate of inflorescence emergence	Rate of leaf emergence	Warm, moist, long days					
Floret no./inflorescence	(?) Inflorescence primordium size	Cool, moist, long days					
Ovule no./floret	(?) Inflorescence primordium size	Cool-mild, moist					
Pollen fertility		Warm (≥ 20°C)					
Ovule fertility		Unknown					
Ovule fertilization	Bee (pollinator) activity	Warm					
Seed weight	Seed size	High nutrient & light					
Seed ripening	Inflorescence exposure	High light, warm, dry					
	Lack of peduncle lodging	Unknown					

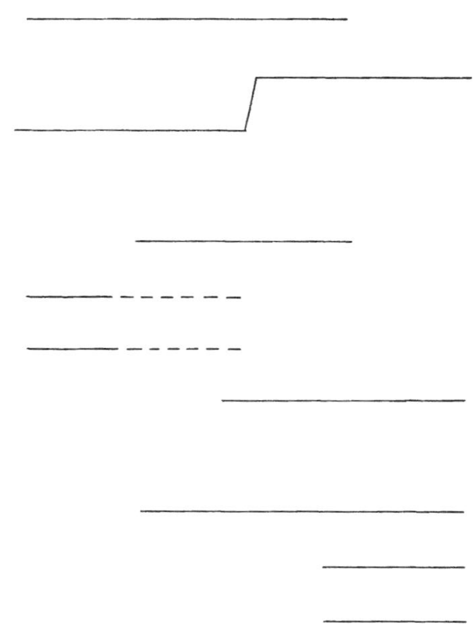

The aim in deciding the time at which to remove grazing animals and shut up or close an area for seed production is to ensure that as many flower heads as possible are already initiated but not yet emerged at the stolon tips at the time of closing. These heads then emerge over a short period of time in the immediate weeks following to give the maximum flower head density and the minimum range of head ages. There must also be adequate time remaining in the growing season for seed ripening to occur. If closing is too early, fewer heads mature at any one time; if closing is too late, even when head density is high, heads are smaller and seed loss at harvest tends to be increased because of the reduced volume of foliage cut along with the seed heads.

The cultivars most widely grown are long-day plants in which inflorescence initiation tends to continue at least up to midsummer, and head emergence continues for a few weeks thereafter. Delayed closing in such crops need not lead to total failure. In low-latitude lines such as Tamar and certain Spanish ecotypes (Figure 3.10), however, initiation declines markedly during spring and very few flower heads would emerge following late defoliation and closing, particularly at high temperatures.

In practice the possibility of changing the time at which flower head initiation and emergence occur naturally is very slight and probably would have little advantage if it were possible. In most situations there is more than adequate time for seed development and ripening available after peak head emergence, although there is little to be gained by attempting to grow cultivars for seed in regions where the longest photoperiods are shorter than the critical daylength for inflorescence initiation. Management practices such as defoliation can be very beneficial in several ways when their timing is based on local knowledge of the plants' developmental behaviour. Other practices which in theory might be beneficial, such as application of mineral nutrient, could have deleterious effects if carried out at the wrong time and more research is required before they can be used with safety.

Having achieved the maximum head production possible per unit area, the main factor limiting seed production by the plant is the low seed set per floret. The causes of this are not yet understood. The developmental sequence from pollination to seed set needs detailed elucidation as a basis for a later full investigation of environmental influences and the role of bees as pollinators. The abundant data on the latter have not been related to floret development and must be considered inadequate for that reason. Priority should now be given to seeking an understanding of post-anthesis floret development if improved seed yields are to be achieved.

References

Adachi, A.; Suzuki, S. (1968) Effect of day length and soil moisture on the flowering, seed maturity and vegetative growth in three types of white clover. *Journal of the Japanese Society of Grassland Science* **14**, 76-85.

Aitken, Y. (1955) Flower initiation in pasture legumes. I. Factors affecting flower initiation in *Trifolium subterraneum* L. *Australian Journal of Agricultural Research* **6**, 212-244.

Aitken, Y. (1960) Axial origin of the terminal inflorescence in the genus *Trifolium. Nature, UK* **187**, 622-623.

Atwood, S.S. (1940) Genetics of cross-incompatibility among self incompatible plants of *Trifolium repens. Journal of the American Society of Agronomy* **32**, 955-968.

Atwood, S.S (1941a) Controlled self- and cross-pollination of *Trifolium repens. Journal of the American Society of Agronomy* **33**, 538-545.

Atwood, S.S. (1941b) Cytogenetic basis of self-compatibility in *Trifolium repens. Genetics* **26**, p. 137 (Abstract).

Atwood, S.S. (1941c) Cytological basis for incompatibility in *Trifolium repens. American Journal of Botany* **28**, 551-557.

Atwood, S.S. (1943) 'Natural crossing' of white clover by bees. *Journal of the American Society of Agronomy* **35**, 862-870.

Beatty, D.W., Gardner, F.P. (1961) Effect of photoperiod and temperature on flowering of white clover *Trifolium repens* L. *Crop Science* **1**, 323-326.

Becker, R. (1941) Untersuchungen über die Blüh- und Befruchtungsverhältnisse und die Blütenmissbildungen des Weissklees (*Trifolium repens* L.) *Futtersaatbau* **1**, 1-84.

Bohart, G.E. (1960) Insect pollination of forage legumes. *Bee World* **41**, 85-97.

Booysen, P. de V.; Laude, H.M. (1964a) Flowering persistence in Ladino clover. *Crop Science* **4**, 518-520.

Booysen, P de V., Laude, H.M. (1964b) Influence of flower initiation and development on internode growth in the Ladino clover stolon. *Crop Science* **4**, 520-524.

Brougham, R.W. (1962) The leaf growth of *Trifolium repens* as influenced by seasonal changes in the light environment. *Journal of Ecology* **50,** 449-459.

Britten, E.J. (1960) Genetic and environmental control of flowering in *Trifolium repens* in the tropics. *Science* **131,** 100-101.

Britten, E.J. (1961) The influence of genotype and temperature on flowering in *Trifolium repens. Agronomy Journal* **53,** 11-14.

Calder, F.W.; Canham, W.D.; Fensom, D.S. (1973) Some effects of Alar-85 on the physiology of alfalfa and Ladino clover. *Canadian Journal of Plant Science* **53,** 269-278.

Chamblee, D.S. (1954) Amount of flowering of Ladino clover lots from different sources. *Agronomy Journal* **46,** 287-288.

Chen, C-C.; Gibson, P.B. (1973) Effect of temperature on pollen-tube growth in *Trifolium repens* after cross- and self-pollination. *Crop Science* **13,** 563-566.

Chubirko, M.M. (1963) Macrosporogenesis and development of the female gametophyte of *Trifolium repens* L. *Ukrainskii Botanichnii Zhurnal* **20,** 47-52.

Chubirko, M.M. (1965) Microsporogenesis and development of the male gametophyte in clover *Trifolium repens* L. *Botanicheski Zhurnal* **50,** 1578-1584.

Clifford, P.T.P. (1977) Cultural methods for 'Grasslands Pitau' white clover seed crops. *New Zealand Journal of Experimental Agriculture* **5,** 147-149.

Clifford, P.T.P. (1979) Effect of closing date on potential seed yields from 'Grasslands Huia' and 'Grasslands Pitau' white clovers. *New Zealand Journal of Experimental Agriculture* **7,** 303-306

Clifford, P.T.P. (1980) Research in white clover seed production. In: Herbage Seed Production. *Bulletin, New Zealand Grassland Association* pp. 64-67.

Cohen, M.M.; Leffel, R.C. (1963) Cytology of pseudo-self-compatibility in Ladino white clover, *Trifolium repens* L. *Crop Science* **3,** 430-433.

Cohen, Y.; Dovrat, A. (1976) Gibberellins and inhibitors in relation to flowering in white clovers (*Trifolium repens* L.) *Journal of Experimental Botany* **27,** 817-826.

Cumming, B.G. (1959) The control of growth and development in red clover (*Trifolium pratense* L.). II. Light, temperature, and the influence of growth regulators. *Canadian Journal of Botany* **37,** 1027-1048.

Darwin, C. (1876) The effects of cross and self fertilisation in the vegetable kingdom. London, UK; Murray, 482 pp.

Darwin, C. assisted by F. Darwin (1880) The power of movement in plants. London, UK; Murray.

Dessureaux, L. (1951) Ovule formation as a factor influencing seed-setting of Ladino white clover. *Scientific Agriculture* **31,** 373-382.

Dunavan, D. (1953) Insect pollination of Ladino clover in South Carolina. *American Bee Journal* **93,** 468-469.

Dunn, G.M; Calder, F.W.; Kilpatrick, R.A. (1962) Variation in white clover in relation to improvement. *Technical Bulletin, University of New Hampshire Agricultural Experiment Station* **104,** 22 pp.

Erith, A.G. (1924) White Clover (*Trifolium repens* L.) London, UK; Duckworth.

Esau, K. (1965) Plant Anatomy (2nd Edition). New York, USA; Wiley.

Evans, L.T. (1959) Flower initiation in *Trifolium subterraneum* L. I. Analysis of the partial processes involved. *Australian Journal of Agricultural Research* **10,** 1-16.

Forster, I.W. (1974) Behaviour and effectiveness of bees in pollinating legumes. *Proceedings of the New Zealand Grassland Association* **36,** 105-110.

Gibson, P.B. (1957) Effect of flowering on the persistence of white clover. *Agronomy Journal* **49,** 213-215.

Green, H.B. (1956) Some factors affecting pollination of white Dutch clover. *Journal of Economic Entomology* **49,** 685-688.

Guignard, M.L. (1880) Recherches d'embryogénie végétale comparée: Légumineuses. *Annales des Sciences Naturelles* **6,**Sér. Bot. 12.

Haggar, R.J. (1961) Flower initiation in Kent Wild White clover (*Trifolium repens* L.) under controlled environmental conditions. *Nature, UK* **191,** 1120-1121.

Haggar, R.J.; Holmes, W. (1963) Wild white clover seed production. 2. A survey on wild white clover seed production in Kent. 1961. *Journal of the British Grassland Society* **18,** 197-203.

Haggar, R.J.; Holmes, W.; Bell, R.D. (1966). Some new techniques for increasing yields of wild white clover seed. *Proceedings of the IX International Grassland Congress,* Brazil, 543-545.

Haggar, R.J.; Holmes, W.; Innes, P. (1963) Wild white clover seed production. I. The effects of defoliation and fertilizer treatment on flowering and seed yields from ryegrass/white-clover swards. *Journal of the British Grassland Society* **18,** 97-103.

Harberd, D.J. (1963) Observations on natural clones of *Trifolium repens* L. *New Phytologist* **62,** 198-204.

Heinrich, B. (1979) Bumblebee economics. Harvard, USA; Harvard University Press.

Huxley, D.M.; Brink, V.C.; Eaton, G.W. (1979) Seed yield components in white clover. *Canadian Journal of Plant Science* **59,** 713-715.

Hyde, E.O.C. (1950) Studies on the development of white clover seed. *Proceedings, 12th Conference of the New Zealand Grassland Association* 101-107.

Hyde, E.O.C. (1954) The function of the hilum in some Papilionaceae in relation to the ripening of the seed and the permeability of the testa. *Annals of Botany, N.S.* **18,** 241-256.

Hyde, E.O.C.; McLeavey, M.A.; Harris, G.S. (1959) Seed development in ryegrass, and in red and white clover. *New Zealand Journal of Agricultural Research* **2,** 947-952.

Jewiss, O.R. (1962) The growth and flowering behaviour of S100 white clover. In: *Experiments in Progress No. 14. Annual Report for 1960-61 of the Grassland Research Institute, Hurley, England* pp. 25-27.

Jewiss, O.R. (1963) The growth and flowering behaviour of S100 white clover. In: *Experiments in Progress 15. Annual Report for 1961/62 of the Grassland Research Institute, Hurley, England* pp. 26-29.

Kawanabe, S.; Yoshihara, K.; Okada, T.; Ueno, M. and Hidaka, M. (1963). Studies on summer depression of pasture crops. 3. Influence of flower bud removal upon vegetative growth of Ladino clover. *Journal of the Japanese Society of Grassland Science* **9,** 31-41.

Knuth, P. (1908) Handbook of flower pollination. Translation by J.R.A. Davis, Oxford, UK; Oxford University Press

Kuester, E. (1942) Zur pathologischen Morphologie der Blüten und Blütenstände 2. Beitrag. Durchwachsung der Blütenstand von *Trifolium repens. Zeitschrift für Pflanzenkrankheit* **52,** 433-449.

Lang, A. (1965) Physiology of flower initiation. In: *Encyclopedia of Plant Physiology Volume 15/1, Berlin, Ruhland, W. editor W.Germany; Springer Verlag,* 1389-1536.

Laude, H.M.; Stanford, E.H.; Enloe, J.A. (1958) Photoperiod, temperature, and competitive ability as factors affecting the seed production of selected clones of Ladino clover. *Agronomy Journal* **50,** 223-225.

Lay, J.K. (1980) White clover seed production with irrigation. In: *Herbage Seed Production. Bulletin, New Zealand Grassland Association* pp. 64-67.

MacLeod, J. (1894) Over de Bevruchtung der Bloemen in het kempischgedeelte van Vlaanderen. *Dodonaea,* 342.

McCloud, D.E.; Cole, W.A. (1951) The effect of photoperiod on the flowering of Ladino and Louisiana white clover. Proceedings, 48th Annual Conference, Association of Southern Agricultural Workers, 137.

Maheshwari, P. (1950) An Introduction to the Embryology of Angiosperms. New York, USA; McGraw Hill.

Mitchell, K.J. (1956) Growth of pasture species. III. White clover (*Trifolium repens*), subterranean clover (*T. subterraneum*) and lotus major (*L. uliginosus*). *New Zealand Journal of Science and Technology* **37A,** 395-413.

Mohamed, Z.A. in B. (1981) Studies on white clover with particular reference to seed yields. Ph. D. Thesis. University College of Wales, Aberystwyth, Wales.

Morse, D.H. (1980) The effect of nectar abundance on foraging patterns of bumble bees. *Ecological Entomology* **5,** 53-59.

Müller, H. (1873) The fertilization of flowers. Translated and edited by D'A. W. Thompson, London. 479 pp.

Oertel, E. (1961) Honeybees in production of white clover seed in the southern states. *American Bee Journal* **101,** 96-99.

Palmer-Jones, T.; Forster, I.W.; Jeffrey, G. L. (1962) Observations on the role of the honey bee and bumble bee as pollinators of white clover (*Trifolium repens* Linn.) in the Timaru district and Mackenzie country. *New Zealand Journal of Agricultural Research* **5,** 318-325.

Penzig, O. (1921) Pflanzenteratologie.

Rembert,D.W. (1977) Ovule ontogeny, megasporogenesis, and early gametogenesis in *Trifolium repens* (Papilionaceae). *American Journal of Botany* **64,** 483-488.

Ridley, J.R. (1967) Effect of temperature on flowering intensity of the Ladino clover stolon. *Dissertation Abstracts* **28,** 13B-14B.

Ridley, J.R.; Laude, H.M. (1968) Temperature and the flowering intensity of Ladino clover stolons. *Crop Science* **8,** 519-521.

Rinderer, T.E.; Harville, B.G.; Lackett, J.J.; Baxter, J.R. (1981) Variation in honey bee morphology, behavior and seed set in white clover. *Annals of the Entomological Society of America,* 459-461.

Roberts, H.M. (1967) Isolation requirements in white clover seed crops. *Report of the Welsh Plant Breeding Station for 1966* pp. 103-105.

Roberts, R.H.; Struckmeyer, B.E. (1938) The effects of temperature and other environmental factors upon the photoperiodic responses of some of the higher plants. *Journal of Agricultural Research* **56,** 633-677.

Roberts, R.H.; Struckmeyer, B.E. (1939) Further studies of the effects of temperature and other environmental factors upon the photoperiodic responses of plants. *Journal of Agricultural Research* **59,** 699-709.

Scullen, H.A. (1952) Ladino seed production in the Pacific Northwest. *American Bee Journal* **92,** 287-288.

Stoddart, J.L. (1959) The effects of gibberellic acid upon growth habit and heading in late flowering red clover (*Trifolium pratense* L.) *Journal of Agricultural Science, UK* **52,** 161-167.

Stoddart, J.L. (1962) Effect of gibberellin on a non-flowering genotype of red clover. *Nature, UK* **194,** 1064-1065.

Stoddart, J.L. (1966) Studies on the relationship between gibberellin metabolism and daylength in normal and non-flowering red clover (*Trifolium pratense* L.). *Journal of Experimental Botany* **17,** 96-107.

Thomas, R.G. (1961a) Flower initiation in *Trifolium repens* L.: a short-long-day plant. *Nature, UK* **190,** 1130-1131.

Thomas, R.G. (1961b) The relationship between leaf growth and induction of flowering in long-day plants (LDP). *Die Naturwissenschaften* **48,** 108.

Thomas, R.G. (1961c) The influence of environment on seed production capacity in white clover (*Trifolium repens* L.) I. Controlled environment studies. *Australian Journal of Agricultural Research* **12,** 227-238.

Thomas, R.G. (1962) The initiation and growth of axillary bud primordia in relation to flowering in *Trifolium repens* L. *Annals of Botany, N.S.***26,** 329-344.

Thomas, R.G. (1972) The role of basal and apical factors in the coordination of growth in the stems of white clover (*Trifolium repens* L.) In: *Plant Growth Substances 1970.* Edited by D.J. Carr, Berlin, W. Germany; Springer-Verlag, 624-632.

Thomas, R.G. (1979a) Inflorescence initiation in *Trifolium repens* L. Influence of natural photoperiods and temperatures. *New Zealand Journal of Botany* **17,** 287-299.

Thomas, R.G. (1980a) Flowering responses of white clover (*Trifolium repens* L.) growing in arctic Norway. *Acta Agriculturae Scandinavica* **30,** 51-58.

Thomas, R.G. (1980b) Growth of the white clover plant in relation to seed production. In: *Herbage Seed Production. Bulletin of the New Zealand Grassland Association, pp. 56-63.*

Thomas, R.G. (1981a) Effect of defoliation on flower initiation in white clover in summer. *Grass and Forage Science***36,** 121-125.

Thomas, R.G. (1981b) Flower initiation in relation to cool season growth of four lines of white clover. *New Zealand Journal of Agricultural Research* **24,** 37-41.

Thomas, R.G. (1981c) The influence of environment on seed production capacity in white clover (*Trifolium repens* L.). II. Responses to the natural environment. *New Zealand Journal of Agricultural Research* **24,** 359-364.

Thomas, R.G. (1981d) Studies on inflorescence initiation in *Trifolium repens* L. The short-long-day reaction. *New Zealand Journal of Botany* **19,** 361-369.

Thomas, R.G. (1982) A comparison of the effects of environment on inflorescence initiation in nine lines of *Trifolium repens* L. *New Zealand Journal of Botany,* **20,** 151-162.

Valle, O. (1963) Breeding and seed production experiences with Finnish white clover. *Annales Agriculturae Fennicae* **2,** 51-58.

van Bogaert, G. (1977) Factors affecting seed yield in white clover. *Euphytica* **26,** 233-239.

Vansell, G.H. (1951) Honeybees activity on Ladino florets. *Journal of Economic Entomology* **44,** 103.

Vasil, H.; West, S.H. (1971) Plantlet formation from flowers of *Trifolium repens. Canadian Journal of Botany* **49,** 327-329.

Ware, W.M. (1925) Experiments and observations on forms and strains of *Trifolium repens* L. *Journal of Agricultural Science, UK* **15,** 47-67.

Watson, D. (1948) Structure of the testa and its relation to germination in the *Papilionaceae* Tribes *Trifoliae* and *Loteae. Annals of Botany, N.S.* **12,** 385-409.

Weaver, N. (1965) Foraging behavior of honeybees on white clover. *Insectes Sociaux* **12,** 231-240.

Williams, R.D. (1931) Self and cross-sterility in white clover. *Bulletin, Welsh Plant Breeding Station, Series H* **12,** 209-216.

Zaleski, A. (1961) White clover investigations. I Effect of seed rates and cutting treatments on flower formation and seed yield. *Journal of Agricultural Science, UK* **57,** 199-212.

Zaleski, A. (1964a) Effect of density of plant population, photoperiod, temperature and light intensity on inflorescence formation in white clover. *Journal of the British Grassland Society* **19,** 237-247.

Zaleski, A. (1964b). Survey of methods used for white clover seed production. *Journal of the National Institute for Agricultural Botany* **10,** 110-115.

Zaleski, A. (1970) White clover seed production. In: White Clover Research Ed. J.Lowe. *Occasional Symposium, British Grassland Society, No.6,* pp. 147-154.

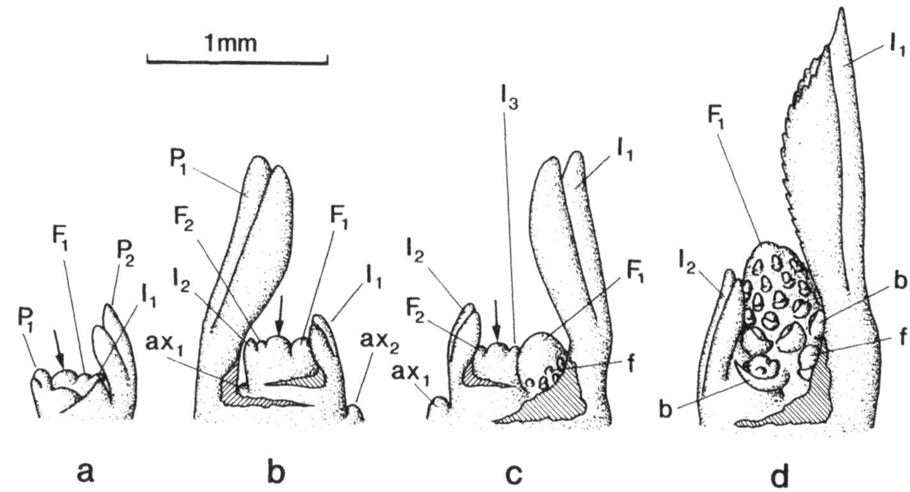

Fig. 3.1 Early stages of development of inflorescence primordia at main stolon apices of plants of Grasslands Huia clover grown in continuous light (CL) at 23°C after pretreatment with warm short days. a, b, c, and d show apical buds dissected out 6, 9, 11 and 13 days respectively after the start of CL. Hatched areas indicate tissues damaged by removal of stipules to expose the organs beneath. The youngest leaf primordium present at the start of CL treatment is labelled P_1 and the next youngest P_2; axillary buds subtended by these are labelled ax_1 and ax_2 respectively. New leaf primordia successively formed after the start of CL treatment are labelled I_1, I_2, and I_3; inflorescence primordia subtended by these are labelled F_1, F_2, and F_3 respectively. Bracts subtending floret primordia are labelled 'b' and the floret primordia themselves 'f'. Arrow heads point to the meristematic dome at each stolon apex. [Original].

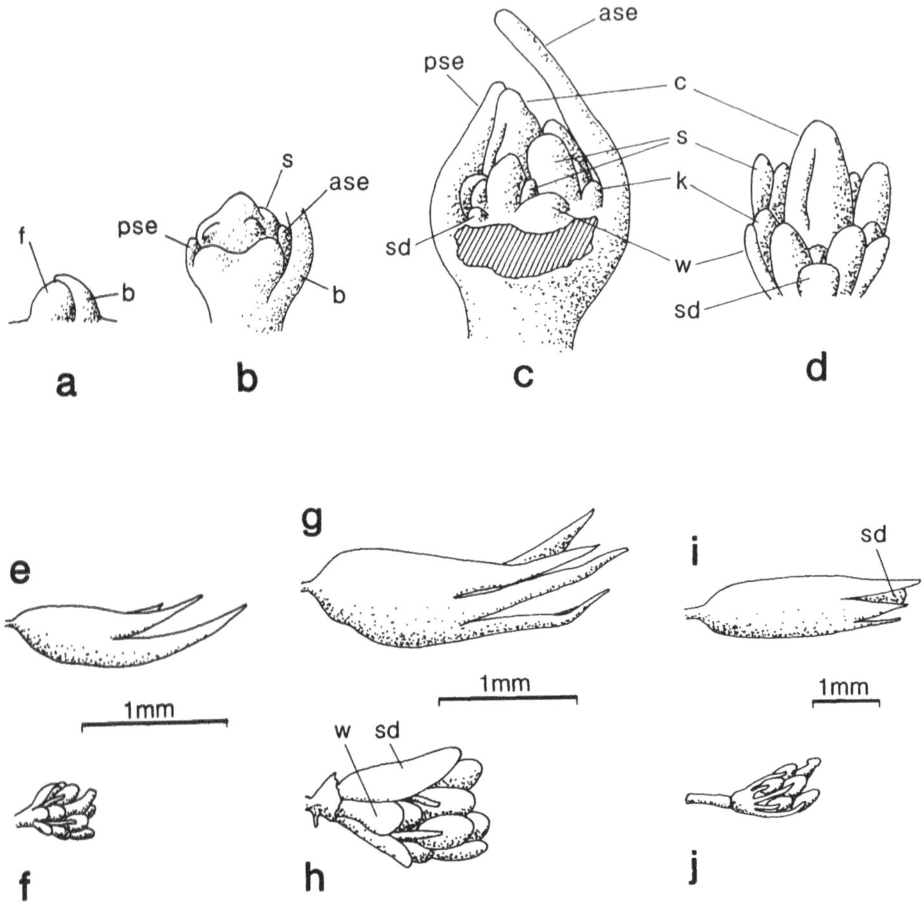

Fig. 3.2 Floret development: (a) newly initiated floret primordium subtended by a bract (as in Fig. 3.1d); (b) intact young floret; (c) slightly older floret with one posterior and one lateral sepal removed (hatching indicates the area damaged by their excision); (d) older floret with calyx removed; (e), (g), and (i) are external views of successively older florets. (f) and (h) respectively show the floral organs of the florets depicted in (e) and (g) with calyces removed; (j) shows the sexual organs of the floret at (i) with both calyx and corolla removed. Florets (c), (e), (g) and (i) respectively are equivalent to the lowermost on inflorescences at nodes 5, 6, 8 and 9-10 in Fig. 3.6. ase - anterior sepal; b - bract; c - carpel; f - floret primordium; k - keel petal; pse - posterior sepal; s - stamen; sd - standard; w - wing petal. [Redrawn, with modifications, from Erith 1924]

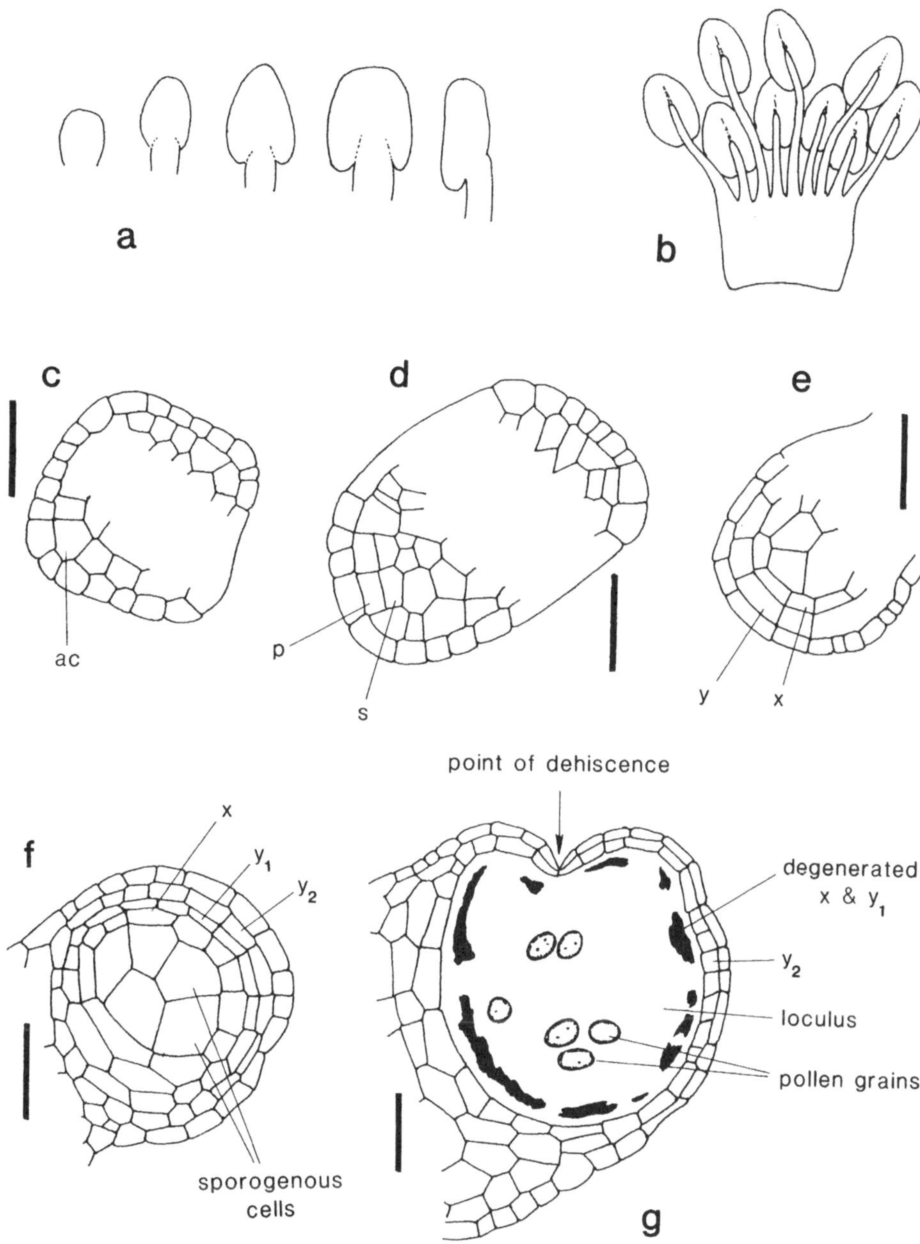

Fig. 3.3 Stamen development: (a) successive stages in stamen development; (b) young stamen 'tube' opened out showing 5 earlier-formed and 4 later-formed anthers; (c)-(g) transverse sections of anthers in successive stages of development. ac - archesporial cell; p - primary parietal cell; s - primary sporogenous cell; x - tapetal cells; y_1 - 'middle layer'; y_2 - endothecium. Vertical scale bars represent 25 μm. [=Redrawn from Erith, 1924].

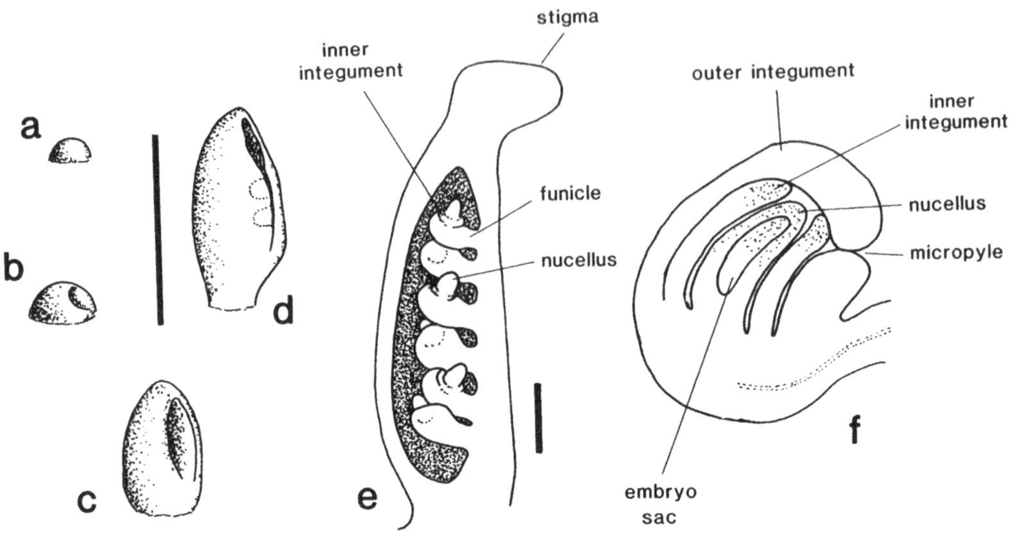

Fig. 3.4 Carpel and ovule development: (a)-(d) successive stages in carpel development from initiation (a) to ovule initiation (d), (e) later stage showing stigma and arrangement of ovules with inner integuments visible (as found in the oldest florets at nodes 7 or 8, see Fig. 3.6), (f) longitudinal section of a nearly mature ovule. (a)-(d) are drawn to the same scale: all vertical scale bars represent 150μm. [Redrawn from Erith, 1924].

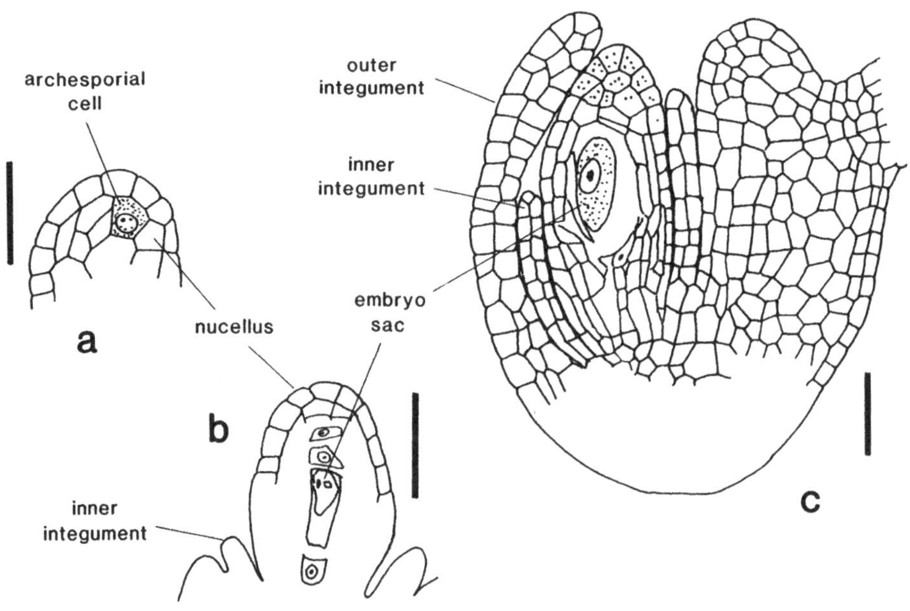

Fig. 3.5 Longitudinal sections of ovules at successive stages of development: (a) at stage shown in Fig. 3.4d, (b) at stage shown in Fig. 3.4e. Vertical scale bars represent 25 μm. [Redrawn from Erith, 1924].

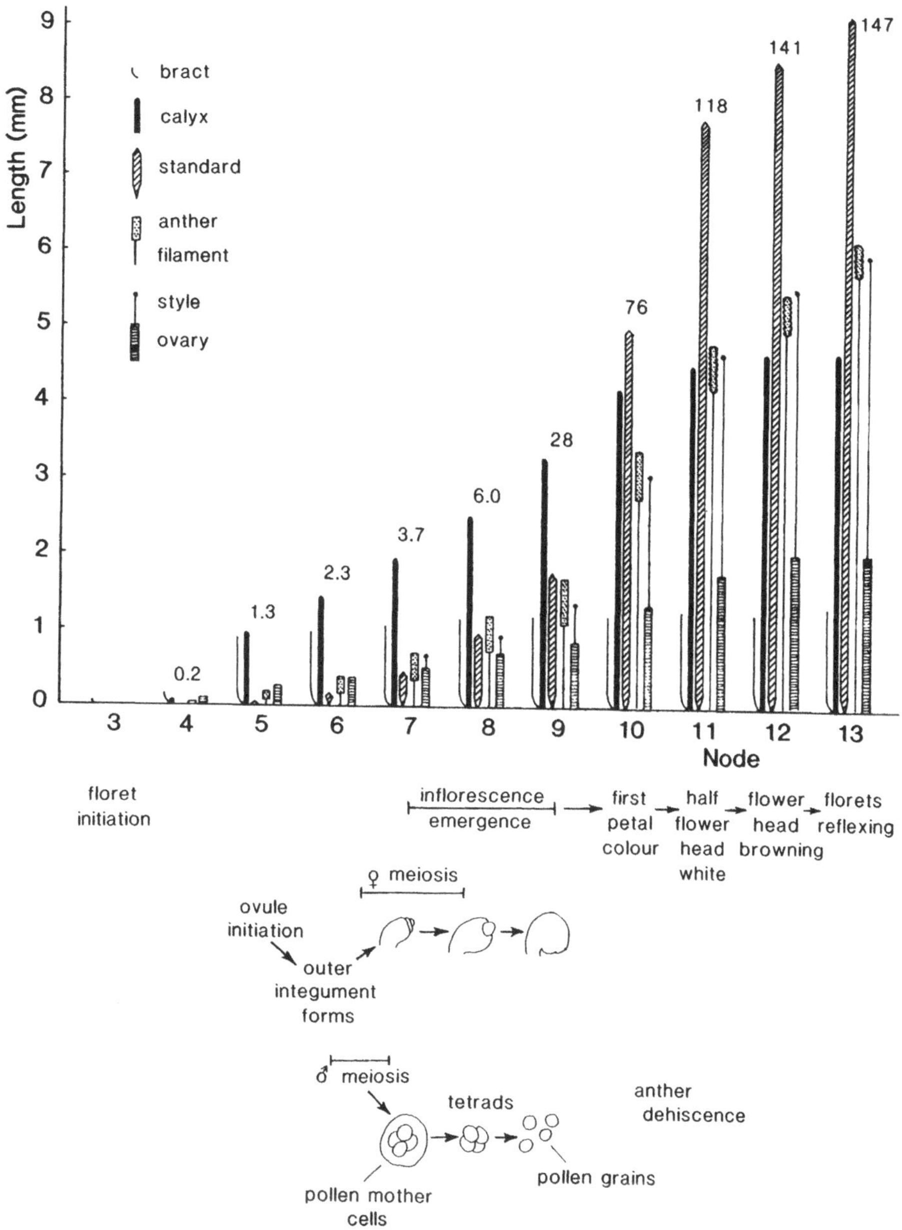

Fig. 3.6 Mean lengths of organs in the oldest florets of inflorescences present at successive nodes of stolons growing in natural early summer conditions at latitude 40°S in New Zealand. The number above each set of floret measurements shows the mean length (in mm) of peduncles at that position. Node 3 was that bearing leaf primordium P_3, node 4 that bearing P_4, etc. (see Chapter 2, Section 2.(A)). The youngest leaf with unfolded leaflets was at either node 7 or node 8. [Original] .

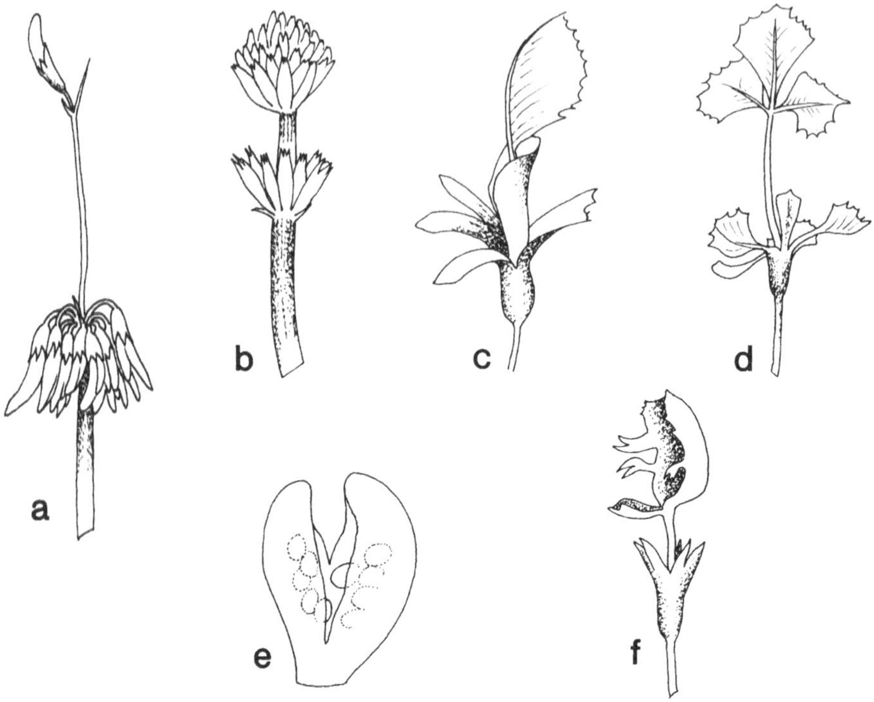

Fig. 3.7 Inflorescence and floret abnormalities (see text explanation). [Redrawn from Erith, 1924].

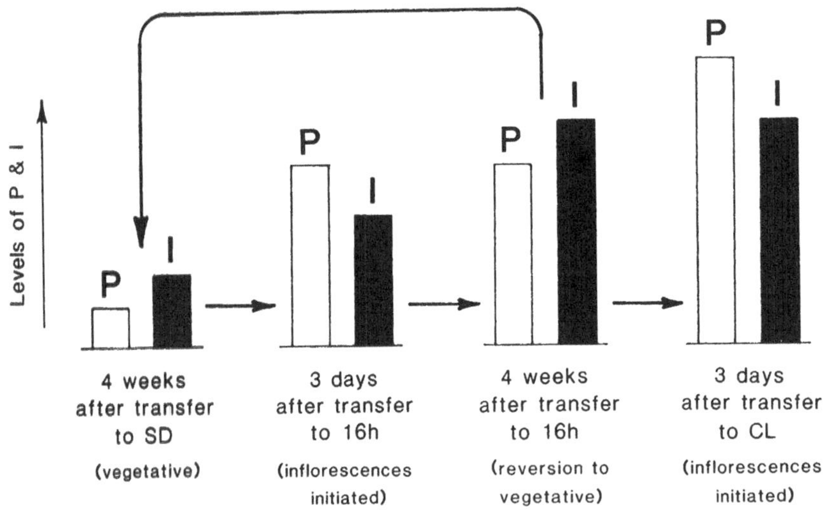

Fig. 3.8 Hypothetical control of inflorescence initiation by the ratio of promoter (**P**) to inhibitor (**I**). [Modified from Thomas, 1981d].

Fig. 3.9 Time of inflorescence initiation in Grasslands Huia grown in the field from May (**M**) to April (**A**) at latitude 40°S. in New Zealand. Mean times of formation of nodes at terminal buds are indicated by positions of vertical bars and dots. Vertical bars indicate the percentage of newly formed nodes which initiated inflorescences. Dots indicate completely non-flowering nodes. [Original].

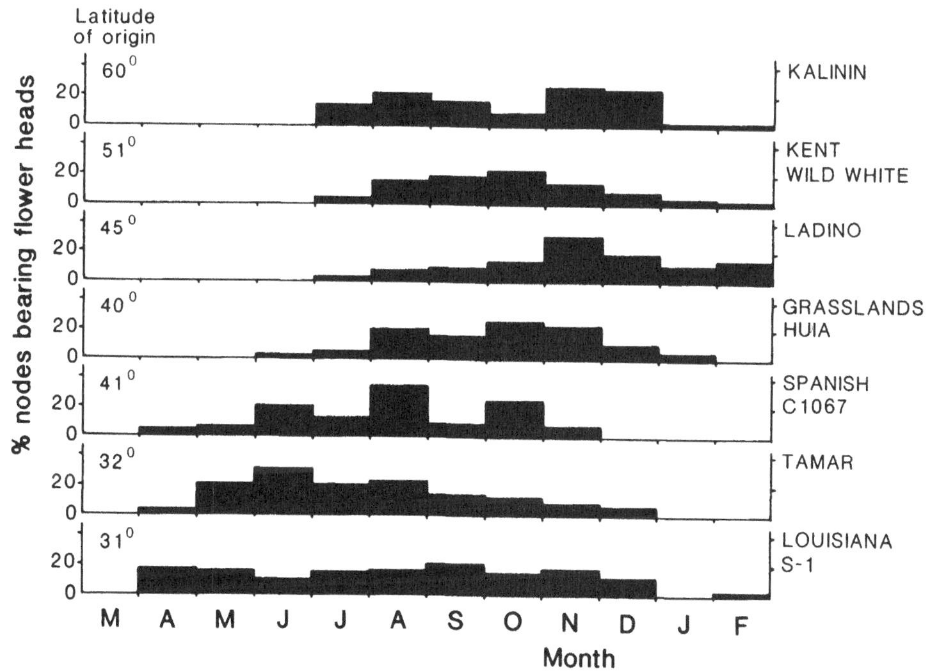

Fig. 3.10 Comparison of inflorescence initiation in apical buds of main stolons of seven lines of white clover grown from March (**M**) to February (**F**) at natural temperatures at latitude 40°S. in New Zealand. Strength of inflorescence initiation is expressed as the mean percentage of new nodes formed each month which bore inflorescences. [Modified from Thomas 1980b] .

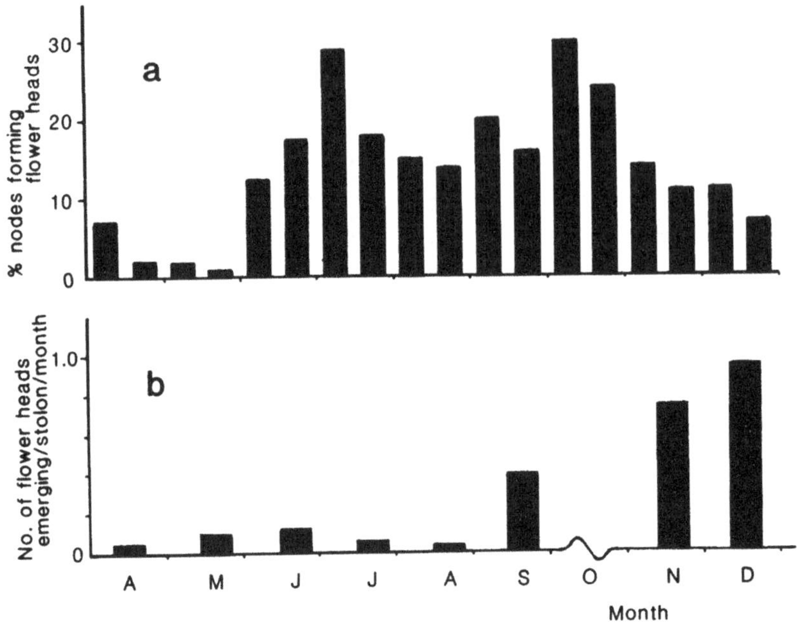

Fig. 3.11 Relationship between time of inflorescence initiation (a) and time of inflorescence emergence (b) in plants of Grasslands Huia clones A and C grown at natural temperatures from April (**A**) to December (**D**) at latitude 40°S. in New Zealand. No emergence data were obtained for October. [Thomas, 1980b].

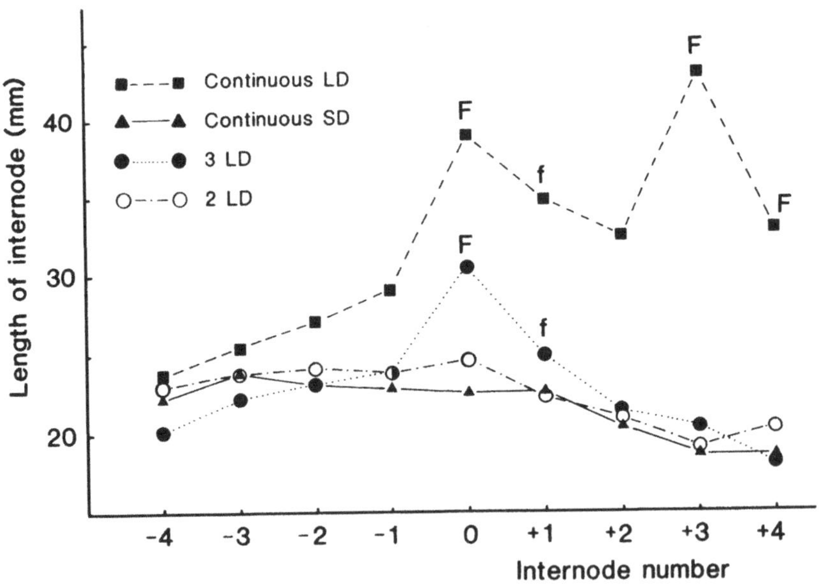

Fig. 3.12 Effect of inflorescences on mean lengths of mature internodes on stolons subjected to different periods of long-day (**LD**) treatments. Internode 0 was the first formed after transfer to LD; those marked + formed later, and those marked − earlier. All plants initiated inflorescences at nodes immediately distal to internodes labelled 'F'; some did so at nodes immediately distal to internodes labelled 'f' [Redrawn from Booysen & Laude, 1964b].

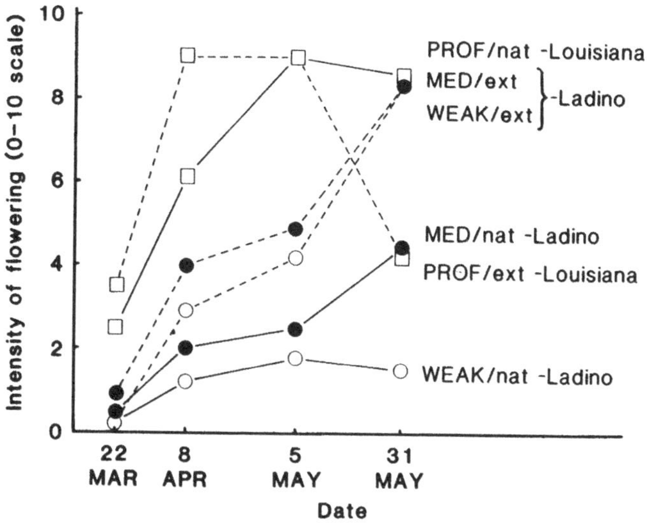

Fig. 3.13 Intensity of flowering (number of inflorescences emerged) in plants of Louisiana and Ladino growing in either a natural environment (nat) or extended photoperiods (ext) at latitude 32°50′N. in the USA. Plants of Louisiana were profuse flowerers (**PROF**); Ladino plants grown were of two lines, medium flowerers (**MED**) and weak flowerers (**WEAK**). See text for details. [Redrawn from Gibson, 1957].

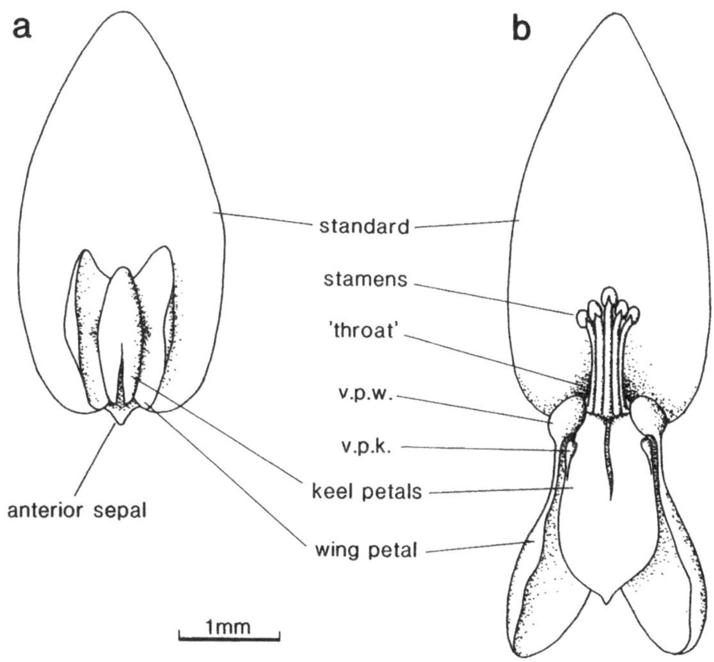

Fig. 3.14 a. 'Bee's eye view' of a floret with wing and keel petals in normal, unvisited position. b. The same floret in which the wing and keel petals have been depressed by a visiting bee, thereby exposing the stamens. v.p.w. and v.p.k. are, respectively, the vesicular protuberances of wing and keel petals; the 'throat' is the space between the standard and the stamens. [Original].

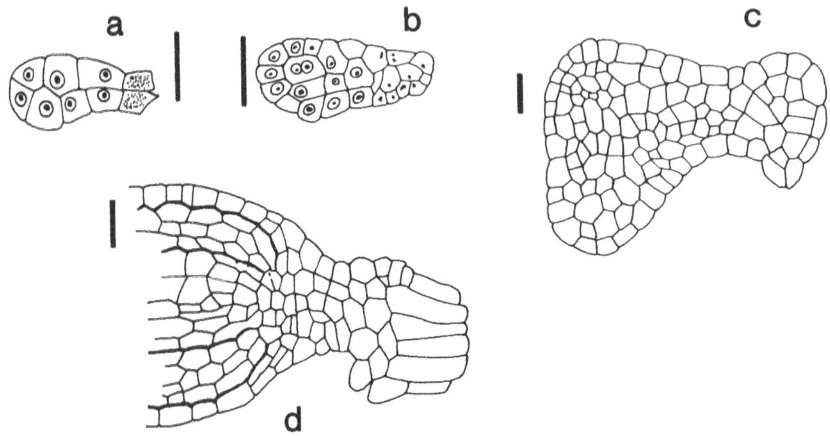

Fig. 3.15 A. Successive stages of development of young embryos. Vertical scale bars represent 25 μm. [Redrawn from Erith, 1924].

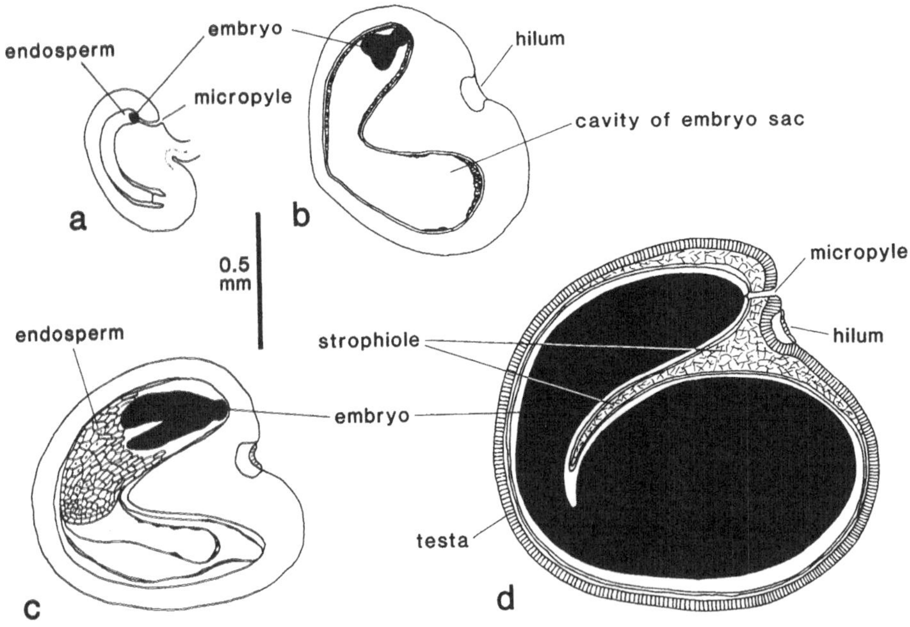

Fig. 3.15 B. Longitudinal sections of ovules at successive stages of development from just after fertilization (a) to maturation into a fully developed seed (d). All are drawn to the same scale. [Redrawn from Erith, 1924].

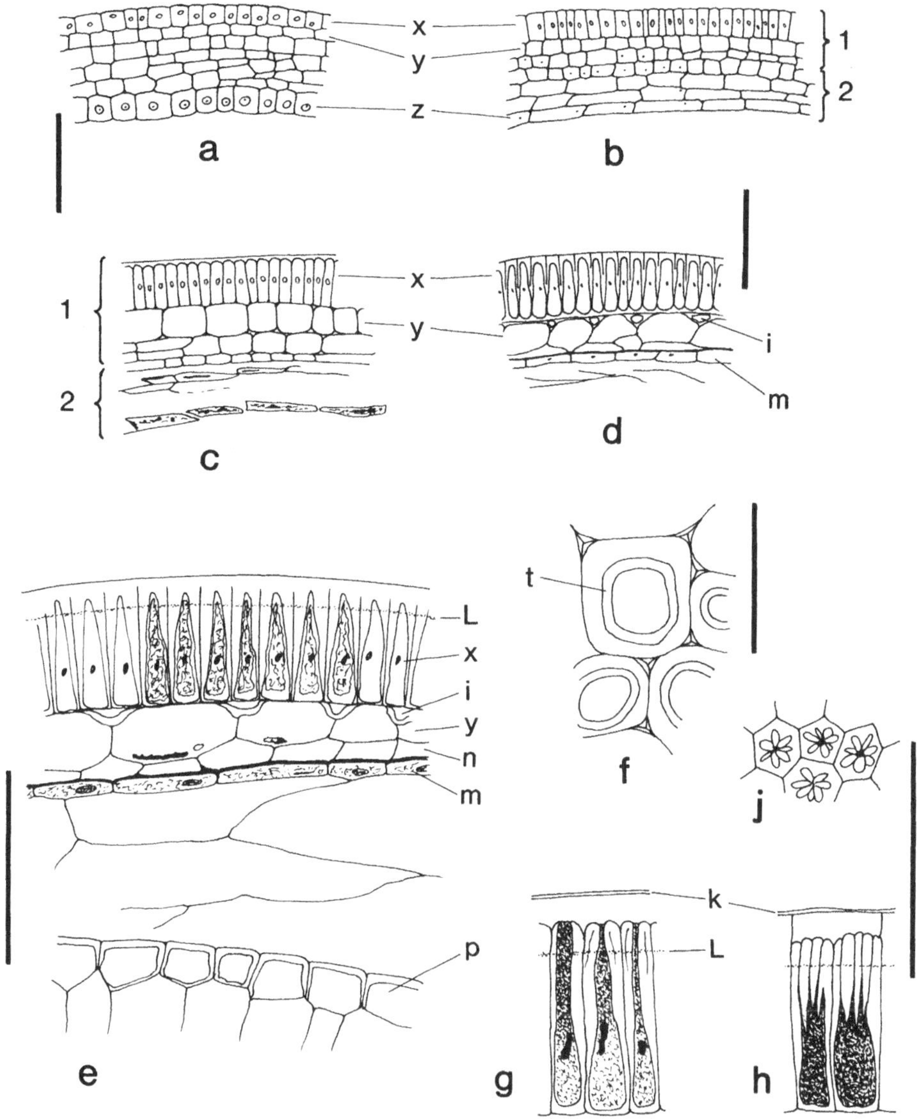

Fig. 3.16 Development of the seed testa: (a) transverse section of the wall of an unfertilized mature ovule; (b)-(d) transverse sections of walls of ovules at successive stages of development after fertilisation; (e) T.S. testa of seed; (f) surface view of subepidermal layer (y); (g) and (h) longitudinal sections of cells of the palisade layer of the mature testa; (j) surface view of ends of cells of the palisade layer. Vertical scale bars represent 50 μm. 1 – outer integument; 2 – inner integument, x – palisade layer; y – subepidermal layer; z – inner epidermis of inner integument; n – 'nutrient' layer; m – 'aleurone' layer; i – intercellular space; t – thick wall of 'pillar cell' 'stalk'; L – light line; k – cuticle. [Redrawn from Erith, 1924]

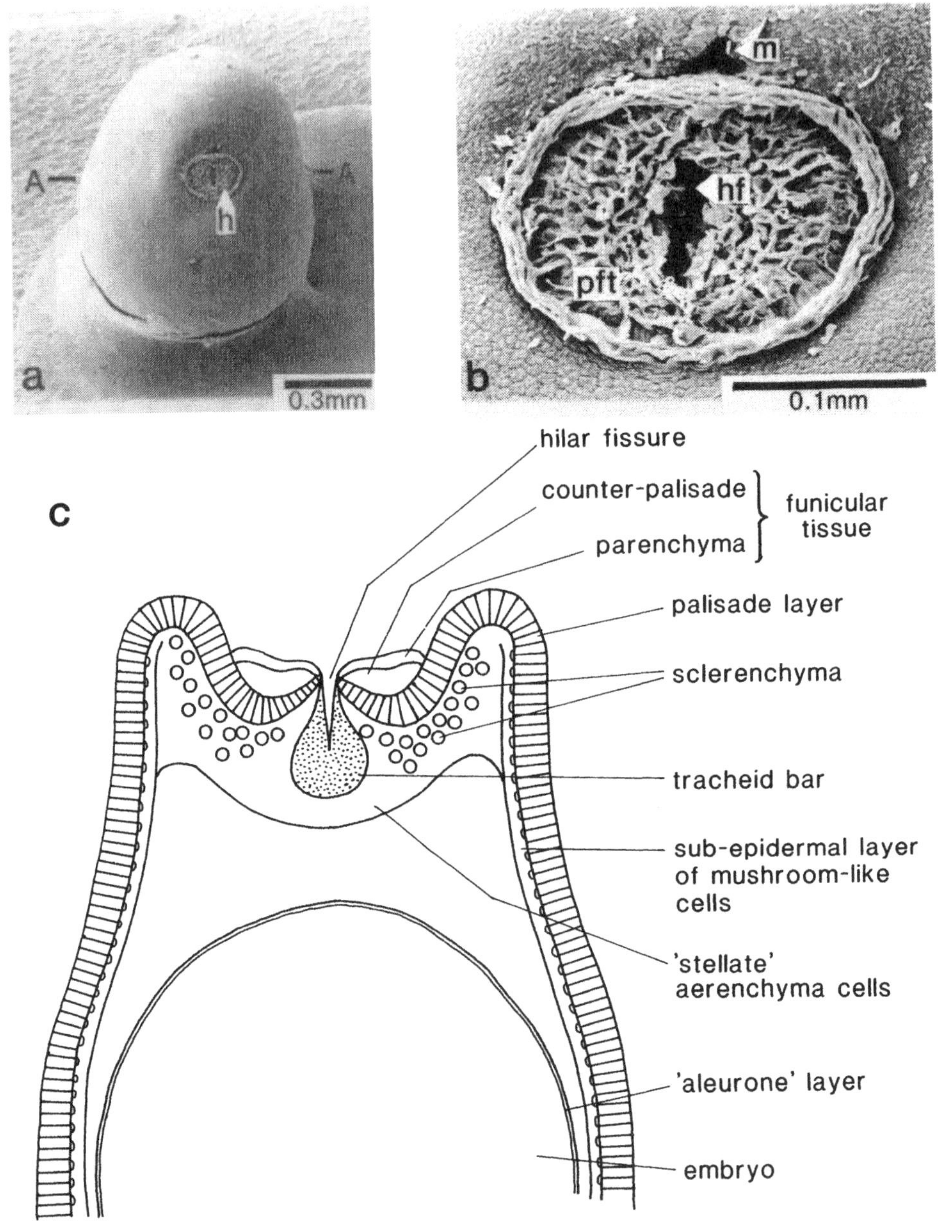

Fig. 3.17 (a) Scanning electron micrograph of seed. h – hilum. [Original]. (b) scanning electron micrograph of hilum of seed in (a). hf – hilar fissure, m – micropyle, pft – parenchymatous funicular tissue. [Original]. (c) Section through seed in region of hilum. Direction of cut indicated by A-A in Fig. 3.17 (a). [Original].

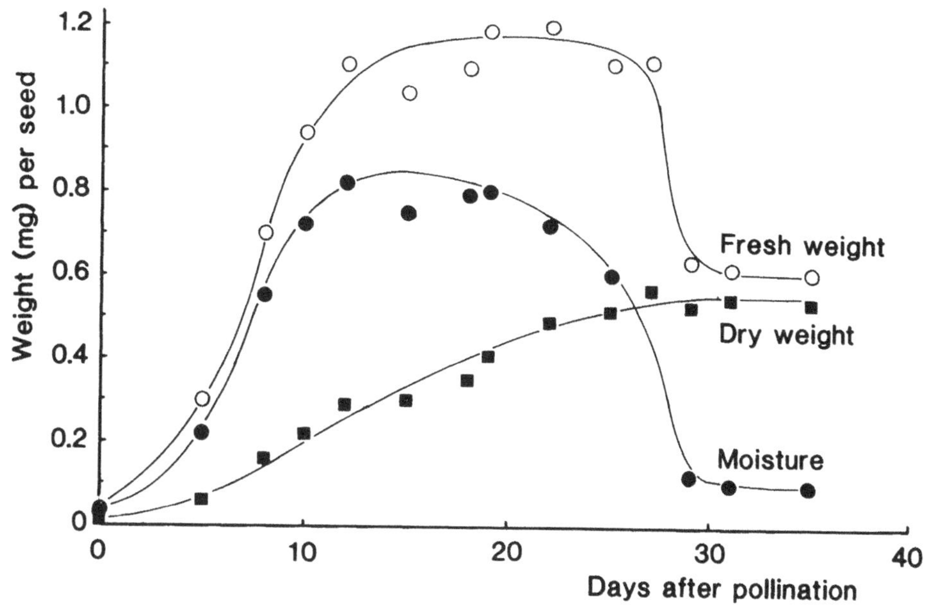

Fig. 3.18 Weight changes in white clover seeds during development. [Redrawn from Hyde et al 1959].

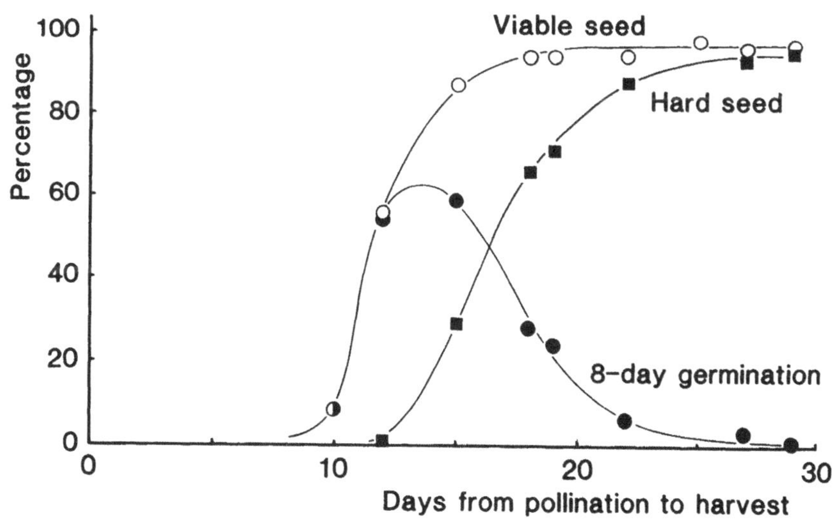

Fig. 3.19 Effect of stage of development of white clover seeds at time of harvesting on percentage germination after 8 days, percentage viability and percentage of hard seeds. [Redrawn from Hyde et al 1959].

Physiology

A. L. Hart
Grasslands Division, DSIR, Palmerston North, New Zealand

Contents

1. Introduction

The plant we know as white clover results from the expression of physiological processes over time and space. In this Chapter an attempt is made to gather together the main findings of physiological research on white clover and, by implication, to show in what aspects of its function data are lacking.

In contrast to some other species which have been used in a great many physiological experiments, white clover is not a common experimental organism. It is certainly not the most convenient plant to work with. Germination and establishment are relatively slow compared with, for example, peas, beans or wheat. Mature plants in the field are likely to have originated from pieces of rooted stolons, whereas plants grown from seed have a primary stem with a tap-root, as well as stolons with adventitious roots. As an outbreeding species it is highly variable and standard errors are often large.

Economics have, perhaps, also played a part in determining how much effort has been expended on white clover. Its herbage, for example does not have the same immediate cash value as wheat grain, soyabeans, maize or rice.

Despite these factors, a fair body of data has been accummulated over the years. More recently, white clover has been attracting more attention with the increased interest in forage systems in which nitrogen is provided by leguminous nitrogen fixation.

2. Light and photosynthesis

(A) Canopy structure

White clover has horizontally placed leaf blades borne at the top of thin erect petioles. When mature, adjacent leaves are sufficiently close to one another to form a distinct canopy.

The leaves of white clover open and close with a circadian rhythm having an endogenous period of 25 to 27 h. Alternate swelling and compression of cells on either side of the pulvinus at the base of each leaflet bring about these leaf movements. The pulvinus has been shown to be the photoreceptor and the site of the endogenous timer (Scott & Gulline 1972). Changes in potassium influx into pulvinules associated with cyclic changes in membrane permeability, and changes in water volume and membrane potential were suggested to be part of the mechanism of opening and closing (Scott & Gulline 1975, Scott et al. 1977) (see Chapter 2, Section 5, (A)).

In a stand with a canopy height of 19 cm, 92% of the incident light penetrated to 17.7 cm, 45% to 12.7 cm and only 5% to 10.1 cm (Mitchell & Calder 1958). In midsummer in New Zealand, 95% of the incident light was intercepted in a pure sward when the leaf area index (LAI) was 3.5 (Brougham 1958a).

Because white clover leaf laminae are horizontal, the critical LAI (LAI at 95% light interception) is lower and the extinction coefficient (k in the equation $\ln I/I_0 = kLAI$ where I/I = fraction of light reaching ground level) is higher than in species with vertically inclined leaves (Brougham 1958a, 1960, Brown et al. 1966). Species with greater critical LAIs and lower extinction coefficients are able to trap more incident light per unit ground area. This is reflected in higher crop growth rates, higher efficiencies of solar energy utilisation and net assimilation rates (Wilfong et al. 1967, Okubo et al. 1969, Agata et al. 1971, Kubota et al. 1971). However, if clover is compared with ryegrass when both are at a low LAI of say 2.0, then it can be shown that, theoretically, clover has a higher rate of photosynthesis than ryegrass. At a low LAI, plants with horizontal laminae can intercept and utilise light better, whereas plants with erect leaves have the advantage at high LAI (Jewiss 1967).

Once the canopy closes in a developing sward, the petioles of subsequent leaves have to elongate sufficiently to bring their laminae above the canopy. New leaves are subjected to lower irradiances of different quality during their development than older leaves, and the rate of leaf senescence increases (Brougham 1958b).

(B) Light modulation by the canopy

The canopy of a pure stand absorbs most visible light but transmits longer wavelengths. Scott et al. (1968) analysed transmitted and reflected radiation in a pure stand of white clover (Figure 1). The leaves in the canopy absorbed most of the visible radiation, while longer wavelengths were transmitted; in clear conditions 40% and 33% of incident radiation at 800 and 1000 nm, respectively, reached ground level. There was a small amount of reflection of incident radiation; as with transmission, reflection of visible wavelengths was less than that of infra-red wavelengths.

(C) Response of photosynthesis to light quality

The response of white clover to different wavelengths of photosynthetically active radiation (PAR) is similar to that in other species.

The relationship between wavelengths and relative quantum yield, relative action spectrum and absorption was determined in square sections of white clover leaves by McCree (1971). Data for white clover from growth chambers are given in Figure 2. Quantum yield expresses the ratio between the rate of photosynthesis and rate of absorption of quanta. Relative quantum yield is normalised to a maximum of 1.00. Relative action (action normalised to a maximum of 1.00) is the ratio between the rate of CO_2 uptake and the rate at which energy is received by the leaf. A leaf with a low absorptance transmits and reflects much of the incident light and vice-versa.

The shape of the curves shown in Figure 2 are typical of all species and conditions tested apart from some differences in the UV and blue up to about 500 nm. The study was done to gain some insight into the best method of measuring PAR. As neither of the curves for relative quantum yield or relative action are flat, neither flux, absorbed quanta, or irradiance are perfect measures of PAR. Both overestimate the effectiveness of blue light relative to red, irradiance (W m^{-2}) more so than quantum flux (μE cm^{-2} s^{-1}).

(D) Carbon dioxide exchange

i. Pathway of CO_2 fixation

White clover is a C_3 plant. It has a number of features typical of C_3 plants: (1) no bundle sheath (Akita et al. 1969); (2) granal chloroplasts in both palisade and spongy mesophyll cells (B.J. Forde & E. Ballantine, personal communication); (3) high chlorophyll a/b ratios in leaves and chloroplasts (Chang & Troughton 1972); (4) high levels of carbonic anhydrase (Atkins 1974); (5) a high CO_2 compensation point and O_2 inhibition of photosynthesis (Akita et al. 1969, Akita & Tanaka 1973, Masterson & Sherwood 1978); (7) low water use efficiency (Tantrum & Mitchell 1972); (8) transpiration rates higher than those of C_4 plants and with a greater response to temperature (Forde et al. 1977); (9) low optimum temperature for growth relative to C_4 plants (Mitchell 1956b, Mitchell & Lucanus 1962).

ii. Effect of irradiance and temperature on CO_2 exchange

In general, when plants are exposed to a range of temperatures and/or irradiances, net CO_2 exchange or gross photosynthetic rate increases with temperature and irradiance (Beinhart 1962, Murata & Iyama 1963, Brown et al. 1966, McCree & Troughton 1966a, Wilfong et al. 1967, Scott et al. 1970, Scott & Menalda 1970, Mächler & Nösberger 1977, Woledge & Dennis 1982).

Response curves for net CO_2 exchange may reach saturation (Wilfong et al. 1967, Woledge & Dennis 1982) or go through a maximum for both these factors (Murata & Iyama 1963, Scott & Menalda 1979) (see Figure 3). Increases in photosynthesis with irradiance are greater at moderate to high than at low temperatures (Beinhart 1962, Scott & Menalda 1970, Woledge & Dennis 1982).

The increase in net photosynthesis with measurement temperature observed in leaves by Woledge & Dennis (1982) was accompanied by a steep decline in diffusion resistance. This was chiefly due to a decline in residual resistance; there was a small decrease in stomatal resistance.

In plants grown at a given irradiance and then subjected to a range of other irradiances the rate of respiration was dependent on the irradiance prevailing during the measurement of CO_2 exchange (McCree & Troughton 1966a). A plant grown at 70 W m^{-2} showed a daily cycle of respiration, with a maximum in the middle of the day and a minimum in the latter half of the night. When the irradiance was lowered to 11 W m^{-2}, respiration fell quickly to establish a new respiratory cycle of lower magnitude. On a return to an irradiance of 70 W m^{-2}, respiration rose as quickly as it had fallen, indicating that the plants had adapted their respiration so as to be proportional to their net daily uptake of CO_2, itself dependent on irradiance. At all light levels there was a positive net CO_2 exchange sufficient to compensate for respiratory losses during dark periods. This work cast doubt on the assumption of a constant respiration rate per unit of leaf area, an assumption which had been used in the development of models of crop growth (Brown & Blaser 1968).

Respiration is stimulated by increases in temperature, possibly more so than by CO_2 uptake. Respiration had a higher Q_{10} (1.39, 10-40°C) than photosynthesis (0.93) in Ladino clover (Murata & Iyama 1963). The Q_{10} for leaf respiration in plants grown at day/night temperatures of 18/14°C was 2.40, and 2.49 for those grown at 8/4°C. Gross photosynthesis of the former had a Q_{10} of 2.5 between 3 and 13°C and 1.4 between 13 and 23°C (these are measurement temperatures) (Woledge & Dennis 1982). The optimum temperature of O_2 uptake by leaf discs was 35°C; that for stolon respiration was 50°C (Beinhart 1962).

CO_2 exchange can undergo temperature acclimation. When tested at a range of temperatures (2-40°C), plants grown at 20°C had the highest net positive exchange rates, while those grown at 10 and 30°C had lower rates. The temperature optima for the plants previously grown at 10, 20 and 30°C were approximately 26, 28 and 30°C respectively (Scott 1970).

Leaves grown at day/night temperatures of 18/14°C and 8/4°C had similar photosynthetic rates when measured at low temperatures, but those from 18/14°C had the highest rates at higher measurement temperatures. Respiration was more affected by growth temperature than was photosynthesis. Leaves grown at 8/4°C had higher rates of respiration than those from 18/14°C when respiration was measured at the same temperature (above 3°C) (Woledge & Dennis 1982).

Altitudinal ecotypes showed both short- and long-term adaptation of CO_2 exchange to light and temperature conditions. Ecotypes from all altitudes grown at low photon flux densities showed higher apparent photosynthesis than those grown at high densities when photosynthesis was measured at low flux density; the reverse was also the case. Similarly, plants grown at low temperatures had higher photosythetic rates than those grown at high temperatures when photosynthesis was measured at low temperature (Mächler & Nösberger 1977). Ecotypes from both high and low altitudes showed greater photorespiratory activity when grown at low temperatures (10/7°C day/night) than at higher temperatures (20/15°C) (Mächler & Nösberger 1978).

With regard to longer-term adaptation, apparent photosynthesis of ecotypes from habitats of low flux density acclimatized best to low flux density during growth and vice versa. Apparent photosynthesis was higher in ecotypes from high altitude especially when grown at low temperatures (Mächler & Nösberger 1977). A high-altitude ecotype also had higher photorespiration when this was measured at 20°C than one from a low altitude. The high photorespiration of the former ecotypes was thought to be compensated for by its ability to photosynthesize at high rates at low temperature (interpreted as an adaptation to a short growing season) (Mächler et al. 1977, Mächler & Nösberger 1978).

iii. Model of respiration in white clover

The idea that the rate of respiration of a plant is not wholly dependent on its size but to a large degree is also proportional to its rate of photosynthesis formed the basis for experiments resulting in equations for respiration (McCree 1970, McCree 1974, McCree & Kresovich 1978, McCree & Silsbury 1978).

The main feature of these equations is the estimation of nighttime CO_2 efflux via the products of a growth coefficient and daytime CO_2 uptake and of a maintenance coefficient and plant size.

iv. CO_2 exchange and LAI

Brown et al. (1966), McCree & Troughton (1966b) and Wilfrong et al. (1967) found a generally linear relationship between net CO_2 exchange and LAI up to LAI values of about 3-3.5 at which net CO_2 exchange reaches a maximum. Net CO_2 exchange remained at or near maximum values as LAI increased to about 5 (Wilfong et al. 1967) and to 7 or 11, depending on light conditions (McCree & Troughton 1966b). These latter values obtained for green leaves under artificial conditions are higher than what seems to have been usually found for white clover in the field, e.g. a pure sward of white clover had a ceiling LAI of 5.5, of which about 3-3.5 LAI units were represented by actively photosynthesizing tissue (Brougham 1958b).

v. LAI and crop growth rate (CGR)

In a similar manner to CO_2 exchange, CGR increases with LAI in an approximately linear fashion up to a LAI of about 3. At higher LAI values, the relationship is curvilinear. The exact form of the relationship is still undecided.

In a pure stand of white clover Brougham (1960) found that the maximum CGR occurred at the critical LAI. Maximum CGR was also correlated with total chlorophyll per m^2 of herbage. It was suggested that the production of dry matter was limited ultimately by the amount of chlorophyll per unit area of land exposed to light levels above the compensation point (Brougham 1960). Similar results were obtained for pure swards of Ladino clover by Okubo et al. (1975), who considered that the chlorophyll content of canopy leaves during the period of optimum LAI could be regarded as an indication of their photosynthetic activity.

It has been postulated that once leaves start to shade one another in the canopy, then the CGR will fall quickly due to respiration increase in relation to photosynthesis, i.e. that there is a sharp optimum LAI for CGR (e.g. Brown & Blaser 1968). In view of the work showing that a positive CO_2 uptake is maintained as irradiation decreases (McCree & Troughton 1966a) and that net CO_2 exchange remains at or near maximum levels as LAI increases above its critical value (McCree & Troughton 1966b, Wilfong et al. 1967), the effects of shading on CGR may not be as deleterious as once thought. CGR may thus remain at a maximum over a fairly wide range of LAI. In any case, it seems likely that the continuous turnover of leaf material (e.g. 1 LAI unit every 5 days; Brougham 1958b) and overtopping of old leaves by new ones would mitigate declines in CGR with canopy closure.

Mature leaves overtopped by new ones do die eventually, and dead material has to be taken into account when CGR is estimated from harvested material (Hunt 1970, Wilman & Martins 1977). Some herbage may disappear because of decomposition or ingestion by microflora (Wilman & Martins 1977). In a ryegrass-white clover pasture after defoliation, 'net' growth rate was initially linear but then became zero. By taking leaf death into account, 'gross' growth rate was shown to continue at positive values, although these were less than the maximum previously attained (Figure 4). The deviation from the maximum was interpreted as being due to an increase in the ratio between respiration and photosynthesis after the critical LAI had been reached (Hunt 1970).

3. Carbon metabolism and nitrogen fixation

Nitrogen fixation is dependent on photosynthesis and respiration for a supply of carbon skeletons and energy.

Defoliation, as expected, causes a decrease in nitrogen fixation (Moustafa et al. 1969, Chu & Robertson 1974, Halliday & Pate 1976, Crush and Tough 1981) (see also Chapter 6, Section 7.(B)). As Figure 5 shows, the effect is rapid (Moustafa et al. 1969, Halliday & Pate 1976). Unless nitrogen fixation is only supported by carbon compounds recently arrived from the shoot, this implies that defoliation may involve more than a simple cessation of carbohydrate supply, such as the disruption of hormonal links.

Nitrogen fixation is not so tightly coupled to photosynthesis that it changes markedly with normal diurnal changes in irradiance. Carbohydrate reserves are sufficient to carry nitrogen fixation through the dark period; any diurnal changes appear to be largely due to variation in temperature (Masterson & Murphy 1976, Haystead et al. 1979, Ryle et al. 1979a). Nitrogen fixation is reduced during the dark when photoperiods are very short, but activity is quickly restored in the next photoperiod (Murphy 1981). Changes in irradiance persisting over a few days can also lead to changes in nitrogen fixation and root respiration (Halliday & Pate 1976, Haystead et al. 1979).

Nodules have respiration rates more than twice those of nodulated roots (Murphy & Murphy 1978). The relative growth rate of nodules is also double that of nodulated roots (Hart et al. 1981).

The magnitude of nodulated root respiration is related closely to nitrogen fixation as measured over one hour (Haystead et al. 1979) or to the rate of accumulation of plant nitrogen during growth (Ryle et al. 1979a). Roots of plants dependent entirely on nitrogen fixation respire a greater proportion of the plant's daily carbon uptake than do roots of plants given mineral nitrogen. Shoot respiration and photosynthesis are unaffected by the source of nitrogen (Table 1) (Ryle et al. 1979a, Haystead et al. 1980). This greater respiratory burden of nodulated plants is thought to be why they yield less than plants utilising combined nitrogen.

Table 4.1. Respiratory losses from the shoot and root of white clover dependent on nitrogen fixation.

A. Percentage of gross photosynthesis respired by								
Shoot			Root			Unrespired		
Nodulated		Nitrate	Nodulated		Nitrate	Nodulated		Nitrate
26	ns	28	34	**	21	40	**	51

From Ryle et al (1979b)

B. Percentage of net daily carbon fixation respired by			
Shoot		Root	
N_2	NH_4NO_3	N_2	NH_4NO_3
14	14	24	18

From Haystead et al (1980)

The generation of hydrogen by nitrogenase represents a loss of respiratory energy unless the hydrogen is retrieved by hydrogenase. Detached nodules of white clover evolve hydrogen (Schubert & Evans 1976, Ruiz-Argueso et al. 1979). According to Ruiz-Argueso et al. 1979 the relative efficiency of nodules ranged from 0.51 to 0.75, depending on the inoculum, and none of the *R. trifolii* strains they tested were able to recycle hydrogen. However, Haystead et al. (1980) arrived at estimates of the ratio between carbon utilised and nitrogen fixed which were lower than those other reported elswhere and from which they suggest that the white clover-*Rhizobium* symbiosis with which they were working had nodule hydrogenase. Field-grown plants of white clover emit hydrogen (Conrad & Seiler 1979, Crush & Tough 1981). The degree to which this hydrogen evolution in the field represents a loss of nitrogen fixing potential is uncertain, but may be quite small (Crush & Tough 1981). Hydrogen evolution rates in these experiments were variable and appeared to be influenced by conditions affecting plant growth.

Estimates of the efficiency of nitrogen fixation in terms of carbon respired per unit of nitrogen fixed fall within the range of values determined for other legumes. Ryle et al. (1979a) found an estimated mean value of 6.6 mg C/mg N, while Haystead et al. (1979, 1980) found values of 3.4 to 4.4 mg C/mg N.

4. Translocation of photosynthate

As in other species, there is translocation of photosynthate in white clover from sources to sinks. The main sinks are those parts of the plant having high rates of metabolism, i.e. root and stolon apices, lateral buds, developing leaves, nodules and flowers (Harvey 1970, Hoshino 1974). The pattern of assimilate distribution from the leaves appears to depend on whether the leaves are on the primary stem (the first-formed stem is characterised by lack of internode elongation) or on stolons.

When the youngest leaves on a primary stem were fed ^{14}C, the main sinks were the stolons (formed from axillary apices) and the roots (Table 2), these sinks receiving 63 and 16.2% of the ^{14}C, respectively. An unexpected finding was that the roots, a sink distant from the terminal leaves, received considerably more assimilate than the apical meristem. Each leaf exported most assimilate to the stolon in its own axil. Assimilate was also exported to all other stolons, with a general tendency for movement to be towards the largest stolons (Ryle et al. 1981a).

Table 4.2. Percentage distribution of ^{14}C-labelled assimilate among plant fractions 24 h after exposure of leaves on the primary shoot of white clover to $^{14}CO_2$ for 30 min.

	Primary shoot leaf exposed to $^{14}CO_2$					
	(E + 2)	(E + 1)	(E)	(E–1)	Mean s.e.	Total
Fed leaf	14.1	18.4	13.3	8.4	±1.60	13.3±0.57
Petiole	2.4	2.2	2.4	2.2	±0.28	2.2±0.11
Other primary shoot leaves	1.1	0.9	1.0	1.0	±0.14	1.0±0.08
Terminal meristem	2.5	3.6	4.3	6.3	±0.94	4.4±0.64
Stolons	67.9	64.6	62.0	58.8	±2.38	63.0±1.41
Roots (+ nodules)	12.0	10.2	17.0	23.4	±2.12	16.2±0.88

E = youngest fully expanded leaf, (E—1) = next leaf younger than E, (E + 1) and (E+2) = next leaves older than E

From Ryle et al (1981a)

The distribution of assimilate from leaves on stolons is more limited, in that leaves may mainly supply those sinks that are close at hand. Hoshino (1974) found that young leaves on stolons exported assimilates chiefly to terminal tissue, while assimilate from older leaves moved to older parts of the stolon or lateral buds and roots. Following defoliation of the primary stem, leaves on stolons being left uncut, nitrogen fixation was reduced in that stolon only. Differential shading had a similar effect (Halliday & Pate 1976). This result implies that assimilate produced by the leaves of a stolon is used only by that stolon together with its roots and nodules, whereas assimilate produced by the primary stem is distributed to all parts of the plant.

The main sources of photosynthate on stolons are leaves 3-6 (counting from the tip). Developing leaves changed from net importers to net exporters of assimilate when they had reached 65-70% of their maximum area. After defoliation, new leaves were supported by reserves until they became self-sufficient and net exporters of assimilate (Hoshino 1974).

Translocation is affected by environmental conditions. Low temperatures, increases in daylength and decreases in irradiance reduce translocation (Hoshino 1974, Ryle et al. 1981b). The source of nitrogen has been shown to change the pattern of distribution in plants with a primary stem (Ryle et al. 1981b). In plants dependent on symbiotic nitrogen fixation, a greater percentage of assimilate from the youngest, fully expanded leaf on the primary stem was transferred to the roots than in plants given mineral nitrogen, at the expense of export to the stolons. When whole shoots were exposed to ^{14}C, more assimilate was transferred to the roots of nodulated plants than to the roots of plants utilising nitrate. Roots of nodulated white clover plants respired more rapidly than those given nitrate (Ryle et al. 1979b), and the greater allocation of assimilate presumably reflects a greater sink strength.

5. Growth and development

(A) Roots

White clover roots range in form from fibrous to taprooted (Caradus 1977). The taproot that develops from the primary seedling root seems to have a life-span of about two years (Westbrooks & Tesar 1955, Ueno et al. 1967). The crown dies soon after loss of the taproot. The cause of taproot death is not known, although it appears to be physiological rather than due to attack by pathogens (Chow 1966). Taproot breakdown may be associated with increasing competition for assimilates by stolons, exacerbated by the long path for assimilate transport between the primary stem and the root (Ryle et al. 1981a).

The spread and establishment of stolons is vital to the persistence of white clover, particularly if flowering and seed production are sparse (Chow 1966). The nodal roots are able to compensate for loss of the taproot (Chow 1966). They have the same ability to support leaves as seedling roots, there being no differences in the allometric relationship between root and lamina dry weight for plants derived from stolon cuttings or seeds (Haydock 1982). Gibson & Trautner (1965) found no differences in height or yield of clover grown with or without primary roots either in the field or in growth cabinets at a number of temperatures. Plants with a primary root system had a primary stem with a radial array of stolons, but there was no visible difference between primary and adventitious root systems.

i. Temperature and root growth

There is a paucity of information on the effect of root temperature on roots and on the plant generally. Usually, in experiments on clover in growth cabinets root temperature has not been controlled independently of shoot temperature. Roots are capable of undergoing adaptive physiological changes at low temperatures, but roots of temperate plants, which in the field may encounter quite low soil temperatures, are often exposed experimentally to relatively high temperatures. The properties of roots at such temperatures may not be relevant to their physiology in the field (Clarkson 1976).

White clover roots are capable of functioning at 5°C (Ollerenshaw & Baker 1981). Nodules are able to overwinter and resume growth in the spring (Pate 1958, Bergersen et al. 1963).

At ambient day/night shoot temperatures of 27°C and a night root temperature of 20°C, daytime root temperatures of 20°C to 40°C had little effect on total dry weight of plants provided with NH_4NO_3 (Small & Joffe 1968). Net assimilation rate decreased with temperature but leaf area ratio increased; relative growth rate was unaffected. Plants in pots where the soil temperature at 1500 h was near 25°C had higher yields than those where the temperature at this time was 30°C (Hogg & Moore 1976).

Plants dependent on symbiotic nitrogen fixation are more sensitive to root temperature than those given combined N. Growth of nodulated plants was reduced more by low root temperature than that of those assimilating nitrate. The magnitude of the effect depended on rhizobial strain (Mytton & Hughes 1980). With regard to higher temperatures, the optimum root temperature for nitrogen fixation was similar to that for uptake of combined N (26°C) but growth of plants dependent on nitrate fixation was severely inhibited in the range 30-45°C. Except for the highest temperature in this range, nodule number seemed unaffected by temperature, so temperature was assumed to have a direct deleterious effect on nitrogen fixation (Small & Joffe 1968).

Plants grown on mineral N were less sensitive to changes in growth cabinet temperatures than those dependent on nitrogen fixation (Arnott & Ryle 1982, Hoglund 1979). The relationships between temperature and nitrogen fixation were not simple. The former authors noted that under a 16-h photoperiod, a decrease in temperature increased the difference in leaf area between N fixers and those given mineral N, but in an 8-h photoperiod the effect was reversed. The latter found that a positive relationship between relative yield of N fixers and temperature was reversed by the passage of time.

Nodulated roots from plants grown in a glasshouse and subjected to a range of assay temperatures showed a broad temperature optimum of 13-26°C for nitrogenase activity (Halliday & Pate 1976).

(B) Root:shoot ratios (see also Chapter 2, Section 4.(D))

The root:shoot ratio of white clover, like that of other plants, is influenced by environmental factors. Davidson (1969 a,b,c) suggested that these variations could be interpreted in terms of an equilibrium between root and shoot activity. In plants growing under favourable conditions the shoot:root ratio is low, but changes with the advent of unfavourable conditions. For example, if the supply of soil nutrients and water is short or unbalanced the ratio may increase (Davidson 1969b, Foulds 1978), while after defoliation or shading it may increase (Mitchell 1956a). These changes maintain the equilibrium between root absorption and photosynthesis.

The relationship between N source and root:shoot ratio is of interest because of the role of white clover as a provider of N in pastures. Ryle et al. (1981b,c) showed that ontogenetic changes in the ratio were not influenced by the source of N, but nodulated plants grown either singly or in swards had higher ratios than those given nitrate. Application of nitrate to nodulated plants shifted the ratios to values intermediate between those of plants fixing N and those utilizing nitrate, largely because of promotion of shoot growth (Ryle et al. 1981c). As the roots of nodulated plants respire more and receive a relatively greater allocation of assimilate than do plants given nitrate (Ryle et al. 1979b, 1981b), the increased shoot growth on application of nitrate to nodulated plants may well involve a redistribution of assimilate away from the root toward the shoot.

(C) Shoots and whole plants

i. *Effect of temperature and irradiance*

Germination of white clover is reduced by low temperature (Scott & Hansen 1977). Growth, though small, occurs at soil temperatures as low as 5°C (Ollerenshaw & Baker 1981). A clinal relationship between growth at low temperature (soil temperature 5°C, air temperature 8-9°C) and winter temperature at the place of origin was found for growth of ecotypes from sites in north Britain (Ollerenshaw & Baker 1981). The cline was less distinct when ecotypes were tested in the field (Ollerenshaw & Baker 1982).

Low day temperatures had a greater effect than low night temperatures on white clover grown under a controlled environment (Mitchell & Lucanus 1960). This is to be expected perhaps, since growth is dependent on material synthesized during the light period.

Growth of shoots and of whole plants increases with temperature (up to an optiumum) and with the amount of photosynthetic energy received (Mitchell 1956a,b, Mitchell & Lucanus 1962, Beinhart 1963, Hart 1976, Jelmini & Nosberger 1978a,b). The optimum temperature for growth is about 24°C, although different parts of the shoot may have different optima and the response curves of the various parts may differ in shape to that of a whole shoot (Mitchell 1956b, Mitchell & Lucanus 1962, Hoshino 1974, Williams & Hoglund 1978). Compared with temperate grasses, Mitchell & Lucanus (1962) showed that white clover was tolerant of high temperature provided the light input was high, growth at 35°C under a 16-h daylength being five times that under a 8-h daylength. On the other hand, white clover is less tolerant of low temperatures than grasses (e.g. Ollerenshaw & Baker 1981).

The growth and development of shoot apices and subsequent branching is affected by temperature and irradiance. Under constant temperature the mitotic index of terminal apices of New Zealand Government Stock white clover showed diurnal fluctuations. When the temperature varied diurnally, the fluctuations became smoother, there being a peak in mitotic index just before the end of the light period (Denne 1966).

Temperature may have an influence on whether apices are vegetative or reproductive, a topic dealt with in greater detail in Chapter 3, Sections 2.(A) and 2.(B) Takeda & Agata (1965a, 1965b) showed that flower formation in Ladino clover increased at high temperatures (25-30° C) and this, together with decreased branch formation, was thought to contribute to a decrease in leaf area.

Increases in temperature promote stolon formation and elongation from apical meristems (Mitchell 1956a, Beinhart 1963, Olusuyi & Raguse 1968). In contrast to apical meristems, Beinhart (1963) found that branching (activation of axillary meristems) decreased as temperatures increased (Figure 6). At low temperatures (10°C), total leaf production was limited by the low activity of the apical meristems which was not offset by the high degree of branching. At high temperatures (30°C) leaf production was limited by the low degree of branching. Maximum leaf production (17° and 23°C) occurred when moderate to high activity of apical meristems was coupled with a high to moderate degree of branching.

Stolon growth and branching are reduced by decreases in irradiance (Mitchell 1956a, Mitchell & Lucanus 1962, Beinhart 1963), although Trautner & Gibson (1966) found slight shading favoured bud formation and development of stolons. There also appear to be interactions between light and temperature in determining shoot growth. Mitchell & Lucanus (1962) found that as temperatures rose under a 16-h daylength, branching increased to a maximum and then declined, whereas under a 8-h daylength the opposite trend was found. Shading nearly halted stolon elongation at 22°C but had almost no effect on elongation at 12°C. Also, the average length of the longest stem in shaded, undefoliated plants at 12°C was nearly the same as that of undefoliated plants in full light at 22°C (Mitchell 1956a).

The contribution to total leaf area made by the leaves on each order of branch has been observed to vary with season, presumably at least in part because of the seasonal changes in light and temperature. In experiments by Beinhart (1963) with spaced plants of Ladino clover, most of the leaves formed between spring and autumn were from stolon branches; this was especially so for leaves formed in spring. Branching was infrequent in summer, presumably because of high temperatures. In the autumn, however, the branches formed in the spring themselves underwent branching, resulting in an increase in LAI, which had fallen over the summer due to the low rates of branching.

Temperature has an effect on leaf development apart from its influence on branching. Leaf appearance rate had a threshold of 2.6°C and lamina size a threshold of 6.3°C (Haydock 1981). Mean leaf dry weight increased in spring in the UK (Haydock 1981). At higher temperatures (18-29°C) leaf dry weight was little affected (Mitchell 1956b). Increases in temperature increased the petiole length (Olusuyi & Raguse 1968, Kishi 1975, Arnott & Ryle 1982), and the rate of progression of leaves through stages of morphological development (Carlson 1966a, Oluyusi & Raguse 1968, Arnott & Ryle 1982). Leaf life span was decreased at high temperatures (Takeda & Agata 1965a).

Leaf area per plant is increased by irradiance. As already noted, this results from the promotion of stolon branching (Beinhart 1963), from increases in rate of leaf appearance (Jelmini & Nosberger 1978b, Arnott & Ryle 1982), and duration of leaf expansion (Arnott & Ryle 1982) and from effects on leaf size (Mitchell 1956a, Arnott & Ryle 1982). Not unexpectedly, irradiance and temperature are interrelated. Mitchell (1956a) found that shading reduced leaf size at 22°C, but at 12°C leaves of shaded plants were larger than those grown in full light. The effects of shading at high temperature may also be exacerbated by the decrease in leaf life span (Takeda & Agata 1965a). Arnott & Ryle (1972) found leaf size to be reduced by a drop in temperature, this effect being greater under an 8-h daylength than a 16-h one. For nodulated plants, total leaf area also decreased with temperature under a daylength of 16-h but the decrease was smaller at a daylength of 8-h. Where plants were given mineral N the effect of daylength on decreases in area with temperature was reduced; in fact, almost all aspects of leaf growth on both the primary axis and stolons, were reduced in plants dependent on symbiotic N fixation (Arnott & Ryle 1982).

In a pure sward of white clover left undisturbed throughout the year, Brougham (1962) found that leaf development was such that the LAI was always above the critical value, i.e. all light was intercepted by actively photosynthesizing tissue throughout the year. There were marked seasonal variations in the height of the canopy above the ground, being lowest in winter and highest in summer. The variation in height was due to variation in petiole elongation. Leaf size also varied, being smallest in winter and greatest in summer. The variation in leaf dimensions was considered to arise from a combination of the influences of incident light and temperature (changing seasonally) and light quality under the canopy.

As suggested by Brougham (1962), petiole elongation and leaf size are affected by light quality (Mitchell 1956a, Olusuyi & Raguse 1968). Light in the red (600-700 nm) and far-red (700-800 nm) end of the spectrum controls morphogenetic processes through its action on phytochrome (e.g. Smith 1975). Warrington & Mitchell (1976) showed that under artificial conditions white clover morphology was determined by the spectral quality of the lamps used. Plants grown under red-biased light (peak irradiance at 600 nm) had long petioles and stolons and large, thin leaves. Clover canopies readily transmit wavelengths greater than 700 nm, but transmission in the 600-700 nm range is quite small (Figure 1) (Scott et al. 1968). While the hypothesis that the spectral quality of the light under the canopy controls leaf morphogenesis seems likely to be correct, it needs to be verified by further experiments.

ii. Defoliation

Defoliation affects root function. Root elongation stops, although less quickly than in some other species, e.g. ryegrass (Evans 1973a,b). Older roots die, nodules are lost, leghaemoglobin is degraded and the rate of acetylene reduction falls, although these effects can be compensated for by the growth and nodulation of new stolon roots (Butler et al. 1959, Moustafa et al. 1969, Sinclair 1973, Chu & Robertson 1974, Halliday & Pate 1976). With defoliated plants, Mitchell (1956a) noted that there was a zone on the roots where nodule development was poorer than either above or below this zone. Nodulation was thought to be occurring in this zone at the time of defoliation.

The size of leaves forming after defoliation is reduced initially, but increases again with time (Mitchell 1956a, Carlson 1966b, King et al. 1978). Severe overall defoliation (leaf area reduced to 16.8%) reduced the rate of regrowth and final leaf size more than a lenient (area reduced to 44%) defoliation (King et al. 1978). The effect of partial (localised) defoliation depended on the position of the excised leaves on the plant. Regrowth of peripheral terminal leaves was affected less by the distribution of defoliation than that of leaves of the crown or lateral nodes. This observation was interpreted in terms of the acropetal movement of assimilates from the primary stem. Peripheral leaves seemed able to use assimilates from all parts of the plant for regrowth, but regrowth of leaves in the centre of the plant depended on assimilates from that zone (King et al. 1978).

The growth of new leaves after defoliation depends on material translocated from any leaves left on the plant and from reserves, probably in the stolons and roots. Carbohydrate levels in the stolons and roots fall after defoliation and then rise again as new leaves develop. Plants high in reserves grow more vigorously than those low in reserves. Repeated defoliation before establishment of new leaves further decreases carbohydrate levels. It appears to take 1-2 weeks before new leaves become self-sufficient and reserves restored to levels existing before defoliation (Moran et al. 1953, Beinhart 1964, Carlson 1966b, Hoshino 1974). Demands are presumably also made on reserves for replacement of lost nodules and regeneration of nitrogenase activity after defoliation.

Defoliation of a sward of white clover which has reached the critical LAI will drastically alter light and temperature conditions above the plants. The cycle of leaf growth and senescence in which increases in petiole length and lamina size cause each new set of leaves to overtop the previous one will be interrupted. The progression from small to larger leaves will begin again as the sward develops towards canopy closure. Light interception by individual leaves from both terminal and axillary buds will improve after defoliation, which may mean that dry matter production soon after defoliation may be higher than in a sward in which the canopy has closed (Brougham 1962, Hunt 1970).

6. Water

The water relations of white clover are complex. Movement of water from soil to plant to atmosphere is influenced by plant factors such as stomatal resistance, leaf water potential and root morphology, and also by external factors such as atmospheric vapour pressure. Assessment of whether water use is controlled predominantly by external factors (Mitchell & Kerr 1966, Forde et al. 1977), internal factors (Shephard 1972, 1975) or some combination of both (e.g. Lawlor & Lake 1976, Burch & Johns 1978) appears to require a careful consideration of each new situation.

One common thread which has emerged relating to the response of white clover to water stress is that the plant does not control leaf water loss efficiently (Sanchez-Diaz & Sanchez-Marin 1974, Burch & Johns 1978, Johns 1978, Aparicio-Tejo et al. 1980). Leaf diffusive resistance increases as water stress increases, but not sufficiently to prevent transpiration continuing. Whether the lack of transpiration control arises from an inability to close stomata or from low cuticular resistance is uncertain. Cox & Boersma (1967) noted that stomata remained open under water stress at high temperatures (although this was compensated for by an increase in mesophyll resistance) and Sanchez-Diaz & Sanchez-Marin (1974) found that stomatal closure during drought cycles was more gradual in white clover than in lucerne. On the other hand, Sanchez-Diaz & Sanchez-Marin (1974) also found a lower cuticular resistance in white clover than in lucerne or *Trifolium subterraneum*.

In dry conditions, white clover leaves senesced rapidly, thus in effect, compensating for poor control of transpiration. From the middle of a dry cycle, LAI values of white clover decreased, until towards the end of the cycle the rate of water loss per unit soil area was similar to that of *Festuca arundinacea*, a species with better control of transpiration (Burch & Johns 1978).

An inability to control water loss can explain the low efficiency of water use of white clover observed in dryland (Johns & Lazenby 1973a,b). Leaf senescence as a means of preventing excessive dehydration was a particularly evident phenomenon and in the hottest months up to 70% of the herbage died in unirrigated swards; even under irrigation 20% of the herbage dried out (Johns & Lazenby 1973a).

A scale for visually estimating the leaf water status of white clover was developed by Shepherd (1975). Relative water content (RWC) over the four ranges 95, 90-80, 70-60 and 50-40% corresponded to the following leaf conditions: (1) turgid, rigid; (2) softer, individual leaflets drooping; (3) all leaflets flaccid, wilted; (4) leaflets shrivelled, inelastic. The estimates were usually within 5% of measured values of RWC.

Root density is important for the ability of white clover to extract soil water (Burch & Johns 1978). White clover was found to have most of its roots within the top 20-cm soil layer, but could extract water to a depth of 90 cm (Evans 1977). The penetration of roots into soils is influenced strongly by soil moisture. Roots of plants in containers did not extend beyond wetted zones (Cullen et al. 1972), root growth was retarded in dry soil (Burch & Johns 1978, Foulds 1978) and a dry soil surface inhibited nodal root formation (Ueno & Yoshihara 1968). Burch (1979) postulated the development of a root-soil contact resistance (equivalent to about 71 times the resistance to radial water flux in the soil) because of root and/or soil shrinkage as the soil dried. In wet soils the major resistance to water flow into and through the roots lies outside the xylem, probably in the endodermis (Burch 1979).

The initial effect of water stress is to reduce extension growth and fresh weight because of decreased turgor, followed by decreases in transpiration, photosynthesis, respiration, nitrogen fixation and dry weight (Peterson & Hagan 1953, Upchurch et al. 1955, Hagan et al. 1957, Engin & Sprent 1973, Johns 1978, Aparicio-Tejo et al. 1980).

In severely wilted leaves of Ladino clover, some 30% of the soluble leaf nitrogen was comprised of proline compared with 1.8% for well-watered leaves (Routley 1966). Aerobic metabolism is necessary for proline formation, and levels are increased by a supply of glucose or light treatment (Routley 1971). Routley (1971) suggested that proline was a storage compound or that its synthesis was a means of turning over ATP and NADPH when anabolic metabolism was low.

Recovery takes place provided stress is not too severe. Upchurch et al. (1955) showed that apparent photosynthesis doubled within 6 h on watering after wilting. Leaf water potential and leaf diffusion potential returned to normal within 3 days after 17 days of water stress. Transpiration also recovered in the same time, though not to quite the levels reached before the drought treatment (Aparicio-Tejo et al. 1980a,b). Flowering is reduced by water stress, as is seed number and weight (Hagan et al. 1957, Foulds 1978).

Germination of clover seeds was inhibited markedly by water potentials of less than -4 bars. The water potentials were established with the use of polyethlene glycol (mol. wt. = 20 000); the resulting osmotic potentials were accepted as producing the same effects as true drought (Sharma 1973).

Clover growth is retarded by flooding (Low & Piper 1960, Finn et al. 1961). Some evidence for variation in reaction to flooding in white clover was found by Hoveland & Mikkelsen (1967). When cv. S-1 Intermediate white clover and Ladino clover were periodically flooded and drained over 4 months, the yield of the intermediate type was reduced more than that of the Ladino type. There were no differences in flooding responses within types.

A lack of oxygen for root metabolism and nitrogen fixation was considered to be the cause of reduced yields under flooding (Low & Piper 1960, Finn et al. 1961, Hoveland & Mikkelsen 1967). Over a 4-month period with flooding 6 days out of every 10, the soil air was found to contain 9.8% 0_2 and 6.0% CO_2, compared with 19.9% 0_2 and 1.1% CO_2 in well-drained soil (Hoveland & Mikkelsen 1967). In flooded soil in which Ladino clover was growing, oxygen diffused to a platinum electrode at a rate of 0.44 x 10^{-7}g cm^{-2} min^{-1} (Finn et al. 1961). This rate of diffusion is below the soil oxygen diffusion rates (1.0-1.5 x 10^{-7}g cm^{-2} min^{-1}) which Gradwell (1969) considered critical for white clover growth. Shoot growth was substantially reduced at less than critical levels of oxygen, although root growth was not depressed to a significant extent.

7. Cold hardiness

Once hardened, white clover is cold hardy, i.e. can withstand temperatures below freezing with minimum subsequent injury.

There is significant genetically based variation in the degree of cold hardiness within the species. Common white clover plants with reticulate stolon growth and protective leaf cover were generally less injured by winter conditions than were Ladino plants, which have a more open, spreading habit. The degree of winter injury varies within white clover type (Ronningen 1953, Ruelke & Smith 1956). Ladino clover sustained greater injury when stored in solid ice at -3°C than did common white clover. Susceptibility to injury within clones varied (Smith 1949). Eagles & Othman (1981) found in studies with a number of cultivars that North European types were more hardy than Mediterranean types.

White clover has been used as a test species in the development of a rapid electrical test for cold hardiness by Greenham & Daday (1957, 1960), who found a correlation between the electrical resistance index of crowns and recovery after exposure to temperatures below 0°C. The magnitude of the index decreased with increasing period or severity of freezing, becoming approximately 100 at death. Differences in mean resistance index, and hence susceptibility to cold injury, were found between ecotypes. Again, North European populations appeared to be considerably more cold hardy than Mediterranean ones (Table 3).

Another measure of resistance to freezing injury is the conductance of distilled water after exosmosis from clover stolons. Conductance was directly correlated with the concentrations of potassium and chloride ions leaked from the tissue and was inversely related to resistance to freezing injury and ability to recover under glasshouse conditions (Flanagan & Sprague 1961).

Table 4.3. Mean resistance index values for ecotypes of white clover with and without cold treatment.

Population	Character	Mean resistance index (n = 6 - 8) Treated			Untreated
		Expt. 1	Expt. 2	Expt. 3	Expt. 4
Mediterranean	Cold susceptible	130**	240**	300**	730*
N. European	Cold hardy	630	690	790	800

*and** = significantly less than the value below at P 0.05 or P 0.01

From Table 1, Greenham & Daday 1957.

In laboratory studies, cold-hardened Ladino clover was injured less severely when cooled slowly than when cooled rapidly (Sprague 1955, Kilpatrick et al. 1966). However, laboratory results may not be entirely applicable to natural conditions, as mid-winter plant temperatures in the open can vary more than laboratory temperatures; also, the rate of cooling in the field was near that found to be lethal in laboratory tests (Sprague 1955).

White clover has been used in some of the many attempts to interpret cold hardiness on the basis of changes in cellular constituents (Wood & Sprague 1952, Ruelke & Smith 1956). It has been pointed out that the wealth of information on changes in cellular constituents 'cannot be fully appreciated however, because of a lack of understanding of what constitutes freezing injury. Unfortunately, the basic premise of such correlative investigations is that an increase (or decrease) in a particular cellular constituent implies a causal relationship to cold hardiness' - which may not necessarily be the case (Steponkus 1978).

8. Plant growth substances

Experiments have been done which implicate growth substances in the morphogenesis of white clover, but no comprehensive study appears to have been undertaken (See also Chapter 2, Section 4).

Indoleacetic acid was thought to influence internode elongation (Thomas 1972). Gibberellic acid has been implicated in secondary thickening of apical internodes (which was also influenced by benzyladenine) (Thomas 1972), and in influencing such morphological characters as leaf size and stolon number (Fletcher & Martin 1962). It has been postulated that morphological differences among varieties of white clover may be based on differences in the levels of endogenous gibberellins (Fletcher & Martin 1962). Gibberellic acid had no effect on the germination of white clover either by itself (Fletcher & Martin 1962) or in combination with 6-benzylaminopurine (Skinner et al. 1958). Kinetin and a number of 6-substituted purines, including 6-benzylaminopurine, were found to stimulate germination (Skinner et al. 1958). Chloroplast development was inhibited by nitrophenols when white clover was used as a bioassay for these compounds (Price & Wain 1976). Ethylene (10 ppm) reduced nodulation, nodule activity and shoot growth when perfused for several weeks through a soil-sand mixture containing white clover (Goodlass & Smith 1979). Various growth substances have been used to control callus and plantlet development (Oswald et al. 1977, Bhojwani 1981).

9. Allelochemicals

Allelochemicals are substances 'by which organisms of one species affect the growth, health, behaviour or population biology of organisms of another species (excluding substances used only as foods by the second species)' (Whittaker & Feeny 1971). White clover produces allelochemical substances and is affected by substances produced by other species.

(A) Extracts and Exudates

Extracts from, and exudates of, white clover have been shown to affect growth and germination of other species (and of white clover) in laboratory and glasshouse studies (Masterson 1962, Fottrell et al. 1964, Grant & Sallans 1964, Hoveland 1964, Masterson 1965, Rice 1972, Tames et al. 1973, Newman & Rovira 1975, Scott 1975, Hale & Mathers 1977, MacFarlane et al. 1982a,b). White clover is 'auto inhibitory'. Its growth was inhibited by leachate collected from sand containing white clover roots (Newman & Rovira 1975). Germination and initial seedling establishment were reduced by water-soluble substances in freeze dried white clover shoots. Experiments with polyvinyl polypyrrolidone suggested that the substances were phenolic compounds (Macfarlane et al. 1982a,b). Exudates from grassland species may affect P uptake by white clover (Newman & Miller 1977). The assumption is that these substances are involved in allelochemical interactions in the field, though evidence for this is only circumstantial.

White clover may also be involved in interactions which affect micro-organisms on plant roots. When white clover was grown with *Lupinus cosentinii* the number of mycorrhizal entry points on the clover roots was increased but the fungus often showed abnormal swellings near the entry points. A substance from the lupin seed coat may have caused this (Morley & Mosse 1976). The abundance of micro-organisms on the roots of grassland plants, including white clover, grown in mixture was different from when they were grown separately, e.g. *Anthoxanthum ordoratum* grown with white clover had more bacteria, fungi and mycorrhizal infection than when grown alone, whereas the microflora on the white clover roots showed the opposite trend (Christie et al. 1978). The growth of white clover was affected by changes in soil micro-organisms brought about by irradiation and inoculation (Newman et al. 1977). The mechanism of such changes is uncertain. There may be direct effects of one plant's exudates on the root flora of another, or there may be changes in the amount of exudate available as a substrate for microbial growth arising from competition among plants for light and nutrients (Christie et al. 1978).

In a number of plants (Walker-Simmons & Ryan 1977) wounding the tissue causes proteinase inhibitor inducing factor (PIIF) to be released. The PIIF moves through the vascular system to nearby and distal tissues where it initiates the accumulation of proteinase inhibitor. Trypsin inhibitor activity was induced in excised leaves of white clover by both 'crude tomato PIIF' and by endogenous PIIF. The levels of trypsin inhibitor activity were significant but not high compared with some other species. This system has been likened to a primitive immunological response to protect the plant from proteinases of insect and microbial pests.

(B) Cyanogenesis

Cyanogenesis (the release of HCN from damaged leaves) is a characteristic of white clover for which the species is polymorphic. The leaves contain two related glucosides, linamarin and lotaustralin. When the leaves are damaged, e.g. on ingestion by animals, the glucosides are hydrolysed by linamarase to produce glucose and an unstable aglycone. The aglycone dissociates to release HCN and either acetone or 2-butanone. The level of the glucosides in the foliage may be sufficiently high for grazing animals to suffer from cyanide poisoning.

Valine and isoleucine are precursors of linamarin and lotaustralin, respectively, in white clover. There are four steps in the synthetic pathway between valine and linamarin. Plants that do not produce linamarin are deficient in at least two of these steps: the production of isobutyradoxime from valine, and linamarin from 2-hydroxyisobutyronitrile. The presence or absence of linamarase is determined by alleles of a single gene (see Section 10.1.3). The presence or absence of both the glucosides appears to be determined by alleles of another independently inherited gene. The biochemistry of cyanogenesis has been covered in greater detail by Tapper & Reay (1973) and Hughes & Conn (1976) among others. Damage arising from causes other than ingestion, e.g. drought, frost, may cause the release of HCN and the inhibition of metabolism. Respiration and photosynthesis of white clover were inhibited by the addition of HCN, respiration more so than photosynthesis (Foulds & Young 1977), but the role of internally generated HCN in environmentally caused damage is uncertain (see Chapter 8, Section 5.(B)).

10. Bloat

White clover is one of the legumes which cause bloat in ruminants. The following section is based on information from Jones et al. (1973), Leng & McWilliam (1973), Jones et al. (1976), Sarker et al. (1976), Howarth et al. (1978a,b, 1982) and Lees et al. (1981). Bloat arises when foams containing gases formed by microbial fermentation stabilise in the rumen. The gases cannot to be dispersed by eructation and cause respiratory and circulatory malfunction.

The foams include substances derived from plant tissue. Colloidal particles consisting of chloroplast thylakoid fragments act as nucleation sites for bubble formation (Howarth et al. 1978b). Soluble proteins (Fraction I and II) are thought to be important but are not the sole constituents of the foams. Plant lipids may act as antifoaming agents but their actual role is uncertain. The role of saponins, which possibly act as foam stabilisers, is also unclear. The onset of bloat may be associated with high rates of fermentation after ingestion of foliage containing high levels of starch and soluble carbohydrate.

Plant properties thought to reduce the risk of bloat are low concentrations of soluble protein, low carbohydrate levels, the presence of tannins and mesophyll cells resistant to mechanical rupture. Tannins are thought to precipitate leaf proteins in the rumen, thus preventing their participation in foam formation. Cells from non-bloat forming legumes have been found to be physically more resistant than those from species causing bloat. A slower release of cell contents may prevent rapid build-up of stable foams, allowing greater opportunity for dispersal of rumen gases.

White clover can have crisp, succulent foliage with easily ruptured mesophyll cells. The leaves by nature contain Fraction I and II protein, starch, soluble carbohydrates and lipids. Saponins have been found in Ladino clover (Bondi et al. 1973). Tannins do not occur in white clover foliage, although they have been found in the flowers.

The risk of bloat could be reduced by breeding a white clover containing tannins or which has tougher mesophyll cells. However, the attainment of either or both of these goals could have undesirable nutritional side effects. *Trifolium repens* var. *giganteum* and *T. semipilosum* cv. Safari (Kenya white clover), do not cause bloat, although the reasons for this are not known.

11. Conclusions

White clover is a temperate, herbaceous dicotyledon and the properties of the individual physiological process, e.g. gas exchange, water movement, reflect this classification. It is also a forage legume, and a physiological interpretation of its agricultural role is desirable but difficult to achieve.

Research on the physiology of white clover has been more sporadic than that on some other species. Further research, or careful extrapolation from knowledge of other plants, into almost every aspect of white clover function could materially help our understanding of its behaviour in agricultural situations. For example, such processes as the control of translocation, the regulation of root and stolon growth, and allelochemical interactions are not fully understood. To reach such an understanding is a difficult goal, as it often requires a holistic explanation of plant function, an explanation as difficult to make for white clover as any other plant.

References

Agata, W.; Kamata, E.; Aoki, K. (1969) Influence of decrease in stolon number caused by low temperature on yield of Ladino white clover swards. *Proceedings of the Crop Science Society of Japan* **38**, 143-148.

Agata, W.; Kubota, F.; Kamata, E.(1971) Dry matter production of forage plants. 3. Influences of light extinction coefficient (k) on dry matter production and optimum cutting frequency in forage plant communities. *Journal of the Japanese Society of Grassland Science* **17**, 235-242.

Akita, S.; Miyasaka, A. (1969) Studies on the differences of photosynthesis among species. II. Effect of oxygen-free air on photosynthesis. *Proceedings of the Crop Science Soiety of Japan* **38**, 525-533.

Akita, S.; Miyasaka, A.; Murata, Y. (1969) Studies on the differences of photosynthesis among species. I. Differences in the response of photosynthesis among species in normal oxygen concentration as influenced by some environmental factors. *Proceedings of the Crop Science Society of Japan* **38**, 507-523.

Akita, S.; Tanaka, I. (1973) Studies on the mechanism of differences in photosynthesis among species. IV. The differential response in dry matter production between C_3 and C_4 species to atmospheric carbon dioxide enrichment. *Proceedings of the Crop Science Society of Japan* **42**, 288-295.

Aparicio-Tejo, P.M.; Sanchez-Diaz, M.F.; Pena, J.I. (1980a) Nitrogen fixation, stomatal response and transpiration in *Medicago sativa*, *Trifolium repens* and *T. subterraneum* under water stress and recovery. *Physiologia Plantarum* **48**, 1-4.

Aparicio-Tejo, P.M.; Sanchez-Diaz, M.F.; Pena, J.I. (1980b) Measured and calculated transpiration in *Trifolium repens* under different water potentials. *Journal of Experimental Botany* **31**, 839-943.

Arnott, R.A.; Ryle, G.J.A. (1982) Leaf surface expansion on the main axes of white and red clovers. *Grass and Forage Science* **37**, 227-233.

Atkins, C.A. (1974) Occurence and some properties of carbonic anhydrases from legume nodules. *Phytochemistry* **13**, 93-98.

Beinhart, G. (1962) Effects of temperature and light intensity on carbon dioxide uptake, respiration and growth of white clover. *Plant Physiology* **37**, 709-715.

Beinhart, G. (1963) Effects of environment on meristematic development, leaf area, and growth of white clover. *Crop Science* **3**, 209-213.

Beinhart, G. (1964) Free sugar concentrations in white clover, *Trifolium repens* L., grown at different temperatures and light intensities. *Crop Science* **4**, 625-631.

Bergersen, F.J.; Hely, F.W.; Costin, A.B. (1963) Overwintering of clover nodules in alpine conditions. *Australian Journal of Biological Science* **16**, 920-921.

Bhojwani, S.S. (1981) A tissue culture method for propagation and low temperature storage of Trifolium repens genotypes. *Physiologia Plantarum* **52**, 187-190.

Bondi, A.; Birk, Y.; Gestetner, B. (1973) Forage saponins. In: Butler, G.W.; Bailey, R.W. (Eds) Chemistry and biochemistry of herbage. London, UK; Academic Press, Vol 1, 511-528.

Brougham, R.W. (1958a) Interception of light by the foliage of pure and mixed stands of pasture plants. *Australian Journal of Agricultural Research* **9**, 39-52.

Brougham, R.W. (1958b) Leaf development in swards of white clover (*Trifolium repens* L.). *New Zealand Journal of Agricultural Research* **1**, 707-718.

Brougham, R.W. (1960) The relationship between the critical leaf area, total chlorophyll content, and maximum growth rate of some pasture and crop plants. *Annals of Botany* **24**, 463-474.

Brougham, R.W. (1962) The leaf growth of *Trifolium repens* as influenced by the seasonal changes in light environment. *Journal of Ecology* **50**, 449-459.

Brown, R.H.; Blaser, R.E.; Dunton, H.L. (1966) Leaf-area index and apparent photosynthesis under various microclimates for different pasture species. *Proceedings of 10th International Grassland Congress*, 108-112.

Brown, R.H.; Blaser, R.E. (1968) Leaf area index in pasture growth. *Herbage Abstracts* **38**, 1-9.

Burch, G.J. (1979) Soil and plant resistances to water absorption by plant root systems. *Australian Journal of Agricultural Research* **30**, 279-292.

Burch, G.J.; Johns, G.G. (1978) Root absorption of water and physiological responses to water deficits by *Festuca arundinacea* Schreb. and *Trifolium repens* L. *Australian Journal of Plant Physiology* **5**, 859-871.

Butler, G.W.; Greenwood, R.M.; Soper, K. (1959) Effects of shading and defoliation on the turnover of root and nodule tissue of plants of *Trifolium repens, Trifolium pratense*, and *Lotus uliginosus. New Zealand Journal of Agricultural Research* **2**, 415-426.

Caradus, J.R. (1977) Structural variation of white clover root systems. *New Zealand Journal of Agricultural Research* **20**, 213-219.

Carlson, G.E. (1966a) Growth of clover leaves - developmental morphology and parameters at ten stages. *Crop Science* **6**, 293-294.

Carlson, G.E. (1966b) Growth of clover leaves after complete or partial leaf removal. *Crop Science* **6**, 419-422.

Chang, F.H.; Troughton, J.H. (1972) Chlorophyll a/b ratios in C_3 and C_4 plants. *Photosynthetica* **6**, 57-65.

Chow, H.S. (1966) The structure and growth habit of white clover in relation to its persistence. *Ph. D. Thesis, University of New Hampshire.*

Christie, P.; Newman, E.I.; Campbell, R. (1978) The influence of neighbouring grassland plants on each others endomycorrhizas and root surface micro-organisms. *Soil Biology and Biochemistry* **10**, 521-577.

Chu, A.C.P.; Robertson, A.G. (1974) The effects of shading and defoliation on nodulation and nitrogen fixation by white clover. *Plant and Soil* **41**, 509-519.

Clarkson, D. T. (1976) The influence of temperature on the exudation of xylem sap from detached root systems of rye (*Secale cereale*) and barley (*Hordeum vulgare*). *Planta* **132**, 297-304.

Conrad, R.; Seiler, W. (1979) Field measurements of hydrogen evolution by nitrogen-fixing legumes. *Soil Biology and Biochemistry* **11**, 686-690.

Cox, L.M.; Boersma, L. (1967) Transpiration as a function of soil temperature and soil water stress. *Plant Physiology* **42**, 550-556.

Crush, J.R.; Tough, H.J. (1981) Hydrogen evolution from white clover. *New Zealand Journal of Agricultural Research.* **24**, 365-370.

Cullen, P.W.; Turner, A.K.; Wilson, J.H. (1972) The effect of irrigation depth on root growth of some pasture species *Plant & Soil* **37**, 345-352.

Davidson, R.L. (1969a) Effects of root/leaf temperature differentials on root/shoot ratios in some pasture grasses and clover. *Annals of Botany* **33**, 561-569.

Davidson, R.L. (1969b) Effects of soil nutrition and moisture on root/shoot ratios in *Lolium perenne* L. and *Trifolium repens* L. *Annals of Botany* **33**, 571-577.

Davidson, R.L. (1969c) Effects of edaphic factors on the soluble carbohydrate content of roots of *Lolium perenne* L. and *Trifolium repens* L. *Annals of Botany* **33**, 579-589.

Denne, M.P. (1966) Morphological changes in the shoot apex of *Trifolium repens* L. 2. Diurnal changes in the vegetative apex. *New Zealand Journal of Botany* **4**, 434-443.

Eagles, C.F.; Othman, O.B. (1981) Growth at low temperatures and cold hardiness in white clover. In: Wright, C.E. (Ed) Plant Physiology and Herbage Production. *British Grassland Society Occasional Symposium* **No. 13**, 109-113.

Engin, M.; Sprent, J.I. (1973) Effects of water stress on growth and nitrogen fixing activity of *Trifolium repens. New Phytologist* **72**, 117-126.

Evans, P.S. (1973a) The effects of repeated defoliation to threee different levels on root growth of five pasture species. *New Zealand Journal of Agricultural Research* **16**, 31-34.

Evans, P.S. (1973b) Effects of seed size and defoliation at three development stages on root and shoot growth of seedlings of some common pasture species. *New Zealand Journal of Agricultural Research* **16**, 389-394.

Evans, P.S. (1977) Root distribution and water-withdrawal patterns of some crop and pasture species. *New Zealand Department of Scientific and Industrial Research Information Series,* **No. 126**, 186-190.

Finn, B.J.; Bourget, S.J.; Nielson, K.F.; Dow, B.K. (1961) Effects of different soil moisture tension on grass and legume species. *Canadian Journal of Soil Science* **41**, 16-23.

Flanagan, T.R.; Sprague, V.G. (1961) Measurements of, and changes in, cold resistance of Ladino white clover. *Crop Science* **1**, 44-47.

Fletcher, W.W.; Martin, D.J. (1962) The effects of gibberellic acid on the germination, growth and development of *Trifolium repens* L. *Journal of Agricultural Science, UK* **58**, 235-241.

Forde, B.J.; Mitchell, K.J.; Edge, E.A. (1977) Effect of temperature, vapour-pressure deficit and irradiance on transpiration rates of maize, paspalum, Westerwolds and perennial ryegrasses, peas, white clover and lucerne. *Australian Journal of Plant Physiology* **4**, 889-899.

Fottrell, P.F.; O,Connor, S.; Masterson, C.L. (1964) Identification of the flavonol myricetin in legume seeds and its toxicity to nodule bacteria. *Irish Journal of Agricultural Research* **3**, 246-249.

Foulds, W.; Young, L. (1977) Effects of frosting, moisture stress and potassium cyanide on the metabolism of cyanogenic and acyanogenic phenotypes of *Lotus corniculatus* L. and *Trifolium repens* L. *Heredity* **38**, 19-24.

Foulds, W. (1978) Response to soil moisture supply in three leguminous species. I. Growth, reproduction and mortality. *New Phytologist* **80**, 535-545.

Gibson, P.B.; Trautner, J.L. (1965) Growth of white clover with and without primary roots. *Crop Science* **5**, 477-479.

Goodlass, G.; Smith, K.A. (1979) Effects of ethylene on root extension and nodulation of pea (*Pisum sativum* L.) and white clover (*Trifolium repens* L.).*Plant and Soil* **51**, 387-395.

Gradwell, M.W. (1969) Soil oxygen and the growth of white clover. *New Zealand Journal of Agricultural Research* **12**, 615-629.

Grant, E.A.; Sallans, W.G. (1964) Influence of plant extracts on germination and growth of eight forage species. *Journal of the British Grassland Society* **19**, 191-197.

Greenham, C.G.; Daday, H. (1957) Electrical determination of cold hardiness in *Trifolium repens* L. and *Medicago sativa* L. *Nature, UK* 180, 541-543.

Greenham, C.G.; Daday, H. (1960) Further studes on the determination of cold hardiness in *Trifolium repens* L. and *Medicago sativa* L. *Australian Journal of Agricultural Research* **11**, 1-15.

Hagan, R.M.; Peterson, M.L.; Upchurch, R.P.; Jones, L.G. (1957) Relationships of soil moisture stress to different aspects of growth in Ladino clover. *Soil Science Society of America Proceedings* **21**, 360-365.

Hale, C.N.; Mathers, D.J. (1977) Toxicity of white clover seed diffusate and its effect on the survival of *Rhizobium trifolii. New Zealand Journal of Agricultural Research* **20**, 69-73.

Halliday, J.; Pate, J.S. (1976) The acetylene reduction assay as a means of studying nitrogen fixation in white clover under sward and laboratory conditions. *Journal of the British Grassland Society* **31**, 29-35.

Hart, A.L.; Jessop, D.J.; Galpin, J. (1981) The response to phosphorus of white clover and lotus inoculated with rhizobia or given KNO_3. *New Zealand Journal of Agricultural Research* **24**, 27-32.

Hart, R.H. (1976) Seedling growth of crownvetch, Ladino clover and alfalfa under shade. *Agronomy Journal* **68**, 683-685.

Harvey, H.J. (1970) Patterns of assimilate translocation in *Trifolium repens* In: Lowe, J. (Ed) White clover research. *Occasional Symposium, British Grassland Society,* **No 6**, 181-186.

Haycock, R. (1981) Environmental limitations to spring production in white clover. In: Wright, C.E. (Ed) Plant Physiology and Herbage Production. *British Grassland Society Occasional Symposium* **No. 13**, 119-123.

Haycock, R. (1982) A comparison of dry matter partitioning in seminal and adventitious plants of *Trifolium repens* L. *Annals of Botany* **50**, 161-165.

Haystead, A.; King, J.; Lamb, W.I.C. (1979) Photosynthesis, respiration and nitrogen fixation in white clover. *Grass and Forage Science* **34**, 125-130.

Haystead, A.; King, J.; Lamb, W.I.C.; Marriott, C. (1980) Growth and carbon economy of nodulated white clover in the presence and absence of combined nitrogen. *Grass and Forage Science* **35**, 123-128.

Hogg, D.E.; Moore, W.A. (1976) Effects of soil temperature on growth and nutrient content of white clover in a glasshouse pot experiment. *New Zealand Journal of Experimental Agriculture* **4**, 463-466.

Hoglund, J.H. (1979) Temperature effects on the nitrogen nutrition of legumes. *Proceedings of the Agronomy Society of New Zealand* **9**, 55-58.

Hoshino, M. (1974) Translocation and accumulation of assimilates and phosphorus in Ladino clover. *Bulletin of the National Grassland Research Institute* (Japan) **5**, 35-84.

Hoveland, C.S. (1964) Germination and seedling vigour of clovers as affected by grass root extracts. *Crop Science* **4**, 211-213.

Hoveland, C.S.; Mikkelson, E.E. (1967) Flooding tolerance of ladino white, Intermediate white, Persian and strawberry clovers. *Agronomy Journal* **59**, 307-308.

Howarth, R.E.; Goplen, B.P.; Brandt, S.A.; Cheng, K.J. (1982) Disruption of leaf tissues by rumen microorganisms: an approach to breeding bloat-safe forage legumes. *Crop Science* **22**, 564-568.

Howarth, R.E.; Goplen, B.P.; Fesser, A.C.; Brandt, S.A. (1978a) A possible role for leaf cell rupture in legume pasture bloat. *Crop Science* **18**, 129-133.

Howarth, R.E.; Goplen, B.P.; Majak, W.; Waldern, D.E.; Chaplin, R.K. (1978b) Chloroplast membranes in the etiology of pasture bloat. *Canadian Journal of Animal Science* **58**, 820-821.

Hughes, M.A.; Conn, E.E. (1976) Cyanoglucoside biosynthesis in white clover (*Trifolium repens*). *Phytochemistry* **15**, 697-701.

Hunt, W.F. (1970) The influence of leaf death on the rate of accumulation of green herbage during pasture regrowth. *Journal of Applied Ecology* **7**, 41-50.

Jelmini, G.; Nösberger, J. (1978a) Einfluss der Temperatur auf die Ertragsbildung, den Gehalt an nichtstruckturbildenen Kohlenhydraten und Stickstoff von *Festuca pratensis* Huds., *Lolium multiflorum* Lam., *Trifolium pratense* L. und *Trifolium repens* L. *Zeitschrift für Acker- und Pflanzenbau* **146**, 143-153.

Jelmini, G.; Nösberger, J. (1978b) Einfluss der Lichtintensität auf die Ertragsbildung und den Gehalt an nichtstruckturbildenen Kohlenhydraten und Stickstoff von *Festuca pratensis* Huds., *Lolium multiflorum* Lam., *Trifolium pratense* L. und *Trifolium repens* L. *Zeitschrift für Acker- und Pflanzenbau* **146**, 154-163.

Jewiss, O.R. (1967) Crop geometry and growth physiology. In: Wilkins, R.J. (Ed) Fodder conservation. *Occasional Symposium, British Grassland Society* **No 3**, 53-65.

Johns, G.G. (1978) Transpirational, leaf area, stomatal and photosynthetic responses to gradually induced water stress in four temperate herbage species. *Australian Journal of Plant Physiology* **5**, 113-125.

Johns, G.G.; Lazenby, A. (1973a) Defoliation, leaf area index, and the water use of four temperate pastures species under irrigated and dryland conditions. *Australian Journal of Agricultural Research* **24**, 783-795.

Johns, G.G.; Lazenby, A. (1973b) Effect of irrigation and defoliation on the herbage production and water use efficiency of four temperate pasture species. *Australian Journal of Agricultural Research* **24**, 797-808.

Jones, W.T.; Anderson, L.B.; Ross, M.D. (1973) Bloat in cattle. XXXIX. Detection of protein precipitants (flavolans) in legumes. *New Zealand Journal of Agricultural Research* **16**, 441-446.

Jones, W.T.; Broadhurst, R.B.; Lyttleton, J.W. (1976) The condensed tannins of pasture legume species *Phytochemistry* **15**, 1407-1409.

Kilpatrick, R.A.; Judd, W.R.; Dunn, G.M.; Rich, A.E. (1966) Agents affecting survival of red and white clovers exposed to low temperatures in a freezing chamber. *Crop Science* **6**, 499-501.

King, J.; Lamb, W.I.C.; McGregor, M.T. (1978) Effect of partial and complete defoliation on regrowth of white clover plants. *Journal of the British Grassland Society* **33**, 49-55.

Kishi, H. (1975) Studies on competition between grass and legume in mixed sward. VI. Effects of the growth stage and the temperature on the upward growth of orchard-grass and Ladino clover in summer. *Proceedings of the Crop Science Society of Japan* **44**, 306-311.

Kubota, F.; Agata, W.; Kamata, E. (1971) Dry matter production of forage plants. 4. Influences of light extinction coefficient (k) on dry matter production in forage plant populations - theoretical analysis. *Journal of the Japanese Society of Grassland Science* **17**, 243-249.

Lawlor, D.W.; Lake, J.V. (1976) Evaporation rate, leaf water potential and stomatal conductance in *Lolium, Trifolium* and *Lysimachia* in drying soil. *Journal of Applied Ecology* **13**, 639-646.

Lees, G.L.; Howarth, R.E.; Goplen, B.P.; Fesser, A.C. (1981) Mechanical disruption of leaf tissues and cells in some bloat-causing and bloat-safe forage legumes. *Crop Science* **21**, 444-448.

Leng, R.A.; McWilliam, J.R. (Eds). (1973) Bloat, Reviews in Rural Science No. 1., University of New England, Armidale, Australia.

Low, A.J.; Piper, F.J. (1960) The influence of water supply on the growth and phosphorus uptake of Italian ryegrass and white clover in pot culture. *Plant and Soil* **13**, 242-252.

MacFarlane, M.J.; Scott, D.; Jarvis, P. (1982a) Allelopathic effects of white clover. 1. Germination and chemical bioassay. *New Zealand Journal of Agricultural Research* **25**, 503-510.

MacFarlane, M.J.; Scott, D.; Jarvis, P. (1982b) Allelopathic effects of white clover. 2. Field investigations in tussock grasslands. *New Zealand Journal of Agricultural Research* **25**, 511-518.

Mächler, F.; Nösberger, J. (1977) Effect of light intensity and temperature on apparent photosynthesis of altitudinal ecotypes of *Trifolium repens* L. *Oecologia* **31**, 73-78.

Mächler, F.; Nösberger, J. (1978) The adaptation to temperature of photorespiration and of the photosynthetic carbon metabolism of altitudinal ecotypes of *Trifolium repens* L. *Oecologia* **35**, 267-276.

Mächler, F.; Nösberger, J.; Erismann, K.H. (1977) Photosynthetic $^{14}CO_2$ fixation products in altitudinal ecotypes of *Trifolium repens* L. with different temperature requirements. *Oecologia* **31**, 79-84.

Masterson, C.L. (1962) Inhibition of *Rhizobium* species by legume seeds. *Irish Journal of Agricultural Research* **1**, 343-344.

Masterson, C.L. (1965) Studies on the toxicity of legume seeds towards *Rhizobium. Annales de l'Institut Pasteur,* (Supplement) **109**, 216-217.

Masterson, C.L.; Murphy, P.M. (1976) Application of the acetylene reduction technique to the study of nitrogen fixation by white clover in the field. In: Nutman, P.S. (Ed) Symbiotic nitrogen fixation in plants. Cambridge, UK; Cambridge University Press, **Vol 7**, p.299-316.

Masterson, C.L.; Sherwood, M.T. (1978) Some effects of increased atmospheric carbon dioxide on white clover (*Trifolium repens*) and pea (*Pisum sativum*). *Plant and Soil* **49**, 421-426.

McCree, K.J. (1970) An equation for the rate of respiration of white clover plants grown under controlled conditions. In: Prediction and measurement of photosynthetic productivity (Proceedings, IBP/PP Technical Meeting, Trebon), Wageningen, Netherlands; Centre of Agricultural Publishing and Documentation (PUDOC) pp. 221-229.

McCree, K.J. (1971) The action spectrum, absorptance and quantum yield of photosynthesis in crop plants. *Agricultural Meteorology* **9**, 191-216.

McCree, K.J. (1974) Equations for the rate of dark respiration of white clover and grain sorghum, as functions of dry weight, photosynthetic rate, and temperature. *Crop Science* **14**, 509-514.

McCree, K.J.; Kresovich, S. (1978) Growth and maintenance requirements of white clover as a function of day length. *Crop Science* **18**, 22-25.

McCree, K.J.; Silsbury, J.H. (1978) Growth and maintenance requirements of subterranean clover. *Crop Science* **18**, 13-18.

McCree, K.J.; Troughton, J.H. (1966a) Prediction of growth rate at different light levels from the measured photosynthesis and respiration rates. *Plant Physiology* **41**, 559-566.

McCree, K.J.; Troughton, J.H. (1966b) Non-existence of an optimum leaf area index for the production rate of white clover grown under constant conditions. *Plant Physiology* **41**, 1615-1622.

Mitchell, K.J. (1965a) Growth of pasture species. III. White clover (*Trifolium repens*), subterranean clover (*T. subterraneum*) and lotus major (*L. uliginosus*). *New Zealand Journal of Science and Technology* **37A**, 395-413.

Mitchell, K.J. (1956b) Growth of pasture species under controlled environment. I. Growth at various levels of constant temperature. *New Zealand Journal of Science and Technology* **38A**, 203-216.

Mitchell, K.J.; Calder, D.M. (1958) The light regime within pastures. *New Zealand Journal of Agricultural Research* **1**, 61-68.

Mitchell, K.J.; Kerr, J.R. (1966) Differences in rate of use of soil moisture by stands of perennial ryegrass and white clover. *Agronomy Journal* **58**, 5-8.

Mitchell, K.J.; Lucanus, R. (1960) Growth of pasture species in controlled environment. II. Growth at low temperatures. *New Zealand Journal of Agricultural Research* **3**, 647-655.

Mitchell, K.J.; Lucanus, R. (1962) Growth of pasture species under controlled environment. III. Growth at various levels of constant temperature with 8 and 16 hours of uniform light per day. *New Zealand Journal of Agricultural Research* **5**, 135-144.

Moran, C.H.; Sprague, V.G.; Sullivan, J.T. (1953) Changes in the carbohydrate reserves of Ladino white clover following defoliation. *Plant Physiology* **28**, 467-474.

Morley, C.D.; Mosse, B. (1976) Abnormal vesicular-arbuscular mycorrhizal infections in white clover induced by lupin. *Transactions of the British Mycological Society* **67**, 510-513.

Moustafa, E.; Ball, R.; Field, T.R.O. (1969) The use of acetylene reduction to study the effect of nitrogen fertiliser and defoliation on nitrogen fixation by field grown white clover. *New Zealand Journal of Agricultural Research* **12**, 691-696.

Murata, Y.; Iyama, J. (1963) Studies on the photosynthesis of forage crops. II. Influence of air-temperature upon the photosynthesis of some forage and grain crops. *Proceedings of the Crop Science Society of Japan* **31**, 315-322.

Murphy, P.M. (1981) Effect of photoperiod on $N_2(C_2H_2)$ fixation and H_2 production by white clover. In: Gibson, A.H.; Newton, W.E. (Eds) Current Perspectives in Nitrogen Fixation. *Proceedings of the IVth International Symposium on Nitrogen Fixation, Canberra 1980*, 463.

Murphy, P.M.; Murphy, M.E. (1978) Relative energy requirements for dinitrogen and nitrate reduction in white clover. *Research Report (Soils), Johnstown Castle Research Station, An Foras Taluntais*, 62.

Mytton, L.; Hughes, D.M. (1980) Symbiotic nitrogen fixation at low temperatures. *Welsh Plant Breeding Station Annual Report* **1979**, 54-56.

Newman, E.I.; Miller, M.H. (1977) Allelopathy among some British grassland species. II. Influence of root exudates on phosphorus uptake. *Journal of Ecology* **65**, 399-411.

Newman, E.I.; Rovira, A.D. (1975) Allelopathy among some British grassland species. *Journal of Ecology* **63**, 727-737.

Newman, E.I.; Campbell, R.; Rovira, A.D. (1977) Experimental aleration of soil microbial populations for studying effects on higher plant interactions. *New Phytologist* **79**, 107-118.

Okubo, T.; Oizumi, H.; Hoshino, M.; Matsumoto, F. (1969) Efficiencies of energy conversion in pasture ecosystem. I. Dry matter production and efficiencies of light utilization in primary canopies of several pasture species. *Journal of the Japanese Society of Grassland Science* **15**, 138-149.

Okubo, T.; Kawanabe, S.; Hoshino, M. (1975) Chlorophyll amount for analysis of matter production in forage crops. II. Seasonal variations in maximum crop growth rate and leaf photosynthesis, and their correlations with chlorophyll content in alfalfa and ladino clover. *Journal of the Japanese Society of Grassland Science* **21**, 124-135.

Ollerenshaw, J.H.; Baker, R.H. (1981) Low temperature growth in a controlled environment of *Trifolium repens* plants from northern latitudes. *Journal of Applied Ecology* **18**, 229-239.

Ollerenshaw, J.H.; Baker, R.H. (1982) Low temperature growth in the field of *Trifolium repens* ecotypes from northern latitudes. *Journal of Applied Ecology* **19**, 519-527.

Olusuyi, S.A.; Raguse, C.A. (1968) Effect of temperature on germination and seedling development of Ladino clover (*Trifolium repens* L.) and Salina strawberry clover (*Trifolium fragiferum* L.) *Crop Science* **8**, 543-544.

Oswald, T.H.; Smith, A.E.; Phillips, D.V. (1977) Callus and plantlet regeneration from cell cultures of Ladino clover and soybean. *Physiologia Plantarum* **39**, 129-134.

Pate, J.S. (1958) Nodulation studies in legumes. II. The influence of various environmental factors on symbiotic expression in the vetch (*Vicia sativa* L.) and other legumes. *Australian Journal of Biological Sciences* **11**, 496-515.

Peterson, M.L.; Hagan, R.M. (1953) Irrigation principles and practices for pastures. *Proceedings of the 6th International Grassland Congress*, 397-403.

Price, D.N.; Wain, R.L. (1976) Studies on plant growth regulating substances. XLI. Structure-activity relationships and metabolism of a group of nitrophenols capable of inhibiting chloroplast development. *Annals of Applied Biology* **83**, 115-124.

Rice, E.L. (1972) Allelopathic effects of *Andropogon virginicus* and its persistence in old fields. *American Journal of Botany* **59**, 752-755.

Ronningen, T.S. (1953) Susceptibility to winter injury and some other characteristics in Ladino and common white clovers. *Agronomy Journal* **45**, 114-117.

Routley, D.G. (1966) Proline accumulation in wilted Ladino clover leaves. *Crop Science* **6**, 358-361.

Routley, D.G. (1971) Proline accumulation in wilting detached Ladino clover leaves. *Advancing Frontiers of Plant Sciences* **28**, 331-339.

Ruelke, O.C.; Smith, D. (1956) Overwintering trends of cold resistance and carbohydrates in medium red, ladino and common white clover. *Plant Physiology* **31**, 364-368.

Ruiz-Argüeso, T.; Maier, R.J.; Evans, H.J. (1979) Hydrogen evolution from alfalfa and clover nodules and hydrogen uptake by free-living *Rhizobium meliloti*. *Applied and Environmental Microbiology* **37**, 582-587.

Ryle, G.J.A.; Powell, C.E.; Gordon, A.J. (1979a) The respiratory costs of nitrogen fixation in soyabean, cowpea and white clover. I. Nitrogen fixation and the respiration of the nodulated root. *Journal of Experimental Botany* **30**, 135-144.

Ryle, G.J.A.; Powell, C.E.; Gordon, A.J. (1979b) The respiratory costs of nitrogen fixation in soyabean, cowpea and white clover. II. Comparisons of the cost of nitrogen fixation and the utilization of combined nitrogen. *Journal of Experimental Botany* **30**, 145-153.

Ryle, G.J.A.; Powell, C.E.; Gordon, A.J. (1981a) Patterns of ^{14}C-labelled assimilate partitioning in red and white clover during vegetative growth. *Annals of Botany* **47**, 505-514.

Ryle, G.J.A.; Powell, C.E.; Gordon, A.J. (1981b) Assimilate partitioning in red and white clover either dependent on N_2 fixation in root nodules or utilizing nitrate nitrogen. *Annals of Botany* **515-523.**

Ryle, G.J.A.; Arnott, R.A.; Powell, C.E. (1981c) Distribution of dry weight between root and shoot in white clover dependent on N_2 fixation or utilizing abundant nitrate nitrogen. *Plant and Soil* **60**, 29-39.

Sanchez-Diaz, M.; Sanchez-Marin, M. (1974) Resistance of leaf tissue to desiccation and stomatal closure in lucerne (*Medicago sativa* L.) and white clover (*Trifolium repens* L.) in relation to water stress. *Annales de Edafologia y Agrobiologia* **33**, 743-754.

Sarkar, S.K.; Howarth, R.E.; Goplen, B.P. (1976) Condensed tannins in herbaceous legumes. *Crop Science* **16**, 543-546.

Schubert, K.R.; Evans, H.J. (1976) Hydrogen evolution: a major factor affecting the efficiency of nitrogen fixation in nodulated symbionts. *Proceedings of the National Academy of Science, U.S.A.* **73**, 1207-1211.

Scott, B.I.H.; Gulline, H.F. (1972) Natural and forced circadian oscillations in the leaf of *Trifolium repens*. *Australian Journal of Biological Science* **25**, 61-76.

Scott, B.I.H.; Gulline, H.F. (1975) Membrane changes in a circadian system. *Nature, UK* **254**, 69-70.

Scott, B.I.H.; Gulline, H.F.; Robinson, G.R. (1977) Circadian electrochemical changes in the pulvinules of *Trifolium repens* L. *Australian Journal of Plant Physiology* **4**, 193-206.

Scott, D. (1970) CO_2 exchange of plants. 3. Temperature acclimatisation of three species. *New Zealand Journal of Botany* **8**, 369-379.

Scott, D. (1975) Allelopathic interactions of resident tussock grassland species on germination of oversown seed. *New Zealand Journal of Experimental Agriculture* **3**, 135-141.

Scott, D.; Hanson, M.A. (1977) Effect of low temperature during initial germination of some New Zealand pasture species. *New Zealand Journal of Experimental Agriculture* **5**, 41-45.

Scott, D.; Menalda, P.H. (1970) CO_2 exchange of plants. 2. Response of six species to temperature and light intensity. *New Zealand Journal of Botany* **3**, 361-368.

Scott, D.; Menalda, P.H.; Brougham, R.W. (1968) Spectral analysis of radiation transmitted and reflected by different vegetations. *New Zealand Journal of Botany* **6**, 427-449.

Scott, D.; Menalda, P.H.; Rowley, J.A. (1970) CO_2 exchange of plants 1. Technique and response of seven species to light intensity. *New Zealand Journal of Botany* **8**, 82-90.

Sharma, M.L. (1973) Simulation of drought and its effect on germination of five pasture species. *Agronomy Journal* **65**, 982-987.

Shepherd, W. (1972) Some evidence of stomatal restriction of evaporation from well-watered plant canopies. *Water Resources Research* **8**, 1092-1095.

Shepherd, W. (1975) Inclusion of plant moisture status in combination method estimates of pasture evaporation. *Journal of Hydrology* **26**, 199-207.

Skinner, C.G.; Talbert, F.D.; Shire, W. (1958) Effect of 6-(substituted) purines and gibberellin on the rate of seed germination. *Plant Physiology* **33**, 190-194.

Small, J.G.C.; Joffe, A. (1968) Physiological studies on the genus *Trifolium* with special reference to the South African species. II. Influence of root temperature on growth, nodulation and symbiotic nitrogen fixation. *South African Journal of Agricultural Science* **11**, 41-56.

Smith, D. (1949) Differential survival of Ladino and common white clover encased in ice. *Agronomy Journal* **41**, 230-234.

Smith, H. (1975) Phytochrome and photomorphogenesis. McGraw-Hill.

Sprague, M.A. (1955) The influence of rate of cooling and winter cover on the winter survival of Ladino clover and alfalfa. *Plant Physiology* **30**, 447-451.

Steponkus, P.L. (1978) Cold hardiness and freezing injury of agronomic crops. *Advances in Agronomy* **30**, 51-98.

Takeda, T.; Agata, W. (1965a) Analysis of the growth in forage crops. I. Effects of high temperature on dry matter and leaf production of Ladino clover. *Proceedings of the Crop Science Society of Japan* **33**, 1-6.

Takeda, T.; Agata, W. (1965b) Analysis of the growth in forage crops. II. Effects of high temperature on photosynthesis, respiration, chemical composition and regrowth of Ladino clover. *Proceedings of the Crop Science Society of Japan* **33**, 1-16.

Tames, R.S.; Gesto, M.D.V.; Vieitez, E. (1973) Growth substances isolated from tubers of *Cyperus esculentus* var. *aureus*. *Physiologia Plantarum* **28**, 195-200.

Tantrum, I.; Mitchell, K.J. (1972) Water loss by crop and pasture species. *New Zealand Agricultural Science* **6**, 7-8.

Tapper, B.A.; Reay, P.F. (1973) Cyanogenic glycosides and glucosinolates (mustard oil glucosides). In: Butler, G.W.; Bailey, R.W. (Eds) Chemistry and biochemistry of herbage. London, UK; Academic Press, Vol 1, 447-476.

Thomas, R.G. (1972) The role of basal and apical factors in the co-ordination of growth in the stems of white clover (*Trifolium repens* L.) In: Carr, D.J. (Ed) Proceedings of the 7th International Conference on Plant Growth Substances. Berlin, W. Germany; Springer-Verlag, p. 624-632.

Trautner, J.L.; Gibson, P.B. (1966) Fate of white clover axillary buds at five intensities of sunlight. *Agronomy Journal* **58**, 557-558.

Ueno, M.; Yoshihara, K. (1967) Studies on the root of herbage legumes. 4. Varietal difference of nodule root growth in relation to the compensation of taproot in white clover. *Journal of the Japanese Society of Grassland Science* **13**, 195-198.

Ueno, M.; Yoshihara, K. (1968) Studies on the root of herbage legumes. 5. Effect of some environmental factors on the penetration of the adventitious roots of white clover to the soil. *Journal of the Japanese Society of Grassland Science* **14**, 100-103.

Ueno, M.; Yoshihara, K.; Hidaka, M. (1967) Studies on the root of herbage legumes. 3. Estimates of the role of the taproot in Ladino clover. *Journal of the Japanese Society of Grassland Science* **13**, 100-104.

Upchurch, R.P.; Peterson, M.L.; Hagan, R.M. (1955) Effect of soil-moisture content on the rate of photosynthesis and respiration in Ladino clover (*Trifolium repens* L.) *Plant Physiology* **30**, 297-303.

Walker-Simmons, M.; Ryan, C.A. (1977) Wound-induced accumulation of trypsin inhibitor activities in plant leaves. Survey of several plant genera. *Plant Physiology* **59**, 437-439.

Warrington, I.J.; Mitchell, K.J. (1976) The influence of blue- and red-biased light spectra on the development of plants. *Agricultural Meteorology* **16**, 247-262.

Westbrooks, F.E.; Tesar, M.B. (1955) Tap root survival of Ladino clover. *Agronomy Journal* **47**, 403-410.

Whittaker, R.H.; Feeny, P.P. (1971) Allelochemics: chemical interactions between species. *Science* **171**, 757-770.

Wilfong, R.T.; Brown, R.H.; Blaser, R.E. (1967) Relationships between leaf area index and apparent photosynthesis in alfalfa (*Medicago sativa* L.) and Ladino clover (*Trifolium repens* L.). *Crop Science* **7**, 27-30.

Williams, W.M.; Hoglund, J.H. (1978) Temperature responses of New Zealand, Spanish, and New Zealand x Spanish white clover populations. *New Zealand Journal of Agricultural Research* **21**, 491-497.

Wilman, D.; Martins, V.M.M. (1977) Senescence and death of herbage during periods of regrowth in ryegrass and red and white clover, and the effect of applied nitrogen. *Journal of Applied Ecology* **14**, 615-620.

Woledge, J.; Dennis, W.D. (1982) The effect of temperature on photosynthesis of ryegrass and white clover leaves. *Annals of Botany* **50**, 25-35.

Wood, G.M.; Sprague, M.A. (1952) Relation of organic food reserves to cold hardiness of Ladino clover. *Agronomy Journal* **44**, 318-325.

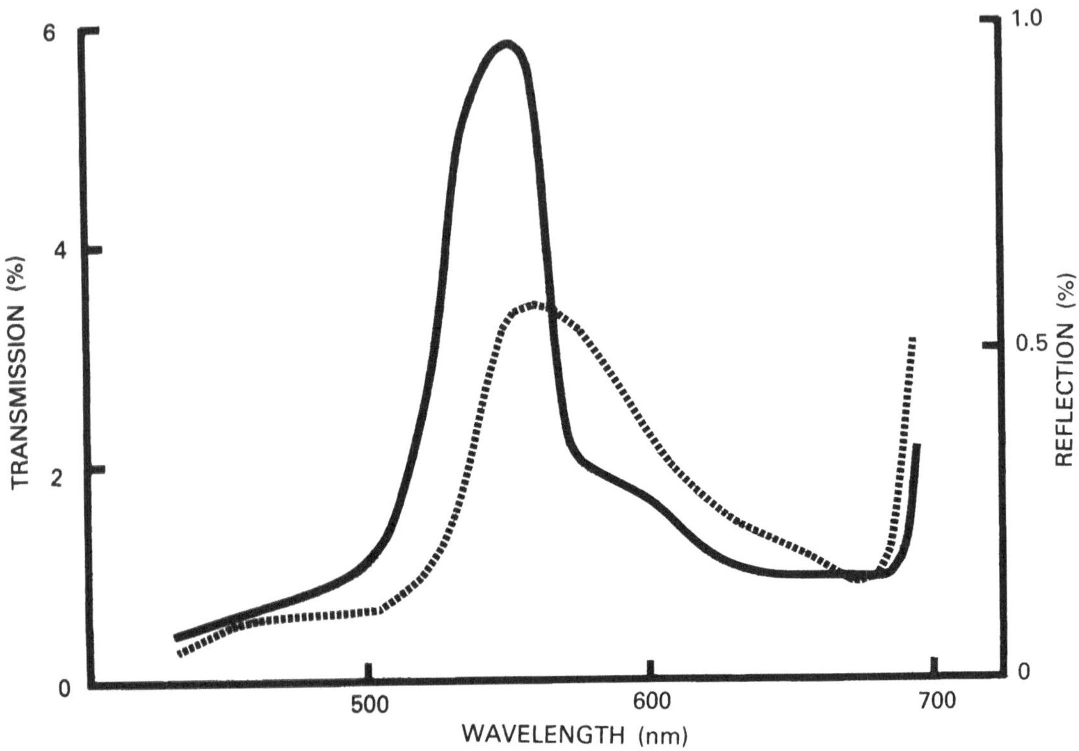

Fig. 4.1 Pattern of transmission (——) and reflection (-------) of visible radiation by white clover. Transmission measurements taken where more than 95% of visible radiation had been absorbed by vegetation. Reflection measurements made with angle of acceptance of instrument limited to 22° and pointed vertically down. From Scott et al. (1968).

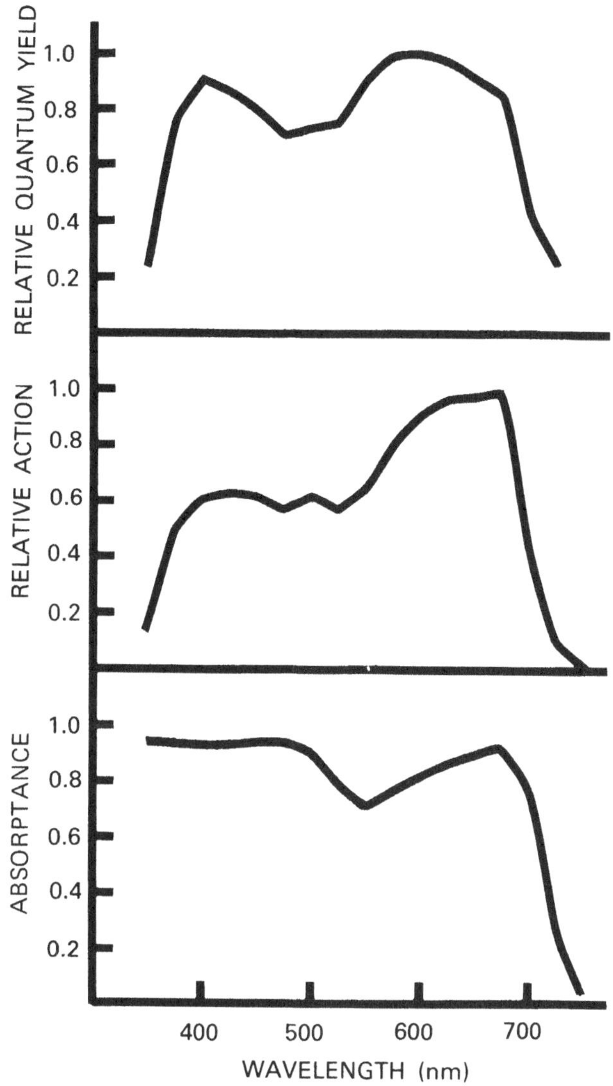

Fig. 4.2 Relationships between relative quantum yield, relative action and absorptance and wavelength for segments of white clover leaves. From Table 3, McCree (1971).

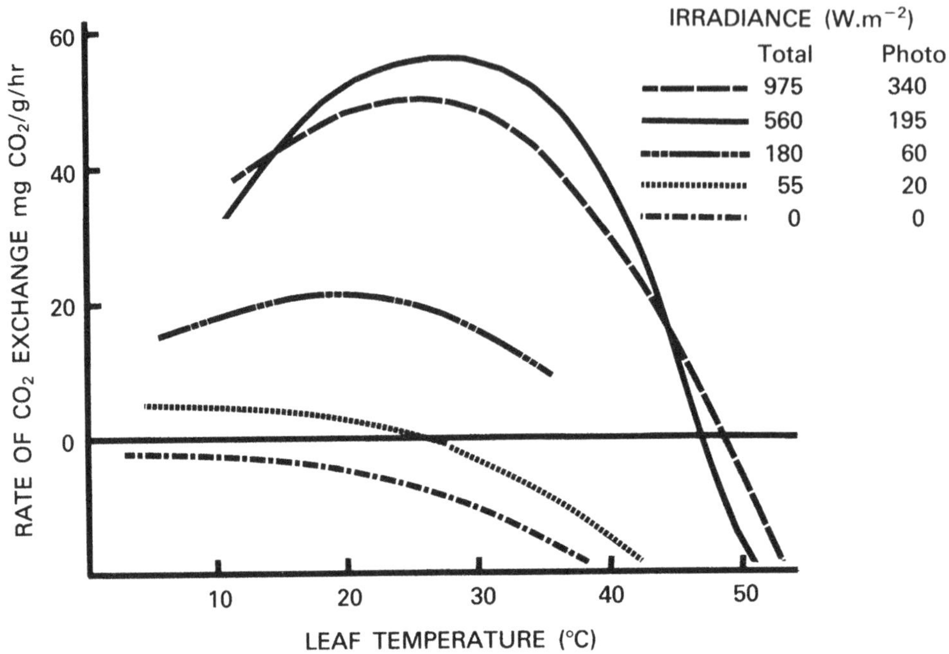

Fig. 4.3 Carbon dioxide exchange of shoots of white clover at various temperatures and irradiances. From Scott & Menalda (1970).

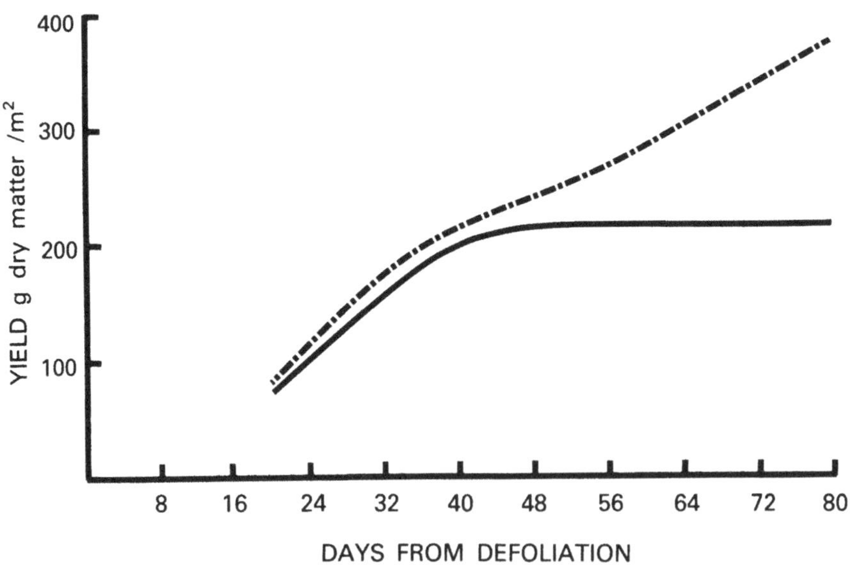

Fig. 4.4 Estimate of the growth of white clover in a ryegrass-white clover sward in spring. Gross growth corrected for leaf death (-------). Net growth (——). From Hunt (1970).

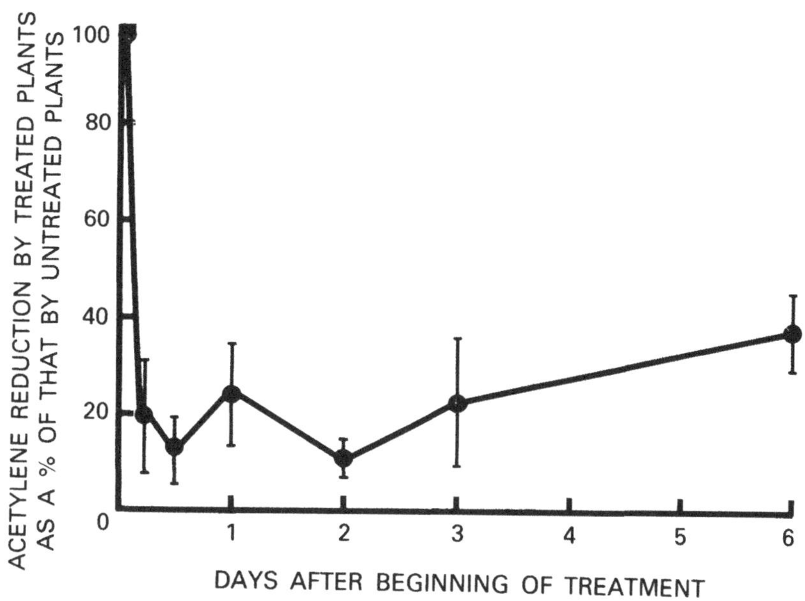

Fig. 4.5 Effects of defoliation on acetylene reducing activity of 12 week old white clover plants. 95% confidence limits are attached to data. From Halliday & Pate (1976).

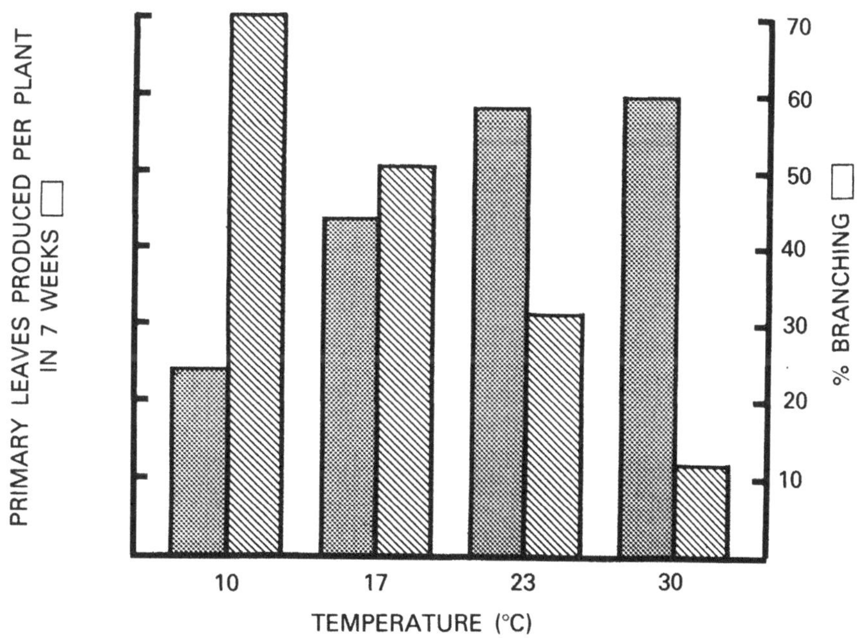

Fig. 4.6 Effect of temperature on primary leaf production and branching (%branching = 100 × no. of primary nodes produced). Data are mean values of 4 clones of white clover. From Beinhart (1963).

Mineral nutrition

J. Dunlop and A. L. Hart
Grasslands Division, DSIR, Palmerston North, New Zealand

Contents

1. Introduction

White clover pastures often need fertilizers to maintain adequate productivity. Consequently, studies of the acquisition and use of nutrients by white clover, and its competition with companion species for them, have a strong practical motivation. This chapter focuses on functional aspects of these processes rather than on data relating productivity to fertiliser application in specific situations.

Nutrients are not absorbed and used independently of each other, and interactions among them are often apparent in the response of growth to various nutrient combinations. In this chapter interactions among nutrients have been interpreted largely in terms of nitrogen metabolism . The presence or absence of the N-fixing symbiosis and pH regulation within and outside the plant provide a useful framework within which to set observations on various aspects of white clover nutrition.

Nutrients applied in fertilisers should be used as efficiently as possible, but attempts to improve efficiency by breeding new white clover varieties have met with little success. Nevertheless, attempts to understand nutrient efficiency, which is reviewed in the final section of this chapter, have served to pinpoint aspects of white clover nutrition on which information is incomplete.

2. The uptake of nutrients

In moving from the soil into a root, nutrient ions have to cross a complex of mineral particles, decaying organic matter, fungi, bacteria, mucilages of plant and bacterial origin, lysed and disintegrating layers of cells at the surface of the roots and pectic ion-exchange material. Across this layer there will be a gradient of pH and fluxes of water, various organic molecules and mineral ions. There may also be a diurnal cycle of drying and wetting. The physical and chemical interactions between nutrient ions and their environment in the rhizosphere are not well understood and are much more complex than is assumed in most models of ion transport by roots. The nature of the rhizosphere is not discussed in detail in this book; in fact, present knowledge of rhizospheres could provide no more than a generalised and fragmented picture of mineral nutrient interactions within this zone and their significance for the mineral nutrition of the plant. These observations are made to emphasize that results of laboratory and glasshouse experiments will seldom be of direct relevance to the field. Research on the mechanisms of ion absorption and transport by roots is an important part of our understanding of the process of mineral nutrition. However, the major challenges facing plant physiologists working in this area are to develop techniques for conducting incisive experiments in the field and to relate the various processes involved in nutrient uptake to each other and to the environment.

(A) Root morphology and anatomy

The movement of ions into the root is often tacitly and occasionally explicitly assumed to be the principal step limiting the rate of uptake in mineral nutrition. However, for nutrients such as phosphate that diffuse slowly in the soil (Nye 1977), the movement of the nutrient to the root will be the rate-limiting step. The plant usually benefits from having a root system which minimises the distance over which nutrients have to diffuse and maximises the surface area across which they can be absorbed. Consequently, the morphology of the root system is ofen important in mineral nutrition.

White clover roots do not possess any unique morphological characteristics. As a seedling matures, its dependence shifts to the roots that originate at stolon nodes (Ueno et al. 1967). In a mature pasture most of the white clover will consist of segments of stolons no longer attached to the primary root system. Nodal root production is greater during the autumn, falling gradually during winter to a minimum in spring and summer (Caradus & Evans 1977). There are significant quantitative differences in root morphology between white clover and some of the grasses with which it grows (Evans 1977). White clover roots are thicker, shorter and less branched than grass roots and in addition its root hairs are shorter and less numerous (see Tables 1 and 2).

Table 5.1 Root characteristics for some pasture species. (Adapted from Evans 1977)

	Length/weight ratio cm mg^{-1}	Diameter mm	Surface area/ weight ratio mm^2 mg^{-1}	Apices/mm
Trifolium repens	27.6	0.26	227	0.51
T. pratense	26.8	0.27	230	0.53
T. subterraneum	22.3	0.28	196	0.61
Lolium perenne	30.6	0.19	181	0.86
Dactylis glomerata	44.4	0.16	220	0.84
Phleum pratense	36.7	0.17	199	0.85
Agrostis tenuis	49.8	0.16	240	0.73
Cynosurus cristatus	35.6	0.20	220	0.73

Table 5.2 Characteristics of roots hairs of some pasture species. (Adapted from Evans 1977).

	Length mm	Percentage with hairs	Area of root hair cylinder mm^2 mg^{-1}†	Volume within root hair cylinder mm^3 mg^{-1}†
Trifolium repens	0.23	68	490	68
T. pratense	0.20	58	420	49
T. subterraneum	0.22	57	380	49
Lolium perenne	0.55	95	1230	411
Dactylis glomerata	0.61	95	1810	623
Phleum pratense	0.77	97	1980	853
Agrostis tenuis	0.68	95	2190	824
Cynosurus cristatus	1.05	98	2510	1467

†Data for the root hair cylinder are expressed on the basis of the dry weight of the root.

Root morphology in white clover ranges from large taprooted plants with few fibrous roots to plants with fine, fibrous roots and no prominent taproots (Caradus 1977). Furthermore, there are differences in the rate of growth, production of apices, and distribution of the roots down the soil profile (Caradus 1981a). Roots of large-leaved types have more root tips at greater depth. Root and shoot morphology appear to be correlated, types with prominent taproots having large leaves and high productivity, and types with fibrous roots having small leaves (Caradus 1981b).

Liming has a large effect on the morphology of white clover roots. Haynes & Ludecke (1981b) found that a shift in soil pH from 4.0 to 4.7 was accompanied by a 2-fold increase in root length per unit weight and a 5-fold increase in the percentage of root length carrying root hairs. Root hair length also increased. Heavier applications of lime tended to produce even larger effects. Apparently, liming stimulated root branching, as the growth of secondary and tertiary lateral roots relative to the taproot and primary laterals was increased. This may have been due to reduced Al toxicity, which is characterised by inhibition of lateral root formation and increase in root diameter. These changes in root morphology were accompanied by increases in the dry matter production and phosphorus content of the plants.

The root anatomy of white clover has been described by Erith (1924) and Soper (1959). It has three xylem poles and normal endodermal development.

In discussions of ion absorption by plants, attention is frequently focused solely on uptake by roots. However, leaves and other organs can also absorb nutrients (Epstein 1972). In white clover plants, significant amounts of stolon are in contact with the soil, both within and on its surface.

Hay (1983) found that at least 29% of stolons are buried and during the late winter the percentage is as high as 90. Wormcasting appears to be important in stolon burial. Soil micro-organisms modify the suface layers of these stolons by first degrading the cuticle and subsequently lysing the cell walls and invading epidermal and subepidermal cells (Hay et al. 1982). Presumably, this removes permeability barriers and allows nutrients to diffuse to absorption sites within the stolons. Hay & Dunlop (1982) demonstrated that white clover plants absorbed phosphate from felt pads moistened with ^{32}P labelled nutrient solution placed against the stolons. The phosphate absorbed in this way was transported to the stolon tips in the same manner as when it is absorbed by nodal roots (see Section 4.(C)). The implication of this result is that stolons can absorb phosphate directly from the soil solution with which they are in contact. The experimental design used in these studies did not enable the quantitative significance of these results to be estimated. However, because of the position of stolons on or just below the soil surface, they are ideally placed to compete for P deposited there through the worm casting activity or by topdressing.

(B) Mycorrhizas and nutrition

In many pastoral soils the absorbing surface of white clover roots is extended by their association with mycorrhizas. The association between vesicular-arbuscular (V-A) mycorrhizas and white clover and its effects on clover mineral nutrition has been examined in numerous studies. Many of these have shown that the presence of V-A mycorrhizas improves growth and P absorption when the plant is grown in soil low in P (e.g. Powell 1976, Crush 1976, Powell 1977b,c, Sparling & Tinker 1978, Hayman & Mosse 1979, Hall et al. 1977). The increases in shoot growth reported range up to 37-fold (Powell 1977b), with 20-fold increases being measured in a number of studies. The large percentage responses are obtained in soils that contain very little available P. In soils higher in P the results are less spectacular, and some workers have obtained growth reductions (Crush 1976, Hall et al. 1977, Hall 1978). Crush (1976) has suggested that as P availability increases, the relationship between the clover host and the mycorrhizal fungi changes from symbiosis to parasitism. Of a number of pasture species studied, white clover appears to be among those sharing most benefit from association with endomycorrhizas (Crush 1974, Powell & Sithamparanathan 1977, Hall 1978, Buwalda 1980).

The stimulation of host growth and increased P uptake is thought to result from the ability of the external hyphae to extend some distance from the root into the soil, thus increasing the volume of soil from which P can be absorbed. In most low-P soils, diffusion of the phosphate ion to the root is the rate-limiting step in absorption. The mycorrhizal hyphae in effect extend the root surface by absorbing the P and translocating it to the host. V-A mycorrhizas can absorb and transport P to a host white clover root 70 mm from the source (Cooper & Tinker 1981). Without V-A mycorrhizas a root would normally deplete the P in a zone extending to less than a millimetre from its surface.

The possibility that V-A mycorrhizas can absorb organic or other forms of P not directly available to the roots has been considered. The current view is that this does not occur to any appreciable extent (Hayman & Mosse 1972), as radioactive labelling experiments suggest that both mycorrhizal and non-mycorrhizal plants use the same pool of soil P. However, the possibility that mycorrhizas absorb P from organic forms remains an open question. Their hyphae probably have extracellular phosphatases capable of hydrolysing organic P compounds at a distance from the root, thus competing positively with other soil micro-organisms (see discussion by Dalal 1977). Mycorrhizas can be particularly beneficial in aiding white clover to absorb P from powdered rock phosphate, a relatively insoluble form of P (Powell & Daniel 1978). However, this probably reflects the ability of mycorrhizas to ramify through the soil rather than any specialised ability to solubilise rock phosphate, as on the micro scale the rock phosphate would not be distributed homogeneously through the soil.

Hyphae can transport P to the white clover host at the rate of 0.3 to 1.9×10^{-9} mol cm^{-2} s^{-1} (Pearson & Tinker 1975). This compares with a rate of approximately 1×10^{-12} mol cm^{-2} s^{-1} for P absorption from nutrient solution by white clover roots (calculated from the data of Bowling & Dunlop 1978 and Evans 1977). The mechanism of transport along the hyphae probably relies on cytoplasmic streaming, as when this is inhibited with cytochalasin, P absorption and transport stop. Mass flow through the hyphae may also contribute to transport, as increased transpiration in the host plant accelerates P transport (Cooper & Tinker 1981).

Mycorrhizal fungi are widely distributed (Powell 1977a) and, except on eroded soils (Hall 1979a), white clover plants growing in natural swards will be mycorrhizal. However, mycorrhizal species appear to vary in their ability to increase growth in P-deficient soil conditions. In a pot trial Powell (1976) found that in 24 of 37 soils, an introduced mycorrhizal fungus, E3, gave an additional stimulation to that produced by the indigenous mycorrhizal fungi. However, in non-sterile soils in the field E3 showed less superiority than in the pot trials. Pastures have a range of mycorrhizal fungi populations, and white clover forms associations with different populations at different stages of the pasture development sequence. The associations predominating at any stage appear to be those which are of greatest benefit to the host under the prevailing conditions (Crush 1978). New mycorrhizal fungal populations can be introduced into non-sterile field soils (Powell 1977c, 1978, Hall 1979b), but the techniques are unsuitable for large-scale inoculation. This problem will have to be solved before mycorrhizal inoculation can become a common practice in pastoral agriculture.

(C) Mechanisms of ion uptake

White clover has been included in studies on various aspects of the movement of ions across membranes into root cells. The aspects include the kinetics of ion transport and coupling of ion transport to energy sources.

Jackman (1965) suggested that the concentration dependence of K absorption by subterranean clover (*T. subterraneum*) and ryegrass (*Lolium perenne*) followed the same trend generally showed by many other species. The Michaelis constants K_m, which described the affinity of the substrate ion for carriers, were similar for the two species. However, the maximum rate of uptake, V_{max}, was greater for ryegrass than for clover. This finding was considered to be significant in explaining why ryegrass competes for potassium better than does clover. These results were corroborated by Dunlop et al. (1979), who also found that for low-salt roots (i.e. roots of plants grown in $CaSO_4$ solutions), white clover and ryegrass had similar values for K_m but that V_{max} for ryegrass was 5 times greater than that for white clover.

Dunlop et al. (1979) also studied the regulation of ion transport. Low-salt roots had proved to be very convenient experimental material which gave results amenable to use in simple models. However, this approach did not offer any suggestions on how plants regulated their ion content. By pretreating seedlings in solutions containing various concentrations of potassium, Dunlop et al. (1979) prepared white clover and ryegrass roots with various tissue contents of K. The values of K_m and V_{max} for K uptake by these roots were then obtained. The K_m indicated a fall in the affinity of carrier molecules for the substrate ions. The values for V_{max} also fell, reflecting a decline in the uptake capacity of the tissue. The percentage change in the K_m was greatest in ryegrass, but the percentage change in V_{max} was greatest in white clover. The extent of the changes was such that as tissue K levels increased, the ability of white clover to compete with ryegrass for K would decrease.

These changes in K_m and V_{max} corroborated an earlier study on barley by Glass (1976), and indicated that the characteristics of the ion transport system were modified to regulate the ion content of the tissue , and also suggest that K_m and V_{max} are not absolute constants. Glass (1976) has proposed a model to explain this regulation of ion transport. He suggested than in addition to the external site on the carrier which binds and transports K into the cell, there are internal sites which also bind K and affect the operation of the carrier molecule. The binding of K to these sites lowers the affinity of the external site for K. Mathematical treatment of the data suggests that in barley (Glass 1976), white clover and ryegrass (Dunlop et al. 1979) there are four internal binding sites for K.

After the development of the chemiosmotic theory (Mitchell 1961, 1966) it has been realised that coupling of ion flows with movement of hydrogen ions across membranes can explain a number of facets of ion transport. The probable mechanism of ion uptake in plant cells is depicted in Figure 1. A membrane-bound ATPase, which uses energy from ATP to transport hydrogen ions from the cell across the plasmalemma to the external medium, is considered to be a prime cause in transport processes. This results in a gradient of pH across the membrane. The process is electrogenic, increasing polarisation of the electropotential across the membrane, with the inside of the cell negative with respect to the medium. This electropotential augments the pH in producing a proton motive force (pmf) for hydrogen ions to flow back into the cell. The passive flux of hydrogen ions is coupled to the movement of many solutes, both organic and inorganic. It is thought that the K^+ ions may be transported into the cell as an integral part of the ATPase-hydrogen ion transport system (Lin 1979) or possibly by a separate carrier which uses energy from the pmf. It is proposed that anions are transported by carriers which simultaniously carry hydrogen ions (symports) into the cell or carry hydroxyl ions from the cell to neutralise the external hydrogen ions (antiports). Neutral organic compounds, e.g. sugars, would be carried with H^+ ions on symports (Pool 1978, Baker 1978).

In white clover it appears that P uptake is coupled to the pmf by a hydrogen ion symport or a hydroxyl antiport (these two alternatives are experimentally indistinguishable). Dunlop & Bowling (1978) demonstrated that when P uptake by the roots was inhibited by eliminating the supply of photosynthetic energy from the shoots, artifically producing a pmf by lowering the external pH produced an immediate increase in the rate of P uptake. Under these circumstances, adding sucrose to the medium also increased the rate of P uptake, but it took more than an hour to produce the same effect as that achieved by the pH shift in 4 minutes. In normal plants the rate of P uptake was also closely related to pH, with a maximum at the low value of pH 4.3. This also suggested the importance of pH for P movement into the plant.

In the same study the relationship of P uptake to the membrane electropotential was examined, and a strong positive linear relationship was found between the two parameters (Bowling & Dunlop 1978). Lowering the supply of photosythates to roots by shading the shoots or removing them had parallel effects on P uptake, and changing the pH of the media also produced a parallel response in electropotential. Removing P from the external media resulted in a substantial 60 mV depolarisation of the electropotential. This pointed to an important role for P in the electrophysiology of root cells and suggested that P transport directly contributed to cell electropotential.

3. Competition for nutrients

White clover is generally a weaker competitor for nutrients than the species with which it grows in pastures. Specifically, it competes poorly with grasses when K (Blaser & Brady 1950, Brown & Rouse 1953, McNaught 1958, Mouat & Walker 1959a), P (Mouat & Walker 1959a, Jackman & Mouat 1972a,b) or S (Walker & Adams 1958) are limiting. The uptake of P and K by white clover grown alone is 4-8 times greater than when grown with ryegrass (Mouat & Walker 1959a), an effect much greater than the inverse effect of white clover on ryegrass. The presence of grasses not only reduces the growth of white clover when P is deficient but as lowers the response of clover to addition of P (Mouat & Walker 1959a, Jackman & Mouat 1972). The effect is shown in Figure 2 which illustrates the large additions of fertiliser necessary to counteract the effect of competition. Jackman & Mouat 1972a also demonstrated that competition from (*Agrostis capillaris*) could reduce the yield of N from white clover by up to 150 kg ha^{-1} in a year. Without adequate P, S and K, productive white clover pastures undergo a plant succession from white clover plus high fertility grasses to high fertility grasses alone and finally to low fertility grasses and weeds (Blaser & Brady 1950, Sears 1960).

The root morphology of white clover contributes to its poor competitive ability. Its thicker and less profusely branched root system explores the soil less thoroughly than the fine, extensively branched root systems of grasses. Furthermore, because of the greater root hair length, the volume of the root hair cylinder and its surface area are several times greater in grasses than in white clover (Tables 1 and 2). Jackman & Mouat (1972b) also measured numbers of root apices of ryegrass and white clover. On the assumption that ion uptake will be most rapid in the zone immediately behind the root tip, they calculated distances between apices and found that the apices on roots of white clover were twice as far apart as those on ryegrass. From this they inferred that a clover root tip penetrating among ryegrass roots would almost always be the subject of direct competition,whereas roots of ryegrass would face much less competition from clover. Whether it is the region immediately behind the root tip, as considered by Jackman & Mouat (1972b), or the total length of the root as is implicit in the measurements of Evans (1977) that is important in nutrient absorption, it is clear that the architecture of root systems of grasses gives them an advantage in competing with white clover for slowly diffusing nutrients.

A common plant adaptation to nutrient stress is a reduction in shoot:root ratio (Davidson 1969), resulting in the production of proportionately more root material for the absorption of nutrients. Mouat (1983) has found than under conditions of P deficiency the shoot:root ratio of white clover is reduced less than is that of browntop. Browntop was able to increase its root growth relative to shoot growth more rapidly in response to a developing nutrient deficiency than the species requiring high fertility. This could be a significant factor in the competition for nutrients by these species.

For a range of plants, including white clover and other pasture species, Mouat (1983) found that the extent of a plant's change in shoot:root ratio in response to nutrient deficiency was related to its root cation exchange capacity (CEC) - plants with a lower CEC had a greater range of shoot:root ratios. This relationship may explain the associations between CEC and competitive ability found in a number of studies (Blaser & Brady 1950, Grey et al. 1950, Mouat & Walker 1959b).

In several studies, interactions between N and competition for K (Blaser & Brady 1950, Brown & Rouse 1953, Mouat & Walker 1959a) and for P (Mouat and Walker 1959a) have been noted. Application of nitrogenous fertiliser intensifies competition from grasses for nutrients. Mouat & Walker (1959b) found no interaction between this effect and cutting treatments in their experiments, and concluded that the applied N was not exerting its effect by increasing competition for light. However, Jackman & Mouat (1972a) argued that the dense mat formed by browntop shades clover stolons. This reduces the development of lateral stolons and terminal buds, thus impairing the general vigour of white clover and its ability to compete for nutrients.

Table 5.3 Critical levels of essential elements in white clover

Element	Critical level in association with ryegrass†	Critical levels in monoculture	
Nitrogen	4.5 -5·5%		
Phosphorus	0.3 -0.4%	0.25%	Jackson et al. (1964)
		0.1 -0.2%	Parks et al. (1967)
		0.15-0.25%	Rendig et al. (1950)
Sulphur	0.25-0.30%	0.10-0.16%	Jordan & Bardsley (1958)
		0.25-0.29%	McNaught & Chrisstoffels (1961)
Potassium	1.8 -2.3%	0.8%	Chandler et al. (1945)
		0.8 -0.9%	Ulrich (1945)
		1.1%	Andrew (1960)
Magnesium	0.12-0.14%	<0.14%	Dorofaeff & McNaught (1962)
Calcium	0.5 -0.8%	1.0%	Andrew (1960)
		0.74%	Andrew & Norris (1961)
Iron	50-70 ppm		
Manganese	25-35 ppm	20 ppm	Gammon et al. (1979)
Boron	25-35ppm		
Zinc	12-18 ppm	12-17 ppm	McNaught & During (1970)
Copper	4- 6 ppm		
Molybdenum	0.05-0.15 ppm		

†These values are from McNaught (1970)

Snaydon (1971) found that roots of white clover plants from calcareous soils competed more effectively when growing on calcareous soils and, conversely, that on acidic soils the plants derived from acidic soils performed best. The effects of root competition generally predominated over the effects of shoot competition.

4. Nutrient Utilization

Despite the agricultural importance of white clover, the literature on this topic is sparse. Only those articles that deal specifically with the assimilation and utilization of nutrients by white clover after uptake are reviewed here. For more general information, the book by Mengel & Kirkby (1978) and the articles by Butler & Jones (1973) and Clarkson & Hanson (1980) are recommended.

(A) Critical levels of essential elements

The critical level of a nutrient is defined as the minimum concentration of the element in a plant necessary to produce a maximum growth (or in some studies 90% or 95% of maximum growth). Values for critical levels of nutrients in white clover, obtained by McNaught (1970) for plants growing with ryegrass in a pasture, are given in Table 3. A single value for the critical level of a nutrient can be misleading, as it will be affected by a number of variables such as the part of the plant amalysed, maturity of the plant, climate, levels of other nutrients and the presence of other plants growing with the clover. Consequently, values reported in the literature for critical nutrient levels in white clover show a considerable range. To avoid some of the problems arising from this variation, critical P and S levels have been reported as ratios to the N concentration (McNaught 1970). This might be especially useful where differences in maturity are invloved. McNaught's (1970) data suggest that in white clover, N:P ratios greater than 18:1 and N:S ratios greater than 14:1 indicate deficiency.

Table 3 also includes values for critical levels of nutrients which are generally lower than McNaught's (1970) and which have been derived from studies in which clover was grown alone in glasshouse, pot or small-plot experiments. These discrepancies may arise from the differences between monoculture and mixed grass-clover swards, but the mechanism by which critical nutrient levels in white clover would be raised by exposing it to competition from ryegrass is not obvious. Almost certainly, some of the factors already listed which affect critical levels also contribute to the differences between the two sets of values.

(B) Visual symptoms of nutrient deficiencies and toxicities

Mineral deficiences and some excesses produce characteristic visual symptoms which can help diagnose the nutritional status of plants. A range of these symptoms has been described by Wallace (1951), Andrew (1956), Millikan (1958) and Dorofaeff & McNaught (1962), and they are summarised in Table 4. Plants not showing symptoms may be still responsive to an added nutrient.

It should be noted that the precise expression of some of the symptoms may depend on the climate and other environmental factors. For example, Millikan (1958) observed that in summer, K deficiency caused numerous small necrotic spots instead of marginal necrotic areas. Consequently visual diagnosis of a nutritional disorder should be confirmed by chemical analysis whenever possible.

Andrew (1956) has reported that Ca and B deficiencies and Al and B toxicities also cause visible effects on the root system, the effect in each case being a reduction in size, with roots thickened and with stubby ends and greatly increased branching. Their colour is also darkened.

Table 5.4 Visual symptoms of mineral disorders in white clover

Element	Site of earliest symptoms	Colouring or chlorosis	Deformities	Advanced symptoms	Anthocyanin pigmentation
Potassium	Older leaves	Necrotic spotting or necrotic areas near leaf margin	–	Necrosis covers whole leaves and death ensues. Deficiency spreads from oldest to youngest leaves	–
Phosphorus	General	Dark green with bluish sheen	Dwarfed prostrate growth	Oldest leaf in turn yellows and dies	On petioles
Nitrogen	Older leaves	General yellowing		Older leaves develop grayish-white necrosis and die	On stolons and possibly leaves
Calcium	Petioles of younger fully formed leaves		Collapse of petiole, with leaf initially remaining green and turgid	Necrotic spotting and interveinal chlorosis near base of mid-rib of older leaves	On leaves of collapsed petioles
Magnesium	Older leaves	Intense pinkish-red (anthocyanin) on leaf margins. General interveinal chlorosis	–	Symptoms extend to younger leaves. Coloration pales and leaf becomes necrotic	On leaf margins
Sulphur	Older leaves	General yellowing. Veins lighter than chlorotic tissues in between		Older leaves become necrotic and die	On stolons
Copper	General	General yellowing	More erect growth habit. Twisting of petioles. Inward cupping of leaflets	Loss of colour in young fully expanded leaves commencing at margins. Loss in colour is diurnal with leaves appearing normal early in the morning but losing colour through the day	

Table 5.4 Visual symptoms of mineral disorders in white clover cont'd

Element	Site of earliest symptoms	Colouring or chlorosis	Deformities	Advanced symptoms	Anthocyanin pigmentation
Manganese	General	Distinct interveinal chlorosis	Outward cupping of leaflets	Necrosis follows curving in severe cases	–
Iron	General	Yellowing	–	Distinct interveinal chlorosis. With increasing severity veins also lose colour. Necrosis, giving dead, papery white tissue	
Boron	General	Intensification of green	Outward curving of leaflets. Leaflets become thick and brittle	Mass of malformed tissue at the base of the plant	In mature plants on margins and underside
Manganese *toxicity*		Necrotic spotting	Leaflets appear crumpled	Necrotic spots coalesce and leaf edges are puckered with mixed necrosis and chlorosis	
Boron *toxicity*		Halo of chlorosis	Extreme edge of leaflet curves upward	With increasing severity halo becomes necrotic, papery white	
Zinc	Generally the young, fully expanded leaves	Light-brown to gray necrosis at the leaf margins which may or may not be preceded by chlorosis	Youngest leaves reduced in size. Inward cupping of leaflets	Sudden collapse of leaf and top of petiole following necrosis. Initiation of numerous small leaves at the base of the plant	

(C) Utilization

i. Nitrogen

The assimilation, transport and partitioning of N by white clover has not been studied as intensively as it has in lupin, pea, cowpea and soybean for example, but its N metabolism appears to be typical of temperate legumes.

Nitrate is reduced by nitrate reductase in the root and shoot. Nitrate reductase activity increases as the supply of NO_3^- increases, activity being greatest in the shoot (Copeland & Pate 1970). N fixation can occur simultaneously with NO_3^- absorption and reduction (Copeland & Pate 1970) provided NO_3^- levels are not too high. Ammonium ions are assimilated in the nodules via the glutamine synthetase-glutamate synthase pathway (Boland et al. 1978). High levels of glutamate dehydrogenase have been found in nodule preparations, but the glutamate dehydrogenase was thought to come from mitochondria and not to be involved in the immediate assimilation of ammonium ions arising from nitrogen fixation (Boland et al. 1978).

Asparagine appears to be the main amino acid resulting from the further metabolism of glutamine formed from ammonium ion assimilation. It is the main free amino acid in white clover plants (Copeland & Pate 1970), herbage (Bathurst & Mitchell 1958) and root nodules (Bulter & Bathurst 1958). The dominant amino compound in nodules is bound aminobutyric acid (Butler & Bathurst 1958).

Asparagine from nodules and nitrate from roots (depending on nitrate availability) are probably the main forms in which N is transported in the xylem, both being found in bleeding sap(Wallace & Pate 1967). The ammonia resulting from nitrate reduction is presumably assimilated by the glutamine synthetase-glutamate synthase pathway (Miflin & Lea 1980).

For transport processes in the nodule, white clover nodules have transfer cells in the pericycle of the vascular bundles (Pate et al. 1969). Transfer cells also occur in the vascular system of the stem and leaves (Gunning & Pate 1969, Gunning et al. 1970).

The operational energy costs of N assimilation *per se* via the metabolism of N in mineral or gas form are probably similar (Gibson 1966, Broughton 1979). However, the yield of white clover that is dependent on N fixation is generally lower than when it is given mineral N for a number of reasons, the relative importance of any particular one depending on the circumstances. Nodulated plants may yield less because their growth is slowed during development of the symbiosis, because they have higher respiration rates, because a greater proportion of assimilate is transported to the roots, and because the particular rhizobium strain-genotype combination is incapable of fixing sufficient N to realise the full potential for growth (Allos & Bartholomew 1955, 1959, Haydock & Norris 1967, Mytton 1976, Murphy & Murphy 1978, Haystead & Marriott 1979, Ryle et al. 1979, 1981, Hart et al. 1981a).

ii. Phosphorus

As with N, there is little published work on the metabolism of P in white clover. After absorption, P is distributed to various parts of the plant according to sink strength (Wilkinson & Gross 1967, Hoshino 1974). If the supply of P permits, considerably more P may be absorbed than is necessary for maximum growth under the particular conditions (Wilkinson & Gross 1965, Hart et al. 1981b).

P is a mobile element (Hill 1980) and may be redistributed. With age, leaves change from being net importers to exporters of P (Bieleski 1973), and this phenomenon may contribute to the decrease in white clover leaf P concentration with age noted by Wilkinson & Gross (1967) and Hart & Jessop (1982). If the P supply is withdrawn, then provided sufficient P has been stored within the plant, growth may continue for some time. The P concentration then falls as it is redistributed and diluted by additional growth. Shoot growth is more adversely affected than root growth, which may even continue unchanged for some time after P withdrawal. Phosphorus is sequestered in the roots to an increasing degree as the supply diminishes (Wilkinson & Gross 1965, Hart et al. 1981b).

The translocation of P appears to be strongly acropetal. Phosphorus absorbed by nodal roots is rapidly translocated to the growing points and young leaves, with a tendency for accumulation in the shoots attached to alternate nodes from the nodal roots to which ^{32}P is applied (Hoshino 1974). Basipetal transport is rare, but occurs when there are growing points of laterals very near to the source of P supply (Hoshino 1974, Hay & Dunlop 1982). The marked unidirectional transport of P appears to be a consequence of apical dominance because on removal of the growing point of the stolon P is rapidly transported basipetally (Hoshino 1974).

As well as translocation to the shoots, there is also translocation to lateral roots developing on nodal roots. Root meristems and primordia and nodules also showed large accumulation of P. Hoshino (1974) found that nodules were a sink for labelled P at all stages of their development.

Phosphorus exists in plants as inorganic P (Pi) and in organic form, the main organic pools being nucleic acids, esters and lipids (Bieleski 1973). Most of the Pi is considered to be vacuolar and a buffer against decreases in P supply (Bieleski 1973). Ester-P may be the source, in the short term, of Pi moving between leaves as they develop (Weste et al. 1974). A limited analysis of the P fractions in leaves of the primary shoot and stolons of white clover has been performed by Hart & Jessop (1982). Total P concentration decreased with leaf age and rose with P supply. The concentration of Pi seemed to be affected mainly by supply, whereas that of the organic fraction (apart from the lipids) changed more with leaf age, indicating that some of the fraction may have been involved in redistribution of P with leaf development. Lipid P showed small changes with P supply and leaf age. This general stability may be a result of there being only a minimum P requirement for structural compounds (Chisholm et al. 1981). In considering the critical concentration of P fractions in white clover tops, Chisholm et al. (1981) also found that the concentration of lipid P was the most stable of those measured.

The distribution of P among the various fractions may be connected in some way with P efficiency (see below). Wilson et al. (1978) reported that *Stylosanthes hamata* cv. Verano, a plant considered to have a greater P efficiency than white clover, has a larger percentage of recently absorbed P in labile pools and a lower percentage in lipid pools than does white clover.

iii. Sulphur

Most organic S in leaves is present in protein or sulpholipid, the ratio between these forms being about 2:1 (Harwood & Nicholls 1979).

Spencer (1959) found that in white clover, protein-S remained the largest S-containing fraction as S supply was increased. Under conditions of low S supply protein-N was low and non-protein N was high, but as the supply increased the balance changed in favour of protein-N (Spencer 1959).

Sulpholipid is part of chloroplast lamellar membranes. Roughan & Batt (1969) isolated it from leaves of white clover where it comprised about 4% of glycerolipids.

Spencer (1959) found only small amounts of SO_4 in white clover. At high levels of S supply, non protein organic S increases rather than SO_4.

iv. Potassium and sodium

Of the four main functions of K, enzyme activation, involvement in transport, anion neutralization and osmotic potential (Clarkson & Hanson 1980), only membrane transport and osmotic potential regulation have been studied in any detail in white clover. There are circadian changes in the content and influx of K in the leaf pulvinules (Scott et al. 1977). Dunlop et al. (1979) have studied the kinetics and regulation of K uptake into roots (see Section 2.(C)).

K transport is strongly acropetal in stolons and it is apparently unable to move basipetally (Nelson & Brady 1953). Most plant K is found in the leaflets and petioles, with petioles and peduncles containing the highest concentations (Wilkinson & Gross 1967).

As noted in Section 5.(B), Na can substitute for K to a large degree without ill effects. White clover roots do not exercise selectivity for K over Na, but K is preferentially transported to the shoot (Whitehead & Jones 1972). Sodium is transported to the shoot to some degree and is accumulated there to higher levels than in other species (Davies et al. 1966, Whitehead & Jones 1969, 1972). White clover has been classified as a natrophile (Smith et al. 1978, Sherrell 1978).

There appears to be genetic variation in the physiology of Na absorption, as Snaydon & Bradshaw (1969) found that plants adapted to calcareous soils had higher concentrations of shoot Na and higher uptake efficiencies than those adapted to acidic soils.

Under high rainfall, Na may be leached from herbage (Whitehead & Jones 1969).

v. Magnesium

Mg appears to be immobile in terms of movement from old to new tissues, there being no redistribution of Mg from mature to younger leaves (Neales 1958, Wilkinson & Gross 1967). The chlorophyll extracted with 80% acetone from leaves contained 12.86% of total Mg, and the 'fibre' fraction 62.47% (Neales 1956). Mg is bonded with the N of the pyrrole rings of chlorophyll. The association with the 'fibre' fraction is less well understood. Mg is thought to contribute to macromolecular structure by virtue of its geometrical bonding, hence it may be involved in cell wall structure Clarkson & Hanson (1980). Alternatively it may have become associated with the fibre during extraction, through binding to electronegative groups (e.g. the O^- of cell wall carboxylate groups).

vi. Manganese

The metabolism of Mn at either adequate or toxic levels appears not to have been studied in white clover. Some of the Mn has been shown to occur in large intracellular inclusions in the root cortex (Walker 1979), at which location it presumably exerts no physiological effects.

vii. Calcium

Populations of white clover adapted to acidic soils have shown a smaller yield response to applied Ca and higher concentrations of Ca in the shoot than those adapted to calcareous soils. The higher Ca concentrations in acidic populations were a result of higher uptake efficiency (shoot Ca content/root weight) but it was not possible to ascertain the extent to which differences in this parameter resulted from differences in translocation and uptake (Snaydon & Bradshaw 1969).

The concentration of Ca in leaves increases with maturity, reflecting its immobility (Wilkinson & Gross 1967). From tests with different substrates on leaf circadian rhythms, Bollig et al. (1978) were led to suggest that Ca^{2+} transport played a role in these rhythms.

viii. Boron

The function of B in plants is poorly understood but it may exert its effects on the apoplasm through its involvement in the maintenance of cell wall and membrane structure (Clarkson & Hanson 1980).

Wilkinson and Gross (1967) found that B accumulated in leaflets of white clover. When B was applied to leaves, most was rendered immobile, probably by absorption to cell walls (Martini & Thellier 1980), but a small proportion remained mobile and was transported to other plant parts. Deficient plants responded quickly to a single foliar application of B, the rate of increase in leaf area approaching that of control plants. This increase in leaf area declined rapidly, however, once the concentration of B again fell due to growth dilution. The pattern of distribution of foliar-applied B was similar in plants that were either sufficient or deficient in B.

ix. Copper

Cu is transported to the shoot (Andrew & Thorne 1962, Wilkinson & Gross 1967) but most is retained in the root (Andrew & Thorne 1962, Jarvis 1980). In studies by Jarvis (1980) roots of plants in flowing nutrient solution retained 71% of that absorbed. Some of the Cu was found to be associated with the cell wall fraction. The percentage of shoot Cu associated with cell walls decreased from 35 to 7% with time, whereas in the roots somewhat more of the root Cu was in the cell wall fraction - 15.6% at the final harvest (Jarvis 1980). Jarvis found that on removing P from the nutrient solution, the association of Cu with the cell walls increased, although there was no change in the overall accumulation of Cu. This observation was unexplained.

(D) Elements essential for nitrogen fixation

Criteria for deciding whether nodulated legumes have a nutritional requirement over and above that of plants depending only on mineral nitrogen have been discussed by Robson (1978). Briefly these criteria are:

1. There may be a negative interaction between the element in question and mineral N.
2. Application of the element increases growth and N concentration in the tops.
3. The element improves nodulation.
4. The element enhances activity of the nodules.

As a general rule none of these criteria should be considered in isolation, although satisfaction of one of them may be sufficient to indicate a requirement for an element.

The number of experiments concerned with whether or not there is a special requirement for a given element by the white clover symbiosis is few. However, the list of elements essential for nitrogen fixation listed by Robson (1978) presumably apply to white clover.

There is a special requirement for molybdenum (Johnson et al. 1952, Walker et al. 1955b, Ouellette 1963, Truong et al. 1971b).

There may be a specific requirement for Ca. Increased Ca availability has been shown to increase the concentration of N in the shoots of plants dependent on symbiotic N fixation. The improved N status was thought to result at least in part from improved nodulation (Andrew & Johnson 1976, Munns & Fox 1977, Munns et al. 1977) (Section 5.5.1).

In white clover seedlings boron has been shown to increase dry matter yield and N accumulation. It has also been suggested that the improved N nutrition arose from an improvement in nodulation (Sherrell 1966).

In that P is required for both host and nodule metabolism, there is no special requirement for P by symbiotic N assimilation. A positive interaction between N and P was observed in white clover by Brock & Hoglund (1974), Hart et al. (1981a) and Hart (1982). Similar conclusions have been reached for *T. subterraneum* by Robson et al. (1981).

The interaction between N and P is not understood, and attempts to explain it can only be speculative. At low levels of P supply when the yield of nodulated plants (-N; no mineral N applied) and those having access to mineral N (+N) are similar (Hart et al. 1981a, Hart 1982), both sets of plants are presumably equally limited by P. As the supply of P increases, the P limitation is overcome and the -N plants are then smaller than their +N counterparts, possibly because of limited N supply and excess respiratory load.

At high levels of P supply when differences in yield between +N and -N plants are significant, the -N plants have a greater proportion of plant P (and N) in their root systems than +N plants (Hart 1982). This is presumably at least partly due to the sink strength of the nodules being very high compared with that in other parts of the plants (Hart 1982). Hart et al. (1981a), for example found up to 19% of the root P in the nodules. The relatively greater distribution of P to the root systems of -N plants may also be associated with the tendancy for -N plants to have higher shoot:root ratios than +N plants, the higher ratios being attributed to poorer N nutrition (Ryle et al. 1981).

There is probably no special requirement for S. McNaught & Chrisstoffels (1961) found a positive interaction between S and N. S increased the percentage N and yield of nodulated plants (Andrew 1977) (in this case percentage N was correlated with percentage S) but this also occurred in plants given mineral N (Spencer 1959). Improved S nutrition increased white clover yield and the activity of the N-fixing system (Masterson 1977), probably because of improved growth rather than specific effects on N fixation.

Potassium has generalised effects on plant growth. Andrew and Bryan (1955) found no interactions between the mode of N nutrition and K on a K-deficient soil. Potassium increased the yield of tops of nodulated plants (Andrew & Robins 1969a), but there was no consistent effect on the concentration of N in the tops (Andrew & Robins 1969b).

Copper is not especially required by the symbiosis, the main effect probably being on plant growth. A positive interaction between Cu and N was found by Hallsworth et al. (1964) and O'Toole & Materson (1970), although Cu increased the percentage N in shoots of white clover on an N deficient medium (Elst et al. 1961).

5. Element interactions and yield

Experiments on the response of white clover to variation in supply of a single nutrient are important in determining how necessary it is for growth or the way in which it is utilised by the plant. However, the results of such experiments are of limited value as indicators of plant response in the 'real' world. Experiments involving nutrient interactions deal with conditions that more closely resemble the 'real' world, as some attempt can be made to learn about the myriad of nutritional effects to which the plant is subject.

Consideration of the mode of N nutrition is a useful starting point in an exposition of the interactions among nutrients affecting the growth of white clover. In plants dependant on N fixation the symbiosis has specific nutrient requirements that differ from those usual for plant growth as discussed above. The fixation of N also entails the extrusion of 0.1-0.2 protons per N atom fixed (Raven & Smith 1976). This may lead to acidification of the rhizosphere with the consequent risk that:

(1) The nodulated plant may be adversely affected by those elements whose effects become manifest at low pH (Israel & Jackson 1978).

(2)The acidity may inhibit the establishment and function of the symbiosis, for which a fairly narrow range of pH is required.

If the plant derives its N from mineral sources the constraints of special nutritional requirements for, and pH sensitivity of, N fixation will be absent. Nutrient effects associated with pH may still occur, however. The assimilation of NH^+_4 requires the extrusion of 1.1-1.25 H^+ per NH^+_4 taken up and consequent acidification of the rhizosphere, whereas the assimilation of NO_3^- entails the extrusion of 0.9-0.05 OH^- per NO_3^- absorbed and the development of a more alkaline rhizosphere. In the field, where NO_3^- and NH^+_4 may be available simultaneously, the effect of one form on the rhizosphere pH may offset that of the other (Raven & Smith 1976).

White clover has not been used in experiments relating rhizosphere pH to the form of N assimilated, but Aguilar & van Diest (1981) have indicated that two other legumes, soybean and alfalfa, acidified their growth medium when they were dependent on symbiotic N fixation, but increased the pH of the medium when given NO_3^-. However, the mode of N nutrition is not necessarily the sole determinant of root zone pH (Foy & Fleming 1978).

Given the importance of N metabolism to growth, its likely effects on rhizosphere pH and the role of white clover as a pasture legume, two broad categories of interactions are especially pertinent to an understanding of white clover nutrition.

1. Those between the element in question and mineral N. The nature of this interaction has been used to determine whether the symbiosis has a special requirement for a particular element (see 5.4.4).

2. Those coming into play a low pH, e.g. the interactions between Al or Mn and P. An understanding of this category has become important for attempts to establish white clover in new agricultural lands. These often have acid, infertile soils, whereas the white clover symbiosis evolved in slightly acid-neutral soils.

(A) Nutrients, pH and lime

The effect of a number of nutrient elements on white clover growth is influenced by soil pH. Lime is often applied to control soil pH, and there is much evidence to show its beneficial effects, one of the most common of which is to reduce the toxic effects of Al and Mn.

White clover yield is inversely proportional to tissue Al concentration (Lowther 1980, Davis 1981a,b,c, Haynes & Ludecke 1981b) and to the level of exchangeable Al in the soil (Shoop et al. 1961, Russell 1978, Khalid & Silva 1979, Davis 1981a,b, Haynes & Ludecke 1981a). Al reduced P and Ca uptake (Andrew et al. 1973, Haynes & Ludecke 1981a) and retards root growth (Haynes & Ludecke 1981b). Al toxicity occurs at pH values below 5.0-5.5 and is generally much reduced by liming to a higher pH, which reduces the level of exchangeable Al in the soil (Shoop et al. 1961, Davis 1981c, Haynes & Ludecke 1981a).

The toxic effects of Al can also be overcome if large amounts of superphosphate are applied. High rates of superphosphate have decreased exchangeable Al in an acid subsoil (Shoop et al. 1961) and a yellow-brown earth subsoil (Davis 1981a,c). In plant shoots the concentration of Al was also reduced by P application (Nordmeyer & Davis 1976, Davis 1981a,b).

In plants provided with mineral N, the 'toxicity threshold value' (where yield was 5% less than the maximum) for Mn concentration in the shoot was 650 ppm (Andrew & Hegarty 1969). The toxic effects of Mn are apparent at low pH where the Mn ions are available for absorption and, as with Al, lime reduced toxicity of Mn by making it unavailable (Lohnis 1951, Walker et al. 1955b, Snaydon 1962b, Truong et al. 1971a, Munns & Fox 1977, Rayment & Verrall 1980, Haynes & Ludecke 1981a). Decreases in Mn availability are similarly reflected in lowered Mn concentrations in plant tissue (Walker et al. 1955b, Dorofaeff & McNaught 1962, Truong et al. 1971a, Widdowson & Walker 1971, Haynes & Ludecke 1981b).

Mn toxicity is usually a problem where soil pH is below 5.5. However, waterlogging can also make Mn^{+2} ions available to plants. Hence, although Mn and Al toxicity often occur together, Mn toxicity can occur in the absence of Al toxicity (Foy & Fleming 1978). This aspect of Mn nutrition has not been studied in white clover.

Lime has beneficial effects other than the alleviation of Al and Mn toxicity. The increase in soil pH *per se* favours rhizobial infection, nodulation and growth (Russell 1966, O'Toole & Masterson 1968, Helyar & Anderson 1970a, Gibson et al. 1975, Munns & Fox 1977, Munns et al. 1977, Jo et al. 1980). The optimum soil pH for nodulated white clover seems to be about 5.5 (McNeur 1954).

Liming increases the concentration of available Ca (Munns & Fox 1977, Rayment & Verrall 1980, Haynes & Ludecke 1981a), thus improving nodulation and increasing tissue Ca concentrations (Andrew 1960, Shoop et al 1961, Andrew & Johnson 1976, Munns & Fox 1977, Munns et al. 1977, Davis 1981c, Haynes & Ludecke 1981b).

There is an interaction between Ca and pH on nodulation. Calcium ions may exchange with H^+ on root and root hair surfaces thus countering inhibitory effects of H^+ on infection and nodulation (Israel & Jackson 1978). On peat soil, an increase in soil Ca permitted nodulation to occur at pH 4.3, 4.7 being the lowest pH at which it occurred without high levels of Ca (O'Toole & Masterson 1968). Ca reduced the inhibitory effect of H^+ ions on nodulation between pH 4.0 and 5.0 for plants growing in sand (Andrew 1976). Israel & Jackson (1978) have suggested that the degree of extrusion of H^+ into the rhizosphere after an excess of cation over anion uptake determines the concentration of Ca required for optimal nodulation.

It is probable that in most temperate agricultural soils white clover is only rarely exposed to Ca deficiency (Walker et al. 1955b), but this may not be so for highly weathered tropical soils (Munns & Fox 1977, Munns et al. 1977), sandy soils grossly deficient in nutrients (Andrew 1960), solodic soils (Russell 1966) and for acid temperate soils (Snaydon 1962a, b). In non-agricultural soils, Ca may be one of the main factors determining the edaphic distribution of white clover (pH and phosphate are others) (Snaydon 1962a, b, c).

Calcium in nutrient solution alleviated Mn toxicity through a reduction of Mn uptake to the shoot (Vose & Jones 1963, Truong et al. 1971a). The reduction in Mn uptake caused by Ca may be offset by concomitant rises in pH, which promoted Mn uptake: '.....substrate pH appears to influence both soil Mn availability and plant processes, the two effects being opposed at all pH levels' (Truong et al. 1971b). The reduction in Mn uptake by Ca is also opposed by P, which tends to increase Mn uptake and exacerbate Mn toxicity. Calcium phosphate applied to soil has also increased the shoot Mn concentration, perhaps indicating that increased Mn availability resulting from the addition of P may prevent responses to phosphatic fertilisers (Truong et al. 1971b).

Lime increases the availability of Mo (Gammon et al. 1955, Walker et al. 1955b, Ouellette 1963, Rolt 1968, Truong et al. 1971b, Widdowson & Walker 1971) which enhances growth through its effect on nitrogen fixation. Walker et al. (1955a, b) noted antagonism between Mn and Mo on acid soils and suggested this arose from the imobilisation of Mo after absorption. Work by Truong et al. (1971b) with acid soils indicated no direct effect of Mo on Mn metabolism. A high Mo supply did not influence Mn phytotoxicity in solution in the presence of combined N. In soil, Mo increased yield in the presence of high levels of exchangeable Mn through promotion of N fixation.

Lime is not always a paragon. Application of lime has reduced the availability of Mn (an essential nutrient though in large amounts it is toxic), Cu, Zn, Fe and B (Andrew & Bryan 1955, Jackman 1961, Russell 1966, Gammon et al. 1979, Davis 1981c, Haynes & Ludecke 1981a). The concentration of Mg and K in herbage may fall after liming of pastures. Plant yield may not be unduly affected but the decreases in Mg may contribute to hypomagnaesemia in cattle (McNaught et al. 1968, 1973a,b, Davis 1981c). The increase in Ca in the soil solution after liming decreases the concentration ratio of Mg (Mg/Mg + Ca) in white clover (Edmeades & Judd 1980), which presumably reduces plant uptake. Lime may even reduce P availability by inducing the formation of complexes containing P and Al (Helyar & Anderson 1970b, 1971, Davis 1981c, Haynes & Ludecke 1981a).

In acid conditions white clover supplied with mineral N shows less response to lime than does clover dependent for N on nitrogen fixation (Helyar & Anderson 1971, Andrew 1976). This may be due to the sensitivity of the N-fixing symbiosis to acidity and to the effect of the alkaline rhizosphere in reducing trace element toxicity. Andrew (1976) reported that the nodulation and growth of N fixing plants was inhibited by low pH, the effect being alleviated by Ca. In contrast, plants completely dependent on mineral N showed only a minor reduction in yield at low pH and no significant effect of applied Ca on yield at pH values between 4.0 and 6.0. Plants provided only with mineral N were relatively susceptible to Al toxicity when the pH of the culture solution was maintained at 4.2 (Andrew et al. 1973).

(B) Cation relationships

The sum of cation concentrations tends to remain constant in plants, consequently inverse relationships between concentrations of individual cations have been observed in white clover. The application of K decreases Na concentration and vice versa in Ladino clover (Cope et al. 1953, Nelson & Brady 1953a). The yield of Ladino clover increased on application of Na alone, probably as a result of Na increasing the availability of soil K (Cope et al. 1953). In other situations Na has replaced K in fertiliser or nutrient solutions with little or no effect on yield (Whitehead & Jones 1972, Schultz et al. 1979). Reciprocal relationships have been observed between K, Ca and Mg (Brown 1957, Andrew 1960, Andrew & Robins 1969b, Andrew & Johnson 1976).

Nodulated plants with no access to NO_3^- ions would be assumed to have a greater content of cations than anions (Israel & Jackson 1978). Whitehead & Jones (1969) found an excess of inorganic cations for such white clover plants, and showed that the ash alkalinity (a measure of organic anions) was approximately equal to the difference between cation and anion content. However, it appears that the cation-anion balance is controlled by factors additional to NO_3^- availability. When K_2SO_4 or KCl was present Noggle (1966) obtained a cation-anion ratio for plants absorbing $Ca(NO_3)_2$ that was quantitatively similar to the value obtained by Whitehead & Jones (1969) for plants fixing their own nitrogen.

(C) Nutrients and water

Adequate water is of such importance to the functioning of the soil-plant system, that there will be a number of interactions between water and nutrients. However, there has been little research on the interactions for white clover. Nitrogen fixation is reduced in white clover by water stress (Engin & Sprent 1973). P uptake is increased by an adequate water supply but decreased by excessive water (Low & Piper 1960). Davidson (1969) showed that water supply, along with N and P, affected the shoot:root ratio. The shoot:root ratio increased with decreasing water availability.

6. Nutrient efficiency

The idea implicit in the concept of nutrient efficiency is that appropriate plant breeding or management programmes may increase yield per unit of plant nutrient used. In particular, the breeding of nutrient efficient varieties has often been described as an attractive way of reducing dependence on fertilizers. However, although this approach was suggested more than 50 years ago by Gregory & Crowther (1928), few if any nutrient efficient crop varieties have appeared. The reason for this may lie partly in the ready availability of fertilisers and their high cost effectiveness, but there are also technical difficulties which have slowed progress towards this goal. Clovers, because of their high P and K requirements in grass/clover pastures, have been seen as species in which it is particularly desirable to improve nutrient efficiency by breeding. The philosophy and progress of this endeavour is discussed in this section.

There are many definitions of nutrient efficiency and it is difficult to see any consensus of opinion emerging at present. Even the word 'efficiency' has been objected to (R. W. Snaydon, personal communication) because it can have widely different connotations for plant physiologists, ecologists and agronomists. Plants that are efficient according to the definitions of one discipline may be inefficient according to those of another. However, the term is retained here because no suitable alternative has been suggested. In this discussion the definition of efficiency used is that of Loneragan (1976), namely, that a plant's efficiency in the use of a nutrient is the relationship between the nutrient present in the soil below a unit area of land and the rate of production of the plant growing thereon. It must be pointed out that white clover plants will almost always be growing in association with a grass. As discussed below this does not complicate the definition so much as the design of the nutrient-efficient species.

(A) Philosophies and strategies

Donald (1968) has pointed out that the aim of most plant breeding has been to increase yield or to eliminate defects. White clover is no exception and because screening work has mostly been carried out on plots of fertile soil, it is unlikely that nutrient efficiency would have been important in the selection of present varieties. Barrow (1978) has pointed out that large-scale screenings for efficiency are unlikely to be successful. Nevertheless, this empirical approach may lead to nutrient-efficient varieties sooner than alternative approaches which are also unproven. Specific attempts to screen for nutrient efficiency are considered below. Donald (1968) has argued that specific breeding objectives are more likely to be reached by the breeding of crop ideotypes, that is, plants with model characteristics chosen to achieve the breeding goal. This appears to be the approach implicit in the work of most researchers working on nutrient efficiency. However, at present most of the reported work is on designing nutrient-efficient ideotypes rather than on breeding.

A useful nutrient-efficient ideotype would have to be theoretically capable of the same production from lower input of fertiliser or greater production from the same input compared with existing cultivars. Furthermore, there would have to be a reasonable likelihood of being able to breed the ideotypes from the material available. To fulfil these requirements there must be good knowledge of the production system, sufficient genetic diversity in chosen characteristics and suitable techniques for identifying the chosen characteristic (Donald 1968).

Cooper (1973) has suggested the following operational sequence for introducing physiological criteria into a breeding programme: (i) establish the baseline of performance by existing material; (ii) identify important components which would be likely to contribute to improved performance' (iii) assess the useful variation avaliable; (iv) introduce screening into the breeding programme. The significance of these last two requirements is illustrated clearly by the formula devised by Cooper (1973) to describe the response to selection (R): $R = Ish^2$, where I=intensity of selection, s=phenotypic standard deviation,and h^2=heritability.

Practicality dictates that the intensity of selection will depend primarily on the speed and ease with which the selection criteria can be measured, as well as their correlation with desired performance. Consequently it is necessary not only to determine the important qualities of a nutrient-efficient plant but also to devise a way of measuring that quality rapidly and preferably at an early stage of the plant's life cycle. Furthermore, that quality will be useful only if it shows some genetic variation.

Because of the importance of grass-cover competition in the mineral nutrition of pastures and the greater competitive ability of the grasses, it has been thought that a nutrient-efficient white clover ideotype would have characteristics approaching those of a grass. However, clovers are a desirable component of pastures because of certain characteristics of their mineral nutrition, e.g. nitrogen fixation,which in turn influence other aspects of their mineral nutrition (Section 4.(D)). Furthermore, while the form of the white clover root system is thought to be a disadvantage in competitive situations (Evans 1977), it may give an advantage under moisture stress, allowing it to compete by occupying a different spatio-temporal niche (M.C.H. Mouat, personal communication). Even so, there are probably gains to be made by identifying the key aspects of the competitive strengths of grasses and incorporating these in the white clover ideotype. Therefore, studies of this type have an important role in programmes aimed at breeding a nutrient-efficient white clover, and are discussed in the following section.

Herbage from grass-clover pastures is only an intermediate product, the marketed product almost always being some animal product. Aspects of animal health do not fall directly within the definition of nutrient efficiency, but the mineral content of the herbage in relation to stock requirements is a valid element in the design of nutrient-efficient ideotypes (Sleper 1979). This design would include consideration of a number of elements which plants require in smaller amounts than animals, such as Na, Mg, Cu and Mo.

(B) Genetic diversity and screening for nutrient efficiency

Snaydon & Bradshaw (1962b) demonstrated that there are substantial intraspecific differences in the ability of white clover to grow under nutrient stress. They found that plant growth in populations from soils high in P declined more when subjected to P deficiency than did growth in populations from low-P soils. The difference between the populations was increased when they were grown in competition with a contrasting population, and was even more marked when grown in competition with grasses adapted to low-fertility conditions (Snaydon & Bradshaw 1962b). The populations also showed different fertiliser responses. It was concluded that a population more suited to the prevailing agricultural conditions could be obtained by selection under conditions similar to those in which the material was later to be used.

Caradus & Dunlop (1978) and Caradus et al. (1980) have screened for nutrient efficiency in white clover. In the first phase of a two-stage glasshouse pot trial, the growth of 460 genotypes at two P levels was measured. The results were examined on a high vs low plot (Caradus 1980) which allowed the performance of each genotype to be readily compared. Two populations were chosen for further study, one group being the most promising, the other with apparently poor performance at low P levels. P response curves were obtained for each genotype of both populations in a second pot trial. There was no consistent pattern between the results of the two stages, but plants with slightly different response curves were obtained (Caradus et al. (1980). The lack of reproducibility between the first and second stage may have arisen from the lack of replication in stage one. However, it may also have been a result of an interaction between climate and relative performance at the P levels. In this trial the experimental treatments introduced differences from those met with in normal pasture in two respects: defoliation by clipping instead of grazing, and absence of a competing species. Extrapolating from the results of Snaydon & Bradshaw (1962b) it would seem that screening experiments might be more productive if the plants were grown in competition with a grass.

Scott (1975) found that the cultivar Tamar yielded more than Huia at all the P levels studied and that it reached maximum growth at a lower P level. However, this advantage depended on climate and on the presence or absence of nematodes. Tamar was found to have a high shoot/root ratio and consequently was less able to cope with repeated defoliation; under this treatment it was less P-efficient than Huia.

Chisholm et al. (1978) measured the rate of P uptake by white clover and *Stylosanthes humilis*, a tropical legume that is highly P-efficient. They found that the initial influx of labelled P was faster for stylosanthes when plants were grown at low P level. However, for plants growing at a high P level white clover had the faster influx.

In a study of the P nutrition of seedlings of eight populations, Spencer et al. (1980) found only minor differences in several aspects of P nutrition. However, they were able to infer some similarities between their results and the findings of Snaydon & Bradshaw (1962b), namely that populations from low-P soils contained higher P concentrations in the shoots and tended to be less responsive to high P. These responses are characteristic of the strategy of tolerance of nutrient stress found in a number of wild plants (Chapin 1980). This strategy includes slow photosynthesis, slow growth rate, and hence low productivity. Consequently, this material may have little value in an agricultural situation where productivity is of prime importance. Therefore, a nutrient-efficient plant of use in agriculture is unlikely to be a product of natural selection and unlikely to be found by surveying existing populations. If so, the required suite of characteristics will have to be brought together by a breeding programme.

(C) Ideotype characters and their genetic variation

It is not yet possible to define all the characteristics required in a nutrient-efficient white clover plant. For many characters of possible importance, conclusive experiments have yet to be performed. Furthermore, most of the more significant experiments have been conducted on *Trifolium* species other than white clover, and of those that have studied white clover, very few have compared different clones or cultivars. In addition, there is little information on nutrient elements other than P. However, there are a few characteristics of a nutrient-efficient white clover plant about which there is some consensus of opinion. These include:

(1) A fine fibrous root system with many long root hairs.

(2) A shoot that maintains a high growth rate at low P contents, i.e. a high utilisation index. (It is important to note, however, that screenng plants for low shoot P content without also measuring amounts of P absorbed would probably be counterproductive).

(3) A shoot form that reduces the number of meristems removed during defoliation by grazing.

(4) A nutrient content that is consistent with stock health.

Fine, fibrous root systems with root hairs would confer an advantage in the uptake of immobile nutrients such as P. Caradus (1977, 1981b) reported that white clover root systems range from those with a very prominent tap root to those without obvious tap roots. However, the fineness of roots of white clover does not approach that of grass root systems. Furthermore, there is a strong positive correlation between prominence of the tap root and larger leaf size, with its associated higher productivity. The morphology of clover roots must also be considered. The greater rooting depth of the tap root is probably associated with the plant's tolerance of water stress, which gives it a competitive edge during the summer. These factors may explain the finding of Caradus (1980) that while differences in P efficiency between pasture legumes and grasses could be related to root characteristics, differences in efficiency within the legume group could not.

Endomycorrhizal associations also increase the volume of soil from which a root can obtain minerals. White clover is almost always mycorrhizal in field conditions and consequently ways of manipulating the association to the advantage of white clover are being sought (Powell 1977c). Crush & Caradus (1980) examined the possibility that the response of white clover types to different endomycorrhizal types may vary. However, they concluded that the lack of interactions between clover-type and mycorrhizal-type suggested that it was probably not worth pursuing mycorrhizal relationships in a breeding programme of selection for nutrient-efficient plants.

The data of Robinson (1942) indicate that there is genetic variation among clones of white clover in their utilisation indices for Ca, P and K. In his experiments, 8 clones were grown in 5 different soils in a glasshouse pot trial. The ranges of Ca, P and K contents on a percentage basis averaged over the 5 soils, were 1.5-, 1.7- and 1.5-fold, respectively, and there was a 3-fold range in the dry matter yields of the tops. The data indicate that the faster growing plants tended to have higher absolute contents of P and K but at lower concentrations than the slow growing plants. This generalisation did not hold true for Ca.

Thomas & Evans (1977) measured the P and Ca content of 23 varieties of white clover and found ranges of 1.4- and 1.44-fold, respectively. The P contents tended to be higher and the Ca contents lower than in Robinson's (1948) study, these differences probably being associated with differences in soil conditions and the fact that Thomas & Evans (1977) used field grown material. However, the usefulness of a high utilisation index and the vigorous growth with which it is frequently associated is diminished because these attributes are, in turn, associated with an open growth form which is not persistent under heavy grazing (Caradus et al. 1980).

Because clover meristems are located at the stolon tips and grass meristems at the stem bases, clover may lose a greater proportion of its meristems through grazing, particularly in the case of the more vigorous white clover varieties which have a more erect form and stolon tips that are particularly vulnerable to removal through grazing. Therefore, high initial plant vigour is unlikely to be found in combination with persistence. Williams et al. (1982) have selected for a shoot form in which the meristems are protected by the rest of the shoot so as to reduce their removal by grazing. These selections are showing promise in hill country pastures where nutrients, especially P, are deficient. It is not known whether this character is beneficial through its relation to the plant's performance under nutrient stress or through some other limitation in the environment. However, conserving the stolon tips with their relatively high nutrient content may reduce the transfer of nutrients to stock camps where other factors prevent their efficient use for herbage production.

Data on 23 white clover varieties (Thomas & Evans 1977) suggest that Ca, Mg, P and Cu would normally be present at adequate levels for stock health, but that Na, Zn and Mn levels would be inadequate. The extent of variation among the 23 varieties in Na, Zn and Mn levels were 2.0-, 1.5- and 1.5-fold respectively. This suggests that there may be scope for increasing the content of these nutrients.

7. Conclusions

The need to improve the mineral nutrition of white clover by breeding is widely acknowledged but despite much research, mainly focussing on P, the goal remains elusive.

Material from sites where soils are low in P appears unlikely to provide useful nutrient efficient agricultural plants because of their low productivity. The alternative approach of designing a model plant, an ideotype, and then selecting and assembling the desired characteristics in a breeding programme, is still at the stage of designing the ideotype. However, because of the economic advantage that nutrient efficiency could confer, this goal is likely to continue to be pursued energetically.

References

Aguilar, A.; Diest, van A. (1981) Rock phosphate mobilization induced by the alkaline uptake pattern of legumes utilizing symbiotic fixed nitrogen. *Plant and Soil* **61**, 27-42.

Allos, H.F.; Batholomew, W.V. (1955) Effect of available nitrogen on symbiotic fixation. *Soil Science Society of America Proceedings* **19**, 182-184.

Allos, H.F.; Bartholomew, W.V. (1959) Replacement of symbiotic fixation by available nitrogen. *Soil Science* **87**, 61-66.

Andrew, C.S. (1956) Visual symptoms of mineral deficiencies and excess in *Trifolium repens* var. Irrigation White. *Australian Plant Nutrition Conference*, 320-325.

Andrew, C.S. (1960) The effect of phosphorus, potassium and calcium on the growth, chemical composition and symptoms of deficiency of white clover in a subtropical environment. *Australian Journal of Agricultural Research* **11**, 149-161.

Andrew, C.S. (1976) Effect of calcium, pH and nitrogen on the growth and chemical composition of some tropical and temperate legumes. I. Nodulation and growth. *Australian Journal of Agricultural Research* **27**, 611-623.

Andrew, C.S. (1977) The effect of sulphur on the growth, sulphur and nitrogen concentrations and critical sulphur concentrations of some tropical and temperate pasture legumes. *Australian Journal of Agricultural Research* **28**, 807-820.

Andrew, C.S.; Bryan, W.W. (1955) pasture studies on the coastal lowlands of subtropical Queensland. I. Introduction and initial plant nutrient studies. *Australian Journal of Agricultural Research* **6**, 265-290.

Andrew, C.S.; Hegarty, M.P. (1969) Comparative responses to manganese excess in eight tropical and four temperate legumes. *Australian Journal of Agricultural Research* **20**, 687-696.

Andrew, C.S.; Johnson, A.D. (1976) Effect of calcium, pH and nitrogen on the growth and chemical composition of some tropical and temperate legumes. II. Chemical composition (calcium, nitrogen, potassium, magnesium, sodium and phosphorus). *Australian Journal of Agricultural Research* **27**, 625-636.

Andrew, C.S.; Norris, D.O. (1961) Comparative responses to calcium of five tropical and four temperate pasture legume species. *Australian Journal of Agricultural Research* **12**, 40-55.

Andrew, C.S.; Robins, M.F. (1969a) The effect of potassium on the growth and chemical composition of some tropical and temperate pasture legumes. I. Growth and critical percentages of potassium. *Australian Journal of Agricultural Research* **20**, 999-1007.

Andrew, C.S.; Robins, B.F. (1969b) The effect of potassium on the growth and chemical composition of some tropical and temperate pasture legumes. II. Potassium, calcium, magnesium, sodium, nitrogen, phosphorus and chloride. *Australian Journal of Agricultural Research* **20**, 1009-1021.

Andrew, C.S.; Thorne, P.M. (1962) Comparative responses to copper of some tropical and temperate pasture legumes. *Australian Journal of Agricultural Research* **13**, 821-835.

Andrew, C.S.; Johnson, A.D.; Sandland, R.L. (1973) Effect of aluminium on the growth and chemical composition of some tropical and temperate pasture legumes. *Australian Journal of Agricultural Research* **24**, 325-339.

Baker, D.A. (1978) Proton co-transport of organic solutes by plant cells. *New Phytologist* **81**, 485-497.

Barrow, N.J. (1978) Problems of efficient fertilizer use. *Plant Nutrition* **1**, 37-52.

Bathurst, N.O.; Mitchell, K.J. (1958) The effect of light and temperature on the chemical composition of pasture plants. *New Zealand Journal of Agricultural Research* **1**, 540-552.

Bieleski, R.L. (1973) Phosphate pools, phosphate transport, and phosphate availability. *Annual Review of Plant Physiology* **24**, 225-252.

Blaser, R.E.; Brady, N.C. (1950) Nutrient competition in plant associations. *Agronomy Journal* **42**, 128-135.

Boland, M.J.; Fordyce, A.M.; Greenwood, R.M. (1978) Enzymes of nitrogen metabolism in legume nodules; a comparative study. *Australian Journal of Plant Physiology* **5**, 553-559.

Bollig, I.; Mayer, K.; Mayer, W.E.; Engelmann, W. (1978) Effects of cAMP, theophylline, imidazole, and 4-(3,4-dimethoxybenzl)-2-imidazolidone on the leaf movement rhythm of *Trifolium repens* - a test of the cAMP hypothesis of circadian rhythms. *Planta* **141**, 225-230.

Bowling, D.J.F.; Dunlop, J. (1978). Uptake of phosphate by white clover. *Journal of Experimental Botany* **29**, 1139-1146.

Brock, J.L.; Hoglund, J.H. (1974) Growth of 'Grasslands Huia' and 'Grasslands 4700' white clovers. II. Effects of nitrogen and phosphorus. *New Zealand Journal of Agricultural Research* **17**, 47-53.

Broughton, W.J. (1979) Effect of light intensity on net assimilation rates of nitrate-supplied or nitrogen-fixing legumes. In: Marcelle, R.; Clijsters, H.; van Packe, M. (eds) Photosynthesis and plant development. The Hague, Netherlands; N.W. Junk, pp. 285-298.

Brown, B.A. (1957) Potassium fertilization of Ladino clover. *Agronomy Journal* **49**, 477-480.

Brown, J.M.; Rouse, R.D. (1953) Fertilizer effects on botanical and chemical composition of white clover-Dallisgrass associations grown on summer day. *Agronomy Journal* **45**, 279-282.

Butler, G.W.; Bathurst, N.O. (1958) Free and bound amino acids in legume root nodules: bound aminobutyric acid in the genus *Trifolium. Australian Journal of Biological Science* **11**, 592-537.

Butler, G.W.; Jones, D.I.H. (1973) Mineral biochemistry of herbage. In: Butler, G.W.; Bailey, R.H. (Eds) Chemistry and Biochemistry of Herbage, London, UK; Academic Press, **Vol. 2**, 127-162.

Buwalda, J.G. (1980). Growth of clover-ryegrass association with vesicular arbuscular mycorrhizas. *New Zealand Journal of Agricultural Research* **23**, 379-383.

Caradus, J.R. (1977) Structural variation of white clover root systems. *New Zealand Journal of Agricultural Research* **20**, 213-219.

Caradus, J.R. (1980) Distinguishing between grass and legume species for efficiency of phosphorus use. *New Zealand Journal of Agricultural Research* **23**, 75-81.

Caradus, J.R. (1981a) Root growth of white clover (*Trifolium repens* L.) lines in glass-fronted containers. *New Zealand Journal of Agricultural Research*. **24**, 43-54.

Caradus, J.R. (1981b) Root morphology of some white clovers from New Zealand hill country. *New Zealand Journal of Agricultural Research* **24**, 349-351.

Caradus, J.R.; Dunlop, J. (1978) Screening white clover plants for efficient phosphorus use. In: Plant Nutrition, Proceedings of the 8th International Colloquium on Plant Analysis and Fertiliser Problems. *DSIR Information Series* **No. 134**, 75-82.

Caradus, J.R.; Dunlop, J.; Williams, W.M. (1980) Screening white clover (*Trifolium repens* L.) plants for different responses to phosphate. *New Zealand Journal of Agricultural Research* **23**, 211-217.

Caradus, J.R.; Evans, P.S. (1977). Seasonal root formation of white clover, ryegrass, and cocksfoot in New Zealand. *New Zealand Journal of Agricultural Research* **20**, 337-342.

Chandler, R.F.; Peech, M.; Chang, C.W. (1945) The release of exchangeable and nonexchangeable potassium from different soils upon cropping. *Journal of the American Society of Agronomy*. **37**, 709-721.

Chapin, F.S. (1980) The mineral nutrition of wild plants. *Annual Review of Ecology and Systematics* **11**, 233-260.

Chisholm, R.H.; Goodwin, D.C.; Wilson, E.J.; Blair, G.J. (1978) Phosphorus efficiency in legumes. In: Plant Nutrition, Proceedings of 8th International Colloquium on Plant Analysis and Fertilizer Problems, *DSIR Information Series*, **No. 134**, 91-92.

Chisholm, R.H.; Blair, G.J.; Bowden, J.W.; Bofinger, V.J. (1981) Improved estimates of critical phosphorus concentration from considerations of plant phosphorus chemistry. *Communications in Soil Science and Plant Analysis* **12**, 1059-1065.

Chisholm, R.H.; Blair, G.J. (1981) Phosphorus uptake and dry weight of stylo and white clover as affected by chlorine. *Agronomy Journal* **73**, 767-771.

Clarkson, D.T.; Hanson, J.B. (1980) The mineral nutrition of higher plants. *Annual Review of Plant Physiology* **31**, 239-298.

Cooper, J.P. (1973) The use of physiological criteria in grass breeding. *Report of the Welsh Plant Breeding Station*, 95-102.

Cooper, K.M.; Tinker, P.B. (1981) Translocation and transfer of nutrients in vasicular-arbuscular mycorrhizas. *New Phytologist* **88**, 327-339.

Cope, J.T.; Bradfield, R.; Peech, M. (1953) Effect of sodium fertilisation on yield and cation content of some field crops. *Soil Science* **76**, 65-74.

Copeland, R.; Pate, J.S. (1970) Nitrogen metabolism of nodulated white clover in the presence and absence of nitrate nitrogen. In: Lowe, J. (Ed) Proceedings of a symposium on white clover research. *Occasional Symposium, British Grassland Society*, **No. 6**, 71-77.

Crush, J.R. (1974) Plant growth responses to vesicular-arbuscular mycorrhiza. VII. Growth and nodulation of some herbage legumes. *New Phytologist*. **73**, 743-749.

Crush, J.R. (1976) Endomycorrhizas and legume growth in some soils of the Mackenzie Basin, Canterbury, New Zealand. *New Zealand Journal of Agricultural Research* **19**, 473-476.

Crush, J.R. (1978) Changes in effectiveness of soil endomycorrhizal fungal populations during pasture development. *New Zealand Journal of Agricultural Research* **21**, 683-685.

Crush, J.R.; Caradus, J.R. (1980) Effect of mycorrhizas on growth of some white clovers. *New Zealand Journal of Agricultural Research* **23**, 233-237.

Dalal, R.C. (1977) Soil organic phosphorus. *Advances in Agronomy* **29**, 83-117.

Davidson, R.L. Effects of soil nutrients and moisture on root/shoot ratios in *Lolium perenne* L. and *Trifolium repens* L. *Annals of Botany* **33**, 571-577.

Davies, W.E.; Griffith, G. ap: Ellington, A. (1966) The assessment of herbage legume varieties. II. In vitro digestibility, water soluble carbohydrate crude protein and mineral content of primary growth of clover and lucerne *Journal of Agricultural Science* **66**, 351-357.

Davis, M.R. (1981a) Growth and nutrition of legumes on a high country yellow-brown earth subsoil. I. Phosphate response of *Lotus, Trifolium, Lupinus, Astragalus*, and *Coronilla* species and cultivars. *New Zealand Journal of Agricultural Research* **24**, 321-332.

Davis, M.R. (1981b) Growth and nutrition of legumes on a high country yellow-brown earth subsoil. II. A comparison of tropical and temperate species. *New Zealand Journal of Agricultural Research* **24**, 333-337.

Davis, M.R. (1981c) Growth and nutrition of legumes on a high country yellow-brown earth subsoil. III. The effect of lime. *New Zealand Journal of Agricultural Research* **24**, 339-348.

Donald, C.M. (1968) The breeding of crop ideotypes. *Euphytica* **17**, 385-403.

Dorofaeff, F.D.; McNaught, K.J. (1962) Magnesium deficiency in white clover (Trifolium repens L.) on a pumice soil. *New Zealand Journal of Agricultural Research* **5**, 310-317.

Dunlop, J.; Bowling, D.J.F. (1978) Uptake of phosphate by white clover. II. The effect of pH on the electrogenic pump. *Journal of Experimental Botany* **29**, 1147-1153.

Dunlop, J.; Glass, A.D.M.; Tomkins, B.D. (1979) The regulation of K+ uptake by ryegrass and white clover roots in relation to their competition for potassium. *New Phytologist* **83**, 365-370.

Edmeades, D.C.; Judd, M.J. (1980) The effect of lime on the magnesium status and equilibria in some New Zealand topsoils. *Soil Science* **129**, 156-161.

Elst, F.H. van der; McNaught, K.J.; Rolt, W.F. (1961) Effect of copper deficiency in white clover on nitrogen fixation. *Nature, UK* **192**, 1315.

Engin, M.; Sprent, J.I. (1973) Effects of water stress on growth and nitrogen-fixation activity of Trifolium repens. *New Phytologist* **72**, 117-126.

Epstein, E. (1972) Mineral nutrition of plants: principles and perspectives. New York, USA; John Wiley and Sons.

Erith, A. (1924). White clover: A monograph. London, UK; Duckworth.

Evans, P.S. (1977) Comparative root morphology of some pasture grasses and clovers. *New Zealand Journal of Agricultural Research* **20**, 331-335.

Foy, C.D.; Fleming, A.L. (1978) The physiology of plant tolerance to excess available aluminium and manganese in acid soils. In: Crop Tolerance to Subtropical Land Conditions. *American Society of Agronomy Special Publication,* **No. 32**, 301-328.

Gammon, N.; Fiskel, J.G.A.; Mourkides, G.A. (1955) Uptake of molybdenum from everglades peat by several grasses and by white clover. *Soil Science Society of America Proceedings* **19**, 488-491.

Gammon, N.; Zakaria, Z.Z.; Ruelke, O.C.; Prine, G.M. (1979) Response of white clover (*Trifolium repens* L.) to manganese rates on an overlimed Myakka fine sand. *Proceedings of the Soil and Crop Science Society of Florida* **38**, 64-66.

Gibson, A.H. (1966) The carbohydrate requirement for symbiotic nitrogen fixation: a whole-plant growth analysis approach. *Australian Journal of Biological Science* **19**, 499-515.

Gibson, D.I.; Hayes, P.; Laidlaw, A.S. (1975) The influence of phosphate and lime on the growth and N fixation of *Lotus uliginosus* and *Trifolium repens* under greenhouse conditions. *Journal of the British Grassland Society* **30**, 295-301.

Glass, A.D.M. (1976) Regulation of potassium absorption in barley roots.*Plant Physiology* **58**, 33.

Grey, B.; Drake, M.; Colby, W.G. (1953) Potassium competition in grass-legume associations as a function of root cation exchange capacity. *Soil Science Society Annual Proceedings* **17**, 235-239.

Gregory, F.G.; Crowther, F. (1928) A physiological study of varietal differences in plants. Part 1. A study of the comparative yields of barley varieties with different manurings. *Annals of Botany* **42**, 757-770.

Gunning, B.E.S.; Pate, J.S. (1969) Transfer cells, plant cells with wall ingrowths, specialised in relation to short distance transport of solutes - their occurrence, structure, and development. *Protoplasma* **68**, 107-133.

Gunning, B.E.S.; Pate, J.S.; Green, L.W. (1970) Transfer cells in the vascular system of stems: taxonomy, association with nodes, and structure. *Protoplasma* **71**, 147-171.

Hall, I.R. (1978) Effects of endomycorrhizas on the competitive ability of white clover. *New Zealand Journal of Agricultural Research* **21**, 509-515.

Hall, I.R. (1979a) Effect of vesicular-arbuscular mycorrhizas on growth of white clover, lotus, and ryegrass in some graded soils. *New Zealand Journal of Acricultural Research* **22**, 479-484.

Hall, I.R. (1979b) Soil pellets to introduce vesicular-arbuscular mycorrhizal fungi into soil. *Soil Botany, Biochemistry* **11**, 85-85.

Hall, I.R.; Scott, R.S.; Johnstone, P.D. (1977) Effect of vesicular-arbuscular mycorrhizas on response of Grassland Huia and Tama white clovers to phosphorus. *New Zealand Journal of Agricultural Research* **20**, 349-355.

Hallsworth, E.G.; Greenwood, E.A.N.; Yates, M.G. (1964) Studies on the nutrition of forage legumes. III. The effect of copper on nodulation of *Trifolium subterraneum* L. and *Trifolium repens*. *Plant and Soil* **20**, 17-33.

Hart, A.L. (1982) The distribution of phosphorus and nitrogen in white clover and lotus inoculated with rhizobia or given mineral nitrogen. In: Scaife, A. (Ed). Plant Nutrition 1982, *Proceedings of the 9th International Plant Nutrition Colloquium.* Farnham Royal, UK; Commonwealth Agricultural Bureaux, 221-226.

Hart, A.L.; Jessop, D. (1982) Concentration of total, inorganic, and lipid phosphorus in leaves of white clover and stylosanthes. *New Zealand Journal of Agricultural Research* **25**, 69-76.

Hart, A.L.; Jessop, D.L.; Galpin, J. (1981a) The response to phosphorus of white clover and lotus inoculated with rhizobia or given KNO_3. *New Zealand Journal of Agricultural Research* **24**, 27-32.

Hart, A.L.; Halligan, G.; Haslemore, R.M. (1981b) Analysis of the response of pasture legumes to phosphorus in a controlled environment. *New Zealand Journal of Agricultural Research* **24**, 197-201.

Harwood, J.L.; Nicholls, R.G. (1979) The plant sulpholipid - a major component of the sulphur cycle. *Biochemical Society Transactions* **7**, 440-447.

Hay, M.J.M. (1983) Seasonal variation in the distibution of white clover (*Trifolium repens* L.) stolons among three horizontal strata in two grazed swards. *New Zealand Journal of Agricultural Research* **26**, 29-34.

Hay, M.J.M.; Dunlop, J. (1982). Phosphate absorption by white clover stolons in pasture. *New Zealand Journal of Agricultural Research* **25**, 211-216.

Hay, M.J.M.; Dunlop, J.; Hopcroft, D.H. (1982) Anatomy and development of the superficial layers in stolons of white clover (*Trifolium repens* L.). *New Zealand Journal of Botany* **20**, 315-324.

Haydock, K.P.; Norris, D.O. (1967) Opposed curves for nitrogen percent on dry weight given by Rhizobium dependent and nitrate dependent legumes. *Australian Journal of Science* **29**, 426-427.

Hayman, D.S.; Mosse, B. (1972). The role of vesicular-arbuscular mycorrhiza in the removal of phosphorus from soil by plant roots. *Revue d'Ecologie et de Biologie du Sol.* 463-470.

Hayman, D.S.;Mosse, B. (1979) Improved growth of white clover in hill grasslands by mycorrhizal inoculation. *Annals of Applied Biology* **93**, 141-148.

Haynes, R.J.; Ludecke, T.E. (1981a) Effect of lime and phosphorus applications on concentrations of available nutrients and on P, Al and Mn uptake by two pasture legumes in an acid soil. *Plant and Soil* **62**, 117-128.

Haynes, R.J.; Ludecke, T.E. (1981b) Yield, root morphology and chemical composition of two pasture legumes as affected by lime and phosphorus applications to an acid soil. *Plant and Soil* **62**, 241-254.

Haystead, A.; Marriott, C. (1979) Effects of rates and times of application of starter dressings of nitrogen fertiliser to surface sown perennial ryegrass-white clover on hill peat. *Grass and Forage Science* **34**, 241-247.

Helyar, K.R.; Anderson, A.J. (1970a) Responses of five pasture species to phosphorus, lime, and nitrogen on an infertile acid soil with a high phosphate absorption capacity. *Australian Journal of Agricultural Research* **21**, 677-629.

Helyar, K.R.; Anderson, A.J. (1970b) Some effects of the soil pH on different species and on the soil solution for a soil high in exchangeable aluminium. *Proceedings of the XIth International Grasslands Congress*, 431-434.

Helyar, K.R.; Anderson, A.J. (1971) Effects of lime on the growth of five species, on aluminium toxicity, and on phosphorus availability. *Australian Journal of Agricultural Research* **22**, 707-721.

Hill, J. (1980) The remobilization of nutrients from leaves. *Journal of Plant Nutrition* **2**, 407-444.

Hoshino, M. (1974) Translocation and accumulation of assimilates and phosphorus in ladino clover. *Bulletin of the National Grassland Research Institute of Japan* **5**, 35-84.

Israel, D.W.; Jackson, W.A. (1978) The influence of nitrogen nutrition on ion uptake and translocation by leguminous plants. In: Andrew, C.S.; Kamprath, E.J. (Eds). Mineral Nutrition of Legumes in Tropical and Subtropical Soils, Melbourne, Australia; CSIRO, 113-129.

Jackman, R.H. (1961) Pasture establishment on a podzolized soil in Northland. II. Plot trials studying nutrient deficiencies. *New Zealand Journal of Agricultural Research* **4**, 361-374.

Jackman, R.H. (1965) The uptake of rubidium by the roots of some graminaceous and leguminous plants. *New Zealand Journal of Agricultural Research* **8**, 763-777.

Jackman, R.H.; Mouat, M.C.H. (1970) The effect of browntop (*Agrostis tenuis* (Sibth.)) and increasing phosphorus deficiency on the growth of white clover (*Trifolium repens* L.). *Proceedings of the XIth International Grassland Congress*, 354-357.

Jackman, R.H.; Mouat, M.C.H. (1972a) Competition between grass and clover for phosphate. I. Effect of browntop (*Agrostis tenuis* Sibth.) on white clover (*Trifolium repens* L.) growth and nitrogen fixation. *New Zealand Journal of Agricultural Research* **15**, 653-666.

Jackman, R.H.; Mouat, M.C.H. (1972b) Competition between grass and clover for phosphate. II. Effect of root activity, efficiency of response to phosphate, and soil moisture. *New Zealand Journal of Agricultural Research* **15**, 667-675.

Jackson, T.L.; Rampton, H.H.; McDermid, J. (1964) The effect of lime and phosphorus on the yield and phosphorus content of legumes in Western Oregon. *Technical Bulletin, Oregon State University Agricultural Experimental Station*, No. 83.

Jarvis, S.C. (1980) The uptake and distribution of copper in perennial ryegrass and white clover grown in flowing solution culture with a controlled supply of copper. *Journal of the Science of Food and Agriculture* **31**, 870-876.

Jo, J.; Yoshida, S.; Kayama, R. (1980) Growth and nitrogen fixation of some leguminous forages grown under acidic soil conditions. *Journal of the Japanese Society of Grassland Science* **25**, 326-334.

Johnson, C.M.; Pearson, G.A.; Stout, P.R. (1952) Molybdenum nutrition of crop plants. II. Plant and soil factors concerned with molybdenum deficiencies in crop plants. *Plant and Soil* **4**, 178-196.

Jordan, H.V.; Bardsley, C.E. (1958) Response of crops to sulfur on Southeastern soils. *Soil Science Society of America Proceedings* **22**, 254-256.

Khalid, R.A.; Silva, J.A. (1979) A study of soil aluminium extraction methods in relation to plant aluminium and yield in tropical soils. *Tropical Agriculture* **56**, 53-63.

Lin, W. (1979) Potassium and phosphate uptake in corn roots. Further evidence for an eletrogenic H+/K+ exchanger and an OH-/Pi antiporter. *Plant Physiology* **63**, 925-955.

Lohnis, M. (1951) Manganese toxicity in field and market crops. *Plant and Soil* **3**, 193-222.

Loneragan, J. F. (1976) Plant efficiencies in the use of stress in problem soils. Cornell University, Agricultural Experimental Station, 193.

Low, A.J.; Piper, F.J. (1960) The influence of water supply on the growth and phosphorus uptake of Italian ryegrass and white clover. *Plant and Soil* **13**, 242-250.

Lowther, W.L. (1980) Establishment and growth of clovers and lotus on acid soils. *New Zealand Journal of Experimental Agriculture* **8**, 131-138.

McNaught, K.J. (1958) Potassium deficiency in pastures. Part 1. Potassium content of legumes and grasses. *New Zealand of Agricultural Research* **1**, 148-181.

McNaught, K.J. (1970) Diagnosis of mineral deficiencies in grass-legume pastures by plant analysis. *Proceedings of the XIth International Grassland Congress* 334-338.

McNaught, K.J.; Chrisstoffels, P.J.E. (1961) Effect of sulphur deficiency on sulphur and nitrogen levels in pastures and herbage. *New Zealand Journal of Agricultural Research.* **4**, 177-196.

McNaught, K.J.; Dorofaeff, F.D.; Karlovsky, J. (1968) Effect of magnesium fertilisers and season on levels of inorganic nutrients in a pasture on Hamilton clay loam. I. Magnesium and calcium. *New Zealand Journal of Agricultural Research* **11**, 533-550.

McNaught, K.J.; Dorofaeff, F.D.; Ludecke, T.E.; Cottier, K. (1973a) Effect of potassium fertiliser, soil magnesium status and soil type on uptake of magnesium by pasture plants from magnesium fertilisers. *New Zealand Journal of Experimental Agriculture* **1**, 329-347.

McNaught, K.J.; Dorofaeff, F.D.; Karlovsky, J. (1973b) Effect of some magnesium fertilisers on mineral composition of pasture on Horotiu sandy loam. *New Zealand Journal of Experimental Agriculture* **1**, 349-363.

McNaught, K.J.; During, C. (1970) Relations between nutrient concentrations in plant tissues and responses of white clover to fertilisers on a gley podzol near Westport. *New Zealand Journal of Research* **13**, 567-590.

McNeur, A.J. (1954) Effect of soil acidity on growth of pasture plants. II. A field trial with white clover (*Trifolium repens*). *New Zealand Journal of Science and Technology* **36A**, 167-176.

Martini, F.; Thellier, M. (1980) Use of an (np) nuclear reaction to study the long distance transport of boron in *Trifolium repens* after foliar application. *Planta* **150**, 197-205.

Masterson, C. (1977) The role of sulphur in legume nitrogen fixation and crop quality. *Irish Grassland and Animal Production Association Journal* **12**, 132.

Mengel, K.; Kirkby, E.A. (1978) Principles of plant nutrition. Worblaufen-Bern/Switzerland; International Potash Institute.

Miflin, B.J.; Lea, P.J. (1980) Ammonia assimilation. In: Miflin, B.J. (Ed) The Biochemistry of plants, Vol 5. New York, USA; Academic Press 169-202.

Millikan, C.R. (1958). Nutrient deficiency symptoms in lucerne and some pasture legumes. *Journal of the Department of Agriculture, Victoria* **56**, 511-531.

Mitchell, P. (1961) Coupling of phosphorylation to electron and hydrogen transfer by a chemiosmotic type of mechanism. *Nature, UK* **191**, 144-148.

Mitchell, P. (1966) Chemiosmotic coupling in oxidative and photosynthetic phosphorylation. Bodmin UK, Glynn Research.

Mouat, M.C.H. (1983) Competitive adaptation by plants to nutrient shortage through modification of root growth and surface charge. *New Zealand Journal of Agricultural Research* **26**, 327-332.

Mouat, M.C.H.; Walker, T.W. (1959a) Competition for nutrient between grasses and white clover. I. Effect of grass species and nitrogen supply. *Plant and Soil* **11**, 30-40.

Mouat, M.C.H.; Walker, T.W. (1959b) Competition for nutrient between grasses and white clover. II. Effect of root cation-exchange capacity and rate of emergence of associated species. *Plant and Soil* **11**, 41-52.

Munns, D.N.; Fox, R.L. (1977) Comparative lime requirements of tropical and temperate legumes. *Plant and Soil* **46**, 533-548.

Munns, D.N.; Fox, R.L.; Koch, B.L. (1977) Influence of lime on nitrogen fixation by tropical and temperate legumes. *Plant and Soil* **46**, 591-601.

Murphy, P.M.; Murphy, M.E. (1978) Relative energy requirements for dinitrogen and nitrate reduction in white clover. In: Research Report (Soils), Dublin, An Foras Taluntais, 62.

Mytton, L.R. (1976) The relative performance of white clover genotypes with rhizobial and mineral nitrogen in agar culture and in soil. *Annals of Applied Biology* **82**, 577-587.

Neales, T.F. (1956) Components of the total magnesium content within the leaves of white clover and perennial ryegrass. *Nature, UK* **177**, 388-389.

Neales, T.F. (1958) The movement of leaf magnesium during the growth of white clover plants. *Australian Journal of Biochemical Science* **11**, 343-348.

Nelson, L.E.; Brady, N.C. (1953a) Some greenhouse studies of certain interactions in Ladino clover using split root techniques. *Soil Science Society of America Proceedings* **17**, 274-278.

Nelson, L.E.; Brady, N.C. (1953b) Translocation of potassium through Ladino clover stolons. *Agronomy Journal* **45**, 513-514.

Nye, P.H. (1977) The rate-limiting step in plant nutrient absorption from soil. *Soil Science* **123**, 292-297.

Noggle, J.C. (1966) Ionic balance and growth of sixteen plant species. *Soil Science Society of America Proceedings* **30**, 763-766.

Nordmeyer, A.H.; Davis, M.R. (1976) Legumes in high country development. *Proceedings of the New Zealand Grassland Association* **38**, 119-125.

O,Toole, M.A.; Masterson, C.L. (1968) Interaction of calcium and hydrogen ion concentration on nodulation of white clover in peat. *Irish Journal of Agricultural Research* **7**, 129-131.

O,Toole, M.A.; Masterson, C.L. (1970) Effects of copper and nitrogen on nodulation and yield of Trifolium repens. In: Lowe, J (Ed) White Clover Research. Occasional Symposium, British Grassland Society, **No. 6**, 251-258.

Ouellette, G.J. (1963) Effects of lime, nitrogen and phosphorus on the response of Ladino clover to molybdenum. *Canadian Journal of Soil Science* **43**, 117-122.

Parks, C.L.; Perkins, H.F.; May, J.T. (1967) A greenhouse study of P and K requirements for ladino clover establishment on kaolin strip mine spoil. *Georgia Agricultural Research* **9**, 2.

Pate, J.S.; Gunning, B.E.S.; Briarty, L.G. (1969) Ultrastructure and functioning of the transport system of the leguminous root nodule. *Planta* **85**, 11-34.

Pearson, V.; Tinker, P.B. (1975) Measurement of phosphorus fluxes in the external hyphae of endomycorrhizas. In: Sanders, F.E.; Mosse, B.; Tinker, P.B. (Eds) Endomycorrhizas. London, UK; Academic Press, 277-287.

Poole, R.J. (1978) Energy coupling for membrane transport. *Annual Review of Plant Physiology* **29**, 437-460.

Powell, C.L. (1976) Mycorrhizal fungi stimulate clover growth in New Zealand hill country soils. *Nature, UK* **264**, 436-438.

Powell, C.L. (1977a) Mycorrhizas in hill country soils. I. Spore-bearing mycorrhizal fungi in thirty-seven soils. *New Zealand Journal of Agricultural Research* **20**, 53-58.

Powell, C.L. (1977b) Mycorrhizas in hill-country soils. II. Effect of several mycorrhizal fungi on clover growth in sterilised soils. *New Zealand Journal of Agricultrural Research* **20**, 59-62.

Powell, C.L. (1977c) Mycorrhizas in hill country soils. III. Effect of inoculation on clover growth in unsterile soils. *New Zealand Journal of Agricultural Research* **20**, 345-348.

Powell, C.L. (1978) Inoculation of white clover and ryegrass seed with mycorrhizal fungi. *New Phytologist* **83**, 81-85.

Powell, C.L.; Daniel, J. (1978) Mycorrhizal fungi stimulate uptake of soluble and insoluble phosphate fertiliser from a phosphate-deficient soil. *New Phytologist* **80**, 251-358.

Powell, C.L.; Sithamparanathan, J. (1977) Mycorrhizas in hill country soils. IV. Infection rate in grass and legume species by indigenous mycorrhizal fungi under field conditions. *New Zealand Journal of Agricultural Research* **20**, 489-494.

Raven, J.A.; Smith, F.A. (1976) Nitrogen assimilation and transport in vascular land plants in relation to intracellular pH regulation. *New Phytologist* **76**, 415-431.

Rayment, G.E.; Verrall, K.A. (1980) Soil manganese tests and the comparative tolerance of kikuyu and white clover to manganese toxicity. *Tropical Grasslands* **14**, 105-114.

Rendig, V.V.; Martin, W.E.; Smith, F.F. (1950) Pasture forage - effect of fertilizers on composition of forage on phosphorus-deficient soils. *California Agriculture* **4**, 6.

Robinson, R.R. (1942) The mineral content of various clones of white clover when grown in different soils. *Journal of the American Society of Agronomy* **34**, 933-939.

Robson, A.D. (1978) Mineral nutrients limiting nitrogen fixation in legumes In: Andrew, C.S.; Kamprath, E.J. (Eds) Mineral nutrition of legumes in tropical and subtropical soils, Melbourne, Australia; CSIRO, 277-293.

Robson, A.D.; O'Hara, G.W.; Abbott, L.K. (1981) Involvement of phosphorus in nitrogen fixation by subterranean clover (*Trifolium subterraneum* L.). *Australian Journal of Plant Physiology* **8**, 427-436.

Rolt, W.F. (1968) Some effects of lime and molybdenum on the growth of white clover in Autea clay. *New Zealand Journal of Agricultural Research* **11**, 193-205.

Roughan, P.G.; Batt, R.D. (1969) The glycerolipid composition of leaves. *Phytochemistry* **8**, 363-369.

Russell, J.S. (1966) Plant growth on a low calcium status solodic soil in a subtropical environment. I. Legume species, calcium carbonate, zinc, and other minor element interactions. *Australian Journal of Agricultural Research* **17**, 673-686.

Russell, J.S. (1978) Soil factors affecting the growth of legumes on low fertility soils in the tropics and subtropics. In: Andrew, C.S.; Kamprath, E.J. (Eds) Mineral nutrition of legumes in tropical and subtropical soils, Melbourne, Australia; CSIRO, 75-92.

Ryle, G.J.A.; Powell, C.E.; Gordon, A.J. (1979) The respiratory costs of nitrogen fixation in soyabean, cowpea, and white clover. *Journal of Experimental Botany* **30**, 145-153.

Ryle, G.J.A.; Powell, C.E.; Gordon, A.J. (1981) Assimilate partitioning in red and white clover either dependent on N_2 fixation in root nodules or utilizing nitrate nitrogen. *Annals of Botany* **47**, 515-523.

Schultz, I.; Turner, M.A.; Cooke, J.G. (1979) Effects of fertiliser potassium:sodium ratio on yield and concentration of sodium, potassium and magnesium in lucerne and white clover. *New Zealand Journal of Agricultural Research* **22**, 303-318.

Scott, B.I.H.; Gulline, H.F.; Robinson, G.R. (1977) Circadian electrochemical changes in the pulvinules of *Trifolium repens* L. *Australian Journal of Plant Physiology* **4**, 193-206.

Scott, R.S. (1975) Nutrient efficiency of some grasses and legumes in relation to environmental stresses. *PhD Thesis, Massey University, Palmerston North*

Sears, P.D. (1960) Grass/clover relationships in New Zealand. Proceedings of the Eighth International Grassland Congress, 130-133.

Sherrell, C.G. (1966) Boron deficiency in white clover (*Trifolium repens* L.) seedlings grown in an organic soil. *New Zealand Journal of Agricultural Research* **9**, 1025-1031.

Sherrell, C.G. (1978) A note on sodium concentrations in New Zealand pasture species. *New Zealand Journal of Experimental Agriculture* **6**, 189-190.

Shoop, G.J.; Brooks, C.R.; Blaser, R.E.; Thomas, G.W. (1961) Differential responses of grasses and legumes to liming and phosphorus fertilisation. *Agronomy Journal* **53**, 111-115.

Sleper, D.A. (1979) Plant breeding, selection and species in relation to grass tetany. *American Society of Agronomy, Special Publication* **No. 35**, 63-77.

Smith, G.S.; Middleton, K.R.; Edmonds, A.S. (1978) A classification of pasture and fodder plants according to their ability to translocate sodium from their roots into aerial parts. *New Zealand Journal of Experimental Agriculture* **6**, 183-188.

Snaydon, R.W. (1962a) Micro-distribution of *Trifolium repens* L. and its relation to soil factors. *Journal of Ecology* **50**, 133-143.

Snaydon, R.W. (1962b) The growth and competitive ability of contrasting natural populations of *Trifolium repens* L. on calcareous and acid soils. *Journal of Ecology* **50**, 439-447.

Snaydon, R.W. (1962c) The response of certain *Trifolium* spp. to calcium in sand culture. *Plant and Soil* **16**, 381-388.

Snaydon, R.W. (1971) An analysis of competition between plants of Trifolium repens L. populations collected from contrasting soils. *Journal of Applied Ecology* **8**, 687-697.

Snaydon, R.W.; Bradshaw, A.D. (1962a) Differences between natural populations of *Trifolium repens* L. in response to mineral nutrients. I. Phosphate. *Journal of Experimental Botany* **13**, 422-434.

Snaydon, R.W.; Bradshaw, A.D. (1962b) The performance and survival of contrasting natural populations of white clover when planted into an upland Festuca/Agrostis sward. *Journal of the British Grassland Society* **2**, 113-118.

Snaydon, R.W.; Bradshaw, A.D. (1969) Differences between natural populations of *Trifolium repens* L. in response to mineral nutrients. II. Calcium meagnesium and potassium. *Journal of Applied Ecology* **6**, 185-202.

Soper, K. (1959) Root anatomy of grasses and clovers. *New Zealand Journal of Agricultural Research* **2**, 329-341.

Sparling, G.P.; Tinker, P.B. (1978) Mycorrhial infection in Pennine grassland. III. Effects of mycorrhizal infection on the growth of white clover. *Journal of Applied Ecology* **15**, 959-964.

Spencer, K. (1959) Growth and chemical composition of white clover as affected by sulphur supply. *Australian Journal of Agricultural Research* **10**, 500-509.

Spencer, K.; Govaars, A.G.; Hely, F.W. (1980) Early phosphorus nutrition of eight forms of two clover species, *Trifolium ambiguum* and *Trifolium repens. New Zealand Journal of Agricultural Research* **23**, 457-475.

Troung, N.V.; Wilson, G.L.; Andrew, C.S. (1971a) Manganese toxicity in pasture legumes. I. Effects of calcium and phosphorus levels in the substrate. *Plant and Soil* **34**, 309-330.

Troung, N.V.; Andrew, C.S.; Wilson, G.L. (1971b) Manganese toxicity in pasture legumes. II. Effects of pH and molybdenum levels in the substrate. *Plant and Soil* **34**, 547-560.

Thomas, T.A.; Evans, D.R. (1977) Mineral and trace element content of legumes. *Annual Report, Welsh Plant Breeding Station* 111-112.

Ueno, M.; Yoshihara, K.; Hidaka, M. (1967) Studies on the roots of herbage legumes. 3. Estimation of the role of the tap root in ladino clover. *Journal of the Japanese Society of Grassland Science* **13**, 100-104.

Ulrich, A. (1945) Critical phosphorus and potassium levels in ladino clover. *Soil Science Society of America Proceedings* **10**, 150-161.

Vose, P.B.; Jones, D.G. (1963) The interaction of manganese and calcium on nodulation and growth in varieties of *Trifolium repens. Plant and Soil* **18**, 372-385.

Walker, G.D. (1979) Occurrence of manganese-rich inclusions in the root cortex of white clover. *New Zealand Journal of Botany* **17**, 1-3.

Walker, T.W.; Adams, A.F.R.; Orchiston, H.D. (1955a) The effects and interactions of sulphur, phosphorus, and molybdenum on the growth and composition of clovers. *New Zealand Journal of Science and Technology* **36A**, 470-482.

Walker, T.W.; Adams, A.F.R.; Orchiston, H.D. (1955b) The effects and interactions of molybdenum, lime and phosphate treatments on the yields and composition of white clover, grown on acid, molybdenum responsive soils. *Plant and Soil* **6**, 201-220.

Walker, T.W.; Adams, A.F.R. (1958) Competition for sulphur in a grass-clover association. *Plant and Soil* **9**, 353-366.

Wallace, T. (1951) The diagnosis of mineral deficiencies in plants by visual symptoms. London, UK; His Majesty's Stationery Office.

Wallace, W.; Pate, J.S. (1967) Nitrate assimilation in higher plants with specieal reference to the cocklebur (*Xanthium pennsylvanicum* Wallr.) *Annals of Botany* **31**, 213-228.

Weste, J.E.; Rowan, K.S.; Chalmers, D.J. (1974) The distribution of phosphorus containing compounds in tomato plants during the development of phosphorus deficiency. In: Bieleski, R.L.; Ferguson, A.R.; Cresswell, M.M. (Eds) Mechanisms of regulation of plant growth, Wellington, New Zealand; The Royal Society of New Zealand, Bulletin, **12**, 151-157.

Whitehead, D.C.; Jones, E.C. (1969) Nutrient elements in the herbage of white clover, lucerne and sainfoin. *Journal of the Science of Food and Agriculture* **20**, 584-591.

Whitehead, D.C.; Jones, L.H.P. (1972) The effect of replacing potassium by sodium on cation uptake and transport to the shoots in four legumes and Italian ryegrass. *Annals of Applied Biology* **17**, 81-89.

Widdowson, J.P.; Walker, T.W. (1971) Effects of lime and molybdenum on the yield and composition of white clover in a sequence of New Zealand zonal soils. *New Zealand Journal of Agricultural Research* **14**, 801-820.

Wilkinson, S.R.; Gross, C.F. (1965) Responses of Ladino clover to phosphorus withdrawal from the nutrient solution. *Agronomy Journal* **57**, 357-360.

Wilkinson, S.R.; Gross, C.F. (1967) Macro and micronutrient distribution with Ladino clover (*Trifolium repens* L.) *Agronomy Journal* **59**, 372-374.

Williams, W.M.; Lambert, M.G.; Caradus, J.R. (1982) Performance of a hill country white clover selection. *Proceedings of the New Zealand Grassland Association* **43**, 188-195.

Wilson, E.J.; Chisholm, R.H.; Blair, G.J. (1978) Phosphorus efficiency in legumes. In: Plant Nutrition: Proceedings of the 8th International Colloquium on Plant Analysis and Fertiliser Problems, Wellington. *DSIR Information Series,* **No. 134**, 575-576.

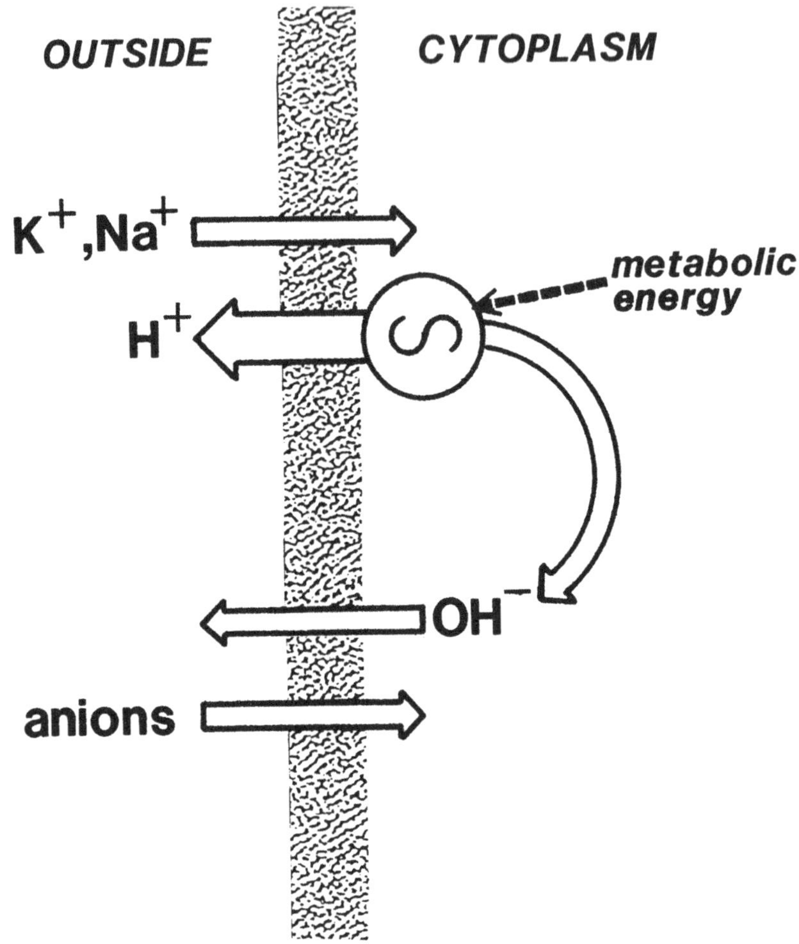

Fig. 5.1 A model to illustrate the coupling of nutrient absorption to the movement of hydrogen ions (after Smith & Raven, 1976).

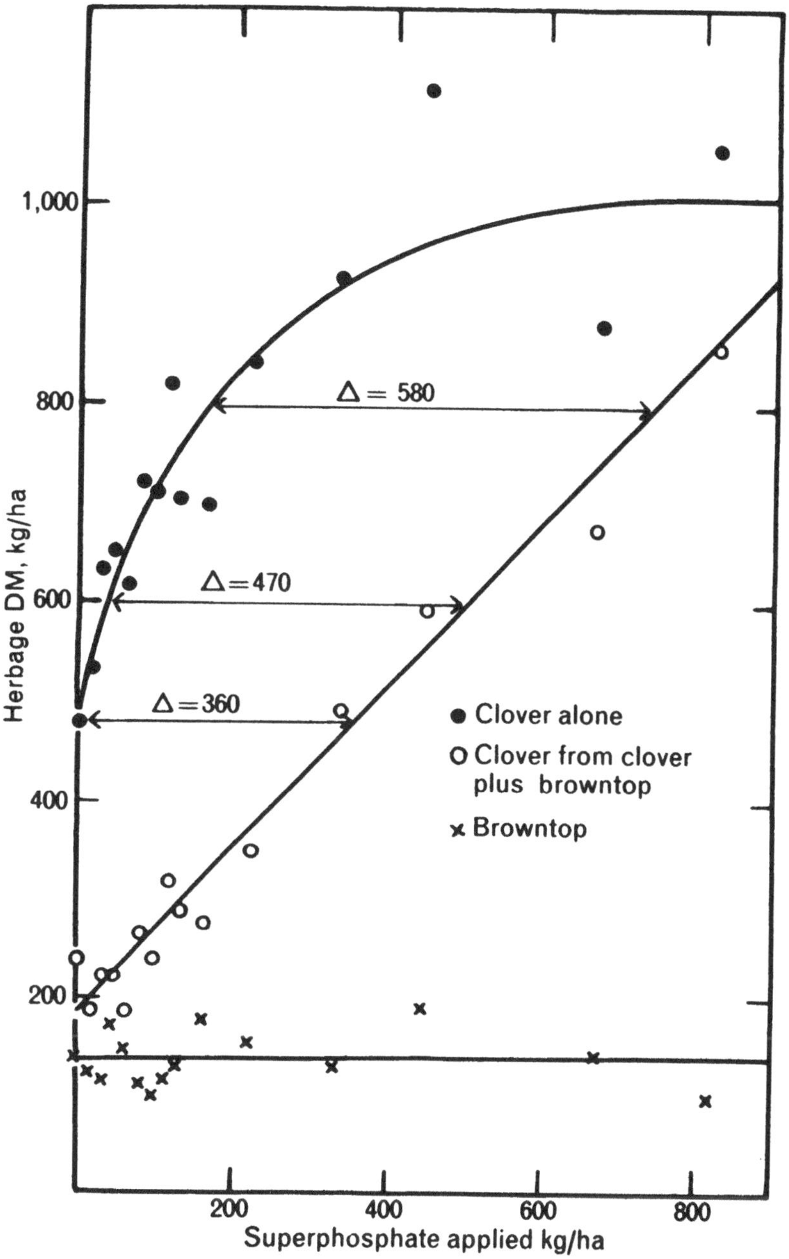

Fig. 5.2 The effect of browntop (*Agrostis capillaris*) on the phosphorus requirements of white clover (from Jackman & Mouat, 1970). Δ =additional superphosphate in kg/ha required by white clover plants growing in competition with browntop to produce the same dry matter yields as white clover growing without competition.

Nitrogen fixation

J. R. Crush
Grasslands Division, DSIR, Palmerston North, New Zealand

Contents

1. Introduction

Despite the importance of white clover nitrogen fixation to temperate agriculture, most research on the development and physiology of the legume nitrogen fixation symbiosis has been on other legumes. However, many of the important processes in infection, nodule development and nodule physiology probably vary only in detail among species.

This chapter discusses research on nitrogen fixation by white clover, with minor references to other species. No attempt has been made to deal critically with *Rhizobium* taxonomy or the biochemistry of nitrogen fixation.

2. Root infection and nodule development

Four distinct stages of infection of *Trifolium* species, including white clover, by *Rhizobium trifolii* have been described by Nutman (1959a). These processes, which are probably common to most legumes, have been reviewed by Bauer (1981). They start with proliferation of rhizobia in the rhizosphere and root hair curling caused by bacterial secretion of two factors (a nucleic acid and a protein or polysaccharide) able to cause root hair deformation (Solheim & Raa 1973). These diffusable factors are apparently absent in non-invasive mutants (Yao & Vincent 1976). Initiation of the infection thread occurs by an invagination process and an open pore is formed by redirection of root hair wall growth (Napoli & Hubbell 1975). Growth of the infection thread remains extracellular within the root hair.

The number of root hairs infected in white clover normally exceeds the number of nodules formed, and development of a lateral meristem in a nodule or lateral root usually inhibits further infection in the vicinity of the meristem (Nutman 1962).

Penetration of a polyploid root cortical cell by an infection thread stimulates the cell into meristematic activity and this, with nearby cells, constitutes the initial nodule primordium (Nutman 1958). White clover nodules are characteristically small (1.5 × 3 mm), pinkish-white, club-shaped to ellipsoidal, but occasionally are larger and fan-shaped. They have an apical meristem capable of enlarging the nodule as nodule cells degenerate progressively from the base. Individual polyploid cells of the nodule are infected by extensive proliferation of the infection thread, which releases bacteria into the cells. Here they multiply rapidly and change into the bacteroid form, each of which is surrounded in white clover by a membrane derived from the cell plasmalemma.

White clover is intermediate among species in rate of root hair infection. Of 12 *Trifolium* species tested by Nutman (1962), 6 had more root hairs infected 11 days after inoculation than did white clover. White clover also showed most variation in time taken to nodulate (12-34 days). Jones (1962) found that 2% of S100 Nomark white clover plants had formed their first nodule within 4 days of inoculation, over 50% by the 7th day and a small percentage required up to 15 days. At each stage from initial infection of root hairs to senescence and degeneration of the nodule, the process can be aborted by a number of factors from different sources of incompatibility and ineffectiveness in the *Trifolium* symbiosis (Nutman 1959b, Mosse 1964).

3. Nodule effectiveness

Both infective and non-infective *Rhizobium* strains may be present in the rhizosphere of white clover. Trifoliin, a lectin produced by white clover root hairs, specifically agglutinates *R. trifolii* and binds to the surface of encapsulated *R. trifolii*. This has been proposed as a mechanism whereby *R. trifolii* strains are selected by white clover (Dazzo & Hubbell 1975, Dazzo & Brill 1977, Dazzo et al. 1978). Wolpert & Albersheim (1976) have shown that legume lectins interact specifically with lipopolysaccharides of their symbionts, although Law & Strijdom (1977) could find no support for this role of lectins with *R. phaseoli* or

R. japonicum. Robertson et al. (1981) concluded that caution is needed in assessing the significance of lectins.

Rhizobium strains capable of nodulating white clovers are widespread in soils, reflecting the environmental tolerance of the host plant. The nodules formed can vary in their effectiveness in fixing nitrogen, depending on the strain of *Rhizobium* involved (Jones 1963). When white clover is exposed to a soil *Rhizobium* population of mixed effectiveness, it tends to select strains effective with itself (Masterson & Sherwood 1974, Brockwell & Katznelson 1976). Host selection for effective strains of *R. trifolii* has also been demonstrated by Robinson (1969a) but Jones & Russell (1972) found that although individual white clovers showed strain preferences, the selection was not necessarily consistent - some plants selected inferior strains. Russell & Jones (1975a) found that of three white clovers, only the S100 variety showed a selection response to a mixed inoculum. There was considerable heterogeneity within any clover variety for selection of particular strains, not surprising in an outbreeding species like white clover.

The selection of *Rhizobium* strains by white clover can be modified by external factors operating on the proportion of different strains in the rhizosphere (Russell & Jones 1975b). Some of these factors are discussed later in relation to soil *Rhizobium* populations. Presumably, selection of the inferior *Rhizobium* strains by white clover would result in elimination of these host genotypes and strains from the pasture under competition from more effective combinations (Robinson 1969b), although Mytton & de Felice (1977) found that competitiveness and effectiveness of *R. trifolii* were not necessarily linked.

4. Biochemistry of nitrogen fixation

Nitrogen fixation consists essentially of the reduction of atmospheric dinitrogen gas to ammonia within the nodule, by action of the nitrogenase enzyme system. This process has a high energy demand. Minchin & Pate (1973) calculated that 32% of photosynthetic carbon is diverted to the nodules where 5% is used in nodule growth, 12% in nodule respiration and 15% is incorporated into amino acid compounds which are translocated to other plant organs. Silsbury (1977) estimated that the energy requirement for symbiotic nitrogen fixation is substantially greater than that required for assimilation of mineral nitrogen.

The biochemistry of symbiotic nitrogen fixation has been reviewed by Bergersen (1971a,b), Postgate (1971), Dilworth (1974), Burris (1974), Burns & Hardy (1975) and Brill (1977). Nitrogenase consists of two components of different molecular weights (220 000 and 55 000 respectively), each containing Fe and the larger molecule also containing Mo. Neither component alone will fix nitrogen. The system requires oxygen, but nitrogenase is inactivated by O_2 at other than low partial pressures. Leghaemoglobin functions to maintain a rapid flux of O_2 into the bacteroids but at low partial pressure (Bergersen et al. 1973). Nitrogen fixation is confined to the bacteroids.

Hydrogen evolution by white clover nodules is dealt with in Chapter 4.

5. Soil *Rhizobium* populations

The distribution of *Rhizobium* species and strains in soil is generally related to the density of suitable host legumes. Therefore, although *R. trifolii* populations can be high ($>$700 000/g soil) in fertile pasture soils (Greenwood 1961, 1964), effective strains may be absent from undeveloped soils of low fertility (Harris 1962, Blair 1967).

R. trifolii is moderately tolerant of low soil pH. For example, lowering soil pH below 6 had little effect on *R. trifolii* numbers or nodulation, but had a major adverse effect on *R. meliloti* (Rice et al. 1977). A further decrease to pH 5 had little effect on *R. trifolii*. Repeated culture of three *R. trifolii* strains at pH 3.5 and 4 resulted in poor growth but the cells were not killed (van Schreven 1972). After 12 subcultures over 300 days, the strains were tested for effectiveness at pH 6.5 and showed only minor evidence of a decrease in nodulation or effectiveness. Although *R. trifolii* is acid-tolerant, soil populations can increase rapidly following application of lime. Mulder & van Veem (1980) showed that in an acid soil with no detectable rhizobia sowing with clover increased the soil population to 7800/g of soil. However, a combination of sowing and liming increased *Rhizobium* numbers to 280 000/g soil. Similar results have been noted during development of infertile soils in New Zealand. In studies by During et al. (1963), the soil *Rhizobium* population increased from <7/g to 7000/g by liming low fertility soils and Lowther (1974a) found that high rates of lime stimulated *R. trifolii* in tussock grassland soils.

R. trifolii is more tolerant of drought than either *R. meliloti* or a *Rhizobium* nodulating *Lotus corniculatus* (Foulds 1971). After 115 days' drought, *R. trifolii* cell numbers in soil had decreased from an initial 10^4/g to 10^3/g, *R meliloti* from 10^3 to 10/g, and 'Lotus rhizobia' from 10^5 to 10/g. Worrall & Roughley (1976) found that moisture stress decreased infection and inhibited nodulation by *R. trifolii* without affecting *Rhizobium* numbers.

The maximum temperature for growth on agar of nine *R. trifolii* strains ranged from 31 to 38.4°C (Bowen & Kennedy 1959). In soil these rhizobia died rapidly at 40°C. High soil temperatures may therefore reduce soil populations of rhizobia in certain areas.

Chemicals released by higher plants or by micro-organisms growing in association with higher plants may be toxic to rhizobia in soil. Parle (1964) demonstrated that the poor nodulation and consequent failure of white clover oversown into some *Notodanthonia*-dominated hill country swards was due to suppression of soil rhizobia populations by a toxin released by the plant or its associated micro-organisms. Complete killing of the *Notodanthonia* was necessary to remove the antibiotic effect. Isolation of an indigenous *Rhizobium* from the problem pasture showed that it was considerably less sensitive to the toxin than a standard commercial inoculant strain. It was not established whether the toxin was produced by the grasses or by some characteristic part of their root microflora, but Parle (loc. cit.) was unable to extract the toxin from *Danthonia* roots and Chatel & Parker (1972) found that cultures of many soil micro-organisms inhibited *R. trifolii.*

Patel (1974) isolated actinomycetes from pasture soils in which satisfactory nodulation of white clover had been difficult to achieve. In the 5 soils examined, 23-70% of the actinomycetes isolated showed some inhibitory activity on *R. trifolii*. However, of the 279 actinomycetes isolated, none inhibited all 12 strains of rhizobia tested.

Bateriophages of *R. trifolii* are isolated readily (Barnet 1972) but their ecological significance is questionable (Marshall 1956a). Schwinghamer et al. (1973) concluded that production of phage-like bacteriocin by some *R. trifolii* strains could give them an advantage in competition.

Agricultural chemicals may influence soil rhizobia populations either directly, or indirectly through effects on host growth. Harmful effects on *R. trifolii* in soil may not be apparent at herbicide concentrations sufficient to affect host growth (Brock 1972, Rolston et al. 1975), but chemicals appear to differ considerably in their toxicity to rhizobia (Tortenson 1975, Fisher 1976). Grossbard (1970) reviewed the effects of herbicides on the symbiotic relationships between *R. trifolii* and white clover and concluded that *Rhizobium* as a genus is relatively tolerant to herbicides. Strains of rhizobia resistant to pesticides and fungicides occur and can be isolated (Gillberg 1971, Odeyemi & Alexander 1977).

It has been shown frequently that if effective *Rhizobium* strains are rare or absent, inoculation of white clover seed with an effective strain is essential for clover establishment on nitrogen-deficient soils. In New Zealand where the indigenous rhizobia are ineffective on white clover (Greenwood & Pankhurst 1977), inoculation is a necessary prerequisite for rapid introduction of white clover into native grassland and scrubland (Sears et al. 1955, During et al. 1962, Harris 1962, Lowther & McDonald 1972, 1973). Inoculation of white clover seed sown into an infertile site at Te Anau increased clover density from 18 to 120 plants/m^2 11 months after sowing, and clover herbage yield by 370% (Cullen 1964). MacKinnon et al. (1977) surveyed legume inoculant usage in New Zealand and concluded that there would be continued demand for inoculants, especially in development of hill country and high country areas which have been devoid of pasture legumes for some years.

Because of the genetic instability of some *R. trifolii* strains, extreme care is required in maintaining cultures for preparation of inocula (Labandera & Vincent 1975). Mortality of *R. trifolii* on inoculated clover seed may occur because of an antibiotic factor in white clover seed diffusates (Vincent 1958, Hale 1976, Hale & Mathers 1977) or storage under adverse environmental conditions, e.g. warm temperatures (Singh & Sharma 1973). Problems of loss of viability of commercially inoculated seed (Thompson et al. 1975) have led to the establishment of official standards for inoculant preparation in Australia (Roughley

1970, 1972), New Zealand (Atkinson 1971) and the USA (Schenberger & Swope 1975). The principles of *Rhizobium* strain selection for use in inocula have been described by Thompson et al. (1974) and Date (1976).

The inoculum may be pelleted onto an inert particle (Fraser 1975) or directly onto the white clover seed coat and covered with a suitable protective coating. Lime coating of legume seed protects the rhizobia as well as having a direct beneficial effect on nodulation (Lowther 1977, Philpotts 1977). The small quantity of lime on pelleted seed can have effects on nodulation similar to those of large soil dressings of lime (Lowther 1974b, 1975). Trace elements required for nitrogen fixation, e.g. molybdenum, can also be pelleted on to legume seed (Graham et al. 1974), but care must be taken to neutralise any toxicity towards rhizobia. Alternatively, a liquid *Rhizobium* culture, either on its own or dispersed in a milled peat medium, is available and can be mixed with the seed directly before sowing. The general procedures involved in legume inoculation have been described by Callaghan (1958), Hastings et al. (1966), Vincent (1970), Roughley (1976), Strijdom & Deschodt (1976) and Burton (1976).

The inclusion of several *Rhizobium* strains in a single commercial inoculant increased the range of soils and climates in which the same inoculant can be used. However in such mixtures where *Rhizobium* growth can be considerable, one strain may eventually completely dominate the culture (Marshall 1956b), even though only mild inter-strain antagonism involving antibiotics, bacteriocides and phage has been observed in *R. trifolii* cultures (Schwinghamer 1971). Similar effects can occur with peat cultures (Marshall 1956b, Waters 1956), but seeding the peat at very high initial rates (10^9 rhizobia/g) prevents much subsequent growth and gives less scope for strain competition. However, the ratio of nodules formed by different strains from a mixed inoculum may not reflect the proportion of strains in the inoculum (Waters 1956), suggesting some selection by the host.

In soils with a high indigenous *Rhizobium* population (10^9/g) the inoculum rate must be of the order of $10^3 \times$ the soil population to establish the inoculant strain in 50% of nodules (Weaver & Frederick 1974). Although *R. trifolii* introduced by seed inoculation into a site with an existing soil *Rhizobium* population could form most of the nodules initially, the influence of the introduced strain may have disappeared by the third season (Roughley et al. 1976). Strains ineffective on white clover will not persist beyond the first season (Brockwell et al. 1972).

At sites with few effective indigenous *R.trifolii*, inoculation allows the host to survive long enough for a mixed soil population to become established. From this the host then selects the most suitable strain for the particular environment (Roughley et al. 1976). Even in soils in which effective rhizobia cannot in the first instance be isolated, maintenance of a host legume by use of fertiliser nitrogen may stimulate multiplication of rhizobia and selection of an effective strain from the microflora (Brockwell et al. 1966a).

Although seed inoculation may have no permanent effects on the rhizobia population of a soil, it is critical to the initial establishment of white clover and the small capital and labour costs of seed inoculation are well justified. Subsequent host selection of *Rhizobium* strains better adapted to the local micro-environment should be regarded as a bonus rather than an alternative to inoculation.

6. Host genotype and *Rhizobium* strain interaction

The volume of nodule tissue developed by white clover can be increased by selection (Jones & Burrows 1968), largely through increase in nodule numbers. However, plant yield is not affected markedly, so that the efficiency of the symbiosis per nodule decreases. Working with S100 white clover, Mytton & Jones (1971) found no increase in plant yield after the first generation of selection. There is an inverse relationship between nodule numbers and nodule size in many effectively nodulated legumes. Nutman (1967) has concluded that plants with few nodules are more efficient than those with many, because abundantly nodulated plants divert a disproportionate amount of their resources into the meristems and cortical regions of their many nodules.

Selection for low seed weight within S100 white clover has resulted in slower nodulation and seedling growth in tube cultures compared with seed fractions of higher seed weight (Mytton 1973). However, these effects were not apparent in larger containers, suggesting that differences in nodule parameters between seed-weight groups in tube cultures may be more a consequence than a cause of plant growth. Culture conditions also affected the results of Connolly et al. (1969) who showed that for white clover plants in sterile test tubes, there was genetic variation in both rate of nodulation and seedling vigour. However, the results could not be confirmed in a non-sterile medium.

Comparison of white clover genotypes grown with *Rhizobium* or mineral nitrogen has shown (Mytton 1976) that performance of young seedlings on agar is not consistent with their subsequent performance in pots. However, vegetative propagules of mature plants did show reasonable agreement between the two culture systems. Under pot conditions, on average, reliance on symbiotically fixed nitrogen reduced yield by 24%. A significant genotype ×nitrogen source interaction suggested that white clover-*Rhizobium* associations vary in their efficiency of nitrogen fixation. Individual shortfalls between potential nitrogen demand and supply of fixed nitrogen were generally related to the growth potential of the plant, so that plants with high potential yields were limited most severely when forced to rely on symbiotic nitrogen fixation. Mytton (loc. cit.) concluded that breeding for increased clover yield by control of plant genotype alone seems likely to increase yield potentials to levels above those which can be supported by natural *Rhizobium* populations, unless plant and *Rhizobium* genotypes can be selected simultaneously. From a comparison of cv. Grasslands Huia and Grasslands 4700 white clovers, Brock & Hoglund (1974) concluded that Grasslands 4700 had a higher growth potential than Grasslands Huia, especially at lower temperatures, but that its potential for symbiotic nitrogen fixation was inadequate and nitrogen supply limited its growth.

When four white clover cultivars (S100, Sabeda, Katrina, Blanca) were inoculated with isolates from the nodules of each *Rhizobium*, strain effects had little influence on plant yield, but there was a highly significant host genotype × *Rhizobium* strain interaction (Mytton 1975). On average a more productive symbiosis resulted when plants were inoculated with isolates from their own nodules than from unrelated plants. This provides further evidence that white clovers will select from heterogeneous *Rhizobium* populations those strains with which they form an effective symbiosis.

7. Effect of environment on nitrogen fixation

The effects of environmental stress on nodulation and nitrogen fixation by legumes have been reviewed by Gibson (1971, 1976), Nutman (1972) and Pate (1976), but much of this work deals with legumes other than white clover. Environmental factors can influence infection, nodulation, nitrogen fixation processes and host growth and nitrogen demand.

(A) Soil pH

White clover is more tolerant of low pH than lucerne but more sensitive than some tropical legumes (Andrews 1976). Optimum soil pH values of 5.3-5.5 for white clover have been suggested by Spurway (1941) and McNeur (1954). Nodulation of white clover seems more sensitive to low pH than is host growth. When white clover was grown with combined nitrogen, growth was reduced by only 20% between pH 6 and pH 4, provided calcium was sufficient (Andrews 1976). When the plants were nodulated, yield was reduced by 55% over the same pH range. Nodulation declined rapidly below pH 5, falling close to a minimum at pH 4.5. The beneficial effects on nodulation of lime-coating inoculated clover seed have already been noted.

White clover and rhizobia from acid soils seem to be adapted to low pH (Holding & King 1963, Lowe & Holding 1970, Masterson & Sherwood 1974). Therefore, an appropriate strain of white clover must be used when assessing the symbiotic effectiveness of soil *Rhizobium* populations (Norris 1965, Sherwood & Masterson 1974). Jones (1963) and Holding & King (1963) found that many *Rhizobium* isolates from soils of pH below 5.5 were inferior in symbiotic effectiveness and suggested that future sowings of commercial white clover might benefit from inoculation.

Most *R.trifolii* strains produce acid in culture and Norris (1956, 1965) considers that acid production is characteristic of rhizobia from advanced legumes, which seem to show a preference for alkaline soils. Acid production was assumed to confer some competitive advantage on rhizobia in the rhizosphere. Brockwell et al. (1966b) examined 94 *R. trifolii* isolates from several sites and found they were all acid producers, irrespective of original soil pH. However, isolates from alkaline soils produced most acid. Identical results were obtained by Jones & Burrows (1969) with 90 *R. trifolii* isolates from 36 sites in Wales. In this trial, final pH of the culture and symbiotic effectiveness on S184 white clover were correlated significantly, prolific acid producers being least effective. However, the results of Sherwood & Masterson (1974) suggest that the relationship between acid production and symbiotic effectiveness may have been different if a range of white clover hosts had been tested.

(B) Light and defoliation

Both shading and defoliation of white clover reduced the number and weight of nodules per plant (Mitchell 1956, Butler et al. 1959, Chu & Robertson 1974). Severe defoliation or shading markedly reduced nitrogen fixation activity, but recovery is noticeable after a few days (Moustafa et al. 1969, Halliday & Pate 1976) and may be complete after 10 days (Chu & Robertson 1974). Yoshida & Yatazawa (1977) showed that the rate of recovery of nitrogen fixation after defoliation is related inversely to the severity of the defoliation. Ladino white clover cut at 6 cm reached maximum nitrogen fixation rates after 16 days' regrowth, whereas plants cut to 1 cm showed little activity after 5 days and took 26 days to recover fully. This suggests that nitrogen fixation will be depressed in closely grazed pastures, particularly if insufficient time is allowed between grazings for nitrogen fixation to reach maximum rates.

Defoliation effects on nitrogen fixation are often interpreted in terms of reduced carbohydrate supply to nodules. Hoglund & Brock (1978) consider that in grazed pasture, increased soil mineral nitrogen from urine, dung and litter, leading to increased uptake of mineral nitrogen by white clover and/or reduced host demand for nitrogen after defoliation, may be the limiting factors. Host demand for nitrogen could be reduced as a result of defoliation because of removal of active meristems, carbohydrate stress on the host, or an internal excess of nitrogen.

(C) Temperature

Incubation of white clover nodules at various temperatures has shown the same broad optimum of 13-26°C in nitrogenase response, with a sharp decline at either extreme (Halliday & Pate 1976), as has been recorded for other legumes (Hardy et al. 1968, Gibson 1971, Pankhurst & Sprent 1976). Thus low soil temperatures probably play a major role in limiting nitrogen fixation by white clover in temperate pastures during spring (Halliday & Pate 1976), although nodules of Ladino white clover will overwinter even at low temperatures and resume growth in very early spring before being replaced by new nodules (Bergersen et al. 1963).

R. trifolii strains differ in sensitivity to temperature extremes. Of four *R. trifolii* strains which nodulated *T. subterraneum* satisfactorily at 22°C, only one formed effective nodules at 30°C (Pankhurst & Gibson 1973). When plants nodulated with a strain sensitive to high temperature were transferred from 22°C to 30°C, bacteroid tissue in the nodules broke down rapidly, whereas nodules formed by a strain tolerant to high temperature were less affected. Northern Scandinavian strains of *R. trifolii* grew better in pure culture at 10°C than strains isolated from more southern soils (Ek-Jander & Fahraeus 1971), but at 20°C there was no detectable difference between the two geographic groups. At 10°C, white clover inoculated with northern *Rhizobium* strains nodulated faster and showed more nitrogenase activity than plants inoculated with strains of southern origin.

(D) Soil nutrient status

Most nutrients affect nodulation and nitrogen fixation indirectly through host growth and demand for nitrogen. The exceptions are minerals such as molybdenum, cobalt, calcium, iron and copper, which are probably all involved directly in nodulation or nitrogen fixation processes (Anderson 1956, Evans & Russell 1971).

Nutman (1962) found that low levels of nitrate (10 and 25 μg N/plant) in the rooting medium delayed initial nodulation by up to 4.8 days compared with nitrogen-free controls. However, after 17 days the number of infections was significantly greater in the nitrate treatments. The increase was assumed to result from the effect of nitrate in prolonging the initial phase of the infection process and suppressing the host plant's ability to inhibit infection.

In mature white clover, mineral nitrogen interacts directly with nitrogen fixation, usually decreasing it in proportion to the amount applied (McAuliffe et al. 1958, Allos & Bartholomew 1959, Lineham & Lowe 1960, Moustafa et al. 1969, Crush et al. 1982).

On a nitrogen-deficient peat soil where white clover showed a linear yield response to fertiliser nitrogen, low rates of fertiliser nitrogen (6, 17, 50 kg N/ha) stimulated nodulation (O'Toole & Masterson 1970), but at 150 kg N/ha, nodulation was reduced. Young (1958) showed that 40 kg N/ha applied to a mixed white clover-grass sward favoured nodulation; higher rates (90-112 kg N/ha) had no effect, but 224 kg markedly reduced the number of large nodules/unit root weight.

(E) Water stress

Engin & Sprent (1973) investigated the effects of the imposition of, and recovery from, water stress in the white clover cultivar S100 Nomark. Water stress depressed nitrogen fixation per plant and per unit weight of nodule. Provided the stress was not too severe, recovery was complete, but longer periods of stress reduced nodule growth and nitrogen fixing efficiency of the nodule tissue. The threshold limit for recovery from stress is a loss of about 20% in nodule fresh weight for soybeans and faba beans (Sprent 1971, 1972) but no data are available for white clover. Meristematic activity in nodules was resumed after watering and acetylene reduction activity increased. Because of its meristematic nodules, white clover recovers faster after water stress than such legumes as soybeans, which have spherical nodules of limited size. However, Foulds (1978) demonstrated that white clover nodules are more sensitive to desiccation than those of *Lotus*, and white clover recovers its nitrogen fixing ability after drought more slowly than subterranean clover (Aparicio-Tejo et al. 1980).

In moist soil 70% of nitrogen fixation activity by white clover occurs in the top 75 mm of soil (Hoglund & Brock 1978, Rumball 1979, Brown et al. 1979, O'Connor et al. 1979, Clark et al. 1979, Brock & Hoglund 1979). As available moisture in the surface soil decreases, nitrogen fixation activity migrates to deeper levels in the soil. As the activity of nitrogen fixation at deeper soil levels is generally low, it is probably mainly only of survival value (Hoglund & Brock 1978). Reduced nitrogen fixation under waterlogged conditions has been attributed to suboptimal oxygen concentrations in the nodules (Minchin & Pate 1975).

8. Nitrogen fixation in pastures

Seasonal patterns of nitrogen fixation by white clover follow fairly closely the clover growth curve. In temperate pastures the limiting factors are low winter temperatures and summer drought. During the growing season high mineralisation rates may reduce fixation per unit of clover grown (e.g. Clark et al. 1979), but this effect may be offset partially by vigorous grass growth. Diurnal patterns of nitrogen fixation have been discussed in Chapter 4.

Annual values for white clover nitrogen fixation recorded in New Zealand and elsewhere are extremely variable (Tables 6.1 and 6.2) and largely reflect differences in clover yields and availability of mineral nitrogen.

The length of the growing season determines potential clover yields, but actual yields are influenced strongly by soil fertility and sward management. In infertile hill pasture, clover yield is limited by competition for scarce nutrients from grasses which tolerate low fertility. However, although annual nitrogen fixation may be as low as 17 kg N/ha, fixation per unit of clover grown is high (Grant & Lambert 1979).

When clovers are sown into a nitrogen-deficient soil, such as after a cropping rotation, competition from grasses is minimum and clovers dominate the sward. Under these circumstances high clover yields and high fixation efficiency combine to give high annual nitrogen totals. For example, in mown plots on a subsoil with clippings removed, Sears et al. (1965) recorded 680 kg N fixed/year from a sward with 56% clover in the herbage. Return of most of the clippings resulted in only 38% clover in the herbage and 410 kg N fixed annually. Nitrogen fixation per unit clover yield dropped from 96 to 73 kg N/t DM.

As pastures develop, soil nitrogen accumulates and grasses eventually dominate (Sears 1962). Consequently, in an established high fertility pasture, clover yields are lower than immediately after sowing. Nitrogen fixation per unit clover grown also declines as clovers make more use of available mineral nitrogen. These two factors mean that annual nitrogen fixation is often less in a developed than in a developing pasture (Hoglund et al. 1979).

Table 6.1 Grass yield, white clover yield, annual nitrogen fixation and fixation per tonne measured clover DM recorded in New Zealand.

Source and System		Grass DM kg/ha/year	Clover DM kg/ha/year	N fixation kg N/ha/year	Fixation efficiency kg N/t DM
Sears (1953), Sears & Evans (1953) Grazed or cut mixed swards, no return of clippings or excreta		3740	6890[1]	620	89
Sears et al. (1965) mown plots					
1. Pure grass, clippings removed		1450	—	15[2]	—
2. White clover clippings removed		—	9420	660	70
5. Grass + clover clippings removed		5580	7060	680	96
7. Grass + clover, 80% of clippings returned		8840	5560	410	73
Brock (1973) mown plots, clippings removed					
White clover, low P fertiliser		—	6480	400	62
White clover, high P fertiliser		—	10 630	570	54
Edmeades & Goh (1978) Farm pastures of different ages	>20 years	4980	1370	45	32.9
	15	4820	2980	101	33.9
	6	2880	2500	88	35.1
	2	5780	4080	142	34.7
Rumball (1979) Grazed pasture		7780	3680	380	103
Brown et al. (1979) Grazed pasture		12 700	1990	110	54
O'Connor et al. (1979) Grazed pasture		5380	1900	160	82
Clark et al. (1979) Grazed pasture		10 700	3030	230	75
Brock & Hoglund (1979) Grazed pasture		8450	4250	210	49
Crouchley (1979) Grazed pasture		9140	1700	160	97
Crush (1979) Grazed pasture		6190	3720	120	34
		9700	4840	190	39
Carran (1979) Grazed pasture		8170	2820	280	99
Grant & Lambert (1979) Grazed pasture		7870	150	17	112
Crush et al. (1982) Grazed pasture		18 000	2200	97	44
		13 350	3000	106	35
		19 000	1200	60	50
		13 500	2800	83	30
		19 200	2100	63	30
		14 800	2500	68	27
Crush et al. (1983) Grazed pasture		13 200	3700	105	41
		12 900	2200	85	39
		12 300	2000	76	38

(1) Clover herbage contained 22% red clover.

(2) Estimated non-symbiotic fixation subtracted by Sears et al. from other values.

Table 6.2 Annual nitrogen fixation by white clover recorded in the UK, the Republic of Ireland, the USA and Australia.

Source	System	Nitrogen fixation kg N/ha/yr.
Halliday & Pate 1976	Pasture	268
Haystead & Lowe 1977	Hill pasture	100
Karraker et al. 1950	Lysimeters	75
Masterson & Murphy 1976	Grazed pasture	83-296
Munro & Hughes 1966	Hill pasture	55
Munro & Davies 1974	Mown hill pasture	81-100
	Grazed hill pasture	90-98
Simpson 1976	Mown plots	474

At any stage of pasture development, stocking rate has an important influence on clover yields. Suckling (1975) showed that at low stocking rates in hill country, rank grass growth suppressed clover. Increasing the stocking rate, especially with cattle, improved clover yields. Alternatively grazing rough hill pasture with goats suppresses weeds and can lead to clover dominance (Lambert et al. 1981).

Annual nitrogen fixation values have only short-term significance on any site. Crouchley (1979) reported 90 and 240 kg N fixed in consecutive years of a grazing trial where management was uniform and soil nitrogen did not change measurably. Summer rain in the second year doubled clover yields. Values for nitrogen fixation per unit clover yield also require careful interpretation. Variation in the ratio of harvested to unharvested material will have a profound effect on this parameter. The proportion of clover harvested varies with root/shoot ratio, the erectness of the clover plant and cutting height. Operator bias in cutting height is also likely to be a major source of variation. However, within an experiment this parameter is useful, especially for expressing seasonal variation in the use of soil mineral nitrogen by clover, e.g. Brock & Hoglund (1979).

Despite the potential for high nitrogen inputs from white clover in pasture, nitrogen deficiency sets the limits to herbage production. It does so because clovers will persist only in a mixed sward under such condition of nitrogen deficiency as limit growth of the taller grasses and admit light to the clover (Ball 1969). There is no evidence that nitrogen fixation processes *per se* limit nitrogen inputs by white clover in pasture. Rather, it is the interaction of host growth (demand for nitrogen) and soil mineral nitrogen availability that determines nitrogen fixation.

9. Conclusions

The robustness of the white clover-*Rhizobium* symbiosis is demonstrated amply by the wide use of the plant under different management systems in temperate agriculture. Rhizobia capable of nodulating white clover effectively seem to survive well in soils where the host plant grows and the nodulation processes are typical of many legumes. In soils where the rhizobia are initially sparse, inoculation and pelleting of clover seed can be vital for successful establishment of white clover seedlings. However, host selection of rhizobia adapted to the local environment may mean eventual replacement of the inoculant strain once white clover plants are established and the soil *Rhizobium* population has expanded and diversified.

In impoverished soils white clover, like other temperate legumes, grows little and consequently fixes little nitrogen. Correction of mineral deficiencies (typically P, S, Mo) leads to clover dominance of the nitrogen-deficient sward and very high rates of nitrogen fixation. As soil nitrogen accumulates, grasses dominate the sward and clover yields and fixation efficiency decline, reducing the total amount of nitrogen fixed.

In a mixed sward the grasses always suffer some degree of nitrogen deficiency, as the lower-growing clovers persist only under conditions of nitrogen deficiency which restrict the growth of taller grasses. Nitrogen deficiency therefore sets the ultimate limit to herbage production from grass/clover pastures unless fertiliser nitrogen is used.

References

Allos, H.F.; Bartholomew, W.V. (1959) Replacement of symbiotic fixation by available nitrogen. *Soil Science* **87**, 61-66.

Anderson, A.J. (1956) Molybdenum deficiencies in legumes in Australia. *Soil Science* **81**, 173-182.

Andrews, C.S. (1976) Effect of calcium, pH and nitrogen on the growth and chemical composition of some tropical and temperate pasture legumes,I. Nodulation and growth. *Australian Journal of Agricultural Research* **27**, 611-623.

Aparicio-Tejo, P.M.; Sánchez-Díaz, M.F.; Pena, J.I. (1980) Nitrogen fixation, stomatal response and transpiration in *Medicago sativa, Trifolium repens* and *T. subterraneum* under water stress and recovery. *Physiologia Plantarum* **48**, 1-4.

Atkinson, J.D. (1971) Legume inoculant service. *New Zealand Journal of Agriculture* **122**, 61.

Ball, Roger (1969) Legume and fertiliser nitrogen in New Zealand pastoral farming. *Proceedings of the New Zealand Grassland Association* **31**, 117-126.

Barnet, Y.M. (1972) Bacteriophages of *Rhizobium trifolii*. I. Morphology and host range. *Journal of General Virology* **15**, 1-5.

Bauer, W.D. (1981) Infection of legumes by rhizobia. *Annual Review of Plant Physiology* **32**, 407-449.

Bergersen, F.J. (1971a) The central reactions of nitrogen fixation. *Plant and Soil, Special Volume* 511-524.

Bergersen, F.J. (1971b) Biochemistry of symbiotic nitrogen fixation in legumes. *Annual Review of Plant Physiology* **22**, 121-140.

Bergersen, F.J.; Hely, F.W.; Costin, A.B. (1963) Overwintering of clover nodules in alpine conditions. *Australian Journal of Biological Science* **16**, 920-921.

Bergersen, F.J.; Turner, G.L.; Appleby, C.A. (1973) Studies on the physiological role of leghaemoglobin in soybean root nodules. *Biochimica et Biophysica Acta* **292**, 271-282.

Blair, I.D. (1967) Studies on *Rhizobium* strains. II. *Rhizobium trifolii*. *New Zealand Journal of Agricultural Research* **10**, 237-252.

Bowen, G.D.; Kennedy, M.M. (1959) Effect of high soil temperature on *Rhizobium* species. *Queensland Journal of Agricultural Science* **16**, 177-197.

Brill, W.J. (1977) Biological nitrogen fixation. *Scientific American* **236**, 68-74.

Brock, J.L. (1972) Effects of the herbicides trifluralin and carbetamide on nodulation and growth of legume seedlings. *Weed Research* **12**, 150-154.

Brock, J.L. (1973) Growth and nitrogen fixation of pure stands of three pasture legumes with high/low phosphate. *New Zealand Journal of Agricultural Research* **16**, 483-491.

Brock, J.L.; Hoglund, J.H. (1974) Growth of 'Grasslands Huia' and 'Grasslands 4700' white clovers. II. Effects of nitrogen and phosphorus. *New Zealand Journal of Agricultural Research* **17**, 47-53,

Brock, J.L.; Hoglund, J.H. (1979) Nitrogen fixation in pasture. VII. Recent alluvium, Palmerston North. *New Zealand Journal of Experimental Agriculture* **7**, 27-29.

Brockwell, J.; Hely, F.W.; Neal-Smith, C.A. (1966a) Some symbiotic characteristics of rhizobia responsible for spontaneous, effective field nodulation of *Lotus hispidus* Desf. *Australian Journal of Experimental Agriculture and Animal Husbandry* **6**, 365-370.

Brockwell, J.; Asuo, S.K.; Rea, G.A. (1966b) Acid production by rhizobia from the genera *Trifolium* and *Lotus*. *Journal of the Australian Institute of Agricultural Science* **32**, 295-297.

Brockwell, J.; Bryant, W.G.; Gault, R.R. (1972). Ecological studies of root-nodule bacteria introduced into field environments. III. Persistence of *Rhizobium trifolii* in association with white clover at high elevations. *Australian Journal of Experimental Agriculture and Animal Husbandry* **12**, 407-413.

Brockwell, J.; Katzelson, J. (1976) Symbiotic characteristics of *Rhizobium trifolii* from Israel in association with 10 species of *Trifolium. Australian Journal of Agricultural Research* **27**, 799-810.

Brown, N.S.; Gray, M.H.; Sinclair, D.P. (1979) Nitrogen fixation in pasture.III. Gisborne Plains, Manutuke. *New Zealand Journal of Experimental Agriculture* **7**, 11-13.

Burns, R.C.; Hardy, R.W.F. (1975) Nitrogen fixation in bacteria and higher plants. Berlin, W. Germany; Springer-Verlag.

Burris, R.H. (1974) Biological nitrogen fixation, 1924-1974. *Plant Physiology* **54**, 443-449.

Burton, J.C. (1976) Methods of inoculating seeds and their effect on survival of rhizobia. In: Nutman, P.S. (Ed) *Symbiotic nitrogen fixation in plants.* IBP Vol 7, Cambridge, UK. Cambridge University Press, 175-189.

Butler, G.W.; Greenwood, R.M.; Soper, K. (1959) Effects of shading and defoliation on the turnover of root and nodule tissue of plants of *Trifolium repens, T. pratense* and *Lotus uliginosus. New Zealand Journal of Agricultural Research* **2**, 415-426.

Callaghan, J.R. (1958) Methods of manufacture and control of legume inoculants prepared in New Zealand. *Proceedings of the New Zealand Grassland Association* **20**, 120-126.

Carran, R.A. (1979) Nitrogen fixation in pasture. Southland, Gore. *New Zeland Journal of Experimental Agriculture* **7**, 39-41.

Chatel, D.L.; Parker, C.A. (1972) Inhibition of rhizobia by toxic soil-water extracts. *Soil Biology and Biochemistry* **4**, 289-294.

Chu, A.C.P.; Robertson, A.G. (1974) The effects of shading and defoliation on nodulation and nitrogen fixation by white clover. *Plant and Soil* **41**, 509-519.

Clark, D.A.; Causley, D.C.; Brougham, R.W. (1979) Nitrogen fixation in pasture. VI. Manawatu Plains, Kairanga. *New Zealand Journal of Experimental Agriculture* **7**, 23-25.

Connolly, V.; Masterson, C.L.; Conniffe, D. (1969) Some genetic aspects of the symbiotic relationship between white clover (*Trifolium repens*) and *Rhizobium trifolii. Theoretical and Applied Genetics* **39**, 206-213.

Crouchley, G. (1979) Nitrogen fixation in pasture. VIII. Wairarapa plains dryland, Masterton. *New Zealand Journal of Experimental Agriculture* **7**, 31-33.

Crush, J.R. (1979) Nitrogen fixation in pasture. IX. Canterbury Plains, Kirwee. *New Zealand Journal of Experimental* Agriculture **7**, 35-38.

Crush, J.R.; Cosgrove, G.P.; Brougham, R.W. (1982) The effect of nitrogen fertiliser on clover nitrogen fixation in an intensively grazed Manawatu pasture. *New Zealand Journal of Experimental Agriculture* **10**, 395-399.

Crush J.R.; Cosgrove, G.P.; Brougham, R.W. (1983) Nitrogen fixation during 1979-81 in 2 pastures on the Manawatu Plains. *New Zealand Journal of Experimental Agriculture* **11**, 17-20.

Cullen, N.A. (1964) Pasture establishment problems at Te Anau. *Proceedings of the New Zealand Grassland Association* **26**, 109-114.

Date, R.A. (1976) Principles of *Rhizobium* strain selection. In: *Symbiotic nitrogen fixation in plants.* Nutman, P.S. (Ed). IBP Vol. 7. Cambridge, UK; Cambridge University Press, pp. 137-150.

Dazzo, F.B.; Hubbell, D.H. (1975) Cross-reactive antigens and lectins as determinants of symbiotic specificity in the *Rhizobium*-clover association. *Applied Microbiology* **30**, 1017-1033.

Dazzo, F.B.; Brill, W.J. (1977) Receptor site on clover and alfalfa roots for *Rhizobium. Applied and Environmental Microbiology* **33**, 132-136.

Dazzo, F.B.; Yanke, W.E.; Brill, W.J. (1978) Trifoliin: A *Rhizobium* recognition protein from white clover *Biochemica et Biophysica Acta* **539**, 276-286.

Dilworth, M.J. (1974) Dinitrogen fixation. *Annual Review of Plant Physiology* **25**, 81-114.

During, C.; Cullen, N.A.; Bennett, G.M. (1962). The establishment of pasture on yellow-brown loams near Te Anau. I. Factors influencing the establishment of clover-grass pasture on cultivated ground. *New Zealand Journal of Agricultural Research* **5**, 278-293.

During, C.; Cullen, N.A.; Mountier, N.S. (1963) The establishment of pasture on yellow-brown loams near Te Anau. II. Factors influencing the establishment of white clover (*Trifolium repens* L.) on uncultivated ground. *New Zealand Journal of Agricultural Research* **6**, 416-431.

Edmeades, D.C.; Goh, K.M. (1978) Symbiotic nitrogen fixation in a sequence of pastures of increasing age measured by a ^{15}N dilution technique. *New Zealand Journal of Agricultural Research* **21**, 623-628.

Ek-Jander, J.; Fahraeus, G. (1971) Adaptation of *Rhizobium* to subarctic environment in Scandinavia. *Plant and Soil, Special Volume* 129-137.

Engin, M.; Sprent, J.I. (1973) Effects of water stress on growth and nitrogen-fixing activity of *Trifolium repens. New Phytologist* **72**, 117-126.

Evans, H.J.; Russell, S.A. (1971) Physiological chemistry of symbiotic nitrogen fixation by legumes. In: *The Chemistry and Biochemistry of Nitrogen Fixation.* Postgate J.E. (Ed). London, UK; Plenum Press, 191-244.

Fisher, D.J. (1976) Effects of some fungicides on *Rhizobium trifolii* and its symbiotic relationship with white clover. *Pesticide Science* **7**, 10-18.

Foulds, W. (1971) Effect of drought on three species of *Rhizobium. Plant and Soil* **35**, 665-667.

Foulds, W. (1978) Response to soil moisture supply in three leguminous species. III. Rate of N_2 (C_2H_2) fixation. *New Phytologist* **80**, 547-555.

Fraser, M.E. (1975) A method of culturing *Rhizobium meliloti* on porous granules to form a pre-inoculant for lucerne seed. *Journal of Applied Bacteriology* **39**, 345-51.

Gibson, A.H. (1971) Factors in the physical and biological environment affecting nodulation and nitrogen fixation by legumes. *Plant and Soil, Special Volume* 139-152.

Gibson, A.H. (1976) Recovery and compensation by nodulated legumes to environmental stress. In: *Symbiotic nitrogen fixation in plants.* Nutman, P.S. (Ed). IBP Vol.7, Cambridge, UK; Cambridge University Press, 385-403.

Gillberg, B.O. (1971) On the effects of some pesticides on *Rhizobium* and isolation of pesticide-resistant mutants. *Archiv für Mikrobiologie* **75**, 203-208.

Graham, P.H.; Morales, V.M.; Zambrano, O. (1974) Seed pelleting of a legume to supply molybdenum. *Turrialba* **24**, 335-336.

Grant, D.A.; Lambert, M.G. (1979) Nitrogen fixation in pasture. V. Unimproved North Island hill country, Ballantrae. *New Zealand Journal of Experimental Agriculture* **7**, 19-22.

Greenwood, R.M. (1961) Pasture establishment on a podzolised soil in Northland. III. Studies on rhizobial populations and the effects of inoculation. *New Zealand Journal of Agricultural Research* **4**, 375-389.

Greenwood, R.M. (1964) Populations of rhizobia in New Zealand soils. *Proceedings of the New Zealand Grassland Association* **26**, 95-101.

Greenwood, R.M.; Pankhurst, C.E. (1977) The *Rhizobium* component of the nitrogen-fixing symbiosis. *Proceedings of the New Zealand Grassland Association* **38**, 167-174.

Grossbard, E. (1970) Effect of herbicides on the symbiotic relationship between *Rhizobium trifolii* and white clover. *British Grassland Society Occasional Symposium* **No. 6**, 47-59.

Hale, C.N. (1976) Some factors affecting the survival of *Rhizobium trifolii* on white clover. *Proceedings of the New Zealand Grassland Association* **38**, 182-186.

Hale, C.N.; Mathers, D.J. (1977) Toxicity of white clover seed diffusate and its effects on the survival of *Rhizobium trifolii. New Zealand Journal of Agricultural Research* **20**, 69-73.

Halliday, J.; Pate, J. (1976) The acetylene reduction assay as a means of studying nitrogen fixation in white clover under sward and laboratory conditions. *Journal of the British Grassland Society* **31**, 29-35.

Hardy, R.F.W.; Holsten, R.D.; Jackson, E.K.; Burns, R.C. (1968) The acetylene-ethylene assay for N_2 fixation: Laboratory and field evaluation. *Plant Physiology* **43**, 1185-1207.

Harris, A.J. (1962) The nutrition of white clover(*Trifolium repens* L.) on Te Anau sandy loam. *New Zealand Journal of Agricultural Research* **5**, 294-301.

Hastings, A.; Greenwood, R.M.; Proctor, M.H. (1966) Legume inoculation in New Zealand. *New Zealand Department of Scientific and Industrial Research Information Series* **No.58**, 37pp.

Haystead, A.; Lowe, A.G. (1977) Nitrogen fixation by white clover in hill pasture. *Journal of the British Grassland Society* **32**, 57-63.

Hoglund, J.H.; Brock, J.L. (1978) Regulation of nitrogen fixation in a grazed pasture. *New Zealand Journal of Agricultural Research* **21**, 73-82.

Hoglund, J.H.; Crush, J.R.; Brock, J.L.; Ball, R.; Carran, R.A. (1979) Nitrogen fixation in pasture. XII. General discussion. *New Zealand Journal of Experimental Agriculture* **7**, 45-51.

Holding, A.J.; King, J. (1963) The effectiveness of indigenous populations of *Rhizobium trifolii* in relation to soil factors. *Plant and Soil* **18**, 191-197.

Jones, D.G. (1962) Variation in nodulation characteristics in S100 Nomark white clover (*Trifolium repens* L.) *Journal of the Science of Food and Agriculture* **13**, 598-603.

Jones, D.G. (1963) Symbiotic variation of *Rhizobium trifolii* with S100 Nomark white clover (*Trifolium repens* L.). *Journal of the Science of Food and Agriculture* **14**, 640-743.

Jones, D.G.; Burrows, A.C. (1968) Breeding for increased nodule tissue in white clover (*Trifolium repens* L.). *Journal of Agricultural Science, UK* **71**, 73-79.

Jones, D.G.; Burrows, A.C. (1969) Acid production and symbiotic effectiveness in *Rhizobium trifolii*. *Soil Biology and Biochemistry* **1**, 57-61.

Jones, D.G.; Russell, P.E. (1972) The application of immunofluorescence techniques to host/plant nodule bacteria selectivity experiments using *Trifolium repens*. *Soil Biology and Biochemistry* **4**, 277-282.

Karraker, P.E.; Bortner, C.E.; Fergus, E.N. (1950). Nitrogen balance in lysimeters as affected by growing Kentucky bluegrass and certain legumes separately and together. *Kentucky Agricultural Experiment Station Bulletin* **No.557**, 16pp.

Labandera, C.A.; Vincent, J.M. (1975) Competition between an introduced strain and native Uruguayan strains of *Rhizobium trifolii*. *Plant and Soil* **42**, 327-347.

Lambert, M.G.; Clark, D.A.; Rolston, M.P. (1981) Use of goats for coarse weed control in hill country *Proceedings, Ruakura Farmers Conference*, pp 167-171.

Law, I.J.; Strijdom, B.W. (1977) Some observations on plant lectins and *Rhizobium* specificity. *Soil Biology and Biochemstry* **9**, 79-84.

Lineham, P.A.; Lowe, J. (1960) Yielding capacity and grass/clover ratio of herbage swards as influenced by fertiliser treatment. *Proceedings of the 8th International Grassland Congress*,113-137.

Lowe, J.F.; Holding, A.J. (1970) Influence of clover source and of nutrient manganese concentrations on the *Rhizobium*/white clover association. *British Grassland Society Occasional Symposium* **No.6**, 79-89.

Lowther, W.L. (1974a) Effect of time and rate of applications of lime, inoculation, and lime pelleting on clover establishment. *New Zealand Journal of Experimental Agriculture* **2**, 145-149.

Lowther, W.L. (1974b) Interaction of lime and seed pelleting on the nodulation and growth of white clover, I. Glasshouse trials. *New Zealand Journal of Agricultural Research* **17**, 317-325.

Lowther, W.L. (1975) Interaction of lime and seed pelleting on the nodulation and growth of white clover. II. Oversown trials. *New Zealand Journal of Agricultural Research* **18**, 357-360.

Lowther, W.L. (1977) Factors affecting the response of clover establishment to inoculation and pelleting. *Proceedings of the New Zealand Grassland Association* **38**, 175-181.

Lowther, W.L.; McDonald, I.R. (1972) Clover response to inoculation and pelleting on Te Anau soils. *Proceedings of the New Zealand Grassland Association* **34**, 122-128.

Lowther, W.L.; McDonald, I.R. (1973) Inoculation and pelleting of clover for oversowing. *New Zealand Journal of Experimental Agriculture* **1**, 175-179.

McAuliffe, C.; Chamblee, D.S.; Uribe-Arango, H.; Woodhouse, W.W. (1958) Influence of inorganic nitrogen on nitrogen fixation by legumes as revealed by N^{15}. *Agronomy Journal* **50**, 334-337.

MacKinnon, P.A.; Robertson, J.G. Scott, D.J.; Hale, C.N. (1977) Legume inoculant usage in New Zealand. *New Zealand Journal of Experimental Agriculture* **5**, 35-39.

McNeur, A.J. (1954) Effect of soil acidity on growth of pasture plants. II. A field trial with white clover (*Trifolium repens*). *New Zealand Journal of Science and Technology* **A36**, 167-176.

Marshall, K.C. (1956a) A lysogenic strain of *Rhizobium trifolii*. *Nature, UK* **177**, 92.

Marshall, K.C. (1956b) Competition between strains of *Rhizobium trifolii* in peat and broth culture. *Journal of the Australian Institute of Agricultural Science* **22**, 137-140.

Masterson, C.L.; Sherwood, M.T. (1974) Selection of *Rhizobium trifolii* strains by white clover. *Irish Journal of Agricultural Research* **13**, 91-99.

Masterson, C.L.; Murphy, P.M. (1976) Application of the acetylene reduction technique to the study of nitrogen fixation by white clover in the field. In: *Symbiotic nitrogen fixation in plants*. Nutman, P.S. (Ed). IBP Vol. 7, Cambridge, UK; Cambridge University Press, 299-316.

Minchin, F.R.; Pate, J.S. (1973) The carbon balance of a legume and the functional economy of its root nodules. *Journal of Experimental Botany* **24**, 259-271.

Minchin, F.R.; Pate, J.S. (1975) Effects of water, aeration, and salt regime on nitrogen fixation in a nodulated legume - definition of an optimum root environment. *Journal of Experimental Botany* **26**, 60-69.

Mitchell, K. (1956) Growth of pasture species III. White clover (*T. repens*), subterranean clover (*T. subterraneum*) and lotus major (*Lotus uliginosus*). *New Zealand Journal of Science and Technology* **37a**, 395-413.

Mosse, B.M. (1964) Electron microscope studies of nodule development in some clover species. *Journal of General Microbiology* **36**, 49-66.

Moustafa, E.; Ball, R.; Field, T.R.O. (1969) The use of acetylene reduction to study the effect of nitrogen fertiliser and defoliation on nitrogen fixation by field-grown white clover. *New Zealand Journal of Agricultural Research* **12**, 691-696.

Mulder, E.G.; van Veem, W.L. (1960) Effect of pH and organic compounds on nitrogen by red clover. *Plant and Soil* **13**, 91-113.

Munro, J.M.M.; Hughes, R. (1966) White clover and hill land improvement. *Journal of the British Grassland Society* **21**, 224-230.

Munro, J.M.M.; Davies, D.A. (1974) Potential pasture production in the uplands of Wales. V. The nitrogen contribution of white clover. *Journal of the British Grassland Society* **29**, 213-223.

Mytton, L.R. (1973) The effect of seed weight on the early growth and nodulation of white clover. *Annals of Applied Biology* **73**, 329-338.

Mytton, L.R. (1975) Plant genotype × rhizobium strain interactions in white clover. *Annals of Applied Biology* **80**, 103-107.

Mytton, L.R. (1976) The relative performance of white clover genotypes with rhizobial and mineral nitrogen in agar culture and in soil. *Annals of Applied Biology* **82**, 577-587.

Mytton, L.R. (1977) The effect of mixtures of rhizobium strains on the dry matter production of white clover grown in agar. *Annals of Applied Biology* **87**, 83-93.

Mytton, L.R.; Jones, D.G. (1971) The response to selection for increased nodule tissue in white clover (*Trifolium repens* L.) *Plant and Soil, Special Volume* 17-25.

Napoli, C.A.; Hubbell, D.H. (1975) Ultrastructure of *Rhizobium*-induced infection threads in clover root hairs. *Applied Microbiology* **30**, 1003-1009.

Norris, D.O. (1956) Legumes and the *Rhizobium* symbiosis *Empire Journal of Experimental Agriculture* **24**, 247-270.

Norris, D.O. (1965) Acid production by *Rhizobium*. A unifying concept. *Plant and Soil* **22**, 143-166.

Nutman, P.S. (1958) The physiology of nodule formation. In: *Nutrition of the legumes*. Hallsworth, E.G. (Ed). London, UK; Butterworth, 87-107.

Nutman, P.S. (1959a) Some observations on root hair infection by nodule bacteria. *Journal of Experimental Botany* **10**, 250-263.

Nutman, P.S. (1959b) Sources of incompatability affecting nitrogen fixation in legume symbiosis. *Society for Experimental Biology Symposia* **13**, 42-58.

Nutman, P.S. (1962) The relationship between root hair infection by *Rhizobium* and nodulation in *Trifolium* and *Vicia*. *Proceedings of the Society of London*, **156**, 122-137.

Nutman, P.S. (1967) Varietal differences in the nodulation of subterranean clover. *Australian Journal of Agricultural Research* **18**, 381-425.

Nutman, P.S. (1972) The influence of physical environmental factors on the activity of *Rhizobium* in soil and symbiosis. In: *Use of isotopes for study of fertiliser utilisation by legume crops. IAEA Publication No. 149*, pp. 55-84.

O'Connor, M.B.; Prime, P.N.; Wilkinson, R.C. (1979) Nitrogen fixation in pasture. IV. Central Island pumice-land, Wairakei. *New Zealand Journal of Experimental Agriculture*, **7**, 15-17.

Odeyemi, O.; Alexander, M. (1977) Resistance of *Rhizobium* strains to Phygon, Spergon, and Thiram. *Applied and Environmental Microbiology* **33**, 784-790.

O'Toole, M.A.; Masterson, C.L. (1970) Effects of copper and nitrogen on nodulation and yield of *Trifolium repens*. *British Grassland Society, Occasional Symposium* **No.6**, 251-258.

Pankhurst, C.E.; Gibson, A.H. (1973) *Rhizobium* strain influence on disruption of clover nodule development at high root temperature. *Journal of General Microbiology* **74**, 219-231.

Pankhurst, C.E.; Sprent, J.I. (1976) Effects of temperature and oxygen tension on the nitrogenase and respiratory activities of turgid and water-stressed soybean and french bean root nodules. *Journal of Experimental Botany* **27**, 1-9.

Parle, J.N. (1964) Some aspects of the nodulation problem occurring in the Wither Hills. *Proceedings of the New Zealand Grassland Association* **26**, 123-126.

Pate, J.S. (1976) Physiology of the reaction of nodulated legumes to environment. In: *Symbiotic nitrogen fixation in plants*. Nutman, P.S. (Ed). IBP Vol. 7. Cambridge, UK; Cambridge University Press, 335-360.

Patel, J.J. (1974) Antagonism of actinomycetes against rhizobia. *Plant and Soil* **41**, 395-402.

Philpotts, H. (1977) Effect of inoculation method on *Rhizobium* survival and plant nodulation under adverse conditions. *Australian Journal of Experimental Agriculture and Animal Husbandry* **17**, 308-315.

Postgate, J.R. (1971) The chemistry and biochemistry of nitrogen fixation. London, UK; Plenum Press, 326pp.

Rice, W.A.; Penny, D.C.; Nyborg, N. (1977) Effects of soil acidity on rhizobia numbers, nodulation and nitrogen fixation by alfalfa and red clover. *Canadian Journal of Soil Science* **57**, 197-203.

Robertson, J.G.; Lyttleton, P.; Pankhurst, C.E. (1981) Pre-infection and infection processes in the legume rhizobium symbiosis. In: *Current perspectives in nitrogen fixation*. Gibson, A.H.; Newton, W.E. (Eds). Canberra, Australia; Australian Academy of Science, 280-291.

Robinson, A.C. (1969a) Competition between effective and ineffective strains of *Rhizobium trifolii* in the nodulation of *Trifolium subterraneum. Australian Journal of Agricultural Research* **20**, 827-841.

Robinson, A.C. (1969b) Host selection for effective *Rhizobium trifolii* by red clover and subterranean clover in the field. *Australian Journal of Agricultural Research* **20**, 1053-1060.

Rolston, M.P.; Chu, A.C.P.; Fillery, I.R.P. (1975) Effect of paraquat on the nitrogen-fixing activity of white clover. *New Zealand Journal of Agricultural Research* **19**, 47-49.

Roughley, R.J. (1970) The preparation and use of legume seed inoculants. *Plant and Soil* **32**, 675-701.

Roughley, R.J. (1972) The Australian inoculants research and control service. *Agricultural Gazette of New South Wales* **83**, 210-211.

Roughley, R.J. (1976) The production of high quality inoculants and their contribution to legume yield. In: *Symbiotic nitrogen fixation in plants.* Nutman, P.S. (Ed). IBP Vol. 7. Cambridge, UK; Cambridge Press, 125-136.

Roughley, R.J.; Blowes, W.M.; Herridge, D.F. (1976) Nodulation of *Trifolium subterraneum* by introduced rhizobia in competition with naturalised strains. *Soil Biology and Biochemstry* **8**, 403-407.

Rumball, P.J. (1979) Nitrogen fixation in pasture. II. Northland warm temperate, Kaikohe. *New Zealand Journal of Experimental Agriculture* **7**, 7-9.

Russell, P.E.; Jones, D.G. (1975a) Variation in the selection of *Rhizobium trifolii* by varieties of red and white clover. *Soil Biology and Biochemistry* **7**, 15-18.

Russell, P.E.; Jones, D.G. (1975b) Immunofluorescence studies of selection of strains of *R. trifolii* by S184 white clover (*T. repens* L.) *Plant and Soil* **42**, 119-129.

Schenberger, L.C.; Swope, A. (1975) Inspection of legume seed inoculants and pre-inoculated seeds. *Purdue University Agricultural Experiment Station, Inspection Report* **No. 106**, 8pp.

Schwinghamer, E.A. (1971) Antagonism between strains of *Rhizobium trifolii* in culture. *Soil Biology and Biochemistry* **3**, 355-363.

Schwinghamer, E.A.; Pankhurst, C.E.; Whitfield, P.R. (1973) A phage-like bacteriocin of *Rhizobium trifolii. Canadian Journal of Microbiology* **19**, 359-368.

Sears, P.D. (1953) Pasture growth and soil fertility I. The influence of red and white clovers, superphosphate, lime, and sheep grazing, on pasture yields and botanical composition. *New Zealand Journal of Science and Technology* **A35**, Supplement 1, 1-29.

Sears, P.D. (1962) Exploitation of high production pastures in New Zealand. *Proceedings of the New Zealand Ecological Society* **9**, 57-63.

Sears, P.D.; Evans, L.T. (1953) Pasture growth and soil fertility. III. The influence of red and white clovers, superphosphate, lime, and dung and urine on soil composition, and on earthworm and grass-grub populations. *New Zealand Journal of Science and Technology* **A35**, Supplement 1, 42-52.

Sears, P.D.; Hyde, E.O.C.; Greenwood, R.M. (1935) Pasture establishment in pumice soils. *New Zealand Journal of Science and Technology* **A37**, 110-140.

Sears, P.D.; Goodall, V.C.; Jackman, R.H.; Robinson, G.S. (1965) Pasture growth and soil fertility, VIII. The influence of grasses, white clover, fertilisers, and the return of herbage clippings on pasture production of an impoverished soil. *New Zealand Journal of Agricultural Research* **8**, 270-283.

Sherwood, M.T.; Masterson, C.L. (1974) Importance of using the correct test host in assessing the effectiveness of indigenous populations of *Rhizobium trifolii. Irish Journal of Agricultural Research* **13**, 101-108.

Silsbury, J.H. (1977) Energy requirement for symbiotic nitrogen fixation. *Nature, UK* **267**, 149-150.

Simpson, J.R. (1976) Transfer of nitrogen from three pasture legumes under periodic defoliation in a field environment. *Australian Journal of Experimental Agriculture and Animal Husbandry* **16**, 863-870.

Singh, A.; Sharma, P.B. (1973) Growth and survival of rhizobia in commercial bacterial inoculants. *Journal of Research of the Punjab Agricultural University* **10**, 95-98.

Solheim, B.; Raa, J. (1973) Characterisation of the substances causing deformation of root hairs of *Trifolium repens* when inoculated with *Rhizobium trifolii. Journal of General Microbiology* **77**, 241-247.

Sprent, J.I. (1971) The effect of water stress on nitrogen fixing root nodules. I. Effects on the physiology of detached soybean nodules. *New Phytologist* **70**, 9-17.

Sprent, J.I. (1972) The effect of water stress on nitrogen fixing root nodules. IV. Effect on whole plants of *Vicia faba* and *Glycine max. New Phytologist* **71**, 603-611.

Spurway, C.H. (1941) Soil reaction (pH) preference of plants. *Bulletin of the Michigan Agricultural Experiment Station* **No. 306**.

Strijdom, B.W.; Deschodt, C.C. (1976) Carriers of rhizobia and the effects of prior treatment on the survival of rhizobia. In: *Symbiotic nitrogen fixation in plants.* Nutman, P.S. (Ed). IBP Vol. 7. Cambridge, UK; Cambridge University Press, 151-168.

Suckling, F.E.T. (1975) Pasture management trials on unploughable hill country at Te Awa. III. Results for 1959-69. *New Zealand Journal of Experimental Agriculture* **3**, 351-436.

Thompson, J.A.; Roughley, R.J.; Herridge, D.F. (1974) Criteria and methods for comparing the effectiveness of *Rhizobium* strains for pasture legumes under field conditions. *Plant and Soil* **40**, 511-524.

Thompson, J.A.; Brockwell, J.; Roughley, R.J. (1975) Preinoculation of legume seeds. *Journal of the Australian Institute of Agricultural Science* **41**, 253-254.

Tortenson, L. (1975) Effect of bentazon and dinoseb on soil microorganisms and on the *Rhizobium*-Leguminosae symbiosis. *Swedish Journal of Agricultural Research* **5**, 177-183.

van Schreven, D.A. (1972) On the resistence of effectiveness of *Rhizobium trifolii* to a low pH. *Plant and Soil* **37**, 49-55.

Vincent, J.M. (1958) Survival of the root nodule bacteria In: *Nutrition of the legumes*. Hallsworth, E.G. (Ed). London, UK; Butterworths, 108-123.

Vincent, J.M. (1970) A manual for the practical study of root nodule bacteria. IBP Handbook No. 15. Oxford, UK; Blackwell Scientific Publications, 164 pp.

Waters, L.M. (1956) Nodulation tests with mixed inoculants. *Journal of the Australian Institute of Agricultural Science* **22**, 141-150.

Weaver, R.W.; Frederick, L.R. (1974) Effect of inoculum rate on competitive nodulation of *Glycine max* L. Merrill. II. Field Studies. *Agronomy Journal* **66**, 233-1236.

Wolpert, J.S.; Albersheim, P. (1976) Host-symbiont interaction. I. The lectins of legumes interact with O-antigen-containing lipopolysaccharides of their symbiont bacteria. *Biochemical and Biophysical Research Communications*. **70**, 729-737.

Worrall, V.S.; Roughley, R.J. (1976) The effect of moisture stress on infection of *Trifolium subterraneum* L. by *Rhizobium trifolii* Dang. *Journal of Experimental Botany* **27**, 1233-1241.

Yao, P.Y.; Vincent, J.M. (1976) Factors responsible for the curling and branching of clover root hairs by *Rhizobium*. *Plant and Soil* **45**, 1-16.

Yoshida, S.; Yatazawa, M. (1977) Nitrogen fixation capacity during the course of cut-recovery in ladino clover. *Journal of Japanese Grassland Science* **23**, 6-13.

Young, D.J.B. (1958) A study of the influence of nitrogen on the root weight and nodulation of white clover in a mixed sward. *Journal of the British Grassland Society* **13**, 106-114.

Population dynamics and competition

W. Harris
* Botany Division, DSIR Lincoln, Christchurch, New Zealand

Contents

* Former Address: Grasslands Division, DSIR, Palmerston North, New Zealand

1. Introduction

White clover is generally sown in association with grasses, which have a higher than average requirement for soil fertility. To explain the success in competing with and maintaining a population in the presence of these other species, it is necessary to examine the physiological responses and performance of white clover in plant communities.

To describe the population dynamics of white clover, factors determining the numbers of units at three stages in the life cycle of the plant will be examined. These stages are defined as: (1) the buried and surface seed populations, (2) the seedling population, and (3) established plants with axillary shoots and/or stolons.

2. The seed population

(A) Surveys

Recent surveys have confirmed Parkinson's (1798) observation that seeds of white clover are abundant in the soil. Champness & Morris (1948), in samples from different types of grassland in England and Wales, observed widespread occurrence of white clover seed and recorded a maximum viable seed population of 19.9 million seeds/ha, equivalent to about 12 kg seed/ha (based on a 1000-seed weight of 0.6 g). They classified white clover among those species with a higher a percentage of buried viable seed than would be expected from the botanical composition of the sward, and observed that under permanent pasture, seeds were confined largely to the top 5 cm of soil.

Hyde & Suckling (1953), in a survey of legume seeds in the top 5 cm of soil of 105 fields in New Zealand, found white clover in all but 20. They recorded a maximum density equivalent to about 296 kg/ha and observed that the population of white clover seed varied with the extent to which pasture management allowed seed to ripen. Densities were lower (average 500 seeds/m^2) for lowland pastures closely grazed by sheep than for similar pastures grazed by cattle (1840 seed/m^2). Similarly, Turkington et al. (1979) observed a buried seed population of 218/m^2 for old permanent pasture in North Wales for which the grazing management provided little opportunity for seed set. An average density of 10 210 seed/m^2 was recorded for fields closed for a seed crop in the previous 5 years. Another discernable cause of variation in the white clover seed population was the incidence of drought as determined by soil type and rainfall, with less white clover seed in drier situations. Charlton (1977) detected no significant site differences in the number of buried seed of white clover on north and south aspects, stock camps, and wet areas. Suckling & Charlton (1978) have reviewed the significance of buried legume seeds in temperate grasslands.

(B) Replenishment of the seed population

The records of buried seed populations indicate the importance of climate and management in determining the replenishment rate of white clover seed populations. For a resident population to set seed, climatic conditions suitable for the initiation of flowers with fertile ovules and pollen, an adequate population of pollinators, and climatic and management conditions that allow seed to mature and ripen will all be necessary. Wheeler & Hill (1957) provide a general description of how climatic factors influence white clover seed production.

Dessureaux (1950, 1951), in Canada, investigated the influence of climatic and genetic variability on seed setting and percentage of seedbearing florets in Ladino clover. The average number of seeds per floret varied during the season from 0.03 to 2.7 and the percentage of seed-bearing florets from 0.8% to 91.9%, depending on the date of flowering. Climatic effects on pollinator activity were considered the main cause of this variation. Genetically based differences in seed set between clones were well defined. (See also Chapter 10, Section 2 (C) (*xv*).

i. Requirements for flowering

Field observations show a wide range of flowering requirements, both within and between varieties, in terms of daylength, temperatures and light intensity (Griffiths et al. 1967). The physiological basis of this variation is considered in Chapter 3.

ii. Pollination

The necessity of cross pollination for complete ovule fertilization was demonstrated by Darwin (1878), who observed the lack of seed set in white clover protected from insects. The honeybee (*Apis mellifera*) and bumble (humble) bees (*Bombus* spp.) are the most frequent insect visitors and generally effect pollination (Davies 1970). Other insects, particularly Diptera and Lepidoptera, visit the flower for honey but do not cross pollinate efficiently (Knuth 1908).

The role of white clover in the honey industry is well documented, as is the relationship between honey bee density, pollen collection activity and seed set in white clover (Harrison et al. 1945, Hammer 1947, Percival 1947, 1950, Vansell 1951, Dunavan 1952, Todd & Vansell 1952, Oertel 1954, Dunham 1957, Weaver 1957, Bohart 1960, Palmer-Jones et al. 1962, Griffiths et al. 1967, Green 1957).

Prolonged overcast and wet weather reduces flowering and bee visits. For this reason most clover seed is produced where clear, bright weather is prevalent, such as the western states of the USA (Gibson & Hollowell 1966), south-east England (Griffith et al. 1967), and drier eastern regions of the South Island of New Zealand (Miller 1944).

iii. Seed maturation

The course of maturation of white clover seed from pollination onward was described by Hyde (1950) and Hyde et al. (1959). Seed development from pollination to full ripening took about 26 ± 5 days depending on weather conditions. For about 10 days after pollination, growth was logarithmic, presumably determined by the rate of cell division in the embryo and seed coat. During this phase moisture content remained near 79%. Seeds were viable from about the 10th day and by the 15th day close to 90% of the seed was viable. Up to the 21st day, growth proceeded uniformly as assimilates were transferred from the plant to the seed. During this second phase, seed dry weight increased 3-fold and water content fell to 63% of the seed weight. Ripening lasted from the 21st to the 25th day. Dry weight remained about the same and moisture content declined to 10%. Hard seed first appeared 12 days after pollination and increased to 96% of the viable seed on the 25th day. Under natural conditions seed devlopment continues until this stage before being shed, so that most naturally reseeded seed is hard. In contrast, at the time of cutting commercial seed crops, various stages of seed development are represented.

A study in Denmark by Bogh (1936) showed that the 1000-seed weight reached a maximum about 20 days from pollination, a result consistent with those cited above.

iv. Plant population density

Plants at low densities produce longer stolons, more stolon branches, and more inflorescences per plant and per unit area (Zaleski 1964). This was shown to be correlated with light intensity at stolon level. Optimal commercial seed crops from first and second year sowings are therefore obtained from low seeding rates (Zaleski 1961, Clifford 1977). Optimal seed yield at low plant density is a generally observed phenomenon described by Holliday (1960).

v. Effects of associate species

Zaleski (1965a) demonstrated that higher yields of white clover seed were obtained from pure sowings than in mixture with grass. However, much white clover seed production is from mixed sowings of grass and legume (Haggar & Holmes 1963, Scott 1973, Zaleski 1965b); undersowing in a cereal is also recommended (Anslow 1962). Zaleski (1965a) found that S23 perennial ryegrass gave more shading, suppressed the clover and reduced the number of inflorescences and seed yield more than did S215 meadow fescue, effects that increased as the grass seed rates increased. However, the suppressive effect of the grass was less where grass and legume were sown in alternate rows than when broadcast together.

Selective suppression of associated species from white clover by herbicides usually increases white clover seed production, a technique most successful where grasses are removed (Zaleski 1967, Wasmuth &

Miles 1972a, Vydrina 1965, Allen et al. 1972). Attempts to remove broadleaved weeds may reduce clover seed yield, but may still be worthwhile because of reduced contamination of the harvested seed by weed seeds.

vi. Defoliation effects

Seed yields are affected by defoliation frequency and timing. Close grazing during spring minimises stolon shading and can lead to high seed yields if grazing is stopped at the correct time. If crops are closed for seed too soon, inflorescence development may be slow and shading from leaves retard it further as the canopy grows. In New Zealand a final defoliation one month before the longest day has resulted in higher seed yields than earlier or later closings (Clifford 1979). Haggar et al. (1963) studied seed production of ryegrass-Kent white clover pastures. In contrast to rotational grazing at approximately 18-day intervals (the traditional management for seed production) continuous sheep grazing during spring ensured that shading of the stolons by associated grass was minimal, so that flower initiation from the stolons was not inhibited. Close defoliation also ensured stolon branching and increased the sites at which flower buds could be initiated.

Zaleski (1961) compared seed production from white clover stands which were either left uncut and the first growth used for seed production, or cut early in spring at the first flower bud, or cut at early flowering, with the seed being harvested from the second growth. Seed production from these cutting treatments varied between each of the three years of the experiment. In a wet season, there was no difference between the number of inflorescences per unit area for the cutting treatments. However, because of larger seeds and more florets per inflorescence, seed yield was highest with the uncut treatment, followed by the early-cut, then the late-cut treatment. In a dry season, the inflorescence number was lowest on the uncut plots and highest on the late-cut plots, but as the latter had exceptionaly short peduncles, much of the seed produced was not harvested. Zaleski suggested that in the dry year, development of the regrowth leaves was limited by water availability which reduced their shading of the stolons and consequent interference with inflorescence development. Thompson et al. (1952) found that in one out of three years, seed production of Ladino clover was greater when the first growth was cut to 15-20 cm before taking seed from the second growth than where seed was taken from the first growth or from more lenient (20-25 cm and 25-30 cm) cuttings of the first growth.

vii. Nutrient and moisture effects

Commercial seed production of white clover requires land of medium fertility with adequate lime, phosphate and potash (Zaleski 1970, Griffith et al. 1967, Anslow 1962). Although in some situations nitrogen application can increase seed production (Schwanbom & Froier 1949), a reduction is more likely if the production is from a grass-legume sward. Haggar et al. (1963) found that application of 78 kg N/ha as sulphate of ammonia to ryegrass-white clover pasture in the month before flower bud appearance more than halved clover seed production compared with control swards.

In recommendations for production of ryegrass and white clover seed under irrigation in New Zealand, Calder (1941) commented that although white clover does not lodge under generous irrigation, much of the plant's energy goes into leaf and stem production. This overgrows the earlier flowers and causes flowering to be sparse and prolonged. He recommended irrigation at the time of closing after a grazing late in October. Thereafter, in most years the natural precipitation would be sufficient to mature a good seed crop.

The requirement for adequate water up to flowering and thereafter a reduced water supply to maximise seed production was confirmed by Zaleski (1966) in an experiment in England.

viii. Losses from predation and disease

Use of insecticides on seed crops has shown that seed losses from insect predation can be large. Thompson et al. (1952) found that the use of DDT on Ladino clover seed crops in Wisconsin increased seed yield by 19, 83 and 36% in three years. They did not specify the damaging insect species.

Strickland (1956) in England reported seed yield increases of 60-75% after insecticide treatment of red and white clover seed crops carrying a high density of the clover weevil (*Apion* spp.). In seed production areas of the South Island of New Zealand, Forster et al. (1962) showed seed damage, mainly from the clover case bearer moth (*Coleophora* spp.), to be widespread, ranging up to 46% of the total seed per head. Thrips (*Halothrips niger* Osborne) caused heavy seed loss in one location. M. H. Campbell (1966) has shown in Australia that seeds on the soil surface are gathered in large numbers by ants and carried to their nests.

Besides these direct effects of insects, other predators and diseases which affect the vegetative parts have indirect effects on seed production (see Chapters 11 and 12).

(C) Dispersal

Methods of dispersal of pasture legume seeds have been summarised by Suckling & Charlton (1978). Long-distance dispersal is principally through the digestive tracts of birds and grazing animals. Seeds can be spread over short distances by dehiscence, by stock trampling, by earthworms and to a small extent by wind.

When ripe, the seed pod is covered by the withered calyx and corolla and soon splits along the ventral suture. The seeds are then easily shaken out (Erith 1924). The average 1000-seed weight for wild white clover is 0.557g (range 0.41-0.75g) and for cultivated white clover 0.612 (range at least 0.60-0.71g) (Adams 1926, Robinson 1947).

Salzmann (1939) found that white clover formed a very large proportion of the seedlings obtained from samples of farmyard manure, and Stebler & Schroeter (1889) cited by Erith (1924) found 5370 white clover seeds with a germination percentage of 62% in 0.454 kg of dried residue of liquid manure.

The importance of hard seeds in dispersal via dung has been shown by Suckling (1952) who observed that sheep consumed seed heads readily, sometimes preferring them to more lush feed. When grazed on swards containing clover the mean daily output of seed per sheep was 1891 mg hard seeds and 676 mg of seeds which absorbed stomach juices. After passage through the digestive tract, 92% germinated normally after scarification and 7% produced abnormal growths, probably as a result of the scarification. Of the seed which had absorbed stomach juices, 11% was viable and a further 9% produced abnormal growth.

In a related experiment (Suckling 1952), hard seed and scarified seed were fed to sheep in capsules and the faeces were collected for 7 days. Significantly, much of the hard seed sample was destroyed in the digestive process as only 27% of the hard seed was recovered. Most of the seed was voided 3 days after feeding, but some continued to appear after 6 days in one of the four sheep used in the experiment. This sheep was killed and 1559 mg hard seed, representing 15% of the seed fed, was found in the digestive tract, most (1307 mg) among the folds of the omasum.

Suckling (1952) also found that in studies with two sheep, 4.3 and 0.76% of the clover seed in herbage consumed was recovered in the dung and that, 76 and 86% respectively was hard. Some seed loss was attributed to trampling by the sheep, but much seed appeared to be destroyed by mastication and cud chewing (to which seed fed by capsule in his experiment would not have been subjected). Sheep grazing an area containing little white clover grazed clover seed heads to the extent that they voided an average of 3 g clover seed per day per animal.

Other workers have demonstrated the passage of white clover seed in viable state through the alimentary tracts of cattle (Lennartz 1957, Yamada & Kawaguchi 1972), goats (Yamada & Kawaguchi 1971) and birds (Krach 1959). Krach experimented with sparrows, pheasants, pigeons and rooks, and concluded that white clover was among plant species with endozoic seed distribution, particularly through the rook.

There is evidence that white clover seed is distributed by earthworms. Bates (1933) observed that the abundance of white clover is correlated with earthworm activity. Chippendale & Milton (1932) considered that earthworm activity could bury seed deeply, and also suggested that seeds may be washed down soil fissures developed during drought. McRill & Sagar (1973) recorded the ingestion of seed of 14 plant species by *Lumbricus terrestris* in Petri dishes over an 18-h period and recovered seed from casts produced in the following 72-h. The earthworms ingested 16% of the white clover seed offered, of which 62.5% was recovered from casts. Non-hard seed of white clover that had 74.7% germination before digestion had 82% germination after passage through the gut; for hard seed the respective figures were 8.3 and 15.4%.

Because of its agricultural uses, white clover seed is spread over large areas to establish new pasture, or oversow existing pasture. The use of stock pre-fed on pasture containing ripe clover seedheads, or the spreading of hay with a high clover seed content on unploughable hill country are recommended practices to introduce clover into such areas (von Hessberg 1932, Madden 1954, Suckling 1952).

(D) Germination requirements

i. *Hardness and longevity*

White clover seed can remain dormant in soil for many years. In storage, viability is retained for long periods at low temperatures, but at higher temperatures, viability can be maintained only in sealed containers. Nearly all seed that ripens under dry conditions is in the dormant, hard condition, but seed ripened in moist environments may be immediately germinable. This failure to form hard seed under moist conditions could be of adaptive significance.

Erith (1924) cited work of Peter (1893) which showed the emergence of white clover seedlings from soil samples taken from under Scots pine and larchwood forests planted respectively 32 and 46 years previously. Lewis (1973) reported an investigation on the viability over 20 years of seed buried at different depths in a mineral soil and in peat. From an initial germination of 90% and 8% hard seed, viability fell to about 1% in the 20-year-old sample. This loss of viability, and the example of Beal's seed viability experiment (Darlington & Steinbauer 1961) in which seed viability was completely lost at a very early stage, probably arose from the fact that the level of hard seed in the samples used was low.

Longevity of seed is influenced by levels of aeration and temperature, as shown by Davies (1956, 1965). Seed of red clover and cv. S100 Nomark white clover was stored at 0°C and at room temperature in tubes sealed under vacuum, under nitrogen after evacuation of air, sealed with air or left unsealed. White clover showed greater longevity than red clover, and after 16 years storage at 0°C viability was close to 90% in all treatments. At ambient temperatures, germination of seed in the unsealed tube declined progressively from the 5th year of storage to about 5% in the 16th year. That of vacuum sealed seed declined to 16% by the 16th year, but the other sealed treatments retained their original viability.

The ecological significance of the hard seed condition is well exemplified by the studies of Hyde (1950, 1954) and Hyde & Suckling (1953). Hyde (1950) concluded that nearly all clover seeds which ripen in the normal manner are in dormant, hard seed condition. He showed (1954) that the proportion of hard seed varies with the intensity of drying during maturation. Mature, unripe seed was ripened at five levels of relative humidity ranging from 30 to 70% and then tested for germination at 20°C over 18 months. The results showed a high negative correlation between the moisture content of the hard seeds and the duration of impermeability, so that by four months all seeds ripened at 70% relative humidity germinated whereas those ripened at 30% relative humidity showed negligible germination after 18 months.

The hilum functions as a hygroscopically activated valve in the otherwise impermeable testa. When relative humidity is high the hilum fissure closes, limiting the absorption of moisture (Hyde 1954). Consequently, hard seeds tend to have a moisture content in equilibrium with the lowest relative humidity to which they have been exposed. The hilar fissure responds to marked fluctuations of humidity, but if humidity increases gradually, the seed absorbs moisture and the hilar fissure remains open.

Both Heinrich (1894) and Chappuzeau (1930) observed that hard seededness was greater in small than in large seeds of white clover. Grant et al. (1964) in a study of seeds of three *T. subterraneum* cultivars showed that dormancy was lowest in large seed. The effect of increased desiccation in increasing hard seededness (Hyde 1954) may explain these observations. An indication of the rate of loss of seed through germination is provided by Suckling (1952) who recorded the germination of seed voided in dung which was placed on soil in pots located in the field. About 5% of seeds germinated every 50 days, so that 15% had germinated after 150 days. Suckling concluded that seasonal rate of germination fluctuated little in the environment of his study, a finding which differs from the observations of Burton (1940) and Blaser & Killinger (1950) which emphasise environmental factors which break dormancy.

ii. Breaking of dormancy

Dormancy in white clover seed appears to be controlled largely if not entirely by the impermeability of the seed coat; scarification is usually sufficient to break dormancy of a considerable proportion of hard seed. Burton (1940) produced evidence that temperature influences the duration of dormancy in populations naturalised in the south-eastern states of the USA. There, white clover behaves as a winter annual, with seed produced in early summer lying dormant until autumn or early winter. Two months after harvest, unscarified and scarified seed was germinated at temperatures of 30, 20 and 10°C. After 28 days the unscarified seed showed 3.1, 3.6 and 64.3% germination, and the scarified seed 70.7, 89.9 and 95.3%, respectively, at 30, 20 and 10°C. These results suggest that seed dormancy in this population was an adaptation to avoid dry summer conditions, the lower temperatures of autumn and winter providing the trigger to break dormancy.

With seed from four white clover lines harvested at the same time, Burton (1940) studied germination under the same range of scarification and temperature treatments after a further 5 months storage during which relative humidity ranged from 60 to 100% and temperatures from 5 to 15°C. After this period of cold storage, germination of unscarified seed at 20° and 30°C was very much higher than that of seed tested at 2 months. However, strains differed in germination response, the scarification requirement for germination at high temperatures being reduced by cold storage much more for one line than another. Such variability provides a basis for adaptation of the life cycle to survival in different environments.

Blaser & Killinger (1950) further defined germination response of Louisiana white clover seed to low temperature exposure. Sods were sampled in autumn after a heavy volunteer spring-early summer seed crop. Germination on these sods was poor at temperatures above 22°C, but pre-exposure to temperatures of 3°C and -18°C followed by higher temperatures increased germination. Alternating cycles of temperature between 3°C and 24-29°C further increased germination. Moist environments were required for the low-temperature responses to be effective, and scarification removed the requirements for low temperature.

Nakamura (1962) investigated the occurrence of hard seed and means of breaking dormancy in samples of Japanese white clover. He noted that potassium nitrate, thiourea, gibberellic acid, alternating temperatures and light had no effect on germination. Additive effects of scarification and pre-chilling on the germination of mature seed at 15°C led him to suggest that both seed coat and embryo share responsibility for dormancy. Wesson & Wareing (1968) presented results suggesting that white clover may be a species which requires a light stimulus to break dormancy after it has lain dormant as a buried seed.

3. The seedling population

(A) Seed size

Within a white clover population, larger seeds show greater emergence than smaller seeds, and swards derived from larger seeds show superior growth in the early stages. However, this effect of seed size is short-lived, and does not carry over into fully established swards. The fate of seedlings from various seed size classes in mixed sowings is unknown, although extrapolation from results with subterranean clover indicates that seedlings from larger seeds would tend to dominate during early establishment. However, unit area yields of swards tend to be independent of seed size and emergence pattern, indicating compensation between plants in the use of environmental resources. There is little information on the importance of seed size *per se* in relation to other factors affecting seedling establishment, e.g. genetic differences in seedling vigour.

i. Emergence

Miller & Pammel (1901) and Chappuzeau (1930) showed that emergence of plants from small seeds of white clover was lower than from large seeds, and that large seeds germinated more quickly than small seeds. Hyde (1950), using scarified white clover seed, showed that the speed of germination increased with advancing age of the seed at harvest. Seed harvested 12 days after flowering took 3.5 days to germinate, whereas mature seed germinated in 2 days at 13°C. After 6 weeks growth in a glasshouse, seedlings from seed harvested at the 12-day stage had approximately half the fresh weight of seedlings from mature seed. The weight of seedlings was found to be directly proportional to the initial seed weight. Hyde (1950) also found that none of seed harvested at 14 days emerged when sown at 4 cm depth at 20°C, whereas 40% of the mature seed emerged.

Black's (1956) work with *Trifolium subterraneum* showed effects of seed size on emergence from sowing depths of 1.25, 2.5 and 5 cm. The smaller seed did not emerge from the 5-cm depth, and seed size was linearly related to potential hypocotyl elongation.

ii. *Seedling size*

Large seed imparts an advantage of both dry weight and leaf area at early seedling growth stages. Evans (1965) in a box experiment in a glasshouse compared growth from seed weight categories of diploid S100 and in a polyploid derivative designated AC9. Growth was observed over three months for plants derived from seed populations with mean seed weights of 0.99 mg (large), 0.63 mg (medium) and 0.45 mg (small). For both varieties the large seeds gave the largest plants, but this difference was more marked for polyploid AC9. The relationship between seed and plant size and the varietal difference were also observed when plants from the seed size populations were grown at 90 x 90 cm spacing in the field. In a third experiment Evans (1965) established plots from broadcast sowings of a range of seed size populations of AC9 and S100. No clear pattern emerged for the effect of seed size on the yield of S100, but for AC9 large (1.22 mg/seed) and medium (0.95 mg/seed) populations outyielded the small (0.62 mg/seed) populations at one cut in the establishment year and the first two cuts in the following year. There were no significant seed size effects for the two cuts later in the second season. Similar trends occurred in mixed sowing with S24 perennial ryegrass.

Mytton (1973) examined effects of seed weight on early growth and nodulation of white clover, at first in agar culture and later after transfer to pot culture. A comparison was made of the effects of fixed N only and of ammonium nitrate supplied at the rate of 1 mg N per plant on the 11th day and 5 mg on the 17th and 24th days. Results (Table 1) show the relationship between seed size and early plant size, development rate, and nodulation rate. The lack of interaction between N source and seed size indicates that the early growth of all seed sizes was not met by symbiotic N fixation. Mytton (1973) found that early differences attributable to seed size did not extend to more mature plants grown in pots.

Table 7.1 Seed size effects on white clover seedling growth and nodulation (Mytton 1972).

	Light Seed (0.36–0.45mg)	Intermediate Seed (0.56–0.65mg)	Heavy Seed (0.76–0.85mg)
Whole plant dry weight after 30 days (mg)	39.6	60.9	69.8
Days to first leaf stage	13.9	12.1	11.8
Days to first trifoliate leaf	18.2	16.4	16.2
Days to nodulate (rhizobium treatment only)	8.93	7.57	7.61
Effect of nitrogen source	Total DM (MG) After 30 days		
Rhizobium only	783	1400	1603
Plus mineral N	1436	2011	2303

The reason for the breakdown of relationships between seed size and yield with time were considered by Black (1957b, 1959) and Taylor (1972) using results obtained with subterranean clover.

iii. Population hierarchy

Seed sown to establish pastures and from natural reseeding will normally vary in size. Direct evidence that variation in seed size may determine the performance of individual white clover seedlings is unavailable, probably because the stoloniferous habit of white clover quickly leads to the loss of identity of individual clover plants in swards, and makes the study of competition between individual plants in this species difficult. In a study of subterranean clover by Black (1958), mixed stands of equal numbers of plants obtained from seeds of two contrasting sizes were established. In the first three months there was no loss of plants obtained from large seeds, but the number of plants obtained from small seeds declined markedly.

Measurements of individual plants in populations of competing plants of various species show that populations of approximately uniform plant size, or a normal distribution of plant size, will move towards an L-shaped skew distribution. Consequently, in the population the proportion of small plants will increase and the population of large plants will decrease. (Koyama & Kira 1956, White & Harper 1970, Harris 1971a). Koyama & Kira theorised that skewness would develop independently of competition as a function of the normal distribution of both seed weight or plant size and relative growth rate. However, Ross & Harper (1972), using *Dactylis glomerata*, showed that the growth of each individual plant was influenced most by the density of previously emerged neighbours and that the spatial distribution of neighbours and effects of seed size were less important. The divergence of relative growth rates between early and late emerging plants showed that early emergers progressively increased their capture of resources at the expense of later emergers, as was demonstrated by Black & Wilkinson (1963) for light interception by subterranean clover. This appears to be an important mechanism by which the dominance hierarchy develops within populations of competing plants, the rate and the time of onset of self thinning being density related. Demonstration of this in subterranean clover was provided by Donald (1963) who cited a thesis study by Stern showing that the variation in growth rate, and consequently plant size, was greatest at high density.

iv. Root size

The effect of seed size on root growth, and their possible consequences for root competition during establishment, were studied by Evans (1973b). In perennial ryegrass cv. Grasslands Ruanui, cocksfoot cv. Grasslands Apanui, timothy cv. Grasslands Kahu and *Agrostis tenuis*, and red clover cv. Grasslands Hamua and white clover cv. Grasslands Huia, seedling size and root length were positively related to seed size, but the clovers had lower relative values than the grasses.

v. Genetic effects

There is little information for white clover on genetic variation in seed size or seedling vigour, and there are no known field studies of varietal mixtures which might indicate the relative importance of seed size *per se* as opposed to genetically controlled seedling vigour. A comparison of *Trifolium fragiferum* and *T. repens* by Harper & Clatworthy (1963) indicated that seed size differences were overridden by other differences between the species. Although *T. fragiferum* seeds were much larger than *T. repens* seeds, no differences were found between the species in their reaction to depth of sowing.

Seedlings of *T. repens* emerged earlier than those of *T. fragiferum*. Although *T. fragiferum* produced larger cotyledons than *T. repens*, this early advantage of *T. fragiferum* was lost because the species had slower rates of leaf expansion and growth. Early root extension of *T. fragiferum* was greater than that of *T. repens*, but hypocotyl extension of *T. repens* was greater than that of *T. fragiferum*. The longer hypocotyl, which showed greater elongation in response to shading, more rapid leaf expansion and a more rapid rate of leaf production in *T. repens* than in *T. fragiferum*, gave *T. repens* an early advantage when the two species were grown in mixed swards.

(B) Effects of environmental factors on seedling development and competition

Except when germination occurs under an existing canopy, seedling growth after germination will proceed for a period with little interference from other plants. Response patterns under non-competitive conditions are considered in detail elsewhere (Chapter 2). Speed of germination and early growth rates of competing species as determined by environment are also important in determining the onset of, and direction of, competition towards dominance or suppression.

McWilliam et al. (1970) examined germination and seedling development in a range of grass and legume species, including Ladino white clover. Seeds were imbibed by immersion in vigorously aerated water at 23°C. The legume seed, which has been scarified, absorbed a considerably greater percentage of water than grasses, white clover more so than lucerne and subterranean clover. Most of the water was absorbed by the legume seed in the first 4 to 8 h, whereas the grass seeds were still imbibing after 24 h, by which time the legumes were germinating. Water loss by dehydration from the fully imbibed seed was more rapid for the legumes in the first 6 h, but after 12 h all species had dried to a moisture content of 25-35%. Ryegrass was superior to all the other species in its ability to germinate under moisture stress, and the lower limit for its germination (-16 bar) was 4 to 6 bar below the limiting potential for the other species, which were all similar in their response pattern to moisture stress.

Sharma (1973) showed drought tolerance at the seedling stage to be in the order *Lolium perenne, Atriplex nummularia, Danthonia caespitosa, Atriplex vesicaria, T. repens*. Although considerable germination of *L. perenne* occurred at a water potential of -15 bar, white clover showed a substantial increase of germination only between -5 to 0 bar.

McWilliam et al. (1970) observed germination rates at temperatures of 5 to 29°C. For all species tested, apart from subterranean clover, temperatures in the range 5 to 30°C did not limit the germination of the group of temperate species used. However, temperature differences in rate of germination were evident. For example, at 10°C the legumes commenced germination in 1-2 days, and were approaching their maximum percentage of 80 to 90 in 4-6 days. For the grasses, germination did not start until the 6th-8th day.

Harper & Clatworthy (1963), comparing *T. repens* variety S100 and a Palestinian variety of *Trifolium fragiferum*, found that both species germinated most rapidly at 25°C but at lower temperatures *T. repens* germinated more rapidly than *T. fragiferum*. At 35°C *T. fragiferum* germinated more rapidly than *T. repens*. Fayemi (1957), comparing germination of five legumes at 6.7 and 25°C, found that germination of Ladino white clover, red clover, alsike clover and lucerne was delayed by the low temperature after imbibition, but low temperature did not reduce viability.

Temperatures below 5°C may limit white clover seedling survival. Mence (1964) showed that at temperatures between -5°C and +5°C, *T. repens* seed died and germination was reduced to about 40%, whereas the effect on *Phleum pratense* was negligible. Rachie & Schmid (1955) compared the response of lucerne, Ladino white clover and a range of *Lotus corniculatus* populations after growth from seed at 22°C for different numbers of days to a 14-day hardening period at 3°C followed by freezing at -10°C for 12 h. The survival of 14-, 19-, 24-, 36- and 44-day-old white clover seedlings was 0, 47, 79, 51 and 16%, respectively.

Nutrient uptake patterns indicated that all the species tested by McWilliam et al. (1970) were likely to benefit from fertilizer application during establishment. Even legumes, and particularly where conditions delay nodulation, would benefit from N applications in the seedling phase if the presence of other species did not interfere (Andries & van Slijcken 1968). The importance of rhizobia for nodulation and an adequate nutrient regime for successful seedling establishment is well illustrated by studies of the effect of seed coating, pelleting and inoculation. As an example, Lowther (1974) examined the effects of application of lime, pelleting with lime or Gafsa-phosphate + dolomite, and inoculation level. On soils of low pH (4.7-5.0) a similar increase of clover performance was obtained by either addition of 2500 kg/ha lime or pelleting. On soils of pH 5.3-5.5, pelleting either had no effect or increased clover growth, whereas liming either depressed or enhanced growth depending on soil type. At pH 6.2, both lime and pelleting decreased growth. On some soils, increasing the inoculation level increased growth in the absence of liming and pelleting, in one example to the extent of replacing the requirements for liming or pelleting for maximum plant growth. Under dry conditions uptake of water by seeds resting on the soil surface is enhanced by absorbent lime or bentonite coatings, resulting in increased germination (Dowling et al. 1971).

(C) Effects of seed placement and sowing method

Establishment studies of white clover generally indicate that a shallow soil coverage of seed (no more than about 2 cm) is beneficial, but that broadcast sowings are more likely to give better results than sowings at depths greater than 2.5 cm.

Insight on how the effects of the strength, tilth and heterogeneity of the soil influence the success of establishment of surface sown seed is provided by Campbell & Swain (1973) and Dowling et al. (1971). Both emphasise the importance of suitable conditions for the entry of radicles into the soil. Campbell & Swain (1973) found that the ability of the radicles of four species to enter the soil declined in the order: perennial ryegrass, phalaris, lucerne and subterranean clover. Factors favouring grass radicle entry were better anchorage of the seedling, better angle of entry and smaller radicle diameter. Dowling et al. (1971) similarly showed poorer radicle entry of legumes, suggesting that an epigeal mode of germination appears to be more closely adapted to subsurface germination. Legumes in particular benefited from some form of restraint applied to the upper surface of the seed. Thus, any restriction placed on the movement of the seed during germination, such as that provided by surface roughness or cover, would aid establishment. Campbell & Swain found that on a level, finely structured surface, slight increases in soil strength markedly reduced percentage radicle entry. At constant soil strength, radicle entry was greater on coarse than on fine tilth, consistent with the generally observed advantage of heterogeneity of the soil surface on establishment. Artificially made depressions in the soil surface increased radicle entry of legumes (Dowling et al. 1971).

Agronomic studies of white clover establishment generally conform to expectations based on the principles given above (Stapledon & Wheeler 1948, Sears et al. 1955, Sonneveld 1958, Yoshihara 1958, During et al. 1962, Cullen 1966a, Herriott 1970, Andries & van Slijcken 1968). Deep sowing (4 cm) was, however, superior to shallow sowing at times of the year when there was moisture at depth but not in the surface layers (Stapledon & Wheeler 1948).

Indication of interaction between seed placement in the soil and species competition was provided by an establishment study of Hart et al. (1968). They found that drill sowing gave better establishment and long-term productivity of tall fescue than broadcast sowing. White clover gave better results from broadcast sowings. Hart et al. considered that these results were attributable to the effect of tall fescue competition on white clover when the two species were placed in close proximity in the drill which was set to sow at 1.3 cm depth in rows 20.3 cm apart.

(D) Time of sowing

The success of white clover establishment may be influenced by the season of seeding. Preference for autumn or spring sowings in different regions is related to differences in the seasonal distributions of temperature and precipitation. Stapledon & Wheeler (1948) in the UK compared the establishment of Italian ryegrass, perennial ryegrass, cocksfoot, timothy and white clover in 2-weekly sowings throughout a year. White clover establishment completely failed during winter, almost entirely because of seedling death. The young seedlings of clover were more vulnerable to winter conditions than seedlings of grasses of equivalent age. Two weeks development under favourable conditions made the difference between survival and death. Suckling (1950), in oversowing studies in hill country in New Zealand, determined percentage establishment from monthly sowings from March through to October. Best establishment was from the early autumn (March) sowings, which ensured longest continuity of moisture supply and sufficient time for establishment before frost damage could occur.

Langille & Warren (1964) in Canada observed establishment and production of Ladino clover in a mixture with *Bromus inermis* and lucerne sown at 2-week intervals from the start of June to early September. Establishment of all three species was successful for sowings up to mid-July, but thereafter there was a decline, particularly for legumes (Figure 1). However, 3 years after sowing there was little difference in composition between swards sown at the different dates.

During et al. (1963), on a bouldery yellow-brown loam near Te Anau, New Zealand, compared oversowing of inoculated white clover seed in July, August, September and November. Best results were obtained from late winter (July and August) sowings, whereas early summer (November) sowing was much inferior. Sowing in November was followed by 2-3 warm, sunny days which possibly reduced the number of effective clover rhizobia on the seed surface. It was considered that more rapid effective nodulation would have reduced the importance of time of oversowing.

Sowing white clover cv. Grasslands Huia in different seasons in New Zealand was studied by Brown (1973a). Emergence from autumn (March) sowing (91%) was greater than from early and late spring sowings. However, late spring sowing produced the highest herbage yield in the first year, which was attributed to the rapid initial development of spring-sown pastures. Similar advantages for late-spring and summer sowings over early-spring and autumn sowings were shown in another experiment in which white clover cv. Grasslands Huia and Grasslands Pitua were compared (Brown 1973b). When mixtures with ryegrass were compared with monoculture sowings, suppression of white clover by ryegrass was greatest in the late-spring sowings.

(E) Disease and predators

White clover seedlings are susceptible to a wide range of diseases and predators and seedling death is often important in determining the percentage establishment in sown swards. A comprehensive coverage of the diseases and pests of white clover is provided in Chapters 11 and 12, and only a few examples are given here. Mence (1964) examined the interaction between soil temperature, soil pathogens and seedling mortality of *T. repens* and *Phleum pratense*. *T. repens* started slow development under adverse temperature conditions (0-10°C) but suffered heavy mortality from attack by soil fungi. The growth of white clover seedlings can be markedly reduced by the effects of the parasitic root nematodes *Heterodera trifolii* (clover cyst nematodes) and *Meloidogyne hapla* (root-knot nematode) (Healy et al. 1973; Yeates et al. 1975).

Radcliffe (1970) observed that young nodulated white clover was particularly susceptible to the third-instar larvae of the scarab *Costelytra zealandica*. Similar susceptibility of white clover seedlings to predation by larvae of the subterranean grass caterpillar (*Wiseana* spp.) has been shown (Harris 1969, Harris & Brock 1972). In the establishment phase, white clover seedlings are particularly susceptible to grazing by slugs (Charlton 1978).

(F) Effects of associated plants on establishment

i. Relative establishment rate

Differences in relative establishment rates can influence the species composition of swards. The growth rates of white clover seedlings and seedlings of other forage plants following spring and autumn sowings were compared by Blaser et al. (1956) in the USA at two locations in Virginia. White clover varieties fell into the categories 'aggressive' (Ladino) and 'non-aggressive' (S100, Virginia and Louisiana) but not into the 'very aggressive' category, which included lucerne, red clover, sweet clover, crimson clover, Italian ryegrass and perennial ryegrass. Ladino and Virginia white clover showed greater seedling growth with autumn sowing than with spring sowing, while lucerne and red clover showed the reverse affect. Grime & Hunt (1975) examined relative rate among 132 species and considered its adaptive significance in the flora of Britain. The measured 'maximum potential relative growth rate' (R_{max}) of white clover was 1.26 ± 0.08, close to the modal value of all species measured. Values of R_{max} ranged from a minimum of 0.22 for the tree species *Picea stichensis* to 2.70 for the annual weed grass *Poa annua*. Examples of values for other species with which *T. repens* is often associated in pastures are *A. tenuis* 1.36, *D. glomerata* 1.31, *Holcus lanatus* 2.01 and *L. perenne* 1.30. Comparison of vegetation samples from a range of habitats showed that in disturbed and/or productive habitats, fast-growing species were predominant, whereas the reverse was true for stable, unproductive habitats. Species such as white clover with R_{max} near to the average were ubiquitous.

ii. Density of plants

The longer term study of density effects on the performance of white clover is made difficult by the vegetative spread of plants, but the indications are that the number of plants per unit area tends to be equalised by the mortality of plants at high seed densities and the greater vegetative spread at low densities. Early yield is generally increased by high sowing density, but after a period of rapid growth these density-related differences generally disappear. For example, Brown (1973a) observed that a density of 1030 seeds/m^2 provided a production advantage over 440 seeds/m^2. However, these effects disappeared once conditions became favourable for rapid clover growth.

Varying the seed rate of other species when establishing swards influences the early yield contribution of white clover but, again, these differences do not usually persist. For example, Harris et al. 1973 varied the seed ratio of ryegrass cv. Grasslands Manawa:white clover in mixtures from 8:1 to 2:1 and 1:1. The contribution of white clover was affected in the first year (Table 2) but the effect had disappeared after 16 months.

Table 7.2 Grass-white clover seed ratio effects on grass and legume herbage yields (kg/ha) in the establishment year. (Harris *et al.* 1973)

	Grass-clover seed ratio		
	1.1	2.1	8.1
Manawa ryegrass	5870	5670	5810
Huia white clover	1440	1160	750

Brougham (1954a) performed an experiment in which short-rotation ryegrass was sown at 0, 11, 17, 22, 34, 45 and 67 kg/ha in mixture with red and white clover at constant rates of 4.5 and 3.4 kg/ha, respectively. Yields for the first 6 months after autumn sowing (Figure 2) show that the linear downward trend was similar for both species as the amount of ryegrass seed increased. The yield of white clover fell from 246 to 67 kg/ha. This was associated with a curvilinear decline in survival of white clover seedlings.

Cullen (1964c) observed similar effects on white clover establishment when a basal mixture of cocksfoot, timothy and clovers was sown with 8 rates of perennial and short rotation ryegrass ranging from 0 to 34kg/ha. The suppressive effects of high sowing rates of ryegrass were more marked on a high fertility soil than on a soil of medium fertility. Unlike the effect on cocksfoot and timothy, the suppression of white clover by high ryegrass seed rates did not persist in the long term.

Parry & Wynn (1954) compared four mixtures in which a 50:50 mixture of New Zealand perennial ryegrass and S23 perennial ryegrass seed was sown at rates of 27 to 18 kg/ha in mixture with white clover at respective rates of 1.1 to 4.5 kg/ha. In the establishment phase the white clover content in the sward increased with increasing clover seed rate and declining ryegrass seed rate. Over a longer period the clover content became similar for all mixtures, but the *Agrostis* content was greater where high ryegrass and low white clover seed rates were combined. Heddle & Herriot (1955) compared eight mixtures in which S23 ryegrass and S143 cocksfoot were sown at either 11 or 22 kg/ha with superimposed seedings of 3.3 or 1.1 kg S100 white clover/ha. In no case did the seed rate of either clover or grass affect the final percentage establishment of clover. Weed suppression by the high seed rate of ryegrass appeared to aid legume establishment.

Herriot & Wells (1960) observed the interacting effects of sowing 11.2 kg S23 ryegrass or S143 cocksfoot/ha with or without fertiliser nitrogen and with 0, 1.1 and 3.3 kg white clover/ha. The white clover seed rate had no measurable effect on grass establishment. Percentage establishment of white clover at two months after sowing was higher with cocksfoot than with ryegrass and was proportional to the seed rate. Seed rate effects were transitory in terms of ground cover, having disappeared at eight months after sowing, but by this time there was a reduction of white clover in the presence of cocksfoot. There were, however, long-term gains of yield from increasing the seed rate of white clover, attributed to the decay of the larger root mass of white clover sown at high rates.

The need to adjust the sowing rates of grasses in grass-legume mixtures to take account of the grass variety was stressed by Green & Corrall (1965). They found that at a given ryegrass seed rate, a larger quantity of clover grew with Irish ryegrass than with the bred varieties S23 or S24 and that Irish ryegrass sown at 22.4 kg/ha was equivalent to S23 or S24 ryegrass sown at about 6 kg/ha in terms of producing herbage with similar clover content.

Experiments in which white clover has been sown with lucerne have shown negligible effects on the performance of lucerne (Garner & Sanders 1940, Fribourg & Kennedy 1953).

In pure and mixed culture, the number of mature plants becomes independent of the number sown above a certain density (Yoda et al. 1963, White & Harper 1970). The ability of plants to survive under high density stress is related to their individual plasticity, as is illustrated by the density related adjustment of plant size, stolon number and length, and flowerhead number in an experiment by Clatworthy (1960). Ten seedlings of *T. repens* were grown in the presence of 0, 5, 10, 20, 40, 80 or 160 seedlings each of *L. perenne, A. tenuis, Trifolium fragiferum* and *T. repens*. Increasing density of each of the associated species reduced the size and number of stolons and flowers of the white clover plants (Figure 3). *L. perenne* reduced the size of white clover plants most, while *T. fragiferum* caused less reduction than did competition from white clover itself.

iii. Cover crops and cover

The interaction between cover crops and undersown pasture plants has been reviewed by Charles (1958) and Santhirasegaram & Black (1965). The main reason for cover crops is to compensate for the low first-year production of pasture by obtaining a cash return from the cover crop. It has also been suggested that the cover crop affords protection to the establishing seedlings, and there is evidence to support this where shelter is provided from extremes of temperature without detrimental effects of competition for light, nutrients and water.

The success with which clover establishes under cover crops may be determined by a variety of factors which are broadly classified by Santhirasegaram & Black (1965). These include the use of crop species or varieties which offer less competition for light, nutrients and water; reduced seed rates of cover species; defoliation of the cover crops at an early stage of growth; and control of the spatial arangement of cover crop and pasture plant, for example, by sowing alternate rows of crop and pasture species.

T.E. Jones (1933) showed a detrimental effect of a barley and oats cover crop on wild white clover. Brougham (1954), in comparisons of barley, short-rotation ryegrass and timothy undersown with red and white clover at 4.5 and 3.3 kg/ha found that the barley cover crop reduced early clover yield to 25% of that in a simple mixture with short rotation ryegrass. However during spring, clover yields increased when conditions favoured rapid development of buried clover seed, and ryegrass yields were reduced by poor nitrogen nutrition.

In the USA, Peters (1961) showed that oats as a cover crop was highly competitive with both undersown legumes and weeds during the period of oats growth. Ladino white clover and birdsfoot trefoil were suppressed more than lucerne and red clover. Recovery of lucerne, red clover and Ladino clover after removal of the oats was rapid under favourable conditions, but under unfavourable conditions the depressing effect of the oats continued into the second year of growth. Davis (1962) in Canada found that irrespective of stage when harvested, oats reduced weed content and yield of cocksfoot and Ladino clover. In the second year reduction of cocksfoot and white clover yields continued under silage and grain management, but where the oats had been sown, yields of grass and clover were equal to or greater than the yields of grass and clover without a companion crop.

In New Zealand, Cullen (1964a) showed that rape (*Brassica napus*) severely suppressed white clover, particularly when sown at a high rate (2.2 kg/ha). This suppression was most severe on a high fertility flat-land site and persisted for a year after sowing, but in a less fertile hill soil, clover suppression was less marked and was similar at all rape seed rates after 7 months. Yields of all pasture components of a ryegrass, cocksfoot, red and white clover mixture were decreased as rape seed rate increased in the range 0 to 2.2 kg/ha (Harris et al. 1973).

Similar suppression of clover and its accentuation under high soil fertility occurred where wheat and oats were the cover crops (Cullen 1964b). Establishment of undersown white clover was unaffected by differences of wheat sowing rate (Scott 1974). Clover sown in winter with wheat established better than that overdrilled the following spring. Irrigation improved the survival of white clover where it was overdrilled late. Charles (1963a, b) in the UK examined the establishment of mixtures with a basal component of S100 white clover. Although white clover made little contribution to total yield in the seeding year, nitrogen application, infrequent defoliation and an oats cover crop reduced yields still further, effects which persisted into the following year. In a related experiment, Charles (1965) examined the effects of an oats cover crop on pasture mixtures based on Italian ryegrass or tall fescue with or without red clover, but in which all contained white clover (Table 3). When sown as the only legume with Italian ryegrass, change from frequent grazing to infrequent cutting increased clover yield, but where red clover was included, white clover was reduced under all regimes, particularly where defoliation was infrequent. In the tall fescue-white clover mixture, management did not affect the legume yield but with red clover added, white clover was reduced by infrequent defoliation and further reduced when undersown with oats.

Table 7.3 Effects of cover-crop, defoliation frequency and companion grass on dry matter yield of white clover (kg/ha) in the year of sowing (Charles 1965).

Associate species	S22 Italian ryegrass	S22 + red clover	S170 tall fescue	S170 + red clover
Management				
5 grazings	767	474	1696	894
Cut at grain harvest grazed once in autumn	1279	337	1671	492
Undersown with oats grazed once in autumn	1365	345	1563	343

During et al. (1963) observed the effects of cover provided by stones and fern on the success of establishment of oversown white clover. Cover afforded protection from frost heave after winter sowing, and when plants were observed in the following summer, the percentage of healthy plants increased with increase of cover. Scott & Archie (1976) found that the proximity of *Festuca novae-zelandiae* tussocks on a dry hygrous soil had little effect on spring germination, but mortality during summer was less on the cooler, shaded side of tussocks. Percentage establishment of white clover, which performed best of several tested, fell from 53% in spring to 0.4% by the end of summer, indicating that stress in that environment was severe.

iv. Defoliation effects

Defoliation may affect the performance of white clover seedlings during establishment by directly removing photosynthetic and reserve tissue from the seedling, by removing herbage of associated species which have overtopped and are shading the seedlings, by modifying other micro-climatic factors such as temperature and humidity, and indirectly by modifying root processes and nutrient uptake.

Blaser & Killinger (1950) in Florida, USA, applied four management treatments to a carpet grass-Louisiana white clover sward after it had been spelled during summer (Table 4). The responses were ascribed to temperature effects, seed coat breakage caused by burning, and defoliation reducing temperatures at the soil level, thus allowing dormancy to be broken (Burton 1940).

Effects of differences in the height of defoliation during the establishment of pasture sown with ryegrass, white and red clover were shown by Sears (1950) in a series of experiments in New Zealand. All experiments involved defoliation to 2.5 cm when the pasture had reached 7.5, 15 or 22.5 cm height. Effects of height before defoliation in all experiments were similar and are illustrated by yields up to October and January for a grazing experiment sown in autumn (March). Increased height before grazing decreased

Table 7.4 Numbers of seedlings of white clover under cutting and burning management of summer-spelled pasture (Blaser and Killinger 1950).

Management Applied November 14	Seedlings/m^2 December 12
Undisturbed	8
Mowed to 7.5 cm	74
Mowed and burned when dry	568
Cut to 1 cm stubble	304

white clover content and these effects persisted into summer under a common grazing regime (Table 5).

Also in New Zealand, Harris et al. (1973) examined variation of height before grazing on the establishment of autumn- and spring-sown, ryegrass-white clover pasture. More lenient grazing increased pasture production from autumn sowing, but there was no effect on yield of clover. For a spring sowing, white clover yield over 18 months was much less under an 18-20 cm to 1 cm than under a 5-6 cm to 1 cm grazing regime. White clover yielded more if grazed less frequently in summer but more if grazed more frequently in autumn.

Adverse effects of inappropriate grazing height on white clover establishment were absent when the sowing rate of short-rotation ryegrass was reduced from 45 to 17 kg/ha (Brougham 1954b).

The foregoing results are explained largely on the basis of inter-species competition for light during establishment where soil fertility is high. At high ryegrass seed rates, clover is shaded earlier and this is intensified if the herbage is not grazed off to a relatively low height. Defoliation during establishment may also affect root competition. Evans (1973b), comparing perennial ryegrass, cocksfoot, timothy, *A. tenuis*, red clover and white clover, found that such defoliation rapidly reduced root elongation; all species then gradually recovered. The depression of root elongation was greater in grasses than in clovers. Defoliation every second day to 25 mm above the base of the shoot almost completely stopped root elongation in all species except white clover, in which it was reduced to about 5% of that in undefoliated plants (Evans 1973a). Similar results were obtained by Tosun (1967) who found that clipping to a height of 15 cm at one-month intervals depressed root growth, but the effect on white clover was comparatively small. Frequent defoliation may therefore improve the competitive ability of white clover both below and above ground.

Table 7.5 Effects of grazing height on white clover yield and plant numbers in November in a ryegrass, white clover and red clover mixture sown in March at Palmerston North, New Zealand (Sears 1950).

Pasture height at grazing up to 21 October	Period ending	Number of grazings	Total yield kg DM/ha	White clover yield kg DM/ha	White clover plants/m^2
7.5 cm	21 October	8	5085	78	197
	30 January	3	6754	1053	
	Total		11838	1131	
15 cm	21 October	4	6530	67	148
	30 January	3	5947	526	
	Total		12477	594	
22.5 cm	21 October	2	6910	11	66
	30 January	3	5914	224	
	Total		12824	235	

v. Disturbance of resident vegetation to aid establishment of oversown white clover

Many reports describe the establishment of white clover oversown into resident vegetation. Charles (1962) concluded that white clover is easier to establish than grasses of economic value. Establishment has sometimes been aided by the presence of cover (During et al. 1963), but most studies indicate the benefit of a partial destruction of the resident vegetation. The rate of establishment of white clover from the buried seed population in a dense old permanent pasture in North Wales was about 0.5 seedlings/m^2 annually, but was much higher in disturbed areas such as molehills (Turkington et al. 1979)

Vegetation may intercept seed before it reaches the soil surface. Edwards et al. (1972) in South Africa oversowed white and red clovers and *Paspalum dilatatum* into three highland sourveld grass species. Up to 40% of oversown seed was intercepted and retained by the crown and leaves of the grass. The percentage establishment of white clover declined with an increase in the density of cover. Recommendations for grazing management to aid establishment of clover in New Zealand hill country by oversowing (Suckling 1954, 1959) are that pastures should be grazed short immediately before sowing to open up the sward and thus allow seed to reach the soil surface. Management should then aim to prevent suppression of clover seedlings by development of long pasture by using either continuous grazing or intensive rotational grazing. Cullen (1966b) similarly recommended reduction of existing vegetation by use of chemicals, fire or close grazing.

Collins & O'Toole (1963) compared the effects of grazing by cattle or sheep on establishment of ryegrass-white clover in oversown bogland. Cattle, by grazing the sward less closely and not selecting the 'finest' herbage, allowed more rapid development of a better balanced ryegrass-white clover sward.

Reference has already been made to the effect of burning resident vegetation on the establishment of natural reseeded white clover (Blaser & Killinger 1950). Collins & O'Toole (1962) found that the burning of native bog vegetation hastened the establishment of white clover where oversown and top dressed with lime, P and K. Pasture establishment studies in Japan by Yamane et al. (1962, 1966) indicated benefits from removing dead mulch by burning to expose the soil surface to broadcast seed and fertiliser. Heavy grazing by sheep after broadcast sowing pressed seed into the soil and reduced competition from native vegetation. Burning mid-altitude tall tussock vegetation in New Zealand considerably increased the area of bare ground, and dry weather for several weeks after created conditions hostile to white clover establishment because of high temperatures and surface dryness (O'Connor & Lambretchsen 1964). Where tussock was not burnt, the inter-tussock areas were covered with a mulch of litter which allowed better clover establishment.

Partial destruction of resident vegetation by machinery generally aids the establishment of white clover. Sprague et al. (1947) found that on areas untilled but which were severely clipped to weaken the existing sod, and limed, fertilized and oversown, Ladino and red clover established satisfactorily, but lucerne, birdsfoot trefoil and sown grasses did not. Hughes et al. (1965) in Wales achieved better white clover establishment on wet *Molinia* pasture after treatment with a forage harvester than with rotary cultivation. Rowlands (1966) compared ploughing and reseeding with surface cultivation using a disc harrow on a Welsh hill-land *Festuca-Agrostis* pasture. Establishment was considerably better under full cultivation, but performance of oversown white clover was considerably better than that of ryegrass on surface cultivated areas. Establishment of oversown white clover was enhanced by surface cultivation of a poor downland pasture at Hurley, England, especially when the treatments were applied in spring rather than summer, and no nitrogen was applied (Norman & Green 1957a).

Modification of resident vegetation by herbicides enhances the establishment of white clover. Jones (1962) compared treatment with dalapon or paraquat together with oversowing, and surface cultivation. Six months after treatment some increase of white clover occurred in plots treated with herbicide alone, but where oversowing was carried out without cultivation the increase was approximately 10-fold. Surface cultivation, however, favoured grass establishment and depressed clover yield. Norman & Green (1957b) found that treatment with 2,4-D amine together with surface cultivation depressed the content of wild white clover but in the year after sowing enhanced the development of cv. S100 Nomark white clover. Other examples of the many experiments which have demonstrated the beneficial role of herbicides in white clover establishment from oversowing are provided by Blackmore (1965) and Allen (1966).

Taylor et al. (1972) in Kentucky, USA, studied the establishment of white clover in a *Poa pratensis* dominant pasture. Subsurface seeding at 0.6 cm depth was better than surface seeding and tillage also improved establishment. Paraquat banding did not aid establishment.

vi. Moisture and nutritional effects

Considerable practical experience has shown the need for adequate mineral nutrition of white clover to ensure successful establishment when oversown into resident vegetation (Suckling 1954, 1959, Norman & Green 1957a, Collins 1961, Cullen et al. 1966, Ludecke 1966, Lowther & McDonald 1973). Stimulation of the non-leguminous components of vegetation by applications of N usually impairs the establishment of white clover (Thompson & Elliot 1950, Collins 1961).

Wilkinson & Gross (1964) examined the establishment of Ladino clover in cocksfoot (*Dactylis glomerata*) sod in a pot experiment using shoot and root partitions (technique of Donald 1958). Five competition treatments compared (i) full competition for mineral nutrients, water and light, (ii) shoot competition only, (iii) root competition only, (iv) 'no competition', and (v) a monoculture with Ladino clover grown alone. Three Ladino clover seedlings were established in each pot, and at 18 days after sowing four moisture level treatments were imposed.

Dry weight (mg) per plant at the end of the 18 days before applying the moisture treatments were: full competition 1.2, shoot only 1.9, root only 1.3, no competition 2.4 and monoculture 3.8. Wilkinson & Gross considered that at this early stage, because of high soil fertility, mineral nutrient deficiency probably did not account for the reduced seedling yield under root competition. Growth of the seedlings at this early stage would have been influenced by cotyledonary reserves, and the extent of shading was less than during subsequent growth periods when shoot competition caused a greater effect than root competition (Table 6).

Wilkinson & Gross (1964) had difficulty in explaining the effect of root competition where nutrients and water were adequately supplied. Plant tissue analyses indicated non-limiting levels of P, K, Ca, Mg, B, Cu, Fe, Zn and Mn. They considered, but largely dismissed, the possible role of allelopathy. However, a review of the effects of associate grasses on white clover content by Chestnutt & Lowe (1970) has suggested a specific suppressive effect of cocksfoot. They concluded that Ladino clover can be readily established in N fertilised cocksfoot sods if light, moisture and nutrients are favourable.

Table 7.6 Dry matter production (g/pot) of Ladino clover seedlings after 10 weeks growth in association with established cocksfoot plants (Wilkinson and Gross 1964).

Competition	Moisture Level				Mean
	Low	Medium	High	Very High	
Full	0.20	0.21	0.27	1.03	0.43
Shoot only	0.50	1.51	1.62	2.64	1.57
Root only	0.83	1.53	2.87	3.82	2.26
No competition	2.01	2.98	7.44	4.66	4.27
Monoculture	1.46	3.80	6.75	7.29	4.83
Mean	1.00	2.01	3.80	3.89	2.67

Clatworthy (1960) sowed white clover at a range of densities in pot culture with 3 water regimes: adequately watered and freely drained, water table maintained close to the surface, and water table maintained at the soil surface (Figure 4). Under freely drained conditions the plants grew rapidly and a response to density was shown. Mean size per plant was reduced by a high water table, but a density response paralleling that in the freely drained treatment developed. The waterlogged plants were so stunted that they did not interfere with each other and no density response was shown.

vii. Allelopathic effects

Newman & Rovira (1975) felt that allelopathy (the influence of plants on others by means of chemical substances) is widespread among grassland species. Because of the difficulties of separating allelopathic effects from the effects of established plants on the micro-environment, most experiments investigating allelopathy have examined the effects of plant extracts on seed germination and the early growth of seedlings.

Hoveland (1964) examined the effect of aqueous extracts of dried and ground roots and rhizomes of 6 grasses on the germination and seedling vigour of 4 clovers. Germination of intermediate white clover (Table 7) was sharply reduced by *Sorghum halepense* extracts, to a less extent by *Paspalum notatum, P. dilatatum, Sorghum almum* and *Cynodon dactylon,* but not by *Festuca arundinacea.* Radicle length of white clover was reduced by all extracts after 2 days, but after 4 days *F. arundinacea* extract stimulated radicle length. Inhibitory effects of the extracts were alleviated when seed were transferred to distilled water. Dilutions of the root extracts of *Sorghum halepense* to concentrations less than 50% did not influence the germination of white clover, but a 12.5% concentration inhibited radicle growth. However, Hoveland (1964) doubted whether concentrations of inhibitors secreted from live roots in undisturbed swards would significantly influence clover.

Table 7.7 Effect of grass root extracts on white clover seedlings (Hoveland 1964).

	Percentage germination		Radicle length mm	
	2 days	4 days	2 days	4 days
Distilled water	84	88	6	15
Cynodon dactylon	21	75	3	4
Paspalum notatum	28	71	3	11
Paspalum dilatatum	23	72	3	7
Sorghum halepense	0	50	0	2
Sorghum almum	4	75	1	3
Festuca arundinacea	80	90	3	17

Grant & Sallans (1964) investigated the influence of aqueous extracts of dried and ground roots and shoots of 4 grasses and 4 legumes on the germination and shoot and root lengths of the same species. Shoot extracts of *Medicago sativa, Trifolium pratense* and Ladino clover significantly depressed the germination of Ladino, but root extracts did not. Legume extracts depressed both shoot and root elongation more than did grass extracts. Depressive effects were shown by both shoot and root extracts of *M. sativa* and by shoot extracts of *T. pratense.*

Ladino clover shoot extracts significantly reduced the germination of all species tested except *Phleum pratense* and *Lotus corniculatus.* Root length in *Bromus inermis* and *T. pratense* was also much reduced. Ladino clover root extracts depressed germination of *Phalaris tuberosa,* and appeared to promote shoot length of the grasses but depressed shoot length in the legumes. Root length of legumes and to a less extent of grasses was reduced.

Campbell (1959), following up the observation that it was difficult to establish white clover after a crop of chou moellier (*Brassica oleracea*), recorded that germination of white clover seeds was reduced from 87% to 10% in the presence of minced chou moellier roots placed on an adjacent filter paper in the same Petri dish. Stem and leaf fractions and unminced roots exerted smaller effects. The effects persisted even after transfer to distilled water treatments. In seedlings from seeds that germinated, radicles became blackened at the tips, root hairs were sparse and within 3 weeks the seedlings had died.

The difficulty of establishing legumes in New Zealand hill country dominated by *Notodanthonia* species may be caused by a toxin secreted by the grass. Parle (1964) suggested that a substance toxic to rhizobia, although associated with living *Notodanthonia,* may be formed by organisms associated with the grass roots. Inhibition of rhizobia and nodulation by *Notodanthonia* was observed by Janson & White (1971).

Scott (1975) recorded the effects of freeze-dried shoot and fresh root-soil materials from several tussock grasslands species on seed germination of white clover. Most of the materials promoted the germination, but shoot material of white clover severely inhibited germination of the test species.

4. The established plant

Productivity of white clover when grown without competition from other species indicates the limits placed on dry matter productivity by environmental factors and genetic potential. Consideration of productivity in monoculture will be followed by examples of the range of variables which have been shown to affect the content of white clover in plant communities. A few experiments in which monocultures and mixtures were grown simultaneously will be considered in detail. Finally, morphological and physiological characteristics of white clover important to its population dynamics and competitive strategies will be examined.

(A) Productivity in monoculture

In temperate climates, annual production of white clover monocultures ranges from about 4 000 to 14 000 kg DM/ha. Reviewing information from experiments in Britain, Spedding & Diekmahns (1972) record a highest annual yield of 11 800 kg DM/ha of S100 white clover under irrigation at Hurley, but some of this yield was from unsown species. Annual yields from unirrigated sowings over eight years varied from 3900 to 8650 kg DM/ha. For weed-free crops of white clover under a 3-4 cut system and low rainfall, production varied from 5000 to 5800 kg DM/ha. With irrigation and monthly cutting during the growing season, annual production was 8700 kg DM/ha. Davies (1970) recorded an annual yield of 9000 kg/ha from Kersey clover at Aberystwyth, Wales.

In New Zealand, Brown (1973b) recorded mean yields under cutting of 6460 and 6040 kg DM/ha from white clover cv. Huia at Gore, Southland (latitude 46°S.), similar to average yields recorded in England and Wales. Further north at Palmerston North (latitude 40°S.) Anderson (1973) obtained annual yields under cutting of 12 490 (84% clover content) and 12 560 kg/ha (96% clover content) from pure sowings of cv. Huia white clover on a silt loam without irrigation. Annual yields of six red clover cultivars in the same experiment varied between 10 000 and 13 500 kg/ha. Brock (1973), also at Palmerston North, obtained weed-free annual yields of cv. Huia under cutting in three different years of 10 110, 12 010 and 9760 kg/ha on a lighter soil with adequate phosphate and irrigation. Respective yields for *Lotus pedunculatus* were 10 360, 10 910 and 8 600 kg/ha and for *Trifolium dubium* 4790, 7639 and 4420 kg/ha. Williams & Barclay (1975) obtained an average of about 8500 kg/ha over two years from pure sowings of New Zealand and Spanish white clover on silt loam at Palmerston North. Annual production, inclusive of some weeds, of 13 000 kg DM/ha on the same silt loam at high level soil fertility and under rotational grazing by sheep, rather than cutting as used in the other cited studies, was recorded by Harris & Hogland (1980) for a pure sowing of Huia white clover. Daily growth rates of white clover reached a peak of 65 kg/ha in mid-summer. By comparison, a pure sowing of Ruanui perennial ryegrass yielded 17 500 kg/ha where fertiliser N was applied, with daily maximum growth rates of 90 kg/ha in late spring. The annual production potential for adequately fertilised, irrigated ryegrass-white clover pasture at Palmerston North has been estimated as 24 000 kg/ha (Brougham 1959a).

(B) White clover content in swards

Consideration of yield of white clover growing with grass in Britain led Spedding & Diekmahns (1972) to conclude that 'any random collection of data from mixed swards will show how variable the actual contribution can be'. Standardised measurements of grazed pastures at 11 geographically dispersed sites (Radcliffe 1974a) provide an assessment of variation in white clover content in swards on farms in New Zealand. Details for pastures in which clover was the main legume are presented in Table 8.

Although the sites span more than 10° of latitude, no clearly defined differences of clover yield related to latitude were found. Neither were there any clear relationships between clover yield and total yield. The Flock House site, which had the lowest total and clover yields, is located on a sandy soil which accentuates summer drought which probably directly limited both total and clover yield. An indication of the range of variation of clover content at a given level of production can be obtained by examining the median range of annual total yields (10 000 to 12 000 kg/ha). Within this range clover content varied from 11.8 to 38.5%.

Peak growth rates span 2 months from 15 November to 25 January. Year-to-year variation at several sites was greater than site-to-site variation. Between-year variation was least for the high rainfall site at Westport and the irrigated site at Winchmore, which suggests that much of the year-to-year variation of clover content in New Zealand may be associated with rainfall.

Table 7.8 White clover contents of several New Zealand pastures

Site	Dargaville	Manutuke	Flock House	Marton	Motueka	Westport	Winchmore	Invermay Taieri Plain	Invermay Hill Site	Winton	Mana Bush
Latitude	35° 55′S.	38° 41′S.	40° 16′S.	40° 05′S.	41° 10′S.	41° 41′S.	43° 48′S.	45° 52′S.	45° 52′S.	46° 11′S.	46° 22′S.
Altitude m.a.s.l.	20	9	9	141	140	60	160	30	100	45	50
Annual rainfall (mm)	1220	998	880	1083	1090	2152	771 + irrigation	687	687	1086	1042
Soil type	Gley estuarine	Recent alluvial heavy silt loam	Alluvial loamy sand	Yellow-grey earth, silt loam	Lowland yellow brown earth	Lowland podzolised yellow brown earth	Silt loam	Recent alluvial	Yellow-grey to yellow brown earth, silt loam	Yellow-grey to yellow brown earth	Silt loam
Number of years clover data	12	14	6	7	6	6	13	12	18	11	6
Total annual yield kg/ha (range)	17 160 (12260–21460)	11 730 (7240–16300)	6 270 (5480–7680)	10 850 (8080–15550)	11 560 (8080–13280)	10 920 (9970–11660)	10 160 (9050–12020)	10 390 (7520–12760)	8 890 (3240–13500)	12 010 (8870–16000)	14 610 (10170–17290)
White clover annual yield kg/ha (range)	2 170 (not stated)	1 380 (460–2500)	1 280 (720–2490)	2 510 (1310–4480)	2 140 (1240–3350)	4 200 (3910–4870)	2 720 (2060–3290)	2 540 (1660–4480)	1 590 (380–2750)	2 690 (1420–4370)	2 280 (1650–3180)
% clover in total yield	12.6	11.8	20.4	23.1	18.5	38.5	26.8	24.4	17.9	22.4	16.1
White clover maximum daily growth rate kg/ha	12.7	9.1	9.3	14.4	16.8	31.2	18.4	18.9	11.7	19.0	24.4
Date of maximum growth rate	28 Dec.	28 Dec.	15 Nov.	25 Jan.	30 Nov.	30 Nov.	11 Jan.	15 Nov.	14 Dec.	25 Jan.	28 Dec.
Source of data	Baars 1976	Radcliffe & Sinclair 1975	Radcliffe 1976	Radcliffe 1976	Radcliffe 1975	Radcliffe 1975	Rickard & Radcliffe 1976	Round-Turner et al. 1976	Round-Turner et al. 1976	Radcliffe 1974	Radcliffe 1974

i. *Influence of weather on white clover content*

It is difficult to distinguish between the effects of light and temperature on the clover content of pasture, as both factors are closely associated with seasonal variation in incoming radiation. In addition, other factors such as rainfall and growth patterns of associated species may be more important in actual site differences of clover content.

Mitchell (1956) demonstrated that the rate of white clover growth was maximum at about 24°C but that the rate differed little over the range 18 to 30°C. Accounting for seasonal light limitations on growth, Mitchell (1956) estimated that the ratio of potential production of white clover relative to ryegrass would be considerably greater in the far north than in the south of New Zealand. Apart from the effects of rainfall, soil factors and management, two further effects confound this temperature based prediction of clover content in New Zealand pastures: in the warmer climate of the north clover would be subjected to more interference from summer active grasses such as *Paspalum dilatatum* (Lambert et al. 1977); and there could be regional adaption of white clover (Hoglund & Brock 1974, Goold & Douglas 1976).

Correlations between temperature, radiation, precipitation and evaporation and white clover percentage were estimated by Kleter (1968) from results obtained from *Lolium perenne, Poa trivialis* and white clover pasture near Wageningen, Netherlands, over six years. Because the correlations were on month-to-month variations, they largely show the relationships between these factors and the seasonal growth of white clover. Both temperature and light levels were shown to be correlated with clover content. The strongest correlation estimated was with mean daytime temperature in April, a 1°C rise of temperature in this month being associated with 1% increase in clover content. This was in a range of clover percentage of 2-10 which compares with the annual mean of about 21% in New Zealand pastures (Table 8). Chestnutt & Lowe (1970) concluded that the normal temperatures in the British Isles may restrict clover growth and, hence, clover content.

Clover content is a function of both clover and the growth rates of the associated grasses. New Zealand results (Table 8) indicate that the highest growth rates of clover in grass-legume mixtures occur at around the longest day (21 December). In Britain (Spedding & Diekmahns 1972) and the Netherlands (Kleter 1968), the clover contibution is greatest after mid-July. These differences in the time of peak clover content between regions could be caused by differences in the time optimum temperature conditions for clover growth are attained, but there is evidence to suggest that grass growth has an important effect on the time of maximum clover growth and content. Spedding & Diekmahns present results showing that yields of clover from monoculture stands were highest before July if water was not limiting.

The direct comparison at Palmerston North, New Zealand, of white clover in monoculture and in mixture with ryegrass shows how the peak of the white clover production curve can be markedly influenced by interference from grass (Figure 5). In this situation Harris & Hoglund (1980) argued that the peak of white clover production in monoculture, which coincided with optimum temperatures of 24°C, was suppressed in mixture with grass by competition for light and water.

Although Kleter (1968) found no correlation between clover and moisture levels as determined by rainfall and potential evaporation, observations in Great Britain suggest that white clover is more abundant in wetter years (Copeman et al. 1960, Cowling 1961, Green & Corrall 1965). In Australia Holford & Gleeson (1976) found that the incidence of annual rainfall in the range 146 to 580 mm on P-deficient granitic soil of the Northern Tablelands of New South Wales had a marked effect on white clover yield and was a controlling factor in the response to P and S. Simpson et al. (1968), in the Upper Sholhaven Valley in southern New South Wales, with an irregular annual rainfall of 760 mm, observed that lucerne and red clover survived summer drought but white clover did not.

ii. *Irrigation effects on clover content*

There is evidence that the white clover content of swards sown with grass is increased by irrigation. White clover also tolerates excess water if drainage is good.

In a 600-mm rainfall zone in the USA, Robinson & Sprague (1947) found that irrigating and clipping to 1 or 3 cm of a *Poa pratensis*-white clover sward resulted in excellent stands of clover. Clover content also remained high under irrigation with 400 kg N/ha and clipping to 6 cm, but without irrigation, high N and lax cutting virtually eliminated white clover. Similar increases of white clover with irrigation of *P. pratensis* and *D. glomerata* based mixtures in the USA, have been recorded by Nelson & Robins 1956, 1957, Robinson & Sprague 1962, Robinson et al. 1952 and Ward et al. 1966.

Irrigation of a riverside meadow on an alluvial soil with approximately 6.35 mm rainfall at Sonning, Berkshire, England, increased the proportion of *Festuca pratensis, Phleum pratense* and white clover at the expense of *Poa* and weed grasses (Munro 1958). Low & Armitage (1959) compared a control and three levels of irrigation and three rates of applied N on white clover-grass pasture at Jealott's Hill, Berkshire, England. With high soil moisture deficits ryegrass continued to grow vigorously under conditions of high N supply, whereas white clover did not. With irrigation, white clover became very competitive with the grasses. The difference in behaviour between grass and clover under the different irrigation regimes could not be explained by the depth of their root system, as grass root development was reduced under the high N rate. However, white clover grown in monoculture produced a greater root weight with deeper wetting of the soil (Cullen et al. 1972), which may conceivably alter the competition between grass and legume for nutrients.

Stiles (1966), in reviewing 10 years irrigation experiments at Hurley, England, comments that where irrigation is not applied, year-to-year variation in white clover is such that it is difficult to see clear effects of fertiliser N on clover content, but with irrigation N application progressively eliminated the white clover. In the Northern Tablelands of New South Wales, Australia, Lazenby & Lovett (1975) observed that in phalaris-white clover mixtures, white clover content under irrigation was much greater than under dryland conditions, although application of high rates of N still led to its disappearance after three years.

Appadurai & Holmes (1964) applied combinations of cutting frequency, cutting height and irrigation treatments to a deep, well drained chalk loam in Kent, England. The ratio of grass to clover was influenced more by moisture than by defoliation. In mid-summer a wet regime gave the highest clover content, but by the end of the season clover content was highest under the medium-wet regime, suggesting that the high moisture regime was excessive at this time. At Ayr, Scotland, Reid & Castle (1965) found that irrigation had no significant effect on white clover content. The annual rainfall at this site (914 mm) and low evapotranspiration probably did not allow soil moisture to fall to a level at which white clover growth was limited. Sato et al. (1967) found that soil moisture content had little effect on the growth of *D. glomerata* in mixture with either Ladino clover or lucerne, but that a moist soil favoured Ladino, while a drier soil favoured lucerne.

Ennik (1960) investigated the effect of ground water levels of 12, 26, 41, 70 and 90 cm below the surface on competitive interactions between ryegrass and white clover. The change of the grass-clover ratio during a 6-week period in the growing season was studied by counting grass tillers and the number of intersections of clover stolons with threads on a frame placed on the cylinders in which the mixtures were grown. The competitive interaction between grass and clover was examined using the concept of the relative reproductive rate, and ratio diagrams (de Wit 1960). It was concluded that at high water levels the grass took up all the N from the soil and that clover fixed a large amount of atmospheric N. At high ground water levels the two species were therefore obtaining N from different sources. At low water levels clover was not efficiently fixing N from the air and was less efficient in obtaining N from the soil than was the grass. Clover had a reduced relative reproductive rate under these conditions.

Available evidence suggests that white clover tolerates excess water better than other legumes. Duthion (1972) observed that white clover and birdsfoot trefoil were less sensitive to excess water than lucerne and sainfoin. Wahab & Chamblee (1972) found that Ladino clover continued to respond positively to irrigation over a 3-year period whereas lucerne declined, probably because of disease.

The effects of drainage on white clover content were demonstrated by Davies & Milton (1945) on upland peat in Wales. Sown grass establishment was successful at all drainage intensities. Clover content remained low on waterlogged areas, which rapidly reverted to the native species, but with reasonable and impeded drainage clover did not fall below 17% over a 5-year period.

iii. *Mineral nutrition and clover content*

(a) General soil fertility.

The complexity of the information on the nutritional requirements and fertiliser responses (Hallsworth 1958; Chestnutt & Lowe 1970, Murphy 1970 (see also Chapter 5) of white clover makes it difficult to review soil fertility effects on clover content. However, in general, but with the exception of N level in mixtures with grass, white clover becomes more competitive with companion species as fertility increases.

Harris (1971b) examined the competitive interaction between white clover, ryegrass and sheep sorrel (*Rumex acetosella*) grown at three levels of soil fertility and three of lime. The species were grown in pots as monocultures and in two- and three-species mixtures. Yield responses were analysed by the two-species competition model of de Wit (1960). Both fertiliser level and lime influenced the competitive power (relative crowding coefficients) of ryegrass and white clover. Ryegrass became more competitive as the fertiliser supply increased. White clover was most competitive at the medium lime level (pH 5.3) at the low fertiliser level. Yield was stimulated by mixing ryegrass and white clover, especially at the lowest fertiliser level.

Sheep sorrel tolerates acid soil, and lime level markedly influenced the competitive interaction between this species and white clover. The effect of fertiliser was not as marked as in the ryegrass-white clover mixture. Where both fertiliser and lime levels were low (pH 4.6), sheep sorrel dominated white clover and yield was stimulated by mixing. Yield was also stimulated where fertiliser level was low and lime level high, but although white clover had a dominant crowding coefficient in this treatment it did not suppress sheep sorrel.

This experiment (Harris 1971b) did not distinguish the separate effects of N, P, K, S, Ca and minor elements on the competitive responses of white clover. Evidence indicates that the availability of N is the most potent soil fertility factor determining the content of white clover in swards.

(b) Nitrogen fertiliser.

Application of fertiliser N to grass-white clover swards increases the competitive effects of the associate grasses and therefore usually suppresses white clover.

A series of experiments by Blackman (1934, 1938) and Blackman & Templeman (1938) involving different N formulations, associate grasses and shading led to the conclusion that the balance between legumes and grasses in swards depends largely on competition for light. As the density and height of grasses was positively correlated with the N supply, additional N depressed white clover by increasing shading by the grasses rather than by any direct effect of N on the clover.

The relationship between the direct response of a grass (*Festuca arundinacea*) to applied N and the associated reduction of white clover yield was shown in a cutting experiment by Frame (1973). At rates above 200 kg N/ha the contribution from clover was negligible. This agrees with the result of a survey of Chestnutt & Lowe (1970), which indicated that when N application exceeds 150 kg/ha little clover normally remains in pastures under British conditions.

Effects of fertiliser N application on the competitive interactions between 6 legume cultivars, including the white clover cultivars Grasslands Huia and Grasslands Pitau, and 9 grass cultivars in a grazing experiment were described by Harris (1980). Huia white clover declined in yield as the effect of grass competition increased. Pitau white clover yielded more than Huia in the presence of the less competitive grasses (*F. arundinacea, D. glomerata* and *Phleum pratense*) and in monoculture. The different behaviour of Pitau was caused by its greater capacity for cool-season growth. Expression of this potential was allowed where competition from associated grasses was low. The effect of applying 180 kg N in the establishment year was to emphasise the competitive effects of the associate grasses.

Analysis of the effect of nitrogen on the competitive interaction between white clover and grass using the theoretical approach of de Wit (1960) was described by Ennik (1970). In a field experiment, perennial ryegrass-white clover mixtures sown at different seed ratios were grown without application of N or with 100 kg N/ha annually. Three cuts were taken in the first year and four in the second. At early cuts, the respective yields of grass and clover were influenced by the initial seed ratios, but by the second year the grass-clover ratio was the same in all mixtures. By the first cut of the first year the grass in the treatment not receiving N was benefiting from the presence of clover, and at this time the clover was not reduced by the increased grass growth. By the third cut of the first year and for all of the second year, the clover in the mixture was lower than the clover yield in the monoculture because of the improved grass growth in the mixture. Nitrogen application increased grass yield and decreased clover yield in the mixture. The role of N fixation and transfer in these relationships is discussed in section 4.(E). Ennik (1970) concluded that the relative competitive abilities of grass and clover suggest that clover growth is limited to the 'space' left by the grass.

Studies using ^{15}N have shown that in pot-grown grass-white clover mixtures given a wide range of fertiliser rates, grasses take up most of the mineral N, with the ryegrass being responsible for about 95% of the uptake of fertiliser N (Walker et al. 1956). Similarly, Akatsuka & Sugihara (1973) found that ^{15}N uptake in a mixture was dominated by *D. glomerata*. Comparisons of several tropical legumes and white clover grown with tropical grasses in south-east Queensland showed that none of the legumes obtained more than about 25% of the annual ^{15}N uptake over a 2-year period (Vallis et al. 1977). However, during spring when growth of the associated tropical grasses was limited by low temperatures, two of the legumes, *Lotononis bainesii* and white clover, were responsible for 70-100% of the total ^{15}N uptake.

(c) Phosphorus.

In general, the competitive power of white clover increases as the level of available soil P increases, although negative efffects of P have also been reported. The relative responses of white clover and associate grasses to P on P deficient soils are probably influenced by the availability of N. In areas that are being developed for temperate pasture production, both deficiencies often occur together. Consequently the initial response to applied P is shown by the legume, which is able to overcome the N deficiency by fixation of atmospheric nitrogen.

Chestnutt & Lowe (1970) suggested that clover has a higher requirement for P than most grasses, but that when P is deficient, clover is better able than grasses to satisfy its requirement for P. However Jackman & Mouat (1972b) refuted the latter conclusion and demonstrated that at suboptimal levels of available P, white clover was less efficient in its response than was *Agrostis tenuis*. Chestnutt & Lowe also suggested that the growth of grasses and clovers and their interrelationship is not unduly affected where levels of available P are above 20 ppm.

There is however, evidence in some situations to suggest that white clover becomes less competitive with associated grasses where fertiliser P is applied. Brockman et al. (1970), in a study of a cut and grazed ley consisting of cv. S215 meadow fescue, S48 timothy and S100 white clover on a moderately low P soil at North Wyke, Devon, found that over a 5-year period P fertiliser increased grass yields in the first four years and decreased clover yields in the last two. Results of Mouat & Walker (1959a) indicated that suppression of white clover by grass, especially ryegrass, was relatively greater where phosphate was applied (Figure 6).

Pasture development sequences involving the responses of white clover to superphosphate applied to P-deficient soils in New Zealand have been thoroughly investigated by Walker (1960), Sears (1953) and others. Pasture development investigations in the Northern Tablelands of New South Wales, Australia, clearly show the responsiveness of white clover to applied superphosphate on P-deficient soils (Wolfe & Lazenby 1973a, b, Holford & Gleeson 1976, Robinson & Lazenby 1976).

Wolfe & Lazenby (1973a) followed the effects of four rates of superphosphate (0, 42, 125 and 375 kg/ha annually) on the development of white clover sown with *Lolium perenne, Festuca arundinacea,* and *Phalaris tuberosa* which were sown at different rates. The effects of grass species and grass sowing rate on total pasture productivity were only minor. The response of white clover, grasses and weeds during four years after sowing in September 1966 is shown in Figure 7. In 1967, white clover showed a linear response to superphosphate applied annually at up to 375 kg/ha, whereas the grasses showed little response. In following years the sown grasses developed a linear response to rates up to 375 kg, and by the fourth year the response of the grass exceeded that of the white clover. Associated soil N analyses indicated that in 1967 the response of the grass to applied P was limited by N deficiency but, in time, N fixed by the white clover became increasingly available to the grass and, by 1970, the improved availability of N, P and S at the highest superphosphate level was reducing legume growth because of competition from sown grass.

In an associated experiment (Wolfe & Lazenby 1973b) the role of N in this succession from swards with high clover to those with high grass content was investigated by annual factorial applications of N fertiliser (0, 56, 112 kg/ha) and superphosphate (0, 188, 375 kg/ha). N fertiliser accelerated the development of a linear response of the sown grass to superphosphate up to the heaviest rate of application.

Without monocultures of the three components it is not possible to be certain which species is the best competitor for P at the different levels of availability, but the responses shown indicate that were P is extremely deficient, the growth of the sown grasses, particularly in the first year, is less restricted than that of white clover. Weeds, mainly *Hypochaeris radicata* and *Eragrostis trachycarpa,* showed the greatest capacity for growth where P was deficient. At higher P levels the relative P response of the components was confounded by N availability and, almost certainly, by competition for light also.

Jackman & Mouat (1972a) encountered difficulty in separating the effects of competition for P between *A. tenuis* and white clover and competition for 'other factors'. White clover was grown in monoculture and in a mixture with *A. tenuis* and given superphosphate at 15 rates in the range 0 to 897 kg/ha on soil deficient in N and P (available P, Olsen's $NaHCO_3$ method, 2-8 ppm). Measures were taken to minimise shading effects by frequent cutting. Results for dry matter yield responses to P are presented in Figure 8. The yield ratio of white clover in mixture to white clover in monoculture indicates that the competitive power of white clover increased as the level of superphosphate application increased.

The response for September 1968 suggests that during establishment, development of white clover was inhibited by competition with *A. tenuis* for P at low phosphate levels. As the swards developed, further fixation of N by white clover allowed it to become 'dominant' at all levels of P, although this dominance was restricted where no P was applied. Removal of the cut herbage probably reduced N transference from the legume to the grass. Potential N transfer would have been greatest where the legume yield was highest, at high phosphate levels. Consequently, at the highest phosphate levels the swards did not progress towards a grass-dominant phase and so differed from the succession to grass dominance shown by Wolfe & Lazenby (1973a). In the 1970-1971 season N was applied to all plots in the experiments of Jackman & Mouat (1973a) and, as a result, reduced the yield of clover in the mixture compared with that of clover in monoculture at all P levels (Figure 8). Apart from the summer of 1970-1971, the yield of *A. tenuis* in the mixture was about the same at all levels of superphosphate, suggesting that frequent cutting prevented the shading and further suppression of *A. tenuis* by clover at high levels of superphosphate. An observed reduction of *A. tenuis* with increased levels of superphosphate in the summer of 1970-1971 was attributed to excessive shading by the clover during an enforced harvest delay.

Jackman & Mouat (1972a, b) considered it likely that shading of clover stolons by *A. tenuis* was important at high P levels even though the results suggested that clover was dominant (Figure 8). As the swards developed, a mat of *A. tenuis* 1 cm thick was formed which covered the white clover stolons and, by shading them, probably inhibited clover bud development. However, as long as *A. tenuis* was able to obtain fixed N, the transfer of which would have been encouraged by repeated defoliation of white clover inducing the sloughing-off of nodules, root breakdown and eventual release of fixed N into the soil, *A. tenuis* would continue to grow and obtain P from the soil.

(d) Potassium.

White clover has a well defined requirement for K (Andrew & Robins 1969, McNaught 1958), probably greater than that of the grasses with which it is most commonly associated (Chestnutt & Lowe 1970). There appear to be few experimental results quantifying the relative competitive abilities of grasses and white clover for K. Competition for K appears to be markedly influenced by the availability of mineral N.

Blaser & Brady (1950) observed the effects of N and K fertilization on a Ladino, *Trifolium pratense, Medicago sativa, Phleum pratense* mixture. In a factorial experiment involving 3 levels of potash (K_2O at 0, 84 and 168 kg/ha) and 3 levels of N (0, 56 and 112 kg/ha), Ladino clover yield and content in the herbage increased with increasing K rate and decreased with increasing N rate. Depression of clover yield was least at the highest K rate (Figure 9). Other reports showing the effect of K on white clover content of swards include Holmes & McLusky (1954), Brown (1957), Linehan & Lowe (1960), Drake et al. (1970),

McNaught & During (1970), Davies & Thomas (1971), and Reith et al. (1973).

Lowe (1966) summarised 15 years' results of field fertiliser experiments on mixtures of grass with white clover in Northern Ireland. In Figure 10 the ratio of clover in the dry matter yield is plotted against rate of N. Where neither N nor K were applied, the average clover content was 15%, but 25.4% where only K was applied. Both with and without applied K, increased N fertilization decreased clover content. Application of high rates of K did not overcome the suppressive effect of high N rates. In Figure 11 clover content is plotted against K_2O rate. With no applied K, clover content was 9.3%, increasing to 19.2% where 250 kg K_2O/ha was applied. Without applied N clover content declined at K_2O rates in excess of 170 kg K_2O/ha. This decline at the highest K rates is similar to that observed for high rates of superphosphate on P-deficient soils (Wolfe & Lazenby 1973a). It is possibly an effect of increased grass competition resulting from the transfer of N from white clover encouraged to grow and fix N by adequate availability of K.

Mouat & Walker (1959a), in a pot experiment, compared the growth of white clover in monoculture and in mixture with *A. tenuis, D. glomerata* and *L. perenne* at 0, 112 and 224 kg K_2O/ha (Figure 12). When the yield of clover in mixture with the grass was expressed as a ratio of its yield in monoculture, white clover maintained its competitiveness as rates of K increased in the mixture with *L. perenne* and *A. tenuis,* but became less competitive with *D. glomerata.*

(e) Sulphur

S is deficient in many extensive grass-legume pastoral systems, especially if they are remote from industrial areas which emit S into the atmosphere. In such pastoral systems the mineralisation of organic S and the presence of available sulphate ion in the soil are insufficient to meet the S requirements of legumes in competition with grasses (Metson et al. 1971).

Sulphur requirements can be met by applying superphosphate, so that the combined application of P and S obscures the response to either element (Metson et al. 1971). Interactions between grass and clover in swards obscure the factors controlling competition between grasses and clovers for S. Where responses to S fertilisers have been measured (other elements except N being non-limiting) it has been difficult to be sure whether S or N was the primary limiting nutrient (Metson 1972, Metson & Collie 1972).

Walker & Adams (1958) applied a 3 x 3 factorial of N and S on a S-deficient soil in Canterbury, New Zealand. The site, which supported a sward dominated by low-fertility grasses, was oversown with *T. repens, T. pratense* and *T. subterraneum.* Without fertiliser N, applied S stimulated clover growth. Without applied S, applied N increased grass growth and suppressed clover. This suppression of clover was reduced where more S was applied. Grasses accounted for 98% of the total uptake of S where S was not applied and in this situation N fixation by the clover was negligible.

(f) Soil acidity.

The increase of white clover content with liming on many soil types has long been recognised, (e.g. Parkinson 1798). Chestnutt & Lowe (1970) concluded that in temperate areas most legumes and their associated root nodule bacteria can persist and increase at soil pH as low as 4 provided Ca is not deficient. Root growth and N fixation are substantially increased and shoot growth moderately increased as pH is raised to 8, though above pH 6 these effects are not marked (McNeur 1953, Andrew 1976). Davies and Milton (1945), in an experiment on upland peat in Wales, showed that responses to liming on peat soils are particularly marked (Table 9). In their experiment, all plots were given basic slag (15% P_2O_5) at 1250 kg/ha and nitrochalk at 125 kg/ha at sowing, and either lime at 2500 kg/ha or no lime. The lime had a marked effect on sward clover content on a reasonably well drained soil and on an area of impeded drainage. Liming was ineffective on waterlogged areas.

The effect of liming on the competitive ability of white clover is influenced by the reaction to lime of those species with which it is associated (Section 4.(B)i, Harris 1971). Snaydon (1962, 1971) and Snaydon & Bradshaw (1962a, b, 1969) also demonstrated differences between white clover populations in their ability to compete and grow in calcareous and acid soils (Table 10). Populations of white clover from calcareous and acid sites were implanted as rooted cuttings in upland *Festuca-Agrostis* swards on a site with

Table 7.9 Percentage botanical composition on seeded areas on peat in the 5th harvest year (Davies & Milton 1945)

Drainage	Treatment	*Lolium perenne*	*Dactylis glomerata*	*Phleum pratense*	*Cynosurus crtistatus*	*Poa trivialis*	*Trifolium repens*	Native species
Unimpeded drainage	No limestone	0.8	8.9	3.1	8.1	0.4	3.8	74.9
	Limestone	7.5	12.8	12.9	22.5	2.9	17.3	24.1
Impeded drainage	No limestone	0.4	trace	8.3	9.1	2.1	6.6	73.5
	Limestone	1.9	1.3	7.9	19.0	3.9	16.6	49.4

acid soil low in Ca and P (Snaydon & Bradshaw 1962a). Populations from acid-upland soils performed better and after two years were the only plants surviving. Populations from calcareous soil had greater root competitive abilities on calcareous soils, while populations from acid soil had greater root competitive abilities on acid soil. Probably because of the larger leaves and longer pertioles of populations from calcareous soils, these had greater shoot competitive abilities than those from acid soils.

(g) Trace elements.

Of the trace elements, probably the most important for practical purposes is Mo. In New Zealand soil groups including moderately to strongly leached yellow-grey and yellow-brown earths, Mo application has reduced the liming requirement from liming to pH 6.5 to liming to pH 5.6-5.8. An example of the effect of rate of sodium molybdate on white clover content on a soil responsive to Mo is shown in Figure 13 (Scott 1963). *Trifolium pratense* was more responsive than white clover at lower rates, but the content of white clover increased relative to red clover as the rate of application exceeded 7 g/ha.

Interactions between responses to micro-elements and between micro-elements and major elements are complex (During et al. 1960, Scott et al. 1963, Cameron 1969, McNaught & During 1970, Cullen & Arnold 1971, Lambert et al. 1974, O'Toole & Masterson 1970). There appears to be no specific study of the effect of the availability of a micro-element on the competitive interaction between white clover and an associated species.

(h) Dung and urine.

Dung and urine affect the botanical composition of pasture by smothering or scorching pasture, by influencing grazing patterns and by altering nutrient concentrations. Nitrogen and K are concentrated in the urine and Ca and P in dung. Dung and particularly urine generally encourage grass growth and can depress clover content.

McDiarmid & Watkin (1971) observed the effect of cattle dung patches 0.3m in diameter on dairy pasture. White clover decayed rapidly beneath the dung patches, and leaving a dung patch on the pasture for 15 days killed 75% of the rooted nodes of clover stolons. At the perimeter of the dung patch, yield of grasses particularly ryegrass, in the sward was stimulated, but there was no change in clover yield.

Weeda (1967) observed the re-establishment of plants on cattle dung patches. With liquid dung, regrowth was comparatively rapid because little of the affected herbage was killed. With thicker dung layers, most plants under the dung patches were killed. The outside 3-6 cm of these patches were covered fairly quickly by tillers of surrounding grasses but the central area often remained sparsely covered for 6-12 months. White clover was usually the first species to recolonise these areas and tended to remain dominant in the dung patch area for 1-1.5 years.

Sears & Newbold (1942) and Sears et al. (1948), in New Zealand on a soil adequately fertilised with lime, found that grazing without return of dung and urine resulted in a sward with a high clover content, with an average annual production of 11 200 kg/ha. Return of urine increased production by 15% and phosphate and lime, found that grazing without return of dung and urine resulted in a sward with a high clover content, with an average annual production of 11 200kg/ha. Return of urine increased production by 15% and increased grass content. Return of dung increased total dry matter production by 18% and although clover content was high in this treatment in the early stages of the trial, grass content progressively increased. Return of both dung and urine produced a grass-dominant sward yielding 33% more than the no-return treatment.

Watkin (1954) found that where sheep grazed with return of urine only, grasses were encouraged and white clover reduced. Wheeler (1958), continuing Watkins' (1954) experiment, gave the effects on clover content of the treatments listed in Table 11. Although urine depressed clover content, reduction of clover by applied fertiliser N was considerably greater. Norman & Green (1958), however, found little effect of cattle urine on botanical composition and concluded that the influence of a single urination was too small to be shown by the method of botanical analysis used. However, cattle dung increased the content of white clover where plots were cut. In grazed plots the dung areas were not defoliated for about 1 year and were grazed to the level of the controls only after 13-18 months. On these ungrazed dung areas, white clover showed no response to dung after 1 year and only a small increase after 2 years.

iv. Light intensity

In reviewing the influence of light intensity on the growth of herbage plants, Black (1957a) concluded that pasture legumes were less tolerant of shade than the grasses. He suggested that the high optimum light intensity requirement of legumes, in addition to their more prostrate growth habit, might be largely responsible for their disappearance from undergrazed or undefoliated grassland associations and for the difficulty of establishing them under cover crops. However, although appreciable variation has been recorded in light saturation level and in maximum photosynthesis in inter- and intra-specific comparisons of temperate species (Cooper & Tainton 1968, Scott et al. 1970), an inherent difference between white clover and other important temperate pasture species in the photosynthetic response of their leaves to light intensity has not been firmly established. Species differences in canopy characteristics, discussed in Section 4(E), are more significant in differentiating the light responses of grasses and clovers.

Reduction of light intensity incident on mixtures containing white clover has most often, but not invariably, reduced the white clover content. Blackman (1938) imposed two light intensity levels (60% and 40% of daylight) on an *Agrostis-T. repens* sward in combination with various N treatments. Shading to 60% of daylight markedly decreased clover content, but shading to 40% daylight had little additional effect. Reduced light intensity in the range 1.0 to 0.33 to 0.11 reduced the yield of both *D. glomerata* and white clover, *Festuca rubra* remained relatively constant in a mixture of the three species grown in a glasshouse, but there were no clear trends in the relative contents of the species (Wilson 1962). Growing a *D. glomerata*-Ladino clover mixture in the field in North Carolina, USA, under slats intercepting about one-third of the incident light virtually eliminated the clover in the second and third year, whereas in full daylight clover produced about 560 kg DM/ha per year (Blake et al. 1966). The yield of *D. glomerata* was not influenced by the shading, but clover declined in all treatments over the three years of the experiment. By contrast, in white clover monocultures in a field experiment in South Carolina, USA, shading in the range from 98% to 27% daylight increased the number of nodes produced, increased the frequency of axillary buds developing and increased the number of nodes that rooted compared with full daylight (Trautner & Gibson 1966). It was suggested that the effect of shade arose from modification of soil surface temperatures and moisture rather than a light effect, supporting the concept that moderate shading by a companion grass may improve white clover persistence.

These shading experiments were with either mixtures or monocultures and it is not possible to distinguish between the direct environmental effects of shading, which, as indicated, may include effects other than a reduction of light intensity, and the effects of plant competition. The effects of light levels of 1.0, 0.5 and 0.25 of daylight on the competitive interactions between white clover, ryegrass and sheep sorrel grown in pots were analysed by Harris (1972) using the de Wit (1960) model of plant competition (Figure 14). Shading markedly reduced the competitive power of white clover, especially when mixed with ryegrass. In full daylight the yields of the mixtures were much greater than the mean of the yields of the monocultures. This effect was greatest for the ryegrass-white clover mixture. Shading reduced this yield stimulation effect, and eliminated it altogether in the ryegrass-white clover treatment shaded to 0.25 daylight. Greater N fixation by white clover in the unshaded treatment accounted for the increased yield of the mixture. Ennik (1960), in a climate chamber study using a range of ratios of ryegrass and white clover in pots, found that during the 3 months of the study the mixtures moved towards an equilibrium ratio. In pots grown under a 'high' light intensity the ratio had a higher white clover content than in pots grown under a 'low' light intensity.

v. Defoliation

The relative contribution of white clover in mixtures is determined by the frequency, intensity and timing of defoliation of swards. Grazing animals selecting white clover as a preferred feed can also introduce a non-uniform intensity effect, depending on the type of animal, the species of associated plants and the genotype of white clover.

Table 7.10 Dry weight yield (g/pot) of two populations of *T. repens* on two types of soil (Snaydon 1962)

	Populations grown alone	
	Calcareous type	Acidic type
Acid soil	18.1	16.6
Chalk soil	14.8	9.8
	Populations grown in 'mixed' plantings	
	Calcareous type	Acidic type
Acid soil	16.0	15.5
Chalk soil	17.5	7.9

(a) Frequency.

In general, more frequent defoliation of a white clover-grass sward increases the content of white clover, e.g. Graham et al. (1961) with Ladino-*D. glomerata* mixtures, Ward et al. (1966) with Ladino-*Poa pratense* mixtures, Bland (1967) with Huia white clover in *L. perenne* and Reid (1968) with a S100 white clover-*D. glomerata* mixture. Brougham (1959b), comparing 4 systems in which the frequency and height of grazing of a ryegrass, white clover and red clover pasture was varied, found that white clover yields and contents were highest under the regime with most frequent grazing. Competition experiments with Ladino clover and cocksfoot by Kishi (1947a, b) and Morita & Fujita (1957a, b) showed that infrequent cutting resulted in the suppression of Ladino.

Wolton et al. (1970) specifically emphasised that the responses of white clover content to defoliation frequency was inconsistent and gave supporting evidence from a series of experiments which showed that white clover was not reduced by increasing the length of the growth period or cutting at a late stage of growth.

The effects of cutting frequencies of 2, 3, 4 and 5 weeks on the yield and composition of mixtures containing ryegrass, cocksfoot and Alta fescue with either lucerne, Ladino clover, broadleaf trefoil (*Lotus pedunculatus*) or all three legumes included, were investigated by Peterson & Hagan (1953) in California. Where Ladino was the only sown legume its content increased with increase of the interval between

Table 7.11 Percentage clover content in plots receiving dung, urine and fertilizer nitrogen (Wheeler 1958).

	Year 1954			
Treatments	†N0	N1	N2	N3
No dung or urine	50	45	23	4
Dung returned	51	47	20	2
Urine returned	43	32	9	–
Dung + urine returned	38	26	4	–
	Year 1955			
Treatments	N0	N1	N2	N3
No dung or urine	34	31	9	1
Dung returned	40	28	18	1
Urine returned	30	24	5	–
Dung + urine returned	12	11	1	–

†Nitrogen treatments applied as nitrochalk (15.5% N).
N0 No fertiliser N applied.
N1 250 kg ha^{-1} applied in March plus 125 kg ha^{-1} at end of August.
N2 375 kg ha^{-1} applied in March plus 94 kg ha^{-1} after each grazing.
N3 375 kg ha^{-1} applied in March plus 188 kg ha^{-1} after each grazing.

defoliations, and at the end of the second growing season Ladino ranged from 56.5% at 2-weekly cutting to 79.1% at 5-weekly. In the mixture containing the three legumes, Ladino decreased with increased cutting interval from 58.1% at 2-weekly cutting to 4.0% at 5-weekly. Lucerne showed the converse response, increasing from 1.1% at 2-weekly cutting to 87.4% at 5-weekly.

Nelson & Robins (1956) found no effect on the percentage content of clover in a Ladino-*D. glomerata* mixture in comparing two cutting frequencies in which the herbage was cut to 3 cm height when it was 13 or 26 cm high. Rousvoal & Gallais (1973) analysed a 5-species diallel including *L. perenne, D. glomerata, H. lanatus, A. stolonifera* and *T. repens*. The yield of white clover remained low under both 3- and 6-weekly clipping and it was less sensitive to effects of associate species than were the other species.

Harris & Thomas (1973) and Harris (1974) analysed the effects of cutting frequency on the competitive interaction between *T. repens, L. perenne, L.perenne* × *multiflorum* and *A. tenuis* by the de Wit (1960) model of competition. Effects of cutting frequency on the competitive interaction between grass and legume varied markedly with the stage of sward development, season and associate grass. In the first autumn and winter both ryegrasses suppressed white clover, especially under infrequent cutting. In the second year the competitive power of white clover grown with *L. perenne* was greater under infrequent than under frequent cutting, whereas the greater suppression of white clover under infrequent defoliation persisted into the second year where hybrid ryegrass was the associate grass. The initial ryegrass dominance was attributed to high soil N availability and greater growth rates of ryegrass than white clover under the low temperatures and light regimes of late autumn and winter. Infrequent cutting at this time allowed fuller development of the ryegrass canopy and increased the shading of clover. Infrequent cutting also increased intraspecific shading between ryegrass plants, reducing tiller numbers per plant, especially for *L. perenne*. The structure of the tiller population of the hybrid ryegrass differed little between cutting frequencies. During summer and early autumn, when light and temperature were near optimum for white clover growth and ryegrass was showing the post-reproductive depression of growth, white clover was able to spread through the swards by stoloniferous growth and occupy available spaces between the ryegrass plants. Similarly, Brock (1971) showed that white clover content increased in response to infrequent grazing by sheep of perennial ryegrass-white clover pasture in summer. In autumn and winter, however, ryegrass responded to infrequent grazing, whereas the clover yield remained about the same under both frequent and infrequent grazing.

In contrast to ryegrass as the associate grass (Harris & Thomas 1973), white clover markedly suppressed *A. tenuis* from establishment under infrequent defoliation compared with frequent defoliation (Harris 1974). *A. tenuis,* a very slow establisher, was overtopped by white clover, an effect reduced by frequent defoliations.

To explain the varied response of white clover content to defoliation frequency, frequent defoliation can be viewed as especially prejudicial to erect grass species, whereas the stoloniferous habit of white clover means that a smaller proportion of its total biomass is removed in any one defoliation. Therefore, where clover is at a competitive disadvantage, frequent cutting may increase clover content. Infrequent defoliation could be expected to increase clover where clover is already at a competitive advantage or at least actively growing and able to take advantage of the longer spells between defoliations, or where infrequent defoliation is predudicial to the associate species growth.

(b) Intensity.

Increased intensity of defoliation of mixed grass-white clover swards has often but not invariably increased the proportion of clover in the sward.

The effects of cutting height on the clover percentage in a *Poa pratensis* sward and the interaction of the cutting treatments with two moisture and two N regimes are shown in Table 12 (Robinson & Sprague 1947). Close cutting increased clover percentage, more so without irrigation and with applied N. Kishi (1973) also found that lenient cutting and N application reduced clover content in a Ladino-*D. glomerata* sward.

On irrigated swards of *L. perenne, Paspalum dilatatum, Bromus uniloides, D. glomerata* and *T. repens* at Werribee, Victoria, Australia, Clark et al. (1974) applied a factorial combination of cutting heights of 3, 6 and 10 cm above ground level and growth heights of 5, 13 and 20 cm above the cutting height.

Table 7.12 Clover percentage in a *Poa pratensis* sod during summer as affected by nitrogen, soil moisture and cutting treatments (Robinson & Sprague 1947)

Cutting treatment	Estimated clover %			
	Not irrigated		Irrigated	
	−N	+N	−N	+N
Cut to 1.25 cm when 10 cm high	57	28	52	48
Cut to 2.5 cm when 10–12.5 cm high	45	24	50	42
Cut to 5 cm when 12.5 cm high	36	1	33	32
Cut to 1.25 cm when 7.5 cm high in early spring, then to 2.5 cm when 10–12.5 cm high	54	20	55	61

Percentage clover content was reduced by increased height of cutting and increased height of growth as follows:

Height of growth (cm)			Height of cutting (cm)		
5	13	20	3	6	10
33.7%	30.8%	27.6%	39.5%	31.0%	23.4%

In another experiment on the same pasture, Clark et al. (1974) imposed 8 treatments in which the number of lenient cuttings in different seasons of the year were varied. Close cutting in all seasons of the year produced 16.1 t DM/ha per year, of which clover contributed 41.4%, and lenient cutting in all seasons produced 13.2 t/ha, of which 26.6% was clover.

Extremely hard cutting can reduce the competitive ability of some cultivars of white clover. Brown (1939) found that the area occupied by white clover cut to height of 1.25 cm was reduced compared with 2.5 cm and that the reduction was particularly marked with Ladino rather than Kent white clover and where *Poa pratensis* was the main associate grass. However, with higher cutting Morita & Fujita (1975a,b) found that varying the cutting height between 5 and 10 cm did not influence the content of Ladino clover in mixture with cocksfoot.

A result similar to the last was observed by Harris & Thomas (1973), who found that the effect of cutting to heights of 2 or 8 cm was negligible compared with the effect of cutting frequency on the competitive inteaction between ryegrass and white clover. However, in *A. tenuis*-white clover mixtures (Harris 1974) *A. tenuis* was more aggressive under frequent cutting to a height of 8 cm compared with a height of 2 cm. The development of a mat of *A.tenuis* under higher cutting impaired the growth of white clover.

(c) Timing.

The effect of intensive defoliation on white clover content varies with season of the year. Generally, frequent intensive defoliations at the start of the clover growing season reduce grass competition, increase sward clover content, but tend to reduce total herbage production. In a series of grazing experiments at Jealott's Hill in the UK (M.G. Jones 1933), four grazing systems were applied to a 2-year-old sward with Kent perennial ryegrass, *Poa trivialis* and Kent white clover fertilised with phosphate and N. The botanical composition after 3 years' grazing treatment is depicated in Figure 15. Hard grazing during spring weather planned (plot 1) or as part of a fixed stocking rate (plot 4) increased clover content. Perennial ryegrass was decreased by this hard spring grazing, whereas the wild white clover suffered little damage from the close grazing and benefited from the reduced competition of the grass. Lenient spring grazing encouraged the full development of ryegrass and checked the growth of white clover for the entire season. Close grazing at monthly intervals from mid-April (plot 3) maintained composition close to that of the original sward. Although N fertiliser substantially increased yield, the different grazing treatments imposed induced similar changes of botanical composition to those in swards receiving phosphate fertiliser only.

The same set of treatments was applied by M.G. Jones (1933) to an old permanent sward dominated by weed grasses and containing a much smaller proportion of wild white clover and ryegrass than the recently sown sward. Botanical composition changed rapidly under the influence of the grazing animal (Figure 16). Hard spring grazing, and to a smaller extent balanced rotational grazing, increased ryegrass and clover content. Grazing at a fixed stocking rate maintained a botanical composition similar to that of the original sward. As in the young sward, fertiliser had a smaller effect on botanical composition than did grazing management. A third experiment with a young grassy sward also showed a marked increase of the white clover content with hard spring grazing.

M.G. Jones (1933) found that autumn management also affected white clover content. An early autumn rest, in which plots were closed on 20 July and given a month's rest before grazing and then rested again until a grazing at the start of November, increased both white and red clover content 5-fold. A late autumn rest, plots being grazed continuously and closely until September and then given a month's rest before the final autumn grazing, increased white clover content 10-fold, with only a small effect on red clover.

Brougham (1960a) imposed 6 grazing treatments, varying the seasonal application of frequent hard grazing (7.5 to 2.5 cm) to a mixed grass and red clover and white clover pasture. Grazing for the remainder of the year was from a height of 18 cm down to 7.5 cm (Table 13). White clover contributed most where grazing was frequent and hard throughout the year. However, white clover also produced a high yield in the system, producing the highest total yield where grazing was frequent and hard in the winter. Brougham (1960a) attributed this to the closer defoliation allowing light to penetrate to ground level, encouraging both tillering of grasses and growth of clovers.

Delay in cutting the first crop of mixtures of Ladino clover with *D. glomerata* or *Bromus inermis* decreased clover yield (Sprague & Garber 1950). Removal of the first crops when 20 to 25 cm high and subsequent growth when it was about 20 cm high maintained the stand of clover better than did cutting later at early head or full bloom stages of the associate grass. Height of cutting had little effect on clover yield where cutting was delayed, but under more frequent cutting, lowering the height of cut from 7.5 to 2.5 cm increased the amount of clover harvested. Ingress of *Poa pratensis* was also enhanced by frequent close cutting.

Table 7.13 Annual yield and white clover content of a mixed grass-clover pasture under six grazing treatments (Brougham 1960).

	Season of hard grazing					
	All	Summer	Autumn	Winter	Spring	None
Total Yield kg DM ha^{-1}	10460	9800†	12746†	14963	10752†	13126
White clover Yield kg DM ha^{-1}	2369	1389	806	1882	1322	1254
White clover content %	22.6	14.2	6·3	12.6	12.3	9.6

†These yields do not include a stubble yield of about 1100 kg DM ha^{-1} obtained during the changeover from the lax to hard grazings.

(d) Uniformity.

In mixed swards, white clover content is increased more under mowing than under grazing (Taylor et al. 1960, Wolton et al. 1970, Scott 1973). There are two likely causes. First, the grazing animal, by returning dung and urine to the sward, maintains soil nutrients at levels favourable to the growth of non-leguminous species. Second, if white clover is a preferred species in a mixture, selective grazing is likely to reduce its content and competitive ability relative to other species present (Norman 1960).

Even when swards are cut to uniform height the defoliation stress placed on individual plants in a sward will vary because of differences in the vertical distribution of plant parts (Sato 1973). Stolons form a large proportion of the biomass of white clover, and as these are largely on the soil surface cutting often leaves the stolons intact. By contrast, tufted grasses will have a higher proportion of their shoot biomass removed by cutting.

However, white clover is a preferred species in the diets of many herbivores (Hodge & Doyle 1967, Lancashire & Keogh 1968, Cowlishaw et al. 1956, Cowlishaw 1960, Archer 1973, Keogh 1974, Hughes 1975, Owen 1976, Murton 1965). Most grazing animals graze selectively and the diet chosen is usually of higher nutritive value than that rejected.

Ivins (1955) concluded that the presence of white clover invariably enhances palatability. Characteristically, white clover has high digestibility, and this declines more slowly with increasing maturity than it does in grasses. Voluntary intake studies have shown that when fed as the sole diet, white clover is consumed in greater quantities than grasses of similar digestibility. Further, the digested or metabolizable energy of white clover is utilised more efficiently for growth and fattening than that in grasses (Thomson & Raymond 1970, Thomson 1977). In some situations, although the grazing animal may prefer the more nutritive legume, overlayers of less preferred herbage may protect the legume. This influenced the grazing of geese (Owen 1976), although they sometimes delved beneath the sward to feed on clover stolons and roots, especially when the soil was wet. In sheep grazing New Zealand hill country the suggested preference of sheep for white clover may have been overriden by the accessibility of the clover in relation to other sward components (Clark et al. 1982). This work showed that dietry legume, as assessed by oesophageal samples, was in direct proportion to pasture legume (Figure 17).

Grazing animals differ in their ability to select out white clover from a sward. Sheep have a precise and small bite and can select out individual plants or plant parts in pasture. This is probably one of the factors accounting for observations of higher clover content in pastures grazed by cattle than by sheep (Clark et al. 1982, Collins & O'Toole 1963). Using rumen samples obtained from fistulated sheep, Cahn & Harper (1976b) concluded that sheep preferred unmarked leaf morphs of white clover, and suggested that the sheep selected the leaf marks visually. There appeared to be some selection by the sheep of larger lamina of white clover, but this character was not obviously linked with variation of leaf morph. Probably a more relevant measure would have been petiole length, which could influence the accessibility of leaves for prehension by sheep, and the finding would have been more convincing if oesophageal samples rather than rumen samples had been used. In contradiction to the results of Cahn & Harper (1976), Charles (1968) found that S100 Nomark formed a greater proportion than leaf-marked S100 of the white clover survivor populations of swards more frequently grazed.

Murton (1965) found that wood pigeons preferred red clover to white clover and attributed this to the larger leaf of red clover which enabled the birds to meet their feed requirements more quickly. Observations (Harris et al. 1983) also suggest that wood pigeons prefer large-leaved varieties of white clover more than small-leaved ones. Polymorphism for cyanogenic glucosides probably has led to the selection of white clover populations able to avoid the preferential grazing by some herbivores. However, Harper (1977) considers this an inefficient defence against predators unless the predator learns to avoid unplatable plants.

In a palatability trial, sheep showed no preference for glucoside-free, medium or high-glucoside strains of white clover grown alone or with perennial ryegrass over 14 months (Corkill 1952). Rabbits definitely preferred glucoside-free plants and damage from slugs (*Agriolimax* spp.) was more severe on glucoside-free white clover. However, Bishop & Korn (1969) found that snail *Helix aspersa* and the slug *Agriolimax reticulatus* did not discriminate between cyanogenic and acyanogenic clovers. Crawford-Sidebotham (1972) offered cyanogenic and acyanogenic *Lotus corniculatus* and *T. repens* to 13 species of slugs and snails. The results suggested that particularly for *L. corniculatus*, the cyanogenic character acted as a defensive mechanism against molluscus although species varied in the extent to which they fed selectively. Angseesing (1974) found selective feeding by the slug species *Arion ater* and *Agriolimax reticulatus* on acyanogenic white clover in preference to cyanogenic clover. This was considered to counterbalance the selection against cyanogenic plants by low temperature and soil water stress.

(e) Defoliation and overwintering.

Available information suggests that in regions with low winter temperatures, lenient infrequent defoliation ensures more successful overwintering of white clover than the frequent close cutting usually considered to maintain clover in swards.

Tesar & Ahlgren (1950) in Wisconsin, USA, grew Ladino clover alone or with *B. inermis, Phleum pratense* or *D. glomerata* and cut the established swards 2, 4 or 6 times a season to 2.5 or 9 cm. Measurements were made from sowing on 12 April 1946 until autumn 1948. Stolon weights of Ladino clover grown alone or with a grass in autumn 1947 or spring 1948 did not differ. Stolon weight was markedly reduced where cutting had been frequent and close (Table 14), and stolons under frequent-close cutting were shorter and smaller in diameter than those under less frequent and lenient cutting.

In these trials, snow cover protected the overwintering stolons for most of the winter, but for 4 days in February when little snow cover was present, minimum teperatures were -14, -17, -16 and -14°C. Stolon weight declined in all treatments over winter, but the percentage reduction of stolon weight, referred to as the percentage killed over winter, was greatest for treatments cut 6 or 4 times to 2.5 cm the previous growing season. This difference could not be attributed to differences of carbohydrate status at the start of winter. Frequent close cutting had reduced percentage available carbohydrates in early autumn to about half that of the less frequent, laxer cut treatments, but a period of recovery of 2 months in the autumn allowed the percentage total carbohydrate to attain similar levels in all cutting treatments (Table 15).

The residual effects of the cutting treatments imposed in 1947 were measured under a common cutting treatment in 1948. This showed that the yield of Ladino clover was closely correlated with the weight of stolon present at the start of the growing season. The percentage ground cover of Ladino clover at the end of the experiment is shown in Table 15.

From this information Tesar & Ahlgren (1950) concluded that defoliating clover closely and frequently produced a high herbage yield, but at the expense of stored food reserves and stolon development. Consequently this treatment, by contrast with more leniently cut treatments, reduced yields in the second year and increased the invasion of Ladino clover stands by *Poa pratensis* and weeds.

Other studies in North America have indicated genotypic differences in the capacity to withstand damage by freezing and have shown the protective influence of cover (Smith 1949, Knight 1953, Sprague 1955, Wood & Sprague 1952). Observations by Harris et al. (1983) in Wales indicated that in that environment both utilisation of carbohydrate reserves and stolon photosynthesis contributed to the overwintering energy requirements of white clover. Cover provided by associated ryegrass reduced the extent of low-temperature damage to stolons compared with that in white clover monoculture.

vi. Pests and diseases

Pests and diseases tend to selectively reduce white clover in mixtures in a manner directly similar to selective defoliation. Pesticides (Blake et al. 1966, White & French 1968) and soil fumigants (Yeates et al. 1976) increased the proportion of white clover in mixtures with grasses. Slugs destroyed white clover in cocksfoot-Ladino clover pastures in Western Washington, USA (Howitt 1961).

Pasture insect pests have been much studied in New Zealand. The grass grub (*Costelytra zealandica*), which in its larval stage feeds on plant roots, and subterranean grass caterpillar (*Wiseana* spp.), which emerges at night from tunnels in the soil to feed on shoot tissue, are two important pests. The effects of these insects on white clover in monoculture and in mixture with ryegrass have been compared by Radcliffe (1970, 1971), Harris (1969) and Harris & Brock (1972). Both insects survived better and made better growth under monocultures of white clover than under ryegrass, and damaged white clover more than the ryegrass. However, mixing with ryegrass appeared to protect white clover against damage from both pests.

Table 7.14 Effects of cutting frequency and intensity on live stolons (g DM per 645 cm^2) of Ladino white clover averaged over all mixtures (Tesar & Ahlgren 1950)

Date	Height of cutting cm	Number of cuttings, 1947 2	4	6
15 September 1947	2.5	10.0	7.4	4.4
	9.0	10.6	9.8	8.7
29 April 1948	2.5	5.4	2.1	1.9
	9.0	5.7	5.8	5.6

Both insects damaged seedlings of white clover more than plants which had formed stolons. Subterranean grass caterpillar stripped leaves off the stolons but left the stolons intact, but by felling whole tillers by eating them through at the base could kill individual grass plants. Areas infested with grass grub and subterranean grass caterpillar are often clovery, as the clover can recover and spread more rapidly than the tufted grasses once the insects pupate and pass into the adult phase.

White clover may increase in response to disease of the associate grass, as shown by Lancashire & Latch (1970). Ryegrass cultivars differing in their susceptibility to crown rust (*Puccinia coronata*) were compared at two cutting frequencies which influenced the development of the disease. Reduction of ryegrass yield by crown rust was compensated for by increased white clover growth. In the treatment cut infrequently and containing the susceptible cultivar Grassland Ruanui, white clover content increased to 80%.

Table 7.15 Cutting frequency and intensity effects on total available carbohydrate content and ground cover of Ladino clover stands sown as pure cultures in 1946 (Tesar & Ahlgren 1950).

Cutting treatments		Total available carbohydrate %			Ground cover %
Number	Height, cm	12 September 1947	2 October 1947	4 November 1947	13 October 1948
2	2.5	24.2	28.0	34.6	70
	9	24.3	29.5	34.1	68
4	2.5	20.6	27.2	33.0	47
	9	25.4	28.3	33.9	68
6	2.5	12.8	23.9	33.8	21
	9	21.5	25.6	34.8	39

vii. Herbicides

Application of selective herbicides rapidly changes botanical composition, and if accompanied by the correct adjustments of management the changes induced can be maintained over a long period. Herbicides that are successful in controlling weed grasses (e.g., paraquat, dalapon, pronamide and carbetamide) usually increase white clover content, although clover growth and N fixation may be checked for a time (Blood 1964, Allen 1965, 1968, Atkinson 1970, Wasmuth & Mills 1972b, Hartley & Atkinson 1972, Paxman & Forgie 1972, Oswald 1976; Rolston et al. 1976). Paraquat has been applied to pasture to increase clover content so that the lamb fattening potential of the pasture is improved (Blackmore 1964, Williams & Palmer 1970).

Control of broad leaved weeds by herbicides usually has a detrimental effect on white clover (Elliot 1961, Thompson & Mitchell 1973, Patterson 1973, Taylor 1973). Elliot (1961) found that the reaction of white clover to applications of 2,4-D, MCPA and mixtures of 2,4-D and MCPB was unpredictable and the extent of damage varied from field to field. The ability of white clover to recover also varied from field to field and full recovery was sometimes delayed until the year after spraying. Recovery was usually most rapid where white clover showed high initial resistance. Picloram used to control woody species has reduced white clover in pasture surrounding the weed species (Amor 1970, Thompson 1973, Meeklah & Mitchell 1973).

Turkington et al. (1979) provide an ecological interpretation for the effect of herbicide on the content of clover in pasture, arguing that part of the fundamental niche of white clover is occupied by grasses and dicotyledonous herbs. Killing of permanent pasture with paraquat increased the growth of transplanted white clover stolon cuttings 65-fold.

viii. Treading

White clover is a characteristic species of trodden areas (Bates 1935, Davies 1938), but within such trodden areas white clover is less tolerant of treading than other species, especially ryegrass. For example, Chappell et al. (1971) found that white clover was one of the species drastically reduced by the public trampling chalk grasslands.

Edmond (1964) examined the effects of sheep treading on sown monocultures of white clover and 9 other species. The observed ranking of tolerance to treading was *L. perenne, Poa pratensis, P. trivalis, L. perenne × multiflorum, A. tenuis, T. repens, Phleum pratense, D. glomerata, T. pratense* and *H. lanatus*. Also, as volunteer white clover appeared in all the plots it was possible to examine the interaction between sown species and treading on the content of white clover (Table 16). The yield of white clover sown as a monoculture was markedly reduced by sheep treading, but in general this reduction was less where another sown species was also present. In particular, the more the yield of white clover was reduced by the presence of a sown species in the untrodden treatment, the smaller was the yield reduction of white clover in the trodden treatments.

Table 7.16 Yield of white clover in association with nine grass species when not trodden and the relative reduction of clover content at four levels of treading (from Edmond 1964).

Sown Species	Not trodden kg DM/ha	Sheep equivalents per ha 0	10	20	40	80
Trifolium repens	448	100	95	92	77	41
Poa pratensis	426	100	97	87	87	39
Poa trivialis	358	100	91	78	88	34
Trifolium pratense	302	100	111	93	89	59
Holcus lanatus	179	100	100	119	94	56
Agrostis tenuis	179	100	138	100	63	81
Phleum pratense	134	100	142	92	108	58
Lolium perenne	134	100	142	117	108	50
Lolium multiflorum × L. perenne	112	100	110	110	100	80
Dactylis glomerata	67	100	133	117	133	50

Of two possible reasons for this effect, the first is that by damaging the canopy of the sown species treading reduced the competitive suppression of white clover. Bates (1935) observed that white clover did not resist treading as well as did the grasses *D. glomerata, Poa pratensis* and *L. perenne*, and concluded that white clover was characteristic of trodden areas mainly because the treading reduced the shading by other species. The second explanation is that the canopy of the associated species protected white clover from the damaging effect of treading. This latter possibility was tested by Brown (1968a), who found that the reduction of white clover yield was less when the trodden herbage was long than when it was cut short. Edmond (1964) observed that white clover was more tolerant of treading in summer than in winter and commented that active growth reduced the intensity of the treading effect.

Other treading experiments on mixed swards have shown reductions of white clover, especially as compared with perennial ryegrass (Brown 1968b, 1971, Edmond, 1958, 1962, 1963, 1966). In Campbell's (1966) study of dairy cow treading during grazing of autumn saved pasture in late winter, no clear effect on white clover content was shown. A comparison of sheep and cow treading by Edmond (1970) suggested that the effects of cow and sheep treading differed little. Edmond (1970) found that particularly heavy sheep treading clearly reduced *H. lanatus* and increased white clover.

(C) Associate species

With some exceptions, white clover yields are generally inversely related to yields of the associated species. Specific cases of antagonism between white clover and other species are sparse. There are indications that white clover can associate more successfully with tufted than with stoloniferous or rhizomatous grasses. Cocksfoot (*D. glomerata*) is suggested to be specifically antagonistic towards white clover, but available results indicate that this could also be a yield effect.

i. Pattern of distribution

In examining the pattern of distribution of *A. tenuis, D. glomerata, L. perenne* and *T. repens* in upland pasture in Wales, Kershaw (1958, 1959) found a change in the association between these species at two scales of sampling. At a small scale, *Lolium* and *T. repens* were positively associated, whereas all the other associations were negative. At a larger scale the *Lolium-T. repens* association remained positive, and positive associations between *D. glomerata* and *T. repens* and *D. glomerata* and *L. perenne* were shown. The negative association between *Argostis* and the other three species remained, although at a reduced level of significance. From this Kershaw concluded that at the smaller scale there was spatial exclusion between *Dactylis* and *T. repens*, but at the larger scale the association reflected a distributional pattern in relation to soil depth. *D. glomerata* and *L. perenne* occupied the deeper soil and *A. tenuis* the shallow soil. Kershaw suggested that the negative association between *T. repens* and *Agrostis* appeared to be a reflection of reduced performance of *T. repens* in areas colonised by *A. tenuis* rather than spatial exclusion of *T. repens* by *Agrostis*. *Agrostis* colonised shallow soil, the shallow soil limiting white clover root development, which in turn reduced stolon growth of *T. repens*. White clover stolons were much longer in areas not occupied by *Agrostis*.

In the succession towards *Agrostis* dominant pasture in New Zealand, the development of positive associations between ryegrass and white clover and negative associations between *Agrostis* and ryegrass and white clover have been observed (Harris & Brougham 1968, Harris 1973a). *Poa* spp., mainly *Poa annua*, also showed positive associations with ryegrass and white clover and negative asociations with *Agrostis*.

More recently, Turkington and co-workers have examined the association of white clover in pasture in detail, using in particular a technique which records the physical contacts between species above ground. In a study of old permanent pasture in North Wales, Turkington & Harper (1979a) showed *T. repens* and *L. perenne* to be positively associated with each other and mostly negatively associated with all other abundant species in the pasture. There was no clear evidence that this association was determined by variation of edaphic factors in the area studied. Analysis of seasonal changes in the pasture (Turkington & Harper 1979b) suggested that the positive association between *T. repens* and *L. perenne* was related to the asynchrony of the growth cycles of the two species.

Turkington et al. (1977) applied similar analysis to four grasslands communities in Ontario, Canada. These communities had been sown 10 or more years before the observations were made and did not contain *L. perenne*. The legumes in the communities were generally negatively associated with other legumes, and grasses were generally negatively associated with other grasses. White clover was the species most commonly positively associated with other species and was positively associated with the grasses *Agropyron repens, Bromus inermis, Festuca* spp., *Poa annua*, and *Poa compressa*, but was negatively associated with *D. glomerata* and the legumes *Medicago lupulina* and *Trifolium pratense*. There was a general trend indicating that where two species were positively related in terms of their total soil environment, they tended to be negatively related in terms of vegetation analysis. For example, *T. repens* and *T. pratense* were positively associated for soil Mg, Ca and pH, but negatively by vegetation analysis. Aarssen et al. (1979)

extended this study to observe changes of association over a 2-year period and categorised associations as being stable, seasonal or temporary. *T. repens* had numerous temporary associations and it was suggested that in being a stoloniferous species and relatively mobile, *T. repens* may come in contact with a greater variety of neighbours than most other species in a community. Turkington & Cavers (1979) then measured change of pattern and association during the two years after sowing of a mixture of five legumes with or without the grasses *D. glomerata* and *B. inermis*. The presence of grass induced positive associations between the legumes, but where grasses were absent the legumes were generally negatively associated except for *T. repens*. This was explained by the stoloniferous growth of *T. repens* allowing it to invade the area of *T. pratense* and *Medicago sativa*, the two other legumes which persisted and which have clumped distributions.

ii. . Content with grasses

Chestnutt & Lowe (1970) comprehensively reviewed the relative amounts of clover in association with *L. perenne, P. pratense, D. glomerata, Festuca pratensis* and *F. arundinacea* and the relative amounts of clover in association with ryegrass cultivars. *D. glomerata* appeared to be the poorest companion grass for white clover, while *L. perenne* and *F. pratensis* were the most compatible (Cowling & Lockyer 1965, Hunt 1956, 1966, Hughes & Davis 1951, Herriot & Wells 1960, Hughes 1951, Reid 1961a, b, Reid & Castle 1965). It was concluded that *D. glomerata* and, to a smaller extent *P. pratense,* showed an aggressiveness towards white clover out of proportion to their yielding ability compared with ryegrass. Chestnutt (1971) found that inclusion of white clover in a mixture had a more favourable effect on the growth of perennial ryegrass than on either *Phleum pratense* or *F. pratensis*. The inclusion of white clover also increased the tiller numbers of *Poa trivialis* and reduced the tiller numbers of *A. tenuis*.

Turkington et al. (1979) observed growth and survival of white clover stolon cuttings transplanted into parts of a permanent pasture in which one of four grass species was dominant. The yield of transplants declined in the order *Cynosurus cristatus* (0.41 g per plant), *L. perenne* (0.21 g) *A. tenuis* (0.17 g) and *H. lanatus* (0.09 g); the percentage survival after 1 year was *C. cristatus* 51%, *A. tenuis* 41%, *H. lanatus* 31% and *L. perenne* 25%.

iii. Habit of associate species

Brown (1939), comparing grass species sown with white clover under frequent close cutting, observed that the content of white clover was much less with the turf-forming grasses *A. tenuis* and *Poa pratensis* than the species with more open stands, e.g. *L. perenne, F. pratensis, Phleum pratense* and *D. glomerata*. Further evidence that *T. repens* does not associate well with stoloniferous species is provided by Haggar (1971) who, in reviewing *Poa trivialis,* suggested that there was an inverse relationship between the content of this grass and white clover. However, the evidence was less conclusive than that for associations with *A. tenuis* described by Kershaw (1958, 1959) and Harris & Brougham (1968).

iv. Yield of associate species

M. G. Jones (1933) observed that wild white clover on the whole seemed to be governed by the ryegrass population - the more ryegrass the less white clover. Where ryegrass was sparse, the competitive effects of *Poa trivialis* came into play. Chestnutt & Lowe (1970) concluded that there were no marked differences between ryegrass cultivars in their compatibility with white clover. Those ryegrass cultivars which suppressed white clover least were of low yield potential or low persistency (Green & Corrall 1965). These observations raise the question as to the extent to which the competitive ability of grasses against white clover is related to their yielding ability.

To answer this question, correlations between clover yield and yield of associate species were calculated for experiments which gave results suitable for this purpose (Hughes 1951, Hughes & Davis 1951, Cowling & Lockyer 1965, Green & Corrall 1965, Harris 1980). The use of high yielding grass cultivars and management that increased the yield of the grass and/or weed component of mixed swards generally reduced the yield of associated white clover. Exceptions to the inverse relationship were observed by Hunt (1956) who found no significant correlation between grass and clover yields, Reid (1961a, b) who obtained a negative correlation only in the first year, and Hunt (1966) who found a positive relationship that strengthened over 6 years.

Camlin (1981) in a 4-year experiment examining the competitive relationship between 3 cultivars of white clover and 10 cultivars of perennial ryegrass given N fertiliser at 200-240 kg/ha annually, specifically used regression analysis to examine the relationship between clover yield and grass yield. In confirmation of the general effect observed in earlier experiments, Camlin (1981) concluded that the compatibility of the ryegrass cultivars with clover was inversely related to the grass component yield, which was determined largely by persistence. The greatest clover contribution was in association with the non-persistent ryegrass cultivars 5321 and Presto (Figure 18).

v. *Cocksfoot effect*

It has often been observed that the white clover content of swards is low when in mixture with cocksfoot. In the experiments of Cowling & Lockyer (1965) and Hughes (1951), cocksfoot was the highest yielding grass and was associated with the lowest content of white clover. Conversely, in the experiment of Harris (1980) where cocksfoot was one of the lowest yielding grasses, it was associated with a high content of clover.

An effect of greater unsown species content in cocksfoot stands on white clover content is also suggested in several experiments (Hughes & Davis 1951, Heddle & Herriot 1954, Reid 1961a, b, Reid & Castle 1965a, b). As previously suggested (Chapter 4, Section 3. (C)), these unsown species, if they are prostrate with a rhizomatous or stoloniferous habit, may interfere more with white clover than would erect grass species. This possibility is illustrated by Heddle & Herriot (1954), who recorded yields of sown grass, white clover and unsown species in swards of S23 ryegrass, Ayrshire ryegrass, S143 cocksfoot and Danish cocksfoot, each sown at three seed rates with white clover. Four cuts were made in the one harvest year and the results are plotted in Figure 19. Yield of white clover was inversely related to the combined yield of sown grass and unsown species. The contribution of the sown grass to this relationship was small, being predominantly a function of the unsown species, mainly *Poa annua* and *Poa trivialis*. These unsown species were most abundant in the cocksfoot swards.

Two further possible explanations are: first, the strong selective grazing of white clover in association with cocksfoot, cocksfoot being less acceptable to grazing animals than ryegrass; second, a greater interference between cocksfoot and white clover caused by similarities in their seasonal growth pattern. This latter possibility is not supported by Hunt (1960), who compared the seasonal growth rates of cocksfoot, timothy, meadow fescue and perennial ryegrass in the west of Scotland, but is a mechanism favoured by Turkington & Harper (1979b).

(D) White clover genotype

Differences between white clover genotypes in their ability to compete with associate grasses and their response to the modifying effects of N fertilisation and defoliation management have been related largely to genotype differences of petiole length and leaf size. Of the three white clover cultivar groups, large, intermediate and low growing, the larger leaved forms are best able to withstand the severe competition from N fertilised grass, whereas the low growing cultivars are considered to be adapted to withstand frequent close defoliation. Clover content of swards has also been shown to be influenced by differences of seasonal periodicity and pest and disease resistance of white clover genotypes. There is also evidence that there are white clover populations adapted for association with specific species.

Low growing, small leaved white clover may be better suited than larger leaved forms to intensive grazing systems (Davis & Cooper 1951). Kent wild (small leaved) and NZ, S100 and Dutch white clovers (all medium leaved) were grown with ryegrass under a rotational grazing system with sheep. Kent contributed most to sward production and, with S100, produced the most animal liveweight gain, whereas Dutch white clover virtually disappeared in the second year of the trial. King (1963) observed smaller leaved, more prostrate white clover genotypes on basic soils compared with acidic soils, attributing this to the greater grazing of vegetation on basic soils. Brock (1974) found the cultivar Grasslands Pitau, which has larger and thicker leaves and less dense stolon formation, to be more responsive to less frequent, lax grazing than grasslands Huia. Adachi et al. (1968) noted that under an infrequent defoliation regime, wild and common white clover failed to persist, but Ladino white clovers maintained their density, presumably because of their greater height.

In Ireland, Connolly (1968) evaluated mixtures of 6 white clover varieties and 3 grass varieties under sheep grazing. Large and intermediate leaved clover varieties yielded more than small leaved varieties, but there was a trend for increased clover yield to be associated with decreased grass yield. Similarly, Davies (1972b), examining correlations between clover content and total and grass yield in a trial assessing 18 clover varieties, observed an inverse relationship between varietal differences of clover yield and grass yield. This form of relationship was also shown by Camlin (1981), comparing S184, S100 and Kersey (Figure 18), grass yield being lowest in association with Kersey.

Reid (1961a, c) compared large leaved Kersey, intermediate leaved S100 and small leaved S184 grown in monoculture and with each of 6 grass cultivars at 3 levels of N fertiliser. As white clover cultivar did not significantly affect total dry matter yield during the trial, Reid concluded that the cultivar of a clover in a seeds mixture was of minor importance compared with the cultivar of grass in determining the response of the mixture to N fertiliser and in determining the clover content of the sward at different levels of N fertiliser. However, Reid (1976) reported better performance of Blanca R.v.P. than S100 at high N levels.

Blanca R.v.P. performed better than 8 other clover cultivars under high N at Trawscoed, Wales, and Cambridge, England (Aldrich 1970a). The ranking of the clover cultivars in their reduction by N fertiliser was generally related to their leaf size, although medium-small leaved S100 showed a smaller reduction and medium-large leaved English Giant a greater reduction in relation to their leaf size. Hoen (1970) also showed the ability of Blanca R.v.P. to maintain itself under high N fertiliser.

Davies (1972a) compared S184, S100 and Kersey sown with ryegrass and fertilised with 0, 190 and 380 kg N/ha annually. Kersey contributed the most clover yield and S184 the least at all N levels. Davies concluded that the longer petioles and larger leaves of Kersey, by allowing the canopy to be raised above that of the grass and by reducing the amount of light passing through to the grass, assisted this cultivar to compete with grass.

Selections of medium-leaved white clover have been obtained which, although similar to S100 in morphological characters, contribute more to yield in grass-clover swards at high levels of N fertiliser (Davies & Evans 1974). It is not clear what character enables these selections, one of which has been released as cultivar Menna, to compete more successfully under high N.

Varietal differences of pest or disease resistance can also influence the clover content of swards, as clearly illustrated by Aldrich (1970b) who reported that *Sclerotinia* reduced the yield of the susceptible cultivar S100, whereas the resistant cultivar Pajbjerg Milka gave good results.

Sowing of the New Zealand bred cultivars Grasslands Huia and Grasslands Pitau, which differ in their capacity for cool-season growth, results in swards of different clover content. The increased clover content obtained in winter by sowing Pitau has been observed to reduce grass and total yield during this season (Barclay 1969, Lambert et al. 1969, Brock 1971, 1974, Goold & Douglas 1976, Suckling 1976).

Agronomic experiments studying interactions between grass and white clover cultivars have not given a clear indication that clover genotypes are specifically adapted for association with particular species. However, work by Turkington and co-workers indicates the existence of biotic ecotypes of white clover adapted for association with particular grass species. Using white clover clones from an old permanent pasture in North Wales, Turkington et al. (1979) demonstrated that the clones differed in their aggressiveness towards a standard clone of white clover. In a related experiment (Turkington & Harper 1979c) white clover plants sampled from areas dominated by either *L. perenne, H. lanatus, C. cristatus* or *A. tenuis* were either transplanted back into areas of each of the four grasses growing naturally in the sward, or into swards of the grasses established in boxes containing a bulk of the soil from the experimental field. Interaction between grass species and clover population was significant, indicating that each clover population tended to yield best when grown with the species of grass from which it was originally sampled. The favoured interpretation of this result was that micro-evolution of *T. repens* had occured in response to different pressures exerted by different species of grass. As the pasture was at least 60 years old, it was considered that there had been sufficient time for this to have occurred.

The possibility that such biotic adaptation could occur within a shorter time was indicated by a study of pasture in Ontario, Canada (Turkington 1979). Ramets of *Medicago sativa, M. lupulina* and *T. repens* from 10-year-old fields of *B. inermis, Phleum pratense* and *D. glomerata* were transplanted into artificial swards of all combinations of legumes and site of origin. Survivorship and dry weight of *T. repens* was greatest when it was transplanted back into swards of the grass species from which it had been sampled. In a second experiment, comparison of *M. sativa* sampled from 1-, 3- and 10-year-old swards of the same three grasses suggested that the micro-evolutionary process of biotic adaption for association with a grass species was rapid.

Besides adaptation arising by selection of the genetic heterogeneity within a population, individual genotypes may be able to adapt to a range of environments by expression of phenotypic plasticity. White clover is particularly plastic in the expression of petiole length and lamina size, and an experiment (Hill 1977, 1980) in which 5 clover genotypes were grown in association with 5 ryegrass genotypes indicated the existence of different phenotypic plasticity for different white clover genotypes. Hill suggested that selection of phenotypic plasticity may produce white clover cultivars better able to adapt to and withstand competition from grass.

(E) Competitive strategies

This section examines those morphological and physiological characteristics of white clover which enable it to successfully survive and grow in plant communities. Characteristics emphasised are (i) rapid attainment of light interception by horizontal leaf arrangement and canopy adjustments arising from changes of petiole length; (ii) the ability to spread widely by stolon growth and explore both the soil and aerial environment; (iii) a different growth periodicity from that of the species with which it is successfully associated; and (iv) the use of fixed atmospheric N rather than mineral N from the soil.

i. *Models defining strategy*

Grime (1973, 1974) suggests that botanical composition is determined by the interaction between disturbance, stress and competition. Species predominating in disturbed habitats have a ruderal strategy characterised by rapid seedling establishment and growth, and diversion of photosynthate to seed production at the expense of vegetative growth. Stress-tolerant species have small and low potential relative growth rates, whereas successful competitors have the potential for tall stature, intensive exploitation of the aerial and soil environment, high potential relative growth rate and a tendency to deposit a dense litter layer. Using indices of competition, stress and disturbance as axes in a triangular ordination model (Grime 1974) *T. repens* was placed with other species characteristic of meadows, enclosed pastures and mown roadsides. These habitats are influenced by approximately equal and interacting effects of competition, stress and disturbance. Compared with 100 other species, the competitive index of white clover was a little above average. Of the agriculturally important grassland species in the same habitat grouping as white clover, *D. glomerata, L. perenne* and *Poa pratensis* were classified as being more competitive than white clover.

Turkington & Cavers (1978) compared the partitioning of dry matter into component parts for the legumes *Medicago lupulina, M. sativa, T. pratense* and *T. repens* and related to this r- and K- selection theory (Harper 1967). Species with a predominantly r-strategy are those characteristic of frequently disturbed habitats and partition a larger proportion of their energy into seed and ramet production to maximise their rate of natural increase. Species with a predominantly K-strategy are characteristic of stable environments and direct their energy towards maintaining competitive ability by elevating their canopies higher to avoid light competition or by having fast growing and extensive root systems. There was clear distinction between *M. lupulina* which rapidly partitioned dry matter to seed production and *M. sativa* which remained vegetative during the 18 weeks of the experiment. *Trifolium pratense* occupied an intermediate position in the r-K continuum between the *Medicago* species. White clover was argued to have either r- or K-strategy depending on whether the development of stolons was classified as reproductive or purely vegetative. Turkington & Cavers (1978) concluded that the formation of stolons by white clover within 7 weeks of germination permitted it to reproduce sooner than *M. lupulina* could reproduce by seed. In this way white clover could behave as an r-strategist in a K-type environment. As will be dicussed later, genotypic variation and phenotypic plasticity of white clover provides individuals which occupy a wide range of the r-K continuum.

ii. Light interception

White clover plants are adapted to re-establish full light interception rapidly after removal of leaves. However, once full light interception is attained, at a relatively low leaf area index (LAI), further leaf growth results in death of overtopped leaves. This is an inefficient partitioning of the plant's energy and reduces its competitive ability after longer intervals of uninterrupted growth.

After defoliation of an established white clover stand, renewal of the leaf canopy is initiated by reserves translocated from the stolons and roots. This phase of dependence on reserves has been demonstrated in Ladino clover (Hoshino et al. 1967, Hoshino & Oizumi 1968, Ochi & Muroga 1967, Sasaki & Kudo 1965, Kumai et al. 1965), though it has been disputed by Carlson (1966). Carlson, however, used seedlings which may not have developed significant quantities of reserves.

Further development of the leaf canopy after the initial phase of dependence on reserves was described by Brougham (1958a). The LAI of a sward of New Zealand white clover grown in late spring reached a ceiling of 5.5 at 20 days after defoliation. Senescence of leaves began when LAI was about 4. As the canopy developed, successive leaves at maturity had longer and heavier petioles and heavier and more expansive laminae. Laminae developed after the ceiling LAI was reached were twice the size of first-formed leaves, and petioles were 2.5 times the weight and 4 times the length of the petioles of leaves formed immediately after defoliation.

Leaves developing after the first layers of leaf are formed are shaded, and petioles of these leaves elongate until the laminae of these leaves penetrate the canopy to reach full light. This process is repeated so that successive crops of leaves have progressively longer petioles until a physical limitation to petiole elongation is reached. Brougham (1958a) observed that in mixtures with grasses, a greater petiole length may be developed because of support by the grass leaf blades and stalks, and noted the genotypically based differences of petiole extension underlying the classification of small, intermediate and large leaved forms.

Brougham (1962) extended his observations to measuring canopy changes in an undefoliated canopy of white clover during a year. The canopy adjusted to seasonal variations in the light environment, so that virtually all incident light was continuously intercepted by leaves with efficient photosynthesis. Shaded leaves not effectively intercepting light quickly senesced and decomposed (Brougham 1958a, Hunt 1970, 1971). This is probably an important pathway in the transfer of legume-fixed N to associated non-leguminous species. White clover does not have the litter accumulating characteristic of highly competitive species (Grime 1973, 1974), instead, the rapid breakdown of its N-rich leaf fall stimulates associated species.

A feature of the white clover canopy which distinguishes it from that of grasses is the orientation of the laminae. The horizontal arrangement of the laminae which, with phototropic movement, orientates them at right angles to the sun, results in a low critical LAI for white clover. The long slender leaves of the grasses, erect for most of their length, allow deeper direct penetration of light into the canopy, whereas light not absorbed by the horizontal laminae of white clover is mostly reflected back into the atmosphere.

Leaf orientation and its effect on LAI was examined by Brougham (1958b) in stands of short-rotation ryegrass (*L. perenne* × *L. multiflorum*), perennial ryegrass, timothy, white clover and a mixture of short-rotation ryegrass and white clover. The LAIs at which 95% of the incident light was intercepted at midday in midsummer were 7.1, 7.1, 6.5, 3.5 and 4.5 for these stands respectively. Maximum values of LAI reached were: short-rotation ryegrass 9.5, perennial ryegrass 8.8, timothy 10.8, white clover 5.5, mixed stand 7.3. Comparing 8 species, Brougham (1960b) recorded critical LAIs of 2.7, 3.1 and 3.2 for white clover which were lower than those of the dicotyledonous species red clover, kale and water cress and considerably lower than for short-rotation ryegrass, perennial ryegrass, cocksfoot and maize.

The light relations between grass and clover in mixed swards have been lucidly described in two experiments of Stern & Donald (1962a, b) with mixtures of *Trifolium subterraneum* and *Lolium rigidum*, and the processes described would apply to white clover-ryegrass mixtures in the establishment year. The mixtures were grown at 4 N levels (Stern & Donald 1962a). At low N clover dominated, at high N grass dominated. Grass yield responded positively to applied N and increased leaf yields of grass disposed above the clover leaf canopy. This reduced the light intensity at the clover leaf canopy, so reducing growth of clover. In the other experiment (Stern & Donald 1962b), roots of grass and clover were segregated in slot-like containers which, when placed together, allowed competition for light but not root competition. As the grass overtopped the clover canopy, the clover, even though it had ample water and nutrients, was shaded so severely as to be eliminated from the sward.

From the light interception characteristics of white clover canopies it can be predicted that prostrate associated species which lie below this canopy will suffer even more severe shading at a given level of LAI than where a grass provides the overtopping canopy. *A. tenuis* was severely depressed where a tall canopy of white clover developed under infrequent cutting in summer (Harris 1974). In another experiment where individual ryegrass genotypes were grown with and without white clover, the more prostrate ryegrass genotypes were overtopped, shaded and suppressed more by white clover than by associated ryegrass plants (Harris 1973b).

Although the body of grass-legume competition studies indicate that white clover is at a disadvantage with the grass in terms of light competition, probably most grasses and white clover are more compatible in their light relations than white clover is with dicotyledonous species of equivalent capacity for leaf elevation. Brougham (1965) compared the regrowth for 60 days after defoliation of a monoculture of white clover with that of white clover mixed with red clover. Thirty days after defoliation, the leaf area of white clover in the mixture was less than 25% of that in the pure stand of white clover (Figure 20). White clover leaves which developed in the mixed stand had longer, thinner petioles and thinner laminae than leaves in the pure white clover stands. Similarly, Morita & Fujita (1971) found that in mixed stands of Ladino and red clover, red clover had the greater leaf area, longer petioles and was the dominant species.

Harper & Clatworthy (1963) measured the growth, development of leaf area and light interception of the closely related species *T. repens* and *T. fragiferum* grown at two densities in monoculture and mixture. Changes in the total leaf area for the species components of the swards recorded from weeks 7 to 21 after sowing (Figure 21) indicated the more rapid development of leaf area by *T. repens* than by *T. fragiferum*, though *T. fragiferum* subsequently developed a higher LAI. Declining light levels later in the experiment probably caused the general decline in LAI (Figure 21). Clover leaves were borne in a broad band which became higher as the swards developed. Early in the experiment, the canopy of *T. repens* was borne slightly higher than that of *T. fragiferum*. Later, because *T. fragiferum* had a longer growth cycle, produced a short vertical stem and had longer petioles than *T. repens*, the canopy of *T. fragiferum* was borne higher than that of *T. repens*. However, because the leaf canopy of *T. repens* developed faster, it gained competitive advantage over *T. fragiferum* which it maintained until the end of the experiment, even though *T. fragiferum* had the potential to gain the advantage in competition for light.

iii. Stolons

The capacity to spread horizontally by growth of stolons is a characteristic important for the competitive ability of white clover. This characteristic allows the species rapidly to attain an equilibrium in plant communities (Ennik 1970, Harris & Thomas 1973) and is the reason why variation in the seed rate of white clover has only a short-term effect on the clover content of swards (3. (F) iii.). White clover growing points proliferate with stolon growth and branching at the soil surface and thus invade unoccupied areas while evading removal during defoliation.

Sanderson (1966), tracing the development of white clover plants, observed that seedlings develop a short primary stem with 5-8 nodes. Primary stolons develop from these nodes to form a rosette and thereafter the plant develops from a series of secondary and tertiary branching stolons. In 2 years or less the proximal ends of the stolons degenerate and individual stolons become independent. Radioactive tracer studies show that absorbed minerals and elaborated metabolites are translocated mainly away from the primary root to the growing tips of the stolons. Hollowell (1966) suggests that after the axillary buds of the primary stems stop developing into stolons, the primary root does not function as a storage organ, becomes weakened and eventually dies. Death may be hastened by disease, as the old tissue is susceptible to infection from soil organisms such as *Fusarium* and *Rhizoctonia.* Root severing experiments have shown that once the nodal root system has been established, further development of the stolons does not rely on minerals and assimilates from the primary root system. Hollowell (1966) suggested that, because in many environments the primary root degenerates within a year, white clover is primarily a winter annual which, under favourable environmental conditions, may behave as a perennial through asexual propagation.

Examination of the effect of complete and partial defoliation on young white clover plants by King et al. (1978) emphasised the acropetal movement of assimilates to support the spread of the plant. They found that defoliation had less effect on the growth of terminal leaves than on leaf growth nearer the centre of the plant. The rate of formation, number and size of internodes was not significantly affected by partial defoliation, suggesting that partial defoliation has little effect on the rate of spread of white clover plants.

Hollowell (1966) observed that abundant flowering and seed setting are associated with early death of the plant. The persistence of white clover is reduced by growing plants in lengthened days which induces profuse flowering (Gibson 1957). White clover populations adapted to regular climatic stress, especially drought, have an annual strategy. Such a strategy is characteristic of the white clover naturalised in the subtropical east coast of Australia which frequently has marked soil water deficits (O'Brien 1970). The ecotype of this region is heavy seeding and has an intermediate leaf-type. In unfertilized swards it behaves mainly as a winter annual, but with adequate soil fertility and favourable seasonal conditions it may perennate. Soft seeds of the ecotype germinate rapidly and the high proportion of hard seed provides a reservoir of buried seed which ensures the long-term persistence of the species.

These observations indicate a range of strategies for white clover survival. Hot, sunny conditions which lead to drought-induced mortality of plants are the same climatic conditions that promote flowering, pollination, seed set and formation of hard seed. By comparison, dull, cool, moist conditions result in low reproductive activity, with less assimilate being diverted away from stolon growth. Moister soil is suitable for root penetration and development of secondary root systems, enabling the stolons more rapidly to become independent of the primary root and shoot system. Further, the opportunity for seedling establishment is limited in permanent pasture in cool moist climates which allow a continuous cover of vegetation to be maintained. For example, establishment of new plants of white clover from seed in permanent pasture in Britain was observed only on molehills (Cahn & Harper 1976a). By comparison, a frequent observation in the southern USA, Australia and New Zealand is the substantial establishment of white clover from seed which occurs in pasture depleted by drought and stress with the onset of conditions favourable for white clover growth.

A measure of the capacity of white clover to spread vegetatively compared with other species is provided by a study by Kishi (1975c) of 4 species grown as pure swards. Tillering of *Dactylis glomerata* ceased when tillers were crowded, whereas *Festuca arundinacea, Poa pratensis* and Ladino white clover all spread indefinitely; Ladino stolons colonised bare ground the most rapidly. How far white clover clones can spread in mixed plant communities is indicated by Harberd (1963), who mapped the distribution of individual clones in an area of fixed dune vegetation. Most of the clones were limited to patches 2 m across, but one clone was distributed in 5 discrete patches which collectively gave a continuous spread of more than 18 m. This clone occupied an ecological zone distinct from a zone occupied by 2 patches of another genotype, suggesting differentiation of ecological tolerance between genotypes.

Harberd (1963) cautioned that the size of the area occupied by a clone should not be regarded entirely as a function of age, as undoubtedly vegetative spread can be limited by ecological factors. Inhibition of stolon formation by white clover seedlings with increased density of associates was described by Clatworthy (1960). The apparent antagonism between white clover and stoloniferous or rhizomatous species possibly results from competition for suitable rooting sites by the prostrate stems of these species as they wander through the sward. Where the associate species are tufted bare soil is ample between the tufts, and in these inter-tuft areas white clover stolons are mostly found.

The pattern of spread of clones can be affected by destruction of stolon sections by pests and diseases, cold damage, treading and grazing. Pascoe (1974) described the fragmentation caused by sheep grazing. On actively growing stolons, new stolons were generally initiated at nodes closer to the base than to the tip of the stolon. Terminal buds on long stolons were more frequently removed by grazing. The removal of the terminal bud stimulated the growth of new stolons from axillary buds, particularly those nearer the grazed end. Grazing often fragmented continuous single plants into several independant growing units. Observations of low-temperature killing of stolons in swards in Aberystwyth, Wales, showed that unprotected sections of stolons were killed (Harris et al. 1983). Large, loosely rooted stolons of the large-intermediate leaved cultivar Olwen mainly died back to the rosette of the primary axis, whereas the dense well-rooted stolons of the small-leaved cultivar S184 became fragmented into lengths of stolon free from the primary root and shoot system.

Genotypes which form canopies of large leaves on long petioles able to avoid shading from associate species differ in stratergy from those genotypes forming dense networks of well-rooted stolons bearing small leaves on short petioles, which are better able to withstand frequent and close defoliation. At the end of the establishment year of an experiment comparing small, intermediate and intermediate-large leaved cultivars of white clover, the weight of stolon was found to be inversely related to the yield of clover herbage harvested from all cuts of the establishment year (Rhodes & Harris 1979). Although the gross yield of all cultivars was similar, the harvestable part of the long petiole-large leaved cultivars was greater. Thus, assimilates appeared to be partitioned to form a tall canopy at the expense of development of the stolon system.

Evidence for the very specific adaptation of white clover genotypes together with phenotypic plasticity (Section 4(D)) suggests that by stolon extension, genotypes may spread over a large area but at the same time be truly successful only in part of the colonised area. As white clover invades a new habitat, it may pass through a phase of rapid growth and vigorous flowering, decline in size as the habitat deteriorates in the presence of other species and then make new vigorous growth after the habitat is disturbed. Indeed, parts of the same clone may be growing vigorously, other parts may be passing through a phase of intensive flower production, while still other parts are dying under some vigorous neighbour. This suggests that a single clone may become dominant over a large area, but in old permanent pastures in England and Wales, white clover populations consisted of a large number of genotypes (leaf morphs) intimately mixed as shown by Cahn & Harper (1976a). Their study was based on 1-m_2 quadrats subdivided into 10 × 10 cm squares to allow the study of pattern. The number of morphs found in a 10-cm square varied from 1.5 to 6.0 with a mean value of 4.2 for old permanent pasture. The corresponding mean for a 5-year-old ley was 2.8. This indicates that a high diversity of clones could be maintained in very old pasture, rather than the number of genotypes present declining as the community aged.

iv. Seasonal growth periodicity

In the UK Blackman (1933) and M.G. Jones (1933) recognised the influence of differences in seasonal growth of grassland species on the botanical composition of swards. Blackman's collection of results from a range of sources showed that the maximum growth of white clover occurs later in the season than most of the grasses with which is it commonly associated in temperate grassland (Figure 22). More recent studies to define processes which allow species diversity in old permanent pastures indicate that species may successfully cohabit because of their different growth periodicities. Turkington & Harper (1979b) observed the markedly different seasonal cycles of leaf production between species in grassland and noted a degree of complementarity in the cycles of forage production in *L. perenne* and *T. repens*. Turkington & Harper (1979a), like Kershaw (1959) and others, noted the association between ryegrass and white clover and it is suggested that these species do not exclude each other because of temporal niche separation. Ryegrass, in the old permanent pasture studied, had a major peak of leaf production in early June and again in August-September, whereas white clover had its peak period of leaf production in July. The other main dimension in the successful cohabitation of the species is that white clover fixes nitrogen, and that ryegrass is among the most responsive of grasses to nitrogen (Chapter 4, Section 5. (F)).

By limiting the period of plant growth, winter cold and seasonal drought can constrain the growth of species in a community to similar temporal niches. The effect of seasonal growth periodicity is most clearly expressed in environments that allow plant growth to continue throughout the year, but in which temperatures are optimum for different species in the plant community at different times of the season. An example of wide differential periodicity in a ryegrass-white clover pasture where water supply was non-limiting was provided by a study at Palmerston North, New Zealand, by Brougham (1959a) (Figure 23). Most of the yield was contributed by ryegrass in winter, spring and early summer months, while white clover produced most of the total yield in late summer and autumn.

Another experiment in the same locality suggested that such a wide separation of growth periodicity of ryegrass and white clover did not occur if water became limiting during summer. Harris & Hoglund (1980) compared *L. perenne, T. repens* and *T. pratense* as monoculture and in two-species grass-legume mixtures. The seasonal distribution of growth of the component species grown as monocultures show the difference of seasonal growth of ryegrass compared with the two clover species (Figure 5). However, the presence of ryegrass markedly changed the growth periodicity of the clovers. White clover contributed less to the yield of the mixtures than ryegrass at all harvests throughout the year, even though the summer yield of white clover in monoculture was clearly greater than that of the ryegrass. By comparison, red clover became the main yield contributor in the mixed sward in summer, although its maximum growth in mixture was delayed until about 6 weeks after it was attained in monoculture. Growth of ryegrass was maximum during November and December when maximum temperatures were in the same range as those optimum for growth of ryegrass (18-21 °C, Mitchell 1956). Also, experimental evidence has suggested that, regardless of environment, growth of ryegrass is depressed in summer after the period of reproductive growth (Leafe et al. 1974). Growth of clovers in monoculture was maximum several weeks before daily temperature maxima were near the growth maxima for clovers (24 °C, Mitchell 1956). Soil water budgeting indicated that water deficits in the surface layers would have limited growth before temperatures had reached optimum for clovers. *Trifolium pratense* was better able than *T. repens* to exploit the niche produced by the reduced summer growth of ryegrass, possibly because its deeper root system allowed it to obtain water from deeper levels in the soil.

Differences in the seasonal growth periodicity of different associate grasses can modify the potential for summer growth of white clover. Vartha (1972) compared growth of white clover sown with perennial ryegrass or *Poa trivialis* alone or with the two grasses together (Figure 24). The greater depression of *P. trivialis* growth in summer provided a larger niche for white clover, which responded by producing greater yields in summer and autumn in the presence of *P. trivialis* than in the presence of ryegrass. If both grasses were sown together, white clover responses was intermediate between that of ryegrass and *P. trivialis* sown separately. By comparing herbage 7.6 cm high cut to 2.5 cm stubble (short) with that 17.8 cm high cut to 5.1 cm (long), the experiment also illustrated how management can influence the expression of seasonal growth periodicity in white clover. Summer clover yield with *P. trivialis* was 39% higher with short cutting than with long cutting, whereas with ryegrass the clover yield was similar with both cutting treatments. Effects of the seasonal timing of grazing in relation to the growth periodicity of species on clover content of permanent pasture have been described previously (Section 7.4.2.5c) with reference to the studies of M.G. Jones (1933).

In regions where summer temperatures are at or near optimum for tropical grass species, results indicate that white clover is forced into a different seasonal niche. Cameron (1969) compared the seasonal growth curve of Ladino grown alone or in association with *Paspalum dilatatum*, *Bromus uniloides* or *Phalaris tuberosa X arundinacea* (Ronpha grass) under irrigation on heavy clay soils in central Queensland, Australia (Figure 25). With paspalum, which has substantially higher summer production, white clover production peaked in the spring months of September, October and November and was depressed throughout the rest of the year, including winter when paspalum was dormant (Figure 25). In mixture with the temperate grasses and in monoculture, white clover production was more evenly distributed throughout the year. Application of various fertiliser and cultural treatments to the paspalum-white clover mixture failed to improve the growth of white clover in winter.

However, in the warm-temperate environment of northern New Zealand where both paspalum and Kikuyu (*Pennisetum clandestinum*) are extensively naturalised and where both grasses have distinctly summer growth patterns, Lambert et al. (1977) found no differences in seasonal growth periodicity of white clover grown with either ryegrass or Kikuyu. Combined sowing of ryegrass and Kikuyu produced higher yields and a more uniform seasonal yield distribution, but reduced the content of white clover (17%) more than where Kikuyu (26%) or ryegrass (27%) were grown separately. Whereas Cameron (1969) irrigated his experiment, that of Lambert et al. (1977) was a dryland experiment, and they indicated that in summer the presence of ryegrass increased stress on clover more than did that of Kikuyu. Possible different, albeit restricted, temporal niches were available where ryegrass or Kikuyu were grown separately. Full exploitation of the spring niche in the presenc of Kikuyu was limited by carry-over effects from the preceding summer growth of Kikuyu. Full exploitation of the summer niche by white clover in the presence of ryegrass was limited by summer drought. Where both grasses were sown together, the period of competitive exclusion of clover by grass was extended.

As well as the effect of seasonal growth periodicity on the ability of a species to capture environmental resources and occupy space at the expense of associate species, seasonal environmental changes may stimulate changes in the habit and hence the competitive ability of species. Seasonal changes in the canopy of the perennial ryegrass sward were described by Leafe et al. (1974). The spring (flowering) and summer (vegetative)canopies differed markedly, the former elevating leaf tissue on the elongating reproductive stems and the latter having an arrangement which caused considerable shading of leaves in the canopy. The leaf development studies of Brougham (1962) showed that as the season advances from winter to summer the mean petiole length and lamina area of white clover leaves increases. Consequently, the two species have their maximum capacity for leaf elevation (capacity to overtop and shade) at different times. Kishi (1974c, 1975a, b) studied the effects of seasonal changes of growth characteristics on competition between Ladino clover and associate grasses. Tall fescue, which produced more top growth in summer-autumn and had high tiller numbers in spring, competed better with clover than cocksfoot and *Poa pratensis*. The latter grew poorly in summer-autumn and their tillering was strongly suppressed by competition from the clover (Kishi, 1974c).

Kishi (1975a) found that the rate of recovery of plant height after cutting in summer was important in determining the outcome of competition for light between cocksfoot and Ladino clover. Clover increased in height more rapidly than cocksfoot in the first 30 days after cutting, but after 60 days cocksfoot overtopped the clover. The decreased vertical growth of cocksfoot in summer was ascribed by Kishi (1975b) to the replacement of reproductive tillers by vegetative tillers. The increased elevation of the canopy of Ladino clover in summer was ascribed to the effect of high temperature on petiole length, but following from Brougham (1962), it seems more likely that temperature exerts an indirect effect of increasing the rate at which a canopy develops, so hastening the shading of developing leaves within the canopy.

Selection from the genotypic variability in seasonal growth periodicity in white clover has produced cultivars with distinctive seasonal growth patterns. Grasslands Pitau has been shown to be an improvement over Grasslands Huia in the north of New Zealand where warmer temperatures in autumn and winter allow it to express its growth potential during those seasons. In the south, where autumn and winter temperatures are lower, it shows no advantage (Goold & Douglas 1976, Lambert et al. 1968, Weeda et al. 1969). The increase in clover content of swards sown with Pitau as compared with Huia is often, particularly in winter, at the expense of the grass component and sometimes the total yield (Brock 1971, 1974, Weeda et al. 1969). Conversely, the presence of a grass with higher than average growth in autumn and winter can completely suppress the cool-season growth potential of Pitau (Harris 1980).

v. Root system

The most important adaptive feature of the white clover root system is its capacity to form nodules hosting the symbiosis with *Rhizobium* which provides a supply of N to the white clover plant independant of that available in the soil. This provides a specific adaptation to soils low in N where competition from other species will be minimal. Hence, in such situations white clover does not need to be an efficient competitor for nutrients. However, white clover is a legume characteristic of high fertility soils and requires a strategy which allows it to survive where levels of P, K and S are adequate. In this respect it differs from legumes such as *Lotus pedunculatus* which dominate white clover by root competition on low-P, high-Al soils (Scott & Lowther 1980) but which fail to persist on soils of high fertility (Brock & Charlton 1978).

Much of the difference in the ability to compete for soil nutrients between white clover and the grasses with which it is commonly associated can be attributed to the more extensive root of the grasses. This has been measured by a number of investigators in several ways (Bennet & Doss 1960, Davidson 1969, Jacques 1943, Haahr et al. 1966, Jackman & Mouat 1972b, Newman 1969, Robinson & Jacques 1958, van Lieshout 1956). Grasses generally have more weight of root per unit weight of shoot, greater length of root per unit of ground surface, deeper root penetration and greater root activity at depth. The beneficial effect of mycorrhiza associated with white clover roots appears to operate by increasing the nutrient uptake surface of the plant (Crush 1974, 1976). Selection for longer root hair length has also increased P uptake, but only where white clover roots were not myccorhizal (Caradus 1981a). Evidence that the size of the root system of white clover is critical to the plant's growth was provided by Bates (1934), who showed that root pruning markedly reduced leaf size. Mouat & Walker (1959b) produced evidence that the lower cation exchange capacity of cocksfoot, ryegrass and particularly *Agrostis tenuis* gives these grasses another advantage in nutrient uptake relative to white clover.

It is suggested that as white clover can meet its N requirement by symbiosis with rhizobia, the species would benefit by using the photosynthate partitioned to the root to support N fixing nodules in priority to supporting extensive root growth. Reaction of the root system to defoliation appears to be less marked for white clover than for ryegrass, cocksfoot and timothy (Evans 1973a, b). The smaller root system of white clover, together with the buffering effect of reserves in the stolon of which by their proximity to the soil are protected from defoliation, may reduce the number of nodules lost as the result of grazing (Butler et al. 1959).

Plants of tufted perennial grasses, such as cocksfoot and ryegrass, develop their nutrient uptake surfaces by extending roots from the restricted area of the base of the plant. In contrast, because of the lateral spread of stolons and development of adventitious roots from the nodes of the stolons, white clover can exploit nutrients in soil well removed from the crown of the plant. Chow (1967) and Ueno et al. (1967) describe how the nutrient requirement of the plant is at first met by the taproot, but later, as the taproot deteriorates, the nodal roots increasingly support the growth of the plant. Ueno & Williams (1967) found that the tap root distributed labelled P to all of the plant, but the ability of the nodal roots to compensate for the loss of the taproot depend on position and size. Large nodal roots close to the taproot distributed P throughout the plant, whereas P taken up by smaller roots further from the taproot was distributed less extensively. The success of nodal roots in supporting the growth of clover plants depends on their frequency, which is greater for genotypes with short internodes (Ueno et al. 1967), and on environmental conditions, particularly wet soil, which enable the nodal roots to penetrate (Cullen et al. 1972, Ueno & Yoshihara 1968).

Studies of the root systems of a range of white clover varieties emphasise the difference between genotypes with root systems dominated by several taproots and genotypes with predominantly fibrous root systems (Caradus 1977, 1981b, c). Strong development of taproots is correlated with the large-leaved habit, and predominantly fibrous root development with the prostrate, small-leaved habit. These variations in root and shoot systems probably have a significant influence on the mineral nutrition and water relations of white clover plants. In sites frequently disturbed or where defoliation or other environmental stresses operate (Grime 1973, 1974), prostrate, fibrous-rooted plants would be better suited to colonise bare soil and exploit nutrients and water in the upper layers of the soil. In sites where environmental factors do not limit the development of the sward canopy, large-leaved white clovers are better adapted to avoid light competition. As these genotypes and conditions restrict stolon formation, the lateral exploration of the environment for nutrients and water by stolon extension and nodal root formation is limited, so that deep penetration by taproots to obtain nutrients and water deeper in the soil becomes important for growth and survival.

vi. Nitrogen relationships

Key factors in the competitiveness of white clover are that it can fix atmospheric N, it is a poor competitor for soil N, high levels of soil N may directly inhibit N fixation (Hoglund 1973), its tissues, which have high N content, rapidly break down on death to release N into the soil, and ryegrass, with which it shows specific positive association (Section 4, (C)i.), is a highly N-responsive species (Bradshaw et al. 1964). Where the main source of plant-available N originates from N recently fixed by clover, species whose active growth depends on continued relatively high availability of mineral N are less likely to completely dominate and exclude clover than species that may respond to N more slowly but which accumulate and recycle N within the plant system. Thus in addition to the niche differentiation within a habitat provided by stoloniferous and tufted habits (Section 4.(C)iii) and seasonal growth periodicity (Section 4.(E)iv), compatibility of white clover and ryegrass is also conditioned by changes in the availability of soil N. These changes may be successional, as in the pasture development sequence (Section 4.(B)iii(C)), or shorter term, within-year oscillations, as in developed ryegrass-white clover pastures (Harris & Thomas 1973). The description of these changes which follows combines notation used in soil-plant N relationships and the plant competition model of de Wit (1960) extended to grass-legume mixtures by Bakhuis & Kleter (1965).

In grass-legume systems the three main sources of supply of N are soil N (S_n), applied fertilizer N (F_n) and clover N (C_n). Walker et al. (1954) and others have applied the equation:

$$G_n = a\,C_n + b\,F_n + c\,S_n \qquad (1)$$

(where n = units of N per unit land area) to relate grass N yield(G_n) to the three sources of N. Although in some situations white clover can take up much of the soil N (Vallis et al. 1977), in many situations it may be assumed that no soil N is absorbed by clover, so that the contribution of soil N to grass nitrogen is constant:

$$G_n = a\,C_n + b\,F_n + K \qquad (2)$$

Cause and effect relationships between G_n, C_n and F_n were considered by Stewart & Chestnutt (1974). General effects indicated were that: (1) clover aided grass yield, but the increased grass yield appeared to inhibit clover yield; (2) fertilizer N reduced the effectiveness of clover in supplying N to the grass, possibly by inhibiting N-fixation; (3) there was delay in the transfer of N fixed by the clover, so that the yield of clover in the previous year and also the decrease in clover yield between successive years was often more highly correlated with grass N yield than the current year's clover yield.

Stewart & Chestnutt (1974) fitted equation (2) to results from a 5-year experiment where 8 grasses and grass mixtures were grown at 4 levels of applied N in the presence or absence of white clover. Combining series and years, the following equation was obtained, with all terms highly significant:

$$G_n = 0.59\ F_n - 0.08\ C_n + 156$$

On the basis of there being strong negative correlation between F_n and C_n owing to depression of clover by N application, an equation giving a more direct measure of the contribution of clover to grass N was defined. This provided the regression for all series and years combined, as follows:

$$G_n = 0.84\ F_n + 0.28\ C_n + 112$$

Changes of the sources of N used by clover and grass plants at different stages in their development when sown in mixture are related to changes in the competitive power of clover and grass as depicted by replacement series diagrams (Figure 26) defined by the following formulae:

$$Og = \frac{k_{gc}\ Z_g}{k_{gc}\ Z_g + Z_c}\ Mg \quad \text{and} \qquad (3)$$

$$Oc = \frac{k_{cg}\ Z_c}{k_{cg}\ Z_c + Z_g}\ Mc \qquad (4)$$

in which Og and Oc represent the yield of grass and clover, Z_g and Z_c the relative frequencies of the components in the mixture at establishment, and Mg and Mc the yield of pure stands of similar density. The constants k_{gc} and k_{cg}, the relative crowding coefficients, are respectively measures of the competitive power of grass to clover and clover to grass.

At first, the N requirements of developing white clover and grass will be met by seed reserves of N (E_n) and uptake of soil N. Once N fixation has begun, N fixed by rhizobia (R_n) will at first be available only to the clover and be incorporated as clover N. Diagrammatic relationships of the N sources to clover and grass N yield at this early stage are as follows:

(a) Utilization of seed reserves

$$E_n + S_n + F_n \xrightarrow{[+]} G_n$$

$$E_n + S_n + F_n \xrightarrow{[+]} C_n$$

(b) N fixation established, clover N derived entirely from Rhizobium- fixed N

$$S_n + F_n \xrightarrow{[+]} G_n$$

$$R_n \xrightarrow{[+]} C_n$$

(c) Clover also taking up soil N and fertilizer N

$$S_n + F_n \xrightarrow{[+]} G_n$$

$$S_n + F_n + R_n \xrightarrow{[+]} C_n$$

At this early stage, inter-specific interference between grass and clover is minimal. The yield of both species in linearly related to the relative seed frequencies (z) and the crowding coefficients (k) for both grass and clover = 1 (Figure 26a). Variation of F_n and S_n at this early stage will modify the growth rates of the species and usually grass yield will increase more than clover N with increased N availability (Figure 26b).

Further development of the mixture will lead to interference between its components, the direction of this being determined by the supply of F_n and S_n and the action of other factors controlling growth of the mixture. The relationship where the relatively greater response of grass to available N increases the adverse effect of grass on clover growth and/or levels of mineral N reduce N-fixation and consequently the amount of N available to clover, is:

(d)

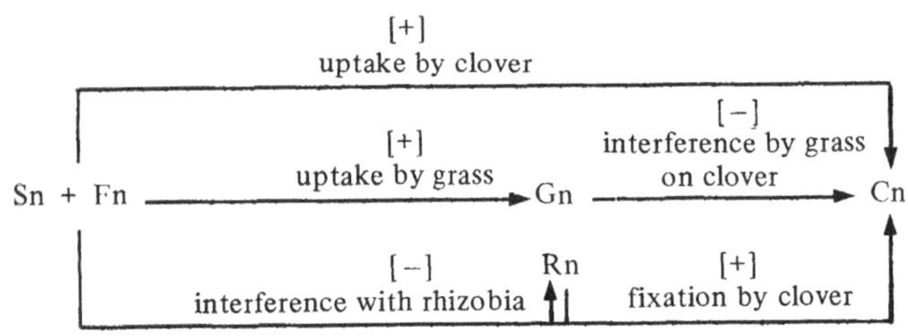

In the extreme case where atmospheric N-fixation is completely inhibited and grass and clover compete for the same source of N, grass, being more efficient than clover in absorbing mineral N, dominates clover ($k_{gc} > k_{cg}$), and the constraint of $k_{gc} = 1/k_{cg}$, which prescibes crowding for the same space (de Wit 1960), applies (Figure 26c). This situation is shown in the early stages of the ryegrass-white clover competition experiments of Ennik (1970) where N was applied, and of Harris & Thomas (1973) where soil N levels were high at sowing.

If high levels of available mineral N are maintained by repeated heavy fertilizer application, the grass will progressively dominate and exclude white clover (Section 4.(B)iii(b)). However, where white clover continues to grow and fix N, death and decay of clover tissue, or consumption of clover by animals and the return of a proportion of the consumed N to soil in excreta, will transfer clover N to grass N. This relationship, where clover obtains its N entirely by fixation, and clover N is transferred to the pool of N available for uptake by grass is:

(e)

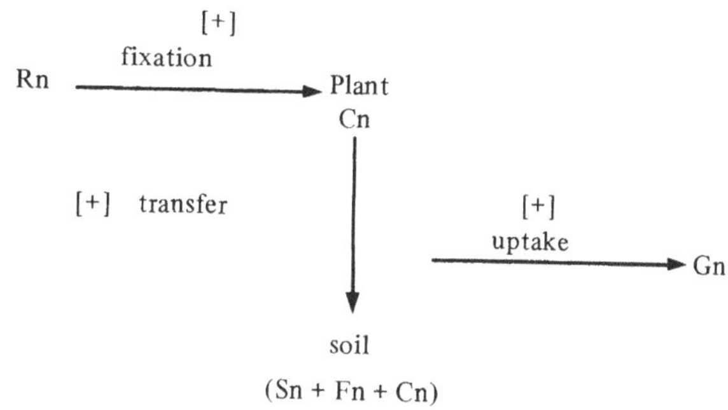

The stage at which transfer of Cn to Gn occurs and influences the competitive interaction between grass and clover varies between systems. It may occur 2-3 months after sowing (Bakhuis & Kleter 1965) or be delayed until the third year after sowing (Bland 1967). To describe the effect of N transfer on grass yield in mixture, Bakhuis & Kleter (1965) modified the equation (3) of de Wit (1960) as follows:

$$Og = \frac{(kgc)_Z \times Zg}{(kgc)_Z \times Zg + Zc} (Mg)_Z$$

which indicates that the (potential) monoculture yield of grass (Mg) varies with the seeding ratio (z). That is, as the clover yield in mixture increases, so does the amount of N available to the grass. $(Mg)_z$ represents the yield of monoculture of grass at a nutrient level similar to that created by clover. Consequently, the yield of grass in mixture with clover may increase to exceed the monoculture yield with decreasing Zg until the density of grass plants is insufficient to capitalise on the improved N supply provided by the clover. These relationships are depicted in Figure 26d where it is assumed that clover yield is unaffected by the presence of grass.

However, N tranfer has other consequential effects on competitive interaction between grass and clover. These are depicted as follows:

(f)

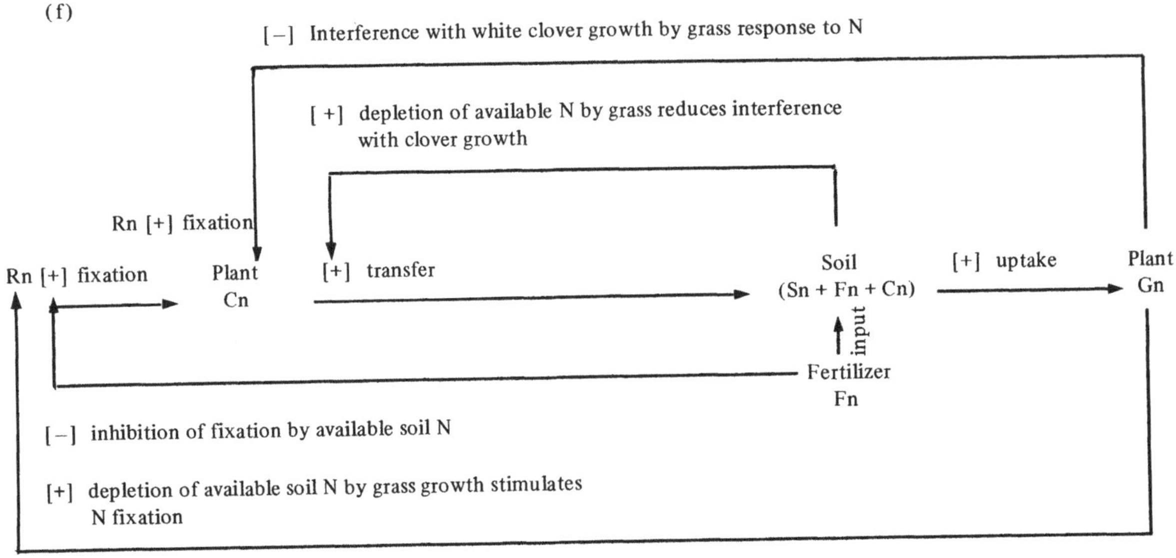

To describe the effect of the grass becoming more competitive with increased amount of clover on the yield of clover, Bakhuis & Kleter (1965) modified the equation (4) of de Wit (1960) as follows:

$$Oc = \frac{(kcg)_z \times Zc}{(kcg)z \times Zc) + Zg} Mc$$

This indicates that the competitive power of clover $(kcg)_z$ increases as Z_c decreases because with decreasing clover content less clover N is transferred to the grass and the grass becomes less competitive. This results in the development of an S-shaped curve for O_c in the replacement diagrams (Figure 26e). An experimental description of the effects of N transfer on the competitive interaction of grass and clover is provided by the experiments of Bakhuis & Kleter (1965), who used root partitions to inhibit N transfer between white clover and ryegrass or cocksfoot, and de Wit et al. (1966), who compared competition between the tropical grass panicum (*Panicum maximum*) and the tropical legume glycine (*Glycine javanica*) in the presence or absence of symbiotic N fixation.

The feed-back mechanisms (f) involved in the grass clover interaction lead to a dynamic equilibrium. Increased grass growth resulting from N-transfer increases shading of clover and improves the ability of grass to compete for other growth factors which are in limited supply. The consequent suppression of clover may at first hasten the senescence and decomposition of clover tissue and promote N transfer, but ultimately reduced clover growth means less fixation and less N being transferred to the soil N pool. Continued growth of the grass will deplete the available soil N. The consequent reduction in grass growth will then reduce the adverse effect of the grass on clover growth.

At high levels of available N there may be a direct inhibition of N-fixation and increased uptake of N by the clover (Hoglund 1971). Together with vigorous uptake of N by grass and reduced replenishment of N by fixation and transfer, the available N will fall to levels less inhibitory to N-fixation. Operation of this mechanism has been indicated by use of the acetylene reduction technique to measure rates of N fixation in simple grass-legume mixtures (Harris & Hoglund 1980). Where fertilizer N was applied, N-fixation per unit of legume yield was reduced, but a greater positive effect on N-fixation was attributed to increased grass growth reducing the amount of soil N available to the legumes.

Establishment of a grass-clover composition equilibrium in mixture is preceded by the S-curvature of the clover yield curve in the replacement diagram becoming so accentuated that the clover yield in mixture is the same at all establishment ratios (Figure 26f). This change is facilitated by the growth of white clover stolons through the sward. As the clover content becomes the same in all mixtures, the stimulus to grass yield by N-transfer can be expected to be the same in all mixtures. If the grass can adjust plant size to compensate for differences in establishment density, grass yield will also become the same in all mixtures. The development of such an equilibrium was shown by the experiment of Ennik (1970) and was approached in the experiment of Harris & Thomas (1973). The latter experiment indicated that, because of winter temperature limitations to N-fixation and/or mineralisation rates and seasonal variations in the utilisation of available N by the grasses, competition between ryegrass and white clover oscillated between situations in which the species were predominantly exclusive, and those in which the grass and clover each stimulated the growth of the other.

The complexity of the interaction between temperature, soil, moisture, N-fixation, mineralisation and grass and clover contents and growth rates in the field is indicated by Hoglund et al. (1979). Interpretation of this complex situation suggests that some of the complementarity of ryegrass and clover is related to the high potential growth rates of grass in late winter and early spring. This depletes soil N at these times, thus providing a soil environment more suited to N-fixation during late spring and summer when temperature and light favour clover growth. This situation emphasises the subtle way in which white clover is adapted to the moist temperate pasture situation. The history of the evaluation of white clover clearly establishes the species as the supreme legume for such pasture.

References

Aarssen, L. W.; Turkington, R.; Cavers, P. B. (1979) Neighbour relationships in grass-legume communities. II. Temporal stability and community evolution. *Canadian Journal of Botany* **57**, 2695-2703.

Adachi, A.; Suzuki, S.; Yamada, T. (1968) Studies on selection method in white clover breeding. 4. Competitive ability of the different types of white clover in the grass swards. *Journal of the Japanese Society of Grassland Science* **14**, 111-115.

Adams, F.M.J. (1926) Some observations on white clover, and a method of distinguishing between the seeds of wild white and dutch clover. *Annals of Applied Biology* **13**, 339-357.

Akatsuka, K.; Sugihara, S. (1973) Uptake of fertilizer nitrogen by pasture plants. *Journal of the Japanese Society of Grassland Science* **19**, 215-221.

Aldrich, D.T.A. (1970a) Interactions between nitrogen fertilizer use and the productivity of white clover. In: White Clover Research. *Occasional Symposium, British Grassland Society* No. 6, 143-146.

Aldrich, D.T.A. (1970b) Clover rot (*Sclerotinia trifoliorum*) in white clover and its influence on varietal performance at different centres. In: White Clover Research. *Occasional Symposium, British Grassland Society* No. 6, 143-146.

Allen, F.C.; Smallridge, T.J.; Walker, I. (1972) Grassweed and sheeps sorrel control in white clover seed crops. *Proceedings of the New Zealand Weed and Pest Control Conference* **25**, 214-219.

Allen, G.P. (1965) The ability of paraquat,dalaponandamitrole-T to kill an *Agrostis/Lolium* pasture in July, October and April. *Weed Research* **5**, 237-251.

Allen, G.P. (1968) The effect of July applications of dalapon on growth and botanical composition of an *Agrostis/Lolium* pasture. *Weed Research* **8**, 309-320.

Allen, H.P. (1966) The role of paraquat as an aid to renewal of grassland. *Proceedings of the 10th International Grassland Congress*, 326-330.

Amor, R.L. (1970) Comparison of granular herbicides for the control of blackberry (*Rubus fruticosus* L.agg.) *Proceedings of the 11th International Grassland Congress*, 673-676.

Anderson, L.B. (1973) Relative performance of the late flowering tetraploid red clover 'Grasslands 4706', five diploid red clovers and white clover. *New Zealand Journal of Experimental Agriculture* **1**, 233-237.

Andrew, C.S. (1976) Effect of calcium, pH and nitrogen on the growth and chemical composition of some tropical and temperate pasture legumes. I. Nodulation and growth. *Australian Journal of Agricultural Research* **27**, 611-623.

Andrews, C.S.; Robins, M.F. (1969) The effect of potassium on the growth and chemical composition of some tropical and temperate pasture legumes. I. Growth and critical percentage of potassium. *Australian Journal of Agricultural Research* **20**, 999-1007.

Andries, A.; van Slijcken, A. (1968) White clover (*Trifolium repens* L.) in grassland. I.W.O.N.L. *Verlagen over Navorsingen. Werkzaamheden van het National Centrum voor Graslanden Groenvoederonderzoek, 1968,* **35**, 11-90.

Angseesing, J.P.A. (1974) Selective eating of the acyanogenic form of *Trifolium repens*. *Heredity* **32**, 73-83.

Anslow, R.C. (1962) The seed production of some herbage species in temperate regions. *Mimeographed Publication, Commonwealth Bureau of Pastures and Field Crops, Hurley,* No. 1/1962, 22-24.

Appadurai, R.R.; Holmes, W. (1964) The influence of stage of growth, closeness of defoliation, and moisture on the growth and productivity of a ryegrass-white clover sward. I. Effect on herbage yields. *Journal of Agricultural Science, Cambridge, U.K.* **62**, 327-332.

Archer, M. (1973) The species preferences of grazing horses. *Journal of the British Grassland Society* **28**, 123-128.

Atkinson, G.C. (1970) Chemical control of barley grass. *Proceedings of the 23rd New Zealand Weed and Pest Control Conference,* 93-96.

Baars, J.A. (1976) Seasonal distribution of pasture production in New Zealand. VIII. Dargaville. *New Zealand Journal of Experimental Agriculture* **4**, 151-156.

Bakhuis, J.A.; Kleter, H.J. (1965) Some effects of associated growth on grass and clover under field conditions. *Netherlands Journal of Agricultural Science* **13**, 280-310.

Barclay, P.C. (1969) Some aspects of the development and performance of 'Grasslands 4700' white clover. *Proceedings of the New Zealand Grassland Association* **31**, 127-134.

Bates, G.H. (1933) The distribution of wild white clover in relation to the activity of earth worms (Lumbricidae). *Welsh Journal of Agriculture* **9**, 195-208.

Bates, G.H. (1934) The relation of leaf size to root structure in *Trifolium repens*. *Journal of Ecology* **22**, 271-278.

Bates, G.H. (1935) The vegetation of the footpaths, sidewalks, cart-tracks and gateways. *Journal of Ecology* **23**, 470-487.

Bennett, O.L.; Doss, B.D. (1960) Effect of soil moisture level on root distribution of cool season forage species. *Agronomy Journal* **52**, 204-207.

Bishop, J.A; Korn, M.E. (1969) Natural selection and cyanogenesis in white clover, *Trifolium repens*. *Heredity* **24**, 423-430.

Black, J.N. (1956) The influence of seed size and depth of sowing on pre-emergence and early vegetative growth of subterranean clover (*Trifolium subterranean* L.) *Australian Journal of Agricultural Research* **7**, 98-109.

Black, J.N. (1957a) The influence of varying light on the growth of herbage plants. *Herbage Abstracts* **27**, 89-98.

Black, J.N. (1957b) Seed size as a factor in the growth of subterranean clover (*Trifolium subterraneum* L.) under spaced sward conditions. *Australian Journal of Agricultural Research* **8**, 335-351.

Black, J.N. (1958) Competition between plants of different initial seed size in swards of subterranean clover (*Trifolium subterraneum* L.) with particular reference to leaf area and light microclimate. *Australian Journal of Agricultural Research* **9**, 299-318.

Black, J.N. (1959) Seed size in herbage legumes. *Herbage Abstracts* **29**, 235-241.

Black, J.N.; Wilkinson, G.N. (1963) The role of time of emergence in determining the growth of individual plants in swards of subterranean clover (*Trifolium subterraneum* L.). *Australian Journal of Agricultural Research* **14**, 628-638.

Blackman, G.E. (1933) Investigations on the intensive system of grassland management. XII. The influence of the grazing factor on botanical composition of intensively managed pastures. *Empire Journal of Experimental Agriculture* **1**, 253-269.

Blackman, G.E. (1934) The ecological and physiological action of ammonium salts on the clover content of turf. *Annals of Botany* **48**, 975-1001.

Blackman, G.E. (1938) The interaction of light intensity and nitrogen supply in the growth and metabolism of grasses and clover (*Trifolium repens*). I. The effects of light intensity and nitrogen supply on the clover content of a sward. *Annals of Botany* **2**, 257-280.

Blackman, G.E.; Templeman, W.C. (1938) The interaction of light intensity and nitrogen supply in the growth and metabolism of grasses and clover (*Trifolium repens*). II. The influence of light intensity and nitrogen supply on the leaf production of frequently defoliated plants. *Annals of Botany* **2**, 765-791.

Blackmore, L.W. (1964) Chemical renovation of pasture on hill country in Southern Hawkes Bay and Northern Wairarapa. *New Zealand Journal of Agriculture* **108**, 122-135.

Blackmore, L.W. (1965) Chemical establishment and renovation of pasture in New Zealand. *Proceedings of the 9th International Grassland Congress*, 307-315.

Blake, C.T.; Chamblee, D.S.; Woodhouse, W.W. (1966) Influence of some environmental and management factors on the persistence of Ladino clover in association with orchardgrass. *Agronomy Journal* **58**, 487-489.

Bland, B.F. (1967) The effect of cutting frequency and root segregation on the yield from perennial ryegrass-white clover associations. *Journal of Agricultural Science, Cambridge, U. K.* **69**, 391-397.

Blaser, R.E.; Brady, N.C. (1950) Nutrient competition in plant associations. *Agronomy Journal* **42**, 128-135.

Blaser, R.E.; Killinger, G.W. (1950) Life history studies of Louisiana white clover (*Trifolium repens* L.) I. Seed germination as related to temperature, pasture management and adaptation. *Agronomy Journal* **42**, 215-220.

Blaser, R.E.; Taylor, T.; Griffeth, W.; Skrdla, W. (1956) Seedling competition in establishing forage plants. *Agronomy Journal* **48**, 1-6.

Blood, T.F. (1964) The effects of low doses of paraquat and dalapon on two sward types based on *Lolium perenne* L. (perennial ryegrass) and *Trifolium repens* L. (white clover). *Weed Research* **4**, 359-361.

Bogh, H. (1936) The seed ripening of the most important herbage legumes and the influence of the harvesting on the size and quality of seed yield. *Årskift, Kongelige Veterinear-og Landbohojskole* 117-120.

Bohart, G.E. (1960) Insect pollination of forage legumes. *Bee World* **41**, 57-64, 85-97.

Bradshaw, A.D.; Chadwick, M.J.; Jowett, D.; Snaydon, R.W. (1964) Experimental investigations into the mineral nutrition of several grass species. 4. Nitrogen level. *Journal of Ecology* **52**, 665-676.

Brock, J.L. (1971) A comparison of 'Grasslands 4700' and 'Grasslands Huia' white clovers in establishing ryegrass-clover pasture under grazing. *New Zealand Journal of Agricultural Research* **14**, 368-378.

Brock, J.L. (1973) Growth and nitrogen fixation of pure stands of three pasture legumes with high/low phosphate. *New Zealand Journal of Agricultural Research* **16**, 483-491.

Brock, J.L. (1974) Effects of summer grazing management on the performance of 'Grasslands Huia' and 'Grasslands 4700' white clovers in pasture. *New Zealand Journal of Experimental Agriculture* **2**, 365-370.

Brock, J.L.; Charlton, J.F.L. (1978) *Lotus pedunculatus* establishment in intensive farming. *Proceedings of the New Zealand Grassland Association* **39**, 121-129.

Brockman, J.S.; Shaw, P.G.; Wolton, K.M. (1970) The effect of phosphate and potash fertilizers on cut and grazed grasslands. *Journal of Agricultural Science, Cambridge, U.K.* **74**, 397-407.

Brougham, R.W. (1954a) Pasture establishment studies. I. The effect of grass seeding rate on the establishment and development of pasture of short rotation ryegrass, red and white clovers. *New Zealand Journal of Science and Technology* **A35**, 518-538.

Brougham, R.W. (1954b) Pasture establishment studies. II. The influence of herbage height at grazing and of grass seeding rate, on the establishment and development of pasture of short rotation ryegaass, red and white clovers. *New Zealand Journal of Science and Technology* **A35,** 539-549.

Brougham, R.W. (1954c) Pasture establishment studies. III. Barley as a cover crop in establishing pastures for the production of late autumn and winter green-feed. *New Zealand Journal of Science and Technology* **A36**, 47-59.

Brougham, R.W. (1958a) Leaf development in swards of white clover (*Trifolium repens*). *New Zealand Journal of Agricultural Research* **1**, 707-718.

Brougham, R.W. (1958b) Interception of light by the foliage of pure and mixed stands of pasture plants. *Australian Journal of Agricultural Research* **9**, 39-52.

Brougham, R.W. (1959a) The effects of season and weather on the growth rate of a ryegrass and clover pasture. *New Zealand Journal of Agricultural Research* **2**, 283-296.

Brougham, R.W. (1959b) The effect of frequency and intensity of grazing on the productivity of a pasture of short-rotation ryegrass and red and white clover. *New Zealand Journal of Agricultural Research* **2**, 1232-1248.

Brougham, R.W. (1960a) The effect of frequent hard grazing at different times of the year on the productivity and species yields of a grass-clover pasture. *New Zealand Journal of Agricultural Research* **3**, 125-136.

Brougham, R.W. (1960b) The relationship between the critical leaf area, total chlorophyll content, and maximum growth of some pasture and crop plants. *Annals of Botany* **24**, 463-474.

Brougham, R.W. (1962) The leaf growth of *Trifolium repens* as influenced by seasonal changes in the light environment. *Journal of Ecology* **50**, 449-459.

Brougham, R.W. (1965) The effect of red clover on the leaf growth of white clover under long spelling during the summer. *New Zealand Journal of Agricultural Research* **8**, 859-864.

Brown, B.A. (1939) Some factors affecting the prevalence of white clover in grassland. *Journal of the American Society of Agronomy* **31**, 322-332.

Brown, B.A. (1957) Potassium fertilisation of Ladino clover. *Agronomy Journal* **49**, 477-480.

Brown, K.R. (1968a) The influence of herbage height at treading and treading intensity on the yields and botanical composition of a perennial ryegrass-white clover pasture. *New Zealand of Agricultural Research* **11**, 131-137.

Brown, K.R. (1968b) Some effects of sheep treading on the herbage and yield of four grasses grown in association with white clover. *New Zealand Journal of Agricultural Research* **11,** 883-890.

Brown, K.R. (1971) Changes in a browntop-perennial ryegrass-white clover pasture trodden by sheep from October to March 1966-1967. *New Zealand Journal of Agricultural Research* **14**, 828-834.

Brown, K.R. (1973a) Some effects of the time of sowing on the first twelve months growth of, 'Grasslands Huia' white clover, sown at two seeding rates or associated with 'Grasslands Manawa' ryegrass. *New Zealand Journal of Experimental Agriculture* **1**, 1-5.

Brown, K.R. (1973b) Effects of four times of sowing on 'Grasslands Huia' and 'Grasslands 4700' white clover. *New Zealand Journal of Experimental Agriculture* **1**, 165-170.

Burton, G.W. (1940) Factors influencing the germination of seed of *Trifolium repens*. *Journal of the American Society of Agronomy* **32**, 731-738.

Butler, G.W.; Greenwood, R.M.; Soper, K. (1959)Effects of shading and defoliation on the turnover of root and nodule tissue of plants of *Trifolium repens, Trifolium pratense* and *Lotus uliginosus*. *New Zealand Journal of Agricultural Research* **2,** 415-426.

Calder, G.G. (1941) Production of ryegrass and white clover seed under irrigation. *New Zealand Journal of Agriculture* **63**, 497-498.

Cahn, M.G.; Harper, J.L. (1967a) The biology of the leaf mark polymorphism in *Trifolium repens* L. I. Distribution of phenotypes at a local scale. *Heredity* **37**, 309-325.

Cahn, M.G.; Harper, J.L. (1976b) The biology of the leaf mark polymorphism in *Trifolium repens* L. II. Evidence for the selection of leaf marks by rumen fistulated sheep. *Heredity* **37**, 327-333.

Cameron, D.G. (1969) Effect of paspalum on the production pattern of Ladino white clover under irrigation on heavy clay soils in Central Queensland. *Queensland Journal of Agriculture and Animal Science* **26**, 373-384.

Camlin M.S. (1981) Competitive effects between ten cultivars of perennial ryegrass and three cultivars of white clover grown in association. *Grass and Forage Science* **36**, 169-178.

Campbell, A.G. (1959) A germination inhibitor and root growth retarder in chou moellier (*Brassica oleracea* var.) *Nature,* **183**, 1263-1264.

Campbell, A.G. (1966) Effects of treading by dairy cows on pasture production and botanical structure, on a Te Kowhai soil. *New Zealand Journal of Agricultural Research* **9**, 1009-1024.

Campbell, M.H. (1966) Theft by harvesting ants of pasture seed broadcast on unploughed land. *Australian Journal of Agriculture and Animal Husbandry* **6**, 334-338.

Campbell, M.H.; Swain, F.G. (1973) Effect of strength, tilth and heterogeneity of the soil surface on radicle-entry of surface-sown seeds. *Journal of British Grassland Society* **28**, 41-50.

Caradus, J.R. (1977) Structural variation of white clover root systems. *New Zealand Journal of Agricultural Research* **20**, 213-219.

Caradus, J.R. (1981a) Effect of root hair length on white clover growth over a range of soil phosphorus levels. *New Zealand Journal of Agricultural Research* **24**, 353-358.

Caradus, J.R. (1981b) Root growth of white clover (*Trifolium repens* L.) lines in glass-fronted containers. *New Zealand Journal of Agricultural Research* **24**, 43-54.

Caradus, J.R. (1981c) Root morphology of some white clovers from New Zealand hill country. *New Zealand Journal of Agricultural Research* **24**, 349-351.

Carlson, G.E. (1966) Growth of white clover leaves, after leaf removal. *Proceedings of the 10th International Grassland Congress,* 134-136.

Champness, S.S.; Morris, K. (1948) The population of buried seeds in relation to contrasting pasture and soil types. *Journal of Ecology* **36**, 149-173.

Chappell, H.G.; Ainsworth, J.F.; Cameron, R.A.D.; Redfern, M. (1971) The effect of trampling on a chalk grassland ecosystem. *Journal of Applied Ecology* **8**, 869-882.

Chappuzeau, B. (1930) Untersuchungen über die Bedeutung von Licht, Feuchtigkeit und Korngrosse bei der Kleekeimung. [Investigation of the significance of light, moisture and seed size in the germination of clover seed.] *Angewandte Botanik* **12**, 99-162.

Charles, A.H. (1958) The effect of undersowing on the cereal cover crop. *Field Crop Abstracts* **11**, 233-239.

Charles, A.H. (1962) Pasture establishment by surface sowing methods. *Herbage Abstracts* **32**, 175-181.

Charles, A.H. (1963a) Establishment studies V. The effect of method of establishment on the behaviour of long-term leys in the seedling year. *Journal of Agricultural Science, Cambridge, U. K.* **60,** 1-10.

Charles, A.H. (1963b) Establishment studies VI. The effect of method of establishment on long term leys in the first harvest year. *Journal of Agricultural Science, Cambridge, U.K.* **60,** 11-17.

Charles, A.H. (1965) Interactions of grass, clover and nurse crop in the seeding year. *Journal of the British Grassland Society* **20**, 241-247.

Charles, A.H. (1968) Some selective effects operating on white and red clover swards. *Journal of the British Grassland Society* **23**, 20-25.

Charlton, J.F.L. (1977) Establishment of pasture legumes in North Island hill country. I. Buried seed population. *New Zealand Journal of Experimental Agriculture* **5**, 211-214.

Charlton, J.F.L. (1978) Slugs as a possible cause of establishment failure in pasture legumes oversown in boxes. *New Zealand Journal of Experimental Agriculture* **6**, 313-317.

Chestnutt, D.B.M. (1971) Effects of white clover on the botanial composition of swards of ryegrass, timothy and meadow fescue and mixtures of these three. *Journal of the British Grassland Society* **26**, 191-213.

Chestnutt, D.B.M.; Lowe, J. (1970) Agronomy of white clover/grass swards. Review in: White Clover Research, Editor J.Lowe. *Occasional Symposium of the British Grassland Society* No. 6, 35-40.

Chippendale, H.G.; Milton, W.E.J. (1932) Note on the occurance of buried seeds in the soil. *Journal of Agricultural Science* **22**, 451-452.

Chow, H.S. (1967) The structure and growth habit of white clover in relation to its persistence. *Dissertation Abstracts International B* No. **10,** 3369.

Clark, D.A.; Lambert, M.G.; Chapman, D.F. (1982) Pasture management and hill country production. *Proceedings of the New Zealand Grassland Association* **43**, 205-214.

Clark, J.; Kat, C.; Santhirasegaram, K. (1974) The effects of changes in heights of cutting and growth on the digestible organic matter production and botanical composition of perennial pasture. *Journal of the British Grassland Society* **29**, 269-273.

Clatworthy, J.N. (1960) Studies on the nature of competition between closely related species. D. Phil Thesis, University of Oxford. In Harper, J.L. *Population Biology of Plants*, 1977.

Clifford, P.T.P. (1977) Cultural methods for 'Grasslands Pitau' white clover seed crops. *New Zealand Journal of Experimental Agriculture* **5**, 147-149.

Clifford, P.T.P. (1979) Effect of closing date on potential seed yields from 'Grasslands Huia' and 'Grasslands Pitau' white clover. *New Zealand Journal of Experimental Agriculture* **7**, 303-306.

Collins, D.P. (1961) The establishment of grass/clover swards on bogland by surface seeding. *Irish Journal of Agricultural Research* **1**, 21-30.

Collins, D.P.; O'Toole, M. (1962) Burning of native vegetation for surface seeding. *Irish Journal of Agricultural Research* **2**, 165-171.

Collins, D.P.; O'Toole, M. (1963) The animal factor in establishment and production from a surface seeded pasture on bogland. *Irish Journal of Agricultural Research* **2**, 225-241.

Connolly, V. (1968) A comparison of white clover and three perennial ryegrass varieties under a system of grazing management. *Irish Journal of Agricultural Research* **7**, 227-242.

Cooper, J.P.; Tainton, N.M. (1968) Light and temperature requirements for the growth of tropical and temperate grasses. *Herbage Abstracts* **38**, 167-176.

Copeman, G.J.F.; Brown, W.J.; Alexander, C.B. (1960) The productivity of herbage varieties 3. Timothy, 1954 to 1957. *Bulletin, North of Scotland College of Agriculture* No. 7, 1-24.

Corkill, L. (1952) Cyanogenesis in white clover (*Trifolium repens* L.). VI. Experiments with high glucoside and glucoside-free strains. *New Zealand Journal of Science and Technology* **A34**, 1-16.

Cowling, D.W. (1961) The effect of white clover and nitrogenous fertiliser on the production of a sward. *Journal of the British Grassland Society* **16**, 281-290.

Cowling, D.W.; Lockyer, D.R. (1965) A comparison of the reaction of different grass species to fertilizer nitrogen and to growth in association with white clover. I. Yield of dry matter. *Journal of British Grassland Society* **20**, 197-204.

Cowlishaw, S.J. (1960) The persistence of white clover under poultry grazing. *Journal of the British Grassland Society* **15**, 6-11.

Cowlishaw, S.J.; Eyles, D.E.; Astbury, J.G. (1956) The effect of two rates of stocking on contrasting swards for poultry. *Journal of the British Grassland Society* **11**, 155-161.

Crawford-Sidebotham, T.J. (1972) The role of slugs and snails in the maintenance of the cyanogenesis polymorphisms of *Lotus corniculatus* and *Trifolium repens*. *Heredity* **28**, 405-411.

Crush, J.R. (1974) Plant growth response to vesicular-arbuscular mycorrhiza 7. Growth and nodualtion of some herbage legumes. *New Phytologist* **73**, 743-752.

Crush, J.R. (1976) Endomycorrhiza and legume growth in some soils of the Mackenzie Basin, Canterbury, New Zealand. *New Zealand Journal of Agricultural Research* **19**, 473-476.

Cullen, N.A. (1964a) The effects of nurse crops on the establishment of pasture. I. Rape (*Brassica napus* L.) forage crops. *New Zealand Journal of Agricultural Research* **7**, 42-51.

Cullen, N.A. (1964b) The effect of nurse crops on the establishment of pasture. II. Cereal nurse crops. *New Zealand Journal of Agricultural Research* **7**, 52-59.

Cullen, N.A. (1964c) Species competition in establishing swards: suppression effects of ryegrass on establishment and production of associated grasses and clovers. *New Zealand Journal of Agricultural Research* **7**, 678-693.

Cullen, N.A. (1966a) Invermay trials show importance of competition between pasture species. *New Zealand Journal of Agriculture* **112**, 31-32.

Cullen, N.A. (1966b) Pasture establishment on unploughable hill country in New Zealand. Factors influencing establishment of oversown grasses and clovers. *Proceedings of the 10th International Grassland Congress,* 851-855.

Cullen, N.A.; Arnold, G.C. (1971) Establishment of pasture on yellow-brown loams near Te Anau X. Major and trace element responses and interactions. *New Zealand Journal of Agricultural Research* **14**, 47-65.

Cullen, N.A.; McNaught, K.J.; Mountier, N.S. (1966) The establishment of pasture on yellow brown loams near Te Anau 4. The effect of lime and phosphate on establishment of clover and grass oversown on uncultivated ground. *New Zealand Journal of Agricultural Research* **9**, 375-387.

Cullen, N.A.; Turner, A.K.; Wilson, J.H. (1972) The effect of irrigation depth on root growth of some pasture species. *Plant and Soil* **37**, 345-352.

Darlington, H.T.; Steinbauer, G.P. (1961) The eighty year period for Dr Beal's seed viability experiment. *American Journal of Botany* **48**, 321-325.

Darwin, C. (1878) Cross and self fertilization, 2nd edition, pp.361. [Cited by Erith (1924)].

Davidson, R.L. (1969) Effect of root/leaf temperature differentials on root/shoot ratios of some pasture grasses and clover. *Annals of Botany* **33**, 561-569.

Davies, R.O.; Milton, W.E.J. (1945) Direct reseeding on upland peat. *Welsh Journal of Agriculture* **18**, 43-47.

Davies, W. (1938) Vegetation of grass verges and other excessively trodden habitats. *Journal of Ecology* **26**, 38-49.

Davies, W.E. (1956) Storage of clover seed. I. First interim report, 1947-1954. *Journal of the British Grassland Society* **11**, 224-229.

Davies, W.E. (1965) Storage of clover seed. II. Second interim report, 1955-1964. *Journal of the British Grassland Society* **20**, 180-181.

Davies, W.E. (1970) White clover breeding. Review in: White clover research. *Occasional Symposium of the British Grassland Society* No. 6, 99-122.

Davies, W.E.(1972a) Breeding white clover (*Trifolium repens*) in relation to companion grass. In: Evaluation of breeding material in herbage crops. *Report of Fodder Crops Meeting, Dublin,* 245-260.

Davies, W.E. (1972b) White clover/grass relationships. *Report of the Welsh Plant Breeding Station, 1972*, 42-44.

Davies, W.E.; Evans, D.R. (1974) White clover collected from high production ryegrass swards receiving large applications of fertilizer nitrogen. *Report of the Welsh Plant Breeding Station, 1974,* 22-24.

Davies, W.E.; Thomas, T.A. (1971) The effect of potassium on forage yields In: Potassium and systems of grassland farming. Henley-on-Thames, UK; Potassium Institute Ltd., 73-78.

Davis, A.G.; Cooper, M.M. (1951) White clover trial. I. Production in terms of liveweight increase. II. Changes in botanical composition. *Journal of the British Grassland Society* **6**, 229-243.

Davis, W.E.P. (1962) Effects of using oats as a companion crop with orchardgrass, *Dactylis glomerata* L. and white clover, *Trifolium repens* L., sown for pasture. *Canadian Journal of Plant Science* **42**, 582-588.

Dessureaux, L. (1950) Variation in the seed setting ability of Ladino white clover, *Scientific Agriculture* **30**, 509-517.

Dessureaux, L. (1951) Ovule formation as a factor influencing seed-setting of Ladino white clover. *Scientific Agriculture* **31**, 373-382.

Donald, C.M. (1958) The interaction of competition for light and nutrients. *Australian Journal of Agricultural Research* **9**, 421-435.

Donald, C.M. (1963) Competition among crop and pasture plants. *Advances in Agronomy* **15**, 1-118.

Dowling, P. M.; Clements, R.J.; McWilliams, J.R. (1971) Establishment and survival of pasture species from seeds sown on the soil surface. *Australian Journal of Agricultural Research* **22**, 61-74.

Drake, M.; Colby, W.G.; Oohara, H. et al. (1970) Five year effects of fertiliser, phosphorus and potassium on yield and composition of alfalfa-orchard grass and Ladino clover-orchardgrass mixtures (1960-1964). *Potash Review* **16/47**, 1-19.

Dunavan, D. (1952) Insect pollination of Ladino clover in South Carolina. *Potash Review* **45**, 124-125.

Dunham, W.E. (1957) Pollination of clover fields. *Gleanings in Bee Culture* **85**, 218-219.

During, C.; Cullen, N.A.; Bennet, G.M. (1962) The establishment of pasture on yellow-brown loams near Te Anau. I. Factors influencing the establishment of clover-grass pasture on cultivated ground. *New Zealand Journal of Agricultural Research* **5**, 278-293.

During, C.; Cullen, N.A.; Mountier, N.S. (1963) The establishment of pasture on yellow-brown loams near Te Anau. II. Factors influencing the establishment of white clover (*Trifolium repens* L.) on uncultivated ground. *New Zealand Journal of Agricultural Research* **6**, 416-431.

During, C.; Lynch, P.B.; Mcnaught, K.J.; Scott, R.S. (1960) A study of the interaction of P, K, S, Mg, Mo, and B fertilisers and ground limestone, applied to grass-clover pasture. *New Zealand Journal of Agricultural Research* **3**, 950-993.

Duthion, C. (1972) The reaction of plants to excess water. *Bulletin Technique d'Information* No. 273/274, 1071-1076.

Edmond, D.B. (1958) The influence of treading on pasture. A preliminary study. *New Zealand Journal of Agricultural Research* **1**, 319-328.

Edmond, D.B. (1962) Effects of treading pasture in summer under different soil moisture levels. *New Zealand Journal of Agricultural Research* **5**, 389-395.

Edmond, D.B. (1963) Effects of treading perennial ryegrass (*Lolium perenne* L.) and white clover (*Trifolium repens* L.) pasture in winter and summer at two moisture levels. *New Zealand Journal of Agricultural Research* **6**, 265-276.

Edmond, D.B. (1964) Some effects of sheep treading on the growth of ten pasture species. *New Zealand Journal of Agricultural Research* **7**, 1-16.

Edmond, D.B. (1966) The influence of animal treading on pasture growth. *Proceedings of the 10th International Grassland Congress,* 453-458.

Edmond, D.B. (1970) Effects of treading on pastures, using different animals and soils. *Proceedings of the 11th International Grassland Congress*, 604-608.

Edwards, P.J.; Booysen, P.de V.; Mapledoram, B.D. (1972) Effects of competition on the establishment of overseeded species in the highland sourveld of Natal. *Agroplantae* **4**, 1-8.

Elliot, J.G. (1961) The reaction of *Trifolium repens* L. in grassland to 2, 4-D, MCPA and mixtures of 2, 4-D and MCPB. *Weed Research* **1**, 184-185.

Ennik, G.C. (1960) De concurrentie tussen witte klaver en Engels raaigras bij verschillen in lichtintensiteit en vochtvoorziening. [The competition between white clover and perennial ryegrass under different conditions of light intensity and moisture supply.] *Mededeling, Institut voor Biologisch en Scheikundig Onderzoek van Landbouwgewassen.* **109**, 37-50.

Ennik, G.C. (1970) White clover/grass relationships: competition effects in field and laboratory. Review in: White Clover Research, Ed. J. Lowe. *Occasional Symposium of the British Grassland Society* No. 6, 165-174.

Erith, A.G. (1924) White clover (*Trifolium repens* L.). A monograph. London, UK; Duckworth and Co., 150pp.

Evans, A.E. (1965) The significance of seed size and weight in white clover. *Report of the Welsh Plant Breeding Station, 1964*, 42-46.

Evans, P.S. (1973a) The effect of the repeated defoliation to three different levels on root growth of five pasture species. *New Zealand Journal of Agricultural Research* **16**, 31-34.

Evans, P.S. (1973b) Effect of seed size and defoliation at three development stages on root and shoot growth of seedlings of some common pasture species. *New Zealand Journal of Agricultural Research* **16**, 389-394.

Fayemi, A.A. (1957) Effect of temperature on the rate of seed swelling and germination of legume seeds. *Agronomy Journal* **49**, 75-76.

Forster, I.W.; Jeffery, G.L.; Palmer Jones, T. (1962) Factors causing losses of clover seed. *New Zealand Journal of Agricultural Research* **5**, 326-330.

Frame, J. (1973) The yield response of a tall fescue/white clover sward to nitrogen rate and harvesting frequency. *Journal of the British Grassland Society* **28**, 139-148.

Fribourg, H.A.; Kennedy, W.K. (1953) The effect of rates of seeding on the yield and survival of alfalfa in meadow mixtures. *Agronomy Journal* **45**, 251-257.

Garner, F.H.; Sanders, H.G. (1940) Studies with lucerne (*Medicago sativa*) - row distances and 'smother' crops. *Journal of Agricultural Science, Cambridge, U. K.* **30**, 182-188.

Gibson, P.B. (1957) Effect of flowering on the persistence of white clover. *Agronomy Journal* **49**, 213-215.

Gibson, P.B.; Hollowell, E.A. (1966) White clover. *Agriculture Handbook, Agricultural Research Service, U.S. Department of Agriculture* **314**, 33 pp.

Goold, G.J.; Douglas, J.A. (1976) A small plot evaluation of three ryegrasses and two clovers, at 5 sites in New Zealand. I. Comparison of 'Grassland Huia' & 'Grassland Pitau' white clovers. *NZ Journal of Experimental Agriculture* **4**, 135-142.

Graham, J.H.; Sprague, V.G.; Zeiders, K.E. (1961) Effect of management and pesticides on persistence of Ladino white clover. *Agronomy Journal* **53**, 356-357.

Grant, E.A.; Sallans, W.G. (1964) Influence of plant extracts on germination and growth of eight forage species. *Journal of the British Grassland Society* **19**, 191-197.

Grant Lipp, E.A.; Ballard, L.A.T. (1964) The interrelation of dormancy, size and hardness in seed of *Trifolium subterraneum* L. *Australian Journal of Agricultural Research* **15**, 215-222.

Green, H.B. (1957) White clover pollination with low honey bee population. *Journal of Economic Entomology* **50**, 318-320.

Green, J.O.; Corrall, A.J. (1965) The testing of grass varieties in swards with clover: The effect of grass seed-rate on comparisons of gross yield. *Journal of the British Grassland Society* **20**, 207-211.

Griffiths, D.J.; Roberts, H.M.; Lewis, J.; Stoddart, J.L.; Bean, E.W.(1967) Principals of herbage seed production. *Technical Bulletin Welsh Plant Breeding Station, Aberystwyth* **1**, pp 135.

Grime, J.P. (1973) Competitive exclusion in herbaceous vegetation. *Nature,* **242**, 344-345.

Grime, J.P. (1974) Vegetation classification by reference to strategies. *Nature,* **250**, 26-31.

Grime, J.P.; Hunt, R. (1975) Relative growth rate, its range and adaptive significance in a local flora. *Journal of Ecology* **63**, 393-422.

Haahr, V.; Knudsen, H.; Neilson, G.; Sandfaer, J. (1966) Root development of some cultivated plants in light sandy soil. *Tidsskrift for Planteavl* **69**, 554-567.

Haggar, R.J. (1971) The significance and control of *Poa trivialis* in ryegrass pastures. *Journal of the British Grassland Society* **26**, 117-121.

Haggar, R.J.; Holmes, W. (1963) Wild white clover seed production. II. A survey on wild white clover seed production in Kent, 1961. *Journal of the British Grassland Society* **18**, 197-203.

Haggar, R.J.; Holmes, W.; Innes, P. (1963) Wild white clover seed production. I. The effects of defoliation and fertiliser treatment on flowering and seed yields from ryegrass/white clover swards. *Journal of the British Grassland Society* **18**, 97-103.

Hallsworth, E.G. Ed. (1958) Nutrition of the Legumes. London, UK; Butterworths Scientific Publications, 359 pp.

Hammer, O. (1947) Er Bierne parat? [Are the bees ready?] *Ugeskrift for* ***Landmaend Kobenhavn*** **92,** 387-389.

Harberd, D.J. (1963) Observations on natural clones of *Trifolium repens* L. *New Phytologist* **62**, 198-214.

Harper, J.L. (1967) A Darwinian approach to plant ecology. *Journal of Ecology* **55,** 247-270.

Harper, J.L. (1977) Population biology of plants. London, UK; Academic Press, 892pp.

Harper, J.L.; Clatworthy, J.N. (1963) The comparative biology of closely related species. VI. Analysis of the growth of *Trifolium repens* and *T. fragiferum* in pure and mixed populations. *Journal of Experimental Botany* **14**, 172-190.

Harris, A.J.; Brown, K.R.; Turner, J.D.; Johnstone, J.M.; Ryan, D.L.; Hickey, M.J. (1973) Some factors affecting pasture growth in Southland. *New Zealand Journal of Experimental Agriculture* **1**, 139-163.

Harris, W. (1969) Some effects of a porina caterpillar (*Wiseana* spp.) infestation on perennial ryegrass, cocksfoot, and white clover. *New Zealand Journal of Agricultural Research* **12**, 543-552.

Harris, W. (1971a) The effect of density, cutting height and white clover (*Trifolium repens* L.) on the structure of a ryegrass (*Lolium* spp.) population. *Journal of Agricultural Science, Cambridge, U. K.* **77,** 385-395.

Harris, W. (1971b) The effects of fertiliser and lime on the competitive interactions of *Rumex acetosella* L. with *Trifolium repens* L. and *Lolium* spp. *New Zealand Journal of Agricultural Research* **14,** 185-207.

Harris, W. (1972) Shading, defoliation, temperature, growth stage and residual fertility effects on competition between *Rumex acetosella, Trifolium repens* and *Lolium (multiflorum x perenne). New Zealand Journal of Agricultural Research* **15**, 687-705.

Harris, W. (1973a) Why browntop is bent on creeping. *Proceedings of the New Zealand Grassland Association* **35**, 101-109.

Harris, W. (1973b) Ryegrass genotype-environment interactions in response to density, cutting height, and competition with white clover. *New Zealand Journal of Agricultural Research* **16**, 207-222.

Harris, W. (1974) Competition among pasture plants. V. Effects of frequency and height of cutting between *Agrostis tenuis* and *Trifolium repens. New Zealand Journal of Agricultural Research* **17**, 251-256.

Harris, W. (1980) An approach to evaluate a large number of mixtures under grazing. *Proceedings 13th International Grassland Congress*, 389-393.

Harris, W.; Brock, J.L. (1972) Effect of porina caterpillar (*Wiseana* spp.) infestation on yield and competitive interactions of ryegrass and white clover varieties. *New Zealand Journal of Agricultural Research* **15**, 723-740.

Harris, W.; Brougham, R.W. (1968) Some factors affecting change in botanical composition in a ryegrass-white clover pasture under continuous grazing. *New Zealand Journal of Agricultural Research* **11**, 15-38.

Harris, W.; Hoglund, J.H. (1980) Influences of seasonal growth periodicity and N-fixation on competitive combining abilities of grasses and legumes. *Proceedings of the 13th International Grassland Congress*, 239-243.

Harris, W.; Rhodes, I.; Mee, S.S. (1983) Observations on environmental and genetic influences on the overwintering of white clover. *Journal of Applied Ecology* **20**, 609-624.

Harris, W.; Thomas, V.J. (1973) Competition among pasture plants. III. Effects of frequency and height of cutting on competition between white clover and two ryegrass cultivars. *New Zealand Journal of Agricultural Research* **16**, 49-58.

Harrison, C.M.; Kelty, R.H.; Blumer, C. (1945) Honeybees and legume seed production. *Quarterly Bulletin of Michigan Agricultural Experimental Station* **28**, 85-89.

Hart, R.H.; Carlson, G.E.; Retzer, H.J. (1968) Establishment of tall fescue and white clover. Effects of seeding method and weather. *Agronomy Journal* **60**, 385-388.

Hartley, M.J.; Atkinson, G.C. (1972) Effect of chemical removal of barley grass on lamb growth rates. *Proceedings of the New Zealand Weed and Pest Control Conference* **25**, 23-28.

Healy, W.B.; Widdowson, J.P.; Yeates, G.W. (1973) Effect of root nematodes on the growth of seedling and established white clover on a yellow-brown loam. *New Zealand Journal of Agricultural Research* **16**, 70-76.

Heddle, R.G.; Herriott, J.B.D. (1954) The establishment, growth and yield of ultra-simple grass seeds mixtures in the south-east of Scotland. *Journal of the British Grassland Society* **9**, 99-110.

Heddle, R.G.; Herriott, J.B.D. (1955) The establishment, growth and yield of ultra-simple grass seeds. Part II. The effects of varying grass and clover seed rates. *Journal of the British Grassland Society* **10**, 157-167.

Heinrich, R. (1894) Die 'harten' Korner der Kleesaaten. [The 'hard' seeds of clover.] *Bericht der Landesversuchsstation zu Rostock* **2**, 97-102.

Herriott, J.B.D. (1970) The influence of seeding method in sward productivity. *Occasional Symposium of the British Grassland Society* No. 6, 239-245.

Herriott, J.B.D.; Wells, D. (1960) Clover nitrogen and sward productivity. *Journal of the British Grassland Society* **15**, 63-69.

Hessberg, von (1932) The natural sowing of white clover in newly sown down pastures. *Mitteilungen, Deutschen Landwirtschaftsgesellschaft* **14**, 885-886.

Hill, J. (1977) Plasticity of white clover grown in competition with perennial ryegrass. *Report of the Welsh Plant Breeding Station, 1976* 24-25.

Hill, J. (1980) Plasticity of white clover grown in competition with perennial ryegrass. *Report of the Welsh Plant Breeding Station, 1979,* 52-54.

Hodge, R.W.; Doyle, J.J. (1967) Diet selected by lambs and yearling sheep grazing on annual and perennial pastures in southern Victoria. *Australian Journal of Experimental Agriculture and Animal Husbandry* **7**, 141-143.

Hoen, K. (1970) Performance in mixed swards of three perennial ryegrass and three white clover varieties at two nitrogen levels. *Irish Journal of Agricultural Research* **9**, 215-223.

Hoglund, J.H. (1973) Bimodal response by nodulated legumes to combined nitrogen. *Plant and Soil* **39**, 533-545.

Hoglund, J.H.; Brock, J.L. (1974) Growth of 'Grasslands Huia' and 'Grasslands 4700' white clovers. I. Effects of temperature and nitrogen. *New Zealand Journal of Agricultural Research* **17**, 41-45.

Hoglund, J.H.; Crush, J.R.; Brock, J.L.; Ball, R. (1979) Nitrogen fixation in pasture. XII. General discussion. *New Zealand Journal of Experimental Agriculture* **7**, 45-51.

Holford, I.C.R.; Gleeson, A.C. (1976) White clover responses to phosphorus and sulphur on granitic soils. *Australian Journal of Experimental Agriculture and Animal Husbandry* **16**, 234-239.

Holliday, R. (1960) Plant population and crop yield. Part I. *Field Crop Abstracts* **13**, 159-167.

Hollowell, E.A. (1966) White clover *Trifolium repens* L. Annual or perennial? *Proceedings of the 10th International Grassland Congress,* 184-187.

Holmes, W.; McLusky, D.S. (1954) The intensive production of herbage for crop drying. Part V. The effect of continued massive applications of nitrogen with and without phosphate and potash on the yield of grassland herbage. *Journal of Agricultural Science, Cambridge, U. K.* **45,** 129-139.

Hoshino, M.; Nishimura, S.; Okubo, T. (1967) Studies on the assimilation of $^{14}CO_2$ in Ladino clover. 3. Uptake and distribution of ^{14}C by the plants in the different stages of regrowth. *Proceedings of the Crop Science Society of Japan* **36**, 269-274.

Hoshino, M.; Olzumi, H. (1968) Studies on the assimilation and translocation of $^{14}CO_2$ in Ladino clover. 4. Utilization and translocation of reserve $^{14}CO_2$ assimilates by the plants in early stages of regrowth. *Proceedings of the Crop Science Society of Japan* **37**, 82-86.

Hoveland, C.S. (1964) Germination and seedling vigour of clovers as affected by grass root extracts. *Crop Science* **4**, 211-213.

Howitt, A. (1961) Chemical control of slugs in orchardgrass-Ladino white clover pastures in the Pacific Northwest. *Journal of Economic Entomology* **54**, 778-781.

Hughes, G.P. (1951) The seasonal output of pastures sown with ultra-simple seed mixtures. *Journal of Agricultural Science, Cambridge* **41**, 203-213.

Hughes, G.P.; Davis, A.G. (1951) The development of swards sown with simple mixtures at different rates of seeding under varying systems of management and manuring. *Journal of the British Grassland Society* **6**, 167-177.

Hughes, J.G. (1975) What sheep eat on developed and undeveloped high country. *Tussock Grassland and Mountain Lands Institute Review* **31**, 20-30.

Hughes, R.; Munro, J.M.M.; Warboys, I.B. (1965) Mechanical treatment for improving upland grazing. *Journal of the British Grassland Society* **20**, 49-53.

Hunt, I.V. (1956) Comparative productivity of 27 strains of grass in the West of Scotland. Part I. Total production for three years. *Journal of the British Grassland Society* **11**, 49-55.

Hunt, I.V. (1960) Comparing the output from mixtures based on perennial ryegrass and on meadow fescue. *Journal of the British Grassland Society* **15**, 308-314.

Hunt, I.V. (1966) The effect of age of sward on the yield and response of grass species to fertiliser nitrogen. *Proceedings of the 10th International Grassland Congress,* 249-254.

Hunt, W.F. (1970) The influence of leaf death on the rate of accumulation of green herbage during pasture regrowth. *Journal of Applied Ecology* **7**, 41-50.

Hunt, W.F. (1971) Leaf death and decomposition during pasture regrowth. *New Zealand Journal of Agricultural Research* **14**, 208-218.

Hyde, E.O.C. (1950) Studies on the development of white clover seed. *New Zealand Grassland Association Proceedings* **12**, 101-107.

Hyde, E.O.C. (1954) The influence of the hilum in some Papilionaceae in relation to the ripening of the seed and the permeability of the testa. *Annals of Botany* **18**, 241-250.

Hyde, E.O.C.; McLeavey, M.A.; Harris, G.S. (1959) Seed development in ryegrass, and in red and white clover. *New Zealand Journal of Agricultural Research* **2**, 947-952.

Hyde, E.O.C.; Suckling, F.E.T. (1953) Dormant seed of clover and other legumes in agricultural soils. *New Zealand Journal of Science and Technology* **A34**, 374-385.

Ivins, J.D. (1955) The palatability of herbage. *Herbage Abstracts* **25**, 73-79.

Jackman, R.H.; Mouat, M.C.H. (1972a) Competition between grass and clover for phosphate. I. Effect of browntop *(Agrostis tenuis* Sibth.) on white clover *(Trifolium repens* L.) growth and nitrogen fixation. *New Zealand Journal of Agricultural Research* **15**, 653-666.

Jackman, R.H.; Mouat, M.C.H. (1972b) Competition between grass and clover for phosphate. II. Effect of root activity, efficiency of response to phosphate and soil moisture. *New Zealand Journal of Agricultural Research* **15**, 667-675.

Jaques, W.A. (1943) Root development in some common New Zealand pasture plants.II. Perennial ryegrass (*Lolium perenne*), cocksfoot (*Dactylis glomerata*) and white clover (*Trifolium repens*). Effect of fertiliser placement on the yield of roots and herbage. *New Zealand Journal of Science and Technology* **A25**, 91-117.

Janson, C.G.; White, J.G.H. (1971) Lucerne establishment studies on uncultivated country.II. A nodulation problem. *New Zealand Journal of Agricultural Research* **14**, 587-596.

Jones, L. (1962) Herbicides to aid pasture renovation. *Journal of the British Grassland Society* **17**, 85-86.

Jones, M.G. (1933) Grassland management and its influence on the sward. *Journal of the Royal Agricultural Society* **94**, 21-41.

Jones, T.E. (1933) The effect of a nurse crop on the establishment and first year yields of various grasses and clovers when sown in pure plots and in mixtures. *Welsh Journal of Agriculture* **9**, 176-190.

Keogh, R.G. (1974) Facial eczema - the importance of litter and control. *Dairyfarming Annual* (1974), 59-63.

Kershaw, K.A. (1958) An investigation of the structure of a grassland community. I. The pattern of *Agrostis tenuis*. *Journal of Ecology* **46**, 571-592.

Kershaw, K.A. (1959) An investigation of the structure of a grassland community. II. The pattern of *Dactylis glomerata, Lolium perenne* and *Trifolium repens*. III. Discussion and conclusions. *Journal of Ecology* **47**, 31-53.

King, J. (1963) Ecotypic differentiation in *Trifolium repens*. *Plant and Soil* **18**, 221-224.

King, J.; Lamb, W.I.C.; McGregor, M.T. (1978) Effect of partial and complete defoliation on regrowth of white clover plants. *Journal of the British Grassland Society* **33**, 49-55.

Kishi, H. (1973) [Studies on competition between grass and legumes in a mixed sward. 1.The growth of two species in a cocksfoot/Ladino clover mixed sward.] *Proceedings of the Crop Science Society of Japan* **42**, 397-406.

Kishi, H. (1974a) [Studies on competition between grasses and legumes in a mixed sward. 3. Influence of cutting frequency on the growth of cocksfoot and Ladino clover mixtures.] *Proceedings of the Crop Science Society of Japan* **43**, 498-504.

Kishi, H. (1974b) [Studies on competition between grasses and legumes in a mixed sward. 4. Influence of light intensity on growth of cocksfoot and Ladino clover mixtures.] *Proceedings of the Crop Science Society of Japan* **43**, 505-509.

Kishi, H. (1974c) [Studies on competition between grasses and legumes in a mixed sward. 2. Trends in botanical composition of several grass/Ladino clover combinations in relation to the growth characteristics of the component grass species] *Proceedings of the Crop Science Society of Japan* **43**, 382-388.

Kishi, H. (1975a) [Studies on competition between grasses and legumes in mixed swards. 5. Effects of regrowth period on the elongation and natural plant height of orchardgrass and Ladino clover after cutting in summer.] *Proceedings of the Crop Science Society of Japan* **44**, 301-305.

Kishi, H. (1975b) [Studies on competition between grasses and legumes in mixed swards. 6. Effects of growth stage and temperature on the upward growth of orchardgrass and Ladino clover in summer.] *Proceedings of the Crop Science Society of Japan* **44**, 306-311.

Kishi, H. (1975c)[Studies on competition between grasses and legumes in a mixed sward. 7. Comparison of the capacity of pasture plants grown in isolation to spread by forming tillers or branches.] *Proceedings of the Crop Science Society of Japan* **44**, 419-424.

Kleter, H.J. (1968) Influence of weather and nitrogen fertilisation on white clover percentage of permanent grassland. *Netherlands Journal of Agricultural Science* **16**, 43-52.

Knight, W.E. (1953) Interrelationships of some morphological and physiological characteristics of Ladino clover. *Agonomy Journal* **45**, 197-199.

Knuth, P. (1908) Handbook of flower pollination 2, 260, 284. [Cited by Erith, (1924)]

Koyama, H.; Kira, T. (1956) Intraspecific competition among higher plants. VIII. Frequency distribution of individual plant weight as affected by interaction between plants. *Journal, Institute of Polytechnics, Osaka City University* **D 7**, 73-94.

Krach, K.E. (1959) Excretion of undigested seeds of clover, grasses and weeds by birds and effect of passage through stomach and intestines on their viability. *Zeitschrift für Acker- und Pflanzenbau* **107**, 405-434.

Kumai, S.; Hirose, M.; Sanada, T. (1965) Studies on the dry matter production of forage crops.1. Experiments on the seasonal growth cycle and growth analysis of ladino clover. *Journal of the Japanese Society of Grassland Science* **11**, 7-13.

Lambert, J.P.; Boyd, A.F.; Brock, J.L. (1974) An evaluation of five varieties of *Lotus pedunculatus* Cav. compared with 'Grasslands Huia' white clover under grazing at Kaikohe. *New Zealand Journal of Experimental Agriculture* **2**, 359-363.

Lambert, J.P.; Rumball, P.J.; Christie, A.F.R. (1977) Comparison of ryegrass and Kikuyu grass pastures under mowing. *New Zealand Journal of Experimental Agricultural* **5**, 71-77.

Lambert, J.P.; Vartha, E.W.; Harris, A.J. (1969) Progress report on 'Grasslands 4700' white clover. *Proceedings of the New Zealand Grassland Association* **31**, 135-142.

Lancashire, J.A.; Keogh, R.G. (1968) Facial eczema and grazing management. *Sheep Farming Annual, 1968*, 29-35.

Lancashire, J.A.; Latch, G.C.M. (1970) The effect of crown rust (*Puccinia coronata* Corda) on the yield and botanical composition of two ryegrass/white clover pastures. *New Zealand Journal of Agricultural Research* **13**, 279-286.

Langille, J.E.; Warren, F.S. (1964) Influence of time of seeding on establishment and persistence of species and yield of forage mixtures. *Canadian Journal of Plant Science* **44**, 78-84.

Lazenby, A.; Lovett, J.V. (1975) Growth of pasture species on the Northern Tablelands of N.S.W. *Australian Journal of Agricultural Research* **26**, 269-280.

Leafe, E.L.; Stiles, W.; Dickinson,S.E. (1974) Physiological processes influencing the pattern of productivity of the intensively managed grass sward. *Proceedings of the 12th International Grassland Congress*, 442-457.

Lennartz, H. (1957) The effect of the passage through the alimentary tract of cattle on the viability of seeds of grassland plants. *Zeitschrift für Acker- und Pflanzenbau* **103**, 427-453.

Lewis, J. (1973) Longevity of crop and weed seeds: Survival after 20 years in soil. *Weed Reserch* **13**, 179-191.

Lieshout, J.W. van (1956) Root development of some field crops. *Verslagen van Landbouwkundigie Onderzoekingen* **62.16,** pp 46.

Linehan, P.A.; Lowe, J. (1960) Yielding capacity and grass/clover ratio of herbage swards as influenced by fertiliser treatments. *Proceedings of the 8th International Grasslands Congress*, 133-137.

Lowe, A.J.; Armitage, E.R. (1959) Irrigation of grassland. *Journal of Agricultural Science, Cambridge, U. K.* **52,** 256-262.

Lowe, J. (1966) Output of pastures under a clover nitrogen regime in Northern Ireland. *Proceedings of the 10th International Grassland Congress*, 187-191.

Lowther, W.L. (1974) Interaction of lime and seed pelleting on the nodulation and growth of white clover. *New Zealand Journal of Agricultural Research* **17**, 317-323.

Lowther, W.L.; McDonald, I.R. (1973) Inoculation and pelleting of clover for oversowing. *New Zealand Journal of Experimental Agriculture* **1**, 175-179.

Ludecke, T.E. (1966) The use of soil sequences in formulating a fertiliser programme when introducing legumes into tussock country in New Zealand. *Proceedings of the 10th International Grassland Congress*, 856-859.

MacDiarmid, B.N.; Watkin, B.R. (1971) The cattle dung patch. 1. Effect of dung patches on yield and botanical composition of surrounding and underlying pasture. *Journal of the British Grassland Society* **26**, 239-245.

McNaught, K.J. (1958) Potassium deficiency in pastures. I. Potassium content of legumes and grasses. *New Zealand Journal of Agricultural Research* **1**, 148-181.

McNaught, K.J.; During, C. (1970) Relations between nutrient concentrations in plant tissue and responses of white clover to fertiliser on a grey podzol near Westport. *New Zealand Journal of Agricultural Research* **13**, 567-590.

McNeur, A.J. (1953) Effect of soil acidicity on growth of pasture plants I. Pot experiments with grasses and clovers in acidified soils. *New Zealand Journal of Science and Technology* **A35**, 311-326.

McRill, M.; Sagar, G.R. (1973) Earthworms and seeds. *Nature, UK* **243**, 482.

McWilliams, J.R.; Clements, R.J.; Dowling, P.M. (1970) Some factors influencing the germination and early seedling development of pasture plants. *Australian Journal of Agricultural Research* **21**, 19-32.

Madden, E.A. (1954) Cattle as distributors of clover seed. *Sheep Farming Annual, Massey Agricultural College, 1954*, 79-82.

Meeklah, F.A.; Mitchell, R.B. (1973) Control of tutu (*Coriaria sarmentosa*) by herbicides. *Proceedings of the New Zealand Weed and Pest Control Conference* **26**, 29-33.

Mence, M.J. (1964) The differential mortality of *Trifolium repens* and *Phleum pratense* seedlings in relation to temperature. *Plant and Soil* **21**, 113-124.

Metson, A.J. (1972) Iron pyrites as sulphur fertilisers. II. Forms of sulphur fertiliser in relation to growth of, and total suphur, sulphate sulphur, and nitrogen/sulphur ratio in a grass-clover pasture. *New Zealand Journal of Agricultural Research* **15**, 565-584.

Metson, A.J.; Blakemore, L.C.; Chittenden, E.T. (1971) Iron pyrites as sulphur fertilisers: field trial with grass-clover pasture on a gley podzol soil at Golden Bay, Nelson. *New Zealand Journal of Agricultural Science* **14**, 104-133.

Metson, A.J.; Collie, T.W. (1972) Iron pyrites as sulphur fertilisers. III. Nitrogen-sulphur relationships in grass and clover separates of pasture herbage in a field trial at Golden Bay, Nelson. *New Zealand Journal of Agricultural Research* **15**, 585-604.

Milne, D.L. (1956) Gall midges affecting seed production in clover. *Annals of Applied Biology* **44**, 669-678.

Miller, F.G.; Pammel, L.H. (1901) A study on the germination and growth of Leguminosae, especially with reference to large and small seed. *Bulletin, Iowa Agricultural College Experiment Station* **62**, 177 pp.

Miller, J.A.S. (1944) Farming in New Zealand. *New Zealand Journal of Agriculture* **69**, 549-557.

Mitchell, K.J. (1956) The influence of light and temperature on the growth of pasture species. *Proceedings of the 7th International Grassland Congress*, 58-69.

Morita, O.; Fujita, T. (1971) [Competition in pasture mixtures. I. Effects of the rate of seeding on competition between Ladino clover and red clover during the vegetative phase]. *Journal of the Japanese Society of Grassland Science* **17**, 48-56.

Morita, O.; Fujita, T. (1975a) [Effects of cutting frequency and cutting height on the growth of Ladino clover and orchard grass in pure and mixed stands]. *Bulletin, Faculty of Agriculture, Mie University* **49**, 55-66.

Morita, O.; Fujita, T. (1975b) [Competition in pasture mixtures. 2. Effects of the cutting height on competition between Ladino clover and red clover]. *Bulletin, Faculty of Agiculture, Mie University* **50**, 11-18.

Mouat, M.C.H.; Walker, T.W. (1959a) Competition for nutrients between grasses and white clover. I. Effect of grass species and nitrogen supply. *Plant and Soil* **11**, 30-40.

Mouat, M.C.H.; Walker, T.W. (1959b) Competition for nutrients between grasses and white clover. II. Effect of root cation exchange capacity and rate of emergence of associated species. *Plant and Soil* **11**, 41-52.

Munro, I.A. (1958) Irrigation of grassland. The influence of irrigation and nitrogen treatments on the yield and utilisation of a riverside meadow. *Journal of the British Grassland Society* **13**, 213-221.

Murphy, W.E. (1970) Effect of fertilizers on clover content of swards. In 'White Clover Research', Ed. J. Lowe. *Occasional Symposium of the British Grassland Society* No. **6**, 221-226.

Murton, R.K. (1965) The Wood Pigeon. New Naturalist Special Volume, Collins. 256pp.

Mytton, L.R. (1973) The effect of seed weight on the early growth and nodulation of white clover. *Annals of Applied Biology* **73**, 329-338.

Nakamura, S. (1962) Germination of legume seeds. *Proceedings of the International Seed Testing Association* **27**, 694-709.

Nelson, C.E.; Robins, J.S. (1956) Some effects of moisture, nitrogen fertiliser, and clipping on yield and botanical composition of Ladino clover-orchardgrass pasture under irrigation. *Agronomy Journal* **48**, 99-102.

Nelson, C.E.; Robins, J.S. (1957) Nitrogen uptake by Ladino clover-orchardgrass pasture under irrigation as influenced by moisture, nitrogen fertilisation and clipping treatments. *Agronomy Journal* **49**, 72-74.

Newman, E.I. (1969) Resistance to waterflow in soil and plant. I. Soil resistance in relation to amounts of root. Theoretical estimates. *Journal of Applied Ecology* **6**, 1-12.

Newman, E.I.; Rovira, A.D. (1975) Allelopathy among some British grassland species. *Journal of Ecology* **63**, 727-737.

Norman, M.J.T. (1960) The relationship between competition and defoliation in pasture. *Journal of the British Grassland Society* **15**, 145-149.

Norman, M.J.T.; Green, J.O. (1957a) The renovation of downland permanent pastures. I. Surface cultivation. *Journal of the British Grassland Society* **12**, 30-38.

Norman, M.J.T.; Green, J.O. (1957b) The renovation of downland permanent pastures. II. Herbicidal treatment, surface cultivation and seeding. *Journal of the British Grassland Society* **12**, 74-80.

Norman, M.J.T.; Green, J.O. (1958) The local influence of cattle dung and urine upon the yield and botanical composition of permanent pasture. *Journal of the British Grassland Society* **13**, 39-45.

O'Brien, A.D.(1970) White clover (*Trifolium repens* L.) in a subtropical environment on the east coast of Australia. *Proceedings of the 11th International Grassland Congress*, 165-168.

Ochi, M; Muroga, T. (1967) Study of regeneration of Ladino clover. Regeneration of Ladino clover according to cutting intervals and its seasonal variation. *Proceedings of the Crop Science Society of Japan* **36**, 145-150.

O'Connor, K.F.; Lambrechtsen, N.C. (1964) Studies on the management of snow tussock grassland. 3. The effect of burning, fertiliser and oversowing on a mid-altitude tall tussock grassland in South Canterbury New Zealand. *New Zealand Journal of Agricultural Research* **7**, 264-268.

Oertel, E. (1954) The value of honey bees to white clover seed growers in Louisiana. *American Bee Journal* **94**, 460-462.

Oswald, A.K. (1977) The effect of seasonal applications of glycophosphates on a mixed sward. *Proceedings of the 1976 British Crop Protection Conference - Weeds.* London, UK; British Crop Protection Council, 961-969.

O'Toole, M.A.; Masterson, S.L. (1970) Effects of copper and nitrogen on nodulation and yield of *Trifolium repens*. In: White clover research, Ed. J. Lowe. *Occasional Symposium of the British Grassland Society* No. 6, 251-258.

Owen, M. (1976) The selection of winter food by white fronted geese. *Journal of Applied Ecology* **13**, 715-729.

Palmer-Jones, T.; Forester, I.W.; Jeffrey, G.L. (1962) Observations on the role of the honey bee and bumble bee as pollinators of white clover (*Trifolium repens* L.) in the Timaru district and Mackenzie country. *New Zealand Journal of Agricultural Research* **5**, 318-325.

Parle, J.N. (1964) Some aspects of the nodulation problem occuring in the Wither Hills. *Proceedings of the New Zealand Grassland Association* **26**, 123-126.

Parkinson, R. (1978) The Experienced Farmer **1**, 95, [Cited by Erith (1924)].

Parry, D.; Wynn, A.J. (1954) Reduced rates for seeding to grass at the College Farm, Aberdeen. *Journal of the British Grassland Society* **8**, 267-272.

Pascoe, W. (1974) The effect of grazing on the growth and development of white clover (*Trifolium repens* L.). M. Agr. Sc. Thesis, Massey University, Summary in: *Proceedings of the New Zealand Grasslands Association* **35**, 305.

Patterson, T.M. (1973) Control of chickweed in pasture. *Proceedings of the New Zealand Weed and Pest Control Conference* **26**, 48-49.

Paxman, R.N.; Forgie, C.D. (1972) Effects of pronamide on various pasture grass species. *Proceedings of the 25th New Zealand Weed and Pest Control Conference,* 56-61.

Percival, M. (1947) Pollen collection by *Apis mellifera*. *New Phytologist* **46**, 142-173.

Percival, M. (1950) Pollen presentation and pollen collection. *New Phytologist* **49**, 40-63.

Peter, A. (1893) Culturversuche mit ruhenden Samen. Nachr. v.d. kg. Ges. d. Wiss., Göttingen, 673. [Cited by Erith (1924)].

Peters, R.A. (1961) Legume establishment as related to the presence or absence of an oat companion crop. *Agronomy Journal* **53**, 195-198.

Peterson, M.L.; Hagan, R.M. (1953) Production and quality of irrigated pasture mixtures as influenced by clipping frequency. *Agronomy Journal* **45**, 283-287.

Rachie, K.A.; Schmid, A.R. (1955) Winter hardiness of birdsfoot trefoil strains and varieties. *Agronomy Journal* **47**, 155-157.

Radcliffe, J.E. (1970) Some of effects of grass grub (*Costelytra zealandica* White) larvae on pasture plants. *New Zeaaland Journal of Agricultural Research* **13**, 87-104.

Radcliffe, J.E. (1971) Effects of grass grub (*Costelytra zealandica* White) larvae on pasture plants. IV. Effect of grass grubs on perennial ryegrass and white clover. *New Zealand Journal of Agricultural Research* **14**, 625-632.

Radcliffe, J.E. (1974a) Seasonal distribution of pasture production in New Zealand. I. Methods of measurement. *New Zealand Journal of Experimental Agriculture* **2**, 337-340.

Radcliffe, J.E. (1974b) Seasonal distribution of pasture production in New Zealand. II. Southland Plains. *New Zealand Journal of Experimental Agriculture* **2**, 341-348.

Radcliffe, J.E. (1975) Seasonal distribution of pasture production in New Zealand. IV. Westport and Motueka. *New Zealand Journal of Experimental Agriculture* **3**, 239-246.

Radcliffe, J.E. (1976) Seasonal distribution production in New Zealand.X. Rangitikei district. *New Zealand Journal of Experimental Agriculture* **4**, 163-170.

Radcliffe, J.E.; Sinclair, D.P. (1975) Seasonal distribution of pasture production in New Zealand. V. Gisborne plains. *New Zealand Journal of Experimental Agriculture* **3**, 247-251.

Reid, D. (1961a) Factors influencing the role of grass-clover leys fertilised with nitrogen at different rates. I. The effects of the variety of companion grass on the yields of total herbage and of clover. *Journal of Agricultural Science, Cambridge* **56**, 143-153.

Reid, D. (1961b) Factors influencing the role of clover in grass-clover leys fertilized with nitrogen at different rates. III. The effects of the variety of companion grass and of the variety of clover on the botanical composition of the sward. *Journal of Agricultural Science, Cambridge, U. K.* **57**, 231-236.

Reid, D. (1961c) Factors influencing the role of clover in grass-clover leys fertilized with nitrogen at different rates. II. The effects of the variety of white clover on yields of total herbage and of clover. *Journal of Agricultural Science, Cambridge, U. K.* **56,** 155-160.

Reid, D. (1968) Studies on the cutting management of grass-clover swards VI. The effects of different closeness and frequency of cutting treatments on the yield and quality of herbage from a cocksfoot/white clover sward. *Journal of Agricultural Science, Cambridge, U. K.* **70,** 59-64.

Reid, D. (1976) The role of white clover in grassland production. *Report of the Hannah Research Institute, 1976,* 66-69.

Reid, D. Castle, M.E. (1965a) The response of grass-clover and pure grass leys to irrigation and to nitrogen treatment. I. Irrigation effects. , *Journal of Agricultural Science, Cambridge, U. K.* **64,** 185-194

Reid, D.; Castle, M.E. (1965b) The response of grass-clover and pure grass leys to irrigation and to fertiliser nitrogen treatment. II. Clover and fertiliser nitrogen effects. *Journal of Agrigultural Science, Cambridge, U. K.* **65,** 109-119.

Reith, J.W.S.; Robertson, R.A.; Inkson, R.H.E. (1973) The nutrient requirements of herbage on deep acid peat. *Journal of Agricultural Science, Cambridge, U. K.* **80,** 425-434.

Rhodes, I.; Harris, W. (1979) The nature and basis of differences in sward composition and yield in ryegrass-white clover mixtures. In: Changes in sward composition and productivity. Ed. Charles, A.H. *Occasional Symposium of the British Grassland Society* No. 10, 55-60.

Rickard, D.S.; Radcliffe, J.E. (1976) Seasonal distibution of pasture in New Zealand. XII. Winchmore, Canterbury Plains dryland and irrigated pasture.*New Zealand Journal of Experimental Agriculture* **4**, 329-335.

Robinson, D.H. (1947) Leguminous forage plants. London, UK; Arnold and Company, 2nd Edition, 119pp.

Robinson, G.C.; Jaques, W.A. (1958) Root development in some common New Zealand pasture plants. 10. Effects of pure sowings of some grasses and clovers on the structure of Tokomaru silt loam. *New Zealand Journal of Agricultural Research* **1**, 199-216.

Robinson, G.G.; Lazenby, A. (1976) Effects of superphosphate, white clover and stocking rate on the productivity of natural pastures, Northern Tablelands, New South Wales. *Australian Journal of Experimental Agriculture and Animal Husbandry* **16**, 209-217.

Robinson, R.R.; Sprague, V.G. (1947)The clover populations and yields of a Kentucky bluegrass sod as affected by nitrogen fetilisation, clipping treatments and irrigation. *Journal of the American Society of Agronomy* **36**, 107-116.

Robinson, R.R.; Sprague, V.G. (1952) Responses of orchardgrass-Ladino clover to irrigation and nitrogen fertilisation. *Agonomy Journal* **44**, 244-247.

Robinson, R.R.; Sprague, V.G.; Lueck, A.G. (1952) The effect of irrigation, nitrogen fertilisation and clipping treatments on the persistence of clover and on total and seasonal distribution of yield in a Kentucky bluegrass sod. *Agronomy Journal* **44**, 239-244.

Rolston, M.P.; Chu, A.C.P; Fillery, I.R.P. (1976) Effect of paraquat on the nitrogen fixing ability of white clover. *New Zealand Journal of Agricultural Research* **19**, 47-49.

Ross, M.A.; Harper, J.L. (1972) Occupation of biological space during seedling establishment. *Journal of Ecology* **60**, 77-80.

Round-Turner, N.L.; Scott, R.S.; Radcliffe, J.E. (1976) Seasonal distribution of pasture production in New Zealand. XI. Otago downland and Taieri plain. *New Zealand Journal of Experimental Agiculture* **4**, 321-328.

Rousvoal, D.; Gallais, A. (1973) [Performance of binary association of five species from permanent grassland.] *Oecologia Plantarum* **8**, 279-300.

Rowlands, A. (1966) Reseeding and surface tratment - comparison of two methods of improving hill grazings. *Journal of the British Grassland Society* **21**, 174-180.

Salzmann, R. (1939) The distibution of viable seeds in farmyard manure. *Schweizerische Landwirtschaftliche Monatshefte* **5**, 40-44.

Sanderson, R.F. (1966) Aspects of the growth of white clover with particular reference to the effects of defoliation and shade. PhD Thesis, University of Reading. [Cited in Spedding and Diekmahns, 1972.]

Santhirasegaram, K.; Black, J.N. (1965) Agronomic practices aimed at reducing competition between cover crops and undersown pasture. *Herbage Abstracts* **35**, 221-225.

Sasaki, N.; Kudo, K. (1965) Physiological and ecological studies on the Ladino clover. 2. On the process of regrowth in relation to respiration and apparent photosynthetic activity. *Bulletin of the Faculty of Agriculture of Hirosaki University* **11**, 39-50.

Sato, K. (1973) [Comparison between four temperate grasses in production structure at each heading stage.] *Journal of the Japanese Society of Grassland Science* **19**, 208-214.

Sato, K.; Nishimura, N.; Ito, M. (1967) Ecophysiological studies on maintenance of density in forage crop sward. 4. The effects of soil moisture and cutting height on the yield and botanical composition in both orchardgrass-Ladino clover and orchardgrass-alfalfa mixtures. *Journal of the Japanese Society of Grassland Science* **13**, 122-127.

Scott, D.(1975) Allelopathic interactions of resident tussock grassland species on germination of oversown seed. *New Zealand Jounal of Agricultural Research* **3**, 135-141.

Scott, D.; Archie, W.J. (1976) Seasonal variation in establishment of oversown seed in relation to tussock proximity. *New Zealand Journal of Experimental Agriculture* **4**, 311-316.

Scott, D.; Menalda, P.H.; Rowley, J.A. (1970) CO_2 exchange of plants. 1. Technique, and response of seven species to light intensity. *New Zealand Journal of Botany* **8**, 82-90.

Scott, R.S. (1963) Long-term studies of molybdenum applied to pasture. III. Rates of molybdenum application in relation to pasture production. *New Zealand Journal of Agricultural Research* **6**, 567-577.

Scott, R.S. (1973) Effects of autumn applied nitrogen and method and frequency of defoliation on a ryegrass, white clover sward. *Proceedings of the New Zealand Grassland Association* **34**, 242-252.

Scott, R.S.; Cullen, N.A.; Davies, E.B. (1963) Long-term studies of molybdenum and lime and their interactions. *New Zealand Journal of Agricultural Research* **6**, 538-555.

Scott, R.S.; Lowther, W.L. (1980) Competition betweenwhite clover'GrasslandsHuia'and *Lotus pendunculatus* 'Grasslands Maku'. I. Shoot and root competition. *New Zealand Journal of Agricultural Research* **23**, 501-508.

Scott, W.R. (1974) Establishment of undersown white clover as affected by wheat sowing rate, time of white clover introduction and irrigation. *New Zealand Journal of Experimental Agriculture* **2**, 151-154.

Schwanbom, N.; Froier, K. (1949) Methods of herbage seed production. *Journal of the British Grassland Society* **4**, 233-261.

Sears, P.D. (1950) The establishment of clovers in pasture. *Proceedings of the New Zealand Grassland Association* **12**, 132-141.

Sears, P.D. (1953) Pasture growth and soil fertility. I. The influence of red and white clovers, superphosphate, lime and sheep grazing, on pasture yields and botanical composition. *New Zealand Journal of Science and Technology* A35, Supplement 1, 1-29.

Sears, P.D.; Goodell, V.C.; Newbold, R.P. (1948) The effect of sheep droppings on yields, botanical composition, and chemical composition of pasture. *New Zealand Journal of Science Technology* **A30,** 231-250.

Sears, P.D.; Hyde, E.O.C.; Greenwood, R.M. (1955) Pasture establishment on pumice soils. *New Zealand Journal of Science and Technology* **A37**, 110-140.

Sears, P.D.; Newbold, R.P. (1942) The effect of sheep droppings on yield, botanical composition, and chemical composition of pasture. I. Establishemnt of trial, technique of measurement, and result for the 1940-1941 season. *New Zealand Journal of Science and Technology* **A24**, 36-61.

Sharma, M.L. (1973) Simulation of drought and its effect on germination of five species. *Agronomy Journal* **65**, 982-987.

Simpson, J.R.; Lipset, J.; Fischer, H.F. (1968) Growth of three pasture legumes on a coarse textured podzolic soil in the Upper Shoalhaven Valley. *Field Station Records, Division of Plant Industry, CSIRO, Australia* **7**, 83-94.

Smith, D. (1949) Differential survival of Ladino and common white clover encased in ice. *Agronomy Journal* **41**, 230-234.

Snaydon, R.W.(1962) The growthand competitive ability of contrasting natural populations of *Trifolium repens* L. on calcareous and acid soils. *Journal of Ecology* **50**, 439-447.

Snaydon, R.W. (1971) An analyis of competition between plants of *Trifolium repens* L. populations collected from contrasting soils. *Journal of Applied Ecology* **8**, 687-697.

Snaydon, R.W.; Bradshaw, A.D. (1962a) The performance and survival of contrasting natural populations of white clover when planted into an upland *Festuca/Agrostis* sward. *Journal of the British Grassland Society* **17**, 113-118.

Snaydon, R.W.; Bradshaw, A.D. (1962b) Differences between natural populations of *Trifolium repens* L. in response to mineral nutrients. I. Phosphate. *Journal of Experimental Botany* **13**, 422-434.

Snaydon, R.W.; Bradshaw, A.D. (1969) Differences between natural populations of *Trifolium repens* L. in response to nutrients. II. Calcium, magnesium and potassium. *Journal of Applied Ecology* **6**, 185-202.

Sonneveld, A. (1958) Sowing grasses and clovers. *Mededeling, Instituut voor Biologisch en Scheikundig Onderzoek van Landbouwgewassen* **38**, 212-215.

Spedding, C.R.W.; Diekmahns, E.C.[Eds](1972)Grasses and legumes in British Agriculture. *Bulletin 49, Commonwealth Bureau of Pastures and Field Crops*, 347-369.

Sprague, M.A. (1955) The influence of rate of cooling and winter cover on the winter survival of Ladino clover and alfalfa. *Plant Physiology* **30**, 447-451.

Sprague, V.G.; Garber, R.J. (1950) Effect of time and height of cutting and nitrogen fertilization on the persistence of the legume and production of orchardgrass-Ladino and bromegrass-Ladino associations. *Agronomy Journal* **42**, 586-593.

Sprague, V.G.; Robinson, R.R.; Clyde, A.W. (1947) Pasture renovation. I. Seed bed preparation, seedling establishment and subsequent yields. *Journal of the American Society of Agronomy* **39**, 12-25.

Stapledon, R.G.; Wheeler, D.E. (1948) An experiment to test the influence of date and depth of sowing on four of the chief grasses and on white clover. *Journal of the British Grassland Society* **3**, 263-272.

Stebler, F.C.; Schroeter, C. (1889) The best forage plants, 138.

Stern, W.R.; Donald, C.M. (1962a) Light relationships in grass-clover swards. *Australian Journal of Agicultural Research* **13**, 599-614.

Stern, W.R.; Donald, C.M. (1962b) The influence of leaf area and radiation on the growth of clover in swards. *Australian Journal of Agricultural Research* **13**, 615-623.

Stewart, D.A.; Chestnutt, D.M.B. (1974) Some mathematical aspects of the clover, grass, and fertilizer nitrogen relations. *Journal of Agricultural Science, Cambridge, U. K.* **82,** 331-342.

Stiles, W. (1966) Ten years of irrigation experiments. *Annual Report of the Grassland Research Institute, Hurley, 1965,* 57-66.

Strickland, A.H. (1956) Problems in estimating insect pest damage to clover-seed crops. *Annals of Applied Biology* **44**, 671-673.

Suckling, F.E.T. (1950) Strike of oversown grass and clover seeds on hill pastures. *New Zealand Journal of Science and Technology* **A32**, No. 2, 1-9.

Suckling, F.E.T. (1952) Dissemination of white clover (*Trifolium repens*) by sheep. *New Zealand Journal of Science and Technology* **A33**, 64-77.

Suckling, F.E.T. (1954) Pasture management trials on unploughed hill country at Te Awa. I. Establishment of experimental area and results for 1949-1951. *New Zealand Journal of Science and Technology* **A36**, 237-273.

Suckling, F.E.T. (1959) Pasture management trials on unploughed hill country at Te Awa. II.Results for 1951-1957. *New Zealand Journal of Agricultural Research* **2**, 485-543.

Suckling, F.E.T. (1976) Evaluation of 'Grasslands Pitau' and 'Grasslands Huia' white clover (*Trifolium repens* L.) varieties on hill country. *New Zealand Journal of Experimental Agriculture* **4**, 143-150.

Suckling, F.E.T.; Charlton, J.F.L. (1978) A review of the significance of buried seeds with particular reference to New Zealand agriculture. *New Zealand Journal of Experimental Agriculture* **6**, 211-215.

Taylor, G.B. (1972) The effects of seed size on seedling growth in subterranean clover (*Trifolium subterraneum* L.) *Australian Journal of Agricultural Research* **23**, 595-603.

Taylor, R.L. (1973) Control of ragwort. *Proceedings of the New Zealand Weed and Pest Control Conference* **26**, 50-52.

Taylor, T.H.; Foote, J.S.; Snyder, J.H.; Smith, E.M.; Templeton, W.C. (1972) Legume seedling stands resulting from winter and spring sowings in Kentucky bluegrass (*Poa pratensis* L.) sod. *Agronomy Journal* **64**, 535-538.

Taylor, T.H.; Washko, J.B.; Blaser, R.E. (1960) Dry matter yield and botanical composition of an orchardgrass/Ladino white clover mixture under clipping and grazing conditions. *Agronomy Journal* **52**, 217-220.

Tesar, M.B.; Ahlgren, H.L. (1950) Effect of height and frequency of cutting on the productivity and survival of Ladino clover. *Agronomy Journal* **42**, 230-235.

Thompson, A. (1973) The effect of herbicides on gorse and pasture species. *Proceedings of the New Zealand Weed and Pest Control Conference* **26**, 13-16.

Thompson, A.; Mitchell, W.J.P. (1973) The control of stinking mayweed in pastures. *Proceedings of the New Zealand Weed and Pest Control Conference* **26**, 40-44.

Thompson, F.B.; Elliot, I.L. (1950) The development of Waikato peat soils. *New Zealand Journal of Science and Technology* **A31**, No. 6 1-16.

Thompson, H.E.; Algren, H.L.; Medler, J.T. (1952) Effect of cutting treatments and the control of injurious insects on seed production capacity in Ladino clover (*Trifolium repens* L.). *Agronomy Journal* **44**, 411-414.

Thomson, D.J. (1977) The role of legumes in improving the quality of forage diets. *Proceedings of the International Meeting on Animal Production from Temperate Grasslands, Irish Grassland and Animal Production Association, Dublin.* 131-135.

Thomson, D.J. .; Raymond, W.F. (1970) White clover in animal production. Nutritional factors review. White clover research. *Occasional Symposium of the British Grassland Society,* No. 6, 277-284.

Todd, F.E.; Vansell, G.H. (1952) The role of pollinating insects in legume seed production. *Proceedings of the 6th International Grassland Congress,* 835-839.

Tosun, F. (1967) The effect of frequency and height of clipping on the top and root growth of some perennial grasses and legumes. *Ataturk Universitesi Ziraat Fakultesi Zirai Arastirma Enstitusu Arastirma Bulteni* No. 23, pp46.

Trautner, J.L.; Gibson, P.B. (1966) Fate of white clover axillary buds at five intensities of sunlight. *Agronomy Journal* **58**, 557-558.

Turkington, R. (1979) Neighbour relationships in grass legume communities. IV. Fine scale biotic differentation. *Canadian Journal of Botany* **57**, 2711-2716.

Turkington, R.; Cahn, M.A.; Vardy, A.; Harper, J.L. (1979) The growth, distribution and neighbour relationships of *Trifolium repens* in a permanent pasture. III. The establishment and growth of *Trifolium repens* in natural and perturbed sites. *Journal of Ecology* **67**, 231-243.

Turkington, R.; Cavers, P.B. (1978) Reproduction strategies and growth patterns in four legumes. *Canadian Journal of Botany* **56**, 413-416.

Turkington, R.; Cavers, P.B. (1979) Neighbour relationships in grass-legume communities. III. Development of pattern and association in artificial communities. *Canadian Journal of Botany* **57**, 2704-2710.

Turkington, R.; Cavers, P.B.; Aarssen, L.W. (1977) Neighbour relationships in grass-legume communities. I. Interspecific contacts in four grassland communities near London, Ontario. *Canadian Journal of Botany* **55**, 2701-2711.

Turkington, R.; Harper, J.L. (1979a) The growth, distribution and neighbour relationships of *Trifolium repens* in a permanent pasture. I. Ordination, pattern and contact. *Journal of Ecology* **67**, 201-218.

Turkington, R.; Harper, J.L. (1979b) The growth, distribution and neighbour relationships of *Trifolium repens* in a permanent pasture. II. Inter- and intra-specific contact. *Journal of Ecology* **67**, 219-230.

Turkington, R.; Harper, J.L. (1979c) The growth, distribution and neighbour relationships of *Trifolium repens* in a permanent pasture. IV. Fine-scale biotic differentiation. *Journal of Ecology* **67**, 245-254.

Ueno, M.; Williams, R.D. (1967) Absorption of radio-active phosphorus from tap- and nodule-roots of white clover (*Trifolium repens.*) *Journal of the British Grassland Society* **22**, 165-169.

Ueno, M.; Yoshihara, K. (1967) Studies on the root of herbage legumes. 4. Varietal differences of nodal root growth in relation to compensation of tap root in white clover. *Journal of the Japanese Grassland Society* **13**, 195-198.

Ueno, M.; Yoshihara, K.; Hidaka, M. (1967) Studies on the root of herbage legumes. 3. Estimation of the role of the tap root in Ladino clover. *Journal of the Japanese Society of Grassland Science* **13**, 100-104.

Ueno, M.; Yoshihara, K. (1968) Studies on the root of herbage legumes 5. Effects of some environmental factors on the penetration of the adventitious roots of white clover to the soil. *Journal of the Japanese Society of Grassland Science* **14**, 100-103.

Vallis, I.; Henzell, E.F.; Evans, T.R. (1977) Uptake of soil nitrogen by legumes in mixed swards. *Australian Journal of Agricultural Research* **28**, 413-425.

Vansell, G.H. (1951) Honeybee activity on Ladino clover florets. *Journal of Economic Entomology* **44**, 103.

Vartha, E.W. (1972) Effects of *Poa trivialis* L. on growth of perennial ryegrass and white clover. *New Zealand Journal of Agricultural Research* **15**, 620-628.

Vydrina, O.S. (1965) [The application of herbicides to white clover grown for seed.] *Nauchnye Trudy Severo-zapadrogo Nauchno-issledovatelsl'skogo Instituta sel'skogo Khozyiaistva* **8**, 207-211.

Wahab, H.A.; Chamblee, D.S. (1972) Influence of irrigation in the yield and persistence of forage legumes. *Agronomy Journal* **64**, 713-716.

Walker, T.W. (1960) Soil fertility and herbage productivity. *Journal of the British Grassland Society* **15**, 74-80.

Walker, T.W.; Adams, A.F.R. (1958) Competition for sulphur in a grass-clover association. *Plant and Soil* **9**, 353-366.

Walker, T.W.; Adams, A.F.R.; Orchiston, H.D. (1956) Fate of labelled nitrate and ammonium nitrogen when applied to grass and clover grown separately and together. *Soil Science* **81**, 339-351.

Walker, T.W.; Orchiston, H.D.; Adams, A.F.R. (1954) The nitrogen economy of grass legume association. *Journal of the British Grassland Society* **9**, 249-274.

Ward, C.Y.; Jones, J.N.; Lillard, J.H.; Moody, J.E.; Brown, R.H.; Blaser, R.E. (1966) Effects of irrigation and cutting management on yield and botanical composition of selected legumes-grass mixtures. *Agonomy Journal* **58**, 181-184.

Wasmuth, A.G.; Miles, K.B. (1972a) The suppression of grass weeds in white clover seed crops with carbetamide. *Proceedings of the New Zealand Weed and Pest Control Conference* **25**, 210-213.

Wasmuth, A.G.; Miles, K.B. (1972b) The control of barley grass with carbetamide. *Proceedings of the New Zealand Weed and Pest Control Conference* **25**, 41-45.

Watkin, B.R. (1954) The animal factor and levels of nitrogen. *Journal of the British Grassland Society* **9**, 35-46.

Weaver, N. (1957) Pollination of white clover. *Gleanings in Bee Culture* **85**, 457-458.

Weeda, W.C. (1967) The effect of cattle dung patches on pasture growth, botanical composition and pasture utilization. *New Zealand Journal of Agricultural Research* **10**, 150-159.

Weeda, W.C.; Cullen, N.A.; Douglas, J.A.; During, C. (1969) Progress report on 'Grasslands 4700' white clover. *Proceedings of the New Zealand Grassland Association* **31**, 143-150.

Wesson, G.; Wareing, P.F. (1969) The role of light in the germination of naturally occurring populations of buried seeds. *Journal of Experimental Botany* **20**, 402-413.

Wheeler, J.L. (1958) The effect of sheep excreta and nitrogenous fertiliser on the botanical composition and production of a ley. *Journal of the British Grassland Society* **13**, 196-202.

Wheeler, W.A.; Hill, D.D. (1957) Grassland seeds. Princeton, New Jersey, USA; D. Van Nostrand Company, Inc., 381-384.

White, J.; Harper, J.L. (1970) Correlated changes in plant size and number in plant populations. *Journal of Ecology* **58**, 467-485.

White, J.H.; French, N. (1968) Leatherjacket damage to grassland. *Journal of the British Grassland Society* **23**, 326-329.

Wilkinson, S.R.; Gross, C.F. (1964) Competition for light, soil moisture and nutrients during Ladino clover establishment in orchardgrass sod. *Agronomy Journal*, **56**, 389-392.

Williams, P.P.; Palmer, P.C. (1970) Progress made with paraquat in relation to pasture management. *Proceedings of the New Zealand Grassland Association* **31**, 96-109.

Williams, W.M.; Barclay, P.C. (1975) Performance of a Spanish white clover (*Trifolium repens* L.) population in New Zealand. *New Zealand Journal of Agricultural Research* **18**, 45-49.

Wilson, D.B. (1962) Effects of light intensity and clipping on herbage yields. *Canadian Journal of Plant Science* **42**, 270-275.

Wit, C.T.de (1960) On competition. *Verslagen van Landbouwkundige Onderzoekingen* **668**, 1-82.

Wit, C.T.de; Tow, P.G.; Ennik, G.C. (1966) Competition between legumes and grasses. *Verslagen van Landbouwkundige Onderzoekingen* **687**, 1-30.

Wolfe, E.C.; Lazenby, A. (1973a) Grass-white clover relationships during pasture development. 1. Effect of superphosphate. *Australian Journal of Experimental Agriculture and Animal Husbandry* **13**, 567-574.

Wolfe, E.C.; Lazenby, A. (1973b) Grass-white clover relationships during pasture development 2. Effect of nitrogen fertiliser with superphosphate. *Australian Journal of Experimental Agriculture and Animal Husbandry* **13**, 575-580.

Wood, G.M.; Spraque, M.A. (1952) Relation of organic food reserves to cold hardiness of Ladino clover. *Agronomy Journal* **44**, 318-325.

Wolton, K.M.; Brockman, J.S.; Shaw, P.G. (1970) The effect of growth at defoliation on white clover in mixed swards. *Journal of the British Grassland Society* **25**, 113-118.

Yamada, T.; Kawaguchi, T. (1971) Dissemination of pasture plants by livestock. I. Recovery and viability of some pasture plants seeds passed through digestive tract of goats. *Journal of the Japanese Society of Grassland Science* **17**, 36-47.

Yamada, T.; Kawaguchi, T. (1972) Dissemination of pasture plants by livestock. II. Recovery, viability, and emergence of some pasture plants passed through digestive tract of dairy cows. *Journal of the Japanese Society of Grassland Science* **18**, 8-15.

Yamane, I.; Iizumu, S.; Kurosaki, Z.; Sato, K.; Sugawara, K. (1962) Pasture establishment by sheep grazing in mountainous grassland. I. Preliminary experiment in 1961. *Journal of the Japanese Society of Grassland Science* **8**, 90-96.

Yamane, I.; Iizumu, S.; Mori, A.; Tsuda, T.; Sato, K.; Sugawara, K. (1966) Pasture establishment by sheep grazing in mountainous grassland. III.Experiment from 1963 to 1965. *Science Report of the Research Institute Tohoku University* **D 17**, 53-84.

Yeates, G.W.; Healy, W.B.; Widdowson, J.P.; Thomson, N.A.; MacDiarmid, B.N. (1975) Influence of nematicides on growth of plots of white clover on a yellow-brown loam. *New Zealand Journal of Agricultural Research* **18**, 411-416.

Yeates, G.W.; Healy, W.B.; Widdowson, J.P.; Thomson, N.A.; MacDiarmid, B.N. (1976) Effect of a soil fumigant on the establishment and growth of a grazed pasture on a yellow-brown loam. *New Zealand Journal of Agicultural Research* **19**, 397-403.

Yoda, K.; Kira, T.; Ogawa, H.; Hozumi, H. (1963) Self thinning in overcrowded pure stands under cultivated and natural conditions. *Journal of Biology, Osaka City University* **14**, 107-129.

Yoshihara, K. (1958) Studies on pasture mixtures. 3. Effects of seeding methods and soil moisture content upon establishment of some pasture crops. *Proceedings of the Crop Science Society of Japan* **27**, 137-140.

Zaleski, A. (1961) White clover investigations. I. Effect of seed rates and cutting treatments on a flower formation and yield of seed. *Journal of Agricultural Science, Cambridge, U.K.* **57**, 199-212.

Zaleski, A. (1964) Effect of density of plant population, photoperiod, temperature and light intensity on inflorescence formation in white clover. *Journal of the British Grassland Society* **19**, 237-247.

Zaleski, A. (1965a) White clover investigations II. Effect of companion grasses and methods of sowing and seed production. *Journal of Agricultural Science, Cambridge, U. K.* **64**, 77-82.

Zaleski, A. (1965b) White clover seed production. *Journal of the National Institute of Agricultural Botany* **10**, 311-316.

Zaleski, A. (1966) White clover investigations. III. Effect of irrigation on seed production. *Journal of Agricultural Science, Cambridge, U.K.* **67**, 249-253.

Zaleski, A. (1967) The use of herbicides in white clover seed crops. *Journal of the British Grassland Society* **22**, 62-65.

Zaleski, A. (1970) White clover in seed production in white clover research *Occasional Symposium of the British Grassland Society* No. 6, 147-154.

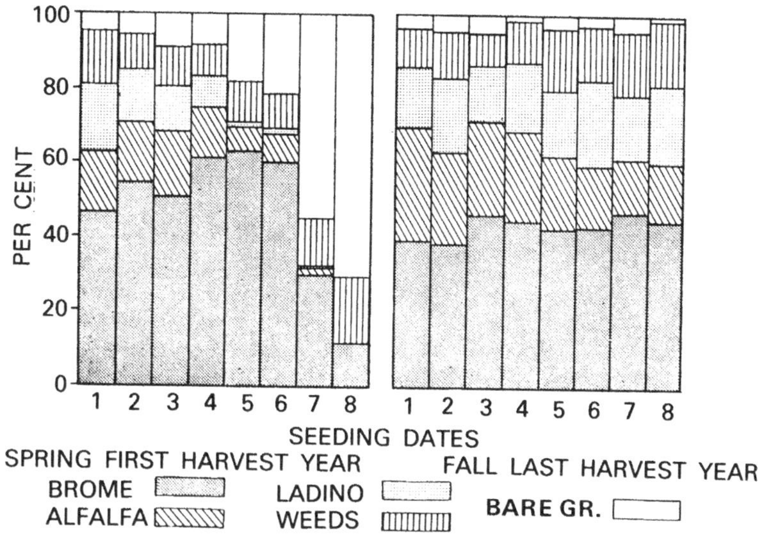

Fig. 7.1 Percentage ground cover by species when seeded at eight dates at successive 2-week intervals from early June in Nova Scotia. (From Langille and Warren (1964) reproduced with permission from *Canadian Journal of Plant Science*).

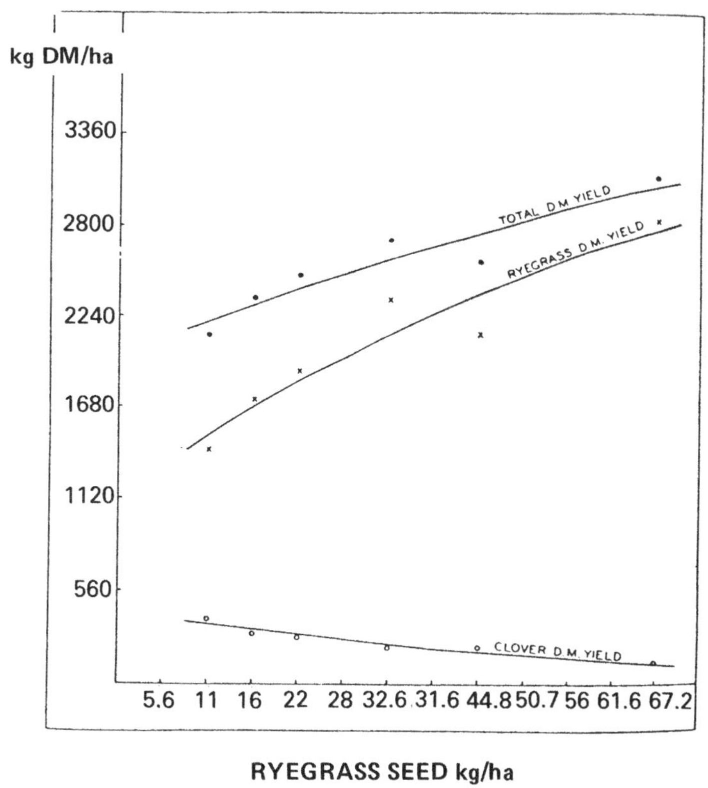

Fig. 7.2 The effect of seeding rate of ryegrass on the total stand yield and yield of ryegrass and clover six months after sowing. (From Brougham (1954a) reproduced with permission from *NZ Journal of Agr. Res.*).

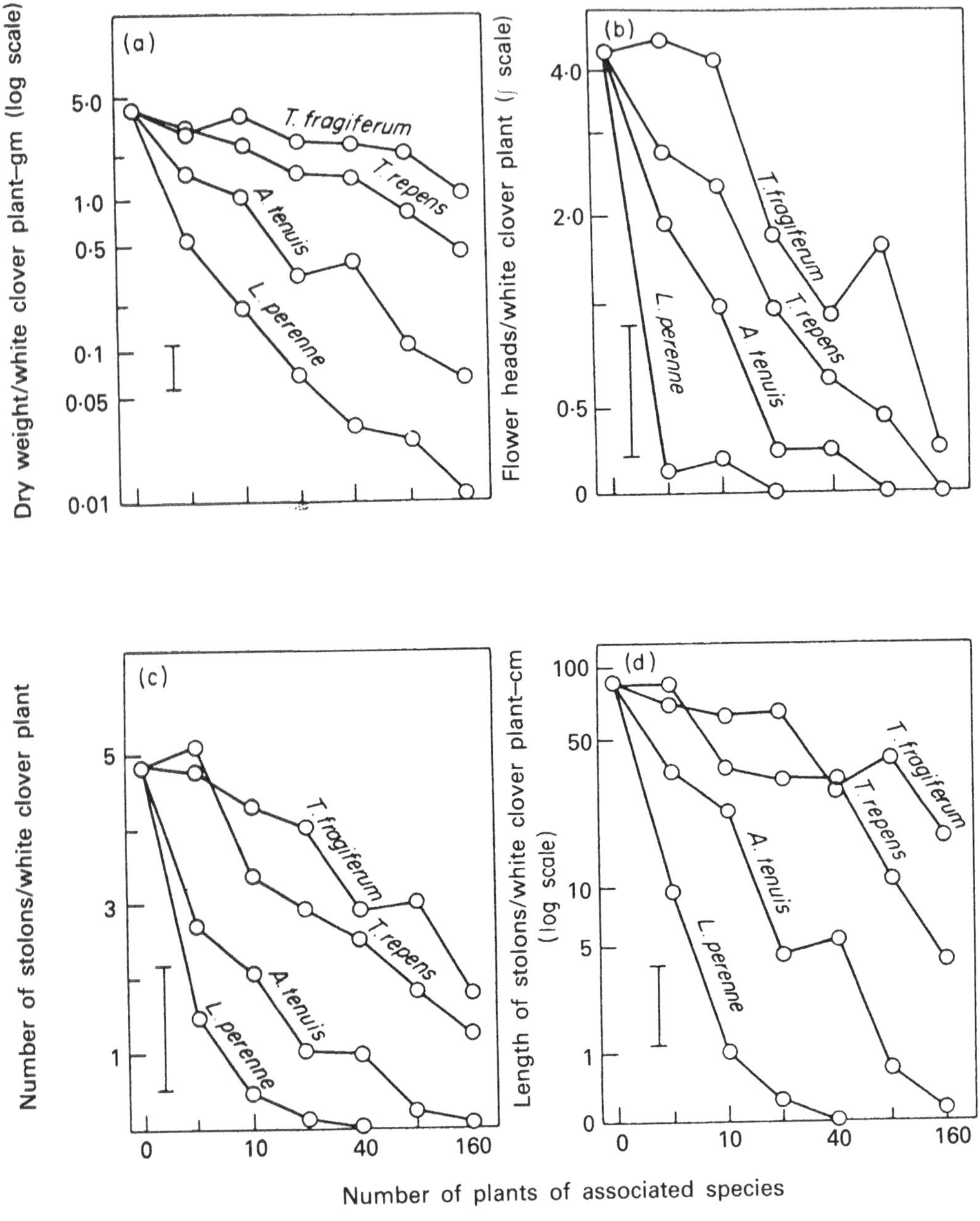

Fig. 7.3 The effect of populations of *Lolium perenne, Agrostis tenuis, Trifolium repens* and *T. fragiferum*, sown at a range of densities, on the growth of *T. repens* sown at the same time and measured after 7 months. Data have been transformed in ways appropriate to homogenize the variance.

(a) Dry weight per plant of *T. repens*
(b) Number of flowerheads produced per plant
(c) Number of stolons per plant
(d) Length of stolons per plant
(From Harper (1977) reproduced with permission from *Population Biology of Plants*)

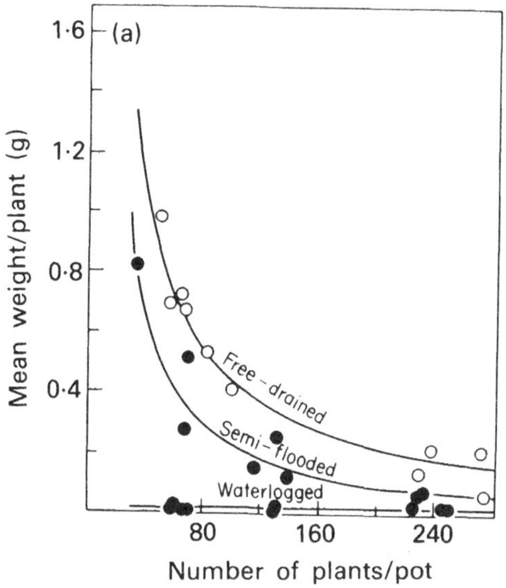

Fig. 7.4 The effect of plant density upon the mean dry weight of plants of *Trifolium repens* grown in pure stands under three water regimes. (From Harper (1977) reproduced with permission from *Population Biology of Plants*).

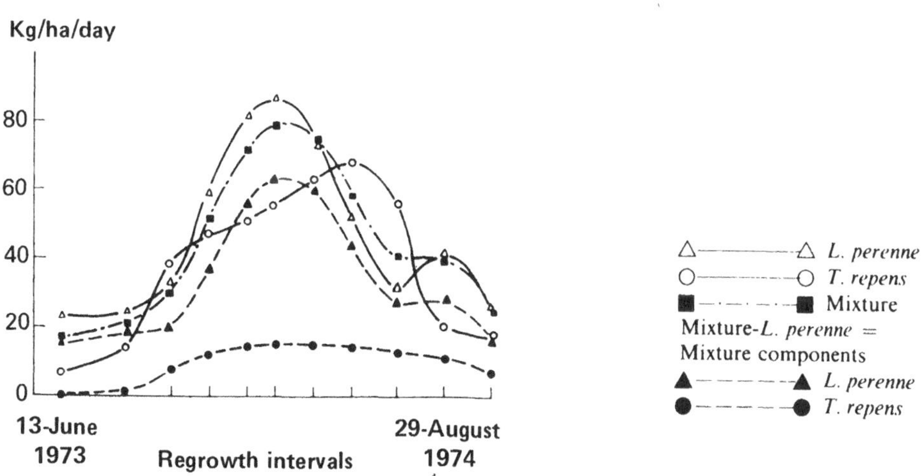

Fig. 7.5 Seasonal growth curves of ryegrass and white clover monocultures and their components in mixture. (From Harris and Hoglund (1981) reproduced with permission from *Proc. XIII Int. Grass. Cong.*).

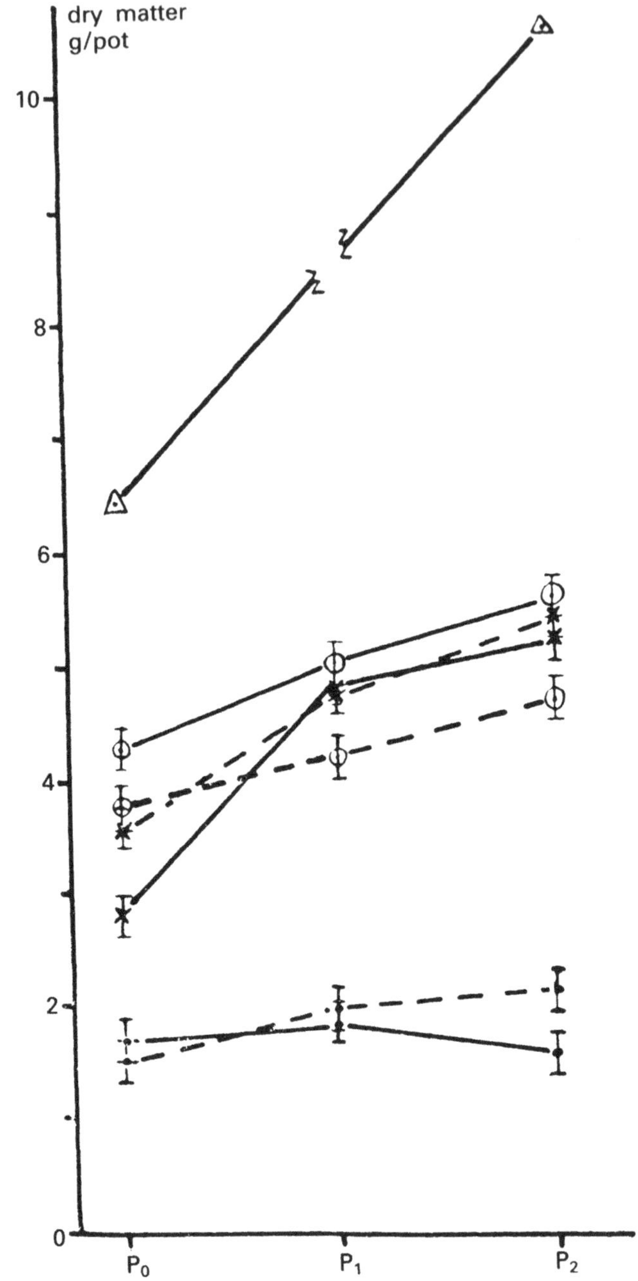

Fig. 7.6 Effect of P treatment averaged over 3 N levels on dry-matter yield of clover grown alone and in association. (From Mouat and Walker (1959) reproduced with permission from Plant and Soil).
P_0 No applied P, P_1 45 kg P_5O_5 ha^{-1}
P_2 90 kg P_2O_5 ha^{-1}

—— Uncut
------ Cut

△ Grown alone
X With Browntop

O With Cocksfoot
• With Ryegrass

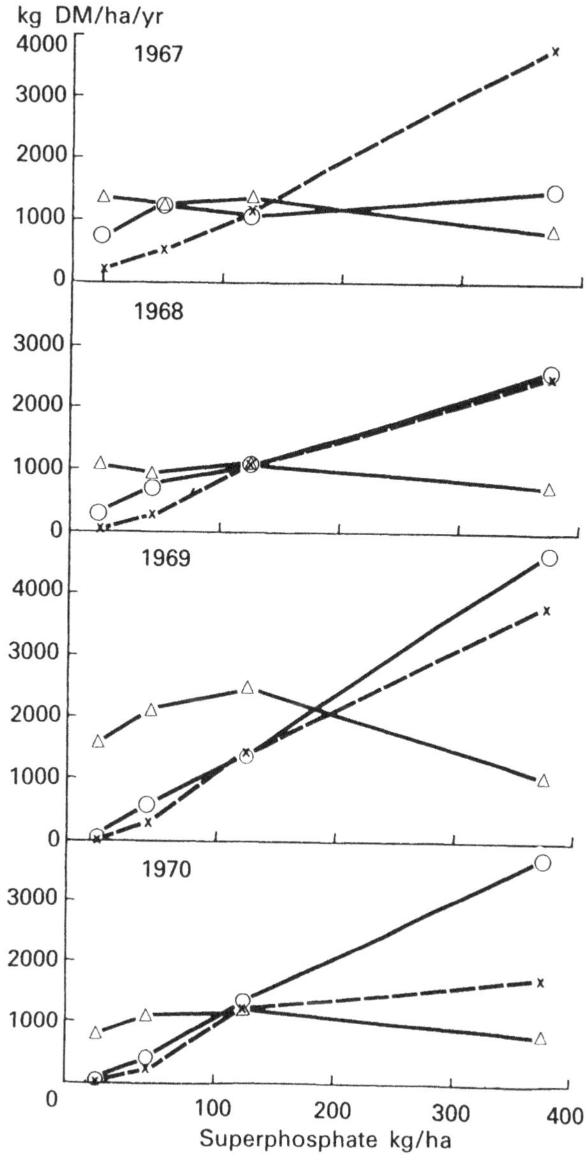

Fig. 7.7 Response of white clover, sown grass and weeds to four rates of superphosphate for four years. X------X White Clover, ○——○ Sown grass, △——△ Weeds. Data obtained from Wolfe & Lazenby (1973a)

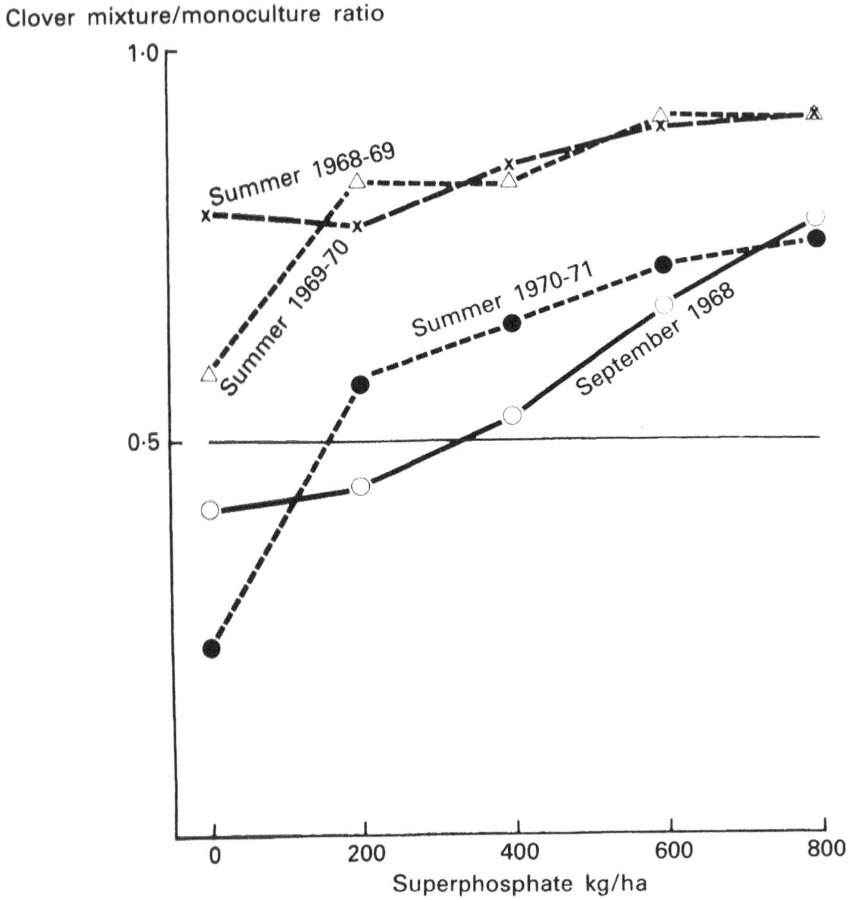

Fig. 7.8 Effect of superphosphate rate on clover yield in mixture with *Agrostis tenuis* relative to clover yield in monoculture. Data obtained from Jackman and Mouat (1982a.)

$$\text{Ratio} = 2\,\frac{\text{Yield of clover in mixture}}{\text{Yield of clover in monoculture}}$$

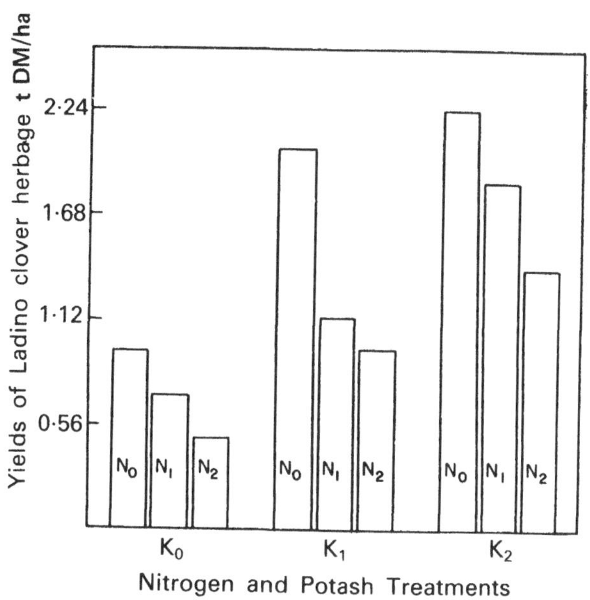

Fig. 7.9 Yield from June to October of Ladino clover in a Ladino-grass mixture at three levels of nitrogen and three levels of potassium. (From Blaser and Brady (1950) reproduced with permission from *Agronomy Journal*).

N_0	0 kg K_2O ha^{-1}	K_0	0 kg K_2O ha^{-1}
N_1	56 kg K_2O ha^{-1}	K_1	84 kg K_2O ha^{-1}
N_2	112 kg K_2O ha^{-1}	K_2	168 kg K_2O ha^{-1}

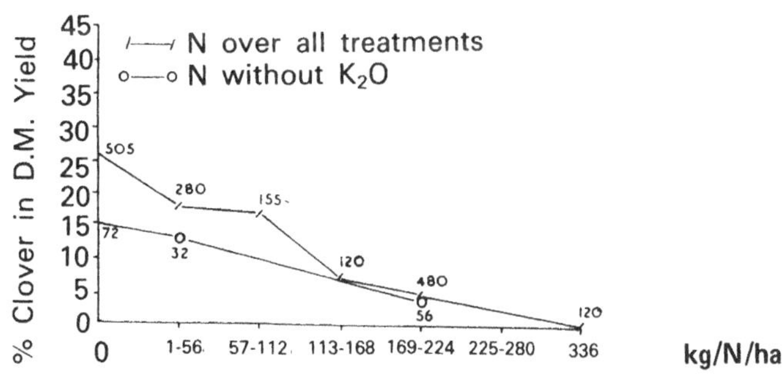

Fig. 7.10 Ratio of clover at various stated rates of N application. Numbers indicate the number of samples for each plotted point. (From Lowe (1966) reproduced with permission from *Proc. Xth. Int. Grass Cong.*).

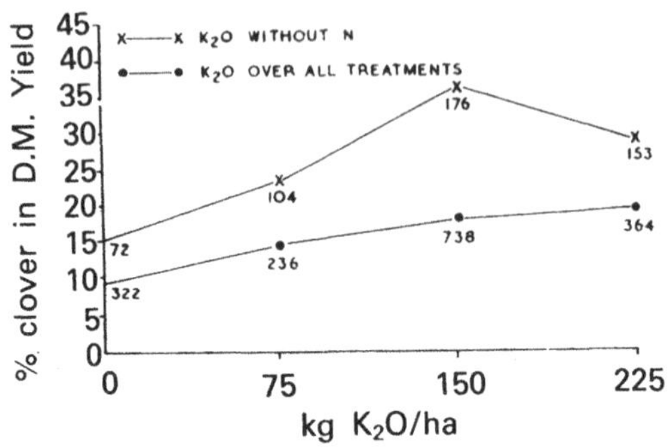

Fig. 7.11 Ratio of clover at various stated rates of K_2O application. Numbers indicate the number of samples for each plotted point. (From Lowe (1966) reproduced with permission from *Proc. Xth. Int. Grass. Cong.*).

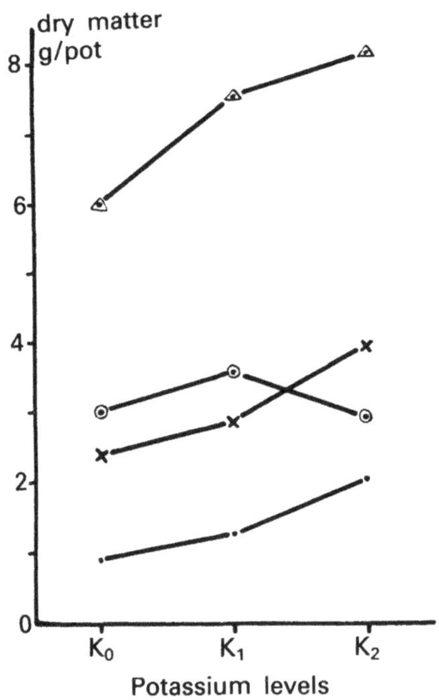

Fig. 7.12 Effect of K level averaged over 3 N levels on dry-matter yield of clover grown in monoculture and in association with grasses. (From Mouat and Walker (1959) reproduced with permission from *Plant and Soil*).

△ Grown alone O With Cocksfoot
X With Browntop • With Ryegrass

K_0 No applied K, K_1 112 kg K_2O ha^{-1}
K_2 224 kg K_2O ha^{-1}

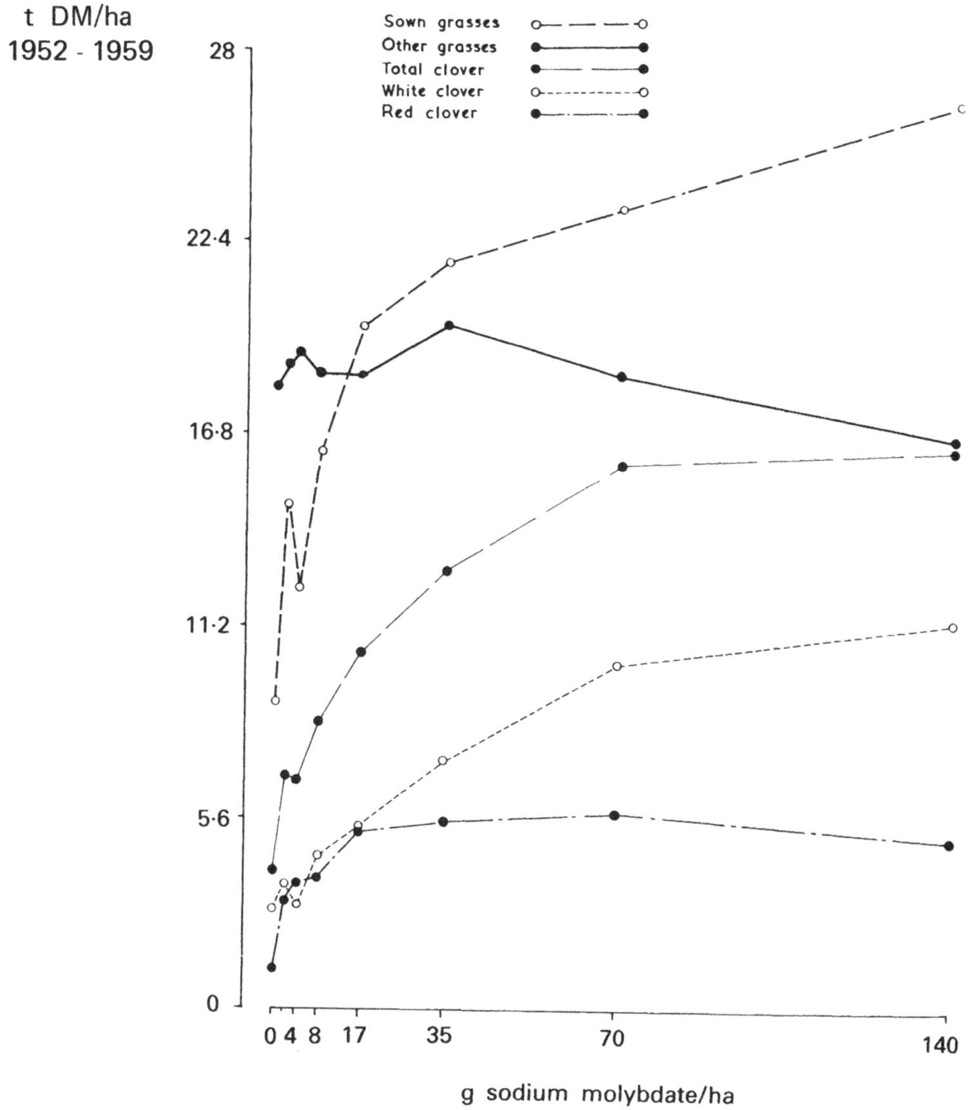

Fig. 7.13 Species response to application of sodium molybdate. (From Scott (1963) reproduced with permission from *NZ Journal of Agr. Res.*).

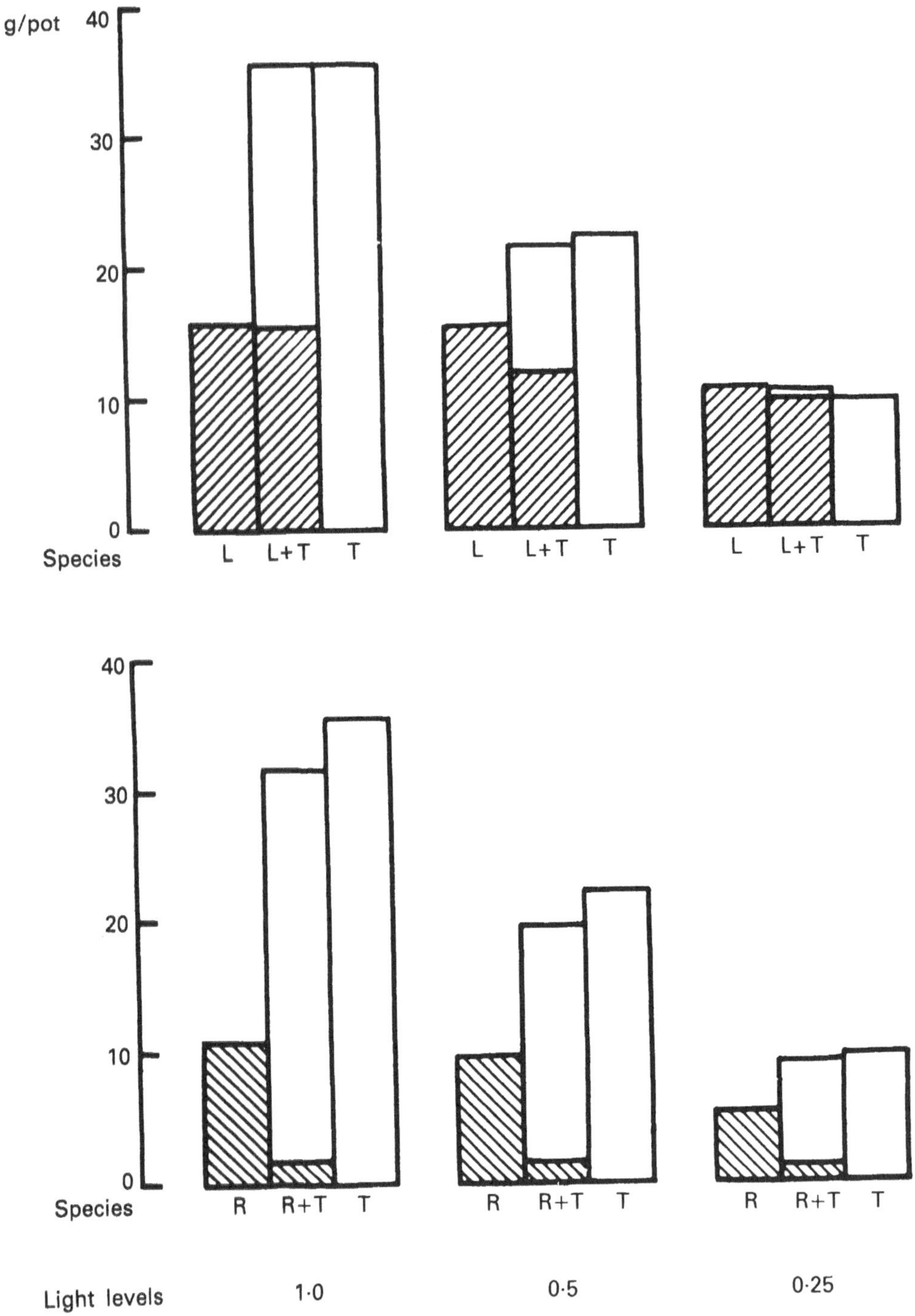

Fig. 7.14 Effect of three light levels on yield of ryegrass (L ▨), white clover (T □) and sheep sorrel (R ▧). Data obtained from Harris (1972.)

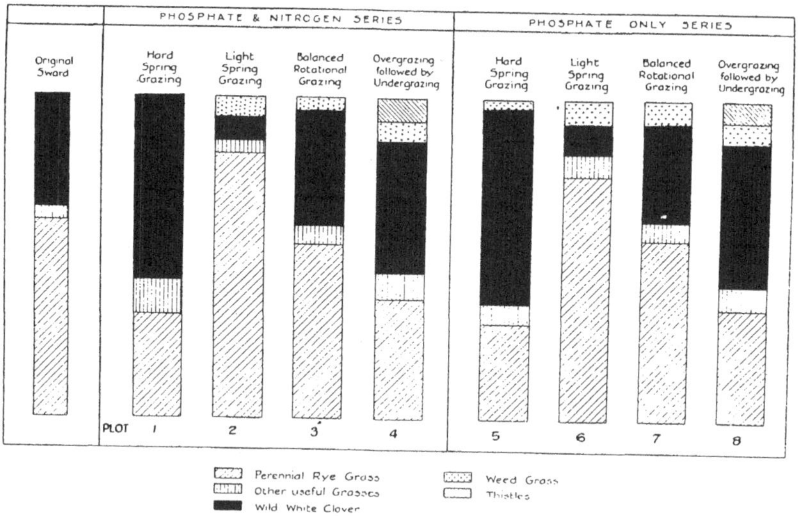

Fig. 7.15 Influence of the timing and method of grazing on the percentage composition of a young clovery sward after three years' treatment. (From Jones, M.G. (1933) reproduced with permission from *Journal Royal Agr. Soc.*).

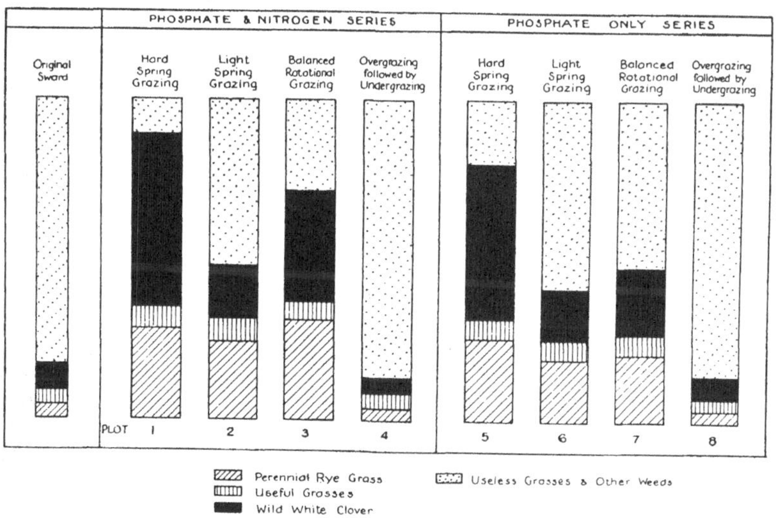

Fig. 7.16 Influence of the timing and method of grazing on the percentage composition of an old permanent sward after two years' treatment. (From Jones, M.G. (1933) reproduced with permission from *Journal Royal Agr. Soc.*).

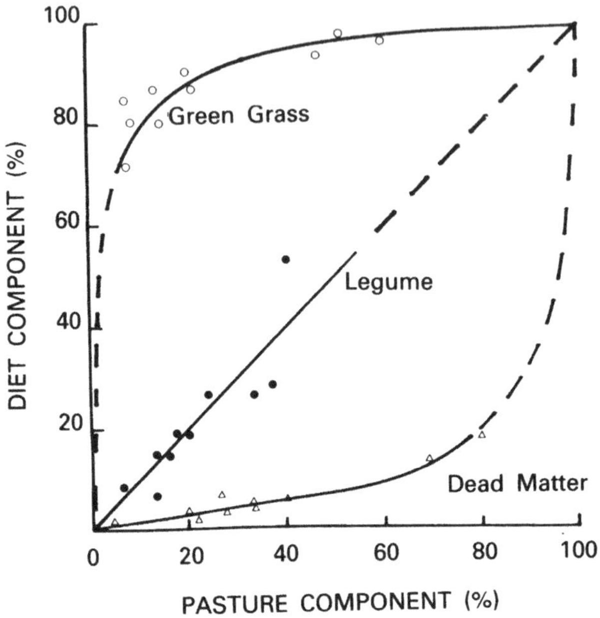

Fig. 7.17 The relationship between diet and sward composition (%) for green grass leaf (○), dead matter (△) and legume (●). (From Clark et al (1982) reproduced with permission from *Proc. NZ. Grass Ass.*).

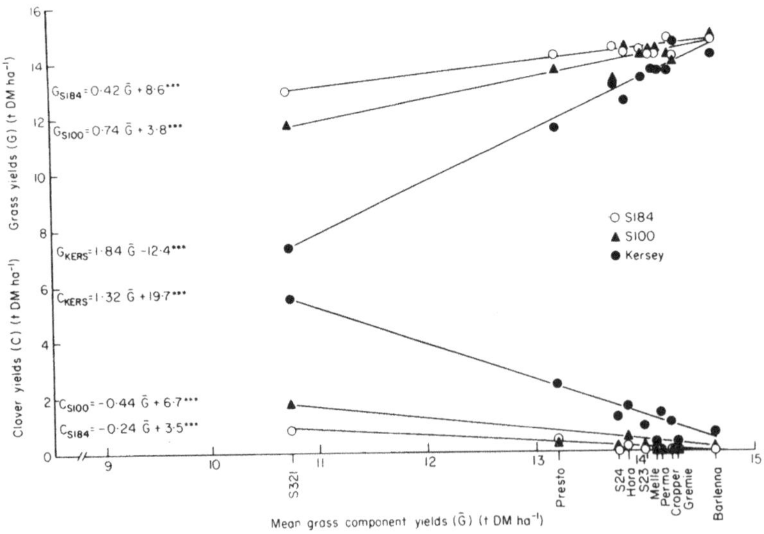

Fig. 7.18 Regressions of grass (G) and clover (C) yields associated with the three clover cultivars Kersey, S100 and S184 against the mean grass component yields (G) of ten ryegrass cultivars in the second harvest year. (From Camlin (1981) reproduced with permission from *Grass & Forage Science*).

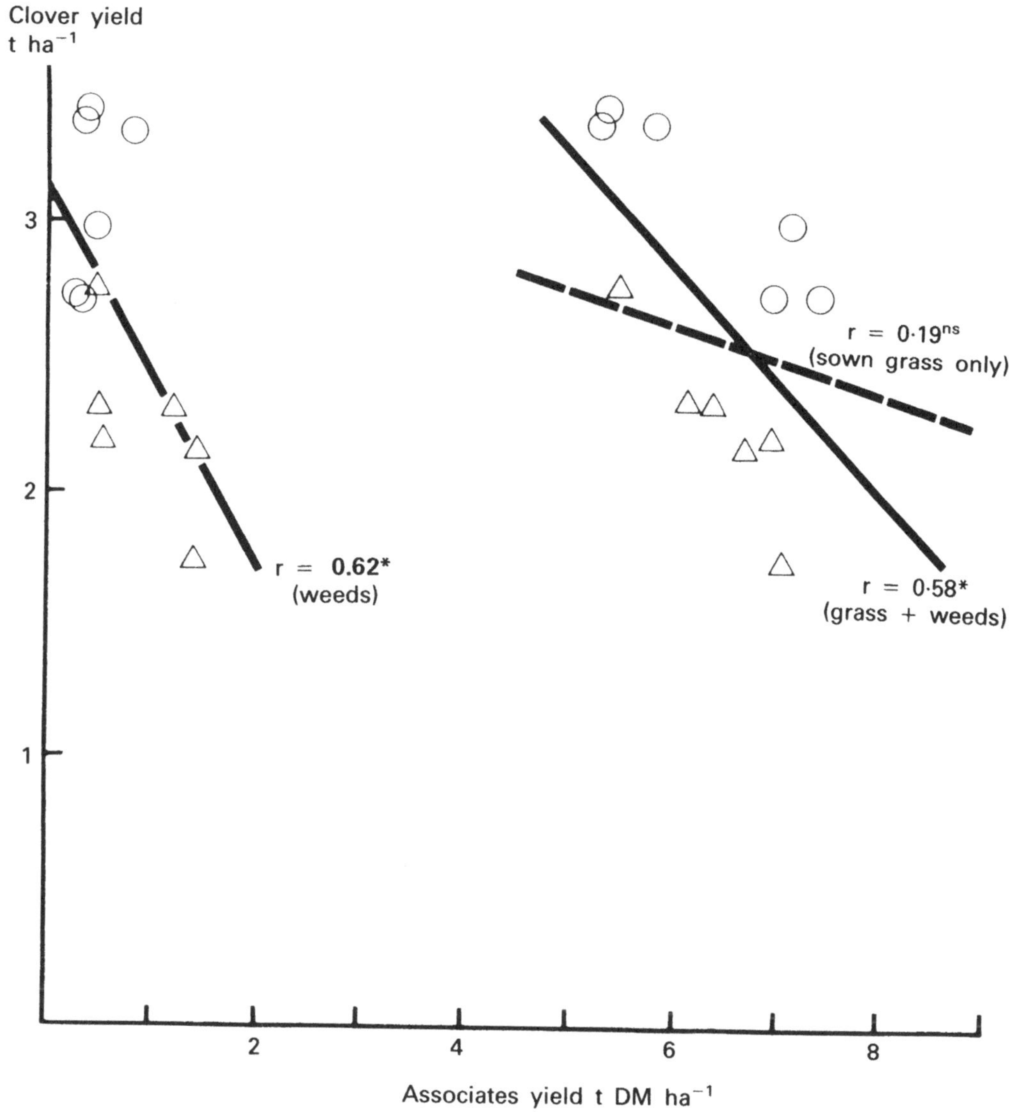

Fig. 7.19 Relationship between clover yield in mixture and the yield of associate sown ryegrass and cocksfoot and weed yield. Data from Heddle and Herriot (1954.)
○ Ryegrass △ Cocksfoot

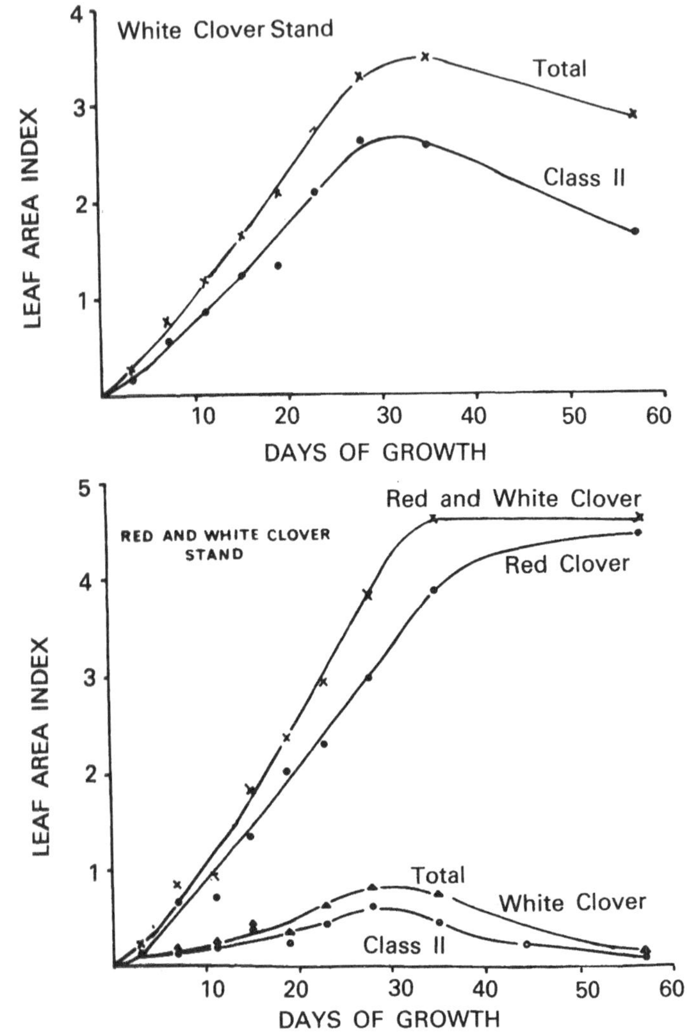

Fig. 7.20 Leaf area index of red and white clover in pure white clover and a mixed white clover-red clover stand at different stages of regrowth. Class II represents fully expanded photosynthetically active white clover leaves. (From Brougham (1965) reproduced with permission from *NZ Journal of Agr. Res.*).

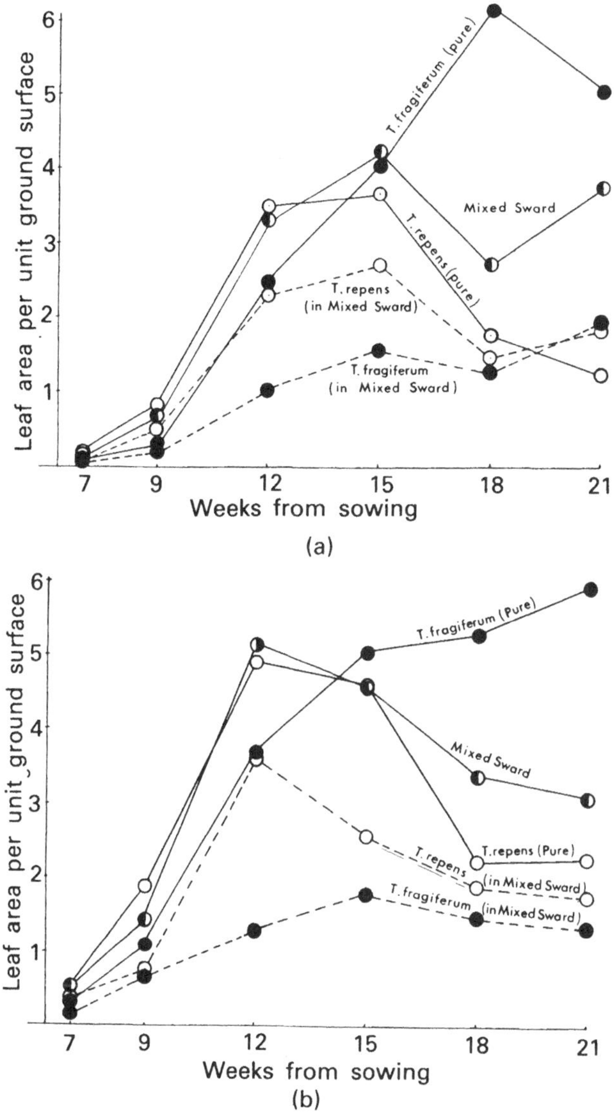

Fig. 7.21 The development of leaf area in pure and mixed swards of *Trifolium repens* and *T. fragiferum* sown at densities of (a) 32 plants per 900 m^2 and (b) 64 plants per 900 cm^2.
○——○ T. repens pure sward ●——● T. fragiferum pure sward
○------○ T. repens mixed sward ●------● T. fragiferum mixed sward
◐——◐ Total mixed sward
(From Harper & Clatworthy (1963) reproduced with permission from *Journal of Experimental Botany*).

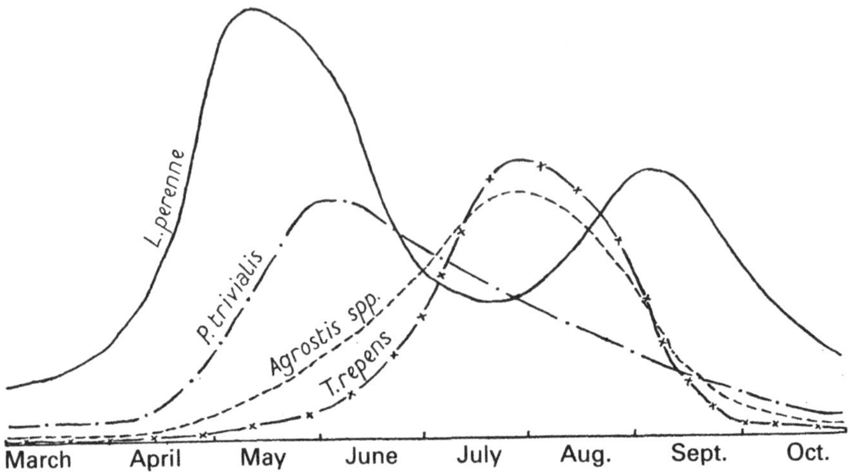

Fig. 7.22 Seasonal productivity curves of *Lolium perenne, Poa trivalis, Agrostis* spp. and *Trifolium repens* in Britain. (From Blackman (1933) reproduced with permission from *Empire Journal of Exp. Agr.*).

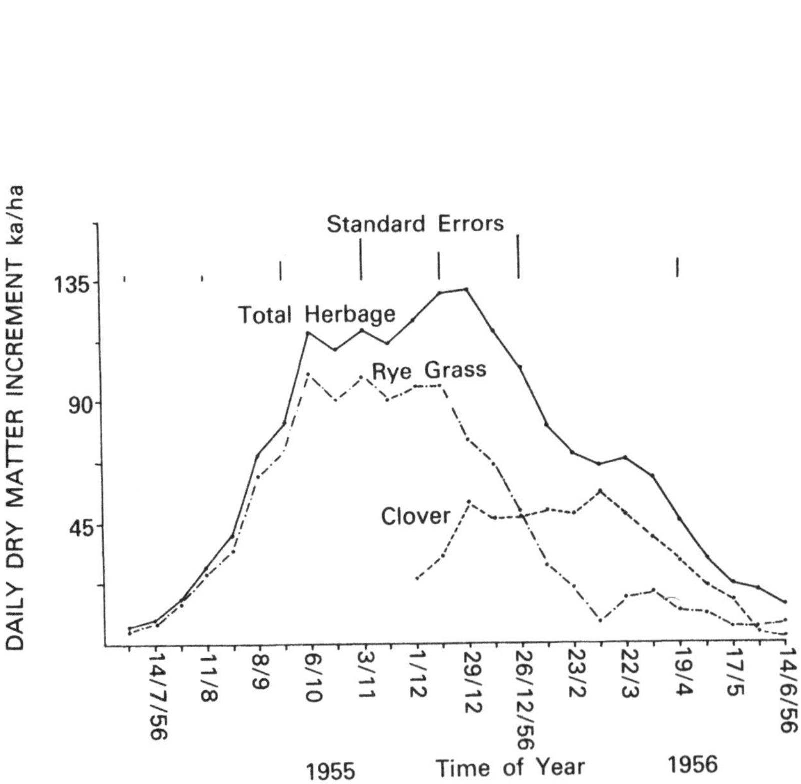

Fig. 7.23 Growth curves of ryegrass, clover and total herbage for an irrigated pasture at Palmerston North, New Zealand. (From Brougham (1959a) reproduced with permission from *NZ Journal of Agr. Res.*).

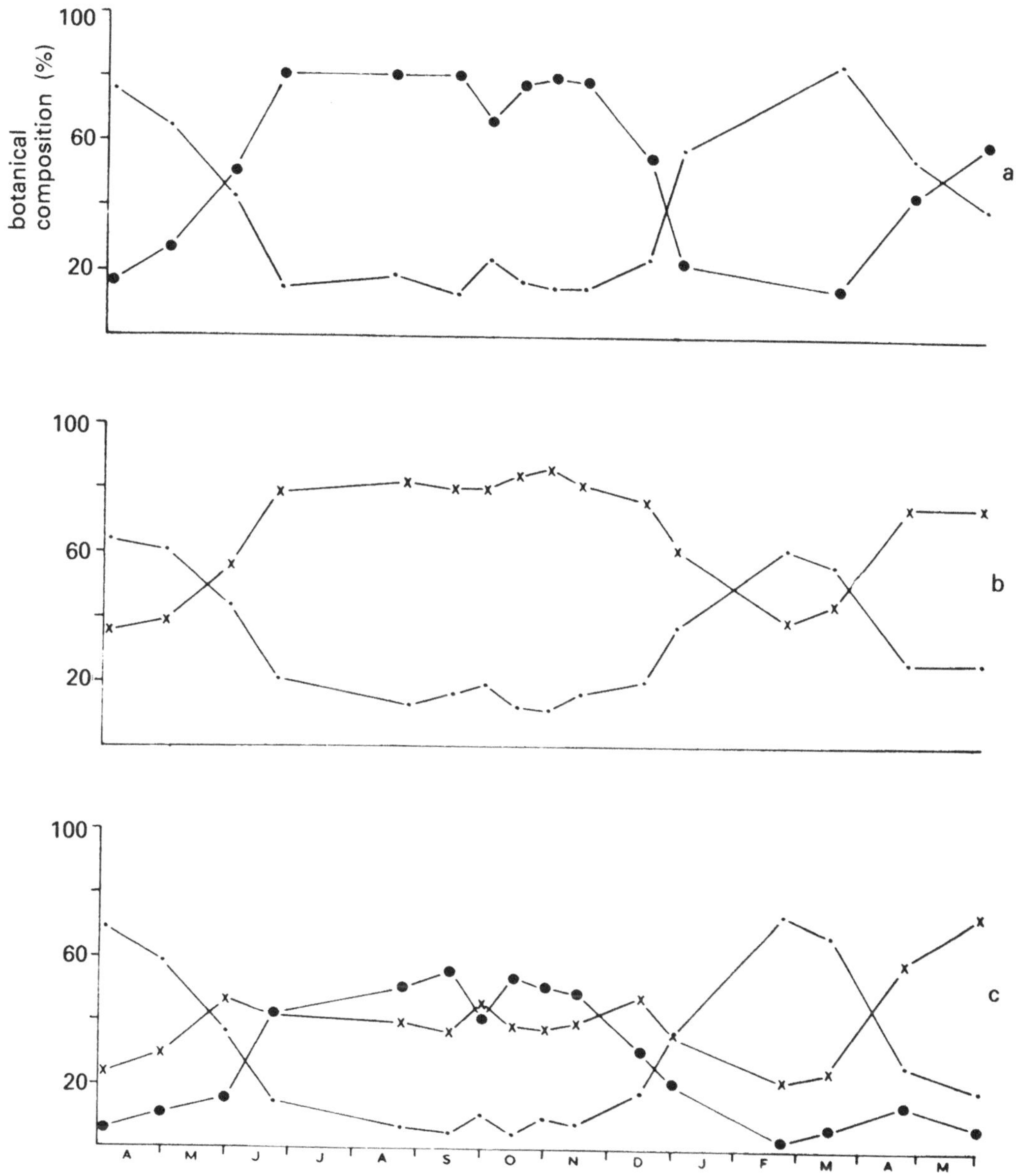

Fig. 7.24 Seasonal changes of botanical composition of mixed ryegrass, *Poa trivialis* and white clover pasture under a long cutting regime.
(From Vartha (1972) reproduced with permission from *NZ Journal of Agr. Res.*).

a. *Poa trivialis*—white clover
b. white clover
c. *Poa trivialis*—ryegrass—white clover

P. trivialis ●——● Ryegrass X——X White Clover ·——·

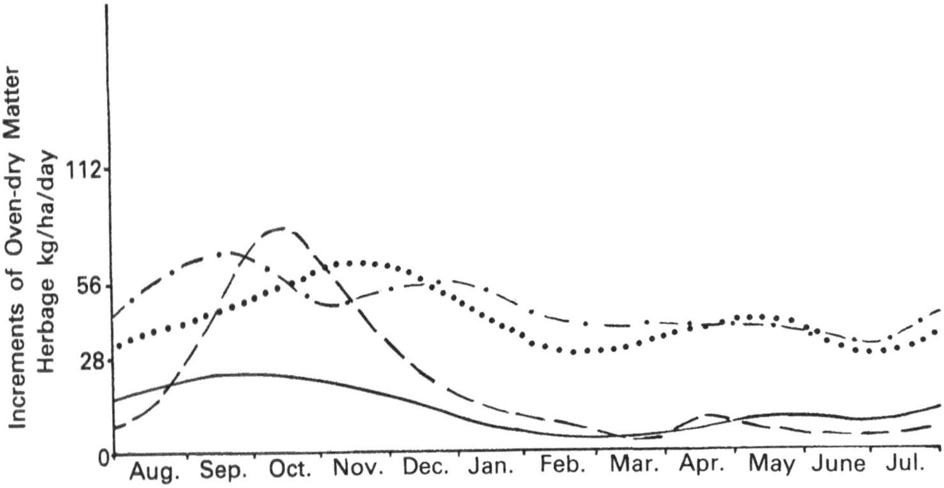

Fig. 7.25 White clover seasonal herbage dry matter production rates in association with different grasses in Queensland, Australia. (From Cameron (1969) reproduced with permission from *Queensland Journal of Agr. & Animal Science*).
...... Clover alone; ------ with *Paspalum dilatatum*
—·—·— with *Bromus unioloides*, —— with *Phalaris tuberosa* x *arundinacea*

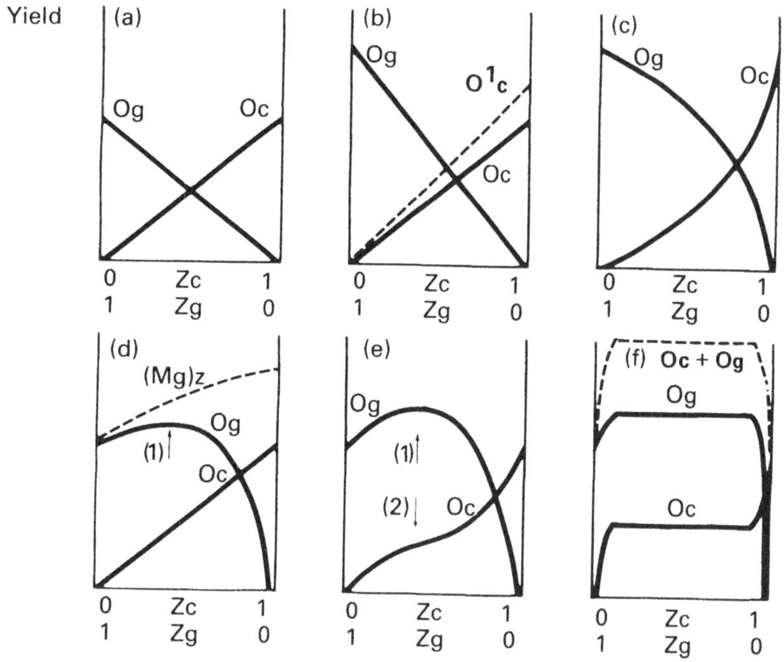

Fig. 7.26 Schematic representation of changes of the competitive interaction between the components of a developing grass and clover sward using replacement series diagrams and the theory of de Wit (1960) and Bakhuis and Kleter (1965). Symbols explained in text.

(a) No competition or N-fixation
$k_{gc} = 1$, $k_{cg} = 1$
$k_{gc} \times k_{cg} = 1$
$M_g = M_c$

(b) No competition or N-fixation. Greater response by grass to available mineral N. O_c^1 = N response by clover.
$k_{gc} = 1$, $k_{cg} = 1$
$k_{gc} \times k_{cg} = 1$
$M_g > M_c$

(c) Competition, no N-fixation
$k_{gc} > k_{cg}$
$k_{gc} \times k_{cg} = 1$

(d) Transfer of N to grass (1). No suppression of clover.
$k_{gc} > 1$, increasing with Zc
$k_{cg} = 1$
$k_{gc} \times k_{cg} > 1$

(e) Transfer of N to grass (1). Adjustment of clover to increased competition from grass (2)
$k_{gc} > 1$, increasing with Z_c
k_{cg} decreasing with Z_c

(f) Equilibrium composition of white clover—grass mixture not fertilised with N.

Adaptive variation

W. M. Williams
* Victorian Crops Institute, Horsham, Victoria 3400, Australia

Contents

* Former Address: Grasslands Division, DSIR, Palmerston North, New Zealand.

1. Distribution

The geographic distribution of *Trifolium repens* is shown in Figure 1. Although a temperate species, found where soil moisture is adequate for growth, white clover is widely adapted to regions from the Arctic to the subtropics and has a wide altitudinal range, being reported from up to 6000 m in the Himalaya Range of India. The likely centre of origin is the Mediterranean region (Vavilov 1951), and the indigenous area includes the whole of Europe and central Asia west of Lake Baikal as well as areas in north Africa (Morocco, Tunisia). It has become naturalised in China, Mongolia, Korea and Japan. Daday (1958) indicated that the first sample reaching Japan was carried by a Dutch vessel which arrived at Nagasaki in 1846. It has also been introduced to southern and east Africa where it is found in cool mountain regions and along water courses.

Early settlers introduced white clover into North America where it has spread over most of the USA and well north into Canada and Alaska. In Central and South America it is restricted to temperate areas, i.e. high altitudes (Guatemala, Costa Rica, Venezuela, Colombia, Peru, Bolivia, Brazil) but becomes common in the more temperate south (Uruguay, Argentina, Chile). In Hawaii, white clover occurs in areas of moderate to high (625-5000 mm) rainfall at 150-2500 m altitude and with mean annual temperatures of 10°-23°C (Britten 1955). Early settlers introduced the species also to Australia and New Zealand. In Australia it is restricted largely to the humid coastal regions but is found in more arid regions wherever water is available. In New Zealand white clover has become the most important pasture legume; being well adapted to the well distributed rainfall and temperate climate, it is found virtually everywhere that the native vegetation has been disturbed.

2. Genetic diversity within white clover populations

As an outcrossing species, white clover is genetically variable. Indeed, white clover populations in permanent grassland are characterised by remarkable genetic diversity within very small areas. Cahn & Harper (1976a) found an average of 4.16 different plants with unique white leaf markings in each 10 × 10 cm square of an old lowland pasture in North Wales. The plant population consisted of many plants in intimate mixture. (The genetic basis of variation in white leaf markings is given in Chapter 10).

Variation among plants in a single field has been studied by Harper (1977) and Burdon (1980a). The latter study showed that in a random sample of 50 plants from one 60-year-old pasture in North Wales, nearly every plant was found to differ significantly from every other plant in at least one character. Significant variation in the following 22 characters was recorded among the 50 plants:

(1) Total dry weight
(2) Shoot dry weight
(3) Root dry weight
(4) Root:shoot ratio
(5) Stolon length
(6) Internode length
(7) Petiole length
(8) Leaf size
(9) Rate of stolon extension
(10) White leaf mark
(11) Red leaf mark
(12) Number of flower heads
(13) Average life of flower head
(14) Average number of florets per head
(15) Peduncle length
(16) Date of appearance of first flower-head
(17) Date of onset of flowering
(18) Date of appearance of last flower-head

(19) Cyanogenic glucoside
(20) Cyanogenic enzyme
(21) *Cyanodothea* resistance
(22) *Pseudopeziza* resistance

Additionally, Burdon & Harper (1980) demonstrated significant variation for relative growth rates among individuals of the same population. Some characters not studied here but also known to vary in populations were: persistence, stolon diameter and stolon number (Knight 1953), self-incompatibility (Corkill 1971), competitive ability (Charles 1968, Turkington 1975), interactions with *Rhizobium trifolii* (Mytton 1975) and resistance to other pests and diseases.

In a population that is at least 60 years old, one or a few superior plants might have been expected to have become dominant. How, then, has such diversity been maintained?

Considerable genetic diversity would be expected if the population were invaded continually by new seedlings arising from interpollinations with other populations. However, diversity is found in white clover populations in the absence of seedling invasion (Cahn & Harper 1976a). Similarly, variation at the micro-scale would be expected on the basis of the demonstration (Snaydon 1962a) that soil contents of various nutrients can alter by several-fold over very short distances (approximately 60 cm). However, genetic diversity is observed even in the absence of such soil variation (Burdon 1980a).

Charles (1968) observed that under sheep grazing, white clover plants without white leaf markings survived better than marked plants. On the other hand, Cahn & Harper (1976b) demonstrated a clear preference by sheep for unmarked leaves where such leaves were more common than marked leaves. It was suggested that genetic diversity for leaf markings might be maintained by preferential selection by grazing sheep of leaves with the most common marking.

There is evidence that certain white clover genotypes are co-adapted to the grass species (perhaps genotype) with which they have been associated. Turkington & Harper (1979c) sampled white clover plants from the same edaphically uniform pasture as that studied by Burdon (1980a). Plants were collected from clumps dominated by each of the four grasses: *Lolium perenne, Holcus lanatus, Cynosurus cristatus* and *Agrostis tenuis*. Ramets of each clover genotype were planted back into areas of pasture dominated respectively by each of the four grasses. Clovers tended to perform better when grown with the grass species from which they were originally collected.

Further evidence for co-adaptation was obtained from a similar glasshouse study in which a standard potting mix was used. Each clover type grew best with the grass with which it was associated when collected, except for that from *H. lanatus,* which also grew well with *A. tenuis* (Figure 2). Such co-adaptation indicates that the variety of associated species in a plant community is itself a source of disruptive selection favouring genetic polymorphism (Harper 1977).

Any given white clover population can be expected to be nodulated by a number of different rhizobium strains. A sample of soil may contain a number of diverse strains of *Rhizobium* bacteria, and various white clover genotypes differ in their abilities to grow with and to select different *Rhizobium* strains. Mytton (1975), for example, grew white clover plants in a John Innes potting mix watered with a suspension of soil from local Aberystwyth fields. He then isolated rhizobia from each of several plants and re-inoculated cuttings of these plants with each rhizobium strain in a factorial series. Plants tended to yield more when nodulated with rhizobium strains from the same or a related plant than with strains from unrelated plants. Thus host plants tended to select from mixed populations those strains with which they formed effective symbioses (Robinson 1969, Jones & Russell 1971, Masterson & Sherwood 1974). Exceptions to this pattern indicated that sometimes the best available strain was not selected and that plant-strain affinities were partly independent of effectiveness. Mytton & de Felice 1975) indicated that genotypes differ in their abilities to select various rhizobial strains out of mixtures (Chapter 10)

Evidence that some fungal diseases may provide a strong selective force in white clover populations was given by Burdon (1980b); selective forces active in maintaining polymorphism at the cyanogenic glucoside and enzyme loci are reviewed in Section 5.

Burdon (1980a) proposed that high genetic diversity in the sampled pasture in North Wales was maintained by strong 'local' selection , i.e. selection exerted on a micro-scale within the field by biotic factors such as identity, age, and size of neighbouring plants, sheep grazing, etc., as discussed above. This 'local' selection occurred against a background of 'general' factors common to the whole area, i.e. physical factors such as rainfall, frost days and soil nutrient availability, which determine whether the species could grow in a particular area. The 'local' environment may change radically from place to place in the field and also from time to time. The result is a complex array of local directional selection forces which, in an otherwise uniform environment, are likely to dominate the immediate fitness of individuals and produce selection on a microscale.

To be successful, a plant must possess not only a suitable set of 'general' fitness characters but also a set of 'local' fitness characters, *and* be lucky enough to establish in a suitable local environment. The nature of the local' environment is complex and ever-changing, so that plants that may be well adapted to one particular location at one time may be less well adapted at a later time or in a different location in the field. The high variability of the biotically-induced 'local' environment exerts a strong disruptive selection force on the clover population and thus maintains a high degree of variability. The stoloniferous nature of white clover may move the plant into contact with a set of factors to which it is less well adapted. Here and there it will be supplanted by other plants better adapted to the local conditions. Thus, by interaction of biotic factors at a 'local' (micro-scale) level, a high degree of diversity of clover types can be maintained even in an apparently physically uniform environment.

3. Adaptation of white clover to different environmental factors

Environmental factors to which there has been evidence of natural selection include:

(a) cultural factors: grazing pressure
(b) edaphic factors: soil fertility and pH
(c) climatic factors: winter cold
summer heat
drought

(A) Grazing pressure

It is well established that white clover populations adapted to intense grazing by sheep tend to be prostrate, small leaved and dense, while those adapted to cattle grazing or hay management systems are more upright, larger leaved and often more open in habit. Experimental evidence for this comes from several sources. An old established pasture in Hawke's Bay, New Zealand was found by Levy & Gorman (1934) to have been divided into two portions, one grazed by cattle and one by sheep. White clover plants transplanted from the cattle grazed pasture were large leaved, tall and highly productive, while those transplanted from the sheep pasture a few metres away were smaller leaved, more prostrate and slightly less productive. As the two populations were of common origin, the morphological differences appeared to have resulted from selection in response to grazing by different types of animal. Parallel results were obtained by Kemp (1937) in the USA and by Knoll & Hermelink (1941) in Germany.

Davies (1962) sowed an area of Welsh hill country at an altitude of 300 m with a mixture of 2 lb of a medium leaved, fairly tall-growing variety (N.Z. certified mother seed) and 0.5 lb of a small leaved, prostrate, wild type variety (S184), along with five different ryegrasses. Samples of the populations were taken after 10 years of hard grazing by sheep, grown as spaced plants at the Welsh Plant Breeding Station and compared with stocks of the original varieties. Results indicated that there had been a general shift toward smaller leaved plants; larger leaved types had been eliminated, only 2% having leaves longer than 25 mm compared with 10% in the original S184 and 47% in the New Zealand variety. There was a corresponding increase of plants with leaflets less than 15 mm long. The possibility that the size differences were entirely the result of loss of vigour in vegetative cuttings was eliminated, as changes in leaf marking frequencies were also recorded.

Larger leaved plants may be under greater selection pressure from grazing animals than smaller leaved plants. Cahn & Harper (1976b) showed that rumen samples from sheep which had grazed pasture contained an excess of larger and a deficiency of smaller leaves. This effect is probably a function of leaf stature (petiole length). Because smaller leaved plants tend to have highly branched stolons with many leaves and roots, they may also be more tolerant of continuous defoliation than large leaved forms.

Adaptation to selective grazing by small animals (rabbits, slugs, snails, etc.) may also have been significant in the evolution of white clover populations. This topic is discussed in relation to the ecology of cyanogenesis (Section 5.).

(B) Edaphic factors

Early attempts to identify specific adapations to soil factors were unsuccessful (e.g. McNeur 1953, 1954). However, the existence of genotypes with different tolerances of soil factors was indicated by Harberd (1963) who determined that the distribution of white clover clones over uneven ground was apparently related to soil characteristics.

In setting out to define the edaphic factors limiting the distribution of white clover, Snaydon (1962a) surveyed soils in the Welsh hills in areas where the species appeared to be near the limits of its edaphic tolerance. Results indicated that white clover has a wide edaphic tolerance, but its microdistribution was associated apparently with variations in edaphic factors over very short distances. Two survey methods were used and, although differing in some aspects, both indicated that levels of calcium, phosphorous and pH influence the microdistribution of white clover in soils marginal for the species. Since levels of these three soil factors tended to be correlated, it was not clear which of them, if any, was the most important variable. Analyses indicated that over very short distances (average 60 cm) soil contents of calcium and phosphate differed by as much as 3-fold, potassium by as much as 2- to 3-fold and pH by up to 0.7 units. Presence of white clover was usually associated with higher levels of soil pH, calcium and, to a lesser extent, phosphorus, but the edaphic tolerance of the species was extremely wide. Although potassium levels differed widely, no overall difference was found between samples with and without white clover.

The wide edaphic tolerance of white clover may be due to individual physiological plasticity or to genetic differentiation of edaphic ecotypes each adapted to a limited range of soil conditions, or a combination of both factors. Research results indicate that edaphic ecotyes, at least in part, contribute to the wide edaphic range of the species. The possible existence of edaphic ecotypes was indicated by Robinson (1942), but the first substantial evidence was produced by Snaydon (1962b). Plants sampled from natural populations from contrasting acid and calcareous soil types were grown in pots containing either their native soil type or the contrasting type. Each population grew relatively better on its original soil type. This effect was magnified if competing mixtures of the two populations were grown in each pot.

Soil phosphate level is apparently associated with differentiation of edaphic ecotypes in white clover. Snaydon & Bradshaw (1962a) obtained a high correlation ($r = +0.96$) between responses of various collected populations to phosphate and the amount of phosphate in the native soil of each population. Growth of ecotypes from high phosphate soils declined more at low phosphate levels than did growth of these from soils low in phosphate. Populations more tolerant of low phosphate levels contained higher concentrations of P in shoots and took up more P per unit weight of root than did less tolerant populations. It appeared therefore that the populations more tolerant of low phosphate were more efficient in extracting phosphate from the soil rather than being better able to metabolise at a lower internal P concentration.

Phosphate is not the only edaphic variable in response to which white clover has differentiated ecotypes. Snaydon & Bradshaw (1969) reported that populations collected from acid and calcareous soils differed in response to calcium, magnesium and potassium. Variation among varieties of white clover in ability to accumulate calcium in the foliage was detected by Vose & Koontz (1959) and discussed by Vose (1960). Growth differences were not closely related to foliar calcium levels: S100 appeared to have a low calcium requirement and uptake, but Dutch and Ladino had high requirements and uptakes. This result agreed with agronomic observations that cv. S100 performed well on hill soils of relatively low calcium content while the other varieties were better adapted to high fertility lowland soils.

The agricultural significance of edaphic ecotypes was demonstrated by Snaydon & Bradshaw (1962b). Five natural populations from widely different soils were inserted into a *Festuca/Agrostis* dominant sward on a heavily grazed acid upland soil low in calcium and phosphorous. Plants from acid soils grew and persisted better than those from lowland calcareous soils, or high-altitude medium fertility soils. After 2 years, only plants from acid soils remained, reduced to 19-25% of the original number. This situation was the reverse of that shown by the same populations grown as spaced plants on a high-fertility lowland soil. Results for plants inserted into the grass sward also contrasted with the results of a trial comparing the same populations as spaced plants at another low-fertility upland site at which lowland and upland populations performed more or less equally well. The spaced-plant trial differed from the sward in lack of competition from native grasses and absence of hard grazing. The importance of competition for magnifying ecotypic differences has been emphasised already (Snaydon 1962b) and in the present case is strengthened by observations that *Agrostis tenuis* and *Festuca ovina* are both strong competitors for phosphate (Mouat & Walker 1959, Bradshaw et al. 1960). Addition of fertiliser improved the performance of the lowland populations in the hill sward (e.g. Figure 3), indicating that adaptation to fertility factors was involved (Bradshaw 1969).

The effect of grazing could have been a reason for differences in ecotype performances, as suggested by King (1963) in a study of 20 populations of white clover from the Scottish hills. A survey showed that populations from higher fertility soils (as indicated by base saturation levels) were significantly smaller leaved than those from lower fertility soils. The latter were, in turn, markedly smaller leaved, more prostrate, less spreading, and earlier flowering than the Kentish-based variety S184. The author suggested that grazing pressures were probably greater on the high fertility than the low fertility hills, resulting in natural selection for smaller-leaved, more prostrate plants at high fertility levels.

Results were consistent with observations indicating that reseeding of upland soils with Kentish and other lowland types frequently failed, while merely changing to a more suitable grazing management, with little addition of fertiliser, often increased the proportion of clover, probably as a result of the slow build-up of plants well adapted to acid upland conditions. If so, there is potential for breeding of plants well adapted for productivity and persistence under low fertility hill conditions.

(C) Climatic factors

Within the indigenous area of the species a number of clear patterns of variation have been found, most relating to climatic factors. Between the Mediterranean and northern Europe, several gradations (clines) of characterisitics occur which can be related to temperature. The most striking of these relates to geographic variation (from south to north) in winter hardiness and winter dormancy, Mediterranean populations usually being frost tender and winter active and northern populations frost hardy and winter dormant. The northern extremes of this south-north cline are represented by populations from Northern Scandinavia, particularly Finland, and northern USSR. To survive winter, plants become dormant and, at the same time, small leaved, prostrate and unproductive. Within Sweden a distinction has been made between natural populations from the north and south, southern material being more luxuriant but less hardy than Norrland types. Thus, southern populations grown in the north may yield well in the first year but later will be outyielded by local material with better winter hardiness (Akerberg & Wiklund 1948).

Experimental evidence for natural selection for winter hardiness has been provided by Swedish work (Nilsson-Leissner 1935, Sylven 1937). One German and two Danish varieties (Morsø and Strynø) were grown at Svalöf in southern Sweden for 2 years before being harvested for seed. Harvested material of two of the varieties was found to yield considerably more than the original varieties, apparently in response to an exceptionally cold winter, survived only by hardier types than the average of the original varieties. After only one generation of being grown for seed in Sweden, Morsø and the German variety gave respective increases in yield of green matter of 9.8% and 33.4% at the first cut and 48.3% and 96.8% in the second cut, compared with the original stocks of these varieties. Strynø was little affected, apparently because it had been fairly highly selected for uniformity (Sylven 1937). Seed production of a second generation at Svalöf further altered Morsø and the German variety but not Strynø (Nilsson-Leissner 1935). Although the intervention of other environmental factors cannot be ruled out, winter hardiness was probably the main character under natural selection. Such results point to the dangers of increasing seed of a bred variety outside its region of adaption.

At the other extreme of the European temperature cline are Mediterranean populations (Spain, Israel, Italy, Morocco, Portugal, etc.) which are often winter-active and may show summer dormancy, presumably in response to high temperatures and drought. Such material has shown considerable promise for cool-season growth in temperate climates. In parts of New Zealand, for example a Spanish population has been shown to have outstanding autumn-winter growth compared with naturalised New Zealand populations (Barclay 1960, Williams & Barclay 1975). In the United Kingdom, Mediterranean populations are autumn-winter active but cold tender in comparison with the winter dormant indigenous materials (Davies & Young 1967). Some Mediterranean populations also have superior heat and drought tolerance, as has been demonstrated in Australia for materials from Israel (Haifa), Lebanon and Portugal (Neal-Smith 1970) and in New Zealand for materials from Algeria, Morocco and Italy (Lodi) as well as from the countries mentioned previously (Williams et al. 1978).

Plants occurring in northern latitudes but from areas with a strong maritime influence, e.g. coastal British Isles, have exhibited frost tolerance associated with an ability to grow at low temperatures (Ollerenshaw et al. 1977, Ollerenshaw & Baker 1981). Such populations are important for plant improvement in regions of middle latitude which have cold winter nights and mild days. New Zealand white clover, which has developed in a maritime climate, also exhibits frost tolerance associated with reasonable winter growth (Barclay 1960).

In parallel with latitudinal adaptive clines in Europe in response to temperature, we would also expect altitudinal clines, and indeed it has been found that populations from high altitudes are relatively uniformly prostrate, small leaved, frost hardy and winter dormant. Reference is made here to the plant habit because accompanying the temperature adaptations is a conspicuous morphological pattern. Mediterranean populations tend to be large leaved, tall (long petioles), and thick stemmed, but northern European populations are prostrate, small leaved and thin stemmed (Daday 1954a). Closer investigation reveals that Mediterranean populations are extremely variable, small leaved types existing under certain conditions (notably when hard grazing predominates and at high altitudes). Such populations were found by Davies & Young (1967) in Sicily, Greece, Turkey and Portugal. Large leaved forms were probably from irrigated areas, e.g. along water courses. Variability falls off as we move northwards, so that natural populations from northern Europe and even the UK tend to be more uniformly small leaved and thin stemmed. Nevertheless, clinal variation of stolon growth in relation to winter temperature was demonstrated by Ollerenshaw & Baker (1981) within a collection of ecotypes from northern Britain (Figure 4). Ecotypes collected from cold sites (high altitude) generally showed little stolon growth at 5°C compared with those from warmer sites.

The apparent concentration of variation in the Mediterranean region is consistent with Vavilov's (1951) concept of the centre of origin of the species. Apparently, the Mediterranean region offers a diversity of suitable habitats for white clover, whereas northern Europe (and high altitude) is marginal for the species and adaptation to certain environmental factors, particularly low winter temperatures, dominates the genetic structure of the populations. White clover populations adapted to adverse conditions of several kinds, e.g. hard grazing, low fertility and cold winters, are generally small leaved and thin stemmed, probably because these characteristics are associated directly with short internodes. Short internodes, in turn, generally are related to multiple branching and many nodes per unit area of land. Since each node is a potential meristem for stem or inflorescence growth as well as root growth, the more nodes there are, the better a plant is able to survive death of part of the plant from any cause.

Accompanying the vegetative pattern of variation is a pattern in flower size, Mediterranean populations generally having large flowers with longer corolla tubes than northern populations. Since northern European bees probably have shorter tongues than Mediterranean species, an important co-adaptation between plant and insect populations may exist (Daday 1954a).

Altitudinal ecotypes of white clover show adaptation to the prevailing temperatures, to the length of the growing season, and to the partial presence of CO_2 in their natural habitats (Maechler & Nösberger 1977,

1978, Maechler et al. 1977). These tend to parallel the short-term adapatations shown by plants at low temperatures. Thus alpine ecotypes and plants grown at 10°/7°C show reduced increases in photosynthetic rate and percentage of activity in C_6-sugars with increasing temperature, and higher photorespiration rates. Adaptation of ecotypes to temperature conditions seems to include a compensation of the dependence on temperature of photosynthesis and a shift of optimal temperature.

A number of European workers have noted the existence of forms with dark green foliage. The incidence of these types appears to increase from south to north, Mediterranean populations often having pale green leaves. Early German studies (Knoll & Hermelink 1941, Boekholt 1933) have indicated that plants with superior persistence and winter hardiness were consistently darker green than inferior types and tended to have no V-marking. Analyses indicated that the darker green plants contained higher protein concentrations. However, no recent studies of this variation are known to the author.

Two other clinal patterns of variation noted by Daday (1954a) were those of the the white V-markings on the leaves and cyanogenesis in the leaves. Nearly every plant from Mediterranean populations was found to have an intense white V-marking, while northern European populations were generally less intensely marked, with fewer types of marking, and often none at all. The nature of the selective forces involved in generation of such clines is unknown, although evidence has been produced for selection of certain leaf marks by grazing sheep (Section 2.). Although these observations do not explain directly the maintenance of geographic clines, they do indicate that selective grazing may be involved in the distributions of various leaf markings in white clover.

Adaptive patterns of variation in cyanogenesis have been widely studied and are discussed later in this chapter.

4. The adaptive significance of flowering date and abundance

In white clover, ecotypes with low intrinsic persistence flower profusely, e.g. Dutch white, while very persistent wild types are often sparse flowering. This inverse relationship with persistence is understandable in view of the physiological relationship between stolon branching and flower initiation, each inflorescence formed blocking the production of a branch stolon (Chapter 3, Section 1.). The trend parallels that seen in other species, e.g. *Lolium perenne,* in which ecotypes adapted to continuous sheep grazing flower sparsely and have high inductive requirements for flowering (Hayward & Breese 1966, Hayward 1967).

In many species flowering time is a highly adaptive character, late flowering being favoured at high altitude and latitude to avoid early spring cold, and early flowering favoured at low altitude and latitude in summer-dry situations to avoid summer stress. However, although Mediterranean material is slightly earlier flowering than northern European material, this trend is not always conspicuous in white clover. Davies & Young (1967) reported the absence of a relationship between flowering time and geographic origin in a range of materials, with a 43-day difference between extremes.

However, although differences in time of appearance of flowers are not always related clearly to geographic source, times and mechanisms of *initiation* of flowers are geographically based. Thus Thomas (1979) has indicated that genotypes obtained from low latitudes initiate inflorescences in early autumn in New Zealand, but genotypes from high altitudes initiate later in winter according to latitude of origin. Furthermore, plants from high latitudes are usually long-day responsive, whereas those from low latitudes tend to lack the photoperiod response. Therefore, Mediterranean plants, for example, stop initiating inflorescences in spring but northern European material continues initiating into summer (see Chapter 3, Section 2.(D) for further details).

Large differences in time of flower appearance do not follow large differences in time of flower initiation, because rate of flower appearance in white clover depends on rate of vegetative growth. Thus inflorescences initiated in early winter do not appear until spring because of slow growth in winter. Those initiated in late winter also appear in spring, because rapid growth in early spring brings them forward quickly.

A further probable reason for the absence of a conspicuous pattern of adaptation in flowering times in white clover is the strong capacity of the species to reproduce asexually. A marked tendency for profuse branching in ecotypes adapted to stress conditions has already been emphasised. Turkington & Cavers (1978) observed that the allocation of dry matter to sexual reproduction in white clover was much lower than in other legume species.

5. Cyanogenesis as an adaptive feature

The foliage of many white clover plants releases hydrogen cyanide (HCN) when damaged. Such plants are said to be cyanogenic. HCN is released because the tissue contains glucosides that contain cyanide (cyanogenic glucosides), and a hydrolytic enzyme (a β-glucosidase) which hydrolyses the cyanogenic glucosides to liberate HCN. White clover contains two cyanogenic glucosides, lotaustralin and its lower homologue linamarin (Melville & Doak 1940). The β-glucosidase in white clover is linamarase (Coop 1940). The genetics of cyanogensis is summarised in Chapter 10, Section 3.

Variation in cyanogenesis and the role of selective forces in maintaining variation in white clover populations have been studied widely. Cyanogenesis is regarded generally as a protection against predatory herbivores, especially insect pests, slugs and snails. If this were the only selection pressure, all populations of white clover would be cyanogenic. However, many populations are acyanogenic or mixed. Studies, reviewed below, have indicated that acyanogenic forms may be favoured under certain climatic or edaphic conditions.

(A) Geographic variation in cyanogenesis

Using picrate paper tests Daday (1954a) surveyed variation in cyanogenic potential of the white clover populations of Europe, and studied the distributions of the glucoside (*Ac*) and enzyme (*Li*) genes on a geographic basis. The distributions of the two genes were found to be correlated with mid-winter (January) average temperatures. In the Mediterranean region some populations contained only cyanogenic (*Ac Li*) individuals and some populations from north-east Europe contained no cyanogenic individuals. In between there was a continuous gradual decrease in frequency of the dominant genes in populations further removed from the Mediterranean. On average, a decrease of 1 °F in January mean temperature was correlated with a reduction of 4.23% in the frequency of the dominant glucoside allele and of 3.16% in the frequency of the dominant enzyme allele. There were no correlations of gene frequencies with rainfall, annual isotherms or mid-summer (July) isotherms. There was a strong positive relationship between the frequencies of the dominant glucoside and enzyme alleles, although the glucoside gene frequency decreased significantly faster than the enzyme gene frequency with fall in January temperature. Quantitative variation in cyanogenic potential paralleled the qualitative differences, i.e. the more cyanogenic plants in the population, the more HCN product per leaf. Picrate scores for cyanogenic plants of north European origin were generally much lower than those of Mediterranean plants.

Parallel results were obtained from a survey of populations from various altitudes in the central European alps (Daday 1954b). Gene frequencies were again found to be correlated with mean January temperatures, the dominant glucoside allele decreasing in frequency from 0.62 to 0.0 and the enzyme allele from 0.57 to 0.02 with increasing altitude. It was again noted that amounts of enzyme and glucoside per leaf were correlated with dominant gene frequencies in the populations. In particular, the activity of linamarase varied quantitatively from sample to sample, many plants with very low activity being sampled. De Araujo (1976) and Brighton & Horne (1977) confirmed that the frequency of cyanogenic plants decreased at high altitudes.

Distributions of the cyanogenic genes outside Europe were described by Daday (1958). In North America and Japan there was a trend for the frequencies of the glucoside gene to decrease from south to north in parallel with the European pattern, but in the Southern Hemisphere (South America, South

Africa, Australia, and New Zealand) gene frequencies had an irregular pattern. A south-north trend was observed also by Portz & Jackobs (1955) in Ladino seed lots harvested in different parts of the USA.

(B) The ecology of cyanogenesis

Daday (1954a, b, 1958) concluded that the pattern of distribution of the cyanogenesis genes was influenced by January temperatures, natural selection favouring acyanogenic plants at cool temperatures. It appeared that natural selection already had produced balanced equilibria in European populations and selection was still in progress in North America and Japan. Absence of similar distribution patterns from the Southern Hemisphere was consistent with this hypothesis, since cline formation in the Northern Hemisphere was most evident where January isotherms were below 40°F[a]. Apart from alpine regions, areas with such low winter temperatures are uncommon in the Southern Hemisphere. Additionally, white clover has been introduced to most parts of the Southern Hemisphere quite recently, and there may have been insufficient time and too much gene flow for patterns to become evident.

Selection against cyanogenic individuals at low temperatures was demonstrated experimentally by Daday (1965). In a field study, acyanogenic individuals were less injured by frost than cyanogenic plants. In a controlled environment study, flowering of acyanogenic individuals was superior at low temperatures (15°/10°C). Calculated fitness values indicated rapid reductions in frequencies of the *Ac* and *Li* genes at low temperatures. Brighton & Horne (1977) similarly showed that cyanogenic plants of *Lotus corniculatus* were damaged by freezing, but acyanogenic plants survived without damage.

The disadvantage of cyanogenic potential at low temperatures was attributed by Daday (1965) to a direct effect of HCN on tissues in which linamarase was activated by low temperatures. Foulds & Young (1977) unsuccessfully attempted to identify the nature of this effect.

A direct effect of HCN on small herbivores feeding on white clover should favour cyanogenic plants. This effect is conspicuous in rabbits and slugs in plant breeding plots in New Zealand (Corkill 1952), and Davies & Young (1967) have reported similar observations from the Welsh Plant Breeding Station. Attempts to confirm these observations under experimental conditions have, however, proved to be surprisingly difficult. The role of cyanogenesis in plant protection has been supported strongly in recent years by D.A. Jones and, because he has reviewed the topic in great detail (Jones 1972, 1973), only a brief account will be given here. Laboratory evidence for selective eating of acyanogenic forms by slugs and snails has been given by Crawford-Sidebotham (1972) who found, however, that the phenomenon was less conspicuous in *T. repens* than in *L. corniculatus*. He also showed that species of mollusc differed in the degree to which they showed discrimination in eating, some species showing none at all. Whitman (1973) obtained qualitive measures (0-3 scale) of herbivore feeding damage for 1000 *T. repens* plants growing in unmanaged grassland in New York State, USA. An association between low feeding damage and cyanogenesis was indicated, although its statistical significance was marginal. Evidence of visible insect feeding damage was found by Dritschilo et al. (1979) to be independent of cyanogenesis. However, aphids significantly favoured acyanogenic white clover. Bishop & Korn (1969) had found no difference in the quantities of cyanogenic and acyanogenic clover eaten by the snail *Helix aspera* or the slug *Agriolimax reticulatus* under glasshouse conditions. Nor was there any difference in growth rate of *H. aspera* fed exclusively on cyanogenic clover. A field study conducted by Miller et al. (1975) at four locations in the USA also failed to detect any selective elimination of acyanogenic seedlings by insects, field mice or rats.

No direct role of cyanogenesis in disease resistance of white clover has been demonstrated.

The relationship between cyanogenesis and resistance to the pepper spot fungus *Stemphylium sarciniforme* was investigated by Wilkinson & Millar (1978, 1979). This pathogen is sensitive to HCN, yet is pathogenic on cyanogenic plants. It was found that although symptom development was faster and

a. For the *Ac* allele in Europe and Asia all populations sampled from areas with a 50°F or higher isotherm were 100% *Ac*. For areas below the 50°F isotherm the gene frequency fell from 100% to 5% at the 30°F isotherm. Cooler areas had 5% or lower frequencies. In North America, dominant gene frequencies below 30% prevailed below the 0°F isotherm.

pathogen development retarded in glucoside-positive plants, tissue death was not related to cyanogenic potential. The fungus was found to carry a β-glucosidase distinct from the plant enzyme, but this appeared to be mycelial in distribution and its role in tissue necrosis was uncertain.

Selection maintains polymorphism for cyanogenesis in white clover populations. However, there is evidence that much of this selection may act indirectly, rather than directly on cyanogenic potential *per se* (Daday 1958, Morley 1959). Daday (1965) reported superior flowering of *Ac Li, Ac li,* and *ac Li* individuals in warm temperatures (30°/25°C) and superior vegetative growth of cyanogenic plants under mild conditions at a coastal site in Australia. The selective advantage of *Ac Li* at high temperatures could perhaps be explained by the protection that HCN affords against predators.

Daday's (1965) controlled environment study indicated selective advantages of *Ac* and *Li* in the absence of predators and also appeared to indicate advantages of either gene in the absence of the other (*Ac li* and *ac Li*, for example, both being at an advantage over *ac li* at high temperatures). Several populations surveyed in Europe showed either one of the genes *Ac* or *Li* at high frequency and the other at low frequency. In Europe the regression coefficients of *Ac* and *Li* frequencies on January temperatures were significantly different, as were the calculated selection responses for *Ac* and *Li* derived from controlled environment observations. It therefore appears that natural selection may be able to act on *Ac* independently of *Li*, and vice versa. Daday (1965) has therefore suggested that in addition to probable direct selection against HCN (by cold temperature), evolution has also linked polygenes enhancing fitness at high or low temperatures with the dominant and recessive alleles, respectively.

Linkage of cyanogenic loci with genes concerned with response to environment was invoked also by Foulds & Grime (1972a, b) to explain results of survey and experimental work on soil moisture supply. In a wild Staffordshire population, experimentally induced severe drought caused *Ac* phenotypes to die three times more frequently than *ac* phenotypes. Surveys indicated that the *Ac* allele occurred at a reduced frequency in droughted habitats in Derbyshire. These effects were apparently independent of whether the *Li* allele was present or absent. Even in moist situtations *ac* phenotypes from the Staffordshire population were higher yielding than *Ac* plants, suggesting possible linkage between *Ac* and genes affecting vegetative vigour.

Associations of either the *Ac* or *Li* dominant alleles with different morphological factors provide scope for selection of either glucoside or enzyme independently. Ennos (1981a, b) found that individuals possessing the *Li* allele had larger leaves than those lacking the enzyme, and that the *Li* phenotype was also associated with total yield in competition experiments. Therefore, the presence or absence of *Li* or other genes in tight linkage disequilibrium had effects not associated directly with cyanogenesis. Dommée et al. (1980) found that plants carrying *Li* grew larger roots than *li* plants. The *Ac* locus showed no association with root growth.

Further evidence for the association of cyanogenesis with different vegetative phenotypes comes from early New Zealand work which found that cyanogenic white clover populations were larger leaved, more productive, and more persistent than acyanogenic populations (Doak 1933, Foy & Hyde 1937). On the other hand, an American study of diverse germplasm showed cyanogenic plants to be larger in diameter but shorter, earlier flowering, and less persistent than acyanogenic plants (Paim & Dean 1976).

The ecological factors affecting distribution of the genes *Ac* and *Li* appear to be complex. The cyanogenic phenotype is apparently selected against at freezing temperatures, but the cyanogenic phenotype could possibly be at a selective advantage where herbivorous predators are likely to wipe out seedling populations. Such predators would be more numerous in warm than in cool climates (Morley 1959). It is also apparent that selection acts not only on the cyanogenic and acyanogenic phenotypes themselves but also independently on *Ac* and *Li* or the chromosome segments carrying them.

How large are the cumulative selection coefficients relating to *Ac* and *Li* (Morley 1959)? If the selection coefficients are large, plant breeders would be advised to include the genes in cultivars bred for warm climates and exclude them from those intended for cooler climates. Estimates of selection coefficients obtained so far have probably been reduced by other factors such as migration and fluctuations imposed by microclimatic variation. The identification of quite strong selection under these circumstances probably indicates that the chromosome segments linked to cyanogenic loci should be considered carefully by plant breeders selecting for widely different environments.

6. Major ecotypes of white clover

Much of the evidence that natural selection results in ecological differentiation of white clover populations has been summarised. The European clines can be considered to have resulted from such selection. Superimposed on the clinal variation patterns, however, is a discontinuous distribution of ecotypes which have evolved as adaptations to local conditions.

(A) Ladino clover (Italy)

Italy has a wide range of variation in its white clover populations, which is consistent with the hypothesis that the Mediterranean is a centre of diversity of the species. The most important ecotype of white clover from Italy is Ladino gigante Lodigiano. This form, most commonly known as Ladino clover or Lodi clover, originated in the Lodi-Cremona district near Milan in the Po River Valley. It is very large in all its organs, including the roots, but is consequently rather open in habit and intolerant of frequent grazing. Better adapted to a hay-type management system, Ladino clover stands are extremely productive in the first year, but tend to thin out later.

Ladino clover was introduced into the USA in 1891 and increased in popularity until by 1950 it was grown in every State except North Dakota (Ahlgren & Fuelleman 1950). This wide distribution reflects its climatic tolerance, its spread being limited only by extremes of cold, drought and heat.

The area of origin of Ladino clover in northern Italy, although in the Mediterranean region, does not have a typical Mediterranean climate. The Po River Valley has a relatively cold winter, mean January temperatures being around 0°C. Thus Ladino clover usually contains very low concentrations of cyanogenic glucosides but populations from the rest of the Mediterranean region are highly cyanogenic (Daday 1954a). Its source of origin probably also explains the wide temperature tolerance of this ecotype; it is hardy enough to withstand the severe winters of the northern USA and Canada, though it also grows well in the western USA under irrigation.

Ladino clover is not well adapted to sustained drought. In south eastern USA it often dies in summer, failing to reseed. However, its large roots and leaves appear to improve its ability to recover faster from temporary drought than small leaved forms (Lovvorn 1948).

Certification of Ladino clover (Lodigiano) in Italy is controlled by the Stazione di Praticoltura di Lodi. However, it has been grown in other countries for so long that natural selection and probably introgression from local material has formed many local substrains. In Quebec, Canada, for example, Dessereaux (1949) reported that the local strain of Ladino was a superior form, and in Australia, Ladino certified has been developed (Barnard 1972). In the USA several distinct varieties have been selected, including Pilgrim and Merit (adapted to the north-eastern States), and Regal and Tillman (adapted to the southeast).

(B) Morsø and Strynø (Denmark)

Two Danish islands were the regions of origin of these ecotypes which have had considerable impact on white clover breeding in Europe. By far the more important of these was Morsø, a wild white clover from the island of Mors in the Limfjord north of Jutland. It came from excellent old permanent pastures, was persistent and somewhat coarser than English wild white. It now forms the basis of many bred varieties of northern Europe.

The other form was Strynø, named from the island of its origin. This was a 'cultivated' white clover used in short leys on the southernmost islands of Denmark. Rather homogeneous in type, it was large leaved and lax in habit.

The great success of Morsø appears to have been its combination of leafiness with considerable winter hardiness. It therefore proved superior to many of the natural ecotypes of the USSR, Norway and Sweden, which are adapted to winter cold and have shown great winter hardiness, but have little potential for productivity. Morsø appears to have met with particular success in Norway and Sweden following seed production in those countries, apparently as a result of genetic shift toward hardier types in the population (e.g. Nilsson-Leissner 1934). In areas where winter hardiness is not a great advantage, e.g. Australia and New Zealand, the winter dormancy of Morsø has been a disability and it has not performed well.

(C) Dutch white clover (Western Europe)

This is the name given by British farmers early this century to all non-wild 'strains' or cultivated forms of white clover (also known as 'ordinary' white clover). The term is now little used, although the type is still characteristic of some European cultivars. Its origin is unknown but its name implies that it was first cultivated in Holland. At the beginning of the 17th century agriculture was more advanced in Holland than in Britain and probably both seed and technology were imported from Holland in about 1700 (Ware 1925b). Early writings indicate that it was short-lived, the first experiment that demonstrated its inferiority in this respect to wild white clover being carried out by the Manchester and Liverpool Agricultural Society in 1886.

This form was characterised by a very lax, open habit, medium-sized leaves and profuse early heading. Most lines were very low or negative for cyanogenic glucosides.

Early this century, seed lines from England, Czechoslovakia, Poland, Russia, Holland, Belgium, Denmark, the USA and New Zealand were predominantly of this form. In most places, however, modern bred varieties have replaced it and, in New Zealand, the Dutch type is now extremely rare.

(D) English wild white clover

This prostrate, small-leaved, dense form of white clover became the centre of much attention in the later 19th century when it was realised that it was extremely hardy and long-lived. Its small leaves, thin stems and short petioles became recognised as deceptively hiding a great bulk of herbage (Ware 1925a). By the beginning of this century it was recognised as an ecological adaptation to very old, continuously grazed sheep pastures in the United Kingdom (especially Kent and the Cotswolds). A certification scheme was started in 1930 for inspection and registration of fields of Kentish wild white. Two grades of certificate were given: Grade A from fields more than 10 years old, and Grade B from fields more recently sown with Grade A.

Comparisons of leaf sizes from Grade A and Grade B seed samples indicated that Grade B gave significantly more large-leaved plants than Grade A, and emphasised the effects of only one generation of seed increase. Uncertified English wild white was also significantly larger in leaf and more variable than certified lines (Hawkins 1960).

The main properties of Kentish wild white have been reviewed by Haggar & Holmes (1963). Its productivity is highest under close grazing where it also shows great persistence. It may also be more tolerant of low fertility soils than larger-leaved clovers. The Aberystwyth variety S184 is a direct selection from Kentish wild white.

An extreme form of wild white was identified by Bradshaw & Pritchard (1955) from Cwm Llefrith in North Wales. This was characterised by small plants, extreme persistence and adaptation to high altitude grazing and climate.

(E) New Zealand ecotypes

White clover was introduced into New Zealand in the mid 19th century by the early settlers. By the 1920s a great deal of variation among populations had developed. The existence of clear-cut categories (strains) rapidly became apparent when New Zealand collections were analysed (Davies & Levy 1931). There were six major types:

Type 1: New Zealand wild white Number 1

A relatively large-leaved, stout stoloned, spreading type with high productivity, dense growth and good persistence. Seasonal spread of growth was relatively even, with substantially more winter growth than ordinary forms. It was ideally adapted to first-class dairy pastures or high fertility soils. It flowered abundantly in the second year, very much later (4 to 5 weeks) than ordinary imported white clover and slightly earlier than Kentish wild white.

Type 2: New Zealand wild white Number 2

A vigorous type, but smaller leaved and more slender stemmed and denser than Number 1 and slightly less productive, especially in winter. It appeared to be better suited than Number 1 to poorer soils and to closely grazed pastures. Flowering behaviour was similar to that of Number 1.

Type 3: Ordinary New Zealand white

A fairly lax type with medium-sized leaves, quite productive in its first summer but lacking winter growth and producing little in its second season. It did not persist more than 2 years and flowered profusely early in the first season. This type tended towards the typical imported Dutch type but appeared to be more persistent, perhaps as a result of crossing with Types 1 and 2. It was found mainly on better soils.

Type 4: Lax early-flowering New Zealand and ordinary European white (also New Zealand 'Dutch')

This type behaved almost as an annual, producing growth in the first spring before going rapidly to seed. Extremely lax in habit and non-productive, it dominated imported seed. It was found in the arable parts of Canterbury where little summer rain in some years was ideal for an annual type.

Type 5: Small-leaved Kentish types

Type 6: Large-leaved Ladino types

Davies & Levy (1931) indicated that the New Zealand Number 1 or Number 2 types were confined largely to old pastures on fertile soils where they appeared to be ecotypic adaptations. They were most prevalent in pastures on the Heretaunga Plains of Hawke's Bay, but were also found in the Sefton-Rangiora district of Canterbury. As already noted (Section 3) Gorman (1934, 1936) showed that the Number 1 type dominated cattle pastures, while the Number 2 type was most prevalent in sheep pastures on the Heretaunga Plains.

After recognition of the strains, attempts were made to improve them by plant breeding. Initially, breeding consisted of mass selection from the best ecotypes. In New Zealand this selection was initiated by Levy & Gorman (1933, 1934). Later selections by pedigree methods culminated in the cultivar Grasslands Huia (Barclay 1960, Corkill 1964).

The New Zealand Number 1 ecotype and selections from it (New Zealand Mother White, Grasslands Huia) show broad adaptation and have been successful in many parts of the world (Gorman 1936, Poole 1946). In the 1966-67 seed year over 80% of certified white clover seed used in Britain was imported from New Zealand (Davies 1970).

The origin of New Zealand Number 1 white clover is unkown. Corkill (1952) suggested that its morphology and high cyanogenic score indicated that it could not have evolved by natural selection from the Dutch or Kentish strains which were the dominant ones introduced to New Zealand in the early days. However, its morphology and cyanogenic behaviour were consistent with a possible French origin.

(F) Australian ecotypes

Australia, like New Zealand, is outside the natural range of distribution of white clover and the species was introduced by the early settlers. As in New Zealand, ecotypic differentiation has occurred in the relatively short time since the introduction of the species.

'Victorian Irrigation' (cv. Irrigation, Barnard 1972) was described as a long-lived spreading type with stout stolons, long internodes and large leaves on long petioles (Drake 1942). It developed in Northern Victoria where it is particularly suited to conditions prevailing in irrigated pastures where the very hot, dry, late summer period would otherwise be too harsh for white clover.

The naturalised ecotype in the Clarence Valley on the sub-tropical northern coast of New South Wales is a free-seeding, early-flowering winter annual. It is found in moister, more fertile areas and is not seen every year. If adequate levels of molybdenum, phosphorus and sulphur are applied, the Clarence Valley ecotype behaves as a perennial. The heavy seed setting of this material is its principal adaptation to ensure regeneration following drought (O'Brien 1970a, b, O'Brien & Cohen 1972).

(G) Ecotypes of the USA

Also outside the indigenous range of white clover, North America has nevertheless developed widespread naturalised populations and accompanying ecotypic differentiation (Ahlgren & Sprague 1940). The early history of white clover introduction into the USA has been summarised by Carrier & Bort (1916). The first English settlers in Ohio and Kentucky, e.g. Boone in 1775, reported that white clover was already there, and it seems that it must have preceded the English into the Ohio Valley. The most reasonable explanation for this is that the French carried it down from Canada, probably around 1700. The limestone soils of the Ohio Valley which are liberally supplied with phosphates were ideal for the rapid spread of the species.

The most distinct ecotypic variation observed in the USA has been the differentiation between winter hardy north-central and north-eastern forms which resemble English wild white and the south-eastern ecotypes (e.g. Louisiana) which are taller, larger-leaved, and somewhat intermediate in type between Ladino and New Zealand Number 1 (Pieters & Hollowell 1937). The Louisiana ecotype has heat and drought tolerance, but lacks winter hardiness and is rather free-seeding. A selected synthetic variety, Louisiana S-1, is not as free-seeding as the unselected ecotype.

Natural populations from New Jersey analysed by Ahlgren & Sprague (1940) proved to be extremely variable, ranging in type from Ladino to wild white but, on the whole, being more or less intermediate. These results suggested that natural selection had not yet produced distinct morphological types in New Jersey. There appeared to be considerable scope for selection and plant breeding from these populations. Midgeley (1938) reported that in Vermont the natural white clover was more persistent than Dutch, English wild or Ladino clover on moist heavy soils. Brown (1939) compared performance under mowing of strains at Storrs, Connecticut, nine of overseas origin and nine collected in the USA. The most persistent were overseas strains from old pastures and the worst were from Denmark, Poland, and commercial Dutch white. US strains were intermediate and showed no evidence of strain differentiation. Dodd (1941) noted strain difference in populations of white clover in Ohio.

(H) The subtropics

Although white clover is not adapted to the subtropical zones, high-altitude regions within those zones have significant populations of introduced white clover. Typical regions are the highlands of Kenya (Bogdan 1965), high altitude Colombia (Crowder et al. 1961), high altitude Hawaii (Britten 1955), sub-tropical Australia (Section 6. (F)) and parts of India. Such areas differ from temperate regions in factors such as daylength, lack of defined seasons, large fluctuations between day and night temperatures and prolonged hot dry periods.

Special requirements for such areas therefore include high temperature and drought tolerance. In Kenya this requirement has been fulfilled to some extent by the variety Louisiana S-1. In Colombia, Crowder (1960) reported that a predominantly medium-leaved, free-flowering form occurs naturally at altitudes of 1500-3000 m. Louisiana S-1 was outstanding in early cuts, but tended to weaken, while Ladino types, slow to start, were later superior. In Hawaii white clover occurs in areas of moderate to high rainfall (600-5000 mm) at 160-2700 m altitude, and at mean annual temperatures of 10° to 25°C (Britten 1955). White clover probably has not been in some of these subtropical areas for long enough to allow ecotypic differentiation. Where differentiation does occur, it is likely to be toward the free flowering annual type.

References

Ahlgren, G.H.; Fuelleman, R.F. (1950) Ladino clover. *Advances in Agronomy* **2**, 207-232.

Ahlgren, G.H.; Sprague, H.B. (1940) A survey of variability in white clover (*Trifolium repens* L.) and its relation to pasture improvement. *New Jersey Agricultural Experiment Station Bulletin* **676**, 42pp.

Akerberg, E.; Wiklund, K. (1948) [Varieties and seed mixtures for pastures in Norrland. I. Results from pasture plant trials performed at the Norrland Branch Station of the Swedish Seed Association in the years 1932-46]. *Sveriges Utsädesförenings Tidskrift* **58**, 31-52 (English summary).

Barclay, P.C. (1960) Breeding for improved winter pasture production in New Zealand. *Proceedings of the 8th International Grassland Congress*, 326-330.

Barnard, C. (Ed) (1972) Register of Australian herbage plant cultivars. CSIRO, Canberra.

Bishop, J.A.; Korn, M.E. (1969) Natural selection and cyanogenesis in white clover, *Trifolium repens*. *Heredity* **24**, 423-430.

Boekholt, K. (1933) [The significance of leaf colour in the breeding of *Trifolium repens* and *T. hybridum*.] *Züchter* **5**, 157-159.

Bogdan, A.V. (1965) Cultivated varieties of tropical and subtropical herbage plants in Kenya. *East African Agricultural and Forestry Journal* **30**, 330-338.

Bradshaw, A.D. (1969) An ecologist's viewpoint. In: Ecological aspects of the mineral nutrition of plants. *British Ecological Society Symposium* **9**, 415-427.

Bradshaw, A.D.; Chadwick, M.J.; Jowett, D.; Lodge, R.W.; Snaydon, R.W. (1960) Experimental investigations into the mineral nutrition of several grass species. III. Phosphate level. *Journal of Ecology* **48**, 631-637.

Bradshaw, A.D.; Pritchard, T. (1955) Ecology of Snowdonia. Wild white clover in natural mountain grassland. *Nature, UK* **176**, 596.

Brighton F.; Horne, M.T. (1977) Influence of temperature on cyanogenic polymorphisms. *Nature, UK* **265**, 437-438.

Britten, E.J. (1955) White clover in Hawaii. *Proceedings of the Hawaii Academy of Sciences 1954-5*, 1 (Abstract).

Brown, B.A. (1939) Some factors affecting the prevalence of white clover in grassland. *Journal of the American Society of Agronomy* **31**, 322-332.

Burdon, J.J. (1980a) Intraspecific diversity in a natural population of *Trifolium repens*. *Journal of Ecology* **68**, 717-735.

Burdon, J.J. (1980b) Variation in disease-resistance within a population of *Trifolium repens*. *Journal of Ecology* **68**, 737-744.

Burdon, J.J.; Harper, J.L. (1980) Relative growth rates of individual members of a plant population. *Journal of Ecology* **68**, 953-957.

Cahn, M.G.; Harper, J.L. (1976a) The biology of the leaf mark polymorphism in *Trifolium repens* L. 1. Distribution of phenotypes at a local scale. *Heredity* **37**, 309-325.

Cahn, M.G.; Harper, J.L. (1976b) The biology of the leaf mark polymorphism in *Trifolium repens* L. 2. Evidence for the selection of leaf marks by rumen fistulated sheep. *Heredity* **37**, 327-333.

Carrier, L.; Bort, K.S. (1916) History of Kentucky bluegrass and white clover in the United States. *Journal of the American Society of Agronomy* **8**, 256-266.

Charles, A.H. (1968) Some selective effects operating on white and red clover swards. *Journal of the British Grassland Society* **23**, 20-25.

Coop, I.E. (1940) Cyanogenesis in white clover (*Trifolium repens* L.). III. A study of linamarase, the enzyme which hydrolyses lotaustralin. *New Zealand Journal of Science and Technology* **22B**, 71-83.

Corkill, L. (1952) Cyanogenesis in white clover (*Trifolium repens* L.). VI. Experiments with high-glucoside and glucoside-free strains. *New Zealand Journal of Science and Technology* **34A**, 1-16.

Corkill, L. (1964) Maori names given to pasture plants bred by Grasslands Division. *New Zealand Journal of Agriculture* **109**, 138.

Corkill, L. (1971) Leaf markings in white clover. *Journal of Heredity* **62**, 307-310.

Crawford-Sidebotham, T.J. (1972) The role of slugs and snails in the maintenance of the cyanogenesis polymorphisms of *Lotus corniculatus* and *Trifolium repens*. *Heredity* 28, 405-411.

Crowder, L.V. (1960) The response of white clover varieties grown at high elevations in Colombia. *Agronomy Journal* **52**, 608-609.

Crowder, L.V.; Ramirez, R.; Chaverra, H. (1961) Performance of temperate zone pasture species within tropical boundaries as represented by Colombia. *Agronomy Journal* **53**, 229-232.

Daday, H. (1954a) Gene frequencies in wild populations of *Trifolium repens* L. I. Distribution by latitude. *Heredity* **8**, 61-78.

Daday, H. (1954b) Gene frequencies in wild populations of *Trifolium repens* L. II Distribution by altitude. *Heredity* 8, 377-384.

Daday, H. (1958) Gene frequencies in wild populations of *Trifolium repens* L. III. World distribution. *Heredity* **12**, 169-184.

Daday, H. (1965) Gene frequencies in wild populations of *Trifolium repens* L. IV. Mechanism of natural selection. *Heredity* **20**, 355-365.

Davies, W.E. (1962) The persistency of sown white clover in a long ley at an upland centre at an altitude of 1 000 ft. *Welsh Plant Breeding Station*, Report for 1961, 45-47.

Davies, W.E. (1970) White clover breeding: review. In: White clover research. *British Grassland Society Occasional Symposium*, **No. 6**, 99-122.

Davies, W.; Levy, E.B. (1931) Strain investigation of grasses and clover. 4. White clover (*Trifolium repens*). *New Zealand Journal of Agriculture* **42**, 75-89, 151-164.

Davies, W.E.; Young, N.R. (1967) The characteristics of European, Mediterranean, and other populations of white clover (*Trifolium repens* L.). *Euphytica* **16**, 330-340.

De Araujo, Aldo, M. (1976) The relationship between altitude and cyanogenesis in white clover (*Trifolium repens* L.). *Heredity* **37**, 291-293.

Dessereaux, L. (1949) Variabilite de populations non selectionnees de trefle blanc Ladino. *Agriculture, Quebec* **6**, 127-131.

Doak, B.W. (1933) A chemical method for the determination of type in white clover. *New Zealand Journal of Science and Technology* **14**, 359-365.

Dodd, D.R. (1941) Some factors affecting the content, fluctuation, and distribution of white clover in permanent seed areas in Ohio. *Proceedings of the Soil Science Society of America* **6**, 288-297.

Dommée, B.; Brakefield, P.M.; MacNair, M. (1980) Differential root growth of the cyanogenic phenotypes of *Trifolium repens* L. *Acta Oecologia, Oecologia Plantarum* **1**, 367-370.

Drake, F.R. (1942) Pasture plant research at Burnley. Report for 1941. *Journal of the Department of Agriculture, Victoria* **40**, 329-336.

Dritschilo, W.; Krummel, J.; Nafus, D.; Pimental, D. (1979) Herbivorous insects colonising cyanogenic and acyanogenic *Trifolium repens. Heredity* **42**, 49-56.

Ennos, R.A. (1981a) Manifold effects of the cyanogenic loci in white clover. *Heredity* **46**, 127-132.

Ennos, R.A. (1981b) Detection of selection in populations of white clover (*Trifolium repens* L.). *Biological Journal of the Linnean Society* **15**, 75-82.

Foulds, W.; Grime, J.P. (1972a) The response of cyanogenic and acyanogenic phenotypes of *Trifolium repens* to soil moisture supply. *Heredity* **28**, 181-187.

Foulds, W.; Grime, J.P. (1972b) The influence of soil moisture on the frequency of cyanogenic plants in populations of *Trifolium repens* and *Lotus corniculatus. Heredity* **28**, 143-146.

Foulds, W.; Young, L. (1977) Effect of frosting, moisture stress and potassium cyanide on the metabolism of cyanogenic and acyanogenic phenotypes of *Lotus corniculatus* L. and *Trifolium repens* L. *Heredity* **38**, 19-24.

Foy, N.R.; Hyde, E.O.C. (1937) Investigation of the reliability of the picric-acid test for distinguishing strains of white clover in New Zealand. *New Zealand Journal of Agriculture* **55**, 219-224.

Gorman, L.W. (1934) Some observations on strains in clovers. *Proceedings of the 3rd Annual Conference of the New Zealand Grassland Association*, 5pp.

Gorman, L.W. (1936) Behaviour of New Zealand certified clovers in New Zealand and abroad. *New Zealand Journal of Agriculture* **53**, 76-85.

Haggar, R.J.; Holmes W. (1963) Kent wild white clover. Ashford Kent, 40 pp.

Harberd, D.J. (1963) Observations on natural clones of *Trifolium repens* L. *New Phytologist* **62**, 198-204.

Harper, J.L. (1977) Population Biology of Plants. London, UK; Academic Press, 892 pp.

Hawkins, R.P. (1960) Investigation on local strains (varieties) of herbage plants. IV. White clover. *Journal of the British Grassland Society* **15**, 28-33.

Hayward, M.D. (1967) The genetic organisation of natural populations of *Lolium perenne*. II. Infloresence production. *Heredity* **22**, 105-116.

Hayward, M.D.; Breese, E.L. (1966) The genetic organisation of natural populations of *Lolium perenne*. I. Seed and seedling characters. *Heredity* **21**, 287-304.

Jones, D.A. (1972) Cyanogenic glycosides and their function. In: Phytochemical Ecology. *Annual Proceedings of the Phytochemical Society, No. 8, 1971*, Academic Press, J.B. Harborne (Ed.), 103-124.

Jones, D.A. (1973) Coevolution and cyanogenesis. In: Taxonomy and ecology. V.H. Heywood (Ed.) Systematics Association Special Volume No. 5; London, UK; Academic Press, 213-242.

Jones, D.E.; Russell, P.E. (1971) The application of immunoflorescence techniques to host plant/nodule bacteria selectivity experiments using *Trifolium repens. Soil Biology and Biochemistry* **4**, 277-282.

Kemp, W.B. (1937) Natural selection within plant species as exemplified in a permanent pasture. *Journal of Heredity* **28**, 329-333.

King, J. (1963) Ecotypic differentiation in *Trifolium repens. Plant and Soil* **18**, 221-224.

Knight, W.E. (1953) Interrelationships of some morphological and physiological characteristics of ladino clover. *Agronomy Journal* **45**, 197-199.

Knoll, J.G.; Hermelink, L. (1941) Beiträge zur Züchtung des Weissklees (vorläufige Mitteilung). *Züchter* **13**, 25-33.

Levy, E.B.; Gorman, L.W. (1933) Strain investigation of white clover. Finalised results of trials at Plant Research Station. *New Zealand Journal of Agriculture* **46**, 78-89.

Levy, E.B.; Gorman, L.W. (1934) Strain investigation in white clover. *New Zealand Journal of Agriculture* **49**, 86-104.

Lovvorn, R.L. (1948) Ladino clover. Italian gift to North Carolina pastures. *Better Crops with Plant Food* **32**, 9-10, and 48-49.

Maechler F.; Nösberger, J. (1977) Effect of light intensity and temperature on apparent photosynthesis of altitudinal ecotypes of *Trifolium repens* L. *Oecologia* (Berlin) **31**, 73-78.

Maechler, F.; Nösberger, J. (1978) The adaptation to temperature of photorespiration and of the photosynthetic carbon metabolism of altitudinal ecotypes of *Trifolium repens* L. *Oecologia* (Berlin) **35**, 267-276.

Maechler, F.; Nösberger, J.; Erismann, K.H. (1977). Photosynthetic $^{14}CO_2$ fixation products in altitudinal ecotypes of *Trifolium repens* L. with different temperature requirements. *Oecologia* (Berlin) **31**, 79-84.

Masterson, C.L.; Sherwood, M. T. (1974) Selection of *Rhizobium trifolii* strains by white and subterranean clover. *Irish Journal of Agricultural Research* **13**, 91-99.

McNeur, A.J. (1953) Effect of soil acidity on growth of pasture plants. I. Pot experiments with grasses and clovers on acidified soils. *New Zealand Journal of Science and Technology* **35A**, 311-326.

McNeur, A. J. (1954) Effect of soil acidity on growth of pasture plants. II. A field trial with white clover (*Trifolium repens*). *New Zealand Journal of Science and Technology* **36A, 167-176.**

Melville, J.; Doak, B.W. (1940) Cyanogenesis in white clover (*Trifolium repens* L.). II. Isolation of the glucosidal constituents. *New Zealand Journal of Science and Technology* **22B, 67-71.**

Midgley, A.R. (1938) Four pasture clovers. *Bulletin, Vermont Agricultural Experiment Station* **No.431**, 16pp.

Miller, J.D.; Gibson, P.B.; Cope, W.A.; Knight, W.E. (1975) Herbivore feeding on cyanogenic and acyanogenic white clover seedlings. *Crop Science* **15**, 90-91.

Morley, F.H.W. (1959) Natural selection and variation in plants. *Cold Spring Harbor Symposia in Quantitative Biology* **24**, 47-56.

Mouat, M.C.H.; Walker, T.W. (1959) Competition for nutrients between grasses and white clover. I. Effect of grass species and nitrogen supply. II. Effect of root cation-exchange capacity and rate of emergence of associated species. *Plant and Soil* **11**, 30-52.

Mytton, L.R. (1975) Plant genotype × rhizobium strain interactions in white clover. *Annals of Applied Biology* **80**, 103-107.

Mytton, L.R.; de Felice, J. (1975) Clover-rhizobium relationships. Welsh Plant Breeding Station Report for 1974, 26.

Neal-Smith, C.A. (1970) Investigation in south-eastern Australia. *Report of the 27th Southern Pasture and Forage Crop Improvement Conference, Athens, Georgia*, 28-31.

Nilsson-Leissner, G. (1934) [White clover strains for seed production in Scania.] (Swedish). *Svensk Frötidning* **3**, 122-124.

Nilsson-Leissner, G. (1935) [Pasture plants and their breeding.] Address delivered at the Annual Meeting of the Swedish Seed Association at Torsta, Tang.] *Sveriges Utsädesförenings Tidskrift* **45**, 274-285.

O'Brien, A.D. (1970a) Evaluating white clover varieties for the Clarence River beef country. *Tropical Grasslands* **4**, 63-69.

O'Brien, A.D. (1970b) White clover (*Trifolium repens* L.) in a subtropical environment on the east coast of Australia. *Proceedings of the 11th International Grassland Congress*, 165-168.

O'Brien, A.D.; Cohen, R.D.H. (1972) Naturalised clover on the north coast. *Agricultural Gazette of New South Wales* **83**, 86.

Ollerenshaw, J.H.; Baker, R.H. (1981) Low temperature growth in a controlled environment of *Trifolium repens* plants from northern latitudes. *Journal of Applied Ecology* **18**, 229-239.

Ollerenshaw, J.H.; Stewart, S.W.; Baker, R.H. (1977) Intra-specific variation within *Trifolium repens* L. in relation to the early spring and late autumn growth of upland and marginal pastures. *International Grasland-kongress, Seckt. Vorträge, Leipzig* **13(1/2)**, 241-250.

Paim, N.R.; Dean, C.E. (1976) Evaluation of cyanogenic and acyanogenic clones and progenies of *Trifolium repens* L. *Agronomy Abstracts* 59.

Pieters A.J.; Hollowell, E.A. (1937) Clover improvement. *Yearbook, United States Department of Agriculture 1937*, 1190-1214.

Poole, A.L. (1946) Trials of New Zealand pasture seeds in Britain, 1941 and 1942 trials. *New Zealand Journal of Science and Technology* **28A**, 145-160.

Portz, H.L.; Jackobs, J.A. (1955)Differences between seed lots of ladino clover in cyanophoric properties. *Agronomy Journal* **47**, 143-145.

Robinson, A.C. (1969) Host selection for effective *Rhizobium trifolii* by red clover and subterranean clover in the field. *Australian Journal of Agricultural Research* **20**, 1053-1060.

Robinson, R.R. (1942) The mineral content of various clones of white clover when grown in different soils. *Journal of the American Society of Agronomy* **34**, 933-939.

Snaydon, R.W. (1962a) Microdistribution of *Trifolium repens* L. and its relation to soil factors. *Journal of Ecology* **50**, 133-143.

Snaydon, R.W. (1962b) The growth and competitive ability of contrasting populations of *Trifolium repens* L. on calcareous and acid soils. *Journal of Ecology* **50**, 439-447.

Snaydon, R.W.; Bradshaw, A.D. (1962a) Differences between natural populations of *Trifolium repens* L. in response to mineral nutrients. I. Phosphate. *Journal of Experimental Botany* **13**, 422-434.

Snaydon, R.W.; Bradshaw, A.D. (1962b) The performance and survival of contrasting natural populations of white clover when planted into an upland *Festuca/Agrostis* sward. *Journal of the British Grassland Society* **17**, 113-118.

Snaydon, R.W.; Bradshaw, A.D. (1969) Differences between natural populations of *Trifolium repens* L. in response to mineral nutrients. II. Calcium, magnesium, and potassium. *Journal of Applied Ecology* **6**, 185-202.

Sylven, N. (1937) The influence of climatic conditions on type composition. *Herbage Bulletin, Imperial Bureau of Plant Genetics*, **No. 21**, 8 pp.

Thomas, R.G. (1979 The effect of temperature on flowering. *Proceedings of the Agronomy Society of New Zealand* **9**, 59-64.

Turkington, R.A. (1975) Relationships between neighbours among species of permanent grassland (especially *Trifolium repens* L.). *Ph. D. Thesis, University College of North Wales, Bangor, University of Wales.*

Turkington, R.A.; Cavers P.B. (1978) Reproductive strategies and growth patterns in four legumes. *Canadian Journal of Botany* **56**, 413-416.

Turkington, R.; Harper, J.L. (1979a) The growth, distribution and neighbour relationships of *Trifolium repens* in a permanent pasture. I. Ordination, pattern, and contact. *Journal of Ecology* **67**, 201-218.

Turkington, R.; Harper, J. L. (1979b) The growth, distribution and neighbour relationships of *Trifolium repens* in a permanent pasture. II. Inter- and intra-specific contact. *Journal of Ecology* **67**, 219-230.

Turkington, R.; Harper, J. L. (1979c) The growth, distribution and neighbour relationships of *Trifolium repens* in a permanent pasture. IV. Fine-scale biotic differentiation. *Journal of Ecology* **67**, 245-254.

Vavilov, N.I. (1951) The origin, variation, immunity, and breeding of cultivated plants. *Chronica Botanica* **13**, No. 1/6.

Vose, P.B. (1960) Intraspecific differences in nutrient uptake and assimilation. *Welsh Plant Breeding Station Report for 1959*, 18-20.

Vose, P.B.; Koontz, H.V. (1959) Uptake of strontium by pasture plants and its possible significance in relation to the fall-out of strontium-90. *Nature, UK.* **183**, 1447-1448.

Ware, W.M. (1925a) Experiments and obervations on forms and strains of *Trifolium repens* L. *Journal of Agricultural Science, UK* **15**, 47-67.

Ware, W.M. (1925b) White clover. *Miscellaneous Publication, Ministry of Agriculture and Fisheries*, **No.46**, 1-26.

Whitman, R.J. (1973) Herbivore feeding and cyanogenesis in *Trifolium repens* L. *Heredity* **30**, 241-244.

Wilkinson, H.T.; Millar, R.L. (1978) Cyanogenic potential of *Trifolium repens* L. in relation to pepper spot caused by *Stemphylium sarciniforme*. *Canadian Journal of Botany* **56**, 2491-2496.

Wilkinson, H.T.; Millar, R.L. (1979) B-glucosidases potentially involved in cyanogenesis during infection of white clover by *Stemphylium sarciniforme*. *Canadian Journal of Botany* **57**, 69-73.

Williams, W.M.; Anderson, L.B.; Cooper, B.M. (1978) Evaluation of clovers on sandy coastal soils. *Proceedings of the New Zealand Grassland Association* **39(1)**, 130-138.

Williams, W.M.; Barclay, P.C. (1975) Performance of a Spanish white clover (*Trifolium repens* L.) population in New Zealand. *New Zealand Journal of Agricultural Research* **18**, 45-49.

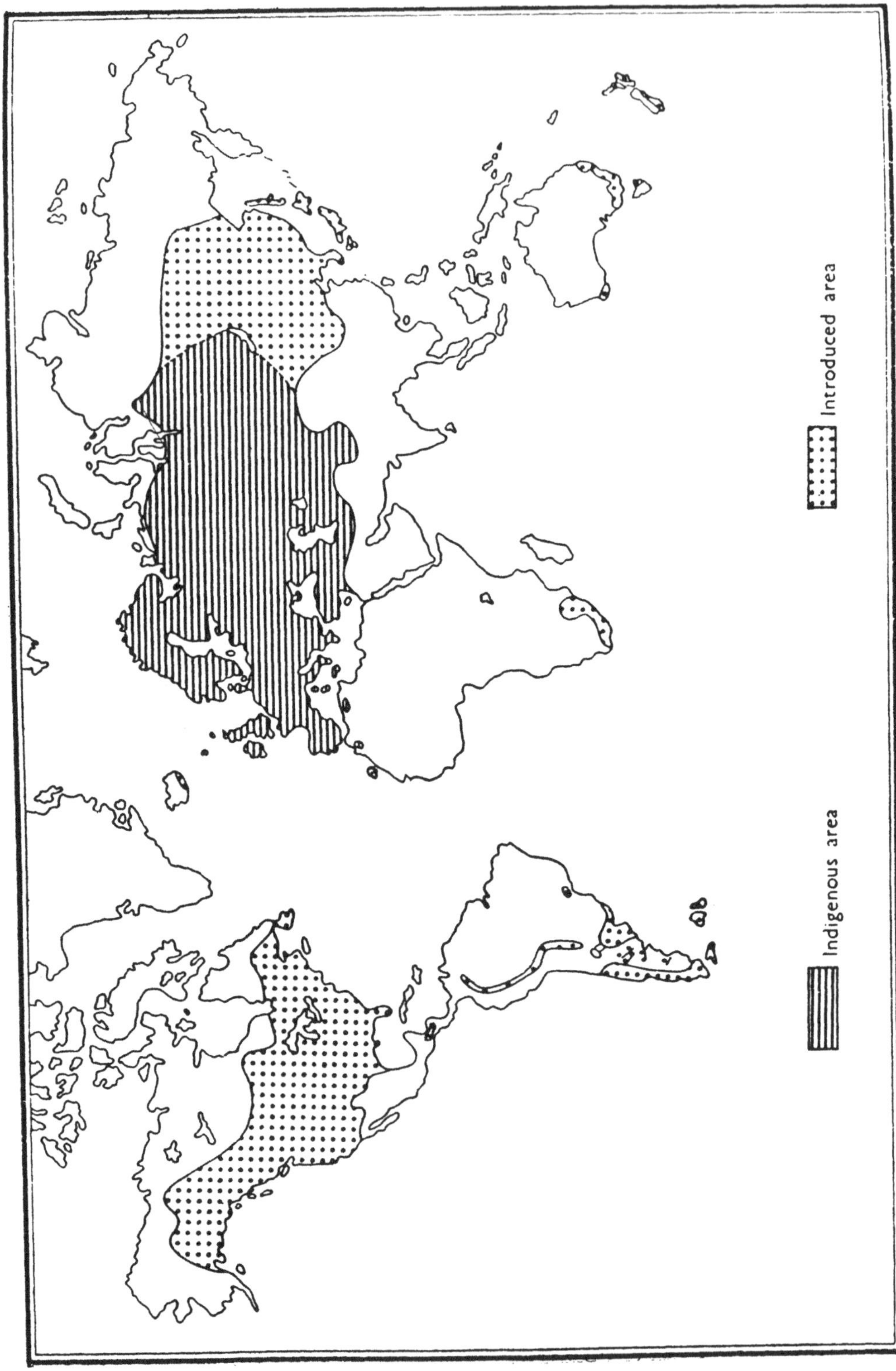

Fig. 8.1 Geographical distribution of *Trifolium repens* L. (From Daday (1958) reproduced with permission from *Heredity*).

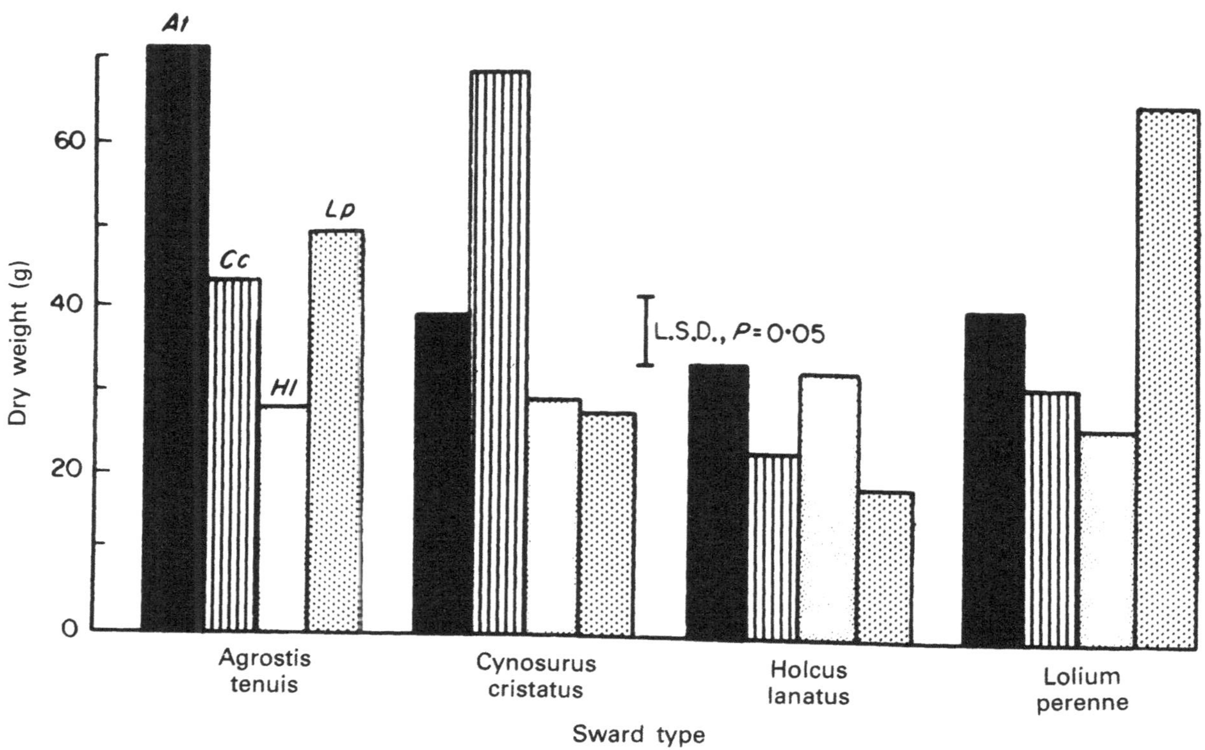

Fig. 8.2 The dry weight of plants of *Trifolium repens* from a permanent grassland sward, sampled from patches dominated by four different perennial grasses and grown in all combinations of mixture with the four grass species. Clover 'types': *At*, *Agrostis tenuis*; *Cc*, *Cynosurus cristatus*; *Hl*, *Holcus lanatus*; *Lp*, *Lolium perenne*.
(From Turkington & Harper (1979c) reproduced with permission from *Journal of Ecology*).

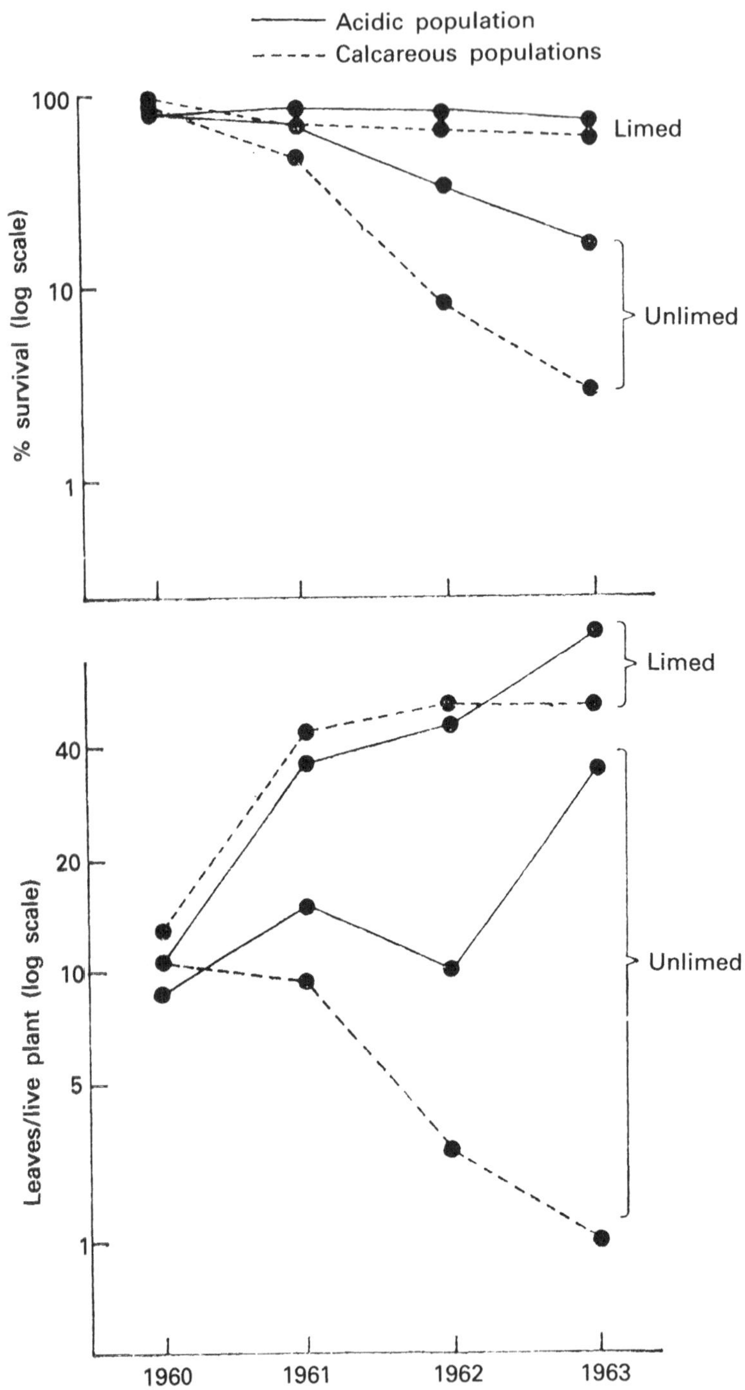

Fig. 8.3 Survival of clover populations in an upland sward near Aber, Caernarvonshire, Wales (data of Snaydon, unpublished). (From Bradshaw (1969) reproduced with permission from *Brit. Ecological Soc. Symp.*)

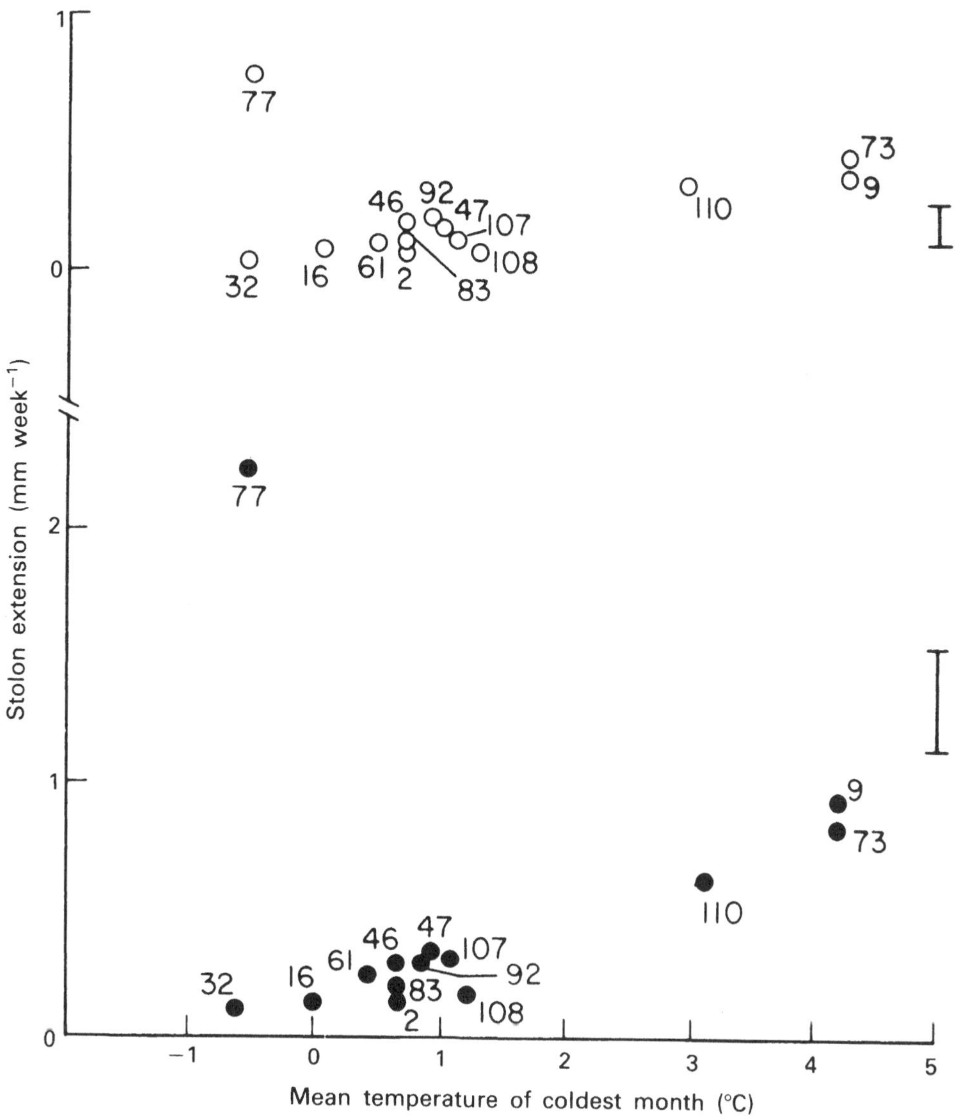

Fig. 8.4 The relationship between main stolon (○), total stolon (●) extension at 5° in a cold cabinet and the winter temperature at the place of origin of 14 *T. repens* ecotypes. The bars are LSDs at P<0·05. The figures are the reference numbers for each ecotype (From Ollerenshaw & Baker (1981) reproduced with permission from *Journal of Appl. Ecology*).

White clover taxonomy and biosystematics

W. M. Williams
* Victorian Crops Institute, Horsham, Victoria 3400, Australia

Contents

1. Introduction

White clover, *Trifolium repens* L., is the most agronomically important of the 250-300 species in the genus *Trifolium*. This genus, together with *Medicago* and *Melilotus*, is classified taxonomically in tribe Trifolieae of subfamily Papilionoideae [a] of the family Leguminosae.

The genus *Trifolium* contains, in addition to *T. repens*, more than 20 other clover species which are cultivated as forages. Among the wild species, about 54 occur only in North America and about 33 exclusively in Africa south of the Sahara. The remainder are distributed mainly in Europe, northern and south-western Asia, and north Africa, with the heaviest concentration around the Mediterranean. About 40 species are confined virtually to the so called Orient (Turkey, Syria, Cyprus, Lebanon, Jordan, Israel, Iran, Iraq) Hossain (1961).

2. Taxonomic relationships

There is no modern world classification of *Trifolium*. Taxonomic treatments have generally been geographically based: European species (Ascherson & Graebner 1908, Coombe 1968) have thus been classified separately from the North American species (e.g. McDermott 1910), Oriental species (Hossain 1961, Zohary 1970), African species (Gillett 1952), Russian species (Komarov et al. 1934-64) and others. Because *T. repens* is native throughout Europe and is now widely distributed throughout the world, it appears in several of these regional treatments, and is classified in slightly different ways by different authors.

The highly detailed review by Gibelli & Belli (1889-1893) has been the basis of most subsequent treatments, such as that by Ascherson & Graebner (1908) dealing with the species of central Europe. Hossain (1961) gives a detailed account of the reproductive structures of the genus, particularly as they affect the protection and dispersal of the seeds, and on this basis recognises eight distinct subgenera, some of which are divided into sections. Some other authors (e.g. Zohary 1970) do not recognise subgenera but classify the genus directly into a number of sections. Most authors agree in placing *T. repens* in section *Amoria* (formerly *Euamoria*), which is placed by Hossain and by Ascherson & Graebner in subgenus *Amoria*. Coombe (1968) designates this same group as section *Lotoidea* of subgenus *Lotoidea*. Species placed in this section by various authors are listed in Table 9.1. *T. glomeratum* and *T. suffocatum*, which are included in section *Amoria* by Ascherson & Graebner and in section *Lotoidea* by Coombe, are segregated as section *Micrantheum* by Hossain and by Zohary, on account of their sessile flower-heads and other floral differences.

3. Cytological relationships

(A) Chromosome number

The first correct publication of chromosome number in *T. repens* was that of Karpechenko (1925) who gave the somatic number as 2n = 32. Erith (1924) had given the number previously as 2n = 16, but her figures showed 16 bodies at metaphases I and II of meiosis, indicating that 32 was the diploid number (Wexelsen 1928). The number 2n = 32 has since been confirmed many times (Britten 1963).

It is evident that 8 is the basic chromosome number in the whole of the tribe Trifolieae, including *Trifolium* (Senn 1938). All closely related species in Section *Amoria* (Table 1) which have been counted so far have a basic chromosome number of 8 (Britten 1963, Pritchard 1962, 1969, Chen & Gibson 1971a). Of 63 species studied so far in subgenus *Amoria* all have 8 as basic number (Pritchard 1969).

a. Subfamily Lotoideae in *Flora Europaea* (Coombe 1968)

Table 9.1 Some *Trifolium* species (mainly European and African) classified into the same section as *T. repens* by various authorities. North American species which belong to subgenus *Amoria* are also listed.

Authorities

1. Ascherson and Graebner (1906–10)	Subgenus Amoria, Section *Euamoria*.
2. Burkart (1939)	Section *Amoria*
3. Coombe (1968)	Subgenus *Lotoidea*, Section *Lotoidea*
4. Gillett (1952)	Section *Amoria*, Subsection *Amoria*
5. Hossain (1961)	Subgenus *Amoria*, Section *Amoria*
6. Zohary (1970)	Section *Amoria*

Species	*Authority*	*Distribution*
acaule Steud.	4	Ethiopia, Kenya
africanum Ser.	4	S. Africa
albopurpureum Torr. & Gray		N. America
ambiguum M. Bieb.	1,3,5,6	Caucasus, Asia Minor
angulatum Waldst & Kit.	1,3	E.C. Europe, S.E. Russia
baccarinii Chiov.	4	Tropical Africa
bifidum Gray		N. America
bilineatum Fres.	4	Ethiopia
bivonae Guss.	3	Sicily
breweri S. Wats.		N. America
burchellianum Ser.	4	Africa (S. Ethiopia southwards)
calocephalum Fres.	4	Ethiopia
carolinianum Michx.		N. America
cernuum Brot.	3	S.W. Europe
ciliolatum Benth.		N. America
euxinum Zoh.	6	Turkey
fistulosum Vaughan		N. America
glomeratum L.	1,3	Europe, Asia Minor
gracilentum Torr. & Gray		N. America
hybridum L.	1,3,5,6	Europe, Asia Minor
isthmocarpum Brot.	1,3,5,6	Mediterranean
masaiense Gillett	4	E. Africa
Mattiriolianum Chiov.	4	Tropical Africa
meneghinianum Clem.$^{+}$	1	S. Europe, Mediterranean
michelianum Savi^{+++}	1,3,5,6	S. Europe, Asia Minor
montanum L.	1,3,5,6	Europe, Asia
nigrescens Viv.$^{\circ}$	1,3,5,6	S. Europe, Mediterranean
occidentale D. Coombe	3	S.W. Britain, Channel Is., N.W. France (coastal)
orphanideum Boiss*	5	Greece, Turkey
pallescens Schreber	1,3	C. & E. Europe (mountains)
palmeri S. Wats		N. America
parnassi Boiss–Spruner	3	Greece (mountains)
petitianum A. Rich	4	Ethiopia
pichiesermollii Gillett	4	Ethiopia
polymorphum Poir**		Chile
pseudostriatum E.G. Baker	4	Africa (Uganda, Zaire, Lake Nyasa)
purseglovei Gillett	4	Uganda, Zaire
radicosum Boiss. & Hohen.	5	Iran, Iraq (high altitude)
reflexum L.		N. America
retusum L.$^{++}$	3,5,6	Mediterranean, Asia Minor
riograndense A. Burkart	2	S. Brazil
ruepellianum Fres.	4	Tropical Africa
semipilosum Fres.	4	S. Arabia, Ethiopia south to Kilimanjaro
stoloniferum Muhl.		N. America

suffocatum L.	1,3	Mediterranean, Asia Minor
tembense Fres.	4	Tropical Africa
thalii Vill.	1,3	Appenines, Pyrenees, N. Spain (high altitude)

+ *T. meneghinianum* is treated as a variety of *T. nigrescens* subsp. *petrisavii* by Hossain (1961) and as a synonym for it by Zohary (1970).

o *T. petrisavii* Clem. is treated as a subspecies of *T. nigrescens* by Coombe (1968), Hossain (1961) and Zohary (1970).

* syn. *T. repens* var. *orphanideum* Boiss (Coombe 1968).

** syn. *T. megalanthum*.

++ syn. *T. parviflorum* (Ascherson & Graebner 1906–10)

+++syn. *T. balansae*.

Most species have the diploid number of 16. Polyploidy is uncommon in *Trifolium*, only 16% of species being polyploid or having polyploid races. However, 70% of known polyploids occur in subgenus *Amoria* which is considered to be one of the more primitive and unspecialised subgenera (Pritchard 1969). *T. repens* is one such polyploid species in the subgenus. Other polyploid species in Section *Amoria* include *T. ambiguum* (2n = 16, 32, 48), *T. burchellianum* (2n = 48, 96), *T. montanum* (2n = 16, 32) and *T. polymorphum* (2n = 32).

Pritchard (1969) indicated that the polyploid species are often indigenous either to regions far removed from the Mediterranean (sometimes considered to be the main centre of origin of the genus [b]) or to alpine regions. In the more specialised sections of the genus, on the other hand, there has been virtually no polyploidy but some reduction of basic chromosome number to x = 7, 6 and 5.

(B) Chromosome morphology

Chromosome size: Chromosomes in *Trifolium* are, in general, small but range in length from an average of 1 μm to about 5 μm in different species. Those of *T. repens* fall at the small end of this range (up to approx. 2 μm, Pritchard 1962). Within *T. repens* Wexelsen (1928) examined chromosomes of three plants of wild white clover and three of Ladino clover. The chromosomes of Ladino clover were approximately 2 μm long and substantially larger than those of wild white clover. Chromosomes in F_1 hybrids were intermediate in size.

Karyotype: Karyotype comparisons within *Trifolium* have been carried out by Wexelsen (1928), Pritchard (1962), and Chen & Gibson (1971a). Wexelsen identified two satellite-bearing chromosomes in *T. repens* and tentatively identified four pairs of small chromosomes. Atwood & Hill (1940) found no consistent criteria for identifying individual chromosome pairs at meiosis.

In Ladino clover, Chen & Gibson (1971a) identified 4 metacentric or nearly metacentric chromosome pairs, 11 submetacentric pairs, and 1 subtelocentric pair bearing satellites. Comparisons with the karyotypes of species which intercross with *T. repens* revealed close similarities among *T. repens, T. occidentale, T. nigrescens, T. petrisavii* and *T. nigrescens* subspecies. *T. meneghinianum,* however, was found to have a dissimilar karyotype from these species, raising some doubt as to the synonymy of *T. meneghinianum* and *T. nigrescens* given by Hossain (1961) and Zohary (1970). Close similarities among the karyotypes of *T. repens, T. nigrescens,* and *T. occidentale* may indicate close phylogenetic relationships among these species. Such observations are supported by studies of chromosome pairing in interspecific hybrids (see Section 3).

(C) Meiosis

T. repens behaves as a regular diploid at meiosis. Atwood & Hill (1940) observed only 5 of 610 microsporocytes to vary from 16 bivalents at metaphase I. The 610 cells were from 11 plants of diverse origin. In none

b. East Africa is also a possible centre of origin - see Section 4. Section 6.

of these cells was multivalent pairing observed. The exceptional cells all showed 15 bivalents and 2 univalents, and some could have been cases of precocious disjunction. This degree of irregularity is exceptionally low even for diploid species.

Chromosome pairing was very regular. Mean chiasma frequencies among 8 different plants ranged from 16.3 to 17.4 per cell. Various cells ranged from 14 to 20 chiasmata, but some of this variation was probably due to the stage of metaphase I observed.

Only 3 cells (0.7%) of 912 anaphase I separations showed other than the regular 16-16 separation. Of the 3, one was a 15-17 and 2 (both from the same plant) were 32-32 separations. No exceptional anaphase II separations were observed among 93 chromosome groups in 60 microsporocytes.

Results obtained by Chen & Gibson (1970a) were consistent with those of Atwood & Hill, except that one quadrivalent was observed.

4. Interspecific hybridisation

Interspecific hybridisation in the genus has proved to be difficult, there being few successful hybridisations among many combinations attempted. This section reviews progress in hybridising *T. repens* with other *Trifolium* species.

(A) Hybridisation with *T. nigrescens*

T. repens and *T. nigrescens* cross with some difficulty, although cetain combinations of plants produce large numbers of hybrids. Williams et al. (1978) reported several hundred putative hybrids after caging plants of each species with a bumble bee and harvesting normal seed. A number of the progeny were confirmed to be triploid (3x = 24) interspecific hybrids showing intermediate morphology and pollen sterility (0.5-6% stainable pollen).

The cross is more successful when *T. repens* is used as female parent. Kazimierski & Kazimierska (1970) obtained 69% of ovules developing into mature flowering hybrid plants with *T. repens* as female, compared with only 7.5% for the reciprocal cross. Poor survival of progeny from the *T. nigrescens* seed parent was apparently due to disturbances in chlorophyll production. Hovin (1962a) obtained 35 seedlings following crosses on to *T. nigrescens* seed parents, but all were albino or chlorophyll deficient. Embryo culture did not greatly improve the success of this cross. On the other hand, reciprocal crosses (*T. repens* x *T. nigrescens*) generally produced normal seeds and seedlings.

Different *T. nigrescens* accessions vary in their effectiveness as pollen parents. Hovin (1962a) found two Italian accessions to be five times more effective than three Turkish accessions in giving more pods set and more developing embryos per pod. Italian pollen parents also gave more vigorous progenies than the Turkish parents.

Hybrids have also been obtained by Trimble (1951), Keim (1953a,b), Evans (1962a), and Trimble & Hovin (1960). Hybrid plants are generally intermediate in morphology between the parent species and show low fertility. Kazimierski & Kazimierska (1970) obtained low seed set (0.6-1.5 seeds per inflorescence) after cross pollination, although the plants were self-sterile. The hybrids, although generally intermediate between the parents, showed larger florets and more elongated leaflets than either parent and tended to be later flowering.

Studies of chromosome pairing at meiosis in *T. repens* x *T. nigrescens* hybrids indicate some homology between the chromosomes of the two species. Hovin (1962a) studied diakinesis in three 24-chromosome hybrids and found an average of 4.2 univalents per cell (range 2-8). Similarly, at metaphase I smeared and sectioned materials averaged 2.9 and 2.5 univalents per cell, respectively. The presence of fewer than 8 univalents per cell may be interpreted as evidence for pairing of *T. nigrescens* and *T. repens* chromosomes. Chen & Gibson (1970b) studied metaphase I in a hybrid plant. Average numbers of trivalents, bivalents and univalents per cell were 4.27, 3.73 and 3.73 respectively. Presence of definite trivalent configurations may be interpreted as further evidence of homology between the chromosomes of the two species. Although tetrads in the hybrid appeared normal, stainable pollen from 10 plants averaged only 7% (range 1% to 16%). These hybrids were self-sterile. Approximately 500 backcrosses to *T. nigrescens* failed to produce seed but 6 viable seeds resulted from a similar number of backcrosses to *T. repens*. These seeds were produced on the hybrid with 16% stainable pollen and gave rise to 6 plants (2n = 28) with up to 50% stainable pollen. The plants resembled the recurrent parent but were less prostrate and had fewer roots at the lower nodes than is typical of *T. repens*. The result suggested that further backcrosses might produce progenies from which monosomic *T. repens* could be derived.

Comparison of meiosis in green and albino sectors of a sectoral chlorophyll chimaeral hybrid showed that meiosis was more regular and pollen viability consequently higher in the green sector (Kazimierski & Kazimierska 1970).

Genetic studies also suggest homology between the chromosomes of the two species. A detailed study of 2 fertile hybrid plants from reciprocal crosses of tetraploid *T. nigrescens* (4n = 32) and tetraploid *T. repens* (4n = 64) was made by Brewbaker & Keim (1953). F_1 and I_1 progeny of these hybrids generally had 48 chromosomes. One of the hybrid plants resulted from a cross between a self-incompatible *T. nigrescens* plant lacking a leaf mark and a self-compatible *T. repens* plant homozygous for a V-mark. The hybrid proved to be self-compatible and, on selfing, segregated non-marked plants, suggesting that the chromosomes of the two species may have paired, i.e. were homologous. The self-compatibility of the hybrid plant also suggested that there was homology among the chromosomes of the two species. Self-compatibility of the tetraploid *T. repens* parent was apparently due to production of 'competition' pollen (Chapter 10, Section 4). If the self-compatibility of the hybrid were also due to this phenomenon, then it was likely that the hybrid was producing pollen carrying both S alleles from the *T. repens* parent after between-species pairing of chromosomes.

T. petrisavii and *T. meneghinianum*

These 'species' were listed by Ascherson & Graebner (1906-10) in section *Amoria* and were given full species rank. They have since been reclassified: Coombe (1968), Hossain (1961) and Zohary (1970) giving *T. petrisavii* subspecies status within *T. nigrescens*. *T. meneghinianum* was treated as a variety of subspecies *T. petrisavii* by Hossain and as a synonym for it by Zohary. It is therefore not surprising that both hybridise with *T. repens* (Leffel & Gibson 1973). Kazimierski et al. (1972) failed to make the cross *T. repens* x '*T. petrisavii*'.

(B) Hybridisation with *T. uniflorum* (Subgenus Amoria, Section Cryptosciadium)

Hybridisation between *T. repens* and *T. uniflorum* (2n = 32) was first achieved by Pandey (1957). Seed set in the cross with *T. uniflorum* as seed parent was 30-50% of that of intra-specific crosses. However, of 30 seeds sown only 2 germinated. One was fully albino and died but the other, although three fourths albino, had a green sector which developed into a mature plant morphologically intermediate between the two parents. Unlike either parent, however, the hybrid was self-compatible and crossed freely with both parents. This, Pandey hypothesised, could have been due to non-homology of the incompatibility loci in the two species allowing for production of gametes lacking incompatibility alleles.

Chen & Gibson (1971b) also had difficulty obtaining hybrids between *T. repens* and *T. uniflorum*. Study of *T. repens* x *T. uniflorum* pollinations indicated that approximately 17% of *T. repens* ovules were fertilised by *T. uniflorum* gametes (*T. repens* x *T. repens* controls gave 68%). Fertilisation was also slower, taking up to 3 days, apparently because of slow pollen tube growth. The hybrid embryo, although growing more slowly than that of *T. repens*, differentiated normally, or nearly so, up to the sixth day after pollination. The hybrid endosperm, on the other hand, began to develop abnormally on the fourth day and failure of embryo development resulted primarily from abnormal development and ultimate disintegration of the hybrid endosperm. Since embryos frequently differentiated cotyledons before disintegration, hybridisation of these two species should be attainable by use of embryo culture. Indeed, Evans (1962a) had obtained 5 possible hybrid plants (the putative hybrids did not flower) by embryo culture. Evans also indicated the presence of pre-fertilisation barriers to hybridisation, pollen tube growth being slower. Williams et al. (1978) reported the use of embryo culture to improve the yield of hybrids from this cross.

Eight hybrid plants between *T. repens* and *T. uniflorum* were obtained following use of another technique by Gibson et al. (1971). Compatible *T. repens* genotypes were initially detected by making large numbers of pollinations of *T. uniflorum* on to *T. repens* plants. If pods enlarged, further pollinations were made to that seed parent. Four of the hybrids had 32 chromosomes while the remainder had 1 to 3 extra chromosomes. The hybrids were vigorous, stoloniferous perennials with floral characteristics intermediate between the parental species.

Crossing *T. repens* and *T. uniflorum* is facilitated by use of *T. occidentale* as a bridging species.

(C) Hybridisation with *T. occidentale*

T. occidentale is a stoloniferous perennial clover indigenous to England, France and the Channel Islands where it grows in relatively dry coastal habitats, on shallow pockets of soil, and in sand dunes (Coombe 1961, Coombe & Morisset 1967). *T. repens* and *T. occidentale* were first crossed artificially by Chou & Gibson (1968), a relatively straight forward procedure.

In crosses of *T. occidentale* with *T. repens* and *T. nigrescens*, differences in cross-compatibility were found to depend on the ploidy of the *T. occidentale* pollen parent. It proved more efficient to use induced tetraploid (4n = 32) *T. occidentale* than the naturally occurring diploid form. Gibson & Beinhart (1969) found that while the cross *T. repens* x *T. occidentale* (2x) produced some hybrids, use of pollen from 4x plants on compatible *T. repens* seed parents produced averages of one or more seeds for every 2 florets pollinated. Crosses in the opposite direction (using *T. repens* as pollen parent) produced seed and plants but hybridity was not confirmed.

The F_1 hybrids were stoloniferous perennials usually intermediate in size between the parents. Fertility of the hybrids varied and appeared to be affected by the environment. Stainable pollen of 10 plants in the greenhouse averaged 19.5%. Sib crosses among these plants generally set seed, as did backcrosses to both parent species.

A few plants were obtained from the cross between tetraploid *T. repens* (4n = 64) x tetraploid *T. occidentale* (4n = 32). The F_1 plants were stoloniferous perennials resembling white clover and had 48 chromosomes.

Meiosis in a 24-chromosome hybrid between *T. repens* and diploid *T. occidentale* included 2-8 trivalents at diakinesis and metaphase I (Chen & Gibson 1970a). All telophase I cells examined had one or more lagging chromosomes. These results indicated that not only was there homology among chromosomes of the two species, but also between chromosomes in the gametic complement of *T. repens*.

Meiosis in 32-chromosome hybrids between *T. repens* and tetraploid *T. occidentale* resembled that of tetraploid *T. occidentale*, i.e. most cells contained 4-6 quadrivalents and 4-8 bivalents with occasional trivalents and univalents. These results strengthen the evidence of homologies indicated by the 24-chromosome hybrid.

T. occidentale has proved useful as a bridging species to facilitate hybridisation of *T.repens* and *T. uniflorum*. Gibson et al. (1971) obtained a hybrid between *T. uniflorum* and tetraploid *T. occidentale* and then obtained many hybrids between this and *T. repens*. They also obtained such three-way hybrids from crosses between *T. uniflorum* and *T. repens* x *T. occidentale* hybrids, and concluded that crossing of *T. repens* or *T. uniflorum* with *T. occidentale* weakens the barrier to combining *T. repens* and *T. uniflorum* germplasm.

i. *Hybridisation of **T. nigrescens** and **T. occidentale***

The cross *T.nigrescens*(2n = 16) x tetraploid *T. occidentale* (4n = 32) is readily achieved, but use of diploid *T. occidentale* as pollen parent has so far not been successful (Chou & Gibson 1968, Gibson & Beinhart 1969). Studies of pollen tube growth in these two crosses revealed that both types of pollen grew well. Many hollow seeds were produced and it appeared that failure of pollen from diploid plants was probably due to a post-fertilisation barrier.

The hybrids obtained were triploids with 24 chromosomes. Meiosis was irregular and, as a result, they were relatively pollen sterile (as determined by I-KI staining). In a study of meiosis, Chen & Gibson (1970b) found trivalents, bivalents and univalents in mean frequencies of 5.69, 2.31 and 2.31 per cell, respectively. From a comparison of trivalent frequencies it was suggested that there is more homology between the chromosomes of *T. nigrescens* and *T. occidentale* than between either species and *T. repens*. However, in view of the rather small differences among trivalent frequencies and the probability that chromosome pairing is under genetic control, such a conclusion seems unjustified.

The triploid hybrids obtained by Gibson & Beinhart (1969) resembled *T. occidentale* more closely than *T. nigrescens*, being perennial, although flowering more profusely than *T. occidentale*, and having procumbent stems rather than stolons. Some had chlorotic leaves. Inflorescence characteristics were intermediate. Backcrosses to both parent species were obtained using the hybrids as seed parents. The backcross progeny crossed with *T. repens*.

(D) Hybridisation with *T. isthmocarpum*

Hybridisation between *T. repens* (2n = 32) and *T. isthmocarpum* (2n = 16), an annual Mediterranean species taxonomically close to *T. repens* has so far produced a single sterile hybrid plant (Kazimierski & Kazimierska 1972). It is apparent from Table 9.2 that the cross was not readily obtained but further attempts might be more successful.

Table 9.2. Results of attempts to hybridise *T. repens* and *T. isthmocarpum* (Kazimierski & Kazimierska 1972)

Species		No. flowers pollinated	No. mature seeds	No. seedlings
♀	♂			
T. repens	*T. isthmocarpum*	77	2	1
T. isthmocarpum	*T. repens*	110	0	—
F1	*T. repens*	103	0	—
F1	*T. isthmocarpum*	86	0	—

The F_1 plant was triploid with 24 chromosomes. Like *T. isthmocarpum* it had an erect stem, and the flowers became reflexed after fertilisation, as in *T. repens*. Somatic mitosis showed frequent lagging chromosomes, bridges and fragments at anaphase. Pollen mother cells had an average of 11.93 univalents (range 7-20), 6.07 bivalents (2-8) and 0.08 (0-2) trivalents per cell at diakinesis and metaphase I. Chromosome separations at anaphase I and II were unequal, with frequent bridges, some chromosomes remaining in the cytoplasm. The pollen was non-viable and the hybrid plant proved to be female sterile in crosses with both parent species.

The karyotype of *T. isthmocarpum* differs from that of *T. repens* (Chen & Gibson 1971a). It carries four satellite-bearing chromosomes and a rather steep gradient in size from largest to smallest, unlike most *Trifolium* species.

Further attempts to achieve hybrids between *T. repens* and *T. isthmocarpum* and among *T. isthmocarpum, T. uniflorum, T. occidentale* and *T. nigrescens* should be attempted in order to confirm the single report and to throw further light on relationships of *T. repens*.

(E) Hybridisation with *T. argutum* Sol (Subgenus Mistyllus)

A report by Kazimierski & Kazimierska (1968) of successful hybridisation between *T. repens* and *T. argutum* (syn. *T. xerocephalum*), an annual Mediterranean species, is of considerable interest, since *T. argutum* has been taxonomically classified as being at a considerable distance from *T. repens* (Hossain 1961).

From the cross *T. repens* (2n = 32) x *T. argutum* (2n = 16) 83 seeds were obtained from which 7 triploid plants developed. These plants showed predominantly *T. argutum* characteristics, although the florets were larger than those of either parent. At pollen mother cell metaphase 1 they averaged 8.68 univalents (range 2-16), 5.28 bivalents (1-8), 1.02 trivalents (0-3) and 0.40 quadrivalents (0-3). The plants were completely sterile.

(F) Hybridisation with *T. ambiguum* (Caucasian or Kura clover)

Although *T. repens* and *T. ambiguum* were reported to be readily grafted (Hely et al. 1952) and fertilisation was known to occur when the species were hybridised (Chen & Gibson 1972, Williams & White 1976, 1977), no hybrid plants were reported until Williams (1978) achieved the cross at the 4x level using embryo culture with nurse endosperm. Fertility of the hybrids was very low but one of the four available had a higher pollen fertility (23%) than the others and gave progeny on selfing and backcrossing to *T. repens* (Williams & Verry 1980).

Analysis of meiosis indicated a high degree of bivalent pairing and regular 16-16 chromosome separation at anaphase I in both the sterile and fertile hybrids. Thus low fertility was apparently not caused by gross meiotic irregularities. F_2 plants produced after selfing the fertile hybrid had 32 chromosomes and most were apparently sterile. However, several had up to 6% stainable pollen and produced seed on selfing and backcrossing to *T. repens*. Two backcrosses of the fertile F_2 hybrid ♀ to *T. repens* were obtained (Williams & Verry 1980), one with 32 chromosomes (6% stainable pollen) and the other with 42 chromosomes (10% stainable pollen).

These results indicate that genetic exchange between *T. repens* and *T. ambiguum* is possible and that the two species are closely related, as indicated by morphological characteristics.

(G) Unsuccessful hybridisations involving *T. repens*

The following species have been unsuccessfully crossed with *T. repens*:

T. alexandrinum L.	*T. hirtum* All.†	*T. pratense* L.
T. alpestre L.†	*T. hybridum* L.*	*T. rubens* L.†
T. arvense L.	*T. incarnatum* L.	*T. scabrum* L.†
T. carmeli Boiss.†	*T. medium* L.	*T. semipilosum* Fres.*
T. desvauxii Boiss. & Bl.†	*T. michelianum* Sav.*	*T. subterraneum* L.
T. fragiferum L.	*T. montanum* L.*	*T. tumens* Stev.†
	T. neglectum Meyer	

Those marked † were given without detail in tabular form by Kazimierski et al. (1972).
*Classified in the same section as *T. repens* (Table 9.1).

i. ***T. alexandrinum* (berseem clover, subgenus *Trifolium*)**

From about 100 pollinations, Trimble & Hovin (1960) obtained 52 seeds from a cross *T. repens* x *T. alexandrinum*. However, the resulting seedlings all died within 2 weeks after germination on agar medium, before their hybrid origin could be determined.

Evans (1962a) attempted reciprocal crosses between *T. repens* and both diploid and tetraploid *T. alexandrinum.* From the diploid cross 9 embryos were excised 12 days after pollination. Two of these grew but remained undifferentiated. Five embryos from the tetraploid cross were excised but all failed to grow further. Kazimierski et al. (1972) obtained no seeds from 94 pollinations of the cross *T. alexandrinum* x *T. repens.*

ii. ***T. arvense* (haresfoot trefoil, subgenus *Trifolium*)**

Trimble & Hovin (1960) obtained 8 seeds from about 100 pollinations with this species. However, no mature seedlings resulted after germination on agar medium. Similarly, Evans (1962a) failed to culture 3 embryos obtained 15 days after crossing *T. repens* and *T. arvense.* No seeds were obtained from 21 flowers pollinated by Kazimierski et al. (1972).

iii. ***T. fragiferum* and *T. neglectum* (strawberry clover, subgenus Galearia (Presl.) Hossain)**

T. fragiferum and *T. neglectum* hybridise readily. Considerable efforts to hybridise these species and their hybrid with *T. repens* were unsuccessful (Kazimierski et al. 1972). *T. fragiferum* was tried reciprocally at both diploid and induced tetraploid levels, and a *fragiferum/neglectum* hybrid was used with self-incompatible as well as self-compatible genotypes of *T. repens.* Kazimierska (1980) has attributed failure to embryo death and not to endosperm failure.

iv. ***T. hybridum* (Alsike clover)**

Extensive attempts to hybridise *T. repens* and *T. hybridum* at both diploid and tetraploid levels have been unsuccessful. As early as 1928, Wexelsen reported failure. Apparently *T. repens* is readily fertilised by T. hybridum pollen (Müller 1960, Chen & Gibson 1972), but developing embryos are either not found (Laczynska-Hulewicz 1965, Kazimierski et al. 1972), or are uncommon and fail to develop fully (Guravich 1949, Trimble & Hovin 1960, Evans 1962a, Kazimierska 1980). Hovin (1962b) attempted to make the cross with several different accessions of *T. repens* and *T. hybridum* at both normal and doubled chromosome number. When *T. hybridum* was used as seed parent 27.3% of pollinated florets set pods when pollinated with *T. repens.* However, the reciprocal cross set only 0.7% pods. Pod set proved to be an unreliable indicator of whether hybrids would result because pods frequently contained stimulated ovules without embryos. From 293 pollinations of *T. hybridum* (2n = 16) x *T. repens* (2n = 32), one embryo was obtained which on embryo culture proved to be the result of self-pollination. Three hundred reciprocal pollinations yielded 2 embryos 12-15 days after pollination. One of these was damaged in transplantation and the other proved to be from a selfing. Similarly from reciprocal crosses at the tetraploid level of both species, embryos were obtained only from accidental self-pollinations. No hybrids were obtained from 747 pollinations, indicating that the barriers to cross-compatibility between these two species are very strong.

Kruse (1971), however, reported that the cross *T. repens* ♀ (2n = 32) x *T. hybridum* (2n = 32) was accomplished relatively easily, yielding a hybrid frequency of 39%. Hybrid embryos required transfer to embryo culture about 14 days after pollination. Three plants resulted, but these were weak and chlorotic.

v. ***T. incarnatum* (crimson clover, subgenus *Trifolium*)**

Wexelsen (1928) failed to obtain any seeds from the cross *T. repens* x *T. incarnatum.* However, only 20 flowers were pollinated.

vi. ***T. medium* (zig-zag clover, subgenus *Trifolium*)**

Pollination of *T. repens* with *T. medium* has resulted in fertilisation and development of hybrid embryos to the 4-5 day stage, when endosperm development ceased (Kazimierska 1978). Only about 29% of *T. repens* ovules were fertilised with *T. medium* pollen (64% lower than intraspecific pollination). *T. repens* pollen failed to grow on *T. medium* stigmas.

vii. ***T. pratense* (red clover, subgenus *Trifolium*)**

No verified cases of mature hybrids between *T. pratense* and *T. repens* have been recorded despite many attempts (e.g., Wexelsen 1928, Kazimierski et al. 1972). By the grafting technique of Evans (1960) followed by emasculation and hand pollination, Lacyzynska-Hulewicz (1965) obtained 0-2 seeds per capitulum in crosses between tetraploid red clover and white clover. Some embryos were excised after 1-2 weeks and cultured. Most proved to be red clover and probably resulted from self-pollination. Three plants with pale pink flowers and large, loose heads were obtained from embryo culture. However, these proved to be completely sterile and their nature is unknown.

Reports by Starzycki (1959, 1961) that *T. repens* and *T. pratense* could be hybridised by the Michurinist grafting technique have yet to be verified.

viii. ***T. semipilosum* (Kenya white clover)**

White & Williams (1976) failed to obtain hybrids from reciprocal crosses between *T. repens* (2n = 32) and *T. semipilosum* (2n = 16). Pollination of *T. semipilosum* with *T. repens* pollen resulted in fertilisation and pod enlargement, but no mature seed was produced. Endosperm failure appeared to be a major cause of seed abortion, development ceasing about 4 days after pollination. Hybrid embryos appeared to develop normally, but were degenerating visibly 10-12 days after pollination, probably as a result of endosperm breakdown. Genotypic variation in pod enlargement was observed, suggesting that more compatible genotypes of the two species might be found. Such combinations might yield embryos sufficiently mature to respond to embryo culture.

ix. ***T. subterraneum* (subterranean clover, subgenus *Trifolium*)**

Evans (1962a) obtained a few embryos 15 days after crossing *T. repens* with diploid and tetraploid subterranean clover, but none grew in embryo culture.

(H) Barriers to hybridisation of *Trifolium* species

Barriers to hybridisation among *Trifolium* species occur both before and after fertilisation, although post-fertilisation barriers tend to be the main causes of failure. Investigations of interspecific *Trifolium* crosses by Müller (1960), Evans (1962b) and Chen & Gibson (1972) revealed no failures of pollen tube germination, but pollen tube growth was generally slower than in intraspecific pollinations. Fertilisation was

Table 9.3. Pollen germination (%) and ovule fertilisation (%) after pollination of *T. repens* stigmas with various related species (Chen & Gibson 1972)

Pollen source	Percent pollen tube germination on *T. repens* stigmas 60 min after pollination	Fertilisation (% ovules) 24 h after pollination
T. repens	100 (after 30 min)	100
T. nigrescens	48	82
T. occidentale (2n)	58	81
T. occidentale (4n)	95	31
T. hybridum	42	61
T. ambiguum	66	28
T. uniflorum	51	18
T. uniflorum × *T. occidentale* (4n)	56	61

observed in the following crosses (Chen & Gibson 1972): *T. repens* (♀) x *T. nigrescens, T. occidentale, T. hybridum, T. ambiguum, T. uniflorum*. Swelling, bursting, coiling and misdirected growth of pollen tubes were common. The amount of fertilisation showed little relationship to extent of pollen tube germination (Table 9.3).

Despite moderate levels of fertilisation, production of hybrid embryos was rare, indicating the importance of post-fertilisation breakdown.

Among both successful and unsuccessful species combinations, different genotypes vary in their ability to combine with those of other species. Evans (1962b), for example, found that in the crosses *T. repens* x *T. alexandrinum* and *T. repens* x *T. hybridum*, pollen tubes penetrated only a short distance into the style before bursting or turning back. Nevertheless differences were apparent in the extent of pollen tube growth among genotypes. In a compatible cross, Hovin (1962a) found that Italian accessions of *T. nigrescens* were significantly more cross-compatible with *T. repens* (Ladino) than were Turkish accessions (Section 3.(A)). A number of techniques have been devised to aid in detection of potentially cross-compatible genotypes. These are discussed in the next section.

Embryo culture has been an important tool in improving the frequency with which successful interspecific hybridisation has been achieved in *Trifolium*. Further species combinations may be obtained from developments and improvements in embryo culture methods, e.g. endosperm transplantation (Williams 1978). However, those which failed because of lack of genetic affinity between genotypes from different species may never be obtaineed.

(I) Techniques for detection of cross-compatible lines

Numerous studies indicate that genotypes differ greatly in their ability to hybridise. A number of techniques have been developed which might assist the detection of species or individuals which are potentially cross-compatible. Those relevant to *Trifolium* are discussed in this section.

i. *Interspecific grafting*

In attempting to improve prospects of interspecific hybridisation in *Trifolium* by grafting, as Michurinist geneticists in the USSR claimed they had done for other distant crosses, Evans & Denward (1955) found a correlation between the success of interspecific grafting and the success of hybridisation (or, at least, interspecific fertilisation). This correlation suggested that grafting could be used to sieve out species combinations which would be expected to fail in hybridisation attempts. Evans (1962a) presented comprehensive data supporting this suggestion. The following crosses mainly involving *T. repens* ($2n = 4x = 32$) were attempted using graft-compatible and graft-incompatible genotypes for each cross. Up to 20 plants of each species were used:

T. repens	×	*T. nigrescens* (2n)
T. repens	×	*T. nigrescens* (4n)
T. repens (4n)	×	*T. nigrescens* (2n)
T. repens	×	*T. hybridum* (2n)
T. repens	×	*T. hybridum* (4n)
T. repens (4n)	×	*T. hybridum* (2n)
T. repens	×	*T. uniflorum* (4n)
T. alexandrinum (2n)	×	*T. repens*
T. alexandrinum (4n)	×	*T. repens*
T. subterraneum (2n)	×	*T. repens*
T. subterraneum (4n)	×	*T. repens*
T. arvense (2n)	×	*T. repens*

Results are summarised in Table 9.4.

Table 9.4. Percentage embryos set after species hybridisation with and without grafting (from Evans 1962a)

	Genotypes compatible as grafts			Genotypes incompatible as grafts
	Stock × Scion	Scion × Stock	Ungrafted controls	Ungrafted plants
Cross in direction shown above	12.0	9.8	11.0	1.1
Cross in opposite direction	8.3	4.7	6.5	0.3
Mean	10.8	7.1	8.7	0.7

Graft-compatible genotypes were clearly more successful than graft-incompatible genotypes. Grafting did not significantly increase crossing, contrary to Michurinist interpretations. Scion x stock results were lower than the reciprocal, apparently because of greater effect of grafting on female than on male fertility.

ii. . Effect of doubling chromosome number on interspecific hybridisation

In some species autotetraploids may cross successfully when diploids of the same species do not cross (Evans 1962a). In *Trifolium*, autotetraploid *T. occidentale* was substantially more effective than the diploid form in hybridisations with *T. repens* and *T. nigrescens* (Chen & Gibson 1968, Gibson & Beinhart 1969). Apart from those examples no strong trend is apparent in the genus. Evans (1962a), for example, obtained similar failure rates in a range of interspecific crosses at diploid and tetraploid levels.

iii. Biochemical methods for detection of cross-compatible combinations

(i) Paper chromatography: Collins and Taylor (1964) used paper chromatography of root tip smears to investigate taxonomic relationships and predict possible interspecific hybridisations within *Trifolium*. Species that had been hybridised and were closely related taxonomically had very similar chromatographic patterns. *T. repens* and *T. nigrescens* showed identical patterns after use of a 4:5:1 n-butanol: distilled water: glacial acetic acid solvent system (the compounds involved were not identified) and thus fell into a group by themselves. *T. semipilosum* gave a unique pattern but one which differed only slightly from that of the *T. repens* group, while *T. ambiguum, T. hybridum* and *T. montanum* were classified together in another group which contained, in addition, species not classified in section Amoria (e.g., *T. medium* - subgenus *Trifolium, T. fragiferum* - subgenus Amoria, section *Vesicastrum*). Thus there were discrepancies between evidence from the chromatographic technique and taxonomic classification.

(ii) Other techniques used in chemo-taxonomy: Other methods have been used successfully to study relationships among species in several genera and to aid in detecting crossable genotypes. Gell et al. (1960), for example, used immuno-electrophoresis for these purposes in *Solanum*. Although not yet attempted in *Trifolium*, the use of serological and other chemo-taxonomic methods could aid considerably in the success of difficult interspecific crosses (Evans 1962b).

iv. Relationship of breeding system to interspecific hybridisation

Lewis (1955) has shown that there is a relationship between self-compatibility and interspecific compatibility:

Style		Pollen	
Self-incompatible (SI)	×	Self-compatible (SC)	Incompatible cross
SC	×	SI	Compatible cross

Interspecific hybridisation in *Trifolium* has been carried out largely with self-incompatible species. Evans (1960) suggested that success might be greater if self-compatible species were used as females or if self-compatibility were induced in self-incompatible species before attempting a cross.

v. Use of bridging species to achieve intespecific hybridisation

Hybrids involving two species normally difficult to cross can sometimes be generated by crossing one of them to a third species (usually one with which both are cross-compatible). Scope for use of this technique in *Trifolium* is limited severely by the small number of successful hybrids obtainable. There is, however, evidence that the bridging technique has helped in at least one case (Section 3.(C)).

vi. Use of pod enlargement to detect compatible genotypes

Genotypes that cross more readily than most can sometimes be detected by performing large numbers of individual crosses and looking for pod enlargement. If pod enlargement occurs further pollinations are made with that combination (e.g., Gibson et al. 1971, Section 3.(B)). However, pod enlargement is not always a reliable indicator that hybrids will result (e.g., Hovin 1962b, Section 3.(G)).

5. Relationships among *Trifolium* species as indicated by *Rhizobium* strain affinities

Norris (1956, 1959) has suggested that studies of *Rhizobium* strain affinities may throw light on relationships among leguminous species. From such studies it was hypothesised that the ancestors of the tribes Trifolieae and Vicieae may have diverged from the main body of the Leguminosae because they exploited the huge tracts of calcareous land left by the retreat of the shallow Cretaceous seas. As a result, many developed obligate calcicole habits and their associated symbiotic bacteria altered from the ancestral cowpea type, which thrives in acid soils, to calcicole forms. The calcicole habit thus sets aside these tribes and their *Rhizobium* strains from the remaining tribes and their *Rhizobium* strains. It was considered that *Trifolium* may be the ancestral type for the *Trifolieae*, having had its origin in mountain regions in the tropics (e.g., East Africa) and from there spread and differentiated into temperate regions.

Norris' studies indicated that there were few affinities between *Trifolium* species from Africa and *Rhizobium* strains from Europe, indicating perhaps that the groups have been long separated. If this is so then groups of *Trifolium* species may have diverged considerably and, in spite of superficial taxonomic similarities, there may be only a limited chance of hybridisation between species from different groups, e.g. between European and African species.

Norris & 't Mannetje (1964) have hypothesised that *Rhizobium* strain affinities can also be used to distinguish ancient and recent species. Recently evolved species or groups of species were expected to have specialised *Rhizobium* strains and ancestral types were expected to have *Rhizobium* strains able to nodulate a wide range of derived types of host. Tests indicated that *Rhizobium* strains from *T. repens* produced no or non-effective nodulation on African species, a result interpreted as support for Norris' (1956) hypothesis that East Africa may be a centre of origin for *Trifolium* from which it has spread and differentiated into the Mediterranean region. However, since *Rhizobium* strains from African clovers were equally ineffective on *T. repens*, the interpretation appears uncertain.

6. The origin of white clover

White clover is a tetraploid species. Polyploidy is uncommon in *Trifolium* and clearly has not been a major mechanism of speciation in the genus. The tetraploid nature of *T. repens* therefore places it apart from most *Trifolium* species. Interspecific hybridisation is also difficult within *Trifolium*, yet *T. repens* hybridises easily with at least three species and with greater difficulty with at least three others. It is tempting to put these two unusual characteristics together and speculate that they were both associated with the origin of the species. Harlan & de Wet (1975) have reviewed the various mechanisms by which polyploid species can arise. Although the origin of white clover is undetermined, this section summarises information which may be relevant to our understanding of its relationships and origin.

White clover has predominantly (perhaps entirely) diploid inheritance, only two satellite chromosomes and regular bivalent formation at meiosis. Presence of trivalents and fewer than 8 univalents per cell in meiotic prophase of triploid hybrids between *T. repens* and the diploid species *T. nigrescens* and *T. occidentale* indicate that: (1) the chromosomes of the diploid species pair with those of *T. repens*, and (2) chromosomes in the gametic complement of *T. repens* can pair with each other, although they do not do so in *T. repens* itself. Such pairing also apparently occurs in *T. repens* x *T. ambiguum* hybrids. Observation (2) indicates that chromosome pairing in *T. repens* may be under genetic control, since bivalent formation is extremely regular, yet on the basis of these results, multivalent formation would be expected. There is also considerable homology between the chromosomes of *T. nigrescens and T. occidentale* as indicated by studies of pairing at diakinesis in hybrids between these species.

There are close similarities among the karyotypes of *T. repens, T. nigrescens, T. occidentale* and *T. ambiguum* (Chen & Gibson 1971a).

White clover is polymorphic for cyanogenic glucosides and hydrolysing enzymes (Chapter 10, Section 3). *T. nigrescens* and its subspecies also contain cyanogenic glucosides and enzyme (Gibson et al. 1972) and *T. occidentale* and *T. ambiguum* contain glucoside but no enzyme.

If white clover is of hybrid origin it could have arisen from a cross between two closely related although divergent populations, perhaps ecotypes, of a single widespread diploid species (Navalikhina 1977). *T. occidentale* and *T. nigrescens* may represent two such ecotypes which have diverged further because of geographic isolation but have nevertheless retained some affinities. On the basis of ease of crossing with white clover, similar karyotypes and chromosome pairing evidence and presence of cyanogenic glucosides, they are diploid species closely related to white clover. *T. ambiguum* is another such closely related species which has evolved into a polyploid series.

Evans (1962a) has suggested that *T. uniflorum* may be an ancestor of *T. repens*. This species is a tetraploid sufficiently different morphologically to be classified into Section *Cryptosciadium* subgenus Amoria and has a distinct karyotype with four satellites (Gibson & Chen 1971). It forms mainly quadrivalents and bivalents at meiosis and is likely to have had an autotetraploid origin. It contains a trace of cyanogenic glucoside and lacks enzyme (Gibson et al. 1972). This evidence suggests that *T. uniflorum* arose from a diploid ancestor closely related to the ancestor(s) of *T. repens*. Its possible role in the origin of white clover is unknown.

T. isthmocarpum and *T. argutum (T. xerocephalum)* have also been suggested as possible diploid ancestors of *T. repens* on the basis of their hybridisation with that species (Kazimierski & Kazimierska 1968, 1972). Both species were crossed with difficulty to *T. repens* and neither gave fertile hybrids. No conclusions can be drawn on the possible significances of these species in the ancestry of *T. repens* until much more research with them is carried out on interspecific hybridisation. *T. isthmocarpum* contains cyanogenic glucosides and hydrolysing enzymes (Gibson et al. 1972), but *T. argutum* was not tested.

7. Classifications of variation within *Trifolium repens*

White clover is an extremely variable species, creating considerable difficulty for taxonomists who have attempted to classify it. No comprehensive summary has been published but one is needed. It is not the purpose of this section to develop a new classification but to review those classifications that have been made.

(A) Common white clover, wild white clover

This has been given various epithets by different authorities, e.g. subspecies *typicum* (Ascherson & Graebner 1906-10) var. *sylvestre* (Erith 1924) subspecies *repens* (Coombe 1968). It is the common form that occurs throughout the entire distribution of the species. Essential features are as follows: stem, petiole, and peduncle glabrous; flowers generally white or pink; flower heads generally 1.5-2.5 cm in diameter, standard petal not more than three times as long as calyx; peduncles usually exceeding the leaves; calyx 10-veined.

Several ecotypes or races of common white clover have been identified. In Britain the term wild white clover is used in a limited sense to denote a small-leafed, densely branching, thin stoloned ecotype indigenous to the UK. Ascherson & Graebner (1906-10) denoted this var. *genuinum* while Zohary (1970) used var. *repens*. Within var. *genuinum* Ascherson & Graebner give *pentaphyllum* and *tetraphyllum* (multiple leaved garden forms) and *roseum* Peterm. (red flowered).

(B) Giant white clover

Giant forms have generally been classified as race or var. *giganteum* Lagr.-Foss (Erith 1924, Ascherson & Graebner 1906-10, Zohary 1970). Such forms occur in several Mediterranean countries: France, Italy, Israel, Lebanon, Morocco. They represent the upper end of the continuous distribution of organ size present in the species. The best known ecotype of this form originated in northern Italy and is cultivated in many parts of the world. It is known variously as Ladino clover, Lodi clover, Italian white clover, giant clover, mammoth white clover. The term, 'Ladino clover', is strictly incorrect. The incorrect name arose from an abbreviation of the correct name, 'Ladino gigante Lodigiano', and has become established by common usage.

Erith (1924) classified var. *giganteum* as a race of var. *sylvestre* (common white clover), while Ascherson & Graebner (1906-10) placed it similarly as a variety of subspecies *typicum* with the same rank as var. *genuinum*. Some early classifications described giant white clover as *T. repens* var. *latum* (e.g., McCarthy & Emery 1894).

Present knowledge indicates that the giant forms are agronomic varieties of *T. repens*. This is reflected in Coombe's (1968) treatment of the species in which giant forms fall into subspecies *repens* along with other common and wild types.

(C) Dutch white clover, cultivated white clover (var. *hollandicum, T. repens cultum* Alef.)

The term Dutch white clover was used by early British agriculturists as the name for all non-wild (cultivated) strains, because most seed sown in Britain from the 17th century onwards was imported from Holland. The name may still be used occasionally to describe forms of common white clover morphologically intermediate between British wild white clover and giant white clover which are also extremely free-flowering, sparsely branching and short lived.

(D) Var. *purpureum* (Bloodwort)

This epithet was used by Erith (1924) to describe a form of common white clover with the upper leaf surface intensively spotted with dark purple or reddish-purple. It was generally white flowered and also existed in multiple-leafed forms. This form is almost certainly identical to the 'red leaf' mutant of white clover conditioned by a single dominant gene, R^1 (see also Carnahan et al. 1955, Corkill 1971, Chapter 10, Section 1).

(E) Var. *rubescens*

Described by Erith (1924) as a widely distributed purple-flowered form of white clover. This form is also a simple mutant (Chapter 10, Section 2).

(F) Var. *grandiflorum* Peterm.

This form with particularly large flowers (1.5 cm long) and flower heads (considerably over 2 cm in diameter) was given by both Erith (1924) and Ascherson & Graebner (1906-10) as a variant of common white clover. It occurs in the Shetland Islands (Erith 1924).

(G) Var. *microphyllum*

Ascherson & Graebner (1906-10) distinguished this very small-leafed form of common white clover on the basis of leaf size (leaflets less than 1 cm long).

(H) Var. *longipes* Peterm.

Also recognised by Ascherson & Graebner, this form was distinguished by its long floret stalks (pedicels), the stalks of the inner flowers being at least three times as long as the calyx tube.

(I) Subspecies *nevadense* (Boiss.) D.E. Coombe.

Described by Coombe (1968) as a small leafed (leaflets less than 1 cm long) form with small white flowers (heads less than 2 cm wide) occurring in southern Spain (Sierra Nevada).

(J) Subspecies *ochranthum* E.I. Nyarady

This strikingly different form, recognised by Ascherson & Graebner (1906-10), Erith (1924) and Coombe (1968) may be a separate species. It has large flower-heads (25-30 mm wide) with the standard petal 3-4 times as long as the calyx. The corolla is yellow or greenish-yellow and later turns dark brown. It has a restricted distribution in Yugoslavia and Romania.

(K) Subspecies *orbelicum* (Velen.) Pawl.

According to Coombe (1968), this form is distinguished from common white clover mainly on the basis of calyx venation, there being only 6 distinct calyx veins. Flowers are cream with the upper calyx teeth 1.5 times as long as the tube. Found in the Carpathians and Balkan Mountains.

(L) *T. biasolettii* Steud and Hochst. (*Biasoletti* Asch. & Graebn.)

A Mediterranean group distinguished mainly on the basis of hairiness of stems, petioles, etc., proportionately smaller plant parts than common white clover, and pink or red flowers. Coombe (1968) gave this form as subspecies *prostratum* Nyman.

(M) Var. *macrorrhizum* Boiss.

Zohary (1970) gave this form as a small-headed, white flowered race with hairy pedicels occurring in Turkey and Iran. Erith (1924) described it as similar to *T. biasolettii* but with white flowers and gave it subvariety rank within variety *Biasoletti*.

(N) *T. orphanideum* Boiss.

This alpine form is distinguished from common white clover mainly by its small size, small flowers and flower heads, very short peduncles (1-2 cm) scarcely taller than the leaves, and its pale pink flowers. It occurs in alpine habitats in Greece, Italy, Israel, Lebanon, Crete and Sicily and is ranked as a subspecies by Coombe (1968) and as a variety by Zohary (1970). According to Erith (1924) it bears sufficient resemblance to var. *biasolettii* to be regarded as a subvariety. Most authorities, however, distinguish var. *biasolettii* on the basis of hairiness of petioles and peduncles, but *T. orphanideum* is glabrous. Hossain (1961) considered *T. orphanideum* to be sufficiently distinctive to be regarded as a separate species within section *Amoria*. This distinction appeared to be based mainly on Hossain's interpretation of stem growth habit - regarding *T. repens* as having creeping stems and *T. orphanideum* matted prostrate caespitose stems. Coombe (1968), however, indicated that *T. orphanideum* was observed to have mixtures of procumbent rooting stems and subcaespitose, non-rooting stems.

Clearly there are discrepancies of interpretation on the significance of variation within *T. repens*. It would be desirable for the ranges of variation within the putative subspecies of varieties, and possible hybridisation among them, to be studied in order to clarify their taxonomic status and perhaps to bring them to the notice of plant breeders.

References

Ascherson, P.; Graebner, P. (1906-10) Synopsis der Mitterleuropäischen Flora, Band IV, Teil 2, 472-617

Atwood, S.S.; Hill, H.D. (1940) The regularity of meiosis in microsporocytes of *Trifolium repens*. *American Journal of Botany* **27**, 730-735.

Brewbaker, J. L.; Keim, W. F. (1953) A fertile interspecific hybrid in *Trifolium* (4n *T. repens* L. x 4n *T. nigrescens* Viv.) *American Naturalist* **87**, 323-326.

Britten, E.J. (1963) Chromosome numbers in the genus *Trifolium*. *Cytologia* **28**, 428-449.

Burkart, A. (1939) *Trifolium riograndense*, nov. spec., endemico en el sur del Brazil. *Darwiniana* **3**, 421-425.

Carnahan, H.L; Hill, H.D; Hanson, A.A.; Brown, K.G. (1955) Inheritance and frequencies of leaf markings in white clover. *Journal of Heredity* **46**, 109-114.

Chen, C. C.; Gibson, P.B. (1970a) Meiosis in two species of *Trifolium* and their hybrids. *Crop Science* **10**, 188-189.

Chen, C. C.; Gibson, P.B. (1970b) Chromosome pairing in two interspecific hybrids of *Trifolium*. *Canadian Journal of Genetics and Cytology* **12**, 790-794.

Chen, C. C.; Gibson, P.B. (1971a) Karyotypes of fifteen *Trifolium* species in section *Amoria*. *Crop Science* **11**, 441-445.

Chen, C. C.; Gibson, P.B. (1971b) Seed development following the mating of *Trifolium repens* x *T. uniflorum*. *Crop Science* **11**, 667-672.

Chen, C. C.; Gibson, P.B. (1972) Barriers to hybridisation of *Trifolium repens* with related species. *Canadian Journal of Genetics and Cytology* **14**, 381-389.

Chou, M.C.; Gibson, P.B. (1968) Cross-compatibility of *Trifolium nigrescens* with diploid and tetraploid *T. occidentale*. *Crop Science* **8**, 266-267.

Collins, G.B.; Taylor, N.L. (1964) Relation of chromatographic patterns and interspecific compatibilities in *Trifolium*. *Crop Science* **4**, 225-226.

Coombe, D.E. (1961) *Trifolium occidentale*, a new species related to *T. repens* L. *Watsonia* **5**, 68-87.

Coombe, D.E. (1968) *Trifolium*. In T.G. Tutin et al. *(Eds) Flora Europaea* Vol. 2. Cambridge, UK; Cambridge University Press, 157-172.

Coombe, D.E.; Morisset, P. (1967) Further observations on *Trifolium occidentale*. *Watsonia* **6**, 271-275

Corkill, L. (1971) Leaf markings in white clover. *Journal of Heredity* **62**, 307-310.

Erith, A.G. (1924) White clover (*Trifolium repens* L.). A monograph. London, UK; Duckworth.

Evans, A.M (1960) Relationship between vegetative and sexual compatibility in *Trifolium*. *Welsh Plant Breeding Station Report for 1959*, 81-87.

Evans, A.M. (1962a) Species hybridisation in *Trifolium*. I. Methods of overcoming species incompatibility. *Euphytica* **11**, 164-176.

Evans, A.M. (1962b) Species hybridisation in *Trifolium*. II. Investigating the prefertilisation barriers to compatibility. *Euphytica* **11**, 256-262.

Evans A.M.; Denward, T. (1955) Grafting and hybridisation experiments in the genus *Trifolium*. *Nature, UK* **175**, 687-688.

Gell, P.G.H.; Hawkes, J.G.; Wright, S.T.C. (1960) The application of immunological methods to the taxonomy of species within the genus *Solanum*. *Proceedings of the Royal Society, Series B*, **151**, 364-383.

Gibelli, G.; Belli, S.C. (1889-93) *Atti della Accademia delle Scienze di Torino, Ser. II.*

Gibson, P.B.; Barnett, O.W.; Gillingham, J.T. (1972) Cyanoglucoside and hydrolysing enzyme in species related to *Trifolium repens*. *Crop Science* **12**, 708-709.

Gibson, P.B.; Beinhart, G. (1969) Hybridisation of *Trifolium occidentale* with two other species of clover. *Journal of Heredity* **60**, 93-96.

Gibson, P.B.; Chen, C.-C. (1971) Reproduction and cytology of *Trifolium uniflorum*. *Crop Science* **11**, 69-70.

Gibson, P.B.; Chen, C.-C.; Gillingham, J.T.; Barnett, O.W. (1971) Interspecific hybridisation of *Trifolium uniflorum* L. *Crop Science* **11**, 895-899.

Gillett, J.B. (1952) The genus *Trifolium* in Southern Arabia and in Africa south of the Sahara. *Kew Bulletin* **No. 3**,367-404.
Guravich, D.A. (1949) Interspecific compatibility within the genus *Trifolium* Ph.D. Thesis, University of Wisconsin.
Harlan, J.R.; DeWet, J.M.J. (1975) On Ö. Winge and a prayer: the origins of polyploidy. *Botanical Review* **41** 361-390.
Hely, F.W.; Bonnier, Ch.; Manil, P. (1952) Effect of grafting on nodulation of *Trifolium ambiguum. Nature, UK* **171**, 884-885.
Hossain, M. (1961) A revision of *Trifolium* in the Nearer East. *Notes from the Royal Botanic Garden, Edinburgh* **23**, 387-481.
Hovin, A.W. (1962a) Interspecific hybridisation between *Trifolium repens* L. and *T. nigrescens* Viv. and analysis of hybrid meiosis. *Crop Science* **2**, 251-254.
Hovin, A.W. (1962b) Species compatibility in subsection *Euamoria* of *Trifolium. Crop Science* **2**, 527-530.
Karpechenko, G.D. (1925) Karyologische studien über die Gattung *Trifolium. Bulletin of the Applied Botany of Genetics and Plant Breeding* **14**, 271-279.
Kazimierska, E. M. (1978) Embryological studies of cross compatibility in the genus *Trifolium* L. II. Fertilisation, development of embryo and endosperm in crossing *T. repens* L. with *T. medium* L. *Genetica Polonica* **19**, 15-24.
Kazimierska, E. M. (1980) Embryological studies of cross compatibility of species within the genus *Trifolium* L. III. Development of the embryo and endosperm in crossing *T. repens* L. with *T. hybridum* L. and *T. fragiferum* L. *Genetica Polonica* **21**, 37-61.
Kazimierski, T.; Kazimierska, E.M. (1968) [Investigations of hybrids of the genus *Trifolium* L. I. Sterile hybrid *Trifolium repens* L. x *T. xerocephalum* Frenzl.] (in Polish) *Acta Societatis Botanicorum Poloniae* **37**, 549-560.
Kazimierski, T.; Kazimierska, E.M. (1970) [Investigations of hybrids of the genus *Trifolium* L. III. Morphological traits and cytogenetics of the hybrid *Trifolium repens* L. x *T. nigrescens* Viv.] (in Polish). *Acta Societatis Botanicorum Poloniae* **39**, 565-591.
Kazimierski, T.; Kazimierska, E.M. (1972) [Investigations of hybrids of the genus *Trifolium* L. IV. Cytogenetics of the cross *T. repens* L. x *T. isthmocarpum* Brot.] (in Polish). *Acta Societatis Botanicorum Poloniae* **41**, 127-147.
Kazimierski, T.; Kazimierska, E.M.; Strzyzewska, C. (1972) Species crossing in the genus *Trifolium* L. *Genetica Polonica* **13**, 11-32.
Keim, W.F. (1953a) An emryo culture technique for forage legumes. *Agronomy Journal* **45**, 509-510.
Keim, W.F. (1953b) Interspecific hybridisation in *Trifolium* utilising emryo culture techniques. *Agronomy Journal* **45**, 601-606.
Komarov, V.L. et al. (Ed.) (1934-1964). Flora URSS Vols. 1-30. Leningrad and Moscow.
Kruze, A. (1971) Interspecific hybrids in *Trifolium*. (Abst. only). *Hereditas* **69**, 298-299.
Laczynska-Hulewicz, T. (1965) Crossing experiments among cultivated clover species *(Trifolium pratense, T. hybridum* and *T. repens). Genetica Polonica* **6**, 1-4.
Leffel, R.C.; Gibson, P.B. (1973) White clover. In Heath, M.E. et al. (Eds.) Forages, the science of grassland agriculture 3rd Ed., Ames, Iowa; Iowa State University Press, pp. 167-176.
Lewis, D. (1955) Sexual incompatibility. *Science Progress* **172**, 593.
McCarthy, G.; Emery, F.E. (1894) Some leguminous plants. *Bulletin, North Carolina Agricultural Experiment Station* 98 pp.
McDermott, L.F. (1910) An illustrated key to the North American species of *Trifolium*. San Francisco, 325 pp.
Müller, G. (1960) Untersuchungen über das Pollenschlanchwachstum bei verschiedenen Trifolium-Artkreuzungen. *Züchter* **30**, 28-30.
Navalikhina, N.K. (1977) [Role of polyploidy in the origin of white clover.] (in Russian). In *Uspekhi poliploidii* Kiev, Ukrainian SSR, 155-165.
Norris, D.O. (1956) Legumes and the *Rhizobium* symbiosis. *Empire Journal of Experimental Agriculture* **24**, 247-270.
Norris, D.O. (1959) *Rhizobium* affinities of African species of *Trifolium. Empire Journal of Experimental Agriculture* **27**, 87-97.
Norris, D.O.; 't Mannetje, L. (1964) The symbiotic specialisation of African *Trifolium* spp. in relation to their taxonomy and their agronomic use. *East African Agricultural and Forestry Journal*, **29**, 214-235.
Pandey, K.K. (1957) A self-compatible hybrid from a cross between two self-incompatible species in *Trifolium. Journal of Heredity* **48**, 278-281.
Pritchard, A.J. (1962) Number and morphology of chromosomes in African species in the genus *Trifolium* L. *Australian Journal of Agricultural Research* **13**, 1023-1029.
Pritchard, A.J. (1969) Chromosome numbers in some species of *Trifolium. Australian Journal of Agricultural Research* **20**, 883-887.
Senn, H.A. (1938) Chromosome number relationships in Leguminosae. *Bibliographia Genetica* **12**, 175-336.
Starzycki, S. (1959) [Variability of hybrids between white clover *(Trifolium repens* L.) and red clover *(T. pratense* L.).] (in Polish). *Hodowla Roslin Aklimatyzacja Nasiennictwo* **3**, 277-319.
Starzycki, S. (1961) [Studies on interspecific clover hybrids, *Trifolium repens* L. x *Trifolium pratense* L.] (in Polish). *Hodowla Roslin Aklimatyzacja Nasiennictwo* **5**, 285-330.
Trimble, J.P (1951) Interspecific hybridisation studies in the genus *Trifolium. M.Sc. Thesis, Pennsylvania State College.*
Trimble, J.P.; Hovin, A.W. (1960) Interspecific hybridisation of certain *Trifolium* species. *Agronomy Journal* **52**, 485.

Wexelsen, H. (1928) Chromosome number and morphology in *Trifolium*. *University of California Publications in Agricultural Science* **2**, 355-376

White, D.W.R.; Williams, E. (1976) Early seed development after crossing of *Trifolium semipilosum* and *T. repens*. *New Zealand Journal of Botany* **14**, 161-168.

Williams, E. (1978) A hybrid between *Trifolium repens* and *T. ambiguum* obtained with the aid of embryo culture. *New Zealand Journal of Botany* **16**, 499-506.

Williams, E.G.; Verry, I.M. (1980) A partially fertile hybrid between *Trifolium repens* and *T. ambiguum*. *New Zealand Journal of Botany* **19**, 1-7.

Williams, E.; White, D.W.R. (1976) Early seed development after crossing of *Trifolium ambiguum* and *T. repens*. *New Zealand Journal of Botany* **14**, 307-314.

Williams, E.; White, D.W.R . (1977) Partially differentiated hybrid embryo obtained after crossing of *Trifolium ambiguum* and *T. repens*. *Proceedings of the 3rd International Congress of SABRAO* **2(b)**, 26-30.

Williams, E.; Williams, W.M.; White, D.W.R.; Pandey, K.K. (1978) Interspecific hybridisation in *Trifolium*. *Progress Report. Clover and Special Purpose Legumes Research* **11**,52-57.

Zohary, M. (1970) Trifolium. In Davis, P.H. (Ed) Flora of Turkey, Vol. 3. Edinburgh, UK; Edinburgh University Press, 384-448.

Genetics and breeding

W. M. Williams

* Victorian Crops Research Institute, Horsham, Victoria 3400, Australia.

Contents

* Former Address: Grasslands Division, DSIR, Palmerston North, New Zealand.

1. Genetics

(A) Leaf marks

i. White V-marks

A characteristic white V-mark occurs on the leaves of most white clover plants. The pale tissue in the pattern is characterized by less-elongated, smaller palisade cells which are irregularly shaped and have more intercellular spaces than the adjacent tissues (Carnahan et al. 1955) and either complete absence of, or a reduced concentration of, chloroplasts (Brewbaker 1955b).

Variation in markings is considerable and is of three main types:

(1) Shape of marking
(2) Position on the leaf
(3) Intensity.

Variations of shape and position are controlled by a complex genetic locus (Corkill 1971) designated *V*. The different shapes and positions may be assumed to represent time zones in leaflet ontogeny at which the *V* genes operate to prevent normal palisade cell development (Brewbaker 1955b). Early attempts to summarize variation controlled by the *V* locus were made by Carnahan et al. (1955) and Brewbaker (1955b). These workers used slightly different symbolism to describe similar variants but later published a joint note (Brewbaker & Carnahan 1956) to unify the nomenclature. This nomenclature is as follows:

v no marking
V^h full V, high on the leaflet
V^i full V, intermediate
V^l full V, low on the leaflet
V^b broken V (point of V absent)
V^{bd} broken V, dominant (i.e. reducing expression of associated V allele)
V^p V point (point of V only)
V^f filled V (V filled in with pale tissue on proximal side)
V^{ba} basal V
V^{by} broken V with yellow tip
V^{fg} filled V with green base (V with pale tissue on distal side) - also called spread V.

Later (Lenoble & Papineau 1970) a further gene designated V^m was described. This has the white tissue distal to the V marking, i.e. is complementary to V^f.

Excellent illustrations of some of these markings have been given by Davies (1963). No marking (*vv*) is recessive to all markings. The majority of the marks exhibit mosaic dominance, i.e. both alleles are expressed in heterozygotes, but overlapping phenotypes sometimes make it difficult to identify both alleles in a heterozygote, and there are some exceptions to the mosaic dominance rule. For example, V point (V^p) is recessive to most other markings and V^m is dominant to V^f and V^l. Partial dominance is also expressed in certain allelic combinations as a loss of intensity of the white marking of the partially recessive allele.

Expression of markings varies greatly with environment and with age and shape of leaves. Segregations of V^h and V^i, or V^i and V^l are virtually impossible to study because of overlaps among phenotypes and V^h, V^i and V^l cannot be identified without reference to standards. Because the variation in position of the markings on leaves is continuous, the V^h, V^i and V^l classification is probably one of convenience. It is more likely that there is a continuous range of variation involving several alleles.

Possible genetic complexity of the V locus was proposed by Brewbaker (1955b) and was demonstrated by Davies & Wall (1959, 1960) who split the V^{by} 'allele' into its two phenotypic components. The phenotype V^{by} is a compound one, consisting of a white broken V and a V-point which is yellow-green. Gamma irradiation of V^{by}/v heterozygotes as defoliated stolons 4-6 inches long produced the following mutations:

Loss of mark 103
Loss of white V 80
Loss of yellow tip 18

There were also numerous cases where the white V was reduced in intensity or reduced to a series of spots on the leaf. Mutant sectors were common. No mutants were found in the unirradiated control (approx. 10 000 leaves).

The phenotype separations indicate structural complexity of V^{by}. The mutant phenotypes obtained are consistent with the interpretation that they are the result of deletions as is common after gamma irradiation, i.e. only losses of genes were recorded. On this basis and from the mutant frequencies obtained, Davies & Wall (1959) concluded that losses of the whole mark and of the white broken V were one-hit events and loss of the yellow tip was a two-hit event. This would place the gene for white V-mark distal on the chromosome to that for yellow-green tip.

The V^{by} system was used by Davies & Wall (1960) in a study of methods of gamma irradiation for production of mutants in white clover. This experiment, which has relevance for mutation breeding, showed that fractionated and chronic doses gave considerably lower frequencies of mutations than acute doses. This was attributed to the multihit nature of some of the events, to an increase in the proportion of resting (and hence more resistant) cells exposed to an initial fraction of radiation, and to a changed sensitivity of irradiated cells. Two full-sib genotypes were used in the study. There was a pronounced difference between them in their sensitivity to mutagenesis despite their close relationship. Apparently sensitivity was under genetic control. Although no mechanism for this was known, the authors suggested possible differences in amounts of protective or sensitizing substances, in respiration rates and oxygen contents, tissue metabolic rates, etc.

Genetic polymorphism and geographic variation in distribution of leaf markings and their possible adaptive significance are discussed in Chapter 8.

ii. Red (purple) leaf marks

A number of red (anthocyanin) pigment patterns occur on the leaves of some white clover plants. These patterns and their inheritance have been described by Carnahan et al. (1955), Davies (1963), Hovin & Gibson (1961), and Corkill (1971).

Red leaf (R^{l}): The upper leaf surface is dark purple except for a narrow margin of green. The lower surface is sometimes slightly coloured. It is conditioned by the single dominant allele of the R locus designated R^{l} (Carnahan et al. 1955). Early taxonomists designated cultivated populations of this form as var. *purpureum*.

Diffuse red leaf (R^{ld}) : A reddish-purple pigmentation is diffused over the entire leaf surface, generally more intensely on the upper surface. It is considerably less intense than the red leaf character and is extremely variable from plant to plant and leaf to leaf. It is controlled by a single dominant gene designated R^{ld} (Corkill 1971).

Red midrib (R^{m}): This dominant character controlled by a further allele of the R locus (Carnahan et al. 1955) has a spiky purple pattern extending along either side of the midrib of each leaflet. A marker strain carrying R^{m} has been produced by Corkill (1963). This character is commonly called 'feathermark'. Davies (1963) has indicated that when R^{m} is associated with V^{by} the intensity of anthocyanin pigmentation inside the V-mark is reduced.

Tests of allelism of R^{l}, R^{ld}, and R^{m} by Corkill (1971) indicated these three genes to be very closely linked, possibly alleles of a single locus but, more probably, separable components of a compound locus. Apparent recombination percentages were:

R^{m}/R^{ld} 0.4
R^{m}/R^{l} 0.5
R^{l}/R^{ld} 0.9 .

However the possiblity of contamination was not ruled out.

Red halo: A number of different forms exist in which the distribution of anthocyanin pigmentation on the leaf is related to the white V-marking. One such form, designated $V^{r}_{2}$by Hovin & Gibson (1961), was described as a strip of red-purple pigmentation along one side of the V mark in young leaves but spreading within the V as leaflets become older, frequently extending to the proximal and distal ends of the midrib.

Another form, designated V^{h}_2 (halo-marking) was described by Corkill (1971). This was similar to V^{r}_2 but did not spread in older leaflets. Both V^{r}_2 and V^{h}_2 appear to be associated frequently with the compound V-allele, V^{by}, but it is not known whether this is a universal relationship. Hoven & Gibson (1961) indicated that V^{r}_2 was a dominant allele, showing an apparent recombination frequency of 0.1% (2 individuals in 3072) with the V allele which they designated V^{b}. Corkill (1971) obtained 1.4% apparent recombination between V^{h}_2 and another V-locus controlled allele, red leaflet (see below). In neither study was contamination ruled out as the explanation for these 'apparent' recombinants. It would appear, however, that the red 'halo' markings are controlled by a gene or genes very closely associated with the V locus.

Red leaflet: The red leaflet character is conditioned by a dominant allele of the V locus, V^{rl}_2 (Corkill 1971). The central leaflet between the V-marking and the attachment of the petiole is pigmented on the upper surface, the colour usually also extending to adjacent tissues in the two subtending leaflets. An intensely white V mark, generally V^{by}, is usually (possibly always) associated with red leaflet. Corkill (1971) failed to obtain any recombinants between leaflet and the V-marking from two isolates involving tests of approximately 1000 gametes.

This character is extremely common and occurs in some populations at high frequency. Davies (1963) described two forms which resemble red leaflet. The first involved pigmentation more or less equally on all three leaflets, often expressed only as a thin red line inside the V mark (always V^{by}). The second, also associated with V^{by}, but uncommon, was a red patch at the point of the V. The inheritance of these forms has not been studied, but they will probably prove to be associated with the V locus also.

Red fleck: In many white clover populations a high proportion of plants have red flecking on their leaflets. There is a wide range of expression of flecking, from an occasional mark on a few leaves of a plant to heavy flecking of all leaves. The degree with which pigmentation occurs on the lower surface of the leaf also varies. In most plants the pigmentation appears to be limited to the upper surface but occasionally the lower surface is also flecked. Carnahan et al. (1955) have shown the character to be associated with the R locus and designated it R^{f}. Pigmentation was shown to be restricted to epidermal cells.

Davies (1963) described an infrequent distinctive form (R^{fa}) with heavy uniform flecking on all leaves on both upper and lower surfaces.

iii. Utility of leaf marks

Most anthocyanin markings are expressed most strongly under cool conditions. In some, especially red leaflet, they can be identified reliably only in cool seasons. Besides their obvious usefulness for genetic studies in white clover the leaf markings also serve as marker genes for plant breeding programmes. They can, for example, be used for investigation of randomness of intercrossing in a polycross and to determine isolation distances for seed production. Varieties can be produced with all individuals carrying the same distinctive mark, e.g. Corkill's (1963) use of red midrib (feathermark). Such varieties can be used by agronomists, ecologists and seed production specialists to study many facets of white clover growth in mixed sowings. Some markers also have potential for characterising bred cultivars. This aspect is dealt with in Section 2 (F) p. 388.

(B) Flower colour

The corolla of white clover florets is rarely pure white. Most commonly a pink venation runs through the petals, particularly the large standard petal. A pink tint in the intervein tissue is also common, particularly at the base. Plants vary greatly in the degree of coloration of the petals, but no study of the genetics of this variation is known to this author.

A number of distinctively coloured flower forms exist within the species, including some with pink flowers resembling those of red clover (*T. pratense* L.) and a reddish-brown form with the petals becoming almost black in older flowers.

Pink flowers also appear to be associated with certain leaf markings. For example, the character R^{ld} (diffuse red leaf) is associated with pink flowers (unpublished observation), but genetic studies to clarify their association are incomplete. Some variants involving flower colour have been listed by taxonomists

(Chapter 9, Section 7). Flower colour variants deserve more attention in genetic studies as they offer some potential as useful varietal markers.

(C) Cyanogenesis

Populations of white clover have long been known to be polymorphic for cyanogenesis (Mirande 1912, Armstrong et al. 1913). Extensive use has been made of the picrate paper or Guignard test for detection of variation in release of HCN. This technique involves placing young leaves in a test tube, adding a drop or two of toluene or chloroform, and introducing a piece of picrate paper which is held in position by a tightly fitting rubber stopper (Corkill 1940, Pusey 1963). Where no HCN is released the picrate paper remains yellow. The presence of HCN is indicated within a few hours by a change in colour to a deep brownish red. The depth of colour is approximately proportional to the amount of HCN liberated (Melville et al. 1940). Using this technique, Corkill (1940) observed that HCN could be generated by mixing foliage from certain pairs of acyanogenic plants. Furthermore, he was able to classify acyanogenic individuals from a polymorphic population into three classes:

1 Those which produced HCN only when linamarase was added to the foliage
2 Those which produced HCN only when linamarin or lotaustralin was added to the foliage
3 Those which failed to produce HCN under any conditions.

He concluded that individuals in the first group were acyanogenic because their foliage lacked the enzyme. Similarly those in the second group lacked the glucosides and the third group lacked both components. Cases where HCN was produced from mixed foliage were mixtures of the first and second groups, absence of either component in one individual being complemented by its presence in the other.

The presence and absence of cyanoglucosides is determined by a single gene difference (*Ac/ac*) (Collinge & Hughes 1984), the cyanogenic character being dominant to the acyanogenic (Williams 1939, Corkill 1942, Atwood & Sullivan 1943). Similarly the presence of enzyme is governed by a single dominant gene (*Li/Li*). The cyanogenic phenotype is therefore dependent on two complementary dominant genes. There is no evidence to suggest that the genes are linked to each other or to the *V*, *R* or *S* loci (see Section 5.).

All acyanogenic plants breed true when self pollinated. However hybridisation of two acyanogenic plants may yield cyanogenic offspring if one contributes glucoside and the other enzyme. Results of genetic tests are consistent with a pattern of disomic inheritance.

Level of cyanogenic activity varies among different cyanogenic plants and follows consistent family patterns, indicating a heritable basis for the variation. Part of this variation is due to dosage effects at the *Li* and *Ac* loci. *Li/li* individuals have enzyme activities intermediate between *Li/Li* and *li/li* (Maher & Hughes 1973). Similarly *Ac/ac* individuals produce amounts of cyanoglucoside between the two homozygotes, although less than half the *Ac/Ac* level (Hughes & Stirling 1982). A further part of the variation may be due to the existence of different *Ac* and *Li* alleles. An *Li* allele conditioning lower levels of enzyme activity than the standard allele was discovered by Maher & Hughes (1973), and Hughes et al. (1984) found evidence that different *Ac* alleles gave quantitative variation in glucoside content.

There is evidence that level of cyanogenic activity is also controlled by modifying genes (Williams 1939). Segregation patterns consistent with modifiers were obtained by Corkill (1942), Atwood & Sullivan (1943), and Hughes & Stirling (1982). An effect of genetic background on cyanogenic phenotype was demonstrated by Hughes (1970) who showed that *Li/li* progeny derived from self pollinating an *Li/li* parent had markedly reduced linamarase activity. Hutton (1957) has demonstrated that doubling the chromosome number in New Zealand certified white clover (a variety with moderate levels of cyanogenic potential) leads to intensification of the cyanogenic reaction.

Two distinct β-glucosidase enzymes in white clover were identified by Hughes (1968), one form being primarily associated with germinating seeds and the other with leaf tissue. The seed form is unlikely to be controlled by the *Li* locus since it occurred in *Li/li* plants at the same levels as in *Li/-* plants. This form may therefore be functionally distinct from the leaf form, functioning in breakdown of storage carbohydrates while the latter form is concerned with cyanogenesis.

i. Environmental variation in cyanogenesis

Relative differences between plants in level of cyanogenesis are usually consistent from test to test, but the absolute results of any one experiment may vary as a result of two factors:

(1) Variation within the plant: Cyanogenic potential varies from organ to organ within the plant. It is absent from the roots, absent or very weak in the stolons, and somewhat weaker in the petioles than in the leaf blades where it reaches maximum expression (Askew 1933). Neither *Ac* or *Li* is expressed in seeds or seedlings before shoot emergence (Collinge & Hughes 1982). The reaction is usually more intense in younger leaves, still folded leaflets giving the best reaction and old leaves often giving negative tests (Williams 1939, Corkill 1942). Linamarase activity shows a similar pattern (Hughes 1970). More detailed analysis by Collinge & Hughes (1982) showed that cyanogenic glucoside content declined during leaf senescence of *Ac/Li* genotypes but not in *Ac/li*.

(2) Variation with season: Observations of many workers (e.g. Askew 1933) indicate that cyanogenic scores vary with season. Corkill (1942) reported that in New Zealand higher reactions were obtained from spaced plants in late spring than in winter. Rogers & Frykolm (1937) in Minnesota noted an increase in intensity of reaction of spaced plants from late autumn through to mid-summer, the increased reaction corresponding with increased growth of plants, rising temperatures, and decrease in moisture supply (culminating in severe drought). Analysis of different *Ac* alleles by Collinge & Hughes (1982) showed variation in the amounts of cyanogenic glucosides produced at different temperatures.

(D) Incompatibility system

Trifolium repens has a well studied gametophytic incompatibility system based on multiple oppositional alleles of the *S* locus. The species also shows frequent pseudo-self-compatibility (breakdown of the strict S allele system). Thus white clover, although usually described as self-sterile, does produce a few seeds when self pollinated. Plants producing less than an arbitrary amount of seed per flower on selfing are described as pseudo-self-fertile and those regularly producing more than that amount are self-fertile. Because pseudo-self-fertility is arbitrarily defined and both genetically controlled and environmentally modified, different authorities have published divergent statistics on its incidence. Williams (1931a) found only 26% of plants to be more or less self-fertile. Most were apparently pseudo-self-fertile, producing fewer than 10 seeds per 100 florets hand selfed, but some gave more seed and were considered to be self-fertile. Low levels of self-fertility were also recorded by Julen (1948) and Zwingli (1956). Nilsson-Leissner (1934), on the other hand, obtained selfed seed from 76.9% and 83.3% of plants selfed in 1932 and 1933 (0.71 and 0.97 seeds per head, respectively). Atwood (1941c) also found that most plants were pseudo-self-fertile. He found great variation among plants but also large variation among ramets of individual clones. This showed that pseudo-self-compatibility was modified by environment and that, under suitable conditions, most white clover plants will produce a few seeds on selfing.

Chen & Gibson (1973) showed that temperature is an important controlling variable in determining success of self-pollination. At 35°C a large percentage of ovules (76.3%) were self-fertilised after 24 hours in genotypes which were highly self-incompatible at 15° or 25°C.

Pollinating technique is also important (Section 2 (D) p. 380).

i. Genetics of self- and cross-incompatibility

Atwood (1940) established experimentally that white clover cross- and self-fertility were genetically controlled by a gametophytic incompatibility system with multiple alleles at the *S* locus. The system was first characterised by East & Mangelsdorf (1925) in *Nicotiana* and named the diploid personate system of multiple oppositional factors. It is characterised by independent action of *S* alleles in both pollen and style. Growth of pollen tubes carrying a given *S* allele is arrested in stigmas bearing an identical allele. A typical diagnostic test for this system, as used by Atwood (1940), is shown overleaf.

Two self-incompatible (or pseudo-self-fertile) parent plants are crossed. The F_1 progeny are also self-incompatible (or pseudo-self-fertile) and every F_1 plant is reciprocally fertile with both parents. The F_1

Parents		S_1S_2 I	x	S_3S_4 II						
F_1	S_1S_3 III	+	S_2S_3 IV	+	S_1S_4 V	+	S_2S_4 VI			
	S_1S_3 (♀) x III	S_2S_3 IV	→	S_1S_2 I	+	S_2S_3 IV				
	S_1S_3 (♀) x III	S_1S_4 V	→	S_1S_4 V	+	S_3S_4 II				
	S_1S_3 (♀) x III	S_2S_4 VI	→	S_1S_2 I	+	S_1S_4 V	+	S_2S_3 IV	+	S_3S_4 II

plants fall into four intra-sterile, inter-fertile groups (designated by roman numerals). By crossing any F_1 group as female with each of the other three groups and performing backcross progeny tests to the F_1 and parent plants, the above pattern is found to be diagnostic for this incompatibility system.

ii. Cytological basis of incompatibility

White clover may have two interference zones active in preventing fertilisation by incompatible pollen - one in the stigma and the other about three-quarters of the way down the style. Atwood (1941b) observed that, between 8 and 20 hours after pollination, incompatible pollen showed a lower rate of germination on the stigma and a generally slower rate of pollen tube growth than compatible pollen. This difference was sufficiently clear to enable the presence of the two classes of pollen to be identified in mixed pollinations. Chen & Gibson (1973) did not detect any difference in pollen grain germination between compatible and incompatible crosses. From the practical viewpoint a stigmatic barrier, if it is common, may be of little significance, because the longest pollen tubes from both compatible and incompatible pollen were found by Atwood (1941b) to grow equally rapidly for the first hour after pollination, i.e. until they had penetrated three-quarters of the style length. Beyond that point compatible pollen tubes continued to grow, although somewhat slowly, and some entered the ovary 3 hours after pollination, effecting fertilisation by 8-10 hours. Observations suggested that the longest tubes showed accelerating growth rates once past the interference zone. A few imcompatible tubes grew slowly into the lower part of the style but none entered the ovary.

The presence of an interference zone about three-quarters of the way down the style was also noted by Pandey (1955). This zone restricts growth of both compatible and incompatible pollen tubes and was interpreted by Pandey (1955) as a physical barrier rather than as a point of exhaustion of the food supply as suggested by Atwood (1941b). However, since imcompatible pollen tubes grew little further than this zone, Pandey considered that the location of the incompatibility reaction was also close to the interference zone.

The cytological basis of pseudo-self-fertility was investigated by Cohen & Leffel (1963). It was found that in self-incompatible plants pollen tubes traversed 75-85% of the style but no further, and in pseudo-self-compatible plants inhibition of pollen tube growth was not complete and some tubes penetrated to the ovary. Attempts to achieve selfing by amputation of stigma and style were unsuccessful.

Chen & Gibson (1973) obtained no evidence for an interference zone in the style. However the point of stopping of incompatible pollen tubes varied with temperature. After self pollination, incompatible pollen tubes made steady growth, comparable to that of compatible pollen tubes. At 15°C the incompatible pollen tubes stopped growth in the upper and middle parts of the style and at 25°C arrest occurred in the lower part of the style and upper part of the ovary. No pollen tubes entered the ovules. However at 35°C cessation of pollen tube growth was less common and a large percentage of ovules (76.3% after 24 hours) was fertilised.

These results suggested that, at low and intermediate temperatures, self incompatibility in *T. repens* is caused by reduced pollen tube growth. Higher temperatures enhance pollen tube growth and break down or circumvent the incompatibility mechanism. There was no evidence of any difference in style structure suggesting an interference zone.

iii. Large numbers of multiple alleles at the S. locus

Gametophytic systems of incompatibility are characterised by large numbers of *S* alleles and the *T. repens* system is no exception. Atwood (1942c) tested two series of plants from his breeding stocks. Results of the first series indicated that, of the 26 alleles successfully tested, 25 (96%) were different. The second sample of 41 alleles contained 34 (83%) unique alleles. Of the others, no more than two were identical.

The methods used in the above studies were extended by Atwood (1944a) to wild populations. An isolated one acre pasture in central Pennsylvania, was sampled at intervals of 4.5 m and one allele from each of 49 plants was tested. Thirty-six (73%) were different. A larger pasture area of less than 40 ha was also sampled at intervals of no less than 100 m, and one allele from each of 49 plants was subsequently tested. Thirty-nine (80%) were different. These results indicate that relatively small natural populations also carry large numbers of *S* alleles. Consequently most members of a population are interfertile.

iv. Self-fertility (S_f) alleles

Naturally occurring self-fertility in white clover was first reported by Atwood (1941a) when a plant averaged over 100 seeds per head when self-pollinated under bags in the field. Extensive genetic tests indicated that self-fertility was conditioned by a self-compatibility (S_f) factor allelic with the *S* series (Atwood 1941a, 1942a). Other such factors occurring naturally in populations were alluded to by Atwood (1942a). It has also proved possible to create such forms very easily by ionising radiation (Pandey 1956).

Plants carrying S_f set as much seed on selfing as on crossing and are cross-compatible with all testers. However, rubbing of heads in the glasshouse is not always a reliable test for S_f because pseudo-self-compatibility can be expressed strongly with this method. Leaving heads to self-pollinate without treatment is also unreliable because although some S_f bearing plants are autogamous (i.e. self-pollinating) as well as self-compatible, this is not always the case.

The S_f allele studied by Atwood (1942a, 1945) was partially dominant to ordinary *S* alleles in the style, i.e. crosses $S_fS_x \times S_xS_x$ gave some seed, although not as much as fully compatible crosses. However, when the cross $S_fS_x \times S_xS_y$ was made only S_xS_y progeny occurred among eight self-incompatible plants tested. Apparently, therefore, the normal oppositional effect was not disturbed in crosses where only a portion of the pollen was inhibited. There was an indication that S_f pollen did not compete as well with S_x on an S_f bearing style. In addition, S_f pollen did not function as well as S_x pollen on S_yS_z stigmas. It was determined that S_f pollen on the style did not stimulate S_x pollen from the same male plant, e.g. the cross $S_4\ S_8 \times S_4\ S_f$ gave no self incompatible progeny. The S_f allele thus did not violate the rule of independent action of *S* alleles in the pollen. Different sources of S_f are expected to differ in their dominance relationships and competition attributes.

The S_f allele would have relevance to white clover breeding methods utilising inbred lines, for although inbreeding has not been used widely, genetic progress in the future may be possible by this method. S_f alleles are therefore worthy of collection and study.

v. Genetics of pseudo-self-compatibility

It is apparent from the work of Atwood that most white clover plants will set some seed on selfing and that the amount of pseudo-self-fertility depends on environment. It has already been shown that the technique used for self pollination affects the results. Variation is also seen when several ramets of a clone are selfed or an individual is selfed in different years or at different temperatures. Atwood (1942b) also demonstrated a genetic component in the variation of pseudo-self-fertility. Two plants differing widely in pseudo-self-fertility were crossed. The first parent showed extremely high pseudo-self-fertility (bordering on S_f) and also high variability from year to year. The second parent produced no seed by manipulation in two seasons and was pratically self-incompatible. Most of the results of Atwood (1942b) support the conclusion that cross-compatibility relationships of the two parent plants conformed to the normal gametophytic system.

Studies of F_1, F_2 and backcross progenies from the original cross produced evidence that pseudo-self-fertility was a quantitative genetic character, probably controlled by several genes. Seven F_1 plants selfed in two successive years gave results ranging widely over the pseudo-self-fertility range. There was nevertheless a significant correlation between the two sets of averages ($r = 0.87$, $P<0.02$). F_2 progenies produced by crossing individual F_1 plants to three others gave selfing figures related to those of the non-recurrent parents. It was evident that the degree of pseudo-self-fertility was not closely under the control of the *S* locus, since plants of relatively high and low pseudo-self-fertility were found among all the *S* genotype classes arising from the original cross.

Julen (1948) showed that it was possible to select for high self-fertility in an inbred line of white clover. The original parent plant showed a self fertility of 4%, increasing to virtual self fertility in I_4. Five other inbred lines showed no increase in self fertility with generations. It was considered unlikely that the *S* locus was directly involved in selection for self fertility but more likely that modifying genes at other loci were progressively accumulated by selection.

Cohen & Leffel (1964) also found that manifestation of pseudo-self-fertility was independent of the *S* genotype. These workers obtained a series of first generation inbred lines from a range of plants varying in pseudo-self-fertility. Each population was divided into its three component genotypes and scored independently for pseudo-self-fertility. There was no significant variation in self fertility among the *S* genotypes tested.

Pandey (1956, 1959), on the other hand, developed a hypothesis which interpreted pseudo-self-fertility as a phenomenon controlled principally by genetic variation at the *S* locus although modified by other minor genes and by the environment. Pandey (1959) suggested that the *S* gene components each consist of a number of elements, variation in which accounts for quantitative variation in degree of self- or cross-fertility. Pseudo-self-fertility was considered to be caused by various degrees of 'partial mutation' at the *S* locus, allowed to express to varying degrees by segregation of modifying genes elsewhere in the genome. A very low number of selfed seeds on a plant may not be due to a partly mutated *S* allele, however, but to new mutations occurring in normal *S* alleles. The main argument for this hypothesis came from work with white and red clovers (Pandey 1956). However, Pandey's experimental evidence did not eliminate other possible explanations and the mutational hypothesis remains inadequately tested.

In view of the potential usefulness of pseudo-self-fertility in plant breeding work (Atwood 1942b), it is desirable that further work be carried out to resolve the genetics of the phenomenon.

vi. Self incompatibility in polyploid white clover.

When the chromosome number of white clover is doubled by the use of colchicine the resulting plants ($4n = 8x = 64$) carry four *S* alleles instead of two. Pollen grains now have two *S* alleles and these are no longer independent in action. The following set of rules governing the behaviour of *S* alleles in tetraploid white clover was devised by Brewbaker and Atwood (1952) following an earlier unsuccessful attempt by Atwood (1944b) to find a satisfactory explanation:

I. Homogenic diploid pollen (e.g. S_1S_1) behaves in the regular oppositional manner; it is inhibited if the allele in the pollen occurs in the style in single or multiple dose and it is uninhibited if the allele is absent from the style.

II. Heterogenic diploid pollen (e.g. S_1S_2) functions or fails depending on the following relationships:

A. Functional

1. Uninhibited - pollen functions if neither allele in the pollen occurs in the style.

2. Interaction - (a) Competition - pollen is *never* inhibited when certain alleles occur together. Competition interaction is said to occur, resulting in the apparent elimination of the functions of both *S* alleles.

(b) Dominance - pollen functions if only one of the alleles in the pollen occurs also in the style and if this allele is recessive to the other. The following is an example:

$$S_2 > S_3 > S_4 > S_1$$

Assume S_2S_4 shows competition interaction.

S_1S_2, S_2S_3	inhibited only by S_2 in the style	
S_1S_3, S_3S_4	inhibited only by S_3 in the style	in single or multiple dose.
S_1S_4	inhibited only by S_4 in the style	

B. Non-functional

1. Inhibited - pollen fails if both alleles in the pollen occur also in the style.
2. Inhibited by dominance - pollen fails if only one of the alleles in the pollen occurs also in the style and if this allele is dominant to the other.

The frequency of occurrence of competition interaction among *S* alleles was found by Brewbaker (1954) to be 8-10% in a population of 11 colchicine-doubled plants. It was predicted that this frequency would effect nearly 50% self-compatibility among plants in advanced generations of seed increase from the 11 clones studied. This would be a serious deterrent to use of tetraploid varieties.

A detailed study of dominance in heterogenic pollen grains was made by Brewbaker (1955a). Dominance relationships of four alleles in six different heterogenic combinations were determined from a massive crossing scheme. The results revealed a non-linear series for the four alleles studied:

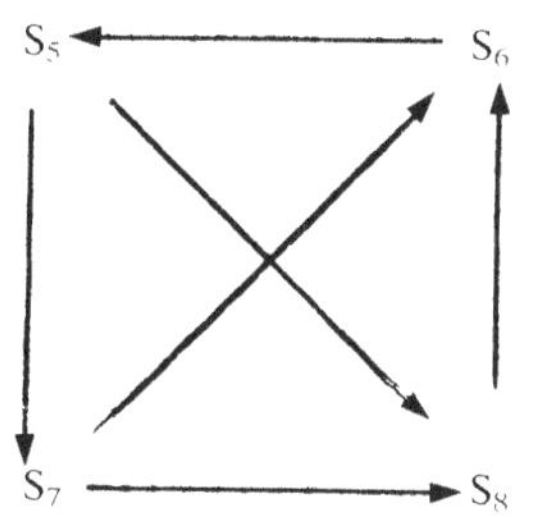

(S_8 dominant to S_6, etc.)

Pollen grains which functioned because of dominance were considerably less effective than uninhibited grains in effective fertilisation, i.e. dominance grains showed a functional disadvantage indicating weak or partial dominance in all cases.

The occurrence of dominance among *S* alleles is of interest to theoretical geneticists since in haploid pollen grains under the gametophytic incompatibility system there is no opportunity for expression of dominance and therefore no opportunity for natural selection to influence the evolution of dominance patterns. This occurrence of well defined *S* alleles dominance patterns in newly induced diploid pollen indicates the development of dominance in the absence of natural selection.

In 32-chromosome white clover plants, seed set is generally little influenced by presence of a proportion of non-functional pollen. In 64-chromosome white clover however, Brewbaker (1953) showed that variations in self- and cross-fertility were directly related to (a) the proportion of functional pollen, and (b) the degree of dominance in the heterogenic pollen. Thus reduced fertility in autotetraploids was shown to be not only due to problems at meiosis but also incompatibility relationships.

vii. Compatibility reactions in interspecific crosses

The problem of incompatibility in making interspecific crosses in *Trifolium* has been dealt with in Chapter 9. However there are a number of recorded studies of self-incompatibility patterns in interspecific hybrids of *Trifolium*. A hybrid between *T. repens* (2n = 32) and tetraploid *T. nigrescens* (4n = 32) (Brewbaker and Keim 1953) was self-compatible, although both parental species carried similar *S* allele systems. Self-compatibility was proposed to arise from competition interaction in heterogenic pollen grains, suggesting homology of the *S*-loci in the two species. Hovin (1963) obtained four allohexaploid hybrids by colchicine doubling of triploid (3x = 24) hybrids between *T. repens* (2n = 32) and *T. nigrescens* (2n = 16). The *T. repens* parent was heterozygous for S_f. One hexaploid was self-compatible and segregated in I_1 and I_2 into self-compatible and self-incompatible classes. Apparently the S_f allele behaved normally in the hybrid. The remaining three hexaploids were self-compatible, although one was pseudo-self-compatible, giving an average of 1 seed per 10 florets. Cross fertility studies suggested that independent action of S alleles and (possibly) dominance occurred, but there was no evidence of competitive interaction of *S* alleles.

Pandey (1957) obtained a self-compatible hybrid between the two self-incompatible species *T. repens* and *T. uniflorum*. Self-fertility of the hybrid was suggested to occur as a result of recombination or independent assortment of *S* genes at different loci, resulting in gametes lacking *S* alleles. Implications of incompatibility patterns in interspecific hybrids for the origin of white clover are discussed in Chapter 9.

(E) Linkage

Only five genetic loci have been identified and studied in white clover - *V* leaf mark, *R* leaf mark, *Li* linamarase, *Ac* linamarin and lotaustralin, and the *S* incompatibility system. The results of Brewbaker (1955) and Corkill (1971) indicate no linkage between any of these loci, although Carnahan et al. (1955) reported possible distant linkage (42% recombination) between the *V* and *S* loci. Independent assortment of *Ac* and *S* was confirmed by Hughes et al. (1984).

Recombination within the complex *V* and *R* loci appears to occur at frequencies up to 1.4% (Corkill 1971).

2. Breeding

(A) Breeding Methods

i. Ecological principles

White clover, like other pasture species, cannot be regarded as a species of intensive cultivation, except perhaps under strict ley management systems, and has been subject to plant breeding and selection for only the last 70 years. Consequently, a great deal of adaptive genetic diversity exists in many grassland regions of the world (Chapter 8). Early white clover breeders exploited this genetic diversity, and the ecological principles which they used to establish their selections continue to be an essential ingredient of white clover breeding today (Williams 1983).

Natural populations consist of individuals which are heterozygous at many genetic loci, and outcrossing, which is almost exclusive, ensures that genetic variability is spread throughout each population. Such a genetic structure enables populations to be adaptable to diverse situations, an important need for a species such as white clover which is grown in extremely variable environments, even within a single field (Burdon 1980, Snaydon 1962a, Turkington & Harper 1979). The white clover breeder is required to achieve improved population performance by accumulation of favourable genes while, at the same time, retaining a high degree of heterozygosity in cultivars.

In white clover the unit of growth and vegetative survival is the relatively short-lived individual stolon

(Beinhart 1963, Chapman 1983, Harberd 1963, Hollowell 1966). However the species also produces long-lived hard seeds which can survive in large quantity in the soil for many years (Champness & Morris 1948, Hyde & Suckling 1953, Suckling & Charlton 1978). White clover therefore has two systems of survival: asexually by stolon profileration, and sexually by seed production. Natural populations vary in the relative degree of development of these systems. In some environments the species survives by natural reseeding and behaves as an annual (Blaser & Killinger 1950, Hutton 1968, Jones & Evans 1977), but in permanent pastures sexual reproduction may be relatively unimportant.

High short-term yield in white clover is associated with large organs (large leaves, long petioles) (Davies & Young 1967a, Williams & Caradus 1979). Vegetative persistence is generally associated with greater density of stolons and is therefore characteristic of small leaved populations (Davies 1962a, Forde & Suckling 1977). High vegetative persistence tends to be correlated with low yield per stolon. Types of white clover with intermediate leaf size (e.g. Grasslands Huia, Aberystwyth S100) frequently yield well and have adequate persistence to cope with a range of environments.

In breeding white clover the primary objective is to improve production from grazing animals. Yield *per se* may be secondary to other characters in approaching this objective. Persistence, complementarity with associated grasses, quality for the grazing animal, compatibility with *Rhizobium* bacteria and compatibility with insect pollinators are all important ingredients in the agronomic success of pasture legumes such as white clover (Davies 1969a).

Species in pasture mixtures must be compatible, and the breeder must therefore take account of the companion grasses when selecting white clover genotypes. Ecological principles are therefore important in determining appropriate evaluation techniques. A large component of genotype × environment interaction in white clover can be ascribed to variation in plant morphology. Small leaved forms predominate as adaptations to cold temperatures, close grazing and low soil fertility, and large leaved forms to warm conditions, light grazing and high fertility. There is a large body of knowledge still to be gained on the ecophysiology of white clover and its relationship to genotype morphology and performance (Williams 1983).

ii. Genetic principles

White clover breeding rests partly on application of sound ecological principles, but improvement of populations by application of genetic principles becomes most important as populations become more closely fitted to a target environment.

The outcrossing nature of white clover and the consequent population structure, based on natural selection of heterozygous genetic combinations, sets the objective for the breeder, who must retain heterozygosity by production of synthetic varieties or development of hybrid varieties.

Genetic analyses of populations of outcrossing pasture species have indicated that genic differences for characters adaptive for vegetative persistence are mainly additive with little evidence for dominance or epistatic effects (Breese & Hayward 1972). On the other hand, characters associated with sexual reproduction, e.g. high seed production, are expected to have reduced genetic variability and show more evidence of dominance and epistasis. It would therefore be expected that populations more dependent on sexual reproduction, i.e. the free-seeding annual populations, would have a different genetic structure from the extremely long-lived persistent populations of permanent pasture. Thus within the species a range of genetic structures can be expected, with strong additive variation in perennial populations leading to high heritabilities for adaptive characters and successful phenotypic selection. However, in short-lived free-seeding populations the expected trend is toward dominance and epistasis with reduced variation for characters associated with survival.

It should be pointed out that although these differences appear likely to apply to white clover, little definitive analysis of population genetic structure has been done in the species. There is no evidence in white clover for maternal effects which might indicate a role of cytoplasmic genetic variation, as has been found in *Lolium* (Breese et al. 1965). Thus, although asexual reproduction is very important in white clover, any effects which the interplay between asexual and sexual reproduction have had in determining the genetic making of populations has not been elucidated. This remains a potentially fruitful area of future research, as do the areas of phenotypic plasticity and genotype-environment interactions and their effects on genetic structure.

Regression techniques are available for the estimation of genotype-environment effects (e.g. Breese 1969). In general it appears that many changes in performance with environment have a strong genetic component and are inherited simply as linear responses through control by genes with additives effects.

Although a tetraploid species, white clover has a diploid breeding system and there are no complications arising from its tetraploid nature. The aim of the breeder is to increase the frequency of favourable genes while avoiding inbreeding depression. This is usually achieved by development of synthetic varieties, most of which are intercrosses (in all possible ways) among several selected individual genotypes. These populations are maintained by random pollination in isolation for a limited number of generations.

Selection of genotypes for population improvement is carried out by phenotypic selection and progeny testing. Base populations may be naturalised populations (e.g. ecotypes) or hybrid populations derived from intercrossing two or more populations. A typical breeding programme would involve evaluation of the base population using spaced plants or small plots. Best plants are phenotypically selected after 2-4 years. These may be allowed to interpollinate in isolation to produce a new cultivar or, more commonly, seed is obtained from individual plants for further selection. This seed is usually grown in a progeny testing nursery for a further 3-4 years and the parent plants with best general combining ability are retained and recombined by interpollination to form a synthetic variety. One or two further cycles of selection may take place before a synthetic variety is finalised, or a phase of hybridisation and selection may be included.

For characters associated with asexual reproduction, phenotypic selection is expected to be effective and, indeed, many examples of successful selection in pasture species are available (Breese & Hayward 1972). Progeny testing is often used because the seed provides not only a genetic test but also may allow an agronomically acceptable sward trial to be carried out and the information to be improved. Breese & Hayward (1972) have warned, however, that care must be taken when using progeny testing in recurrent selection programmes. Outcrossing methods may tend to be biased in favour of genes showing dominance and epistasis to the detriment of other favourable genes, e.g. favourable recessives. For recurrent selection of pasture plants such as white clover, it may therefore be more efficient to use progeny testing methods which involve some inbreeding as these give better estimates of homozygous variation (Latter 1964).

White clover breeding is usually a 10-20 year process. It generally takes 1-2 years to set up a base population, 5-8 years to carry out initial phenotypic selection and one progeny testing period of 1-2 years for seed multiplication of the synthetic before wide scale testing is possible. Seven to 10 years is therefore a minimal period from the beginning of a programme until the breeders testing is completed and widespread agronomic evaluation begins. Because evaluations must be done in grazed pastures in a diversity of environments, the testing phase can take 5-10 years, and periods of 15-20 years from start to release of a new cultivar are common.

iii. Phenotypic selection

Individual phenotypic selection is used extensively in the early stages of programmes where it is necessary to reduce large heterogeneous populations to smaller groups. As expected from theoretical considerations, phenotypic selection is more successful for characters of higher heritability such as leaf size and disease resistance than for more complex characters such as yield.

The low effectiveness of phenotypic selection for yield in white clover was demonstrated by Davies (1960). Plants from six F_1 families were classified as 'good' and 'poor' on the basis of vigour and were checked by clonal propagation. A series of crosses, good × good, good × poor, poor × poor was made within each family. The poor × poor crosses produced F_2 families which were no different in vigour from the good × good or good × poor crosses, even though the mean vigour of the poor parents was only 27% of that of the good parents. Thus selection within F_1 families made no difference to the yield of F_2 families. The process was repeated by selecting good and poor plants within each F_2 family and making a similar series of crosses as well as outcrosses to unrelated tester plants. Although F_3 families, on average, showed a 10% depression relative to outcrossed families, the F_3s nevertheless indicated that selection in the F_2 was successful. The good × good (F_2) crosses yielded 103.8% of the yield of outcrossed families while the poor × poor (F_2) crosses yielded only 78.5%. No effect of selection in the F_1 carried through to F_3. Outcrossing good and poor F_2 plants to unrelated tester plants restored the F_1 situation, mean yields of good and poor outcrosses being 99.2 and 101.7, respectively.

These results are consistent with the finding of Suzuki et al. (1958) that heritability for yield (in Ladino clover) is very low. Greater success in the F_2 may have been a consequence of inbreeding (full-sib mating) in F_1 families. Suzuki et al. (1958) analysed heritabilities of 10 characters in a Ladino clover population by investigation of offspring-parent regressions. The lowest estimate obtained was for yield and it was concluded that phenotypic selection for yield had little theoretical validity.

iv. Clonal selection

White clover is particularly amenable to clonal subdivision by cuttings and therefore suited to clonal selection procedures. Cloning allows assessment of the average performance of a genotype over a range of environments, studies of threshold values for extreme environmental stresses, responses to different defoliation regimes, and studies under simulated sward conditions. It is most likely to be useful for characters which cannot be studied on a single individual or for single plant characters of low heritability with no non-additive genetic effects (Latter 1964, Rusmini 1965). Davies (1960) noted that evaluation of genotypes as clones frequently contradicted impressions gained from individual evaluations. A breeding scheme suggested by Gibson et al. (1963) placed heavy reliance on clonal evaluation. This approach was based on a study which indicated that clonal evaluations of parent plants for yield and persistence were sufficiently correlated with their polycross progeny performances to be useful in speeding selection programmes and reducing plant numbers before progeny testing. Coefficients of determination of progeny performance on clonal performance were 0.40 and 0.45 respectively for yield and persistence. Dijkstra & de Vos (1972) also obtained reasonable correlations between parental clones and their open-pollinated progenies for survival and plant spread under grazing in mixtures with grass. Zimmerman (1958), on the other hand, found no relationship between yields of parent clones and their polycross progenies.

Clonal evaluations could play an important role in white clover breeding. Studies of possible gains in efficiency over individual selection and progeny testing methods are justified.

v. Seedling selection

White clover populations often show large variability for seedling vigour both among and within seed lines. This variability could be useful for selection of lines with rapid and vigorous establishment and for selection of higher yielding cultivars if it could be demonstrated that selection of vigorous seedlings led to vigorous mature plants. This has indeed been reported by Zimmerman (1958) who selected within a wild population for seedlings with rapid early growth and obtained lines with growth superior to two commercial cultivars. Heritabilities for seedling yield and seedling survival under cutting have been shown to be moderately high (Connolly 1972).

vi. Family selection

Latter (1964) has discussed the use of family selection schemes and sib testing for selection of either individual plant characters or sward performance. In these schemes selected individuals are pair-crossed, polycrossed, top-crossed with a standard cultivar or simply open-pollinated to produce a maternal line of seed which is then usually grown as a progeny and subjected to selection by a number of schemes (maternal line selection). Schemes involving selection of individuals based on performance of their full-sib and half-sib relatives as well as the individuals themselves seem to be well suited to white clover breeding and constitute the most frequently used selection strategy. Selection programmes involve one or more cycles of selection of the best individuals from the best progenies, culminating in a final progeny testing phase to arrive at the parent plants for a new cultivar.

White clover cultivars are usually synthetic cultivars based on polycrosses of three or more parent plants or bulks of three or more seed lines. Polycross progeny testing is therefore an appropriate procedure in the last stages of a breeding programme. As well as giving information on the general combining ability (GCA) of the parent plants the plant replication involved also provides an opportunity for the breeder to collect sufficient seed for sward trials. Successful utilisation of polycross progeny testing in white clover has been reported by Zimmerman (1958) and Barclay (1960, 1969). It should be noted, however, that recurrent use of polycross progeny testing may fail to utilise desirable recessive genes (Hayward & Breese 1972).

In a single cycle of selection Barclay (1960) reduced the number of parent plants of New Zealand pedigree white clover from 57 to 7 on the basis of polycross progeny testing and obtained yield increases of 53% for winter growth and 36% for summer growth.

Diallel progeny testing has rarely been used in white clover. Barclay (1960) analysed GCA and specific combining ability (SCA) effects among the seven parent plants of New Zealand pedigree white clover selected by the polycross procedure. Both GCA and SCA were significant for summer growth, whereas only GCA was significant for winter growth. This result was interpreted as indicating that selection had been more effective for summer growth than for winter growth. Investigation of the contributions of each parent plant to the estimates of GCA indicated that further gains in winter growth without loss of summer growth could be obtained by selection of appropriate plants.

vii. Inbreeding

Direct effects of self-fertilisation of white clover plants can be severe. Atwood (1938) found that first generation inbred plants were 30% less vigorous than their parents while four second generation inbred lines were on average 40% less vigorous than their non-inbred parents.

Inbreeding may be of concern in the production of synthetic cultivars from small numbers of parent clones. Carnahan (1960) has shown that mean forage yields (clover + brome grass) of syn 2 populations from six two-clone synthetics (pair-crosses) averaged approximately 10% less than the syn 1 yields. In this study potentially more drastic effects of inbreeding may have been ameliorated by compensation provided by grass growth and by selective establishment of more vigorous genotypes in dense sowings. The results nevertheless showed a decline from syn 1 to syn 2 large enough to be of concern and to indicate that more than two clones should be used for a synthetic cultivar, although it may be possible to select genotypes that tolerate some inbreeding.

Davies (1960) obtained a yield depression of 10% (relative to outcrossing) after two generations of full-sib mating. The vigour of two clone synthetics at the 64-chromosome level also fell in advanced generations (Davies & Evans 1976). Mean green yields of five F_1 families grown as spaced plants were 40% better than the 64-chromosome lines from which the parent plants were selected. However, by the F_2 the advantage had disappeared and the F_3 families yielded only 75% of control values. Morphological characters (leaf size, petiole length, and spread of stolons) showed similar trends in advanced generations.

Production of inbred lines for potential use in generating hybrid cultivars is not completely restricted by the self-incompatibility system, as high temperatures or self-fertility (Sf) alleles (Section 4) can be used. Pseudo-self-fertility can be used as a method of generating S-allele homozygotes which may be utilised for production of hybrid cultivars. Progeny testing based on inbreeding methods (Latter 1964) may be a more efficient breeding strategy than outbreeding methods for pasture species because it identifies a larger range of the genetic variation available (Hayward & Breese 1972) .

viii. Hybrid vigour

Hybrid vigour has been detected in white clover in crosses between inbred lines (Atwood 1946) and between open-pollinated lines (Ellerström 1969, Semenov et al. 1973) but is not always obtained (e.g. Barcikowska 1976). Use of hybrid vigour could be made in a number of ways:

Single cross hybrid between inbred lines
Single cross hybrid between two selected genotypes
Double cross hybrid among inbred lines

Double cross hybrid among selected genotypes
Varietal hybrid between open-pollinated lines
First generation synthetic cultivar
Advanced generation synthetic cultivar

Methods at the top of this list provide for high utilisation of available hybrid vigour but at high cost and sacrifice to variability while those lower down the list make use of hybrid vigour at low cost and provide variability at the same time.

Production of commercial cultivars based on early generation hybrids cannot be realistically contemplated at present, although it is technically feasible. Clonal propagation provides the easiest method, but self-fertility genes are available which would allow production of inbred lines, and male sterile forms have been collected (Atwood 1944, Connolly 1970) which may allow for production of hybrid seed. The expense involved in production of cultivars by such methods is unlikely to be compensated for by large yield gains. The uniformity of a hybrid cultivar is also unlikely to be a desirable feature.

The synthetic cultivar provides the most efficient method of cultivar production because it provides maximum variability, and fits conveniently at the end of a selection programme utilising polycross or diallel progeny testing without need for expensive manipulations involving extensive cloning, self-fertilisation or male sterility. The principal use of hybridisation in white clover is likely to remain the use of crossing among diverse lines for the purpose of increasing available genetic variability.

By using pseudo-self-fertility it is possible to devise schemes of variety building which may be cheap alternatives to the synthetic. An example is given below:

Step 1 Generate genotypes homozygous for at least four different S alleles by using pseudo-self-fertility.

Step 2 Evaluate genotypes for specific combining ability according to the classical double cross scheme:

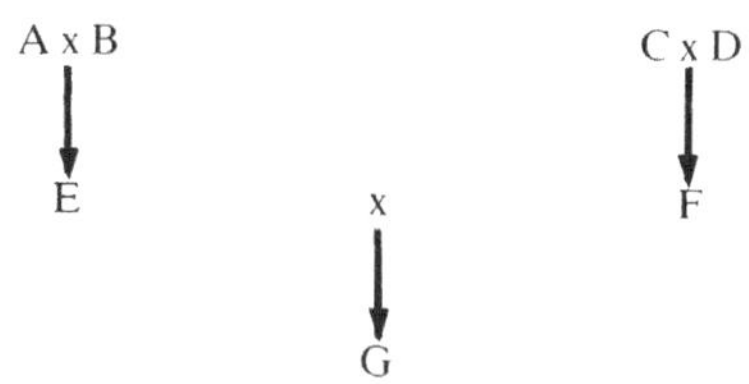

Step 3 Produce commercial seed of line G as follows:

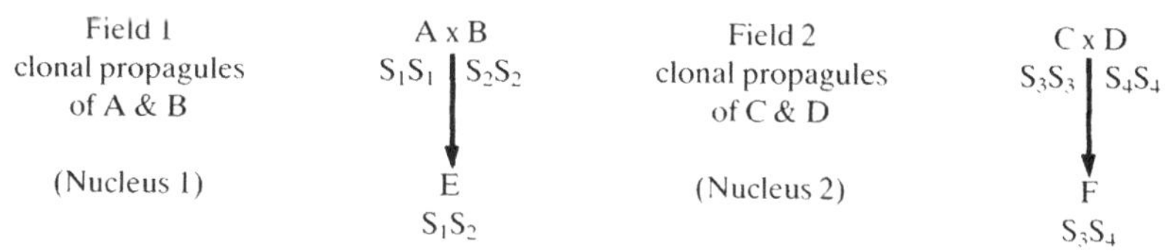

Seed of E and F could be produced in quantities similar to that produced for nucleus seed (Syn 1) of a synthetic cultivar. Nucleus 1 + nucleus 2 (lines E and F) are then mixed and sown to produce Breeder's seed which will be 100% hybrid and essentially equivalent to a first generation synthetic cultivar but will be available in commercial quantities. Subsequent multiplication would be equivalent to current synthetic cultivar production.

Such a scheme was suggested by Atwood (1942) while a similar one was devised for lucerne by Tysdal et al. (1942). It has been developed in red clover by Anderson et al. (1972) who have, however, found that a number of questions remain to be answered before the method can be effectively used for commercial breeding. In particular, problems associated with methods of selection of inbred lines, methods of inbred maintenance, stability of S alleles as influenced by inbreeding and stability of seed production require attention. More importantly, it has yet to be determined that the method produces a cultivar better than a synthetic of four parents, A,B,C and D.

ix. Mutation breeding

White clover, like other diverse species in which the natural genetic variation has been little exploited by plant breeding has been subjected to almost no mutation breeding. The mutation approach can be useful where single changes are required without alteration of the essential characteristics of a cultivar, e.g. addition of foliar tannins to established cultivars for bloat control, or generation of male sterile forms of established cultivars.

Davies & Wall (1959, 1960) determined that acute doses of gamma-irradiation were more efficient for mutagenesis than chronic or fractionated doses (Section 1). X-irradiation of 10-20 Kr was most suitable for inducing variation in quantitative characters such as leaf size and number of inflorescences in Ladino clover (Palenzona & Ferrari 1965, 1966). W.T. Jones and W.M. Williams in New Zealand, using chemicals and ionising radiation, and D.R. Marshall in Australia, using ionising radiation, have attempted unsuccessfully to induce foliar mutations for condensed tannins in large populations of white clover.

Self-fertility mutants can be created easily by use of ionising radiation (Pandey 1956) or chemical mutagenesis (Ezakova & Piskovackaya 1977).

Recent development of techniques for the exploitation of genetic mutations arising with high frequency in tissue culture (Larkin & Scowcroft 1981) may soon lead to success in this area with white clover.

x. Induced polyploidy

White clover is a tetraploid species with 2n = 4x = 32 chromosomes. Many plant breeders have investigated the effects of doubling the chromosome number with colchicine to produce plants with 8x = 64 chromosomes. Results have indicated that induced polyploidy shows little promise as a method for the improvement of the agronomic performance of the species.

Induction of polyploidy has been relatively straightforward. Methods have been compared by Atwood (1944b), De Roo (1952), Brewbaker (1952), Evans (1955), Tiemann (1966) and others. The most successful technique has involved placement of a drop of colchicine solution on the growing tip between the cotyledons of very young seedlings. Treatments with the most severe effects on plant morphology were found to be most effective by Atwood (1944b).

Polyploid plants or sectors have been identified by observing leaf and stolon sizes, flower sizes or seed sizes, the polyploids usually being larger in these characteristics. Evans (1955) and Frandsen (1945) showed that polyploid white clover had increased stomatal lengths and pollen grain sizes. Brewbaker (1955) identified polyploid sectors on the basis of larger pollen grains. Numbers of pollen grain pores have also been used. Najceveska & Speckmann (1968) determined that while normal 4x plants showed no more than 25% of polymerous grains, 8x plants always showed more than 50% polymerous grains. The same study showed that 8x plants could be distinguished on the basis of number of chloroplasts in the leaf stomata, 4x plants averaging 12.6 (range 9.4-14.7) and 8x plants 20.4 (14.5-26.6).

Inferiority of the polyploid form of white clover has been demonstrated by Atwood (1944b), Turesson (1962), De Roo (1968), Williams (1968) and Davies (1969). When colchicine has been used in the doubling process some plants have produced mixtures of 32-chromosome and 64-chromosome stolons. Atwood (1944b) used this phenomenon to compare growth potentials of 4x and 8x cuttings from the same genotype by taking paired sets of cuttings from 10 such chimaeral plants. The cuttings were grown to maturity and compared for vigour in the second year of growth (scale 1-10). The 32-chromosome cuttings averaged 8.38 while their 64-chromosome counterparts averaged only 4.53. Differences were statistically significant for 9 of the 10 selected pairs.

Turesson (1962) obtained generally higher green matter yields from polyploid white clover lines, but part of this apparent increase over the normal form may have been due to ease of harvesting, the polyploid form being taller. The polyploids were reported to be of disappointing type, being slow spreading and unsuitable for agriculture. De Roo (1968) reported that an experimental octoploid white clover produced in

Belgium yielded 9.6% more green matter than the original material but 1.9% less dry matter. Octoploids were less resistant to clover phyllody than tetraploids (De Roo 1976).

Williams (1968) obtained clover DM yields from two polyploid selections (ex S100) which were markedly inferior to the original cultivar. Davies (1969) obtained similar results from a wider range of polyploids. A Finlay-Wilkinson analysis indicated that the polyploid forms did not respond to improved conditions to the same extent as the diploid cultivars.

Polyploid populations may show some useful characteristics. Tiemann (1966) found among 12 polyploid lines studied that two combined good forage yields and quality with satisfactory seed yields. On the whole the polyploids compared well with their parental plants, five being superior in green and dry weights although containing 1.4% less DM and generally being poor in seed setting. There was a correlation between parent and polyploid for seed yield but not for green weight or dry weight. The polyploids were later flowering, showed less diminution in yield with successive cuts and generally improved in seed yield in successive years.

Rynska (1967) reported that although polyploid forms were slower developing they were more winter-hardy than the normal form, apparently because of a higher carbohydrate content. In Japan, under conditions of severe summer drought, Adachi et al. (1968) showed that whereas Ladino clover faded out in the later part of the second year, polyploid Ladino persisted and contributed about 40% of the yield in mixed swards.

Some improvement in nutritive quality of polyploids has been indicated. Davies et al. (1968) reported that polyploid S100 (AC 9) contained a higher concentration of phosphorus than the normal cultivar. As phosphorous levels in white clover are inadequate for lactating cows and only just adequate for beef animals, this improvement may be significant. AC 9 was also superior in contents of nitrogen, potassium and sodium but lower in calcium than S100. However, none of these elements was at levels which are limiting to the needs of grazing animals. In general the effect of chromosome doubling was to increase concentrations of intracellular substances (water, protein, fats, sugars, carotenes, phosphorous) but not increase cell wall materials (crude fibre, cellulose, calcium). There was no effect on *in vitro* digestibility. Reduced fibre content was also reported by Sullivan (1944), and Yoshihara et al. (1957) and Tiemann (1966) failed to detect any differences between ploidy levels in protein content. Tiemann also reported no significant differences in levels of Ca or P.

Reduced fertility and seed set are common in induced polyploids (Levan 1942) and polyploid *T. repens* is no exception, although variation is considerable (Tiemann 1966). Forty-one families of polyploid white clover investigated by Frandsen (1948) averaged yields of 71.6 kg seed per hectare compared with 275.8 for normal white clover. Bijok et al. (1968) also obtained reduced seed set in polyploids, only 3% of pods containing two seeds compared with 16% in normal white clover. Studies of possible causes of reduced seed set have revealed no meiotic irregularities (Bijok et al. 1970, Mackiewica 1970a) or any irregularities in development of embryo sacs or ovules (Bijok et al. 1970, Mackiewica 1970b). No degeneration of pollen grains was indicated by Bijok et al. (1970), but Mackiewica (1970b) reported that polyploid plants had higher percentages of aborted pollen grains and lower pollen viability. Polyploidy causes some breakdown of the gametophytic incompatibility system, resulting in increased self fertility. This is reviewed in detail in Section 4.

Results indicate that the octoploid chromosome number exceeds the optimum for *T. repens*. The hexaploid ($6x = 48$) level appears to be even less desirable. Although Turesson (1962) reported that 6x plants were rather attractive, the material became very heterogeneous in advanced generations and was found to be reverting to the diploid level (Tiemann 1980). Tiemann & Schreiter (1981) showed that fertility and seed yield of hexaploids were markedly lower than those of tetraploid or octoploid forms. Aneuploidy occurred at a frequency of 72% in the progeny of 6x plants.

(B) Genetic resources and their use

The use of genetic resources in breeding white clover was reviewed by Williams (1983). Genetic resources available to the white clover breeder are: (1) Local populations (wild or bred), (2) Introduced populations (wild or bred), (3) Closely related species.

i. Local populations

Local ecotypes may be multiplied directly for use, or used as base populations for selection of improved cultivars. Direct multiplication generally involves a seed certification scheme to ensure that quality standards are met. Local ecotypes still in use commercially include Kent wild white (England), Ladino gigante Lodigion (Italy), Victorian Irrigation (Australia) and several others.

Improved cultivars that have largely replaced the original ecotypes include New Zealand wild white (Davies & Levy 1931a,b, Doak 1933, Hadfield 1929, Levy & Gorman 1933, 1934) replaced by Grasslands Huia (Barclay 1960, Corkill 1949, 1964), and Louisiana white clover (Pieters & Hollowell 1937) replaced by Louisiana S-1 (Owen 1953).

The value of selection of improved cultivars from ecotype populations was questioned by Williams (1950). However, the approach has been very successful (e.g. Corkill 1949) and many examples of its use are to be found in the list of cultivars at the end of this Chapter.

ii. Introduced populations

Introduced germplasm can be multiplied directly for release as a cultivar, or used as a base population for selection or in hybrids with local material. Examples of cultivars developed by direct selection from introductions include Haifa (Barnard 1972), Kivi (Julen 1963), Hero, Robusta and Nora (Nilsson-Leissner & Nilsson 1940) and Sabeda and Olwen (Davies 1972) (see Index of Varieties, Germplasm and Cultivars at the end of this Chapter).

Davies & Young (1967) hypothesised that the most likely source of suitable introduced germplasm for any region is an area one step back towards the Mediterranean centre of origin. Populations from such areas have sufficient winter-hardiness for most winters and also have higher yields and longer growing seasons than local material. Similarly populations taken from lowland to hill regions were suggested to be ideal breeding material. This hypothesis is supported by observations in Britain that use of French germplasm has led to several cultivars, and development of other European cultivars follows a similar pattern.

When introduced material is grown for long periods natural selection and introgression from local populations leads to some local adaptation. Experimental evidence for this was obtained by Nilsson-Leissner (1935) and Sylvén (1937) for Danish and German populations grown in Sweden. Differentiation of Ladino populations has been reported from Canada (Dessureaux 1949), USA (Chamblee 1954, Portz & Jackobs 1955, Jackobs & Hittle 1958) and Australia (Barnard 1972). Selections from local Ladino populations in various parts of the USA include Pilgrim (Leffel & Gibson 1973), Merit (Wilsie 1964), Regal (Johnson et al. 1970) and Tillman (Gibson et al. 1969). New Zealand wild white apparently differentiated into two forms, one adapted to cattle grazing (No. 1) and the other to sheep grazing (No. 2) (Levy & Gorman 1934). ML-48-65 is a Canadian line developed after about 16 years of natural selection for winter-hardiness in introductions from the USSR (Cooke & Sonmor 1977).

Introductions are often used in hybrids with local populations to provide properties lacking in the local material. Examples include Blanca R.v.P. developed in Belgium from local wild white x Kersey (UK), and S100 which was composited from genotypes of New Zealand, Dutch and English wild white origin for use in the UK. When hybridisation involves an introduction carrying deleterious characters in addition to desirable ones it may be necessary to backcross to the locally adapted material. Barclay (1960, 1963) bred Grasslands Pitau white clover by one generation of backcrossing to New Zealand wild white after hybridising a large leaved Spanish ecotype with New Zealand material. Selection was carried out in the backcross populations for plants carrying the seasonal growth characteristics of both parents, but closer in leaf size to the New Zealand parent.

Hybrid populations are usually intermediate in type between the parents, although where selection is made for increased yield, leaf size is likely to move towards that of the larger leaved parent. Reciprocal effects arising from direction of crossing are generally unimportant, although Barcikowska (1976) found a small maternal effect for grain weight and thickness of flower stalk in one cross.

iii. Interspecific hybridisation

As noted earlier (Chapter 9), *T. repens* can be hybridised with a few other *Trifolium* species. *T. occidentale, T. nigrescens, T. uniflorum* and *T ambiguum* all produce fertile hybrids with white clover and therefore justify investigation for agronomic characters which might contribute to varietal improvement.

T. occidentale

This species resembles *T. repens* in general appearance, although it is cytologically, ecologically, and geographically distinct and lacks vegetative vigour. It has a relatively narrow distribution, being found in dry, warm places, on rocks, in dry pastures and sand dunes, always near the sea, and geographically restricted to the Channel Islands and the Channel coasts of southern England and north west France (Coombe 1961, Coombe & Morisset 1967). Its relatively short stems and very small leaves render it morphologically suited to dry, windy, coastal habitats. It is probably resistant to salty soils. Coombe (1961) has indicated that it is not a highly variable species and, being self-compatible, is probably highly homozygous.

Gibson et al. (1971) have reported that *T. occidentale* may be a useful source of virus resistance, showing no symptoms of clover yellow vein, white clover mosaic, or peanut stunt viruses. The species may be useful for improvement of drought tolerance in white clover, although its small stature suggests low yield potential. *T. occidentale* may also facilitate hybridisation of *T. repens* and *T. uniflorum* (Gibson & Chen 1975).

T. uniflorum

This is a perennial species with a deep, woody taproot which could be useful in providing resistance to root pests and diseases as well as drought tolerance. It also has a large seed which could aid in the improvement of seedling vigour, and is reported to have some resistance to sooty blotch (*Cymadothea trifolii*), powdery mildew (*Erisyphe polygoni* DC) and to viruses, only clover yellow mosaic virus causing symptoms on it (Gibson et al. 1971). Gibson & Chen (1975) have released germplasm of *T. uniflorum* x *T. occidentale* hybrids (designated SC-2) and SC-2 x *T. uniflorum* hybrids (designated SC-3). The frequency of obtaining viable hybrids from crosses of *T. repens* x SC-2 and SC-3 was reported to be higher than for crosses with *T. repens* and *T. uniflorum*. SC-2 and SC-3 therefore facilitate the combining of germplasm of *T. repens* and *T. uniflorum*. Embryo culture also produces high numbers of *T. repens* x *T. uniflorum* hybrids (Williams et al. 1982, Williams & Williams 1983).

T. nigrescens

This Mediterranean species is extremely variable but is a free seeding winter annual. Hybrids with white clover are weak perennials which root infrequently at the nodes and may be more suceptible to viruses than *T. repens* (Gibson et al. 1971). It does not, at present, appear to be a promising source of germplasm for improvement of *T. repens*.

T. ambiguum

T. ambiguum has a large rhizomatous root mass which may be useful for drought tolerance and insect resistance and is reported to be resistant to seven different viruses (Barnett & Gibson 1975). This species is taxonomically very closely related to *T. repens* (Chapter 3) and has given fertile hybrids with *T. repens* (Williams & Verry 1980, Williams & Williams 1983).

Other species

A number of other *Trifolium* species, which have not yet been successfully crossed with *T. repens*, carry characteristics which breeders would like to transfer to white clover.

T. semipilosum has shown some promise in Australia (Jones 1973), and in New Zealand has shown potentially useful drought tolerance but little growth potential, especially in winter (Williams et al. 1978). The possiblity of hybridisation with *T. repens* has been considered but attempts, so far, to cross these species are not encouraging (White & Williams 1976).

A character which may be desirable in white clover foliage is a moderate level of condensed tannin for control of bloat (Jones & Lyttleton 1971) and protection of proteins from rumen digestion. Two species, *T. arvense* (haresfoot trefoil) and *T. affine* contain tannins in the leaves but neither species is closely related to *T. repens* and prospects of interspecific hybridisation are negligible. In such cases more sophisticated methods such as the genetic 'transformation' technique reported by Pandey (1975) may ultimately provide an answer.

(C) Breeding Objectives

i. Plant type

Plant 'type' in white clover can be described by use of three parameters: spread (diameter), height (petiole length and, perhaps, petiole angle) and density. All three parameters are affected by the environment and vary with e.g. season, soil fertility, soil moisture and plant age. Gibson & Trautner (1965), for example, observed that plants of Louisiana S-1 were taller than plants of Ladino at low temperatures, whereas the reverse applied at high temperatures (Table 1). Density is, in turn, a complex concept involving stolon branching frequency, leaf size, and stolon thickness. Breeding for yield and persistence involves manipulation of these morphological parameters. Leaf size, plant height or petiole length, and green weight have been shown to be highly correlated in plants of diverse geographic origins (Davies & Young 1967, Barcikowska 1976, Table 2).

Table 10.1. Petiole lengths (cm) of Ladino and Louisiana S-1 white clovers at four temperatures

	Temperature (°F)				
	50	62	74	86	Cultivar means
Ladino	8	11	18	19	14
Louisiana S-1	12	14	16	17	14

From Gibson & Trautner (1965).

Broad sense heritabilities of leaf size, petiole length, and stolon thickness have been shown to be around 0.50 when calculated from parent-offspring regressions (Suzuki et al. 1958, Barcikowska 1976).

$$H = \frac{\sigma^2 p}{\sigma^2_p + \sigma^2_p}$$

where $\sigma^2 e + \sigma^2 p$ = total variance between plants

and $\sigma^2 e$ = variance among leaves within plants

Broad sense heritability estimates for leaflet length by Davies & Young (1967) varied considerably from year to year. In one year most populations ranged from 0.2 to 0.5, two populations reaching heritability values of 0.7. In another year some populations had heritability values below 0.2. Heritability values for leaflet shape (length/width) were generally under 0.2, with a few ranging to 0.4.

Table 10.2. (a) Correlation coefficients (r) for spaced plant characters of white clover populations (means of correlations for 3 years). From Davies & Young (1967). All quoted r values highly significant ($P<0.01$).

	Leaf height	Leaflet Length	Leaflet Width	Leaflet Length/width	Plant Spread	Green weight
Leaflet length	0.89					
Leaflet width	0.88	0.98				
Leaflet length/width	–*	–	–			
Plant spread	0.76	0.64	0.59			
Green weight	0.78	0.80	0.80	–	0.78	
Flowering date	–	–	–	–	–	–

*Non-significant and/or heterogeneous coefficients.

(b) Phenotypic correlation coefficients (r) of clones. Second year of growth modified from Barcikowska (1976).

	Plant height	Leaf surface area	Rosette Diameter	Length of stolons
Leaf surface area	+0.38**			
Rosette diameter	+0.60**	−0.16		
Length of stolons	+0.34**	+0.48**		
Petiole length		+0.46**		
Green weight	+0.61**	+0.41**	+0.69**	+0.07
Stolon thickness	+0.42**	+0.70**	+0.03	+0.53

**r significant at $P=0.01$.

ii. Root morphology

There have been few studies of variation among white clover genotypes for root characteristics. Ueno & Yoshihara (1967) compared the growth habits of adventitious roots of white clover cultivars after disappearance of the tap root. The cultivars best able to develop nodal roots into suitable substitutes for the tap root were those with short stolon internodes and better root growth at the plant base. Caradus (1977, 1981) observed that large-leaved, large-organed plants had thicker roots than smaller-leaved plants. This applied to both the seedling tap root and to adventitious roots formed by mature plants.

A great deal more research on white clover root systems is required before genetic manipulations of root morphology can be expected to contribute to improved yield or persistence of white clover cultivars.

iii. Yield

White clover shoot yield can be divided into a distinct horizontal component immediately above the ground surface (the stolons) with a vertical component (the leaves) above. 'Harvestable' yield is made up almost entirely of leaf. The horizontal stolon component is largely unharvested and, with the root system, provides for regrowth, plant spread, and persistence through stolon branching, i.e. a system of asexual reproduction. Few data are available on variation among white clover genotypes for the relative proportions of leaf and stolon. Williams & Hoglund (1978) have shown that such variation does exist. In a cool temperature (12.5°/6°C, day/night) a Mediterranean population (from Spain) produced approximately 48% of its shoot dry matter in the leaves and a New Zealand population produced only 40% as leaf. At moderate temperatures (17.5°/10° and 22.5°/14°C) this situation was reversed. This is an aspect of white clover growth which, with studies of root growth and ratio of shoot to root, deserve more attention by plant breeders.

Under conditions optimum for white clover growth, larger leaved forms are more productive than smaller leaved forms (e.g. Davies 1958a, Lehle 1959). However, in pastures, conditions are optimum for only short periods and in many environments the main morphological feature for persistence under sub-optimal conditions (dense stolon branching; Beinhart et al. 1963, Chapman 1983) becomes important for much of the time. Dense stolon branching is found in populations adapted to survival under continuous sheep grazing as in Kent wild white (Ware 1925) and in populations adapted to other stress factors such as winter cold (Daday 1954a). Densely branched forms are more numerous in populations of all types, from the giant-leaved Ladino to the small-leaved forms indigenous to Northern Europe. They tend to show agronomic advantages over their less-branched counterparts, as has been shown, for example, in the large-leaved forms used in the south-eastern USA (Gibson et al. 1963, Beinhart et al. 1963, Knight 1953a).

Rhodes & Harris (1979) have shown a strong inverse correlation between harvested yield and unharvested stolon stubble for cultivars grown under four treatments (Figure 1). Recent cultivars bred for increased stature, e.g. Olwen, appear to have achieved high harvest yields in the first year but at the expense of unharvested stolon. This may lead to reduced persistence and competitive ability in subsequent years.

Taken over the species as a whole, the most densely branched types occur among less-productive small-leaved forms. Usually the plant breeder must balance the characteristics for high productivity under 'ideal' conditions (large leaf, rapidly elongating thick stolons) with those for survival under adverse conditions (dense branching, usually associated with smaller leaf, thinner stolons, and lower productivity). Some of the most successful ecotypes for grazing lands are of the 'intermediate' type, combining medium-sized leaves for quite high productivity with medium stolon branching capacity which provides adequate persistence for many stresses (e.g. Haycock & Ollerenshaw 1982). From such ecotypes some excellent commercial cultivars have been developed. Outstanding examples include the New Zealand No. 1 ecotype (Levy & Gorman 1934) from which New Zealand certified white clover was selected (Barclay 1960) and Denmark's Morsö ecotype which forms the basis of several European cultivars.

White clover breeders have not used sophisticated index selection methods to combine attributes for high yields and persistence, as it has been possible to make good progress by empirical methods. The continuing requirement for broadly based, variable cultivars containing a range of variation around a given plant type will probably be met in the future by selection methods which partition yield into components and base selection on observations of separate components, e.g. vertical and horizontal (Rhodes & Harris 1979). This approach will require a greater understanding of stolon branching and the partitioning of assimilates into stems and leaves (see Chapter 4).

iv. Seasonal yield

White clover grows best in temperate environments in late spring and summer and is poorest in winter (Chapter 7, Section 3. Haycock 1984). The Mediterranean region has proved to be a rich source of autumn and winter-growing, but frost tender, ecotypes of pasture species including white clover (Eagles & Othman 1981). These winter growing forms have been most successful in warm-temperate climates without severe winter cold. In cool temperate climates the best that the plant breeder can hope to achieve is a cultivar which closes the winter feed gap, i.e. grows later into autumn and begins growth earlier in spring. The discovery by Ollerenshaw et al. (1978) and Ollerenshaw & Baker (1981) that some populations from regions with a cold winter climate also show good growth at low temperatures may provide plant breeders with new scope for selection in this direction.

Use has been made of Mediterranean germplasm for improvement of cool season productivity in New Zealand (Barclay 1960, 1963, Williams & Barclay 1975), Australia (O'Brien 1970a,b) and in Britain (Davies & Evans 1968, Connolly 1970).

In New Zealand, a population of Spanish origin was found to grow better in the cool season than the indigenous material (Barclay 1960). Comparisons of Spanish and New Zealand white clovers at several locations over several years indicated that the Spanish was superior in autumn and early winter in the warmer North Island but in the cooler South Island advantages of the Spanish were less consistent and restricted to autumn. At other times of the year the Spanish was inferior to indigenous New Zealand material. The Spanish material often began to show its 'cool-season' superiority in late summer when temperatures were high and moisture sometimes limiting. Furthermore the Spanish material was usually at its poorest relative to New Zealand material in spring when temperatures were similar to those prevailing in autumn when the Spanish was at its best. These observations indicated that the cool-season growth of Spanish white clover was not controlled closely by temperature *per se*. More likely the Mediterranean material had a completely different seasonal growth rhythm from the maritime New Zealand population. The maritime material showed restricted autumn-winter growth with a spring-early summer peak, but the Mediterranean material tended to show restricted winter-spring growth with a late summer-autumn growth peak (Williams & Barclay 1975). Hybridisation of Spanish and New Zealand white clovers gave rise to F_1 material combining the seasonal growth characteristics of the parents. Improved cool-season growth was apparent in the F_1 (Barclay 1963) but mainly in the warmer North Island (Williams & Sedcole 1978). The cultivar Grasslands Pitau (formerly Grasslands 4700) was developed from selected F_1 material by back-crossing to New Zealand plants to improve plant density followed by further selection for cool-season growth (Barclay 1969). This cultivar continues to reflect the behaviour of its Spanish parent, showing improved cool-season growth regularly in the warmer North Island but only sporadically in cooler South Island regions (Goold & Douglas 1976, Lambert et al. 1969).

In Australia the cultivar Haifa (Barnard 1972) has been developed by selection within a population introduced from Israel. In addition to its cool-season growth this cultivar shows superior drought tolerance, another attribute of Mediterranean material. Other Mediterranean populations reported to do well in Australia have been from Portugal, Algeria and Lebanon (Neal-Smith 1970).

Ecotypes from Israel have featured strongly in the British Isles where interest has lain not only in improvement of cool-season growth but also in enlarged leaf sizes for improvement of competition with grasses. Mediterranean populations are frequently large leaved (Chapter 8). Davies & Evans (1968) reported that F_1 and F_2 hybrids involving Kersey × Israeli ecotypes and Kersey × Turkish ecotypes showed improved summer and autumn growth at Aberystwyth and maintained satisfactory winter hardiness. The cultivars Kerry and Katrina have been released by the Welsh Plant Breeding Station as a result of this programme. Connolly (1970), working at Carlow, Ireland observed that four ecotypes from Israel showed better autumn growth than Blanca (the best cultivar available for that region) but were inferior at all other times. F_1 progenies from crosses of the best Israel ecotype with Blanca performed well in mown grass-clover plots, the F_1 mean being better than the parental controls in all seasons except early spring. However, the difference reached statistical significance only in late autumn. The F_1 also showed less severe growth depression than Blanca when high N levels were applied. In that environment improved tolerance to high N would probably be more important than autumn growth, which constitutes only a small proportion of annual yield.

v. Persistence

White clover is usually described as a perennial species, but its 'perenniality' is the result of renewal of plant parts rather than the existence of a single long-lived meristem (Hollowell 1966). Individual stolons have relatively short life-spans. Close observation of white clover plants growing in fixed dune vegetation in Britain indicated that no part of a stolon survived for more than two growing seasons (Harberd 1963). In the summer-dry environment of South Carolina individual stolon segments 5 months old have been observed to show approximately 75% death in the presence of insects and fungal pathogens (Beinhart et al. 1963). Under New Zealand hill country conditions Chapman (1983) showed that less than 10% of stolons survived for longer than a year.

The seedling tap root also has a limited life-span. Studies of Ladino clover in Michigan by Westbrook & Tesar (1955) showed that some tap roots were already dead late in the first production year, and by the middle of the second summer after sowing, all tap roots were dead in four separate trials. Neither grazing management nor improved soil fertility prolonged the life of the tap roots beyond the second summer, although root death in the first season could be slowed by reduced severity of defoliation and topdressing with both P and K.

Tap root death does not appear to be associated strongly with lack of persistence of white clover. Knight (1953b) found no association between root characteristics and persistence of Ladino clover genotypes. James et al. (1980) improved persistence of 'Tillman' Ladino clover in North Carolina by treatments with insecticide, but this effect was not associated with improved tap root survival. Gibson & Trautner (1965) demonstrated that tap root death *per se* was not associated with loss of production of individual white clover plants because adventitious roots were able to enlarge and fulfil the functions of a tap root.

Persistence of white clover has the following elements:

1 A requirement for continual renewal of the stolon population by asexual reproduction.

2 A requirement for survival of the stolon roots.

3 Ability to produce seed when environmental stress limits capacity for asexual reproduction.

Continual renewal of the stolon population is best achieved by dense stolon branching. Knight (1953b) demonstrated that persistent clones of Ladino clover produced large numbers of stolons with relatively small stolon diameters. Other evidence for association of persistence with dense stolon branching is summarised in Chapter 4.

The importance of maintaining the stolon adventitious roots was demonstrated by Janes et al. (1980). Ground cover of Ladino clover after 3.5 years was 12% in control plots. Treatment with combined fungicide (benomyl) and insecticide (carbofuran) improved ground cover to 88% by reducing stolon and root damage. Gibson & Trautner (1965) suggested a possible association between plant death and failure of the adventitious root system in the period immediately following tap root death which often coincides with a summer stress period.

The possibility of breeding for persistence by adjusting the season when a cultivar will lose its seedling taproot has not been investigated. It is not known whether populations possess repeatable variation in longevity of the seedling taproot, i.e. inbuilt physiological variation, or variation related to season of growth, rather than to disease susceptibility. It is not known how much genotypes vary in the ease or rapidity with which they form thickened roots under different environmental conditions or, indeed, whether thickened roots are required for maximum productivity.

There is, to some extent, an inverse relationship between vegetative persistence and profuseness of flowering. This is illustrated by a comparison between the very profuse flowering 'Dutch' type which is normally annual or biennial, and the very persistent 'English wild' type which is sparsely flowering. Stolon nodes can produce either a branch stolon or a flower head but not both. Each flower head is therefore produced at the expense of a potential branch stolon, and the more flower heads the fewer branch stolons are available to contribute to vegetative persistence. Gibson (1957) demonstrated that profuse flowering reduced persistence (stolon survival) of white clover during a very hot Alabama summer. Genotypes from northern latitudes which normally flowered poorly in Alabama were induced to flower profusely by application of long days from February to June, inclusive, and compared with the same genotypes given normal daylengths. Average numbers of live stolons transecting a one metre line by late July were 28 and 8 for the normal and extended daylength treatments, respectively.

Flowering and seeding play an important role in maintaining stand persistence when environmental stress renders vegetative persistence through stolon proliferation unsuccessful, e.g. in sub-tropical Australia (Jones & Evans 1977). Studies of farmers' fields in Michigan by Westbrook & Tesar (1955) indicated that natural reseeding of Ladino clover was helpful in maintaining stands. Suckling (1952) showed that natural dissemination of white clover seed in sheep droppings can be an important method of spreading seed in hill regions. Some flowering is therefore important for persistence and a compromise between vegetative persistence and flowering under stress for reseeding is usually a desired breeding objective. Variation in timing and prolificacy of flowering are important in this regard and are discussed in Section *xv*. p. 378.

vi. Cold tolerance

White clover populations vary widely in cold tolerance. Much of this variation is adaptive (Chapter 8), populations from regions with freezing winters being cold hardy and winter dormant and those from regions with mild winters being winter active and often frost tender. Thus populations from low altitude Mediterranean climates are usually winter active and frost tender (Davies 1970; Greenham & Daday 1957) compared with those from north west Europe. In the severe winter climate of northern Europe even material of cool temperate maritime origin is unable to survive (Akerman 1942) and in the northern USA material from southern states will not survive (Aamodt et al. 1939, Smith 1949, Ronningen 1953) but locally adapted material will.

Considerable variation in the frost tolerance of naturally occurring genotypes within northerly latitudes of 52° to 68° N was observed by Ollerenshaw & Haycock (1984). Analysis suggested that exposure was of importance in determining frost tolerance with plants from exposed sites being more frost tolerant than those from sheltered or snow-covered sites at the same latitude. In collecting frost tolerant plants, factors such as wind exposure, snow cover and solar radiation early in the day were more important. Few genotypes were identified which combined frost tolerance and ability to grow at low temperatures, indicating that scope exists for breeding for improved winter survival and growth at sub-optimal temperatures.

Ladino clover is adapted to a wide range of environments and is widely used in many parts of the world. Its survival in the severe winters of the northern USA is generally adequate unless ice sheets form, but this phenomenon is frequent enough to limit Ladino stand life in some localities. There has been considerable interest in development of cold-hardy selections from Ladino clover. Flanagan (1952) and Ronningen (1953) detected genetic variation for cold tolerance within Ladino clover and indicated that selection of hardier lines with desirable agronomic characteristics should be possible. In general, winter injury in Wisconsin was related to plant vigour and density of growth, the most vigorous, dense plants being less winter-injured (Ronningen 1953).

Laboratory methods for screening plant material for cold tolerance have been described by Smith (1949), Flanagan (1952), Wood & Sprague (1952), Greenham & Daday (1957, 1960), Berezovskii & Agaserdiev (1979) and Eagles & Fuller (1983). Advantages of laboratory screening over field evaluations include avoidance of variations in degree and duration of ice cover and the possible occurrence of lethal air temperatures in the field.

Wood & Sprague (1952) compared a Ladino clone selected for cold resistance in the field with a clone selected for moderate cold resistance and unselected Ladino. Freezing to 10°F distinguished the two selected clones from the unselected material but did not separate the good from the moderate one. Freezing to 5°F caused more stolon death and virtually killed both the moderate and unselected lines, leaving the highly resistant clone less injured (Table 3). Vegetative vigour during the growing season appeared to be inversely related to cold hardiness, the cold resistant clone being less productive than unselected material. Autumn dormancy was also apparently associated with greater cold hardiness. Analysis of the clones for metabolic factors associated with cold hardiness are reviewed by Hart (Chapter 4, Section 7).

Harris et al. (1983) observed large differences between cultivars in changes of leaf, stolon and root biomass, stolon length and reserve carbohydrate during winter in the UK. Early spring growth was correlated with the weight of stolon which overwintered. Cultivars such as S184 both entered and completed winter with more biomass than large-leaved cultivars such as Olwen. Furthermore, S184 was able to support a greater weight of leaf and, in general, this type of small-leaved white clover was better adapted to persistence under adverse conditions and to early season growth in Britain than types with the ability to grow at low temperatures.

vii. Drought and heat tolerance

White clover is best adapted to moist temperate environments, but nevertheless its ability to survive drought varies greatly. In the south eastern USA, for example, populations adapted to Mississippi and Louisiana show considerable relative drought and heat tolerance. Selection of genotypes from these populations has led to the breeding of heat tolerant lines which show improved survival in very hot, dry summers (Owen 1949). One such selection culminated in the release of Louisiana S.1 white clover (Owen 1953) which has shown good summer survival in Queensland (Schroder 1963).

Within Ladino clover, Yamada (1959) has demonstrated the existence of variability for drought tolerance, dense, leafy ('polyphyllous') plants being superior to open ('oligophyllous') types. Many Mediter-

Table 10.3. Percentage recovery of Ladino stolons following exposure to 10° and 5°F temperatures in a cold chamber. Results determined after 25 days growth in a glasshouse. (Expressed relative to unfrozen control samples).

	Treatment			
	Frozen to 10°F in 17.5 hours		*Frozen to 5°F in 21.5 hours*	
	Stolons alive (%)	Dry weight of leaves (%)	Stolons alive (%)	Dry weight of leaves (%)
Unselected material	76.9	52.8	3.0	0.4
Moderately cold resistant clone	86.8	50.7	0	0
Very cold resistant clone	104.0	74.1	42.0	14.4
Means	89.2	59.2	15.0	4.9

Adapted from Wood and Sprague (1952).

ranean populations have adapted to survival in hot, dry summers by development of a seasonal growth rhythm, giving maximum growth in autumn-winter, flowering in early spring and then undergoing a period of relative dormancy from late spring until late summer, presumably associated with physiological attributes conferring drought and heat tolerance. Such a growth cycle is exhibited by the Spanish population studied in New Zealand by Barclay (1960), Williams & Barclay (1975) and Williams & Hoglund (1978). In eastern Australia, ecotypes from Israel, Lebanon and Portugal have shown better heat and drought tolerance than other lines of white clover (O'Brien 1970a,b, Neal-Smith 1970). The cultivar Haifa was developed in eastern Australia by selection from material collected in Israel for superior heat tolerance, productivity, persistence and seedling ability (*CSIRO Plant Introduction Review* 7, 1970). In northern New South Wales the Clarence Valley ecotype appears to exhibit extreme Mediterranean adaptations, i.e. good cool season growth, early flowering and free seeding. In years of adequate summer rainfall this ecotype behaves as a perennial, but where summer rainfall is poor it is a well adapted winter annual (O'Brien & Cohen 1972). Drought resistance is also a requirement of some European regions, including Scandinavia and the Balkan States. Swedish studies (e.g. Akerman et al. 1939, cf. Jones 1940) indicated the existence of superior drought tolerance of deep rooting, large leafed forms with long petioles. Savlof released the cultivar Hero in 1939 as a more drought tolerant cultivar than Morsö.

viii. Nitrogen tolerance

Large quantities of fertiliser nitrogen are applied to mixed pastures in some regions. Under these conditions white clover may be completely eliminated from the sward. This elimination has little effect on the nitrogen economy of the pasture, but without white clover the feed is of poorer quality, and there is a good case for trying to retain white clover in the sward. Several workers have indicated the existence of variation in nitrogen tolerance among white clover cultivars. Reid (1961) found Kersey to be more persistent in mixed swards than S100 or S184 when nitrogen fertiliser was applied but noted no differences in the absence of applied nitrogen. Curran & Reilly (1965) showed differences both among varieties (Kentish, S100, New Zealand mother white, and a synthetic composite) and within cultivars for growth at three levels of nitrogen. The Belgian cultivar Blanca R.v.P. was superior to all other cultivars tested, including Kersey, when grown with high nitrogen fertiliser applications in Ireland, although it was not markedly better without fertiliser nitrogen (Hoen 1970, Connolly 1970a). In English trials Blanca also performed well both with and without nitrogen, although it tended to be better with nitrogen (Aldrich 1970a).

A major programme began in 1949 at the Welsh Plant Breeding Station to develop larger leaved forms better able to compete with grasses boosted with nitrogen fertiliser (Jones & Davies 1966). Initially the large leaved Ladino clover was hybridised with S100. Ladino itself can be outstanding in the sowing year in Britain (and elsewhere) but usually fails to persist. F_1 Ladino x S100 families were generally outstanding in vigour and had the desired leaf size. F_2 families were slightly closer to S100 in leaf size but showed the expected range of variation (Davies 1958c). F_3 and F_4 families were established in 1956 and 1957 in a sward trial which lasted until 1961 (Davies 1962). Many progenies proved superior to S100, Nomark especially in dry periods in late summer. Recovery from grazing was also quicker. The programme was extended to include large-leaved materials from other Mediterranean countries combining morphology similar to Ladino with high cyanogenesis to eliminate the problem of slug damage. Material from Turkey and Israel was used, the latter showing frost tenderness. When crossed with Kersey the Mediterranean materials gave hybrids with large leaves and satisfactory winter hardiness (Davies & Evans 1968). A further selection from large leaved material collected in France produced an experimental cultivar in about 1964. Cultivars released by the Welsh Plant Breeding Station from this programme include:

Sabeda (from material collected in Normandy, north west France recommended for use under high nitrogen conditions).

Olwen (from material collected in south west France).

Kerry (from crosses between Kersey and Israel material).

Katrina (from crosses between Kersey and ecotypes from Turkey and Israel).

High nitrogen tolerance may be a characteristic of some Mediterranean white clover populations. Hoglund & Brock (1974) in a growth cabinet study observed that high nitrogen applications were more suitable for Grasslands Pitau (New Zealand x Spanish) than for New Zealand white clover, particularly at high temperatures.

Davies & Evans (1975) investigated the nitrogen tolerance of 12 collections from pastures with a history of high nitrogen application and high stocking rates. Ages of the pastures ranged from 5 to 200 years. Most of the collection lines were small-leaved types inferior to S184 at both low N (nil) and high N (200-275 kg/ha/year). However, two collections were outstanding, exceeding the yield of S100 in mixed ryegrass-clover swards by 70% at high N. They were also superior at low N. Total yields of the mixed swards at high N were reduced by the presence of these stronger growing clovers as a result of competition. It would therefore be necessary to determine whether increased animal performances arising from a better balanced sward would compensate for loss of total yield. Neither line exhibiting superior performance at the high N levels was large leaved; both were similar to S100 in morphology and leaf size. However Wilman & Asiegbu (1982a,b) have confirmed that larger leaved cultivars are more tolerant of applied N than small leaved cultivars.

ix. Pest resistance

(a) Nematodes

Stem eelworm (*Ditylenchus dipsaci* (Kühn) Filipjev). Cultivar differences in resistance to stem eelworm have been demonstrated by Danish tests (Bingefors 1969) and in New Zealand (Williams 1972, Williams & Barclay 1972). The most resistant materials screened have been of Ladino stock, including cultivars Regal and Merit (US), Espanso (Italy), NFG Gigant (Germany), European Giant, and the original Ladino ecotype.

Because of the sprawling nature of white clover, a moderate proportion of resistant plants appears to be adequate to maintain full production of native stands (Williams & Barclay 1972). Such a level is present in Crau (France), Blanca R.v.P. (Belgium), Aberystwyth S184 (UK) and Grasslands Huia (New Zealand) (Williams 1972).

However, yield losses in the first year may be considerable even with a moderate level of resistance. For example, Williams & Barclay (1972) found that first year monoculture yields of Huia were reduced 36% by eelworm attack. With grass competition during establishment eelworm attack could lead to poor clover content in new pastures and therefore to poor pastures.

Successful selection for resistance in the cultivar Grasslands Pitau was reported by Williams (1977). A resistant selection gave over 2.5 times the yield of the unselected cultivar under spaced plant conditions after eelworm attack.

Clover cyst nematode (*Heterodera trifolii*) Goffart). Differences among plants in resistance to clover cyst nematode have been reported by Kuiper (1960) and Norton (1967). Kuiper identified resistant plants in populations of diverse origin, including Ladino, Kersey, S100, New Zealand and Hungarian material. Yeates et al. (1973) screened six populations of white clover for resistance and found all to be susceptible hosts. However the data suggested that some plants may have been more resistant than others. Dijkstra (1971) noted that seedling clover plants were very sensitive to cyst nematodes, but mature plants showed little or no sensitivity to large numbers of cysts. Selection and breeding from resistant plants led to progenies with fewer susceptible seedlings, fewer cysts on progeny plants, and smaller cysts, indicating that it was possible to select populations giving lowered nematode multiplication. A population of diverse origin selected for resistance gave only 5.5% very severely infected plants (7400 cysts per plant) compared with 100% for a susceptible standard cultivar (Witte cultivarklavar CB) when inoculated with a particular strain of nematode. However, only 5.5% of the resistant population were virtually uninfected, 89% having 20-400 cysts per plant. Selection for resistance is apparently successful but rather slow and it may take many generations to achieve high frequencies of resistant plants in a cultivar.

Root knot nematode (*Meloidogyne* species). Little success has been achieved in selection for tolerance or resistance to this pest. Gibson (1973) reported the release of SC-1 germplasm selected for tolerance to root knot nematode (*M. incognita*) and has described the technique used. Yeates et al. (1973) screened for gross differences among six lines and found all six to be susceptible to *M. hapla*.

(b) Chewing insects

White clover is affected by many chewing insects (Chapter 12), including root feeders (e.g *Costelytra zealandica*), foliage feeders (e.g. *Wiseana* spp.) and seed feeders (e.g. *Coleophora* spp.). However, documented examples of genotype differences in resistance or tolerance are few. There is no direct evidence from white clover that the cyanogenic glucosides in the foliage of some plants influence chewing insects, although the high cyanogenic potential of populations in warm climates suggest that there may be some protective role for these compounds. Nayer & Fraenkel (1963) showed that linamarin in glucose solution checks feeding by the Mexican bean beetle (Jones 1973). Harris & Brock (1972) found no difference in feeding damage by *Wiseana* caterpillars to cyanogenic or acyanogenic populations.

Cultivar differences in damage by *Coleophora* spp. were demonstrated by Pearson (1980) who showed that Grasslands Pitau had 24% seed loss and Grasslands Huia 34% loss.

Variation among white clover populations for tolerance of *Costelytra zealandica* was shown by Wilson (1978a,b) and Wilson & Farrell (1978), but this was apparently related directly to plant vigour and there was no evidence for resistance. Consistent low preference feeding and reduction in larval weight were observed for a few of 96 genotypes attacked by clover root curculio (*Sitona hispidulus*) (Powell et al. 1983, Powell & Campbell 1983).

(c) Slugs and snails

Evidence for differential resistance of cyanogenic and acyanogenic genotypes to feeding by slugs and snails has been briefly summarised in Chapter 4 and in more detail by Jones (1971, 1973). Demonstration of differential resistance has been complicated by effects of species of pest, alternative food sources and environment, but it is apparent that cyanogenesis is not a simple tool for clear cut prevention of pest damage.

x. *Disease resistance*

(a) Viruses

There are few reports of selection for resistance or tolerance to virus diseases. Cope et al. (1978) screened 120 clones in 5 replicated field trials and found 3 clones virus-free after 2 years. Viruses infecting susceptible clones were AMV, PSV and CYVV (see Chaper 11). Smrz et al. (1977) screened 51 cultivars and new selections for immunity to AMV and BYMV. No immune cultivars were found but G66 and Milka Octoploid showed the lowest infection. Six clones were inoculated with a range of 6 viruses by Barnett & Gibson (1975). Apparent resistance to AMV (one clone), BYMV (6 clones), CYVV (5 clones) and PSV (2 clones) was found. No resistance to BYMV or WCMV was observed. Harville & Derrick (1978) found that Nolins Improved white clover was apparently less affected by PSV and CYVV than Louisiana S.1 which was severely infected. Differential tolerance to PSV and CYVV was demonstrated by Ragland & Campbell (1983). Clark & Barclay (1972) inoculated 4900 plants with WCMV and found none to be immune and Scott (1982) obtained a similar result after inoculation of 1344 plants. In the latter study clonal differences occurred in virus concentration but these were not related to yield losses.

A search for resistance to several viruses among the species closely related to *T. repens* was made by Barnett & Gibson (1975). *T. ambiguum* was infected only by CYMV, a virus for which no resistance was discovered, and was resistant to AMV, BYMV, CYVV, PSV, RCVMV and WCMV. *T. montanum* showed resistance to AMV, BYMV and CYVV and *T. thallii* to these and PSV. Other species showed resistance to one of the seven viruses used. Of these species, only *T. ambiguum* has been successfully hybridised with white clover (see Section 8.(C)).

(b) Fungal and bacterial diseases

Clover rot (*Sclerotinia trifoliorum* Eriksson). Differential resistance to clover rot has been shown by Hanson et al. (1953), Aldrich & Doling (1967), Aldrich (1970b), and several other workers (Chapter 11). Pajbjerg Milka is fairly resistant and Aberystwyth S100 and Grasslands Huia susceptible. Kersey and Kent are intermediate. Other cultivars with resistance to clover rot include Pronitro, Sonja, Menna and Blanca (Aldrich 1978, Davies 1978).

Stolon and root rot complexes. Various stolon, crown and root rot symptoms are attributed to pathogen complexes involving several species of fungi (Chapter 11), but there is little published effort to select for resistance to these complexes. Cope (1978) reported severe crown rot in old experimental fields and used these for screening populations for resistance.

In subtropical Australia, Irwin & Jones (1977) found some apparent resistance to the stolon rot fungus *Pythium middletonii* in naturalised white clover populations. Locally collected seed from two locations showed 14% and 32% survival in seedling tests compared with complete death of commercial cultivars. Plant breeding for resistance to rot fungi is likely to be difficult where a pathogen complex involves several species and simultaneous resistance is required for them all.

Leaf spot diseases. Yield losses from attacks by leaf spot diseases (Chapter 11) have not been determined. However leaf spot diseases cause white clover to become oestrogenic (Wong & Latch 1971, Wong et al. 1971). Therefore breeding of resistant cultivars could have dual significance in improvement of herbage yield and quality.

Wilkinson & Millar (1978, 1979) have studied the fungus *Stemphylium sarciniforme* on white clover in relation to the effect of cyanogenic potential of the host on symptom development. This fungus is susceptible to HCN but exhibits little difference on cyanide positive or negative plants.

Williams & Skipp (unpublished) have obtained evidence for resistance of some white clover plants to rust (*Uromyces* spp.), *Pseudopeziza* leafspot, and sooty blotch (*Cymadothea trifolii*).

xi. Tolerance of low soil phosphorus levels

Attempts are being made to develop populations able to survive and produce in soil low in phosphate. Genotypes showing differential phosphate nutrition can be identified (Caradus & Dunlop 1978, Caradus et al. 1980) and selection for longer root hairs has been successful (Caradus 1979), although the significance of the latter character to phosphate nutrition is not yet determined. Robinson (1942) showed that there was no relationship between the phosphorus or calcium content of a clone and its ability to grow at low levels of lime and phosphate.

Caradus & Williams (1981) carried out selection on low fertility hill soils for ability to survive and yield. Selected plants, derived from indigenous hill populations, performed better than lowland cultivars such as Grasslands Huia on low-fertility soils but were inferior on high-fertility lowland soils. A glasshouse study showed that the selected plants, which had small leaves and dense stolon production, had a lower total P content and lower growth rates than larger leaved plants. They also partitioned a higher proportion of P into stolons and roots than into leaves and therefore exhibited a low harvest index for P. It was suggested that the tendency for the selected hill country plants to have more of their P below harvest height may provide them with a mechanism for survival and growth in grazed swards on soils low in P. Such plants growing in hill country pastures should also help to alleviate the serious problem of P loss from slopes and its transfer and concentration on flat areas (stock camps).

Analysis of eight seminatural populations of white clover from the United Kingdom and two cultivars (Huia and Kent wild white) showed that there were large differences among populations for total P uptake (Caradus 1983). Most of this variation was accounted for by differences in root length rather than uptake per unit root length, suggesting that improvement of performance on low P soils might be achieved by selection for large root systems.

Endomycorrhizas stimulate P uptake and growth of white clover in infertile soils. If significant interactions between mycorrhizal fungi and plant genotype occur, then it would be necessary for the breeder to select in the presence of the appropriate fungus. Although Powell (1982) did obtain evidence for genotype x mycorrhizal fungus interaction in white clover, Crush & Caradus (1980) did not and generally found mycorrhizal infection to be beneficial only at very low P levels. A case can be made when selecting for low fertility soils for carrying out selection in the presence of indigenous mycorrhizal fungi.

xii. Nitrogen fixation

Improvement of the symbiotic effectiveness of white clover is of great interest and importance. Agronomists and microbiologists have developed formulae for correction of nutrient deficiences and provision of more effective bacterial strains which have had large effects on clover growth and nitrogen (N) fixation. Plant breeders have attempted to select for improved nodulation (and therefore improved N fixation), improved tolerance of the white clover plant and its N-fixing system to additions of fertiliser N, and the possibility of improving the N-fixing potential of the species in marginal areas (e.g. low fertility acid soils at low temperatures) by combining selections of host plants and bacterial strains. Much of this work has been carried out at the Welsh Plant Breeding Station, Aberystwyth (Jones & Davies 1966).

Recent New Zealand studies (Hoglund & Williams 1984) showed that genetic variation existed within the progenies of seven Grasslands Huia plants for symbiotic N sufficiency, i.e. the ratio of growth with symbiotic N to growth with non-limiting mineral N. N sufficiency varied with temperature, being close to 1.0 at high temperatures and markedly less at low temperatures. Over all temperatures, plants showed wide genetic variation and generally improved when crossed with a Spanish line. A selection strategy was suggested that would simultaneously breed for improved yield and N sufficiency. This involved initial field selection for yield under non-limiting N conditions followed by laboratory screening for symbiotic N sufficiency.

xiii. Host plant variation in nodulation

Plants with high productivity in the absence of high levels of soil N must be well nodulated and capable of fixing adequate atmospheric N to sustain high growth rates. Selection and breeding of such plants might be expected to improve nitrogen fixation. This traditional approach looks at the end-product of a complex association involving several factors, e.g. nodulation rate, total weight of nodules per plant, number of nodules per plant, nodule size, strain variation among nodules on a single plant. An alternative approach involves analysis of these components of nodulation and their genetic variation so that optimal combinations of all important components of N fixation can be synthesised. Such a study was undertaken by Jones (1962) who analysed variation among plants of the variety S100 Nomark for nodulation parameters when inoculated with an effective strain. The experiment, conducted on agar slopes in test tubes, revealed variation in times of primary nodulation, rates of nodule production, nodule numbers, sizes and weights. Earliness of nodulation, i.e. time between inoculation and appearance of nodules, was correlated with harvest dry weight at 70 days. Connolly et al. (1969) obtained a similar result. Selection of a cultivar with uniformly early nodulation might therefore be expected to improve plant establishment especially in marginal situations such as low fertility hill soils where competition from grasses is strong. Such selections in red clover, although successful, have not improved nodulation effectiveness (Nutman 1959).

The genetics of nodulation rate in white clover may be rather complex. Connolly et al. (1969) analysed a diallel series of crosses among plants selected for differences in nodulation rate, finding reciprocal effects as well as additive and non-additive effects. However, estimates of general combining ability were interpreted as being large enough for selection to be effective.

In Jones' (1962) work, nodule numbers did not correlate with harvest dry weights of the plants but total nodule weights did show a significant positive correlation with plant dry weight. Nodule sizes and abundance were variable and inversely related. The close relationship between total nodule weight and total plant weight suggested that selection for high nodule weight might improve N fixation.

In studying the genetics of nodulation characteristics, it is important to specify the bacterial strains involved. The strength of plant genotype x bacterial strain interactions has been illustrated by Mytton (1975) who, in a diallel cross inoculation study with eight genotypes and bacteria isolated from each of the eight plants, found that most plants yielded better when nodulated with their own bacteria or bacteria from a plant of the same cultivar. This has broad implications for plant breeding, development of commercial inoculants and application of laboratory research to field conditions. Thus in plant breeding, selection of both hosts and rhizobial strain simultaneously should lead to improved growth and N fixation. Commercial inoculants suitable for one white clover cultivar may not necessarily suit another (Bonish 1980). Experimental results obtained in the laboratory on agar probably will not correlate with field results using the same host genotypes unless care is taken to ensure that the bacterial strains are the same (Connolly et al. 1969, Mytton & Jones 1971).

Occurrence of cultivar x strain interactions in the field were demonstrated by Young & Mytton (1983). In a hill land study, S184 outyielded Grasslands Huia when inoculated with one strain, while the reverse applied with a different rhizobial strain.

Successful selection for increased nodule volume in S100 Nomark using *Rhizobium* strain W6 was carried out on agar slopes by Jones & Burrows (1968). Increase in nodule volume was accompanied by an increase in nodule number. Plant size was not significantly increased by two generations of selection so that nodule efficiency (i.e. plant growth per unit of nodule volume) fell. A similar selection experiment with S100 Nomark was reported by Mytton & Jones (1971) who carried out selection through three generations. As in the previous study, selection for increased nodule volume was successful for the first two generations but it regressed in the third generation. Plant yields, on the other hand increased in the first generation but regressed to the original size by the third generation. Increase in nodule volume was again accompanied by an increase in numbers of nodules, which were smaller on average than in the original population. Symbiotic efficiency again declined with selection. The authors noted that the base population was seven plants and an indication of inbreeding was obtained when crosses between third generation plants selected for large and small nodule volume were heterotic.

These failures to increase symbiotic effectiveness by selection for high nodule volume appeared to indicate that factors other than total nodule volume were important. One possible factor is individual nodule size. Nutman (1967) suggested that plants with fewer, larger nodules may be more efficient than those with many small nodules, because large quantitites of metabolites are not diverted to as many growing points and cortical regions. Mytton & Jones (1971) noted an apparent relationship between plant and nodule sizes and have suggested that large individual nodule size should be included as a selection parameter.

A potentially fruitful area of plant selection is bacterial strain preference by individual genotypes of white clover. Jones & Russell (1972) observed that within S184 there was considerable variation among plants for selection of effective and ineffective strains from a 1:1 mixture. Selection can be modified by factors such as pH (Jones & Morley 1981) and the relative concentrations of the two strains in the growth medium (Russell & Jones 1975a). In one study with two pure strains and a 1:1 mixture, S100 was preferentially nodulated by one strain while Kersey and Pajbjerg did not exhibit strain preference (Russell & Jones 1975b). All three cultivars showed interplant variation in strain preference resulting in large variances in shoot weight among plants inoculated with pure strains. Jones & Hardarson (1979) studied cultivar differences in strain preference using mixtures of two antibiotic resistant mutants. The highly selected cultivar S100 was nodulated almost exclusively by one strain (75 *str*) but S184 and Pajbjerg Smalbladet were more variable (Figure 2). Ninety-three per cent of the total variance in strain preference was genotypic, 68% due to cultivar and 25% to genotypes within cultivars. Genetic control of preference by the host was confirmed when clones derived from individual seedlings showed preference similar to the seedling. Further, Hardarson & Jones (1979b) found that preference was highly heritable: progenies from crosses between plants with 90-100% preference for strain 75 *str* showing 85% preference, while progenies from parents with less than 50% preference had only 13% of their nodules inhabited by strain 75 *str*. Intermediate values were found for progenies from other crosses. Inheritance was additive with no evidence of dominance or maternal effects. Strain preference was also found to be affected by temperature and soil type (Hardarson & Jones, 1979a) and by pH (Jones & Morley 1981). Therefore selection of effective *Rhizobium* strains should involve not only compatability with the host cultivar but also competitiveness with indigenous strains under appropriate soil and temperature conditions.

Reduction of variation in strain preference by plant breeding could contribute to improved overall N fixation by reducing the numbers of plants poorly adapted to inoculated strains and also by reducing the proportions of nodules contributed by poor native strains. Thus selection of cultivars uniformly highly selective for one strain of *Rhizobium* may increase N-fixing potential. Such an approach would require the simultaneous selection of plant and bacterial genotypes as suggested by Mytton (1975).

A major deterrent to nodulation and N fixation is a high level of available soil N. It is conceivable that white clover cultivars or *Rhizobium* strains could be selected which continue to fix N at soil N levels which are normally inhibiting, and thus allow the legumes to raise soil N levels to a new threshold. The 'supernodulating' mutant *nts* 382 of *Glycine max* appears to have the potential to achieve this (McNeil et al. 1985, Caroll et al. 1985).

xiv. Herbage quality

White clover is of very high feeding value for ruminants (Thomson & Raymond 1970), being rated in New Zealand work as 86% higher than perennial ryegrass for liveweight gain of lambs (Ulyatt 1970). There are few disadvantages of white clover as a feed for ruminants. With the exception of bloat in cattle, which is important, other problems (cyanosis, goitrogenic effects, oestrogen effects, photosensitisation, and some minor feed flavours in dairy products) are minor. Very few genotypic differences in quality factors have been found.

(a) Feeding value

Differences among three related white clover varieties for *in vitro* digestibility, dry matter content, and percentage water soluble carbohydrates were found to be very small by Davies et al. (1966). The only significant difference was a reduced DM content for octoploid S100 (AC9) relative to the normal form. No differences in palatability to cows feeding on young pasture containing different cultivars was found by Vez (1959). In older regrowth (more than six weeks) cultivars with dense foliage were favoured.

(b) Mineral content

Mineral content varies greatly with environment. However genetic variations in mineral content between populations have been demonstrated by Robinson (1942) for Ca, P and K; Snaydon & Bradshaw (1962a) for P; Vose & Jones (1963) for Ca; Snaydon & Bradshaw (1969) for Ca and Na but not Mg or K; and Davies et al. (1966) for P, K and Na. This last result came from differences between AC9, an octoploid form of S100, and tetraploid cultivars S100 and AC8 (S100 x Ladino).

White clover, and legumes generally, are considerably higher in content of Ca, Mg and P than grasses. Mg is important for control of hypomagnesaemic tetany and scope is indicated for manipulation of Mg levels in herbage by selection. However, the main problem appears to be grass dominance in early spring, the main danger period, and selection for early spring growth would probably be more effective than selection for high Mg content in white clover.

When cyanogenic plants are fed, the conversion of cyanide into thiocyanate in the animal inhibits accumulation of iodine by the thyroid gland. This can be offset by high iodine levels in the herbage (Johns 1956). Although cyanogenic white clover contains higher than normal iodine levels, it nevertheless was found to be goitrogenic to guinea pigs, rats and sheep (Flux et al. 1956). A short (43 days) sheep grazing trial with cyanogenic and acyanogenic white clovers was carried out by Butler et al. (1957), but serious goitrogenic effects were not observed. Although this potential problem can be overcome by selection of low cyanide cultivars, it is apparent that the problem is not a major one and highly cyanogenic cultivars are in common use without causing problems.

(c) Protein quantity and quality

An early study by Boekholt (1933) showed that protein content varied among plants of white and alsike clovers and that this variation was related to leaf colour, dark-leaved plants being 30% higher in leaf protein than plants with light-green leaves. Furthermore, leaf colour was found to be highly heritable and therefore suitable as a tool for selecting high protein cultivars. No significant differences in crude protein concentrations among three selected white clover cultivars were measured by Davies et al. (1966). Similarly Vacek & Novotny (1975) and Vacek & Zapletalova (1982) found no significant differences in protein content among three cultivars, although differences in individual amino acids did occur.

Possible benefits of reducing herbage protein solubility for the N efficiency of ruminant nutrition have been suggested by several workers (e.g. Clark 1975, Broderick 1975, Mertens 1977). However, so far no attempts have been made to achieve this in white clover by plant selection.

(d) Oestrogen content

White clover is usually non-oestrogenic, but it can become potent if infected with leaf diseases (Wong & Latch 1971, Wong et al. 1971, Newton et al. 1970). The literature on incidence of oestrogens in white clover was reviewed by Todd (1970). Selection of plants resistant to leaf diseases (see Section *x*.) should result in non-oestrogenic white clover cultivars.

(e) Cyanogenic potential

Many white clover populations contain cyanogenic glucosides. The variation in this characteristic is reviewed in Section 1 (C) and Chapter 8, Section 5. The genetics p. 348 indicate that selection of cultivars with high or low levels of cyanogenic glucosides is straightforward (e.g. Corkill 1956). Low cyanide cultivars are often considered to be desirable because of links between cyanosis and goitrogenic effects (see Section (b) above). The presence of some cyanogenic potential is considered by some workers to be desirable to prevent excessive grazing by pests (slugs, snails, rabbits, etc, Corkill 1952, Davies 1970, Jones 1973). Although there are reports of cyanide poisoning of sheep (Evans & Evans 1948) and cattle (Gurnsey et al. 1977) by direct introduction of white clover through a rumen fistula there are no substantiated reports of poisoning of animals grazing naturally. Coop & Blakely (1949, 1950) and Corkill (1952) tested white clover strains with high glucoside levels and found no deleterious effects on sheep. Tetley (1953) obtained some evidence to suggest that cyanogenic white clover might help to control intestinal parasites of sheep.

No studies of genetic variation in photosensitisation effects (Todd 1970) or feed flavours (Riddet et al. 1937) are known.

xv. Flowering and seed set

The agronomy of white clover seed production will not be reviewed here, as it has been covered by Haggar et al. (1963), Haggar & Holmes (1963) and Zaleski (1970) for the UK, Marble et al. (1970) for California and Scott (1973) and Clifford (1980) for New Zealand. However, improvement of seed yields can be an important breeding objective.

Seed yields are closely related to numbers of seed heads produced. Brigham & Wilsie (1955) found that visual ratings of numbers of heads per plant were correlated with seed set and gave enough information to improve seed by selection in Ladino clover. High parent-offspring correlations for seed set of open pollinated progenies and single cross progenies were observed. There were slight, although statistically non-significant, negative correlations between seed set and forage yields, and seed set and plant density among parental clones. A significant positive correlation was found between seed setting and forage yield of progenies from a diallel crossing scheme, indicating that high yields of both forage and seed are possible.

Seed head number is an important component of seed yield, but plants have the ability to compensate for reduced head number. Binek (1983) found that removal of up to 48% of inflorescences from the cultivar Podkowa did not reduce seed yield. Compensatory increases of 58-78% occurred in both number and weight of seed per head, although 1000-seed weight altered little (less than 8%). From a plant breeding viewpoint, reduction to a standard number of heads per plant (20 or 40) improved repeatability of 1000-seed weight and seeds per head and therefore improved selection efficiency.

Large differences among plants for seed set per flower do occur, as observed by Atwood (1940, 1944a). Studies revealed that number of ovules per flower was the main factor governing differences among F_2 lines. Dessureaux (1951) analysed the seed setting abilities of Ladino clover clones into components i.e. number of ovules per floret, seeds set per floret, and percentage of seed bearing florets. Clones were found to differ significantly in all three characters. Selection was successful in changing the numbers of ovules and seeds in the direction of high and low values without reducing genetic variation in early generations. A high number of ovules and a high degree of pollination appeared to be the main factors favouring high seed setting. Selection based on seed set often increased ovule number and generally improved efficiency of pollination. Breeding for a greater number of ovules, a higher percentage of seed-bearing florets and a higher proportion of ovules developing into seeds was recommended to improve seed setting in Ladino clover. Supporting results have been obtained more recently by Cebrat et al. (1978, 1982). In a study of 18 clones from 5 origins, Van Bogaert (1977) found that number of florets per head was variable and highly heritable (H = 0.75). Similarly, number of seeds per pod (determined by X-ray photography) was highly heritable. Numbers of ovules per pod differed markedly from those obtained by Dessureaux:

	Number of ovules per floret (total of 100 florets)					
	8	7	6	5	4	<4
Van Bogaert	2	13	62	21	2	0
Dessureaux	0.1	0.6	21.1	26.1	47.7	4.3

Wide differences between the materials used by the two workers are indicated. It is apparent that selection for greater numbers of florets per head and seeds set per floret, in addition to more heads per plant, should improve potential white clover seed yields. Ujj Meszaros (1981) successfully selected for increased numbers of florets and seeds per head in mutagen-treated material.

Amount of flowering is affected by several environmental factors. Reference has already been made

(Chapter 2) to the differential enhancement of flowering among genotypes by cold treatment. Davies (1962) showed that low light intensities under glasshouse conditions generally diminished flowering, but certain genotypes were more seriously affected than others. Van Bogaert (1977) obtained a significant difference in average number of florets per head in two consecutive, although climatically different, years. However, the difference was only 2.4 florets per head in favour of the drier, sunnier year (75.3 versus 77.7 florets per head). Average numbers of seeds per pod were 2.52 in the wetter of the two years and 3.90 in the warmer, drier year, percentage seed-bearing florets being 82.2 and 97.6 respectively.

Seed size varies considerably in white clover, but it is not considered an important agronomic character. Large seed does not necessarily establish better than small seed (Evans 1965). Octoploid lines have substantially larger seed than normal white clover. Evans (1965), for example, indicated that S100 had a mean seed weight of 0.6202 mg and AC9, a colchicine-doubled form of S100, had a mean seed weight of 0.933 mg. The inheritance of seed size in white clover has apparently not been studied.

Seed coat colour in white clover varies from bright yellow to dark red. There is a definite genetic component to this variation, some parent plants producing seed of one colour or the other, but there is also a strong environmental influence and ageing generally increases the proportion of red seed. Germination capacity of red seed has been shown to be slightly lower than that of yellow seed (West & Harris 1963).

White clover is relatively unspecialised in the insects needed for pollination. World wide, honey bees are the main pollinators, the small corolla size being suitable for them to gather nectar without difficulty. In many locations bumble bees or small halictid or andrenid bees are also significant pollinators (Free 1970, Davies 1971). Variation in nectar production has been reported, low nectar levels resulting in poor seed set as a result of poor bee attraction (Kropacova & Haslbachova 1969). Strength of flower scent may also be successfully selected for.

Variation in effectiveness of fertilisation has been indicated by Dessureaux (1951), Van Bogaert (1977) and Cebrat et al. (1978). Van Bogaert (1977) estimated fertilisation values averaging 42.1% in a cool, wet year and 65.1% in a warmer, drier year.

xvi. Herbicide tolerance

Very little has been reported on this aspect of white clover breeding. Cultivar differences in tolerance to paraquat at the seedling stage were demonstrated by Faulkner (1980). The small-leaved cultivar Pronitro was most tolerant but other small-leaved cultivars were susceptible. Selection for improved tolerance was unsuccessful in Pronitro, but a pasture response was obtained in a mixture of cultivars.

(D) Techniques used in breeding

i. Controlled cross pollination

(a) Hand pollination

Hand pollination involves the placement of foreign pollen on the stigma of a floret using a small brush, a toothpick or even a dehiscing anther held in tweezers. Atwood (1941c) indicated that neither tripping [a] nor emasculation are necessary to reduce selfing, even for plants with reasonably high levels of pseudo-self-fertility, unless special precautions are justified (e.g. for critical genetic studies). However, both emasculation and tripping facilitate application of pollen to the stigma.

a. The style and stamens are enclosed in a tense position within the keel. Tripping is the release of tension by application of downward pressure, making the stigma available for pollination.

In a study of the amount of seed set and selfing occurring during hand pollination, Atwood (1941c) compared three methods: (1) emasculated and tripped, (2) tripped only and (3) neither emasculated nor tripped. Six plants with different degrees of pseudo-self-fertility and showing recessive solid green leaves were crossed with pollen from plants homozygous for white V markings. All methods gave similar numbers of seeds per head but in method (1) it was easiest and in (3) most difficult to apply pollen. Seed was germinated and numbers of solid green seedlings observed. Method (1) gave 3/557, method (2) 0/214, and method (3) 0/260.

Emasculation can be achieved by a number of methods. Use of suction by attachment of a glass capillary tube to a suction pump is relatively simple and damage to the stigma can be reduced with practice (Atwood 1941c). A forceps method has been described by Williams (1954) in which the entire corolla, staminal tube and anthers are removed by gripping the corolla on the underside midway between the tip of the calyx and the tip of the standard and pulling gently. As a precaution against self pollination, the pistil can be sprayed immediately with water.

(b) Bee pollination

Isolation of two or more plants in a bee cage results in a high percentage of cross seed unless highly self fertile plants are involved. Atwood (1943) crossed eight clones differing in degree of pseudo-self-fertility and carrying recessive alleles at two loci with other clones homozygous for the dominant marker genes. Pair crosses were performed in bee cages over two seasons. Percentage crossing ranged from 85% to 100% among cages and differed among clones. Only a slight inverse relationship existed between degree of pseudo-self-fertility and amount of crossing. However, where a self fertile clone was used percentage outcrossing was only 18.8%. It is apparent therefore that all plants being used in later stages of a breeding programme should be investigated for their degree of self fertility before they are used in controlled crosses in bee cages.

Honey bees are the main pollinators of white clover (Valle 1936), but controlled crossing in bee cages is often easier to manage with bumble bees, as these large insects may be easily caught in the wild and rinsed in luke-warm water to remove foreign pollen before being introduced to bee cages. Honey bees are more difficult to handle because they are smaller and more mobile.

ii. Controlled self pollination

Under suitable conditions most white clover plants will produce a few seeds on selfing. The success of selfing, however, depends on the environment (especially temperature) and on the technique employed. In the glasshouse, Atwood (1941c) obtained seed from pseudo-self-fertile plants only when entire heads were rubbed by hand every second day throughout flowering. When selfed by placement of pollen on stigmas with or without emasculation, the same plants failed to set seed.

In the field, Atwood (1941c) compared the effects of hand manipulation of bagged heads with non-manipulation. Without manipulation 190 heads on 76 plants produced an average of 0.5 seeds per head (Williams (1931b) obtained only about one eighth of this level). When heads were gently rolled between thumb and fingers within the bags every day for six days after bagging, seed set increased to 8.5 per head. In a later experiment, manipulations of about four heads on each of 615 plants gave an average of 5.7 seeds per head. Ten percent of plants gave no seed, 73% 1-9 seeds per head, 16% 10-15, and 1% gave more than 50 seeds per head. Six manipulations, three manipulations (every second day after bagging) and one manipulation on the third day after bagging were compared. Results were not all conclusive, but one manipulation was generally inferior and three sufficient.

Pseudo-self-fertility in bee pollination cages fell between that for hand pollination (i.e. zero) and that for rubbed heads. Relative rankings of genotypes for degree of pseudo-self-fertility were the same with either bee pollination or hand rubbing of heads. The reasons why hand rubbing was more successful than bee pollination or hand application of pollen to stigmas are unknown, although with manipulation very few pollen grains stick to the stigma and the incompatibility reaction may consequently be weak.

Use of high temperatures to improve pseudo-self-fertility has been discussed in Section 1 (D) p. 350.

iii. Pedicel reflection as an indicator of successful pollination

Atwood (1940) discovered that the success of a pollination could be predicted from the behaviour of the flowers after pollination. Within 20-24 hours of compatible pollinations the standards and wings folded tightly around the keel, but after incompatible pollinations they remained unchanged. Even more indicative of a successful pollination was the degree of pedical reflection 60-72 hours after pollination. After a compatible pollination the pedicels became completely reflexed and the petals began to wither and turn brown. After an incompatible cross the pedicels did not move and petals did not turn brown. When part of a head was pollinated with compatible pollen and the remainder with incompatible pollen the difference was evident within the head.

iv. Screening for agronomic value

The agronomic value of white clover genotypes is best determined under grazing in mixed swards with associated grasses because this is how most white clover is used. For various reasons plant breeders often screen their material by other methods, and it is critical to know how closely the alternative methods relate to the pasture situation. Methods include: (1) widely spaced plants, (2) rows or closely spaced plants, (3) sown monoculture swards, (4) mixed grass-clover swards. Any of these systems may be either defoliated by grazing animals or mown.

(a) Widely spaced plants

The growing of widely spaced (60 cm or more) plants without associated grass has been by far the most common method used for evaluation of breeding material. The system requires fewer seeds than most others, labour input is low because weeds can be controlled with machinery, and individual plants are easily observed, thus reducing scoring times and facilitating identification and recovery of selected plants.

Spaced plants, however, give limited information about probable performance in mixed swards. Several studies have shown that plants of dense type under spaced conditions perform better than those of open type when grown in mixed grass-clover swards, but this result has not been invariable. Atwood & Garber (1942), for example, observed that best sods were formed by tall, spreading and dense plants but several cases were observed in which two plants very similar in morphological type performed very differently in mixed grass-clover associations. Poor mixed swards were usually formed by bunchy (non-spreading) types with short internodes, by very prostrate plants which appeared to be smothered by the grass, and by open types which maintained that habit in the mixtures. Mixed swards mown high (5 cm) and low (2.5 cm) were highly correlated, but some low-grazing genotypes performed relatively better under low clipping. Gibson et al. (1963) indicated that plants with widely scattered leaves and a low frequency of stolon branching (open or 'viney' types) were inferior to denser ('non-viney') types in both spaced plant nurseries and in mixed swards. Dijkstra & de Vos (1972) obtained best performances in mixed swards under grazing from the so-called 'culture' types (intermediate leaf size, dense). These types also performed well as mown spaced plants, whereas fine-leaved lines performed poorly as spaced plants but better in grazed mixed swards. Large leaved (Ladino) types were poor as spaced plants and very poor when grazed in mixed swards. The data of Andries & van Slijcken (1968) also indicated that fine-leaved 'pasture' types tended to perform better in mixtures than as monocultures, although they failed to achieve very high rankings relative to intermediate-coarse and coarse types.

Low correlations between results for widely spaced plants and those from mixed swards have been obtained. Davies & Tyler (1961) compared spaced plants with an equivalent set planted at 30 cm spacings and oversown with grass. Correlations were significant in the sowing year for leaf size, height and clover yield, but in the second year correlations for yield were not significant in any of the three experiments reported. Leaf sizes were correlated in each experiment, but correlations for plant height were also lost in the second year.

Better correlations between performances in monocultures (or as widely spaced plants) and in mixed swards have been reported for better performing lines than for poorer lines (Dijkstra & de Vos 1972, who also cite data of Dijkstra & van Dijk 1966 and Andries & van Slijcken 1968 for similar observations).

Athough this phenomenon is clearly of great importance to the plant breeder who retains only the best lines, it cannot be inferred that widely spaced plants provide adequate information on performances in mixed swards for selections to be evaluated solely by that technique. If density of growth is taken into account in spaced plants, estimation of performance in swards can be improved. Area of plant spread is of considerably less use than density, and height and type (as indicated by leaf size) are of virtually no value (Dijkstra & de Vos 1972).

A number of workers have investigated the possibility of using growth under low light intensities to screen for competitive ability. This work was reviewed briefly by Dijkstra & de Vos (1972) and indicated that selection at low light intensity is not appropriate for assessing ability to grow in associtation with grass.

(b) Rows, closely spaced plants and sown monocultures

Rows of white clover in the absence of associated grass are subject to the same reservations which apply to widely spaced plants. Rows sown directly into cultivated soil allow observation of seedling establishment of families, a phase of growth often not as easily studied by other methods.

Monocultures developed either from closely spaced (10-30 cm) plants or from sowings allow observation or measurement of productivity without the difficulties associated with the presence of a grass companion. Use of closely spaced plants is advantageous if seed supplies are limited. However, monoculture swards probably provide no real advantages over widely spaced plants and have the disadvantage that performances of individual genotypes cannot be followed unless plots are constituted from cuttings obtained from a single individual. Such plots, although demanding of labour, may have a place in some breeding situations where clonal replication can lead to significantly improved genetic gains. Where such expense is to be employed, however, greater benefit would probably be obtained from similarly constituted mixed clover-grass associations.

(c) Mixed clover-grass swards

Although more expensive and difficult to handle than widely spaced plants, the mixed clover-grass plot is usually the most suitable evaluation unit available to the breeder, as it combines competitive ability of progenies or genotypes with other selection criteria. The benefits of such mixtures can be partially wasted if plots are mown regularly, as there is clear evidence that frequent mowing is more conducive to white clover growth than to grasses (e.g. Hudson 1933, Brown & Munsell 1956). Without clipping, however, total yield (grass + clover) and clover yield are measured less accurately because the breeder is required to estimate yield by eye or to use some system of alternating grazing and mowing (Hudson 1933). Developments in the field of electrical or mechanical devices which estimate pasture production may improve such measurements.

Mixed clover-grass swards can be set-up from small quantities of clover seed by use of either widely spaced or closely spaced plants oversown with or alternating with grass. Such systems have been used by Atwood & Garber (1942), Davies (1958b), Gibson et al. (1963), Gibson (1964) and Dijkstra & Vos (1972).

Gibson (1964) observed that yields of forage obtained by planting at 15 cm spacings were equal to or higher than those obtained from broadcast sowings. In addition, the CV was lower for the 15-cm spacings, giving more sensitivity in comparisons of breeding lines. It was recommended that 15-cm spacings be used except when individual plant evaluations are required, where 30 cm spacing was recommended. Individual genotypes can also be evaluated at close spacings by clonal propagation as used by Gibson et al. (1963).

Selection of white clover for competitive ability with grass has been of considerable importance in regions where N applications to grasslands are high (e.g. Western Europe, UK). Use of mixed clover-grass swards has been a fundamental part of programmes to breed white clover cultivars able to withstand such conditions. Such a programme has been in progress at the Welsh Plant Breeding Station, Aberystwyth since 1948 (Davies 1958) (see Section 2 (C) *viii*. p. 370. for details).

v. *Tissue culture*

Tissue culture methods are available in white clover for a diversity of applications. Embryo and ovule culture have been used extensively to aid successful interspecific hybridisation. Callus culture is available for application of somaclonal variation and generation of cell suspension cultures. Cell suspensions provide an opportunity for *in vitro* genetic manipulation procedures such as transformation and variant selection, and protoplasts offer these prospects as well as potential for protoplast fusion. In addition, tissue culture may offer new methods for the long-term storage of special genetic material such as parental clones for synthetic cultivars.

(a) Embryo and ovule culture

The technique of embryo culture with transplanted nurse endosperm has recently produced several legume hybrids (Williams & Williams 1983). The method, described by Williams (1978) involves removal of pods 14-16 days after pollination and surface sterilisation of them with a hypochlorite:cetrimide solution. After rinsing, the pods are dissected under sterile conditions on filter paper discs. Any dissectable embryos are removed and placed on 'nurse endosperm'. Nurse endosperm is obtained from the female parent by normal pollination several days after the interspecific cross was made. Each hybrid embryo and its nurse endosperm is transferred on its filter paper disc to a sterile nutrient medium in a test tube. Williams (1978) has given a medium successfully use to culture embryos of *T. ambiguum* x *T. repens* hybrids. Plugged or sealed tubes are kept in plastic bags at room temperature without supplementary lighting for 4-6 months before transplantation to soil. Nakajima et al. (1969) obtained mature seeds from ovules excised as early as 1-2 days after anthesis and cultured on media containing cucumber or melon extract.

(b) Callus culture

Callus cultures are relatively easily established from stem and root tissues of white clover (Gresshof 1980), but there are some genetic differences in ability to callus. Regeneration of plantlets from callus appears, however, to show large effects of genotype, certain genotypes expressing good capabilities of regeneration while others fail (Mohapatra & Gresshof 1982). In general, plant regeneration from white clover callus is obtained on media with a high (1:2) cytokinin to auxin ratio (Oswald et al. 1977, Gresshof 1980). White (1984) was able to select a genotype with long-term regenerative capacity in cell suspension by selecting for continued regeneration from callus on solid medium. Callus tissues of Ladino clover reflected the salt susceptibility of the plants when grown in cultures containing 62.5 mM NaCl. Callus culture is thus a potential tool for selection for salt tolerance (Smith & McComb 1981). Plants regenerated from callus were cytologically stable ($2n = 32$) in Gresshof's (1980) study, although Pelletier & Pelletier (1971) obtained polyploid and aneuploid regenerants.

(c) Cell suspensions

Successful regeneration of plantlets from cell suspensions of Regal white clover were obtained by Oswald et al. (1977) using an auxin/cytokin ratio of 5:1 to initiate cell suspensions from callus and to generate callus from cells. Plantlets were obtained from callus as outlined in the previous section. White (1984) selected a single genotype (of Spanish origin) which could be induced to regenerate after 24 months in suspension culture by coordinated removal of 2,4-D and picloram from the medium and an increase in ammonium concentration. This treatment induced meristemoid formation in the suspension culture. The meristemoids developed into plantlets when planted onto solid medium.

(d) Protoplasts

Regeneration of plantlets from protoplast cultures has been achieved by Gresshof (1980) and Ahuja et al. (1983). In both cases, protoplasts formed callus from which regeneration took place, three cultivars of white clover (Ac32, S184 and Olwen) showing different plating abilities. Regenerated shoots tended to fuse into plate-like structures but nevertheless developed into plantlets. Bhojwani & White (1982) obtained a reproducable plating efficiency of approximately 37% provided that a controlled environment was used, but different genotypes within Grasslands Huia showed different morphogenic responses. Shoot buds were obtained from one type of protoplast-derived callus but these did not develop into plantlets. The other types developed only roots.

(e) Organ culture and somatic embryogenesis

The use of tissue culture for the long-term maintenance of genetic stocks requires a method that is not associated with production of spontaneous genetic variation (somaclonal variation, Larkin & Scowcroft 1981). Current knowledge suggests that callus culture must be avoided because it does give rise to variation. However cultures in which organ integrity is maintained or where somatic embryogenesis occurs from whole organs are possible candidates for genetically stable preservation. Long term research is required to establish the genetic stability of such cultures.

Plantlet regeneration was obtained from shoot tip cultures of two cultivars of white clover (S100 and S184) by Cheyne & Dale (1980), and Maheshwaran & Williams (1984) obtained plantlets by somatic embryogenesis from immature embryos.

(E) Maintenance of cultivars

Two methods are available for maintaining parental material of white clover cultivars which have selected parents.
(1) Maintenance of the parent plants for many years.
(2) Long-term storage of seed derived directly from the parents.

It is recommended that both methods be employed for all cultivars. Maintenance of plants is a demanding operation requiring frequent replanting to prevent build-up of diseases and pests and contamination by other plants. Continual checking of leaf markings and other diagnostic characters is essential.

Seed storage on the other hand is less laborious and less prone to error and was recommended by Davies (1956, 1965). Investigation of seed longevity of S100 showed that germination remained unimpaired at 0°C for more than 16 years, irrespecitve of packaging method (Davies & Young 1965). Plants from this seed showed a slight but insignificant drop in green yield and seed set (Davies & Young 1967). Storage of seed at room temperature in sealed tubes retained high germination for 15 years. In open tubes, however, seed germination started to decline after 5 years and was below 10% by the 14th year (Davies & Young 1965).

Subfreezing storage (-15°C, 60% relative humidity) maintained germination values over a 20-year period (Rincker 1983), mean drop in germination percentage being 3.5% over 27 samples from 14 cultivars.

Tissue culture (Page 383) offers potential for the future maintenance of genetic material. The most suitable method is likely to be low temperature storage of whole shoot cultures. As far as is known, such cultures are less subject to genetic mutation than are other forms of cell or tissue culture such as callus, cell suspensions or protoplasts. However, it will be a number of years before experiments started in the late 1970s can be evaluated and tissue culture confirmed as a relatively safe method of cultivar maintenance.

i. Genetic shift in cultivar multiplication

Most white clover cultivars are synthetic cultivars produced by intermating a number of selected parent plants. Maintenance of a constant genetic composition requires the following:
(1) Similar flowering among the parent plants each time a synthesis is made.
(2) Similar seed set of each progeny line in subsequent generations.
(3) No contamination from outside sources.

During commercial seed production these conditions are rarely satisfied and variation among seed lots of cultivars does occur as demonstrated for Ladino clover in the USA (Chamblee 1954, Portz & Jackobs 1955, Jackobs & Hittle 1958). Problems are usually greater when seed is increased outside the region of adaptation of a cultivar and natural selection removes a proportion of the plant population before flowering. An extreme example of such selection was given by Sylven (1937) showing that cultivars of German and Danish origin grown for seed in Sweden became more winter hardy as a result of natural selection for 2-3 years. Gradual selection for persistent types is common in white clover seed production (Schwanbom & Froier 1949), but can be avoided by restricting stand life. Norgaard (1967) found no shift in persistence of Morso Otofte IIK grown in Denmark unless extreme conditions were imposed.

Genetic shift can also occur as a result of differential flowering and seed set among surviving plants. A comprehensive study of this was made in cultivar Pilgrim at Davis, California (Laude & Stanford 1960). This cultivar showed marked shifts in flowering time and intensity when grown for seed in California. Studies of the 21 parent clones showed that these varied by 8-10 weeks in flowering date at Davis, and duration of flowering varied from a few days to 14 weeks. Large differences in vigour of the clones were also noted (Laude et al. 1958). Variation in seed contribution of individual clones occurred with change of location, time of closing and time of harvest (Stanford et al. 1960, Booysen & Laude 1964). Table 4 shows the differences recorded for seed contributions of the parent clones at two harvests. Different clones were the best seed producers at different harvests, and at different locations. Genetic shifts were minimised by changing closing dates to achieve seed production from flowers initiated in the longest photoperiod available. This was attributed to the adaptation of Pilgrim to higher latitudes (Stanford et al. 1962).

Table 10.4. Effect of harvest time on seed contribution (%) of parental clones of Pilgrim white clover. From Stanford et al (1960).

	Harvest 1 (July)	Harvest 2 (October)
Range in seed contribution per parent clone	0-23	0-23
Contributions of the four highest yielding clones to total seed yield	73	59
Contributions of the seven lowest yielding clones to total seed yield	1.75	7.2

Bula et al. (1964) grew six selected clones at Clemson, South Carolina and at four locations in Western USA. It was concluded that syn 1 progenies produced outside the region of adaptation of the clones (Clemson) did not differ much from locally produced progenies. However, seed contribution of each clone varied with location. Daylength and temperature differences at the various locations influenced flowering date and intensity of flowering of the clones.

A study at Weihenstephan comparing seed lots of a German composite multiplied for two generations in Germany or the USA was reported by Simon et al. (1974). Leaf size, plant diameter, height, regrowth and flowering density of most of the USA-grown populations were somewhat lower than for the German-grown populations, but were similar to the means of the original seed. No patterns emerged relating population changes to harvest year or generation. It was apparent that genetic shift was as likely to occur during seed production in Germany as in the USA and that plant characteristics and yield were retained when seed was produced in the USA. Similar results and conclusions were obtained by Chisci & Ferrari (1967) from a study of Ladino clover cv. Espanso, selected at Lodi, Italy and increased in the USA.

Aberystwyth S184 has proved to be stable during seed increase outside the UK. Kelly & Boyd (1966) and Boyd (1971) reported very few shifts in flowering date among many samples received from Canada, USA and Finland, and no shifts in leaf size were noted. Some differences in petiole and stolon thickness were observed with advancing generations.

In some cultivars not all plants flower in the sowing year, and seed harvested that year will probably show greater population shift than in later years (Dovrat & Waldman 1966). Where ripening conditions are ideal, closing date can be delayed until late-flowering plants can produce viable seed in time for harvest. Thus choice of locations with ideal ripening conditions and proper choice of closing and harvest dates can help to prevent genetic shifts towards earlier flowering (Kelly & Boyd 1966).

(F) Identification of cultivars

The large number of white clover cultivars available on world markets has resulted in a requirement for distinguishing characteristics in order that cultivars can be identified and so that contaminants can be detected. Distinguishing characteristics can be grouped into two categories:
(1) Morphological factors (including flowering).
(2) Genetic markers.

i. *Morphological factors*

Genetic uniformity of morphological characteristics in white clover cultivars, as with all open-pollinated populations, is never achieved. Trueness-to-type is a statistical feature of the whole population, not a characteristic of individual plants. Mean values for population samples are compared statistically. Such data are useful for separation of cultivars and may be used for statistical detection of contamination.

(a) Leaf size

The leaf size of a white clover plant varies with position on the plant and with environment. Hawkins (1959) has determined leaf size variations within individual plants (Table 5) and shown a large effect of stolon type and leaf position.

Table 10.5. Variation in length of leaflet (mm) on one plant.

	Position of leaf counting from apex of stolon									
	2	3	4	5	6	7	8	9	10	Mean
Primary stolon										
Kersey	24	28	30	30	27	32	35	34	–	30.0
Kent		18	21	20	20	20	19	24	23	20.6
Secondary stolon										
Kersey	14	18	15	14						15.2
Kent		18	18		14	10				15.0

Variation from year to year is illustrated by the data in Figure 3, extracted from Hawkins (1959). The data indicate that leaflet length varies greatly from year to year and that variety x year interactions are such that, even given a reference cultivar, distinctions among cultivars may be unreliable and only large differences can be consistently recognised. Similar variation occurred for leaflet widths. Therefore, comparisons of leaf sizes, even if carried out at a given time, from replicated plots, on leaves from selected positions, are still open to substantial variation relative to one another.

(b) Leaflet shape

Leaflet shape can be described by the ratio leaflet length:width. This ratio is apparently little affected by environment (Hawkins 1959) and was the primary character used to distinguish among seven cultivars:

Leaflet length:width >1.27

Late flowering	Ladino
Intermediate flowering	Kersey
Early flowering	Dutch

Leaflet length:width <1.30

Intermediate flowering

Smallest leaves	Kent
Small leaves	S184
Intermediate leaves	New Zealand, S100

(c) Stolon and petiole thickness

These characters show similar variation to leaf size and the same rules apply to their application.

(d) Flowering time

Flowering time is only of limited used in varietal identification because only markedly early and late cultivars can readily be distinguished from the majority which fall in the 'intermediate' category (Hawkins 1959).

(e) Number of vascular bundles in petioles

Determination of numbers of vascular bundles in petioles may be suitable for separation of widely divergent types. However, it is subject to environmental variation and therefore is of limited application. Gausman & Fuelleman (1952) found very consistent differences between Ladino clover samples and Illinois indigenous white clover populations in a three-year study. Seven Ladino populations from diverse sources averaged 7.65 ± 1.57 bundles/petiole (range 6.70-9.82) while seven Illinois indigenous populations averaged 4.22 ± 0.86 (range 3.91-4.57). Gibson & Schultz (1953) determined average numbers of vascular bundles per petiole in 10 petioles on each of 12 clones of Ladino and an intermediate type grown in three environments. Although the two types could be distinguished on average within any one environment, some of the Ladino plants were no different from intermediate plants. It was also found that intermediate clones in cool field conditions had as many bundles as Ladino under hot, dry conditions.

	Greenhouse	*Cool moist field*	*Hot dry field*	*D**
Ladino	6.7	7.6	5.6	0.76
Intermediate type	5.0	5.6	4.1	0.60

*Difference required for significance (1% level)

ii. Genetic markers

(a) Cyanogenesis

The genetics of cyanogenesis in white clover has been described in Section 1 (C). Natural populations vary greatly in ability to generate HCN, much of this variation being adaptive in nature (Chapter 8, Section 5). It is therefore not surprising that commercial cultivars also vary greatly in cyanogenic reaction (Daday 1955). Picrate paper tests (Corkill 1940) can be used in different ways to achieve increasing resolution in detecting variable differences.

Bulk samples of seedlings or foliage

Bulk samples can be scored quantitatively for HCN production using the picrate paper colour scale in which depth of colour change is proportional to the amount of HCN liberated (Melville et al. 1940). This test was shown to correlate well with agronomic type in New Zealand populations (Foy & Hyde 1937) and has been used since that time for certification of white clover seed. The test has low resolution for detection of acyanogenic contaminants. Corkill (1952) showed that high scoring samples could nevertheless contain up to 30% acyanogenic individuals and that even with 64% contamination the colour change was slight. However, in a reverse contamination experiment one positive seedling in 50 was detectable. Therefore the bulk test is very sensitive for detection of positive contaminants in a cyanide-free strain but insensitive in the reverse situation.

The bulk test is probably still sufficiently correlated with agronomic type in New Zealand to be of use in detecting populations of poor agronomic value. However, separate assays of individual plants will probably provide more sensitive measures for detecting small differences among populations, including contamination levels, in situations where time and labour permit this sort of analysis.

(b) Analyses of individual plants

Corkill (1952) determined the frequencies of the four possible phenotypes (glucoside + enzyme, glucoside only, enzyme only and neither glucoside nor enzyme) in several white clover populations and also included evaluation of the level of HCN emitted by each plant. Daday (1954, 1955) used these methods to demonstrate large differences among white clover cultivars for frequency of plants containing glucoside (e.g. Weibulls tetraploid 0, S100 Nomark 100%). Enzyme frequencies were slightly less varied and less characteristic for each cultivar. A strong positive correlation was found between the frequencies of plants containing glucoside and the intensity of the picrate reaction. Although determination of the frequency of glucoside-carrying plants was most useful, the simpler test giving frequency carrying glucoside and enzyme was almost as good. Chesneaux (1972) included percentage plants as one of three characters used for distinguishing among the 19 cultivars of white clover included in the French Official Catalogue.

(c) Leaf markings

V-markings

Presence or absence of white V-marking can be scored very easily in seedlings a few weeks old and statistics on frequencies of non-marked plants are useful for cultivar identification (Carnahan et al. 1955, Davies & Young 1963, Chesneaux 1972). Aberystwyth S100 Nomark was produced by the Welsh Plant Breeding Station as a characteristically non-marked cultivar. Detection of contamination of this cultivar by pollen from unrelated wild plants depends on the frequency of the recessive non-marked gene in the wild population. Carnahan et al. (1955) found that *vv* individuals occurred in natural populations in the USA at a frequency of approximately 9%, a figure consistent with the findings of Davies & Young (1963) for a range of commercial cultivars in the UK. The frequency of the gene *v* could then be calculated as approximately 0.3. Therefore about 70% of contamination of S100 Nomark would be detectable from observation of leaf marks alone.

Most other available leaf marks generally show mosiac dominance. Use of such genes for cultivar identification is complicated by two factors:

(1) Presence of similar genes in natural populations.

(2) Masking of contamination by recessive genes.

Therefore there is very limited scope for use of available leaf marks as varietal 'labels'. Probably the less common alleles provide the greatest scope. Cultivars homozygous for filled V (V^{f}), basal V (V^{ba}) or V-point (V^{p}) would allow identification of a high proportion of natural contaminants, although variability of expression of these genes may be a problem. V^{by} (broken V with yellow point) occurs at low frequency in many varieties (0-3.7%, Davies & Young 1963) but is dominant and would mask contamination from foreign pollen sources.

Anthocyanin markings

Anthocyanin leaf markings were used by Chesneaux (1972) as one of three characters for distinguishing cultivars. Frequencies of plants showing markings ranged from 0% (Major) to 63.4% (Kersey). However, expression of anthocyanin markings is poor at high temperatures and the red-fleck character, in particular, is unreliable.

Use of any of the dominant anthocyanin markers (red midrib, red halo, diffuse red leaf, red leaf, or red flowers) would be subject to the objection that their presence would cover contamination from foreign pollen sources.

Index of Varieties, Germplasm and Cultivars

The list below is neither complete nor consistent. It contains names of white clover cultivars known to be in commerce, and those bred but not in commerce, as well as experimental lines, ecotypes and cultivars which have appeared in the recent English language scientific literature. Some of the designations may be synonymous. Many cultivar names appearing in the early literature referred to local populations and have been omitted.

Coverage of non-English literature is minimal. Therefore many omissions of cultivars etc, can be expected, and there is very little information on some. A comprehensive checklist of varieties was produced by Caradus (1986).

Aberyswyth S100

Bred at the Welsh Plant Breeding Station by R.D. Williams.

Released in 1935.

Developed from material of diverse origin - New Zealand, Dutch, English wild white.

Cyanogenic (92.5%, Daday 1955; 85-95.3%, Hawkins 1959).

Selected for medium sized leaves on long petioles, relatively short stolons which readily root at the nodes. Internodes are short and plants consequently dense and leafy. Suitable for both sheep and cattle pastures, persistence is intermediate between English wild white and Dutch types (Williams 1945).

Some references: Aldrich et al. (1958), Andries & van Slijcken (1968), Davies (1958), Davis & Cooper (1951, 1953), Harkess et al. (1970), Hunt et al. (1965), Laissus & Teilhard de Chardin 1962), Munro et al. (1975), Stapledon (1947), Troughton (1955a), Vacek (1970), Williams (1972), Ziegenbein (1959), Zwingli (1956), Zwingli & Grundbacher (1958).

Aberystwyth S100 Nomark

Bred at the Welsh Plant Breeding Station by selection from S100 but not released into commerce. It retains the essential characteristics of S100 but lacks the white V-marking on the leaf.

Cyanogenic (98.1%, Daday, 1955; higher than S100, Nasinec & Mika, 1979).

Some references: Connolly (1968), Davies (1958), Troughton (1955a), Vacek (1970).

Aberystwyth S184

Bred at the Welsh Plant Breeding Station as a selection from English wild white clover.

Released in the 1930's.

Medium cyanogenesis (69.7%, Daday, 1955; 59.3-67.1%, Hawkins, 1959).

Similar in appearance to English wild white (i.e. small leaved) but with slightly larger leaves, darker green, denser, and less variable. Persistent and hardy (Williams 1945).

References: Davies (1958), Hunt et al. (1965), Munro et al. (1975), Troughton (1955a), Williams (1972).

Ala Lu

Selected from Ladino clover at Tuskegee, Alabama, US on the basis of high seed yields. Adapted to southern USA (Crops & Soils 5: 24, 1952).

References: Craigmiles & Crowder (1960).

Angeliter Milka

Milka Pajbjerg K & V is marketed under this name in Germany.

Arcadia

Selected by Northrup King (USA) from REGAL.

Aria

Maintained by Van Engelen Zaden, B.V., Vlijmen, Holland.

Selected for low HCN ecotypes.

A low growing (grazing) type with good persistence and winter-hardiness. Released in 1968. Formerly designated VERTAS wittee weideklaver.

Reference: Andries & van Slijcken (1968).

Armada

Maintained by Van Engelen Zaden B.V., Vlijmen, Holland.

Selected for low HCN from ecotypes.

A tall growing (hay-type) with good persistence and good resistance to *Sclerotinia trifoliorum* in NIAB trials (UK).

Released in 1965.

Formerly designated VERTAS kultuurklaver.

References: Andries & van Slijcken (1968), Nasinec & Mika (1978), Williams (1972).

Babinska
Polish cultivar of Ladino type selected from crosses between Ladino and Dutch types.
Barbian
Maintained by Barenbrug Holland B.V., Arnhem, Holland.
Selected from old Dutch pastures and released in 1955 under the name Witte weidklaver BARENZA.
Registered as BARBIAN in 1972.
Low in HCN (15% with glucosides, French Variety List 1975).
Small leaved and fairly prostrate, strongly creeping and well adapted to grazing (resembles ABERYSTWYTH S184 and KENT wild white). Good resistance to clover rot.
Barenza
See BARBIAN.
Beta
Swedish cultivar selected from Morsö.
References: Manner (1958), Vacek (1970).
Blanca RvP
Developed at the Rijksstation voor Plantenveredeling, Belgium, from crosses between indigenous Belgian material and Kersey from the UK.
Released in 1954 and has become one of the most successful European cultivars.
Cyanogenesis very low (Nasinec & Mika 1978).
Selected for productivity, persistence and competitiveness with companion grasses. Small-medium leaves with medium-thick stolons.
Marketed in West Germany as TRIBLA.
References: Andries & van Slijcken (1968), Connolly (1968, 1970a), Hoen (1970), Kasper (1976), Vacek (1970), Williams (1972).
Brasov
Maintained by Institutul Agronomic Dr Petru Groza, Cluj, Romania.
C.B. Pasture and C.B. Hay
See WITTE WEIDELAVER C. B., CULTURA and PERTINA.
Clarence Valley
An intermediate-leaved ecotype adapted to subtropical and temperate areas on the east coast of Australia. Mainly a winter annual with heavy seeding capability but can behave as a perennial given adequate fertility and moisture.
References; O'Brien (1970a,b), Ostrowski (1972), Thomas (1961), Wilson (1965).
Cocarde
Maintained by Cebeco-Handelsraad, Rotterdam, Holland.
Combi Maaiklaver and Combi Weideklaver
Dutch cultivars listed by Andries and van Slijcken (1968) as very good in grazed mixtures with perennial ryegrass.
Crau
Maintained by INRA, France.
Medium-large leaved with low HCN, some resistance to clover rot and stem eelworm.
References: Laissus & Teilhard de Chardin (1962), Williams (1972), Williams & Caradus (1979), Rhind et al. (1979).
Cultura
Maintained by Cebeco-Handelsraad, Rotterdam, Holland.
Selection started in 1937 and the cultivar was added to the Dutch list in 1950. Formerly designated C.B. HAY.
Developed from Morsö and Morsö Otofte.
Cyanogenesis low.
Medium-large leaved, high yielding and winter hardy.
References: Adries & van Slijcken (1968), Harkess et al. (1970), Nasinec & Mika (1970), Vacek (1970).

Daeno
Maintained by Dachnfeldt, Odense, Denmark.
Also marketed as DAENO K (K = suitable for short leys).
Acyanogenic.
Diana
Maintained by Cebeco-Handelsraad, Rotterdam, Holland.
Donna
Bred at the Welsh Plant Breeding Station, Aberystwyth from material collected in the Scilly Isles. Leaves medium-small, susceptible to clover rot.
Duron
An old Canadian cultivar no longer marketed.
Selected in Ontario from introductions from the UK and Scandinavia.
Reference: Hall (1948).
Dutch White
Name given by common usage to a local cultivar of small-medium leaved, profuse flowering, annual-type white clover grown in Eastern England. (Hawkins 1960, Williams 1945). See Chapter 8.
English Giant
Derived from Dutch white in England.
Acyanogenic (Nasinec & Mika 1978).
Esko
Selected from TAMMISTO at the Kesko Co. farm at Länsi-Hahkiala, Finland.
Released in 1968 but did not reach the market because of seed production difficulties.
Espanso
Bred and maintained by Instituto Sperimentale per le Colture Foraggere, Lodi, Italy.
Selection from Lodigiano 107/66.
Acyanogenic and resistant to *Ditylenchus dipsaci*.
References: Rhind et al. (1979), Williams (1972), Ferrari (1965-66).
European Giant
Fairly resistant to *Ditylenchus dipsaci* (Williams 1972).
Feathermark
A line of GRASSLANDS HUIA developed by Corkill (1963), homozygous for the gene 'red midrib' (R^m). Maintained by Grasslands Division, DSIR, Palmerston North, New Zealand.
Fries-Groninger
Dutch landrace.
Acyanogenic and was susceptible to *Ditylenchus dipsaci*.
References: Andries & van Slijcken (1968), Nasinec & Mika (1978), Williams (1972).
G66
Bred by F. Grubner at the Institute of Irrigation Research, Svarvas, Hungary.
Large-leaved Ladino type developed for mowing under irrigation.
Giant
Polyploid cultivar reported by Crowder (1960) to be drought and disease resistant.
Gigant
A recently developed cultivar designated GIGANT was reported from the USSR by Sulga (1966, 1969). An older cultivar GIGANT BELYI (or BELIJ) was described by Martyaenko (1935, 1939) and shown to be acyanogenic (Nasinec & Mika 1978). GIGANT ULADOVSKIJ was also tested by Nasinec & Mika (1978) and shown to be acyanogenic.
Grasslands Huia
Bred by Grasslands Division, DSIR, Palmerston North, New Zealand.
Early selections known as New Zealand Mother White and New Zealand Permanent Pasture were released

in 1930. Final selection in 1957 and named Grasslands Huia in 1964. Also marketed as HUIA.
Developed from New Zealand ecotypes (Chapter 8).
Cyanogenic (84.4%, Daday, 1955; 85.2-93.7%, Hawkins, 1959).
A medium-small leaved cultivar with long petioles and densely branching habit it is highly productive and very persistent. Widely adapted to maritime regions in many parts of the world, Huia grows mainly in spring, summer and autumn, but also shows winter growth in mild conditions.
References: Aldrich et al. (1958), Barclay (1960), Corkill (1949, 1964), Davies (1958), Davis & Cooper (1951, 1953), Andries & van Slijcken (1968), Esser (1960), Harkess et al. (1970), Munro et al. (1975), Rasmussen (1962), Stapledon (1947), Vacek (1970), Williams (1932), Williams (1945), Williams (1972), Williams & Barclay (1972, 1975), van Slijcken (1960), Laissus & Teilhard de Chardin (1962).

Grasslands Pitau

Bred at Grasslands Division, DSIR, Palmerston North, New Zealand by P.C. Barclay.
Former designation GRASSLANDS 4700. Also marketed as PITAU.
Released in 1972.
Developed by hybridisation of New Zealand material (mainly HUIA) with a winter-active Spanish introduction. The hybrid was backcrossed to New Zealand plants to improve plant type.
Cyanogenic.
A medium-leaved cultivar, larger leaved than HUIA, especially in autumn and winter, thicker stemmed and slightly less dense than HUIA. It expresses its superior cool season production best in high fertility situations where grazing pressure is moderate, but can tolerate hard grazing. It grows better in northern New Zealand than in the south.

References: Barclay (1960, 1963, 1969), Brock (1971, 1974), Brown (1973), Brock & Hoglund (1974), Clifford (1977), During & Weeda (1974), Goold & Douglas (1976), Lambert et al. (1969), McDonald & Cullen (1973), Oram (1982), Williams (1972), Williams & Barclay (1972, 1975), Williams & Hoglund (1978), Williams & Sedcole (1979), Weeda et al. (1970).

Haifa

Developed by the New South Wales Department of Agriculture, Australia from material collected in Israel.
Large leaved, resembling Ladino in form, although more stoloniferous, not as tall and more dense. Leaves pale green with very conspicuous white V-markings.
Better adapted to heat and low rainfall than most other cultivars, also winter-active and early flowering.
References: Barnard (1972), O'Brien (1970a,b), Scarsbrick (1977), Wilson (1955-64, 1960, 1965).

Hero

Bred at Svalöf, Sweden by G. Nilsson-Leisner and released in 1939.
Selected from Morsö for tolerance to summer drought by selection for deep roots, tall growth with large leaves but relatively dense habit.
Adapted to use for grazing or silage in southern and central Sweden.
References: Hall (1948), Nilsson-Leissner & Nilsson (1940).

Huia

See GRASSLANDS HUIA.

Irrigation

An ecotype selected by the Victorian Department of Agriculture, Australia from the irrigation districts of northern Victoria.
First certified in 1936.
Also known as VICTORIAN IRRIGATION. Medium leaved, resembling HUIA although slightly more prostrate. Performs well under irrigation at high summer temperatures and low humidity.
Cyanogenic.
References: Archer & Gresser (1979), Barnard (1972), Drake (1942), Elliott (1940), Martin (1954), Nasinec & Mika (1978), Vacek (1970).

Iygeva 4

Cultivar with superior yield and resistance to *Sclerotinia* in Estonia, USSR.

Reference: Jaagus (1971).

K30

Bred by K. Ujj Meszaros at the University of Agricultural Sciences, Keszthely, Hungary. More prostrate than LOVASZAPTONAI, dark green, winter hardy and mycoplasma resistant.

Karina

Maintained by P.H. Petersen, Lundsgaard, Germany.

Released in 1971.

Derived from crosses between STEINACHER and MILKA and selected for vigour.

Lacks a white leaf marking.

Katrina

Bred by the Welsh Plant Breeding Station, Aberystwyth from crosses between Kersey and ecotypes from Turkey and Israel. Medium-large leaved.

Kent Wild White

An ecotype from Kent, England maintained by the Kent National Farmers Union and certified under a scheme adopted in 1930.

A typical English wild white with slender stolons, a dense branching habit, small leaves and great persistence. Slow growing in early spring and late flowering.

Moderately cyanogenic.

References: Aldrich et al. (1958), Davis & Cooper (1951, 1953), Davies (1958), Haggar & Holmes (1963), Harkess et al. (1970), Hawkins (1960), Ware (1925b).

Kerry

Bred at the Welsh Plant Breeding Station, Aberystwyth from crosses between Kersey and an ecotype from Israel.

Large-leaved and thick stemmed with early flowering.

Kersey

Discovered in 1924 growing in a lucerne field in Suffolk, England. Reached commercial production in about 1946.

Medium-large leaved with early spring growth and quick recovery from defoliation, even in dry seasons. Thick stolons with rapid spread and good nodal rooting. Persistent and high yielding.

Cyanogenic and susceptible to *Ditylenchus dipsaci*.

References: Andries & van Slijcken (1968), Barrows (1949), Connolly (1968, 1970), Daday (1955), Davies (1958), Harkess et al. (1970), Hawkins (1960, 1967), Hunt et al. (1965), Nasinec & Mika (1978), Williams (1972).

Kivi

Selected at Svalöf, Sweden from New Zealand material.

Previously known as SVO414.

Very productive and sufficiently winter hardy for southern Sweden.

Cyanogenic but very susceptible to *Ditylenchus dipsaci*.

References: Akerberg & Halling (1959), Andries & van Slijcken (1968), Julen (1963), Munro et al. (1975), Williams (1972).

Krus 33

See RADI.

Ladino Gigante Lodigiano

An ecotype from the Po River Valley in the Lodi-Cremona area near Milan, Italy. First noticed in 1834-45 and since has become spread around the world. Naturally selected strains exist in several countries, including USA, Canada and Australia. Several cultivars have been selected from it.

Seed of original material is harvested in Italy and certified under a scheme administered by the Stazione Sperimentale di Praticoltura, Lodi, Italy (Chisci, 1963).

Characterised by giant leaves and very thick stems, it is more tap-rooted than smaller leaved white clovers (Caradus 1976). It has a wide climatic tolerance and is probably most limited in its usage by lack of tolerance to grazing and characteristic lack of persistence. Extremely productive.

Virtually acyanogenic but susceptible to damage from slugs.
Resistant to *Ditylenchus dipsaci.*
References: Ahlgren & Fuelleman (1950), Andries & van Slijcken (1968), Barnard (1972), Davies & Levy (1931), Davies (1958), Laissus & Teilhard de Chardin (1962), Nasinec & Mika (1978), Rudorf & Schoene (1936), Vacek (1970), Ware (1925), Williams (1972), Zwingli (1956), van Slijcken (1960).

Lena
Maintained by Swedish Seed Association, Svalöf. Developed at the Ultura Station by selection from Morsö.
Small-leaved, formerly designated UO148.
References: Akerberg et al. (1961), Rhind et al. (1979).

Liganta
Maintained by Deutsche Saatveredelung, Lippstadt-Bremen, Germany.
A large-leaved Ladino type listed in 1975.
Bred from crosses between Ladino and Morsö.
Cyanogenesis low.
Tall growing with large leaves, persistent, high yielding, and late flowering.

Lirepa
Maintained by Deutsche Saatveredelung, Lippstadt-Bremen, Germany.
Listed in 1976.
Medium-large leaved with only a small marking, winter-hardy, high yielding and moderately resistant to clover rot. Early-medium flowering.

Lodi Ötofte
Produced by the Ötofte station of the Dansk Planteforedeling, Denmark from Ladino clover.
Probably the first cultivar bred from Ladino (Whyte et al. 1953).
Acyanogenic. Resistant to *Ditylenchus dipsaci* (Bingefors 1969).
Reference: Rasmussen (1962).

Louisiana Improved
See NOLINS IMPROVED.

Louisiana S.1
Selected at the Louisiana Agricultural Experiment Station, USA by C.R. Owen and released in 1950.
A five-clone synthetic selected from local Louisiana indigenous clover (Chapter 8) for heat and drought tolerance and improved yield.
Medium-large leaved, erect, open and coarse in habit, early flowering. Produces seed over most of the year and can flower in short days in tropical areas. Tends to behave as an annual.
Cyanogenic.
References: Barnard (1972), Craigmiles & Crowder (1960), Blaser & Killinger (1950), Jamieson, (1957), Grylls & Day (1966), Killinger & Harris (1959), Nasinec & Mika (1978), O'Brien (1970), Owen (1953), Rhind et al. (1979), Schroder (1963), Williams (1972).

Lovaszpatonai
Developed at the Agricultural University of Keszthely, Hungary by M. Horn, J. Lukocki, and K. Ujj Meszaros from old pastures. State approved in 1967.

Luclair
Maintained by Institut National de la Recherche Agronomique, France and bred by Lenoble at Lusignan.
A persistent, cold tolerant cultivar of Dutch type, somewhat sensitive to clover rust (*Uromyces*).
Produced by multiplication of a single hybrid.

Lune de Mai
Bred by Lenoble at Lusignan, France and maintained by the Institut National de la Recherche Agronomique, France.
A giant type selected from six local populations, a 20-clone synthetic.
Acyanogenic.

Major
Maintained by Institut National de la Recherche Agronomique, France.

Makibashiro (Makibashiroshiro)
(White pasture). A synthetic Japanese cultivar based on 8 clones selected from Tregor (France), Pertina (Netherlands) and line 25849 (Australia). It is suitable for grazing, is similar in disease resistance to New Zealand white, has excellent seed yields and is tolerant of snow and cold.
Reference: Sato (1980).

Mecklenburger
German cultivar listed by Nasinec & Mika (1978) as acyanogenic.
Reference: Vacek (1970).

Mega
Maintained by Dansk Planteforaedling, Borkop, Denmark.
Also marketed as MEGA TRIFOLIUM K & V.

Menna
Bred by the Welsh Plant Breeding Station, Abersystwyth from plants collected in a field in Surrey, England with a history of high N application and high stocking rates.
Grows well at both high and low soil N levels.
Medium-small leaved. Fair resistance to *Sclerotinia trifoliorum*.
Reference: Davies & Evans (1975).

Merit
Released in 1960 by Iowa State University, USA as a selection from Ladino clover. Adapted to the central corn belt and north-east USA.
A 30-clone synthetic selected for drought resistance, winter-hardiness and leaf-hopper resistance. Resistant to *Ditylenchus dipsaci*.
Acyanogenic.
References: Leffel & Gibson (1973), Nasinec & Mika (1978), Vacek (1970), Williams (1972), Wilsie (1963, 1964).

Milka
Also marketed as ANGELITER MILKA (Germany), MILKA PAJBJERG K & V, PAJBJERG MILKA, PAJBJERG MILKA K & V.
Maintained by Dansk Planteforaedling, Borkop, Denmark.
Medium leaved cultivar of high performance and resistant to clover rot (*Sclerotinia*). Virtually acyanogenic. Reported to be resistant to *Ditylenchus dipsaci* by Bingefors (1969) but found to be susceptible by Williams (1972).
References: Aldrich (1970), Andries & van Slijcken (1968), Connolly (1970a), Harkess et al. (1970), Nasinec & Mika (1978), Vacek (1970), Williams (1972).

Milkanova
Maintained by Dansk Planteforaedling, Borkop, Denmark. Also marketed as MILKANOVA K & V.
Medium-large leaved cultivar with a high degree of resistance to clover rot (*Sclerotinia trifoliorum*).

Millay
Ladino type cultivar from Chile.
References: Avendono & San Juan (1967), Rhind et al. (1979).

Minamioha
(Norin GO3). A 9 clone Ladino type synthetic with good yield distribution, autumn yield, heat and drought tolerance and improved resistance to *Leptosphaerulina trifolii* and *Cercospora zebrina*.
Reference: Sakai (1983).

Minn. A
Not a cultivar but a germplasm release from the Minnesota Agricultural Experiment Station, USA.
Composite of seed from crosses among 7 PILGRIM plants which can be traced to collections made in Minnesota.
Selected for winter survival and large plant type, it is smaller leaved and lower yielding than Ladino clover but may be useful for breeding for cold hardiness.
Reference: Hovin & Thomas (1975).

Mira
Maintained by Dansk Planteforaedling, Borkop, Denmark. Also marketed as MIRA ÖTOFTE K & V.
ML-48-65
A very winter-hardy line developed at Melfort, Saskatchewan, Canada from progeny of three introductions from USSR which survived several winters from 1949-1952. Released in 1966.
Able to withstand hard grazing or mowing, it is medium-leaved and very persistent.
Reference: Cooke & Sonmor (1977).
Morsö
Old Danish ecotype forming the basis of several European cultivars (Chapter 8).
References: Akerberg (1934, 1936), Hansen (1942), Munro et al. (1975), Nilsson-Leissner (1934), Rudorf & Schoene (1936), Schneider (1932), Vacek (1970), Ware (1925), Whyte (1938), Williams (1972).
Morsö Otofte
Bred at Otoftegaard, Denmark in 1927 as a selection from MORSÖ, this was the first Danish bred cultivar.
Small leaved, early flowering, persistent and high yielding in Denmark but used mainly for annual or short duration leys rather than permanent pastures (Frandsen 1940).
Slightly cyanogenic (Andries & van Slijcken 1968, Hawkins 1959).
References: Hansen (1937).
Nesta
Bred at the Welsh Plant Breeding Station, Aberystwyth by hybridisation of Kersey and S100 Nomark and backcrossing to S100.
Medium-leaved but petioles and stolons medium thin. Fair resistance to *Sclerotinia*. Good performances when N fertiliser applied.
New Zealand Certified Mother Seed
See GRASSLANDS HUIA.
New Zealand Certified Permanent Pasture
See GRASSLANDS HUIA.
NFG Gigant
Also GIGANT or NFG GIGANTE.
German cultivar maintained by Deutche Saatveredelung, Lippstadt-Bremen.
Bred by crossing Italian Ladino x Morsö and released into commerce in 1953.
Large-leaved, tall, high yielding, persistent and winter hardy.
Cyanogenesis low.
Fairly resistant to *Ditylenchus dipsaci* (Williams 1972).
References: Ziegenbein (1959), Zwingli & Grundlacher (1958).
Nolins Improved
Local cultivar in South-east USA developed by W.T. Nolin, Hamburg, Louisiana by selection from the Louisiana strain.
Cyanogenic (Nasinec & Mika 1978).
References: Davis (1951), Nolin (1950), Killinger & Harris (1959), Winchester & Kretschmer (1959).
Nora
Bred by Weibullsholm Plant Breeding Institute, Landskrona, Sweden by selection of a Norwegian population from Indahl.
Released in 1947.
Small-medium leaved, resistant, frost and drought tolerant with vigorous early spring growth.
Cyanogenesis low (9.1% cyanogenic, Daday 1955).
Very susceptible to *Ditylenchus dipsaci* (Williams 1972).
Also known as WEIBULL'S NORA, WEIBULL'S V322.
Reference: Schwanbom (1947).
Nordic
Ladino type cultivar (Hovin & Thomas 1975).
Olwen

Bred by the Welsh Plant Breeding Station, Aberystwyth from plants collected in South-west France.
Medium-large narrow leaf and thick stolons, early flowering.
Best performances in the absence of *Sclerotinia*

Osceola

Ladino type selected for summer persistence at the University of Florida, USA. Synthetic cultivar with 35 parents.
Produces twice the number of inflorescences produced by other Ladino types and outyields most other cultivars in Florida (Baltensperger et al. 1984).

Otofte

Maintained by Dansk Planteforaedling, Borkop, Denmark.
Low in cyanogenesis (Nasinec & Mika 1978).
Reference: Vacek (1970).

Ovcak

Czechoslovakian cultivar selected from a southern Bohemian ecotype, released in 1963. Similar to VIGLASSKA with good seed but slow regrowth and low forage yields.
Cyanogenesis very low.
References: Nasinec & Mika (1978), Vacek (1970), Vacek & Novotny (1975).

Palupin

Czechoslovakian cultivar listed by Nasinec & Mika (1978) as low in cyanogenesis.

Pastevec

Czechoslovakian cultivar selected from an ecotype from the Bohemian-Moravian upland and released in 1961.
A typical pasture type, late flowering and sparse flowering with poor seed yield.
Cyanogenesis very low.
References: Nasinec & Mika (1978), Vacek & Novotny (1975).

Pathfinder

An old Canadian cultivar selected from a collection of wild white clover at Nappan, Nova Scotia.
Intermediate type adapted to moderate rainfall and tolerant of acid soils.
Selected for spread, leafiness, persistence and abundant flowering (Hall 1948).
References: Zwingli & Grundbacher (1958).

Pergamino El Lucero Mag

Argentinian cultivar (Pergamino morning star MAG).
Reference: *Idia 1956 106-108:* 303. (Plant Breeding Abstracts 28: 182).

Pertina

Maintained by Cebeco-Handelsraad, Rotterdam, Holland.
Selected from old pastures, using clonal evaluation, beginning in 1935.
Added to the Dutch list in 1940 as C.B. PASTURE.
Small-medium leaved, winter hardy, persistent and suitable for permanent pastures.
Cyanogenesis low.
References: Harkess et al. (1970), Nasinec & Mika (1978), Williams (1972).

Pilgrim

Released in 1953 in the USA by Northeast Stations, USDA. A Ladino-type 21-clone synthetic cultivar adapted to Northeast USA and Canada.
Acyanogenic.
References: Booysen & Laude (1964), Gervais (1961), Laude & Stanford (1960), Laude et al. (1958), Leffel & Gibson (1973), Seed World 73: 23 (1953), Stanford et al. (1960, 1962).

Pitau

See GRASSLANDS PITAU.

Podkowa

Polish cultivar of Dutch type selected from a local population in Central Poland.
Formerly the main Polish cultivar but no longer recommended because of disease susceptibility (Staszewski,

personal communication).
References: Barcikowska (1968), Staszewski (1966).
Probstheider
German cultivar listed by Nasinec & Mika (1958) as medium-high in cyanogenesis.
Reference: Vacek (1970).
Pronitro
Small leaved cultivar maintained by Zwaan en de Wiljes B.V., Scheema, Holland.
Selected from material collected in old Friesland pastures.
Small leaved (similar to S184) and resistant to *Sclerotinia trifoliorum.*
Radi
Polish cultivar of Ladino type selected for frost resistance from American and Byelorussian strains.
Able to withstand temperatures below -20°C, drought resistant, and productive.
Marketed in West Germany as KRUS 33 by Kruse & Co, Bielefeld.
Regal
A Ladino type cultivar selected at the Auburn University Station, Alabama, USA and released in 1962.
Selected for high yield, summer production and improved persistence.
Made up of five clones. Adapted to Southeastern USA.
Acyanogenic and resistant to *Ditylenchus dipsaci.*
References: Johnson et al. (1970), Nasinec & Mika (1978), Ward (1965), Williams (1972).
Rema
Polish cultivar of Dutch type adapted to low fertility grasslands.
Retor
Dutch cultivar maintained by D.J. van der Have B.V., Holland.
Medium-large leaved.
Acyanogenic (Nasinec & Mika, 1978).
Robusta
An old Swedish cultivar bred at Weibullsholm by E. Akerberg and released in 1939. No longer listed.
Selected from STRYNO (Chapter 8), it was fairly tall and open, with good drought resistance. Adapted for southern Sweden.
Also known as WEIBULL'S ROBUSTA.
References: Hall (1948), Nilsson-Leissner & Nilsson (1940).
Sabeda
Bred at the Welsh Plant Breeding Station, Aberystwyth and released in 1973.
Selected from plants collected in the Normandy region of France.
Medium-large leaves, late flowering, and persistent at high N levels.
Reference: Munro et al. (1975).
Sacramento
Sacramento Valley Milling Co., Ordberd, California, US.
Saga
See SONJA.
SC-1
Germplasm released in 1972 by the South Carolina Agricultural Experiment Station, USDA, Clemson.
SC-1 is the first generation recombination among 145 genetically diverse white clover clones selected for tolerance to root-knot nematode (*Meloidogyne incognita*). Each clone selected both at the seedling and mature plant stage.
Ladino type with large leaves and thick stolons.
Reference: Gibson (1973).
Siral
Maintained by Pasture Research Unit, New South Wales Department of Agriculture, Berry, NSW, Australia. Registered in 1976.
Derived from an Algerian collection (CPI 19434).

Survives severe moisture stress.
Medium-large leaves and good cool season production.
Cyanogenic.
References: MacKay (1977).

Skultuna

Old Swedish cultivar highly susceptible to *Ditylenchus dipsaci* (Williams 1972).

Smalbladet Pajbjerg

Maintained by Dansk Planteforaedling, Borkop, Denmark.
Small-medium leaves and low in cyanogenesis.
References: Andries & van Slijcken (1963), Harkess et al. (1970).

Sonja

Maintained by Weibullsholm Plant Breeding Institute, Landskrona, Sweden.
Formerly designated SAGA or WW V 10. first listed in Sweden in 1976.
Originates from crossing NORA x DUTCH.
Medium leaved with light green colour and large flower heads.
Good resistance to clover rot (*Sclerotinia trifoliorum*).
Reference: Jönsson (1983).

Star

Selected by R.D. Ensign and maintained by the University of Idaho, Moscow.
Star is an extremely free-seeding selection from Idaho common white clover.
Yields well in northern Idaho. (Ensign 1981).

Steinacher

Maintained by BayWa AG, München, German Federal Republic.
Developed from material originating in the Rhon region and registered about 1945.
Medium leaved and fairly resistant to winter-kill.
References: Vacek (1970), Ziegenbein (1959).

Stryno

Danish ecotype (see Chapter 8). Akerberg (1934, 1936), Rudorf & Schoene (1936), Ware (1925).

Svea

Bred at Swedish Seed Association, Svalof, southern Sweden by B. Johansson, released in 1934.
Selected from Swedish wild white which is characterised by small leaves, prostrate habit, winter-hardiness, persistence and late spring growth.
References: Nilsson-Leissner & Nilsson (1940), Whyte (1938).

Svolder

Maintained by Swedish Seed Association, Svalof, Sweden.
Medium leaved.

Szarvasi 4

Bred by Z. Nagy at the Institute of Irrigation Research, Szervas, Hungary. State approved in Hungary, 1967.
Medium-large leaved, few but large flowers, persistent.
References: Kapas (1967), Rhind et al. (1979).

Tamar

Cultivar of Israeli origin (Dr A. Dovrat, Volcani Centre, Bet-Dagan).
Marketed in Holland and Germany (Zwaan and de Wiljes' Zaadteelt and Zaadhandel GmbH).
Large leaved annual type, tall growing, profuse flowering with pale green leaves and heavy white leaf markings.
Reference: Scott (1976).

Tammisto

Selected at the Tammisto Plant Breeding Institute, Finland in the 1930s from southern-Finnish pastures, but never released because of seed production problems.
Selected for winter hardiness and yield.

References: Valle (1963), Whyte (1938).
Tillman
Released in 1965 by the South Carolina Agricultural Experiment Station, USDA, Clemson, South Carolina, USA.
A six-clone Ladino based synthetic cultivar selected for profusely branched stolons, sparse flowering, persistence, general disease resistance, and improved yield. Tested as S.C. EXPT. VAR.
Adapted to southern parts of USA suitable for Ladino clover. Seed World 98: 16 (1966).
References: Gibson et al. (1969), Jones (1966).
Titan
Maintained in France by Northrup King Semences, Verfeil.
Tongala Irrigation
A medium leaved cultivar from Victoria, Australia available commercially in South Africa.
Reference: Edwards & Mappledoram (1966).
Tregor
Maintained by UCOPAC Vilmorin Grandes Cultives, 77-Verneuil l'Etang, France.
Trevise
Maintained by L. Clause, S.A., Bretigiy-sur-Orge, France.
Medium leaved, cyanogenic and susceptible to *Ditylenchus dipsaci*.
References: Nasinec & Mika (1978), Williams (1972).
Tribla
See BLANCA.
Undrom
Maintained by the Swedish Seed Association, Svalof, Sweden.
Small leaved.
Vaccares
Maintained by UCOPAC Vilmorin Grandes Cultives, 77-Verneuil l'Etang, France.
Medium leaved.
Vertas Kultuurklaver
See ARMADA.
Vertas Witte weideklaver
See ARIA.
Viglasska
Maintained by the Research and Breeding Institute for Fodder Plants, Troubsko, Czechoslovakia, and released in 1950.
Selected from a local ecotype near the village of Viglas in Central Slovakia.
A medium leaved pasture type with relatively slow recovery after cutting. Small flower heads but good seed set.
Early flowering.
References: Vacek (1970), Vacek & Navotny (1975).
Von Kameke
Maintained by Deutsche Saatveredelung Lippstadt-Bremen, Germany.
In commerce since 1937.
Small-medium leaved pasture type with good persistence and resistance to *Sclerotinia trifoliorum*.
Reference: Vacek (1970).
Wildenelsk Otofte
(Wild English Otofte). Dansk Planteforaedling, Borkop, Denmark.
Low in cyanogenesis.
Wilkla
Selected from a local ecotype by D.J. van de Have, Holland.
Characterised by finely branched stolons and dense foliage.
Cyanogenic and very susceptible to *Ditylenchus dipsaci*.
References: Hall (1948), Williams (1972), Zwingli & Grundbacher (1958).

Witte Weideklaver C.B.

Dutch landraces marketed as intermediate type strains (Whyte et al. 1953). See CULTURA and PERTINA.

Zero Pajbjerg K & V

Bred by the Dansk Planteforaedling, Borkop, Denmark.

Listed by Andries & van Slijcken (1968) as a good cultivar under grazing.

Cyanogenic and fairly susceptible to *Ditylenchus dipsaci* (Williams 1972) although reported resistant in Danish tests (Bingefors 1969).

Reference: Harkess et al. (1970).

References

Aamodt, O.S.; Torrie, J.H.; Smith, O.F. (1939) Strain tests of red and white clovers. *Journal of the American Society of Agronomy* **31**, 1029-1037.

Adachi, A.; Suzuki, S.; Yamada, T. (1968) [Studies on selection methods in white clover breeding. 4. Competitive ability of the different types of white clover in grass swards] (Japanese). *Journal of the Japanese Society for Grassland Science* **14**, 111-115.

Ahlgren, G.H.; Fuelleman, R.F. (1950) Ladino clover. *Advances in Agronomy* **11**, 208-230.

Ahuja, P.S.; Lu, D.Y.; Cocking, E.C.; Davey,M.R. (1983) An assessment of the cultural capabilities of *Trifolium repens* L. (white clover) and *Onobrychis viciifolia* Scop. (sainfoin) mesophyll protoplasts. *Plant Cell Reports* **2**, 269-272.

Akerberg, E.(1934) [Some results of breeding and seed production of pasture plants at Weibullsholm] (Swedish). *Svensk Frötidning* **3**, 3-10.

Akerberg, E. (1936) [Experiments with seed mixtures for pasture leys conducted at Weibullsholm during the period 1933-35] (Swedish). *Landtmannen-Svensktland* **20**, 426-429.

Akerberg, E.; Bingefors, S.; Manner, R.; Umaerus, R.E.; Valcuaxter, M. (1961) [Herbage crops, The Swedish Seed Association, 1886-1961] (Swedish). *Sveriges Utsadesförenings Tidskrift* **71**, 473-495.

Akerberg, E.; Halling, S. (1959) Svalofs Kivi vitklöver [Svalofs Kiwi white clover] (Swedish) *Svalöf Arsbok, 1959*, 72-75.

Akerman, A. (1942) [The hardiness of agricultural plants during three severe winters] (Swedish). *Landtmannen-Svensktland* **26**, 655-659.

Aldrich, D.T.A. (1970a) Interaction between nitrogen fertilizer use and the productivity of white clover. *British Grassland Society Occasional Symposium* **No.6**, 227-232.

Aldrich, D.T.A. (1970b) Clover rot (*Sclerotinia trofoliorum*) in white clover and its influence on varietal performance at different centres. *British Grassland Society Occasional Symposium* **No.6**, 143-146.

Aldrich, D.T.A. (1978) The way ahead with white clover - the choice of variety. *NIAB Fellows Conference Report* **No.4**, 1-3.

Aldrich, D.T.A.; Doling, D.A. (1967) Varietal resistance to clover rot in white clover. *Nature, UK* **214**, 946-947.

Aldrich, D.T.A.; Davis, A.G.; Cooper, M. McG. (1958) White clover trials 6. Residual soil fertility. *Journal of the British Grassland Society* **13**, 147-150.

Anderson, M.K.; Taylor, N.L.; Kirtuavip, R. (1972) Development and performance of double cross hybrid red clover. *Crop Science* **12**, 240-242.

Andries, A.; van Slijcken, A. (1968) [Witte klaver (*Trifolium repens* L.) in grassland] (German) *Verslagen over navorsingen IWONL* **No. 35**, 11-90.

Archer, K.A.; Gresser, A. (1979) White clover in New South Wales. *New South Wales Department of Agriculture Plant Industry Bulletin* p. 2.5.10, 16 pp.

Armstrong, H.E.; Armstrong, E.F.; Horton, E. (1913) Herbage studies: 2 Variation in *Lotus corniculatus* & *Trifolium repens*. *Proceedings of the Royal Society, London* **B86**, 262-269.

Askew, H.O. (1933) Determination of hydrocyanic acid in white clover. *New Zealand Journal of Science and Technology* **15**, 227-233.

Atwood, S.S. (1938) A 'one-leaved' white clover. *Journal of Heredity* **29**, 239-240.

Atwood, S.S. (1940) Genetics of cross-incompatibility among self-incompatible plants of *Trifolium repens*. *Journal of the American Society of Agronomy* **32**, 955-968.

Atwood, S.S. (1941a) Cytogenetic basis of self-compatibility in *Trifolium repens* (Abstract). *Genetics* **26**, 137.

Atwood, S.S. (1941b) Cytological basis for incompatibility in *Trifolium repens*. *American Journal of Botany* **28**, 551-557.

Atwood, S.S. (1941c) Controlled self- and cross-pollination of *Trifolium repens*. *Journal of the American Society of Agronomy* **33**, 538-545.

Atwood, S.S. (1942a) Genetics of self-compatibility in *Trifolium repens*. *Journal of the American Society of Agronomy* **34**, 353-364.

Atwood, S.S. (1942b) Genetics of pseudo-self-compatibility and its relation to cross-incompatibility in *Trifolium repens*. *Journal of Agricultural Research* **64**, 699-709.

Atwood, S.S. (1942c) Oppositional alleles causing cross-incompatibility in *Trifolium repens*. *Genetics* **27**, 333-338.

Atwood, S.S. (1943) 'Natural crossing' of white clover by bees. *Journal of the American Society of Agronomy* **35**, 862-870.

Atwood, S.S. (1944a) Oppositional alleles in natural populations of *Trifolium repens*. *Genetics* **29**, 428-435.

Atwood, S.S. (1944b) Colchicine induced polyploids in white clover. *Journal of the American Society of Agronomy* **36**, 173-174.

Atwood, S.S. (1945) Behaviour of self-compatibility factor and its relation to breeding methods in *Trifolium repens*. *Journal of the American Society of Agronomy* **37**, 991-1004.

Atwood, S.S.; Garber, R.J. (1942) The evaluation of individual plants of white clover for yielding ability in association with bluegrass. *Journal of the American Society of Agronomy* **34**, 1-6.

Atwood, S.S.; Brewbaker, J.L. (1953) Incompatibility in autoploid white clover. *Cornell University Memoir* **319**, 1-47.

Atwood, S.S.; Sullivan, J.T. (1943) Inheritance of a cyanogenetic glucoside and its hydrolysing enzyme in *Trifolium repens*. *Journal of Heredity* **34**, 311-320.

Avendono, T.R.; San Juan, H.A. (1967) [Ladino clover: Millay is an excellent forage resource] (Spanish). *Investigacion Progreso Agricola* **1**, 28-29.

Baltensperger, D.D.; Dean, C.E.; Horner, E.S. (1984) Registration of Osceola white clover. *Crop Science* **24**, 1211.

Barcikowska, B. (1968) [Problems of breeding white clover] (Polish). *Postapy Nauk Rolniczych* **15**, 3-11.

Barcikowska, B. (1976) Studies on some utility features of the crosses *Trifolium repens* L. form *cultum* and on relationships between clones and their generative progeny. *Genetica Polonica* **17**, 191-210.

Barclay, P.C. (1960) Breeding for improved winter pasture production in New Zealand. *Proceedings of the 8th International Grassland Congress*, 326-330.

Barclay, P.C. (1963) Extending season of growth of pasture. *Sheepfarming Annual*, 103-109.

Barclay, P.C. (1969) Some aspects of the development and performance of 'Grasslands 4700' white clover. *Proceedings of the New Zealand Grassland Association* **31**, 127-134.

Barnard, C. (1972) Register of Australian Herbage Plant Cultivars. CSIRO, Canberra.

Barnett, O.W.; Gibson, P. B. (1975) Identification and prevalence of white clover viruses and resistance of Trifolium species to these viruses. *Crop Science* **15**, 32-37.

Barrow, G.E. (1949) Kersey white, a new strain of white clover. *Journal of the Ministry of Agriculture* **55**, 522-523.

Beinhart, G. (1963) Effects of environment on meristematic development, leaf area and growth of white clover. *Crop Science* **3**, 209-213.

Beinhart, G., Gibson, P. B.; Halpin, J.E.; Hollowell, E.A. (1963) Selection and evaluation of white clover clones III. Clonal differences in branching in relation to leaf area production and persistence. *Crop Science* **3**, 89-92.

Berezovskii, Yu. V.; Agaverdiev, A. Sh. (1979) [Determining the frost resistance of herbage crops by means of recording very weak luminescence.] (Russian). *Doklady Vsesoyuznoi ordena Lenina akademii selsko-khozaistvennykh nauk im V.J. Lenina* **no.8**, 15-16.

Bhojwani, S.S.; White, D.W.R. (1982) Mesophyll protoplasts of white clover: isolation, culture and organogenesis. *Plant Science Letters* **26**, 265-271.

Bijok, K.; Goral, M.;Goral, S. (1968) [Cytological studies, pollination, and seed setting in polyploid forms of *Trifolium repens* L.] (Polish). *Hodowla Roslin, Aklimatyzacjai Nasiennictwo* **12**, 423-435.

Bijok, K.; Goral, M.; Goral, S. (1970) [Embryological studies on polyploid forms of *Trifolium repens* L.] (Polish). *Acta Agrobotanica* **23**, 259-277.

Binek, A. (1983) [The structure of seed yield in clones of white clover (*Trifolium repens* L.) after the reduction of inflorescences to a standard number per plant] (Polish). *Acta Agraria et Sylvestria, Agraria* **22**, 21-29.

Bingefors, S. (1969) The use of nematode resistant varieties of grasses and legumes. *Herbage Abstracts* **39**, 107-111.

Blaser, R.E.; Killinger, G.W. (1950) Life history studies of Louisiana white clover (*Trifolium repens* L.) I. Seed germination as related to temperature, pasture management and adaptation. *Agronomy Journal* **42**, 215-220.

Boekholt, K. (1953) [The significance of leaf colour in the breeding of *T. repens* and *T. hybridum*] (German). *Züchter* **5**, 157-159.

Bonish, P.M. (1980) Nodulation of white clover: plant influences on the effectiveness of *Rhizobium trifolii*. *New Zealand Journal of Agricultural Research* **23**, 239-242.

Booysen, P. DeV.; Laude, H.M. (1964a) Influence of flower initiation and development on internode growth in the Ladino clover stolon.

Crop Science **4**, 520-524.

Booysen, P. DeV.; Laude, H.M. (1964b) Flowering persistence in Ladino clover. *Crop Science* **4**, 518-520.

Boyd, M.M. (1971) The stability of cultivars of grass and clovers when grown for seed in differing environments. Part 2. *Journal of the National Institute of Agricultural Botany* **12**, 267-278.

Breese, E.L. (1969) The measurement and significance of genotype-environment interactions in grasses. *Heredity* **24**, 27-44.

Breese, E.L.; Hayward, M.D. (1972) The genetic basis of present breeding methods in forage crops. *Euphytica* **21**, 324-336.

Breese, E.L.; Hayward, M.D.; Thomas, A.C. (1965) Somatic selection in perennial ryegrass. *Heredity* **20**, 367-379.

Brewbaker, J.L. (1952) Colchicine induction of tetraploids in *Trifolium* species. *Agronomy Journal* **44**, 592-594.

Brewbaker, J.L. (1954) Incompatibility in autotetraploid *Trifolium repens* I. Competition and self-compatibility. *Genetics* **39**, 307-317.

Brewbaker, J.L. (1955a) Incompatibility in autotetraploid white clover II. Dominance and double reduction. *Genetics* **40**, 137-152.

Brewbaker, J.L. (1955b) V-leaf markings of white clover. *Journal of Heredity* **46**, 115-123.

Brewbaker, J.L.; Atwood, S.S. (1952) Incompatibility alleles in polyploids. *Proceedings of the 6th International Grassland Congress* **1**, 267-272.

Brewbaker, J.L.; Carnahan, J.L. (1956)Leaf marking alleles in white clover - uniform nomenclature. *Journal of Heredity* **47**, 103-104.

Brewbaker, J. L.; Keim, W., F. (1953) A fertile interspecific hybrid in *Trifolium* (4n *T. repens* L. x 4n *T.nigrescens* Viv.) *American Naturalist* **87**, 323-326.

Brigham, R.D.; Wilsie, C.P. (1955) Seed setting and vegetative vigour of Ladino clover (*Trifolium repens* Leyss.) clones and their diallel crosses. *Agronomy Journal* **47**, 125-127.

Brock, J.L. (1971) A comparison of 'Grasslands 4700' and 'Grasslands Huia' white clovers in establishing ryegrass-clover pastures under grazing. *New Zealand Journal of Agricultural Research* **14**, 368-378.

Brock, J.L. (1974) Effects of summer grazing management on the performance of 'Grasslands Huia' and 'Grasslands 4700' white clovers in pastures. *New Zealand Journal of Experimental Agriculture* **2**, 365-369.

Brock, J.L.; Hoglund, J.H. (1974) Growth of 'Grasslands Huia' and 'Grasslands 4700' white clovers II. Effects of nitrogen and phosphorus. *New Zealand Journal of Agricultural Research* **17**, 47-53.

Broderick, G.A. (1975) Factors affecting ruminant responses to protected amino acids and proteins. In Protein nutritional quality of foods and feeds Volume 1, Part 2. M. Friedman (Ed.), Marcel Dekker Inc., New York, pp. 211-259.

Brown, B.A.; Munsell, R.I. (1956) *Connecticut (Storrs) Agricultural Experimental Station Bulletin* **313**, 1-42.

Bula, R.J.; May, R.G.; Garrison, C.S.; Rincker, C.M.; McAllister, D.R. (1964) Growth response of white clover, *Trifolium repens* L., progenies from five diverse geographic locations. *Crop Science* **4**, 295-297.

Burdon, J.J. (1980) Interspecific diversity in a natural population of *Trifolium repens*. *Journal of Ecology* **68**, 717-735.

Caradus, J.R. (1977) Structural variation of white clover root systems. *New Zealand Journal of Agricultural Research* **20**, 213-219.

Caradus, J.R. (1979) Selection for root hair length in white clover *(Trifolium repens* L.). *Euphytica* **28**, 489-494.

Caradus, J.R. (1980) Distinguishing between grass and legume species for efficiency of phophorus use. *New Zealand Journal of Agricultural Research* **23**, 75-81.

Caradus, J.R. (1981) Root morphology of some white clovers from New Zealand hill country. *New Zealand Journal of Agricultural Research* **24**, 349-351.

Caradus, J.R. (1983) Genetic differences in phosphorus absorption among white clover populations. *Plant and Soil* **72**, 379-383.

Caradus, J.R. (1986) World checklist of white clover varieties. *New Zealand Journal of Experimental Agriculture* **14**, 119-164.

Caradus, J.R.; Dunlop, J. (1978) Screening white clover plants for efficient phosphorus use. *Proceedings of the 8th International Colloqium on Plant Analysis and Fertiliser Problems, Auckland, New Zealand*. A.R. Ferguson et al. (Eds). *New Zealand DSIR Information Series* **No. 134**, 75-81.

Caradus, J.R.; Dunlop, J.; Williams, W.M. (1980) Screening white clover for different responses to phosphate. *New Zealand Journal of Agricultural Research* **23**, 211-217.

Caradus, J.R.; Williams, W.M. (1981) Breeding for improved white clover production in New Zealand hill country. Plant Physiology and herbage production, 7-9 April 1981, meeting held at the University of Nottingham. Occasional Symposium no. 13 (ed. C.E. Wright) Lurgan, UK, L.M. Press Ltd, 163-168.

Carnahan, H.L.; Hill, H.D.; Hanson, A.A.; Brown, K.G. (1955) Inheritance and frequencies of leaf markings in white clover. *Journal of Heredity* **46**, 109-114.

Carroll, B. J.; McNeill, D. L.; Gresshof, P. M. (1985) Isolation and properties of novel soybean (*Glycine max* (L.) Merr.) mutants that nodulate in the presence of high nitrate concentrations. *Proceeding of the National Academy of Sciences, USA*, **82**, 4162-4166.

Cebrat, J.; Sobotka-Mioduszewska, A.; Kobiezynska-Golab, Z. (1978) [Preliminary embryological investigation of the causes of low fertility in white clover (*Trifolium repens* L.)] (Polish). *Hodowla Roslin, Aklimatyzacjai Nasiennictwo* **22**, 311-320.

Cebrat, J.; Kobierzynska-Golab, Z.; Ramenda, S. (1982) [The variability of quantitative characters which affect fertility in five varieties of white clover (*Trifolium repens* L.)] (Polish). *Hodowla roslin, Roslin, Aklimatyzacjai Nasiennictwo* **26**, 11-34.

Chamblee, D.S. (1954) Amount of flowering of Ladino clover lots from different sources. *Agronomy Journal* **46**, 287-288.

Champness, S.S.; Morris, K. (1948) The population of buried viable seeds in relation to contrasting pasture and soil types. *Journal of Ecology* **36**, 149-173.

Chapman, D.F. (1983) Growth and demography of *Trifolium repens* stolons in grazed hill pastures. *Journal of Applied Ecology* **20**, 597-608.

Chen, Chi-Chang; Gibson, P. B. (1973) Effect of temperature on pollen tube growth in *Trifolium repens* after cross- and self-pollinations. *Crop Science* **13**, 563-566.

Chesneaux, M.T. (1972) [Possibilities of controlling varietal identity in white clover (*Trifolium repens* L.) under artificial conditions] (French). *Annales de l'Amelioration des Plantes* **22**, 311-319.

Cheyne, V.A.; Dale, P.J. (1980) Shoot tip culture in forage legumes. *Plant Science Letters* **19**, 303-309.

Chisci, G.C. (1963) La Coltivazione del Trifoglio Ladino. *Estratto da L'Informatore Agrario Verona* **52.**

Chisci, G.C.; Ferrari, A. (1967) [Trials to compare generations of Ladino clover (*Trifolium repens* var. *giganteum*) multiplied in the original environment and the U.S.A.] (Italian). *Genetica Agraria* **21**, 148-162.

Clark, J.H. (1975) Nitrogen metabolism in ruminants: protein solubility and rumen bypass of protein and amino acids. In 'Protein nutritional quality of foods and feeds.' Volume 1, Part 2, M. Friedman (Ed), Marcel Dekker Inc., New York, pp. 261-304.

Clark, M.F.; Barclay, P.C. (1972) The use of immuno-osmophoresis in screening a large population of *Trifolium repens* L. for resistance to white clover mosaic virus. *New Zealand Journal of Agricultural Research* **15**, 371-375.

Clifford, P.T.P. (1977) Cultural methods for 'Grasslands Pitau' white clover seed crops. *New Zealand Journal of Experimental Agriculture* **5**, 147-149.

Clifford, P.T.P. (1980) Research in white clover seed production. In 'Herbage seed production'. Grassland Research and Practice Series No. 1. *New Zealand Grassland Association*, pp. 64-67.

Cohen, M.M.; Leffel, R.C. (1963) Cytology of pseudo-self-compatibility in Ladino white clover, *Trifolium repens* L. *Crop Science* **3**, 430-433.

Cohen, M.M.; Leffel, R.C. (1964) Pseudo-self-compatibility and segregation of gametophytic self-incompatibility alleles in white clover, *Trifolium repens* L. *Crop Science* **4**, 429-431.

Collinge, D.B.; Hughes, M.A. (1982) Developmental and physiological studies on the cyanogenic glycosides of white clover, *Trifolium repens* L. *Journal of Experimental Botany* **33**, 154-161.

Collinge, D.B.; Hughes, M.A. (1984) Evidence that linamarin and lotaustralin, the two cyanogenic glucosides of *Trifolium repens* L., are synthesised by a single set of microsomal enzymes controlled by the *Ac/ac* locus. *Plant Science Letters* **34**, 119-125.

Connolly, V. (1968) A comparison of six white clover and three perennial ryegrass varieties under a system of grazing management. *Irish Journal of Agricultural Research* **7**, 227-242.

Connolly, V. (1970a) White clover breeding. *An Foras Talúntais, Plant Sciences and Crop Husbandry Division Research Report 1970*, 28-31.

Connolly, V. (1970b) Variation and seasonal productivity of populations of white clover (*Trifolium repens*) and their potential in breeding improved varieties. *British Grassland Society Occasional Symposiun* **No.6**, 133-138.

Connolly, V. (1972) Genetic analysis of seedling growth in white clover. *An Foras Talúntais, Plant Science and Crop Husbandry Division Research Report* 1972, 22.

Connolly, V.; Masterson, C.L.; Coniffe, D. (1969) Some genetic aspects of the symbiotic relationship between white clover (*Trifolium repens*) and *Rhizobium trifoli*. *Theoretical and Applied Genetics* **39**, 206-213.

Cooke, D.A.; Sonmor, L.G. (1977) Registration of ML-48-65 white clover germplasm (Reg. No. GP8). *Crop Science* **17**, 189.

Coombe, D.E. (1961) *Trifolium occidentale*, a new species related to *Trifolium repens* L. *Watsonia* **5**, 68-87.

Coombe, D.E.; Morisset, P. (1967) Further observations on *Trifolium occidentale*. *Watsonia* **6**, 271-275.

Coop, I.E.; Blakley, R.L. (1949) The metabolism and toxicity of cyanides and cyanogenetic glucosides in sheep I. Activity in the rumen. *New Zealand Journal of Science and Technology* **30**, 277-291.

Coop, I. E.; Blakley, R.L. (1950) The metabolism and toxicity of cyanides and cyanogenetic glucosides in sheep. III. The toxicity of cyanides and cyanogenic glucosides. *New Zealand Journal of Science and Technology* **A31**, 44-58.

Cope, W.A. (1978) Effect of inbreeding on field performance of white clover selections. *Crop Science* **18**, 144-146.

Cope, W.A.; Walker, S.K.; Lucas, L.T. (1978) Evaluation of selected white clover clones for resistance to viruses in the field. *Plant Disease Reporter* **62**, 267-270.

Corkill, L. (1940) Cyanogenesis in white clover (*Trifolium repens* L.) I. Cyanogenesis in single plants. *New Zealand Journal of Science and Technology* **B22,** 65-67.

Corkill, L. (1942) Cyanogenesis in white clover (*Trifolium repens* L.) V. The inheritance of cyanogenesis. *New Zealand Journal of Science and Technology* **B23**, 178-193.

Corkill, L. (1949) Pasture improvement in New Zealand. *Empire Journal of Experimental Agriculture* **17**, 157-169.

Corkill, L. (1952) Cyanogenesis in white clover (*Trifolium repens* L.) VI. Experiments with high-glucoside and glucoside-free strains. *New Zealand Journal of Science and Technology* **A34**, 1-16.

Corkill, L. (1956) Breeding for quality in pasture plants. *Sheepfarming Annual* **1956**, 47-51.

Corkill, L. (1963) A white clover strain with a distinguishing leaf mark (a note). *New Zealand Journal of Agricultural Research* **6**, 457-459.

Corkill, L. (1964) Maori names given to pasture plants bred by Grassland Division. *New Zealand Journal of Agriculture* **109**, 138.

Corkill, L. (1971) Leaf markings in white clover. *Journal of Heredity* **62**, 307-310.

Craigmiles, J.P.; Crowder, L.V. (1960) Comparison of Ladino and Louisiana white clovers in three grass sods. *Agronomy Journal* **52**, 379-381.

Crowder, L.V. (1960) The response of white clover varieties grown at high elevations in Colombia. *Agronomy Journal* **52**, 608-609.

Crush, J.R.; Caradus, J.R. (1980) Effect of mycorrhizas on growth of some white clovers. *New Zealand Journal of Agricultural Research* **23**, 233-237.

Curran, P.L.; Reilly, M.L. (1965) Varietal and clonal differences in white clover. *Scientific Proceedings of the Royal Dublin Society, Series B* **1**, 163-172.

Daday, H. (1954a) Gene frequencies in wild populations of *Trifolium repens*. I Distribution by latitude. *Heredity* **8**, 61-78.

Daday, H. (1954b) Gene frequencies in wild populations of *Trifolium repens*. II Distribution by altitude. *Heredity* **8**, 377-384.

Daday, H. (1954c) Gene frequencies in strains of *Trifolium repens* L. *Nature* . **174,** 521-522.

Daday, H. (1955) Cyanogenesis in strains of white clover. *Journal of the British Grassland Society* **10**, 266-274.

Davies, D.R.; Wall, E.T. (1959) Separation by gamma irradiation of two elements at the V locus in *Trifolium repens*. *Nature* **184**, 1957-1958.

Davies, D.R.; Wall, E.T. (1960) Induced mutations at the V locus of *Trifolium repens*. I Effects of acute, chronic and fractionated doses of gamma irradiation on induction of somatic mutations. *Heredity* **15**, 1-15.

Davies, H. (1978) Variety and nitrogen interaction in white clover. *NIAB Fellows Conference, Report* **4**, 4-6.

Davies, W.; Levy, E.B. (1931) Strain investigation of grasses and clover 4. White clover (*Trifolium repens*). *New Zealand Journal of Agriculture* **42**, 75-89, 151-164.

Davies, W. Ellis, (1956) Storage of clover seed. I First interim report, 1947-54 *Journal of the British Grassland Society* **11**, 224-229.

Davies, W. Ellis (1958a) The yields of pure sown plots of eight white clover strains under cutting. *Journal of the British Grassland Society* **13**, 34-38.

Davies, W. Ellis (1958b) Breeding better clovers. *Agricultural Review, London* **3**, 12-18.

Davies, W. Ellis (1958c) Legume breeding. *Welsh Plant Breeding Station Report for 1950-56*, 51-65.

Davies, W. Ellis (1960) The effect of F1 selection on the subsequent yield of clover families. *Welsh Plant Breeding Station Report for 1959*, 29-30.

Davies, W. Ellis (1962a) The persistency of sown white clover in a long ley at an upland centre at an altitude of 1000 ft. *Welsh Plant Breeding Station Report for 1961*, 45-47.

Davies, W. Ellis (1962b) The breeding of a new large-leaved white clover. *Welsh Plant Breeding Station Report for 1961*, 44.

Davies, W. Ellis (1962c)The flowering of white clovers under glasshouse conditions. *Welsh Plant Breeding Station Report for 1962*, 50-51.

Davies, W. Ellis (1963) Leaf markings in *Trifolium repens*. In 'Teaching Genetics', (Eds) C.D. Darlington & A.D. Bradshaw. Oliver and Boyd, Edinburgh and London, pp. 94-98.

Davies, W. Ellis (1965) Storage of clover seed. II Second interim report 1955-64. *Journal of the British Grassland Society* **20**, 180-181.

Davies, W. Ellis (1969a) Herbage legumes - special considerations. *Occasional Symposium of the British Grassland Society 'Grass and Forage Breeding'* **No. 5**, 21-27.

Davies, W. Ellis (1969b) The assessment of herbage legume varieties. IV The potential of 64-chromosome white clover. *Journal of Agricultural Science, Cambridge* **73**, 139-144.

Davies, W. Ellis (1970) White clover breeding: review. In Occasional Symposium of the British Grassland Society 'White Clover Research', 99-102.

Davies, W. Ellis, (1971) Host-pollinator relationships in the evolution of herbage legumes in Britain. *Science Progress* **59**, 573-589.

Davies, W. Ellis (1972) New varieties, herbage legumes. *Welsh Plant Breeding Station Report for 1971*, 17-18.

Davies, W. Ellis; Evans, D.R. (1968) Improving competitive ability in white clover. *Welsh Plant Breeding Station Report for 1967*, 43-44.

Davies, W. Ellis; Evans, D.R. (1975) White clover collected from high production ryegrass swards receiving large applications of fertiliser nitrogen. *Welsh Plant Breeding Station Report for 1974*, 22-23.

Davies, W. Ellis; Evans, D.R. (1976) Loss of vigour during seed multiplication of 64-chromosome white clover lines based on two individuals. *Welsh Plant Breeding Station Report for 1975*, 20-21.

Davies, W. Ellis; Griffith, G. Ap.; Ellington, A. (1966) Assessment of herbage legume varieties. II In vitro digestibility, water-soluble carbohydrate, crude protein and mineral content of primary growth of clover and lucerne. *Journal of Agricultural Science, Cambridge* **66**, 351-357.

Davies, W. Ellis; Thomas, T.A.; Young, N.R. (1968) Assessment of herbage legume varieties, III Annual variation in chemical composition of eight varieties. *Journal of Agricultural Science, Cambridge* **71**, 233-241.

Davies, W. Ellis; Tyler, B.F. (1961) The assessment of breeders' material. *Welsh Plant Breeding Station Report for 1960*, 31.

Davies, W. Ellis; Young, N.R. (1963) The use of genetic markers for the identification of white clover cultivars. *Welsh Plant Breeding Station Report for 1962*, 33-35.

Davies, W. Ellis; Young, N.R. (1965) The long term storage of small samples of clover seed. *Welsh Plant Breeding Station Report for 1964*, 41-42.

Davies, W. Ellis; Young, N.R. (1967a) The characteristics of European, Mediterranean, and other populations of white clover (*Trifolium repens* L.). *Euphytica* **16**, 330-340.

Davies, W. Ellis; Young, N.R. (1967b) Long term storage of white clover seed. *Welsh Plant Breeding Station Report for 1966*, 59-60.

Davis, J.N. (1951) Louisiana improved mother white. *Southern Seedsman* **14(19)**, 24,50.

Davis, J.N.; Cooper, M.M. (1951) White clover trial. I Production in terms of live-weight increase. II Changes in botanical composition. *Journal of the British Grassland Society* **6**, 229-243.

Davis, J.N.; Cooper, M.M. (1953) White clover trial. III Production in the third and fourth harvest years. IV Botanical composition of the swards. V Earthworm population. *Journal of the British Grassland Society* **8**, 115-130.

Dessureaux, L. (1949) [Variability in non-selected populations of Ladino white clover] (French). *Agriculture, Montreal* **6**, 127-131.

Dessureaux, L. (1951) Ovule formation as a factor influencing seed setting of Ladino white clover. *Science in Agriculture* **31**, 373-382.

Dijkstra, J. (1971) Breeding for resistance to *Heterodera* in white clover. *Euphytica* **20**, 36-46.

Dijkstra, J.; De Vos, A.L.F. (1972) The evaluation of white clover (*Trifolium repens* L.) in monoculture and in mixture with grass. *Euphytica* **21**, 432-439.

Doak, B.W. (1933) Quantitative HCN determinations of white clover types. *Imperial Bureau of Plant Genetics: Herbage Plants, Bulletin* **11**, 43.

Dovrat, A.; Waldman, M. (1966) Changes in flowering responses and seed yields in some northern forage varieties multiplied in a southern latitude. *Proceedings of the 10th International Grassland Congress*, 787-792.

Drake, F.R. (1942) Pasture plant research at Burnley. Report for 1941. *Journal of the Department of Agriculture, Victoria* **40**, 329-336.

During, C.; Weeda, W.C. (1974) A comparison of 'Grasslands Pitau' and 'Huia' white clovers in the Waikato under two intensities of winter grazing. *New Zealand Journal of Agricultural Research* **2**, 23-25.

Eagles, C.F.; Fuller, M.P. (1983) Evaluation of cold hardiness in forage grasses and legumes. In *The utilisation of genetic resources in fodder crop breeding, 3-16 September 1982, Aberystwyth, UK*. Welsh Plant Breeding Station, 265-270.

Eagles, C.F.; Othman, O.B. (1981) Growth at low temperature and cold hardiness in white clover. Plant physiology and herbage production, 7-9 April 1981, held at the University of Nottingham. *Occasional symposium no. 13*. Ed. Wright, C.E. 109-113.

East, E.M.; Mangelsdorf, P. (1925) A new interpretation of the hereditary behaviour of self-sterile plants. *Proceedings of the National Academy of Sciences, USA* **11**, 166-171.

Edwards, P.J.; Mappledoram, B.D. (1966) Select the clover variety to suit your area. *Farming in South Africa* **41**, 13-15.

Ellerström, S. (1969) [Heterosis in crosses between Swedish and exotic white clover material] (Swedish). *Sveriges Utsadesforenings Tidskrift* **5**, 269-281.

Elliott, H.G.(1940) White clover. *Journal of the Department of Agriculture, Western Australia* **17**, 186-192.

Ensign, R.D. (1981) Registration of Star white clover (Reg, No. 5). *Crop Sceince* **21,** 631-632.

Esser, J. (1960) [Results from tests made in 1955-58 with imported seed of lucerne and white clover] (German). *Saatgut Wirtschaft* **12**, 273-275.

Evans, A.M. (1955) The production and identification of polyploids in red clover, white clover and lucerne. *New Phytologist* **54**, 149-162.

Evans, A.M. (1965) Investigations on diploid and polyploid white clover. The significance of seed size and weight in white clover. *Welsh Plant Breeding Station Report for 1964,* 42-46.

Evans, E.T.R.; Evans, W.C. (1948) Studies of the biochemistry of pasture plants. 2 The pharmacological properties of certain pasture plant juices, and their possible significance in the aetiology of some disorders of the grazing animal. *Journal of the British Grassland Society* **3**, 249-261.

Ezakova, O.F.; Piskovackaja, R.G. (1977) [The effect of chemical mutagens on self-fertility in white clover] (Russian). *International Grassland Kongress Section-Vorträge, Leipzig* **13(1/2)** 381-385.

Faulkner, J.S. (1980) Variation in paraquat tolerance in seedlings of *Trifolium repens. Record of Agricultural Research* **28**, 27-31.

Ferrari, A. (1966) *Relazione sull'attivita della Stazione sperimentale di Praticoltura, Lodi 1965-66,* 14-29.

Flanagan, T.R. (1952) Cold resistance of Ladino clover and its measurement. *Abstracts of Doctoral Dissertations, University of Pennsylvania 1951* **14**, 36-39.

Flux, D.S.; Butler, G.W.; Johnson, J., M.; Glenday, A.C.; Peterson, G.B. (1956) Goitrogenic effects of white clover (*Trifolium repens* L.). *New Zealand Journal of Science and Technology* **A38**, 88-102.

Forde, M.B.; Suckling, F.E.T. (1977) Sampling white clover (*Trifolium repens* L.) from New Zealand hill-country pastures. *Proceedings of the 3rd International Congress of the Society for the Advancement of Breeding Researches in Asia and Oceania (SABRAO)*, Canberra, Australia, February 1977, 14(6)-15-18.

Foy, N.R. Hyde, E.O.C. (1937) Investigation of the reliability of the 'picric-acid test' for distinguishing strains of white clover in New Zealand. *New Zealand Journal of Agriculture* **55**, 219-224.

Frandsen, H.N. (1940) Improvement of herbage plants in Denmark. In 'The breeding of herbage plants in Scandinavia and Finland.' *Imperial Agricultural Bureaux Joint Publication* **no. 3**, 71-92.

Frandsen, K.J. (1945) [Observations on polyploid forms of some cultivated plants] (Danish, English summary). *Tidsskrift for Planteavl* **49**, 445-496.

Frandsen, K.J. (1948) [Observation on polyploid forms of cultivated plants (*T. pratense, T. hybridum* and *T. repens*] (Danish). *Beretning om Nordiske jordbrugsforskeres forenings Kongres* 1947, **1-3**, 508-527.

Free, J.B. (1970) Insect pollination of crops. Academic Press, London.

Gausman, H.W.; Fuelleman, R.F. (1952) Distinguishing Ladino clover from white clover plants. *Agronomy Journal* **44**, 96-97.

Gervais, P. (1961) Performance of white clover varieties under simulated conditions at Lennoxville, Quebec. *Forage Notes* **7**, 28.

Gibson, P. B. (1957) Effect of flowering on the persistence of white clover. *Agronomy Journal* **49**, 213-215.

Gibson, P. B. (1964) A technique requiring few seed for evaluating white clover strains. *Crop Science* **4**, 344-345.

Gibson, P. B.; Beinhart, G.; Halpin, J.E. (1969) Registration of Tillman white clover. *Crop Science* **9**, 522.

Gibson, P. B.; Beinhart, G.; Halpin, J.E.; Hollowell, E.A. (1963) Selection and evaluation of white clover clones. I Basis for selection and a comparison of two methods of propagation for advanced evaluations. *Crop Science* **3**, 83-92.

Gibson, P. B.; Chen, Chi-Chang; Gillingham, J.R.; Barnett, O.W. (1971) Interspecific hybridisation of *Trifolium uniflorum* L. *Crop Science* **11**, 895-899.

Gibson, P. B.; Chen, Chi-chang (1973) Success in hybridising and selfing *Trifolium repens* at different temperatures. *Crop Science* **13**, 728-730.

Gibson, P. B.; Schultz, E.F., Jr (1953) The effect of environment on the number of vascular bundles per petiole in white clover. *Agronomy Journal* **45**, 123-124.

Gibson, P. B.; Trautner, J. L. (1965) Growth of white clover with and without primary roots. *Crop Science* **5**, 477-479.

Gibson, P. B.; Chen, Chi-Chang (1975) Registration of SC-2 and SC-3 clover germplasms (Reg. no. GP4 and GP5). *Crop Science* **15**, 605-606.

Goold, G.J.; Douglas, J.A. (1976) A small-plot evaluation of three ryegrasses and two clovers at five sites in New Zealand. *New Zealand Journal of Experimental Agriculture* **4**, 135-141.

Greenham, G.C.; Daday, H. (1957) Electrical determination of cold hardiness in *Trifolium repens* L. and *Medicago sativa* L. *Nature, UK* **180**, 541-543.

Greenham, G.C.; Daday, H. (1960) Further studies on the determination of cold hardiness in *Trifolium repens* L. and *Medicago sativa* L. *Australian Journal of Agricultural Research* **11**, 1-15.

Gresshof, P.M. (1980) In vitro culture of white clover: callus, suspension, protoplast culture, and plant regeneration. *Botanical Gazette* **141**, 157-164.

Gurnsey, M.P.; Jones, W.T.; Merrall, M.; Reid, C.S.W. (1977) Cyanide poisoning in cattle; two unusual cases. *New Zealand Veterinary Journal* **25**, 128-130.

Hadfield, J.W. (1929) Certification of grass and clover seeds. System established by the Department of Agriculture. *New Zealand Journal of Agriculture* **39**, 289-295.

Haggar, R.J.; Holmes, W.; Innes, P. (1963) Wild white clover seed production. I The effects of defoliation and fertiliser treatment on flowering and seed yields from ryegrass/white clover swards. *Journal of the British Grassland Society* **18**, 97-103.

Haggar, R.J.; Holmes, W. (1963) Kent wild white clover. Wye College, Kent, 40 pp. (illustrated).

Hall, M. (1948) Five hundred varieties of herbage and fodder plants. *Commonwealth Bureau of Pastures and Field Crops, Aberystwyth, Bulletin* **39**, 258-264.

Hansen, J. (1937) [Trials with strains of white clover 1934-1937 and 1928-1934] (Danish). *Tidsskrift for Planteavl* **42**, 689-698.

Hansen, J. (1942) [Trials with strains of white clover, 1934-1940] (Danish). *Tidsskrift for Planteavl* **46**, 642-660.

Hanson, A.A.; Graham, J.H.; Kreitlow, K.W. (1953) The isolation of ladino clover plants resistant to *Sclerotinia trifoliorum*. *Canadian Journal of Agricultural Science* **33**, 84-90.

Harberd, D.J. (1963) Observations on natural clones of *Trifolium repens* L. *New Phytologist* **62**, 198-204.

Harkess, R.D.; Hunt, I.V.; Frame, J. (1970) The effects of variety and of companion grass on the productivity of white clover. *British Grassland Society Occasional Symposium* **No. 6**, 175-180.

Hardarson, G.; Jones, D. G. (1979a) Effect of temperature on competition amongst strains of *Rhizobium trifolii* for nodulation of two white clover varieties. *Annals of Applied Biology* **92**, 229-236.

Hardarson, G.; Jones, D. G. (1979b) The inheritance of preference for strains of *Rhizobium trifolii* by white clover (*Trifolium repens*). *Annals of Applied Biology* **92**, 329-333.

Harris, W.; Brock, J.L. (1972) Effect of porina caterpillar (*Wiseana* spp.) infestation on yield and competitive interactions of ryegrass and white clover varieties. *New Zealand Journal of Agricultural Research* **15**, 723-740.

Harris, W.; Rhodes, I.; Mee, S.S. (1983) Observations on environmental and genotypic influences on the overwintering of white clover. *Journal of Applied Ecology* **20**, 609-624.

Harrison, J.E. (1936) Permanent pasture mixtures. *Journal of Agriculture, Victoria* **34**, 521-528.

Harville, B.G.; Derrick, K.S. (1978) Identification and prevalence of white clover viruses in Louisiana. *Plant Disease Reporter* **62**, 290-292.

Hawkins, R.P. (1959) Botanical characters for the classification and identification of varieties of white clover. *Journal of the National Institute of Agricultural Botany* **8**, 675-682.

Hawkins, R.P. (1960) Investigations on local strains (varieties) of herbage plants. IV White clover. *Journal of the British Grassland Society* **15**, 28-33.

Hawkins, R.P. (1967) A note on Kersey white clover. *Journal of the National Institute of Agricultural Botany* **11**, 226-227.

Haycock, R.; Ollerenshaw, J.H. (1982) Winter survival and spring growth of white clover cultivars at an upland site in northern England. *Journal of Agricultural Science, Cambridge* **98**, 471-473.

Haycock, R. (1984) Dry-matter distribution and seasonal yield changes in five contrasting genotypes of white clover. *Journal of Agricultural Science, Cambridge* **102**, 333-340.

Hoen, K. (1970) Performance in mixed swards of three perennial ryegrass and three white clover varieties at two nitrogen levels. *Irish Journal of Agricultural Research* **9**, 215-223.

Hoglund, J.H.; Brock, J.L. (1974) Growth of 'Grasslands Huia' and 'Grasslands 4700' white clovers. I Effects of temperature and nitrogen. *New Zealand Journal of Agricultural Research* **17**, 41-45.

Hoglund, J.H.; Williams, W.M. (1984) Genotypic variation in white clover growth in response to temperature and nitrogen. *New Zealand Journal of Agricultural Research* **27**, 19-24.

Hollowell, E.A. (1958) Registration of varieties and strains of white clover (*Trifolium repens* L.) Louisiana S1 (Reg. no. 1). *Agronomy Journal* **50**, 692.

Hollowell, E.A. (1966) White clover (*Trifolium repens* L.), annual or perennial? *Proceedings of the 10th International Grasslands Congress*, 184-187.

Hovin, A.W. (1963) Compatibility reactions of interspecific hybrids between *Trifolium repens* L. and *T. nigrescens* Viv. *Crop Science* **3**, 487-489.

Hovin, A.W.; Gibson, P.B. (1961) A red leaf marking in white clover. *Journal of Heredity* **52**, 295-296.

Hovin, A.W.; Thomas, H.L. (1975) Registration of Minn A white clover germplasm (Reg. no. GP6). *Crop Science* **15**, 606.

Hudson, A.W. (1933) Measurement of pasture production: extension and demonstration field trials. *Imperial Bureau of Plant Genetics: Herbage Plants, Bulletin* **11**, 21-35.

Hughes, M.A. (1968) Studies on the β-glucosidase system of *Trifolium repens* L. *Journal of Experimental Botany* **19**, 427-434.

Hughes, M.A. (1970) Cyanogenesis in white clover. In *Occasional Symposium British Grassland Society, White Clover Research*. **No. 6**, 123-131.

Hughes, M.A.; Stirling, J.D. (1982) A study of dominance at the locus controlling cyanogenesis cyanoglucoside production in *Trifolium repens* L. *Euphytica* **31**, 477-483.

Hughes, M.A.; Stirling, J.D.; Collinge, D.B. (1984) The inheritance of cyanoglucoside content in *Trifolium repens* L. *Biochemical Genetics* **22**, 139-151.

Hunt, I.V.; Harkess, R.D.; Martin, T.W. (1965) Comparison of production of four varieties of white clover. *Research Bulletin, West of Scotland College of Agriculture*, **no. 34**.

Hutton, E.M. (1957) Some effects of induced autopolyploidy in white clover, barrel medic, and 'Wimmera' ryegrass. *Journal of the Australian Institute of Agricultural Science*, **23**, 227-231.

Hutton, E.M. (1968) Australia's pasture legumes. Farrer memorial oration, June 1968. *Journal of the Australian Institute of Agricultural Science* **34**, 203

Hyde, E.O.C.; Suckling, F.E.T. (1953) Dormant seeds of clovers and other legumes in agricultural soils. *New Zealand Journal of Science and Technology* **A34**, 375-385.

Jaagus, M. (1971) [On preliminary results of trials on evaluation of basic breeding material of white clover at the Iygev Breeding Station] (Estonian). *Lead, Loode Kogmik. Eesti Maaviljel. ja Maaparand. Lead. Uurimise Inst.* **24**, 97-102.

Jackobs, J.A.; Hittle, C.N. (1968) Variations among seedlots of certified white clovers. *Agronomy Journal* **50**, 327-330.

James, J.R.; Lucas, L.T.; Chamblee, W.V. (1980) Influence of fungicide and insecticide applications on persistence of ladino clover. *Agronomy Journal* **72**, 781-784.

Jamieson, T.V., Jr. (1957) Louisiana S-1 white clover moves west to Texas. *Rice Journal* **60**, 38-39.

Johnson, W.C.; Donnelly, E.D.; Gibson, P.B. (1970) Registration of Regal white clover. *Crop Science* **10**, 208.

Jones, D.A. (1966) On the polymorphism of cyanogenesis in *Lotus corniculatus*. I Selection by animals. *Canadian Journal of Genetics and Cytology* **8**, 556-567.

Jones, D.A. (1972) Cyanogenetic glycosides and their function. In Phytochemical ecology, J.B. Harborne (Ed.), Academic Press, London, pp. 103-124.

Jones, D.A. (1973) Coevolution and cyanogenesis. In Taxonomy and ecology, V.H. Heywood (Ed.), *Systematics Association Special Volume* **No. 5**, Academic Press, London, pp. 213-242.

Jones, D.G. (1962) Variation of nodulation characteristics in S100 Nomark white clover (*Trifolium repens* L.). *Journal of the Science of Food and Agriculture* **13**, 598-603.

Jones, D.G.; Burrows, A.C. (1968) Breeding for increased nodule tissue in white clover (*Trifolium repens* L.). *Journal of Agricultural Science, Cambridge* **71**, 73-79.

Jones. D.G.; Russell, P.E. (1972) The application of immunofluorescence techniques to host plant/nodule bacteria selectivity experiments using *Trifolium repens*. *Soil Biology and Biochemistry* **4**, 277-282.

Jones, D.G.; Morley, S.J. (1981) The effects of pH on host plant 'preference' for strains of *Rhizobium trifolii* using fluorescent ELISA for strain identification. *Annals of Applied Biology* **97**, 183-190.

Jones, D.G.; Davies, W.E. (1966) The improvement of symbiotic effectiveness in white clover (*Trifolium repens* L.). *Proceedings of the 10th International Grassland Congress*, 648-652.

Jones, D. G.; Hardarson, G. (1979) Variation within and between white clover varieties in their preference for strains of *Rhizobium trifolii*. *Annals of Applied Biology* **92**, 221-228.

Jones, R.J. (1973) The effect of cutting management on the yield, chemical composition and *in vitro* digestibility of *Trifolium semipilosum* grown with *Paspalum dilatatum* in a sub-tropical environment. *Tropical Grasslands* **7**, 277-284.

Jones, R.M.; Evans, T.R. (1977) Soil seed levels of *Lotononis bainesii*, *Desmodium intortum* and *Trifolium repens* in sub-tropical pastures. *Journal of the Australian Institute of Agricultural Science* **43**, 164-166.

Jones, R.P. (1940) Meadow plant day at Svalöf. *Herbage Reviews* **8**, 25-31.

Jones, W.T.; Lyttleton, J.W.(1971) Bloat in cattle XXXIV A survey of legume forages that do and do not produce bloat. *New Zealand Journal of Agricultural Research* **14**, 101-107.

Jönsson, H.A. (1983) White clover (*Trifolium repens*) Sonja. *Agri. Hortique Genetica* **41**, 73-82.

Julén, G. (1948) Inbreeding in herbage plants. In Svalöf 1886-1946. *Carl Bloms Boktryckeri A.-B., Lund*, 211-236.

Julén, G. (1963) Svalöf's Kiwi white clover (Swedish). *Sveriges Utsädesförenings Tidskrift*, **73**, 241-250.

Kapas, S. (1967) The work of the National Institute for the evaluation of varieties and production in 1967, Budapest, Hungary. *Plant Breeding Abstracts* **40**, 108.

Kasper, J. (1976) The utilisation of simple grass and clover mixtures based on native and foreign varieties of white clover (*Trifolium repens* L.) (Czech). *Pol'nohospodarstvo* **21**, 334-344.

Kelly, A.F.; Boyd, M.M. (1966) The stability of cultivars of grasses/clovers when grown for seed in differing environments. *Proceedings of the 10th International Grassland Congress*, 777-782.

Killinger, G.B.; Harris, H.C. (1959) Pasture grass and legume responses to various fertiliser and management practices. *Report of the Florida Agricultural Experiment Station 1959*, 48-49.

Knight, W.E. (1953a) Breeding Ladino clover for persistence. *Agronomy Journal* **45**, 28-31.

Knight, W.E. (1953b) Interrelationships of some morphological and physiological characterisitics of Ladino clover. *Agronomy Journal* **45**, 197-199.

Kropacova, S.; Haslbachova, H. (1969) A study of the activity of the honey bee (*Apis mellifera* L.) on white clover (*Trifolium repens* L.) (Czech). *Rostlinna Vyroba* **15**, 1123-1134.

Kuiper, K. (1960) Resistance of white clover varieties to the clover cyst eelworm, *Heterodera trifolii* Goffart. *Nematologica (Suppl. 2)*, 95-96.

Laissus, R.; Teilhard de Chardin, B. (1962) [Increase in yield of leys due to white clover] (French). *Fourrages* 1962 **11**, 47-57.

Lambert, J.P.; Vartha, E.W.; Harris, A.J. (1969) Progress report on 'Grasslands 4700' white clover. *Proceedings of the New Zealand Grassland Association* **31**, 135-142.

Larkin, P.J.; Scowcroft, W.R. (1981) Somaclonal variation - a novel source of variability from cell culture for plant improvement. *Theoretical and Applied Genetics* **60**, 197-214.

Laude, H.M.; Stanford, E.H.; Enloe, J.A. (1958) Photoperiod, temperature and competitive ability as factors affecting the seed production of selected clones of Ladino clover. *Agronomy Journal* **50**, 223-225.

Laude, H.M.; Stanford, E.H. (1960) Environmentally induced changes in gene frequency in a synthetic forage variety grown outside the region of adaptation. *Proceedings of the 8th International Grassland Congress*, 180-184.

Latter, B.D.H. (1964) Selection methods in the breeding of cross-fertilised pasture species. In: Grasses and Grasslands, C. Barnard (Ed.), Macmillan, Melbourne, 168-181.

Leffel, R.C.; Gibson, P.B. (1973) White Clover. In: Heath, Metcalfe and Barnes (Eds.) Forages, the science of grassland agriculture, 3rd edition, Iowa State University Press, Ames, Iowa, pp. 167-176.

Lehle, H. (1959) [A contribution to knowledge of the yield characters of white clover and their importance for breeding] (German). *Zeitschrift für Acker und Pflanzenbau* **108**, 384-412.

Lenoble, M.; Papineau, J. (1970) [Note on a new leaf markings in *Trifolium repens*]. *Annales Amélioration des Plantes* **20**, 485-487.

Levan, A. (1942) Plant breeding by induction of polyploidy and some results in clovers. *Hereditas, Lund.* **28**, 245-246.

Levy, E.B.; Gorman, L.W. (1933) Strain investigation of white clover. Finished results of trials at Plant Research Station. *New Zealand Journal of Agriculture* **46**, 78-89.

Levy, E.B.; Gorman, L.W. (1934) Strain investigations in white clover. *New Zealand Journal of Agriculture* **49**, 86-104.

McNeil, D. L.; Carroll, B. J.; Gresshof, P. M. (1985). A supernodulation and nitrate-tolerant symbiotic (*nts*) soybean mutant. *Plant Physiology* **78**, 34-40.

Mackiewicz, T. (1970a) Microsporogenesis and heterochromatin grains in octoploid *Trifolium repens* L. *Genetica Polonica* **11**, 37-44.

Mackiewicz, T. (1970b) Female gametophyte development in octoploid white clover (*Trifolium repens* L.) *Genetica Polonica* **11**, 241-247.

Mackiewicz, T. (1970c) Pollen grain morphology, viability, and pollen tube development *in vitro* and *in situ* in octoploid white clover (*Trifolium repens* L.) *Genetica Polonica* **11**, 249-256.

Maher, E.P.; Hughes, M.A. (1971) Isolation of linamarin-lotaustralin from *Trifolium repens*. *Phytochemistry* **10**, 3005-3007.

Maher, E.P.; Hughes, M.A. (1973) Studies on the nature of the Li locus in *Trifolium repens* L. II. The effect of genotype on enzyme activity and properties. *Biochemical Genetics* **8**, 13-26.

Maheshwaran, G.; Williams, E.G. (1984) Direct somatic embryoid formation on immature embryos of *Trifolium repens, T. pratense,* and *Medicago sativa*, and rapid clonal propagation of *T. repens. Annals of Botany* **54**, 201-211.

Manner, R. (1958) [Hammenhogs Beta white clover] (Swedish) *Varen 1958*, Hammenhogs, Otto J. Olsen and Sons AB, 12-25.

Marble, V.G.; Jones, L.G.; Goss, J.R.; Jeter, R.B.; Burton, V.E.; Hall, D.H. (1970) Ladino clover seed production in California. *California Agricultural Experiment Station Extension Service Circular* 554.

Martin, G.J. (1954) Species and strains of pasture plants. *Tasmanian Journal of Agriculture* **25**, 315-321.

Martynenko, I. (1935) ['Gigant Belyi'] (Russian) *Semenovodstvo* **No. 7**, 50-52.

Martynenko, I. (1939) [Breeding and agrotechnique of Ukrainian clovers] (Russian) *Mater. Sovesc. korm. Trav.*, Jan 1938, 16-19.

Melville, J.; Coop, I.E.; Doak, B.W.; Reifer, I. (1940) Cyanogenesis in white clover (*Trifolium repens* L.) IV. Methods of determination and general considerations. *New Zealand Journal of Science and Technology* **22B**, 144-154.

Mertens, D.R. (1977) Importance and measurement of protein insolubility in ruminant diets. *Proceedings 1977 Georgia Nutrition Conference*, 30-41.

Mirande, M. (1912) [On the presence of hydrocyanic acid in white clover (*Trifolium repens* L.)]. *Compte rendu de l'Academie des Sciences, Paris* **155**, 651-653.

Mohapatra, S.S.; Gresshof, P.M. (1982) Ecotypic variations of *in vitro* plantlet formation in white clover (*Trifolium repens*). *Plant Cell Reports* **1**, 189-192.

Munro, J.M.M.; Davies, D.A.; Morgan, T.E.H. (1975) Hill upland research. Evaluation of varieties: clover persistency. *Welsh Plant Breeding Station Report for 1974*, 48-49.

Mytton, L.R. (1975) Plant genotype x rhizobium strain interaction in white clover. *Annals of Applied Biology* **80**, 103-107.

Mytton, L.R.; Jones, D.G. (1971) The responses to selection for increased nodule tissue in white clover (*Trifolium repens* L.) *Plant and Soil, Special Volume*, 17-25.

Najcevska, C.M.; Speckmann, G.J. (1968) Numbers of chloroplasts and pollen grain pores in diploid and tetraploid varieties of some *Trifolium* species. *Euphytica* **17**, 357-362.

Nakajima, T.; Doyama, T.; Matsumoto, H. (1969) *In vitro* culture of excised ovules of white clover, *Trifolium repens* L. *Japanese Journal of Breeding* **19**, 373-378.

Nasinec, J.; Mika, V. (1978) [Varietal differences in the content of cyanogenetic glycosides in Dutch clover (Trifolium repens L.)] (Czech). *Rostlinna Vyroba* **24**, 67-74.

Nayer, J.K.; Fraenkel, G. (1963) The chemical basis of the host selection in the Mexican bean beetle *Epilachna varivestis* (Coleoptera, Coccinellidae). *Annals of the Entomological Society of America* **56**, 174-178.

Neal-Smith, C.A. (1970) Investigations in south-eastern Australia. *Report, 27th southern pasture and forage crop improvement, Athens, Georgia*, 28-31.

Newton, J.E.; Betts, J.E.; Drane, H.M.; Saba, N. (1970) The oestrogenic activity of white clover. *British Grassland Society Occasional Symposium* **No. 6**, 309-313.

Nilsson-Leissner, G. (1934) [White clover strains for seed production in Scania] (Swedish). *Svensk Frotsidning* **3**, 122-124.

Nilsson Leissner, G. (1935) [Investigations of self-sterility in white and alsike clover] (Swedish). *Nordisk Jordbrugsforskning* **16**, 168-173.

Nilsson-Leissner, G.; Nilsson, F. (1940) Herbage plant breeding in Sweden. In: The breeding of herbage plants in Scandinavia and Finland. *Imperial Agricultural Bureaux, Joint Publication* **No. 3**, 15-51.

Nolin, W.T. (1950) It's evergreen. A new clover that reseeds, gives hot season pasturage and rivals ladino in growth. *Southern Seedsman* **13(11)**, 15, 55.

Norgaard, H.S. (1963) [Natural selection in red and white clover under different methods of seed production] (Danish). *Ugeskrift for Landmaend 1963*, **No. 3**, 273-275.

Norton, D.C. (1967) Relationships of *Heterodera trifolii* to some forage legumes. *Phytopathology* **57**, 1305-1308.

Nutman, P.S. (1959) Sources of incompatibility affecting nitrogen fixation in legume symbiosis. *Symposium of the Society for Experimental Biology* **13**, 42-58.

Nutman, P.S. (1967) Varietal differences in the nodulation of subterranean clover. *Australian Journal of Agricultural Research* **18**, 381-425.

O'Brien, A.D. (1970a) Evaluating white clover varieties for Clarence River beef country. *Tropical Grasslands* **4**, 63-69.

O'Brien, A.D. (1970b) White clover (*Trifolium repens* L.) in a subtropical environment on the east coast of Australia. *Proceedings of the 11th International Grassland Congress, Australia, 1970*, 165-168.

O'Brien, A.D.; Cohen, R.D.H. (1972) Naturalised clover on the north coast. *Agricultural Gazette of New South Wales* **83**, 86.

Ollerenshaw, J.H.; Baker, R.H. (1981) Low temperature growth in a controlled environment of *Trifolium repens* plants from northern latitudes. *Journal of Applied Ecology* **18**, 229-239.

Ollerenshaw, J.H.; Haycock, R. (1984) Variation in the low temperature growth and frost tolerance of natural genotypes of *Trifolium repens* L. from Britain and Norway. *Journal of Agricultural Science, Cambridge* **102**, 11-21.

Ollerenshaw, J. H.; Stewart, W. S.; Baker R. H. (1978) Intraspecific variation within *Trifolium repens* L. in relation to early spring and late autumn growth of upland and marginal pastures. *Proceedings of the 13th International Congress, Section* **1-2**, 241-249.

Oram, R.N. (1982) *Trifolium repens* L. (white clover) cv. Grasslands Pitau (Reg. no. B-la-7). *Journal of the Australian Institute of Agricultural Science* **48**, 191-192.

Ostrowski, H. (1972) White clover (*Trifolium repens*) in subtropical Australia - a review. *Tropical Grasslands* **6**, 97-106.

Oswald, T.H.; Smith, A.E.; Phillips, D.V. (1977) Callus and plantlet regeneration from cell cultures of ladino clover and soybean. *Physiologia Plantarum* **39**, 129-134.

Owen, C.R. (1949) Breeding for heat tolerance in Louisiana white clover. *Proceedings of the Association of Southern Agricultural Workers 1949*, 52.

Owen, C.R. (1953) Louisiana S1 white clover. *Louisiana Agricultural Experiment Station Bulletin* **479**, 15 pp.

Palenzona, D.L.; Ferrari, A. (1965) [Factors that modify the effect of ionising radiations in ladino clover] (Italian). *Genetica Agraria* **19**, 199-208.

Palenzona, D.L.; Ferrari, A. (1966) [Correlations between characteristics in ladino clover (*T. repens* var. *giganteum*)] (Italian). *Genetica Agraria* **20**, 307-312.

Pandey, K.K. (1955) Pollen-tube behaviour in *Trifolium* species. *Science and Culture* **20**, 504-505.

Pandey, K.K. (1956) Mutations of the self-incompatibility alleles in *Trifolium pratense* and *T. repens. Genetics* **41**, 327-343.

Pandey, K.K. (1957) A self-compatible hybrid from a cross between two self-incompatible species in *Trifolium. Journal of Heredity* **48**, 278-281.

Pandey, K.K. (1959) Mutations of self-incompatibility gene(s) and pseudo-compatibility in angiosperms. *Lloydia* **22**, 222-234.

Pearson, W.D. (1980) Management practices reduce case bearer (Lepidoptera: Colephoriade) damage to white clover seed crops. *New Zealand Journal of Experimental Agriculture* **8**, 71-74.

Pelletier, G.; Pelletier, A. (1971) [Culture *in vitro* of tissue of white clover (*Trifolium repens*); variability of regenerated plants *Annales de l'Amélioration des Plantes* **21**, 221-233.

Pieters, A.J.; Hollowell, E.A. (1937) *Clover Improvement. Yearbook, United States Department of Agriculture 1937*, 1190-1214.

Portz, H.L.; Jackobs, J.A. (1955) Difference between seed lots of ladino clover in cyanophoric properties. *Agronomy Journal* **47**, 143-145.

Powell, C. Ll. (1982) Phosphate response curves of mycorrhizal and non-mycorrhizal plants. III. Cultivar effects in *Lotus pedunculatus* Cav. and *Trifolium repens* L. *New Zealand Journal of Agricultural Research* **25**, 217-222.

Powell, G.S.; Campbell, W.V.; Cope, W.A.; Chamblee, D.S. (1983) Ladino clover resistance to the clover root curculio (Coleoptera: Curculionidae). *Journal of Economic Entomology* **76**, 264-268.

Powell, G.S..; Campbell, W.V. (1983) Adult *Sitona hispidulus* feeding preferences among ninety-six genotypes of Ladino white clover. *Journal of the Georgia Entomological Society* **18**, 294-300.

Pusey, J.G. (1963) Cyanogenesis in *Trifolium repens*. In: C.D. Darlington and A.D. Bradshaw (Eds.), *Teaching Genetics*, Oliver and Boyd, Edinburgh and London, 99-104.

Ragland, C.K.; Campbell, C.L. (1983) Effect of clover yellow leaf vein and peanut stunt viruses on yield of white clovers in a clover-tall fescue stand. *Phytopathology* **73**, 505.

Rasmussen, F. (1962) [Trials (in Denmark) with foreign varieties and ecotypes of clover and grass species, 1955-59] (Danish). *Tidsskrift for Planteavl* **66**, 1-28.

Reid, D. (1961a) Factors influencing the role of clover in grass-clover leys fertilised with nitrogen at different rates. I. The effects of the variety of companion grass on the yields of total herbage and of clover. II. The effects of the variety of white clover on the yields of total herbage and of clover. *Journal of Agricultural Science* **56**, 143-160.

Reid, D. (1961b) Factors influencing the role of clover in grass-clover leys fertilised with nitrogen at different rates. III. The effects of the variety of companion grass and of the variety of clover on the botanical composition of the sward. *Journal of Agricultural Science, Cambridge* **57**, 231-236.

Rhind, J.M.L.C.; Smith, A., Goodenough, D.C.W.; Morrison, A.R.J.; Ellis, J.I. (1979) Cultivar evaluation of forage crops. Cedara Press, Natal Region, Department of Agricultural Technical Services, Pietermaritzburg.

Rhodes, I.; Harris, W. (1979) The nature and basis of differences in sward composition and yield in ryegrass-white clover mixtures. *British Grassland Society Occasional Symposium* **10,** 55-60.

Riddet, W.; Hodgson, J.N.; Levy, E.B.; Sears, P.D. (1937) Investigation of feed flavour in cream and butter. Part I. Feed flavour in cream. *Proceedings of the New Zealand Grassland Association* **5**, 96-111.

Rincker, C. M. (1983) Germination of forage crop seeds after 20 years of subfreezing storage. *Crop Science* **23**, 229-231.

Robinson, R.R. (1942) The mineral content of various clones of *T. repens* when grown on different soils. *Journal of the American Society of Agronomy* **34**, 933-939.

Rogers, C.F.; Frykolm, O.C. (1937) Observations on the variations in cyanogenetic power of white clover plants. *Journal of Agricultural Research* **55**, 533-537.

Ronningen, T.S. (1953) Susceptibility to winter injury and some other characteristics in ladino and common white clovers. *Agronomy Journal* **45**, 114-117.

De Roo, R. (1952) [Colchicine induced tetraploidy in some fodder crops] (French). *Mededelingen van de Landbouwhogeschool en de Opzoekingsstations van de Staat te Gent* **17**, 281-292.

De Roo, R. (1968) [Experience gained on polyploid grasses, clovers and fodder beet] (French). *Revue de l'agriculture* (Brussels) **21**, 1721-1739.

Rusmini, B. (1965) [A method of improving Ladino clover by means of clonal lines] (Italian). *Sementi Elette* **11**, 152-155.

Russell, P.E.; Jones, D.G. (1975a) Immunofluoresence studies of selection of strains of *R. trifolii* by S.184 white clover (*T. repens* L.) *Plant and Soil* **42**, 119.

Russell, P.E.; Jones, D.G. (1975b) Variation in the selection of *Rhizobium trifolii* by varieties of red and white clover. *Soil Biology and Biochemistry.* **7**, 15-18.

Rudorf, W.; Scheone, G. (1936) [*Trifolium repens* and *T. hybridum* trials of bred strains from different sources in mixtures with grasses; and test of productivity in *Phalaris arundinacea* 1932-35] (German). *Pflanzenbau* **13**, 1-15.

Rynska, A. (1967) [Winter survival and generative development of di- and polyploid clovers (*Trifolium pratense* L., *T. hybridum* L. *T. repens* L.)] (Polish). *Roczniki Nauk Rolniczych, Seria A* **93**, 25-77.

Sakai, K. (1983) [New summer crop cultivars (white clover)] (Japanese). *Japanese Journal of Breeding* **33**, 504-505.

Sato, H. (1980) [Introduction of new varieties of summer crops (white clover)] (Japanese). *Japanese Journal of Breeding* **30**, 409-410.

Scarsbrick, B. (1977) Haifa - a persistent, vigorous white clover, *Agricultural Gazette of New South Wales* **88**, 33-35.

Schneider, G. (1932) [Morso white clover, a little known grassland plant which gives good yields] (German). *Mitteilungen der Deutschen Landwirtschaft Gesellschaft* **47**, 686-687.

Schroder, C.A. (1963) Louisiana clover makes a hardy pasture. *Queensland Agricultural Journal* **8**, 1-3.

Schwanbom, N. (1947) [Weibulls new Nora white clover] (Swedish, English summary). *Agri Hortique Genetica* **5**, 10-15.

Scott, R.S. (1976) The phosphate nutrition of white clover. *Proceedings of the New Zealand Grassland Association* **38**, 151 159.

Scott, S.W. (1982) Tests for resistance to white clover mosaic virus in red and white clover. *Annals of Applied Biology* **100**, 393-398.

Scott, W.R. (1973) Pasture seed production. In: R.H.M. Langer (Ed.), *Pastures and Pasture Plants*, Reed, Wellington, New Zealand.

Semenov, A.L.; Vlasova, K.S.; Grushevskii, M.P. (1973) Efficiency of hybridisation in white clover breeding (Russian). Abstract in *Plant Breeding Abstracts* **45**, 7426.

Simon, U.; Kastenbauer, A.; Garrison, C.S. (1974) Growth type and yield comparisons of forage species after seed multiplication in Germany and in the United States. I. Red clover, alfalfa, and white clover. *Crop Science* **14**, 682-686.

Smith, D. (1949) Differential survival of Ladino and common white clover encased in ice. *Agronomy Journal* **41**, 230-234.

Smith, M.K.; McComb, J.A. (1981) Use of callus cultures to detect NaCl tolerance in cultivars of three species of pasture legumes. *Australian Journal of Plant Physiology* **8**, 437-442.

Smrz, J.; Musil, M.; Vacek, V. (1977) Virosis occurrence in the assortment of white clover (*Trifolium repens* L.). (Czech). *Ochrana Rostlin* **13**, 7-12.

Snaydon, R.W. (1962a) Micro-distribution of *T. repens* L. and its relation to soil factors. *Journal of Ecology* **50**, 133-143.

Snaydon, R.W. (1962b) The growth and competitive ability of contrasting natural populations of *T. repens* L. on calcareous and acid soils. *Journal of Ecology* **50**, 439-447.

Snaydon, R.W.; Bradshaw, A.D. (1962a) Differences between natural populations of *Trifolium repens* L. in response to mineral nutrients. I Phosphate. *Journal of Experimental Botany* **13**, 422-434.

Snaydon, R.W.; Bradshaw, A.D. (1962b) The performance and survival of contrasting natural populations of white clover when planted into an upland *Festuca/Agrostis* sward. *Journal of the British Grassland Society* **17**, 113-118.

Snaydon, R.W.; Bradshaw, A.D. (1969) Differences between natural populations of *Trifolium repens* L. in response to mineral nutrients. II. Calcium, magnesium, and potassium. *Journal of Applied Ecology* **6**, 185-202.

Stanford, E.H.; Laude, H.M.; Enloe, J.A. (1960) Effect of harvest dates and location on the genetic composition of the syn 1 generation of Pilgrim ladino clover. *Agronomy Journal* **52**, 149-152.

Stanford, E.H.; Laude, H.M.; Booysen, P. de V. (1962) Effects of advance in generation under different harvesting regimes on the genetic composition of Pilgrim ladino clover. *Crop Science* **2**, 497-500.

Stapledon, R.G. (1947) The precise value and application of the Aberystwyth and other strains of herbage plants. *Journal of the British Grassland Society* **2**, 113-118.

Staszewski, Z. (1966) [Studies on the white clover - *Trifolium repens* L. v. *typicum* Aschers et Graebn. F. *giganteum* Lagr. et Fosset] (Polish). *Hodowla Roslin Aklimatyzacja i Nasiennictwo* **10**, 151-173.

Suckling F.E.T. (1952) Dissemination of white clover (*Trifolium repens*) by sheep. *New Zealand Journal of Science and Technology* **33A**, 64-77.

Suckling, F.E.T.; Charlton, J.F.L. (1978) A review of the significance of buried legume seeds with particular reference to New Zealand agriculture. *New Zealand Journal of Experimental Agriculture* **6**, 211-215.

Šul'ga, P.M. (1966) [The white clover variety Gigant] (Russian). *Meadow and Pasture* 1966, **No.3**, 31-32.

Šul'ga, P.M. (1969) [New legume and grass cultivars] (Russian). *Breeding and Seed Growing 1969* **No.6**, 67-68.

Sullivan, J. T. (1944) Further comparisons of plants with different chromosome numbers in respect to chemical composition. *Journal of American Society of Agronomy* **36**, 537-543.

Suzuki, S.; Adachi, A.; Yamada, Y. (1958) [Studies on the technique of selection in white clover breeding. III. On the heritability of some characters in Ladino clover] (Japanese). *Bulletin of the National Institute of Agricultural Science, Tokyo (Series G)* **14**, 169-178.

Tetley, J.H. (1953) Inhibition of populations of *Haemonchus contortus* in sheep fed on white clover (Trifolium repens) high in lotaustralin. *Nature (London)* **171**, 311.

Thomas, R.G. (1962) The initiation and growth of axillary bud primordia in relation to flowering in *Trifolium repens* L. *Annals of Botany* **26**, 329-344.

Thomson, D.J.; Raymond, W.F. (1970) White clover in animal production: nutritional factors. *British Society Occasional Symposium* **No. 6**, 277-284.

Tiemann, H. (1966) [Comparative investigations on the influence of genome doubling on productivity in white clover (*Trifolium repens* L.)] (German) *Zuchter* **36**, 302-313.

Todd, J.R. (1970) White clover in animal production: animal health factors: review. *British Grassland Society Occasional Symposium* **No. 6**, 297-307.

Troughton, A. (1955a) A comparison of three strains of white clover. I. Herbage production. *Journal of the British Grassland Society* **10**, 197-212.

Troughton, A. (1955b) A comparision of three strains of white clover. II. Observations on bloat. *Journal of the British Grassland Society* **10**, 297-305.

Turesson, G. (1962) Results of colchicine doubling in red, alsike, and white clover. *Agri Hortique Genetica* **20**, 111-135.

Turkington, R.; Harper, J. L. (1979) The growth, distribution and neighbour relationships of *Trifolium repens* in a permanent pasture. I. Ordination, pattern and contact. *Journal of Ecology* **67**, 201-218.

Tysdal, H. M.; Kiesselbach, T. A.; Westover, H. L. (1942) Alfalfa breeding. *Nebraska Agricultural Experiment Station, Research Bulletin* **124**.

Ueno, M.; Yoshihara, K. (1967) [Studies on the roots of herbage legumes. 4. Varietal differences of nodal root growth in relation to the compensation of tap root in white clover] (Japanese, English Summary). *Journal of the Japanese Society for Grassland Science* **13**, 195-198.

Ujj Meszaros, K. (1981) [Effect of crossing extreme variants on number of florets per inflorescence in white clover (*Trifolium repens* L.)] (Hungarian). *Novenytermeles* **30** , 169-176.

Ulyatt, M.J. (1970) Evaluation of pasture quality under New Zealand conditions. *Proceedings of the New Zealand Grassland Association* **32**, 61-68.

Vacek, V. (1970) [An evaluation of the fodder crop collection of white clover (*T. repens* L.)] (Czech). *Rostlinna Vyroba* **16**, 997-1005.

Vacek, V.; Novotny, F. (1975) [Content of amino acids in Persian clover (*Trifolium resupinatum* L.)] (Czech). *Sbornik Vedeckych Praci* 1975, 49-61.

Vacek, V.; Zapletalova, I. (1982) Contrasting of foreign varieties of white clover (*Trifolium repens* L.) from the aspect of forage quality.

Pol'nohospodarstvo 28, 757-765.
Valle, O. (1936) Pollination and seed formation of the clover species. *Herbage Reviews* **4**, 71-77.
Valle, O. (1963) Breeding and seed production experiences with Finnish white clover. *Annales Agriculturae Fenniae* **2**, 51-58.
Van Bogaert, G. (1977) Factors affecting seed yield in white clover. *Euphytica* **26**, 233-239.
Van Slijken, A. (1960) [The behaviour of two varieties of white clover in association with six types of perennial ryegrass under conditions of grassland farming] (French). *Revue de l'Agriculture (Brussels)* **13**, 117-120.
Vez, A. (1959) [An examination of the palatability of different varieties of white clover] (French). *Landswirtschaftliches Jahrbuch der Schweiz* **8**, 735-741.
Vose, P.B.; Jones, D.G. (1963) The interaction of manganese and calcium on nodulation and growth in varieties of *Trifolium repens. Plant and Soil* **18**, 372-385.
Ward, C.Y. (1965) New white clover variety superior in station tests. *Mississippi Farming Research* **28**, 6.
Ware, W.M. (1925a) Experiments and observations on farms and strains of *Trifolium repens* L. *Journal of Agricultural Science, Cambridge* **15**, 47-67.
Ware, W.M. (1925b) White clover. *Ministry of Agriculture and Fisheries, Miscellaneous Publication* **No.46**, 1-26.
Weeda, W.C.; Cullen, N.A.; Douglas, J.A.; During, C. (1970) Progress report on 'Grasslands 4700' white clover. *Proceedings of the New Zealand Grassland Association* **31**, 135-142.
West, S.H.; Harris, H.C. (1963) Seedcoat colours associated with physiological changes in alfalfa and crimson and white clovers. *Crop Science* **3**, 190-193.
Westbrook, F.E.; Tesar, M.B. (1955) Tap root survival of ladino clover. *Agronomy Journal* **47**, 403-410.
White, D.W.R. (1984) Plant regeneration from long-term suspension cultures of white clover. *Planta* **162**, 1-7.
White, D.W.R.; Williams, E. (1976) Early seed development after crossing of *Trifolium semiplosum* and *T. repens. New Zealand Journal of Botany* **14**, 161-168.
Whyte, R.O. (1938) Herbage plant improvement in Finland. *Herbage Reviews* **6**, 263-270.
Whyte, R.O.; Nilsson-Leissner, G.; Trumble, H.C. (1953) *Legumes in agriculture*, FAO, Rome.
Wilkinson, H.T.; Millar, R.L. (1978) Cyanogenic potential of *Trifolium repens* L. in relation to pepper spots caused by *Stemphylium sarciniforme. Canadian Journal of Botany* **56**, 2491-2496.
Wilkinson, H.T.; Millar, R.L. (1979) β-Glucosidases potentially involved in cyanogenesis during infection of white clover by *Stemphylium sarciniforme. Canadian Journal of Botany* **57**, 69-73.
Wilman, D.; Asiegbu, J.E. (1982a) The effects of clover variety, cutting interval and nitrogen application on herbage yields, proportions and heights in perennial ryegrass-white clover swards. *Grass and Forage Science* **37**, 1-13.
Wilman, D.; Asiegbu, J.E. (1982b) The effects of variety, cutting interval and nitrogen application on the morphology and development of stolons and leaves of white clover. *Grass and Forage Science* **37**, 15-27.
Williams, E. (1978) A hybrid between *Trifolium repens* and *T. ambiguum* obtained with the aid of embryo culture. *New Zealand Journal of Botany* **16**, 499-506.
Williams, E. G.; Verry, I. M. (1980) A partially fertile hybrid between *Trifolium repens* and *T. ambiguum. New Zealand Journal of Botany* **19**, 1-7.
Williams, E. G.; Verry, I. M.; Williams, W. M. (1982) Use of embryo culture in interspecific hybridization. In Vasil, I. K.; Scowcroft, W. R.; Frey, K. J. (Eds) *Plant Improvement and Somatic Cell Genetics*. Academic Press, New York. pp 119-128.
Williams, I.G. (1968) White clover appraisal and evaluation. *Report of the Welsh Plant Breeding Station for 1967*, 55-56.
Williams, R.D. (1931a) Self and cross-sterility in white clover. *Welsh Plant Breeding Station, Series H, Bulletin* **No. 12**, 209-216.
Williams, R.D. (1931b) Methods and technique of breeding red clover, white clover, and lucerne. *Imperial Bureau of Plant Genetics, Herbage Plants Bulletin* **No. 3**, 46-76.
Williams, R. D. (1932) White clover trials. 1926-30 *Journal of the Ministry of Agriculture* **39**, 708-718.
Williams, R.D. (1939) Genetics of cyanogenesis in white clover (*Trifolium repens*). *Journal of Genetics* **38**, 357-365.
Williams, W. (1945) Varieties and strains of red and white clover - British and foreign. *Welsh Plant Breeding Station, Series H, Bulletin* **No. 16**, 26pp.
Williams, W. (1950) The present position in the breeding of herbage legumes. *Annals of Applied Biology* **37**, 310-313.
Williams, W. (1954) An emasculation technique for certain species of *Trifolium. Agronomy Journal* **46**, 182-184.
Williams, W.M. (1972) Laboratory screening of white clover for resistance to stem nematode. *New Zealand Journal of Agricultural Research* **15**, 363-370.
Williams, W.M. (1977) Selection for resistence to stem eelworm in white clover. *Proceedings of the 3rd International SABRAO*

Congress **2 (14b),** 19-23.
Williams, W.M. (1983) Exploitation of the genetic resource through breeding In: J.G. McIvor, R.A. Bray (Eds.), 'Genetics Resources of Forage Plants'. Commonwealth Scientific and Industrial Research Organisation, Melbourne, 261-274.
Williams, W.M.; Anderson, L.B.; Cooper, B.M. (1978) Evaluation of clovers on sandy coastal soils. *Proceedings of the New Zealand Grassland Association* **39**, 130-138.
Williams, W.M.; Barclay, P.C. (1972) The effect of clover stem eelworm on the establishment of pure swards of white clover. *New Zealand Journal of Agricultural Research* **15**, 356-362.
Williams, W.M.; Barclay, P.C. (1975) Performance of a Spanish white clover (*Trifolium repens* L.) population in New Zealand. *New Zealand Journal of Agricultural Research* **18**, 45-49.
Williams, W.M.; Caradus, J.R. (1979) Performance of white clover lines on New Zealand hill country. *Proceedings of the New Zealand Grassland Association* **40**, 162-169.
Williams, W.M.; Hoglund, J.H. (1978) Temperature responses of New Zealand, Spanish, and New Zealand x Spanish white clover populations. *New Zealand Journal of Agricultural Research* **21**, 491-497.
Williams, W.M.; Sedcole, J.R. (1979) Regional variation in performance of New Zealand x Spanish white clover (*Trifolium repens* L.) hybrids in New Zealand. *New Zealand Journal of Agricultural Research* **22**, 77-81.
Williams, W.M.; Williams, E.G. (1983) Use of embryo culture with nurse endosperm for interspecific hybridization in pasture legumes. *Proceedings of the XIV International Grassland Congress, Lexington, Kentucky* 163-165.
Wilsie, C.P. (1963) Merit: a new ladino clover. *Iowa Farm Science* **17**, 5-6.
Wilsie, C.P. (1964) Registration of Merit ladino clover. *Crop Science* **4**, 448.
Wilson, E.R.L. (1978) White clover and grass grub. I. Screening technique for resistance and tolerance. *New Zealand Journal of Agricultural Research* **21**, 715-721.
Wilson, E.R.L. (1978b) White clover and grass grub. II. Screening 74 overseas lines for resistance and tolerance. *New Zealand Journal of Agricultural Research* **21**, 723-726.
Wilson, E.R.L.; Farrell, J.A.K. (1978) Some problems encountered in screening white clover for resistance to grass grub. *Proceedings of the Second Australasian Conference of Grassland Ecology*, 91-94.
Wilson, G.P.M. (1955-64) Annual pasture species improvement reports. Grafton Experiment Farm, NSW Department of Agriculture.
Wilson, G. P. M. (1960) Report of research in the Bellingen district (1957 to 1960). NSW Department of Agriculture.
Wilson, G. P. M. (1965) Seed yield of white clover varieties at Grafton. *Agricultural Gazette, New South Wales.* **76,** 308.
Winchester, J.A.; Kretschmer, A.E., Jr. (1959) Investigation of agronomic crops for forage, cover and special uses. *Report of the Florida Agricultural Experiment Station 1959*, 291.
Wong, E.; Flux, D.S.; Latch, G.Ç.M. (1971) The oestrogenic activity of white clover (*Trifolium repens* L.). *New Zealand Journal of Agricultural Research* **14**, 639-645.
Wong, E.; Latch, G.C.M. (1971) Effect of fungal diseases on phenolic contents of white clover. *New Zealand Journal of Agricultural Research* **14**, 633-638.
Wood, G.M.; Sprague, M.A. (1952) Relation of organic food reserves to cold hardiness of ladino clover. *Agronomy Journal* **44**, 318-325.
Yamada, T. (1959) [Studies on drought tolerance of ladino clover] (Japanese, English summary). *Bulletin of the National Institute of Agricultural Sciences, Series D* **No. 17**, 143-169.
Yeates, G.W.; Healy, W.B.; Widdowson, J.P. (1973) Screening of legume varieties for resistance to the root nematodes *Heterodera trifolii* and *Meloidogyne hapla. New Zealand Journal of Agricultural Research* **16**,81-86.
Yoshihara, K.; Kawanabe, S.;Ueno, M. (1957) [The yield and chemical composition of polyploid ladino clover] (Japanese). *Proceedings of the Crop Science Society of Japan* **26**, 65-67.
Young, N.R.; Mytton, L.R. (1983) The response of white clover to different strains of *Rhizobium trifolii* in hill country reseeding. *Grass and Forage Science* **38**, 13-19.
Zaleski, A. (1970) White clover seed production. *British Grassland Society Occasional Symposium* **No. 6**, 147-154.
Ziegenbein, G. (1959) [The agronomic value of different ladino varieties] (German). *Mitteilungen der Deutschen Landwirtschaftsgesellschaft* **74**, 426-429.
Zimmerman, K.F. (1958) [A contribution to methods of breeding white clover] (German). *Züchter* **28**, 17-25.
Zwingli, W. (1956) [Fertility relationships in Swiss white clover populations, with supplementary cytological studies] (German). *Zeitschrift für Pflanzenzuchtung* **36**, 237-288.
Zwingli, W.; Grundbacher, F. (1958) [White clover trials with some well known bred varieties] (German). *Mitteilungen der Schweiz Landswirtschaft* **6**, 155-159.

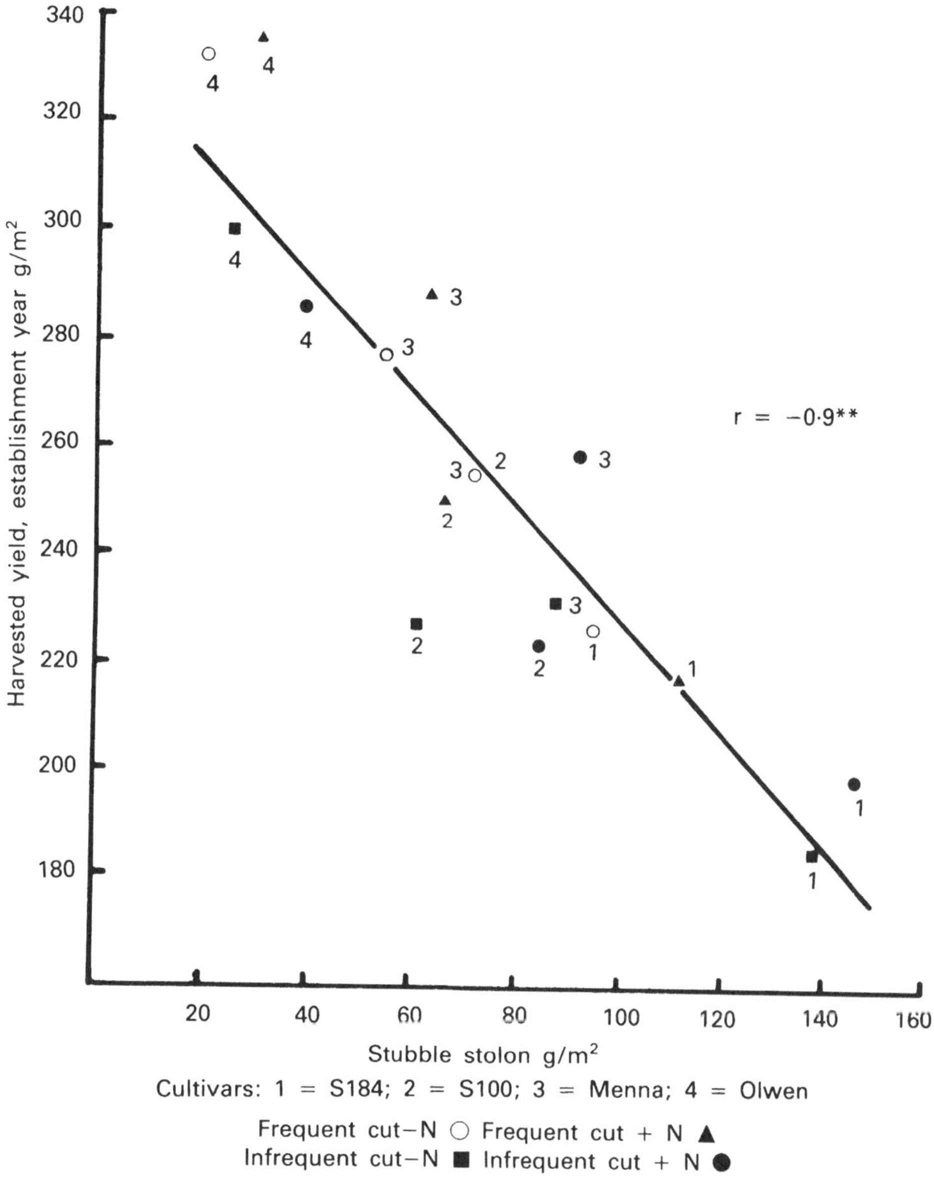

Fig. 10.1 Harvested yield and stubble stolon weight in four white clover cultivars. (From Rhodes and Harris (1979) reproduced with permission from *British Grasslands Soc. Occ. Symp.*).

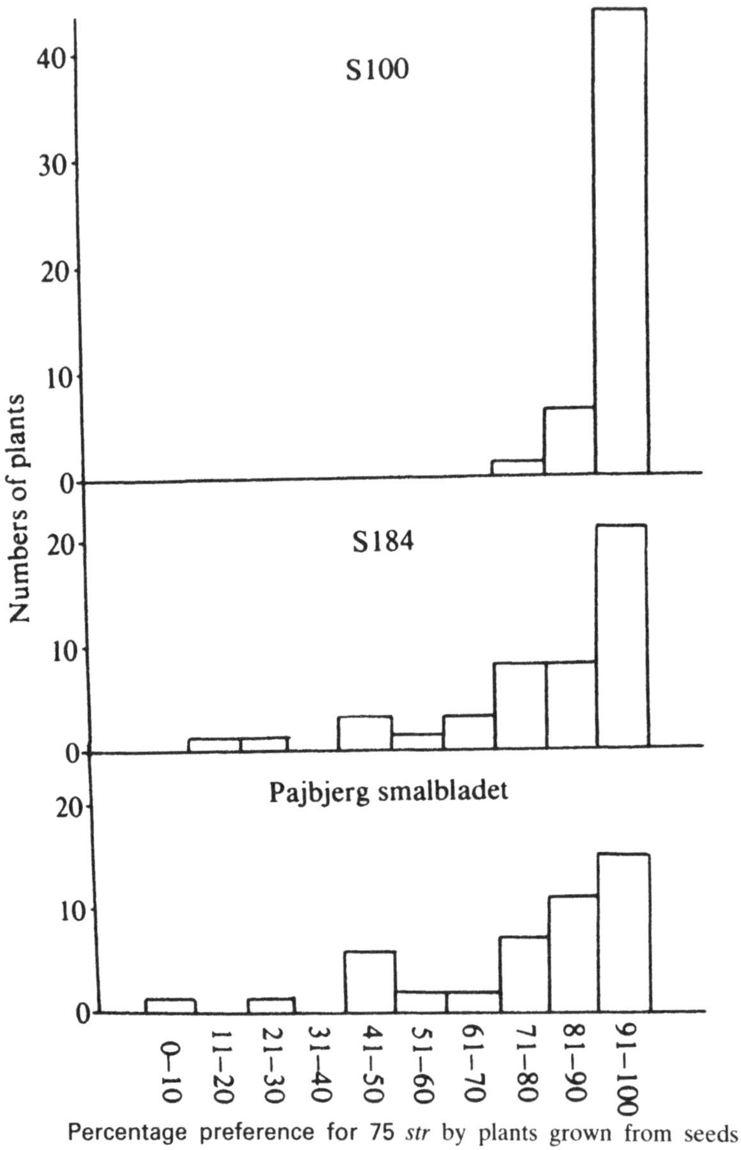

Fig. 10.2 Frequency distribution of the preference for 75 *str* of 50 plants of the white clover cultivars S100, S184, and Pajbjerg Smalbladet. (From Jones & Hardarson (1979) reproduced with permission from *Annals of Appl. Biology*).

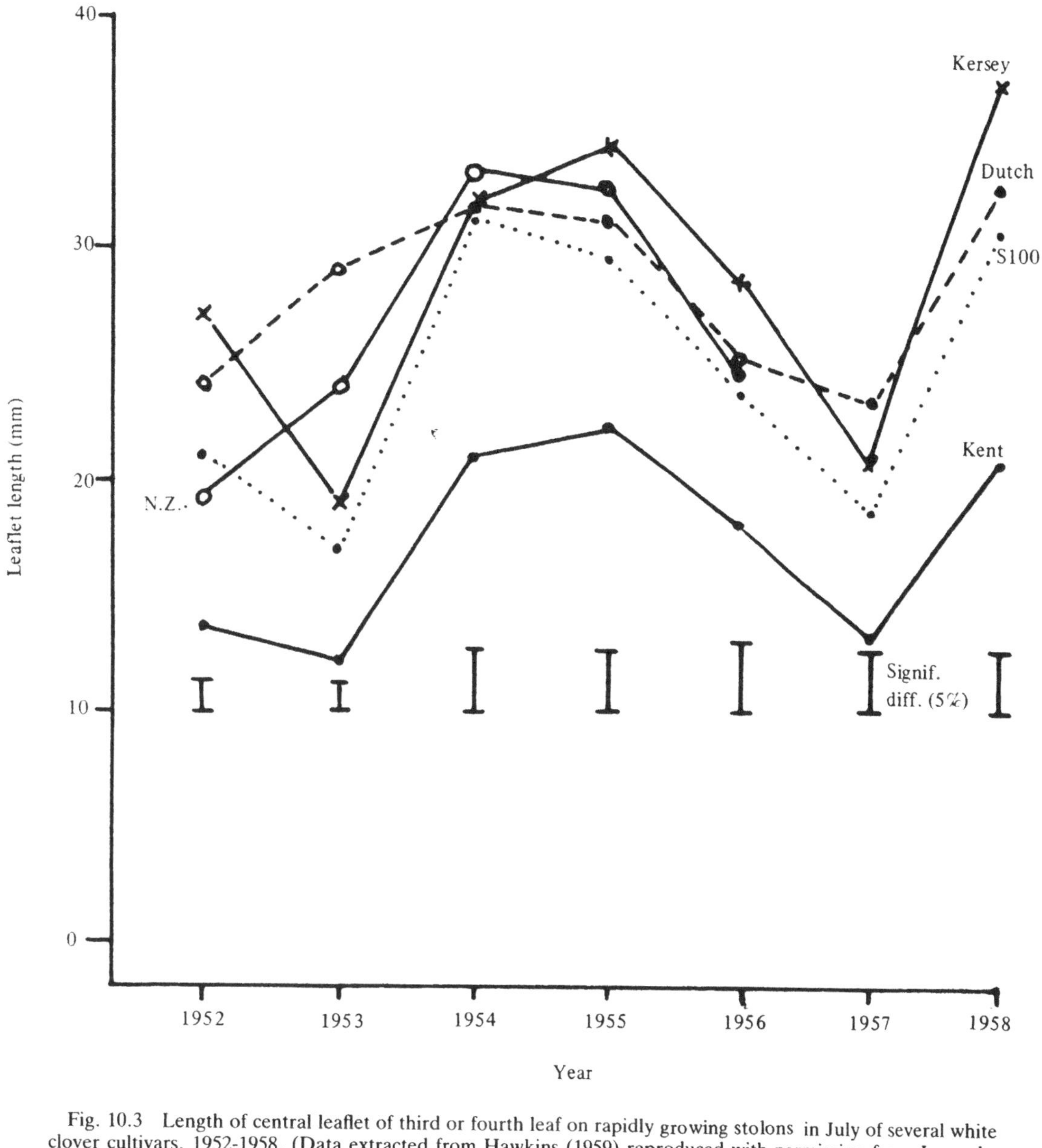

Fig. 10.3 Length of central leaflet of third or fourth leaf on rapidly growing stolons in July of several white clover cultivars, 1952-1958. (Data extracted from Hawkins (1959) reproduced with permission from *Journal Nat. Inst. Agr.*).

Diseases

G. C. M. Latch and R. A. Skipp
Plant Diseases Division, DSIR, Palmerston North, New Zealand

Contents

1. Introduction

The impact of white clover diseases in grazed pastures is seldom dramatic enough to alarm farmers or pasture agronomists. However, white clover is a host for numerous pathogenic microorganisms, and many facets of its growth and its role in livestock production and soil fertility can be affected by disease. For instance, the establishment of new pastures can be hindered by damping-off diseases which kill young seedlings. Foliage diseases, fungal root rots and systemic virus infections can reduce clover herbage production. Viruses, clover phyllody and other diseases may affect nodulation and nitrogen fixation dramatically. Some diseases affect the nutritional quality of clover herbage. Others cause oestrogenic coumestans to accumulate in diseased leaves, compounds that can reduce animal fertility if ingested in quantity at critical times in the breeding cycle. Diseases such as rust and clover rot, which kill or severely impair the growth of mature plants, can hinder breeders in their evaluation of white clover seed lines in field plots.

This chapter considers specific diseases of white clover caused by fungi, bacteria, viruses, and mycoplasma-, spiroplasma- and rickettsia-like organisms. It must be remembered, however, that limitations to the growth of white clover in grazed pasture probably result from a complex interaction of the effects of one or more diseases with other constraints such as moisture stress, grazing pressure and competition. The literature review and the writing for this chapter we completed in 1982.

2. Fungal and bacterial diseases

(A) Seedling diseases

i. Damping-off

Some soil-inhabiting fungi, particularly species of *Pythium*, *Fusarium*, *Rhizoctonia*, *Phoma* and *Gliocladium*, kill emerging white clover seedlings and thus can cause poor or uneven establishment of newly sown pasture (Hanson & Kreitlow 1953, Graham et al. 1957). Infected roots and hypocotyls rot, causing seedlings to fall over as if pinched at soil level (post-emergence damping-off). Sometimes seedlings die before they have emerged from the soil. The pathogens must be isolated from diseased tissue and grown in pure culture before they can be identified. Keys for identification of *Pythium* and *Fusarium* species are provided by Robertson (1980) and Booth (1971).

The severity of damping-off depends on enviromental factors such as temperature and soil moisture. Graham et al. (1957) found that more seedlings died at temperatures higher and lower than that optimum for clover growth. Mence (1964) showed that losses of white clover seedlings from a mixed sward of *Trifolium repens* L. and *Phleum pratense* L. increased when soil temperatures fell below 10°C. White clover appeared particularly vulnerable under these conditions because it was more susceptible to fungal attack, and clover growth was retarded more than grass growth by low temperatures. Soil moisture content affects damping-off fungi differently. Graham et al. (1957) found that *Pythium debaryanum* Hesse was most damaging in wet soils, whereas damping-off by *Fusarium* species was more severe in dry soils.

Seedlings appear to be most vulnerable to fungal attack during the first few days of growth and become highly resistant thereafter. Halpin & Hanson (1958) reported that isolates of *Pythium debaryanum, P. ultimum* Trow and *P. irregulare* Buisman, which proved to be highly pathogenic to clover seedlings when applied to soil in pots at the time of seeding, caused little damping-off when inoculation was delayed for 2-3 days. Temperature, soil type and soil moisture content influenced the rate at which lucerne and red clover seedlings lost their susceptibility to damping-off fungi in pot experiments (Chi & Hanson 1962).

Sowing clover seed when conditions are most favourable for rapid germination and growth will minimise seedling loss from damping-off fungi by reducing the period during which seedlings are susceptible to infection. Dusting seed with fungicides protects lucerne seedlings from damping-off (Falloon & Skipp 1982), and may be found suitable for use on white clover. The fungicides must not be antibacterial if seed is to be inoculated with *Rhizobium trifolii*

(B) Aerial rots and blights

i. Clover rot

This important disease, caused by *Sclerotinia trifoliorum* Eriksson, has been reported from most countries where clover is grown. Although *S. trifoliorum* attacks many legumes (Kreitlow et al. 1953, Cappellini 1960), most studies have been concerned with its effect on red clover. Recently, severe attacks of clover rot have been reported on white clover (Aldrich & Dolling 1967, Dixon 1975, Scott & Evans 1980) and consequently interest in the effects of this disease on *T. repens* has been renewed. The disease can be damaging in mixed grass-white clover pastures but its effect is most dramatic on pure stands or spaced plants in breeders' plots.

Clover rot first appears in autumn. Initially, isolated plants are affected, but as the disease progresses, large bare patches can develop in a clover field. The disease spreads faster in warm, moist conditions; dry cold weather will check its spread temporarily. Damage is often most serious in late winter and early spring. The outer leaves of infected plants appear water-soaked and olive brown. These wilt and become covered with white mycelium which spreads by contact to adjacent leaves. Mycelium also grows down the petioles and stems to the plant crown. Leaves and stems may die within a few days but the crown is more resistant, which explains why clover fields with large bare patches often recover rapidly when weather conditions become unfavourable for growth of the pathogen.

The fungus forms hard, black, irregularly shaped resting structures (sclerotia) 2-10 mm in diameter, most commonly around the crowns of dead plants, but sometimes in or near the host. Usually, sclerotia within 5 cm of the soil surface germinate in autumn and form light-brown, disc-like fruiting bodies (apothecia) 4-8 mm in diameter borne at the apex of slender stalks. Deeply buried sclerotia can remain dormant in soil for at least 5 years (Khaleeva & Petrov 1970). The formation of apothecia apparently occurs in response to decrease in temperature in autumn (Honda & Yunoki 1973), although apothecia are occasionally found in spring (Kreitlow 1949, Williams & Western 1965). The upper surface of each apothecium is lined with cylindrical asci, each of which produces 8 ascospores. Ascospores are forcibly ejected from asci and are dispersed by wind.

The main source of inoculum was thought by some workers (Wadham 1925, Frandsen 1942) to be mycelium which developed from germinating sclerotia. However, others (Dijkstra 1964, 1966, Loveless 1951, Williams & Western 1965) considered that the disease was disseminated both within and between crops by ascospores rather than by mycelial strands from sclerotia. Observational evidence supports this view. Loveless (1951) showed that even though ascospores were short-lived, they could infect healthy clover leaves if the surface remained moist. Ascospores were discharged in autumn and infections gave rise to small black spots on the leaves. These lesions remained inconspicuous until, in response to warm weather and long periods of high humidity, the fungus became aggressive, causing water-soaking of leaves and plant death. The dead tissues became covered with *S. trifoliorum* mycelium. Honda & Yunoki (1975) also found that the minute ascospore lesions would enlarge to form typical clover rot symptoms. Ascospore infections of white clover became active about one month after the plants were covered by snow and sclerotia formed about one month later, infections remaining dormant for more than 3 months under unfavourable conditions.

The fungus normally survives during summer as sclerotia, but in the UK mycelium has also been reported to persist in the crowns of clover plants (Justham & Ogilvie 1950). In New Zealand, clover rot can be found in long, dense pastures on dairy farms during summer, but damage is seldom severe.

The fungus may be spread by infected seed or in seed contaminated by sclerotia. Justham & Ogilvie (1950) reported the isolation of *S. trifoliorum* from surface-sterilised red clover seed. Leach (1958) surface-sterilised seed of four different clover species, including white clover, and placed it on a nutrient medium to check for fungal infection. Although 8 of 19 samples of *T. incarnatum* L. were infected by *S. trifoliorum*, seed from the other clover species was healthy. Pape (1931) found up to 200 sclerotia of *S. trifoliorum* per kg of commercial red clover seed. Modern seed cleaning machinery screens out contaminants more efficiently and there have been few reports of sclerotia in commercial white clover seed from seed testing stations in the UK and New Zealand. Scott(1981) detected mycelium of *Sclerotinia* sp. (thought to be *S. trifoliorum*) on white clover seed harvested from a field heavily infected with the fungus. Thus it appears that seed-borne infection may occur, but infection in a new crop of white clover probably arises from other sources.

Control of clover rot is difficult because of the longevity of sclerotia in the soil, the dispersal of ascospores over long distances by wind, and the wide host range of the pathogen. Repeated sprays with benomyl or iprodione controlled clover rot (Jenkyn 1980), increasing dry matter production in field plots by up to 14%. However, the development of resistant varieties probably offers the best approach for reducing losses from clover rot.

Plant breeders have had moderate success in developing varieties of red clover resistant to clover rot. Some white clover cultivars are more resistant to the disease than others (Zwingli & Grundbacher 1958, Åkerberg 1958, Niemann 1962, Dixon 1975, Scott & Evans 1980). Screening white clover plants in the glasshouse for resistance to clover rot has proved unsatisfactory because results do not correlate closely with results from field trials (Hanson & Graham 1955). Attempts are being made to improve glasshouse techniques, using ascospores as inoculum (Macfarlane 1979). Dixon (1975) evaluated three field assessment techniques for clover rot and concluded that visual assessment with a diagrammatic key was the simplest method available. Breeding programmes are made more difficult by the genetic variability of the pathogen; hybridisation between different strains of *S. trifoliorum* can occur after hyphal anastomosis (Bjorling 1942, Frandsen 1942).

ii. Southern blight

In the south eastern USA (Halpin et al. 1963) and Japan (Kitani et al. 1968, Iida et al. 1969), during very hot humid weather, pure stands of white clover can be damaged severely by *Sclerotium rolfsii* Sacc., the vegetative state of the basidiomycete fungus *Athelia rolfsii* (Curzi) Tu & Kimbrough. This fungus causes the southern blight disease of white clover and also attacks a wide range of important crops in tropical and warm temperate climates (Brooks 1953, West 1961).

Clover plants are infected by mycelial strands which develop either after germination of resistant, thick-walled sclerotia, or from reservoirs of fungal mycelium within plant debris. Under favourable conditions *S. rolfsii* grows very rapidly, but is restricted to the soil surface because the fungus cannot compete well with antagonistic micro-organisms within the soil (Watkins 1961). Mycelium from the soil surface grows over stolons, petioles and leaflets which are killed by the action of extracellular toxins, possibly a combination of pectic enzymes and oxalic acid (Watkins 1961). Plants may be completely killed within a few days of mycelium appearing on the soil surface, or, when only the older stolon tissue is destroyed, assume a ringed, 'doughnut-like' form. Stands can be damaged so badly that reseeding is necessary the following year. A description of the fungus is given by Mordue (1974a).

Control of *S. rolfsii* on clover by the fungicide quintozene, or an antagonistic fungus, *Trichoderma lignorum* Tode ex Harz, has been recommended in Japan (Kitani et al. 1968). Plants with an erect habit, and those not suffering from a summer die-back of stolons (see Section 2.(C), are reported to show least damage from southern blight (Halpin et al. 1963).

(C) Stolon rots

In the moist, temperate climate of New Zealand and parts of Europe, mixed permanent pastures may retain constant populations of white clover for many years. Plant losses are compensated for adequately by natural reseeding and vegetative propagation from rooting stolons (Crowder & Craigmiles 1960). However, in areas with harsh winters (e.g. Canada and the northeastern USA) or very hot summers (e.g. the southern USA and Queensland, Australia), this balance between clover destruction and growth is easily upset, and white clover in pastures can suffer rapid decline (Stuckey 1962, Schillinger & Leffel 1964, Blake et al. 1966, Gibson & Hollowell 1966). Rotting of stolons and roots by pathogenic fungi has been found to be a major cause of the poor persistence of white clover in these areas (Hanson 1953, Westbrooks & Tesar 1955, Kilpatrick & Dunn 1956, Halpin et al. 1963, Willis 1965, Irwin & Jones 1977).

White clover stolons may often be found dying during hot weather, particularly when exposed on bare soil, a condition known as summer die-back in the southeastern USA (Halpin et al. 1963). A rot appears in the crown of the affected plant and progresses from old into young stolon tissue, causing necrosis and defoliation. Severely affected plants often have a ringed, doughnut-like appearance because the centre of the plant has died. No single fungus has been found to be responsible for this disease, but a number of those most frequently isolated are considered to constitute a pathogen complex, including species of *Fusarium* and *Rhizoctonia*, *Curvularia trifolii* (Kauff.) Boedijn, *Colletotrichum destructivum* O'Gara, *Macrophomina phaseolina* (Tassi) Goidanich, *Leptodiscus terrestris* (Gerdemann), Ostazeski and *Sclerotium rolfsii* (Garren 1955, McGlohon 1959, Halpin & McCarter 1961, Hansen & Curl 1964).

McCarter & Halpin (1962) showed that fungi isolated from stolons were pathogenic to seedlings over a wide range of temperatures, but caused greatest damage at high temperatures. Kilpatrick (1960) found that after exposure to hot sun, stolons developed localised blistered and bleached areas which served as sites for invasion by pathogens, particularly when humidity was high. *Fusarium* and *Rhizoctonia* species, *Colletotrichum destructivum* and *Trichoderma viride* Pers. ex S.F. Gray were isolated from rotting tissue. Heat damage and rot were less in mixed swards, and in shaded stolons (Kilpatrick 1960, Stuckey 1962). A self-shading effect was probably also responsible for an observed greater summer persistence of compact rather than spreading clones of white clover (Halpin et al. 1963).

Curvularia trifolii, *Colletotrichum destructivum*, *Bimuria novae-zelandiae* D. Hawksw., Chea & Sheridan and species of *Fusarium*, *Rhizoctonia*, *Phoma*, *Phomopsis* and *Cylindrocarpon* have been found in stolons rotting under cool conditions in New Zealand (R.A Skipp & M.J. Christensen unpublished data). In the USA, Hanson (1953) isolated similar fungi from stolons of Ladino clover which had been injured by low temperatures during winter. Hanson found that these fungi were weakly pathogenic and unable to attack vigorously growing stolons. Damaging rots therefore occurred only when plants had been weakened by exposure to excessive heat or cold.

Some of the species that constitute this complex of stolon-rotting fungi also cause root and stem rots, or leaf spots of clover and other legumes. Inoculum for stolon infection comes from diseased plants or the soil.

Colletotrichum destructivum is often isolated from minute black spots on the petioles and stolons of clover or from aerial parts of other forage legumes such as lucerne and sainfoin. It can often be recognised by the spore-forming cushions (acervuli) with dark, septate hairs (setae) which form on rotting tissue. Spores are readily dispersed in water droplets from rain splash. A feature of many *Colletotrichum* species is their ability to cause latent, symptomless infections and *C. destructivum* can often be isolated from tissue lacking any visible damage. Other species of *Colletotrichum* sometimes found in rotting stolons are *C. trifolii* Bain & Essary and *C. graminicola* (Ces.) Wilson (Garren 1955, Halpin & McCarter 1961). Tiffany & Gilman (1954) describe the diagnostic features of *Colletotrichum* species from legumes.

Rhizoctonia solani Kuhn has frequently been named as the cause of root and stolon rots of white clover, but since the taxonomy of *Rhizoctonia* has been uncertain until recently, other species of *Rhizoctonia* may have been mistaken for *R. solani*. *R. solani* typically has wide hyphae (up to 12 μm in diameter) with multinucleate cells; its perfect (sexual) stage, *Thanatephorus cucumeris* (Frank) Donk, is described by Mordue (1974b). Other *Rhizoctonia* species have narrow hyphae and binucleate cells; their perfect stages are species of *Ceratobasidium*. The *Rhizoctonia* states of both *Thanatephorus* and *Ceratobasidium* are abundant in soils and are frequently pathogenic on a wide range of plants. Species of *Fusarium*, including *F. oxysporum* Schlect., *F. solani* (Mart.) Sacc. and *F. roseum* Link ex S.F. Gray, have frequently been isolated from rotting stolons. *F. roseum* was originally described as one species in the taxonomic system of Snyder & Hansen (1945), but is now considered to be a group of species which includes the plant pathogens *F. avenaceum* (Corda ex Fr.) Sacc., *F. acuminatum* Ellis & Everhart, *F. sambucinum* Fckl. and *F. culmorum* (W. G. Smith) Sacc.

Sclerotium rolfsii, as described in Section 2.(B) is the cause of southern blight, a disease that often follows summer die-back of the stolons. Rotting stolons may also be invaded by *Trichoderma viride*, an antagonist to *S. rolfsii*.

Curvularia trifolii is primarily a pathogen of clover leaves (see Section 2.(E)) which will invade stolons.

Macrophomina phaseolina, the charcoal rot fungus, is a warm-temperature pathogen of the roots of many plants, and persists in the soil as small (less than 1 mm diameter), hard, black, sclerotia (Brooks 1953).

Leptodiscus terrestris, not normally considered a pathogen of white clover, causes a root and crown rot of lucerne, red clover and birdsfoot trefoil (Gerdemann 1953, McVey & Gerdemann 1960). *L. terrestris* produces spores and sclerotia which aid its persistence in the soil.

Irwin & Jones (1977) reported that a serious decline in clover stolon cover which occurs in pastures in parts of Queensland, Australia in the late summer was caused by *Pythium middletonii* Sparrow, possibly in association with plant-parasitic nematodes. Root and stolon rots typical of those found in the field developed after inoculation with pure cultures of the fungus. Field applications of fungicide (captafol) and nematicide (fenamiphos) reduced stolon death.

(D) Root rots

Root rots of white clover have been studied extensively in the USA, where many perennial legumes, including lucerne, and red, white and alsike clover often lack sufficient persistence to maintain good production for more than 2 years. A *Fusarium* root rot complex, which causes death of secondary roots and a progressive necrotic discoloration of cortical and/or vascular tissues of taproots, is regarded as a major cause of this poor persistence (Fulton & Hanson 1960, Leach et al. 1961, Moody et al. 1967, Leath et al. 1971, O'Rourke 1976). Pathogenicity studies have shown that isolates from one legume could damage other legume species (Carroll & Elliott 1964, Chi 1965, Aubé 1966, Leath & Kendall 1978).

The species of *Fusarium* (*F. oxysporum*, *F. solani* and *F. culmorum*) most commonly isolated from diseased roots of forage legumes are the same as those associated with stolon rots of *T. repens*. These fusaria are abundant in all kinds of soil as either resistant resting spores (chlamydospores) or growing as soil saprophytes (Booth 1971). *F. oxysporum* has been found to be the major component of the rhizosphere-rhizoplane flora of lucerne, and white and red clover, and may readily be isolated from young, apparently disease-free roots of these plant species (Thornton 1965, O'Rourke & Millar 1966, Smith & Peterson 1966, Skipp & Christensen 1981). Infected seed may also serve as a source of inoculum; McGee & Kellock (1974), for example, found *F. oxysporum* and *F. avenaceum* in seed of subterranean clover.

The role of *Fusarium* spp. in root rots, persistence problems and poor clover productivity has been difficult to establish, partly because vigorously growing plants may not show symptoms. Plants in the field usually succumb to root rot gradually, often after they have been exposed to stress. The cause of root rot in forage legumes is discussed in detail by Fulton & Hanson (1960), Leath et al. (1971), Menzies (1973a) and Fezer (1961), who comments that, in diseases caused by weak pathogens, information on the conditions necessary for pathogenicity is as important as knowledge of the relative virulence of the different causal agents.

Much work has been concerned with aspects of environment and management which predispose plants to debilitating root rots. Westbrooks & Tesar (1955) found that Ladino clover plants with low levels of available carbohydrates in their roots in winter died from root rot during their second production year. After studying red clover, Colville & Torrie (1962) also concluded that the level of root carbohydrate stored during winter greatly influenced plant survival.

McCarter & Halpin (1961a), Menzies (1973b) and Moody et al. (1967) showed that severe clipping, simulating hard grazing, greatly increased the severity of root rot on white clover plants grown in unsterilised field soil in the glasshouse. With one exception (Leath & Kendall 1978), studies with other legumes have also shown that frequent and severe clipping increases damage by root-rotting fungi (Fulton & Hanson 1960, Fezer 1961, O'Rourke & Millar 1966, Siddiqui et al. 1968). Defoliation reduces carbohydrate supply to roots and causes depletion of root reserves (Lukezic et al. 1969, Nelson & Smith 1969).

Shading has a similar effect on root reserves and root growth. Butler et al. (1959) and Evans (1973) showed that both shading and defoliation greatly reduce the rate of elongation of white clover roots. Artificially shaded plants rapidly disappeared from a mixed white clover-*Dactylis glomerata* L. sward (Blake et al. 1966), but whether these died from root rot or from some other cause was not known. High temperatures also depress root growth while favouring shoot growth (Menzies 1973a). This decline in root vigour may partly explain why *Fusarium* spp. and other fungi cause more root rot at high temperatures (McCarter & Halpin 1962).

Stress caused by insects feeding on growing shoots can affect the health of roots. Leath & Byers (1977) found that infestation of shoots of lucerne, and white and red clover by pea aphid, *Acyrthosiphon pisum* (Harris), increased root rot caused by *F. roseum*. Similarly, survival of plants inoculated with *F. roseum* was decreased when they were infested with potato leafhopper, *Empoasca fabae* (Harris), in field cages. Importantly, significant differences in stand density between infested and non-infested plots were not detected while insects were present in the cages but became apparent only in the following year after the plants had been exposed to winter conditions. These results suggest that the effects of stress on perennial legumes can be long-term and cumulative (e.g. insect damage plus winter injury). The physiological effects of insect feeding are probably complex, but a decreased export of carbohydrates from leaves for accumulation in the roots would be expected (Willard 1930).

Leaf diseases also seem to predispose plants to greater damage from root-rotting fungi. O'Rourke & Millar (1966) showed that infection of lucerne leaves with *Ascochyta imperfecta* Peck increased the severity of root rot. Red clover died from *Fusarium* root rot more rapidly in plants systemically infected with red clover vein mosaic virus (Denis & Elliott 1967), white clover mosaic virus or clover yellow mosaic virus (Watson & Guthrie 1964) than in plants not infected with these viruses. McCarter & Halpin (1961b), however, found that infection of white clover with bean yellow mosaic virus did not increase root rotting.

Little is known about the effect of mineral nutrition on root rots of white clover. However, work with other legumes, summarised by Leath et al. (1971), indicates that fungal attack is favoured by nutrient imbalance, particularly potassium deficiency.

As well as weakening the roots physiologically, insect and nematode damage can provide infection sites in the thickened tap root (Leach et al. 1963, Leath et al. 1971, Willis & Thompson 1979). Clover root curculio larvae of the weevil *Sitona hispidula* (F), which feed on root nodules and small secondary fibrous roots and gnaw deep cavities in tap roots of white, red and alsike clovers (Leach et al. 1961), can also harbour pathogenic species of *Fusarium* (Kilpatrick 1961a). Graham & Newton (1960) have shown in glasshouse experiments that invasion by larvae of *S. hispidula* increases the severity of rot in white clover taproots infected with *F. roseum*.

Field observations also suggest a link between the severity of *Fusarium* root rot and larval infestation (Kilpatrick & Dunn 1958, 1961). However, in a field trial by Dunn et al. (1964) in which larval populations and insect injury were controlled by a pre-planting application of dieldrin, the severity of root rot in treated and control plots did not differ significantly; winter injury was particularly severe in this experiment. These workers also reported the appearance of an internal breakdown of taproot tissue which, in its early stages, was not associated with the presence of pathogenic fungi. Cressman (1967) considered that a similar condition of red clover taproots was physiological in origin. Poor persistence was thought to result from the destruction of taproots by insects and secondary fungal pathogens after this breakdown in internal tissues.

A more convincing demonstration of the role of insects in root disease development in the field was provided by Leach et al. (1961, 1963) who showed that alsike clover plants, protected from clover root curculio injury by application of organochlorine insecticides, developed significantly less rot in the vascular tissues of their taproots than did control plants. James et al. (1980) found that carbofuran increased yield and persistence of Ladino clover. Both stolon rotting and root damage caused by larvae of *S. hispidula* decreased in insecticide-treated plots. The fungicide benomyl reduced root rot particularly when combined with insecticide. (Further information about the clover root curculio is given in Chapter 12).

Rotting roots may be more attractive to nematodes and insects than healthy roots. Edmunds & Mai (1967) found that lucerne roots were invaded more readily by *Pratylenchus penetrans* Cobb. when infected by *F. oxysporum*, while the root borer *Hylastinus obscurus* Marsh. is also attracted to fungal lesions on red clover roots (Leath & Byers 1973).

Fungi other than *Fusarium* spp. may cause root rots of white clover. Menzies (1973b) showed that the roots of clover, taken from some warmer areas of New Zealand where clover persistence was poor, were invaded by *Codinaea fertilis* Hughes & Kendrick. Of the fungi isolated from diseased tissue, 40% were *C.*

fertilis and only 19% were species of *Fusarium*. Laboratory tests confirmed that *C. fertilis* was a primary pathogen of white clover roots, and showed that the disease was favoured by high temperatures and frequent clipping of the foliage. Apart from a record of *C. assamica* (Agnihothrudu) Hughes & Kendrick on the roots of tea (Agnihothrudu 1962), *Codinaea* spp. are not known to be pathogens but are found on dead tissue, mainly of deciduous trees (Ellis 1976). In New Zealand, *C. fertilis* had previously been recorded only on the dead leaf bases of the indigenous palm *Rhopalostylis sapida* Wendl. & Drude (Hughes & Kendrick 1968), which makes this demonstration of pathogenicity to an introduced species, *T. repens*, particularly interesting. *C. fertilis* has recently been found to cause a root rot of Ladino clover in North Carolina in the USA (Campbell 1980). Attempts are being made to screen white clover for resistance to *C. fertilis* in New Zealand (S.A. Menzies, personal communication).

Kilpatrick (1959a) isolated a number of pathogenic fungi from white clover roots in the USA, including *Colletotrichum destructivum, Gliocladium roseum* Bainier, *Leptodiscus terrestris, Sclerotinia trifoliorum, Macrophomina phaseolina* and species of *Pythium, Rhizoctonia* and *Thielaviopsis*. We have found that roots of field-grown white clover and of seedlings grown in several New Zealand pasture soils in the glasshouse, were invaded by several *Fusarium* species (including *F. oxysporum*), *Codinaea fertilis, Ceratobasidium cornigerum* (Bourd.) Rogers, *Thielaviopsis basicola* (Berk. & Br.) Ferraris, species of *Cylindrocarpon, Phoma, Chrysosporium* and *Pythium* and dark, sterile mycelium (Skipp & Christensen 1981, Skipp et al. 1982). A newly described ascomycete, *Bimuria novae-zelandiae* (Hawksworth et al. 1979), was also common in white clover roots. Pure cultures of these fungi were pathogenic to seedling roots of *T. repens* (Skipp & Christensen 1982). However, root rots did not appear to seriously damage the pastures studied.

Other micro-organisms that invade and possibly damage white clover roots cannot be isolated on general laboratory culture media and must be studied by other techniques. Selective media containing antibiotics may be needed to detect species of *Pythium* and *Phytophthora* (Johnson & Keeling 1969, Stovold 1974). Endomycorrhizal fungi, which extensively colonise the secondary rootlets of white clover plants in most natural soils (Crush 1976), must be studied by direct microscopic observation (Phillips & Hayman 1970) because they fail to grow on artificial media. Techniques used to clear and stain roots for microscopy should preserve root-invading nematodes (Skipp & Christensen 1981), since these can affect root growth and the activities of pathogenic fungi.

Effective control of root rot is difficult, but some general principals have emerged. The conclusions of Fulton & Hanson (1960) from a study of red clover root rot apply equally to white clover: "The best method of controlling root rot is to keep plants growing vigorously. This cannot be done at all times but sturdy adapted varieties, suitable liming and fertilisation, proper crop rotation and good management should help. In attempting to develop resistance to this disease complex, first consideration must be given to selecting or developing plants with superior inherent vigour and the ability to tolerate unfavourable conditions." It will probably be necessary to make selections from plants exposed both to root pathogens and to stress. The physiological mechanisms involved in the resistance of plants to root pathogens are poorly understood; more knowledge could provide principles upon which to base methods of plant selection.

Present attempts to introduce white clover more widely into infertile hill country and other unfavourable areas should not overlook the possibility that the increased stresses on plants imposed by these environments could lead to the development of root rot and persistence problems from this cause. These may not be apparent in early trials because of an initial lack of pathogen inoculum in the soil.

Leath et al. (1973) showed that a soil drench with the fungicide benomyl increased the yield and persistence of red clover. A similar treatment might be considered for white clover pastures as an alternative to renovation or re-seeding, but much work needs to be done to determine if the application of fungicide to pasture would be economic.

(E) Leaf diseases

i. Sooty blotch

Sooty blotch or black blotch, caused by *Mycosphaerella killianii* Petrak (formerly called *Cymadothea trifolii* (Killian) Wolf), is a distinctive and widespread disease of many species of *Trifolium*. Crimson clover suffers most damage (Killian 1923, Hanson & Kreitlow 1953). Its significance on white clover may have been underestimated, because the disease is inconspicuous in pasture, symptoms being confined to the underside of leaves. White clover yield is reduced by stunting and partial defoliation (Booth & O'Rourke 1973). Toxins (Amelung 1966) and oestrogens (Newton et al. 1970, Wong & Latch 1971a,b, Wong et al. 1971) accumulate in diseased herbage and may affect livestock health and performance. In New Zealand, this is the predominant disease of white clover during the critical period before mating of sheep (R.A. Skipp & M.G. Lambert, unpublished). Sheep fertility may be reduced if large amounts of infected material are ingested.

The characteristic sooty blotch symptoms first appear in spring and may continue until autumn when, in some areas of New Zealand, they can be very severe indeed. Dark green to black, cindery-textured fungal outgrowths up to 1 mm in diameter appear on the lower surface of infected leaves. Leaves initially remain green and appear healthy but eventually become discoloured, wither and die. Descriptions of the fungus and its life cycle are given by Killian (1923), Bayliss-Elliott & Stansfield (1924) and Wolf (1935).

The conspicuous leaf symptoms are caused by an asexual conidial stage of the fungus, *Polythrincium trifolii* Kunze ex Ficinus & Schubert. Each spot consists of a subepidermal mycelial mat (stroma) from which arise clusters of conidiophores. These produce light brown, two-celled, pear-shaped conidia at their tip.

Growth of the fungus on artificial media ceases within a few hours of conidial germination, but in living leaf tissue an extensive intercellular mycelium develops after cuticular or stomatal penetration (Camp & Whittingham 1972). The conidial stage of the fungus thus behaves as an obligate parasite. Conidia are regarded as the means of localised re-infection during summer and autumn. Inoculation of plants in the field is reported to give good infection (Killian 1923). However, in the glasshouse, artificial inoculation is frequently unsuccessful, although accidental infection from outside sources can cause severe epidemics. Thus, more information about conditions necessary for infection is needed if experimental work is to be carried out under controlled conditions.

The fungus also develops a sexual stage which assists survival during winter. Flask-shaped fruiting bodies (perithecia) begin to develop during autumn-winter in leaves showing the *Polythrincium* stage of the disease. Maturation of perithecia is usually delayed until spring, by which time the infected leaves have entered the herbage litter. Club-shaped sacs (asci), each containing eight, 2-celled, spindle-shaped ascospores form within the perithecia, and ascospores are discharged under favourable conditions. These provide the main source of spring infection. Killian (1923) showed that the conidial stage of the fungus developed in plants inoculated with ascospores.

A third spore stage produces flask-shaped cavities within shiny black stromata which develop among and between the stromata of the conidial stage. During autumn, in humid weather, fine white tendrils containing spores ooze from small pores in these cavities. The function of these spores is disputed. Wolf (1935) regarded the flasks as spermogonia producing male gametes required for fertilisation and the development of asci, whereas Killian (1923) and Baylis-Elliott & Stansfield (1924) stated that they produce asexual pycnidiospores capable of infecting clover leaves.

Selection of white clover plants resistant to *M. killianii* appears to offer a practical means of control. O'Rourke (1971) showed that cv. Blanca was resistant, whereas New Zealand mother white clover and the Welsh S100 cultivars were susceptible. Selection of resistant individuals from within adapted cultivars is also possible (R.A. Skipp & W.M. Williams unpublished). If selection and breeding for resistance is to be more extensive, it will be necessary to find out whether physiological races of the fungus exist which might overcome resistance incorporated into new lines.

ii. Pepper spot

Pepper spot caused by *Leptosphaerulina trifolii* (Rost.) Petrak occurs on white clover leaves during cool wet periods of winter and spring (Garren 1955). Significant damage has been reported in pastures in the northeastern USA (Kreitlow et al. 1953) and Ireland (O'Rourke 1970). In New Zealand, where this is the most common disease in early winter (R.A. Skipp & M.G. Lambert, unpublished), severe infections have killed single-spaced plants in breeders' plots. Disease severity is increased by frost (Kilpatrick & Dunn 1956), virus infection (Creelman 1967) and sprinkler irrigation (Curl & Weaver 1958).

Minute (less than 0.3 mm diameter), black, pin-prick spots on petioles and on both surfaces of the leaf lamina are characteristic of the disease. The spotting is most intense on older leaves, which can appear grey and die prematurely (Kreitlow et al. 1953). 'Green islands' may be seen around spots on leaves becoming yellow as they senesce. Peduncles, flowers and seeds may also become infected; Kilpatrick (1958b) isolated *L. trifolii* in trace amounts from seed of a naturally infected crop.

Fungal fruiting bodies (pseudothecia) usually develop in senescent and dead leaves but may occasionally be found within exceptionally large spots in leaves on infected plants. The pseudothecia are globose and are packed with irregularly shaped, two-walled asci each containing 8 ascospores. Ascospores of *L. trifolii* are elliptical, with 3 or 4 transverse and up to 2 longitudinal septa (Booth & Pirozynski 1967a). They are forcibly discharged from the asci. Pseudothecia are very resistant to unfavourable environmental conditions and can resume ascospore discharge upon thawing and wetting (Elliot & Wilcoxon 1964).

Each spot arises from an infection by a single ascospore. Clover leaves are penetrated by short germ tubes arising from one or more cells of the ascospore. Penetrated host cells collapse rapidly, possibly being killed by an extracellular fungal toxin (Sundheim & Wilcoxon 1965). The spots become visible to the naked eye within one day of infection, but thereafter expand very slowly.

The fungus can be grown in culture but light is required for sporulation. Short days (10 h, Kilpatrick 1962), and light of 310-330 nm wavelength (Leath 1971) favour maturation of the pseudothecia and ascospore discharge.

Barrière et al. (1974) considered pepper spot of lucerne to be caused by another species, *L. briosiana* (Poll.) Graham & Luttrell, which they distinguished from *L. trifolii* by characteristics of growth in culture and the size and degree of septation of ascospores (Graham & Luttrell 1961). However, the clover type has been found on lucerne (Carr 1957) and isolates from lucerne will infect white clover (O'Rourke 1972). Latch (unpublished) found that *Leptosphaerulina* isolates from several species of *Vicia, Medicago* and *Trifolium* would cross-infect all hosts in glasshouse tests. Booth & Pirozynski (1967a) reported that morphological differences disappeared when clover and lucerne isolates were grown in pure culture. They preferred to regard *L. trifolii* as a cosmopolitan species, which can infect plants from several families in addition to members of the leguminosae.

iii. Pseudopeziza leafspot

A leafspot disease very common in the cooler, more humid areas of white clover production such as Ireland (O'Rourke 1976), the UK (Carr 1971), New Zealand (Dingley 1969) and the northern USA (Kreitlow et al. 1953) is caused by *Pseudopeziza trifolii* (Biv.) Fckl. Severe infection may kill leaves, particularly during autumn, winter and early spring, but the disease is generally regarded as less damaging than similar leafspots on lucerne and red clover. Oestrogenic coumestans accumulate in diseased leaves (Wong & Latch 1971a,b). In New Zealand, where the disease occurs predominantly in late autumn and winter (R.A. Skipp & M.G. Lambert, unpublished), high levels of coumestans induced by this fungus are not likely to affect the breeding performance of ewes which are usually mated in autumn.

The presence of *Pseudopeziza* leafspot in pasture can often be detected by the cupping of clover leaflets, which exposes the undersufaces of leaves and gives the plants a silvery appearance. Symptoms of infection first appear as minute chocolate-brown spots on the upper surface and very occasionally the lower leaf surface also. These spots enlarge to form buff-coloured, roughly circular lesions up to 2 mm in diameter. Mature lesions often have darker, minutely fringed margins. Under conditions of high humidity, a central fruiting body (apothecium) forms and swells to give a conspicuous, jelly-like cushion composed of sterile hair cells and spore-forming asci.

Each ascus contains 8 ascospores. Ascospores are forcibly discharged from the asci, frequently in pairs, and become impacted on nearby leaf tissue or are carried further by air currents. The fungus probably survives unfavourable conditions on infected debris, and resumes spore discharge after rehydration of the apothecium (Carr 1971). Infection of leaf tissue requires a relative humidity approaching 100% (Schmiedeknecht 1964a). The impacted ascospores produce fine penetration pegs at their point of contact with the host cuticle; these enter epidermal cells directly, separate infections being effected by both spores of a pair. Kristinsson (1967) has suggested that a similar clumping of the ascospores of *Pseudopeziza medicaginis* (Lib.) Sacc. favours the association of compatible mating types necessary for the formation of sexual fruiting bodies. Infected cells are not immediately killed by the fungus and visible symptoms often do not appear on leaves for 2-3 weeks after infection.

Schmiedeknecht (1964b,c) has shown that white clover is not attacked by *P. trifolii* isolated from red clover, and has defined the *Trifolium* host range of two specialised forms of *P. trifolii*. The 'repentis' form attacks *T. repens, T. hybridum, T. fragiferum, T. resupinatum* and *T. dubium*, whereas the 'pratensis' form infects *T. pratense* and *T. incarnatum*. *Pseudopeziza medicaginis*, which is closely related to *P. trifolii*, causes a serious leafspot of lucerne but will not infect white clover. *P. trifolii* has smaller ascospores than *P. medicaginis* (Jones 1919) and requires higher humidity and lower temperatures for optimum infection (Schmiedeknecht 1964a).

Pseudopeziza species will not readily grow from infected tissue on to artificial media, but pure sporulating cultures can be obtained on agar media when discharged ascospores are used as inoculum (Frosheiser 1960). Alternatively, the fungus can be maintained on white clover plants in the glasshouse. Adequate infection for maintenance and screening purposes can be obtained by suspending leaves bearing mature apothecia over experimental plants for 2-3 days at 100% relative humidity.

iv. Rust

White clover is the natural host for two species of rust fungi, *Uromyces trifolii-repentis* Liro and *Uromyces trifolii* (Hedw. f. ex DC.) Fckl. Both species occur throughout the world and *U. trifolii-repentis* has been recorded frequently in the USA, Canada (Kreitlow et al. 1953, Kilpatrick & Dunn 1956, Willis 1965), Australia (Latter 1952) and Japan (Hiratsuka 1973), whereas *U. trifolii* is more common in Europe (Wilson & Henderson 1966, Jørstad 1967, O'Rourke 1976) and New Zealand (G.F. Laundon personal communication). It is not clear how much rust diseases damage white clover in grazed pastures, but in New Zealand separately spaced plants in breeders' evaluation plots can die after severe infection by *U. trifolii*. There has been some selection for resistance to *U. trifolii-repentis* in the USA where rust can severely damage stands used for seed production (Kreitlow et al. 1953).

Uromyces trifolii-repentis

This rust is macrocyclic, i.e. its life cycle is complex and involves several distinct spore forms produced in different types of pustule. In late summer, small (up to 0.5 mm diameter) red-brown pustules termed uredinia can be found on the under-surface of infected leaves, and petioles. Each pustule produces a mass of dry powdery urediniospores which erupt from the surface of infected tissue. The uredinial, brown rust stage is most conspicuous in the USA (Hanson & Kreitlow 1953) but is considered to be only transitory elsewhere (Latter 1952, Jørstad 1967). Severe infection causes leaf distortion and yellowing.

Uredinia are superseded in autumn by darker brown pustles (telia) of similar size which also form on petioles and the lower surfaces of leaves. The stalked, thick-walled teliospores produced within telia probably assist survival of the fungus during winter. Germinating teliospores produce basidiospores; these give rise to the infections forming the other two types of spore-producing structure, the pycnia and aecia.

Pycnia that develop in autumn or early spring are inconspicuous, flask-shaped bodies embedded in leaf tissues. Pycniospores are discharged through a minute pore in the pycnium. Aecia, in clusters up to 5 mm in diameter, are found on leaves containing pycnia. The cup-shaped, creamy yellow aecia produce chains of aeciospores which in Australia may be the most important spore stage for fungal dispersal (Latter 1952).

The different types of spores are described in detail by Wilson & Henderson (1966). Urediniospores of *U. trifolii-repentis* are cinnamon brown and covered with minute spines; there are 2-4 approximately equatorial germ pores. Teliospores are chestnut brown and their walls are smooth, bearing the remains of a stalk at the base, and a single apical germ pore which is covered by a prominent colourless papilla. Pycniospores are smooth walled and colourless. Aeciospores are pale yellow and their walls have a finely warted appearance.

A number of *Trifolium* species have been found to be susceptible to *U. trifolii-repentis* after artificial inoculation (Kobel 1920, Davis 1924, Mains 1936). *Uromyces fallens* (Arthur) Barth., the rust that attacks red clover, is morphologically similar to *U. trifolii-repentis* and infects some of the same species of *Trifolium* (Kobel 1920). Arthur (1934) suggested that the similarities between these two rusts are so great that they are best regarded as a single species. However, the rust from white clover has never been found to infect red clover and *vice-versa*; thus retention of the two species is probably still justified (Laundon 1973, 1975). The diagnostic characteristic of *U. trifolii-repentis* is that its urediniospores have 2-4 equatorially arranged germ pores, whereas urediniospores of *U. fallens* have 4-7 scattered germ pores.

Uromyces trifolii (*sensu* Walker 1978; syn. *U. flectens* Lagerheim; ? syn. *U. nerviphilus* (Grognot) Hotson) is a microcyclic rust. Thus its life cycle is simple and only one spore-forming stage (telia) occurs on the host plant. Individual telia are often about 1 mm long but may coalesce to produce erupting pustules up to 15 mm long. The largest pustules occur on petioles, peduncles and leaf veins, causing gross distortion. Pustles are also found on the leaf lamina and occasionally on stolons. This rust is seen most often during autumn and winter.

The affinity of *U. trifolii* for petiole and leaf vein tissue distinguishes it from *U. trifolii-repentis* which forms small pustles mainly on the leaf lamina. However, the teliospores of the two species are identical, and difficulties in identification may arise when plants have mixed infections, particularly when the uredinial stage of *U. trifolii-repentis* is absent. The identity of the species forming telia can be determined by inoculating clover plants with teliospores from the pustules. Infection by teliospores of *U. trifolii* results only in the production of telia, whereas aecia and pycnia but no telia are formed after teliospore infection by *U. trifolii-repentis* (Davis 1924).

By repeated inoculations with teliospores, Mains (1935) confirmed that in the USA a true microcyclic form of *U. trifolii* which produces only telia exists. Reports by other workers (Arthur 1934, Roane & Miller 1976) of a North American form of this rust which also produces aecia are probably erroneous and are possibly examples of mixed infections of both *U. trifolii* and *U. trifolii-repentis* (Jørstad 1967, Laundon 1975).

Latter (1952) has shown that a number of species of *Trifolium* are susceptible to *U. trifolii* from white clover, but another form of the fungus which is commonly found on strawberry clover does not infect *T. repens*. The possibility of physiological races within the white clover form of *U. trifolii* cannot be ignored when selecting for resistance. Effective selection appears to be possible because of the considerable variation in reaction to *U. trifolii* shown by clovers of different genotype (Latch & Skipp, unpublished data).

An important feature of infection by *U. trifolii* is that the fungus becomes systemic within the plant. These infections probably arise from mycelium within the stolons (Kobel 1920). Thus, rusted plants brought in from the field and grown in a glasshouse often continue to develop pustules on newly formed leaves.

Basidiospores are the primary source of infective inoculum. These spores are released from short promycelial hyphae produced by germinating teliospores. Germ tubes produced by basidiospores penetrate the leaf cuticle and initiate the infection.

Rust pustules frequently contain dark, flask-shaped bodies (pycnidia) formed by the fungus *Darluca filum* (Biv.) Cast. which grows parasitically on the rust. It is not known whether the presence of this fungus reduces rust growth or increases plant damage.

v. *Curvularia leaf blight*

Ladino white clover is particularly susceptible to infection by *Curvularia trifolii* and suffers significant damage in Japan (Takahashi & Iida 1969) and the eastern USA (Lehman 1951, Kilpatrick & Dunn 1956). The feed quality of infected clover can be affected. Isawa & Nishihara (1975) found that crude protein levels were lower and lignin content was higher in diseased than in healthy clover. Kilpatrick et al. (1966)

suggested that infection with *C. trifolii* during autumn reduces cold hardiness and thus affects persistence.

Wet, warm (23-26°C) periods, frequent in the southeastern USA from autumn to spring, favour infection by *C. trifolii* (Kreitlow et al. 1953, Lehman 1951), but under sprinkler irrigation the disease may be seen in mid-summer (Curl & Weaver 1958). Infection often occurs at the leaf tips, and lesions spread inwards in a wedge-shape which is at first chlorotic but becomes water-soaked and brown; the leading edges of infections remain yellow (Lehman 1951). Lesions may expand to kill entire leaflets, and the infection may then progress into the petioles which wilt and collapse. Flowers, peduncles and developing seeds may also develop extensive lesions (Kilpatrick 1961b). Stolons are not normally damaged during leaf blight epidemics (Kreitlow et al. 1953), but *C. trifolii* may cause stolon die-back which occurs when plants are exposed to high temperatures (Halpin & McCarter 1961) (see Section 2.(C)).

The fungus sporulates profusely on leaf tissue maintained under high humidity, but less readily on media containing agar (Kilpatrick 1958b). The conidia are often curved, club- to pear-shaped, brown, usually three septate, with a pronounced hilum marking the site of attachment to the conidiophore. Illustrated descriptions of *C. trifolii* are given by Ellis (1971) and Falloon (1976).

White, red, alsike, crimson and subterranean clover are natural hosts for *C. trifolii* (Kreitlow & Yu 1955, Nishihara 1966) but a range of *Trifolium* and *Medicago* species were found to be susceptible after conidial inoculation in the glasshouse (Kilpatrick 1958b). Falloon (1976) has shown that *C. trifolii* damages some turf grasses at high temperatures (25-35°C),but it is not known if the isolates that infect turf grasses also infect white clover. Specialised forms of the fungus with a limited range of *Trifolium* hosts have been described by Nishihara (1966). Kilpatrick (1958b) found 32 resistant seedlings among 14 000 which had been inoculated with the fungus, indicating that selection of resistant material for breeding should be possible. About 2% of inoculated red clover plants were also found to be resistant to *C. trifolii*.

vi. Cercospora leafspot

Leafspots, and sunken, elongated petiole lesions caused by *Cercospora zebrina* Pass. can often be found on white clover during summer and autumn. Damaging outbreaks in parts of the USA have been reported (Hanson 1953, Gibson & Hollowell 1966).

The leaf lesions are light brown to purplish-black, circular (up to 5 mm in diameter) or, when confined by veins, rectangular. Under humid conditions, colourless, multicelled, needle-shaped conidia are produced abundantly on both surfaces of the leaf. The conidia are dislodged easily from the conidiophores and are dispersed by air currents. They produce germ tubes which grow over the leaf surface and penetrate through stomatal openings. Infected tissue dies rapidly (Berger & Hanson 1963a).

Summer black-stem disease of lucerne and red clover is also caused by *C. zebrina*. Although isolates of *C. zebrina* from these plants show host specialisation on a range of *Trifolium* and *Medicago* species (Berger & Hanson 1963a), there appear to be sufficient morphological and biochemical similarities to retain them within a single species (Peterson & Latch 1969). The fungus does not sporulate well on artificial media (Berger & Hanson 1963b), which makes it difficult to produce the large quantities of inoculum needed for screening white clover seed lines for resistance to *C. zebrina*.

vii. Other diseases of leaves and flowers

Downy mildew caused by *Peronospora trifoliorum* de Bary is of potential importance in the UK (Carr 1971) but causes little damage elsewhere. Infection often occurs in wet weather during late spring and is accompanied by the development of chlorotic patches on leaflets and leaf bases. Sporangia are produced singly on each pointed tip of dichotomously branched sporangiophores which emerge through stomata (Gäumann 1923). The undersurface of the leaflets often appears violet-grey where there is profuse sporulation of *P. trifoliorum*. Infections can become systemic and thus damage a succession of developing leaves, sometimes killing the entire plant (Carr 1971). Lucerne, red clover and alsike clover are infected by different host-specialised forms of *P. trifoliorum*.

A specialised form of *Erysiphe trifolii* Grev., which causes a powdery mildew disease of red and crimson clover, may also be found on white clover (Sampson & Western 1954). The fungus was previously known as a form of *E. polygoni* DC. ex St-Amans (Kapoor 1967). The disease is recognised easily by the white, powdery appearance of the upper leaf surface because of the covering of superficial mycelium and chains of spores. Leaves may eventually become chlorotic and die. An account of the host-parasite relations of the form of *E. trifolii* on red clover is given by Carr (1971).

Stemphylium trifolii Graham is found on white clover in Japan (Iida et al. 1969) and the USA (Graham 1957). Irregular light brown spots up to 6mm across are formed on infected leaflets, usually at their edge. Sporulation occurs on leaves kept under high humidity. The conidia are light brown and the smooth wall is constricted at the site of origin of transverse septa. Longitudinal septa are also present. *Stemphylium sarciniforme* (Cav.) Wiltsh., which causes a damaging leafspot of red clover, and *S. botryosum* Wallr., which is pathogenic to lucerne, have also been recorded on white clover (Chilton et al. 1943, Kilpatrick & Dunn 1956). Illustrated descriptions of these two species are given by Ellis (1971). *S. botryosum* is the conidial state of an ascomycete, *Pleospora herbarum* (Pers. ex Hooker) Rab. (Booth & Pirozynski 1967b). White clover seed may become infected by *Stemphylium* spp. (Kilpatrick 1958a).

A leafspot caused by *Stagonospora meliloti* (Lasch) Petr. can be damaging in some parts of the USA (Albrecht 1942, Kilpatrick & Dunn 1956) and is often seen in Europe and New Zealand during cool, wet weather. The lesions are oval, about 3 × 5 mm, off-white, with a darker or watersoaked margin. Minute black specks are often visible in the pale centre of the lesion. These are the flask-shaped, spore-containing pycnidia which are embedded in the dead leaf tissue. In this species of *Stagonospora*, they have a characteristic fringed beak around their opening at the leaf surface (Jones & Weimer 1938). The spores have up to 3 septa. Lucerne and sweet clover are the most severely affected hosts of *S. meliloti*. A related fungus, *S. recedens* (Massal.) Jones and Weimer, infects red clover (Jones & Weimer 1938) but has also been recorded on white clover in Japan (Narita 1972). Other fungi that produce spores in pycnidia may invade necrotic lesions caused by *Stagonospora* or may themselves cause similar leafspots. *Ascochyta imperfecta* causing spring black stem of lucerne and red clover has been recorded on white clover (Moore 1959, Dingley 1969), and species of *Phoma* are also found. Identification can be difficult because a diagnostic characteristic, degree of septation in the spore, is variable within these genera. *Phoma* is predominantly aseptate, *Ascochyta* uniseptate and *Stagonospora* two or more septate (Brooks 1953).

Rhizoctonia leguminicola Gough and Elliott, the cause of black patch, can damage white clover and other legumes in parts of North America during warm humid weather (Weimer 1950, Leach & Elliott 1951, Gough & Elliott 1956, Berkenkamp 1977). The disease is more serious on red clover, sometimes causing considerable losses in seed production, mainly through seed infection and subsequent damage. The fungus produces a mycotoxin which causes cattle fed on infected red clover hay to salivate excessively (Rainey et al. 1964). Infected leaves develop dark brown, irregular, concentrically zonate spots resembling the so-called target spot of red clover caused by *Stemphylium sarciniforme*. No spores are formed and the fungus spreads by growth of the dark, coarse, superficial mycelium which can become lodged in the seed coat after flower infection.

Seed of *T. repens* can also be infected with *Sclerotinia spermophila* Noble which causes discoloration and reduces germination (Noble 1948). In culture the fungus produces a conidial stage similar to that of *Botrytis anthophila* Bond., which frequently infects red clover flowers but has only once been recorded on white clover (Hardison 1948).

Nishihara (1968) reported that the relatively weak and unspecialised parasite *Choanephora curcurbitarum* (Berk. & Rav.) Thaxter causes a leaf blight of Ladino clover in parts of Japan where summer heat damage occurs. Stolon, petiole and leaf tissue may be affected and killed, eventually becoming covered by a superficial white mycelium. Sporangiophores that emerge through the epidermis have swollen tips covered with elliptical, red-brown, minutely striate sporangia.

The yeast *Rhodotorula glutinis* (Fres.) Harrison var. *rubescens* (Saito) Lodder can cause a leafspot and seed pitting of white and other clovers during prolonged periods with high night humidity (Kilpatrick 1959b). Infected leaves develop chlorotic spots which turn brown but retain an expanding watersoaked margin. Infected seeds tend to be shrivelled and non-viable.

A bacterial leafspot caused by *Pseudomonas syringae* van Hall is found frequently in the USA on white and red clover during wet weather, but causes little damage (Hanson & Kreitlow 1953). Leaf lesions first appear as minute, translucent flecks which enlarge to become irregularly-shaped, blackish-brown spots with a diffuse yellow-green margin. Dead tissue may be shed from drying leaves. Dark, sunken lesions also appear on petioles. The bacteria are exuded from the spots and may be spread by rain, insects or by seed contamination (Jones et al. 1923). In Japan, *P. andropogonis* (Smith 1911) Stapp causes a leafspot on clover which can spread to sorghum and maize, causing a serious leaf streak disease (Goto & Okabe 1965).

3. Virus diseases

(A) White clover mosaic virus (WCMV) Group Potexvirus

White clover in pastures throughout the world may be infected by WCMV (Pratt 1961). The virus was found in 6 of 19 pastures in the southern states of the USA examined by Barnett & Gibson (1975), 6 of 30 examined in eastern Canada (Pratt 1968) and 11 of 30 examined in the UK (Gibbs et al. 1966). In New Zealand, Fry (1959) found few infected plants in young pastures, but in those more than 4 years old often more than 50% of white clover plants were infected.

Information on the effect of WCMV on the productivity of clover in pastures is sparse. In greenhouse tests, Fry (1959) showed that the virus reduced yield by 25% and Tapio (1970) found that one strain of WCMV reduced yield by up to 47%. In caged field plots of healthy and WCMV-infected clones, virus infection reduced yield of Ladino clover by an average of 56% in each of two successive seasons (Kreitlow 1964). Considerable reductions in flower production and seed yield have been reported in WCMV-infected clones used for the production of breeders' seed (Barnett & Gibson 1977). One important aspect of white clover infection by this virus is the effect it has on nodulation by *Rhizobium trifolii*. Guy et al. (1980) report a 71% decrease in the number of nodules in virus-infected plants, although nodule size was unaltered.

Symptoms of WCMV on leaves of naturally infected white clover plants may appear as chlorotic markings along veins or in interveinal areas. Symptoms are most noticeable on vigorously growing plants in spring. However, many infected plants show no symptoms.

The virus can be transmitted easily by sap inoculation, but it is debatable whether aphids are involved in natural transmission. Goth (1962) reported that 10% of white clover clones became infected by WCMV from the aphids *Acyrthosiphon pisum* (Harris), but this has not been confirmed by others (Tapio 1970). Hampton (1963) reported that WCMV in red clover can be transmitted by seed, but there is no evidence of seed transmission in white clover (Barnett & Gibson 1977).

Many species of the Leguminosae are susceptible to WCMV and plants from several other families have been artificially infected. Using small numbers of test plants, Barnett & Gibson (1975) found that only 2 of the 22 species of *Trifolium* they inoculated with WCMV did not become infected. Clark & Barclay (1972) inoculated 4900 white clover plants in the field and found that none was immune to WCMV.

White clover mosaic virus particles are flexuous filaments, 480 nm long and 13 nm wide (Bercks 1971). Several strains of the virus have been noted (Bancroft et al. 1960, Tapio 1970).

(B) Clover yellow mosaic virus (CYMV) Group Potexvirus

This virus was named by Pratt (1961) who recovered two serologically distinct viruses (WCMV and CYMV) from naturally infected white clover showing mosaic symptoms. Johnson (1942) had previously shown that mosaic symptoms on white clover from Washington State were caused by a complex of two viruses which he called pea mottle and pea wilt. Pratt (1961) concluded that pea wilt virus was identical to WCMV as recorded in other countries (Fry 1959, Bos et al. 1960) and that pea mottle, which he renamed CYMV, was a separate virus common only in the western USA. The limited distribution of this virus was confirmed by Agrawal et al. (1962). Clover yellow mosaic virus was not detected in a survey of clover viruses in pastures in the southern USA (Barnett & Gibson 1975), nor was it seen in surveys of Canadian clovers by

Pratt (1968) and Gates & Bronskill (1974). The geographic localisation of this virus may reflect the distribution of its vector, which is as yet unknown.

Systemic symptoms of CYMV infection appear in newly developing leaves of white clover plants 5-10 days after inoculation. Irregular chlorotic flecks and streaks develop parallel to the lateral veins and the leaves sometimes become distorted. Field symptoms are most noticeable in spring and autumn. The severity of symptoms varies with the strain of CYMV present.

This virus is readily transmitted mechanically and by the parasitic dodder plant. No insect vector is known. Hampton (1963) reported that the virus may be transmitted from severely infected red clover plants to their seed, but there is no information about the possibility of seed transmission in white clover.

The effect of CYMV on the yield and persistence of white clover was studied by Pratt (1967). Plants infected with severe strains of CYMV were more prone to winter injury than healthy clovers. Summer hay yields from Dutch white clover plots inoculated 9 months earlier with mild or severe strains of CYMV were 46% and 76% less, respectively, than from virus-free plots.

Few plant species have been found to be naturally infected with CYMV. Pratt (1961, 1967) found the virus in lucerne, sweet clover, *Chenopodium album* L., *Stellaria media* L. as well as several species of *Trifolium*. It has also been found in peas (Ford & Baggett 1965) and apples (Welsh et al. 1973). Clover yellow mosaic virus has been mechanically transmitted to a number of other plant species (Pratt 1961), and Barnett & Gibson (1975) were able to infect all 22 *Trifolium* species they inoculated.

Clover yellow mosaic virus particles are flexuous filaments 539 nm in length (Bercks & Brandes 1963). Several strains differing in host range and severity of symptoms have been found (Pratt 1967, Welsh et al. 1973).

(C) Clover yellow vein virus (CYVV) Group Potyvirus

Hollings & Nariani (1965) first found and described CYVV in white clover in England. Infected plants were sometimes stunted and showed a mild veinal yellowing, especially when they contained other viruses such as WCMV and RCVMV. However, Gibbs et al. (1966) found that in British pastures most white clover plants infected with CYVV were symptomless. Although this virus is common in many pastures, the first experiments determining its effect on herbage yield were not conducted until 1980. Gibson et al. (1980) and Gibson et al. (1981) found that infection with CYVV reduced growth of white clover leaves, stolons and roots, with fewer nodules being formed on infected than on healthy roots. Detrimental effects of CYVV on white clover seed production have been demonstrated by Barnett & Gibson (1977) who found that diseased plants produced fewer flowers.

The survey of viruses in British white clover pastures by Gibbs et al. (1966) revealed that CYVV was the second most common virus, occurring in 18 of 26 fields and in 9% of the plants sampled. However, a recent small survey of white clover plants growing in upland sites in the UK (Scott & Hughes 1980) indicated that CYVV was present in 51 and WCMV in 53 of the 68 plants tested. Clover yellow vein virus was also the most common one infecting white and alsike clovers in Sweden (Lindsten et al. 1976). In the USA Barnett & Gibson (1975) found CYVV in 8 southern States, the highest infection being 47% of plants in one field. Harville & Derrick (1978) also detected CYVV in most white clover fields around Baton Rouge, Louisiana. Pratt (1968, 1969) recovered the virus from chlorotic white clover in eastern and western Canada. He suggested that previous reports of bean yellow mosaic virus in white clover (Houston & Oswald 1953, Kreitlow et al. 1957) were probably infections with CYVV. Both CYVV and BYMV appear to be distantly related, as shown in microprecipitin tests by Varma et al. (1970).

The virus can be mechanically transmitted by sap inoculation and also by the aphids *Acyrthosiphon pisum*, *Aulacorthum solani* (Kltb.), *Macrosiphum euphorbiae* (C. Thomas) and *Myzus persicae* (Sulz.) in a non-persistent manner (Hollings & Nariani 1965, Singh & Lopez-Abella 1971). Seed transmission has not been reported.

Clover yellow vein virus has also been recovered from naturally infected red clover (Pratt 1968) and coriander (Singh & Lopez-Abella 1971). A number of other species have become infected after artificial inoculation. Barnett & Gibson (1975) were able to infect 19 of 22 *Trifolium* species and found resistance to this virus in some lines of white clover. Apparent resistance to CYVV and other viruses was also found by Harville & Derrick (1978) in 'Nolin's Improved' white clover in Louisiana. They thought it likely that natural selection had occurred, as seed used in replanting was always harvested late in the year, after hot weather and viruses had eliminated susceptible plants.

The virus particles are filamentous and flexuous, 767 nm × 12 nm. Many particles are smaller (590-700 nm), probably because they fragment easily (Hollings & Stone 1974). Several strains of CYVV have been reported (Singh & Lopez-Abella 1971). There are no consistent differences in host ranges and symptoms among the many described strains of potyviruses on clovers. Serological tests provide the most reliable way of distinguishing CYVV from other members of this group (Hollings & Stone 1974).

(D) Bean yellow mosaic virus (BYMV) Group Potyvirus

There are many reports of this virus infecting clovers throughout the world and its effects on herbage yield have been considerable (Gibson et al. 1979, Hutton & Peak 1954). Bean yellow mosaic virus has been reported on white clover in the USA by Houston & Oswald (1953) and Kreitlow et al. (1957). However, Pratt (1969) argues that their reports were not substantiated by serological or particle size data. He believes it possible that the virus they were investigating was clover yellow vein virus, which has a distant serological relationship with BYMV (Gibbs et al. 1966). In their survey of pastures in 8 southern States of the USA for viruses in white clover, Barnett & Gibson (1975) were unable to find BYMV infected clover, nor were they able to infect white clover with BYMV, although several other species of clover proved susceptible. However, new techniques for virus detection have led recently to two new reports of BYMV in white clover. McLaughlin et al. (1978) used the ELISA technique to detect the virus in the USA and Scott & Hughes (1980) found BYMV in a few upland-grown white clover plants in the UK when they used a rapid immune electron microscopy method.

The virus is readily transmitted mechanically and by a number of aphid species in a non-persistent manner. Virus particles are long and flexuous, about 750 × 15 nm. A number of strains of the virus have been noted (Bos 1970).

(E) Red clover vein mosaic virus (RCVMV) Group Carlavirus

Red clover vein mosaic virus was described first by Osborn (1937). Infected red clover plants show a vein chlorosis or vein mosaic and this type of symptom has been described for white clover infected with RCVMV in the USA (Roberts 1957, Hagedorn et al. 1959). Gibbs et al. (1966) remarked that white clover plants in the UK naturally infected with RCVMV were symptomless unless also infected with CYVV.

Red clover vein mosaic virus has been reported to occur in white clover in the USA, Canada, Holland, Germany and the UK. In their survey of viruses in white clover from UK pastures, Gibbs et al. (1966) found RCVMV to be the most common disease and that it was present in 62% of pastures and in 12% of all white clover plants surveyed. In contrast, Scott & Hughes (1980) found that only 6% of white clover plants in upland sites in the UK and France were infected with RCVMV. Possibly, the climate was less favourable for the aphid vectors of RCVMV in these high-country sites. Surveys by Pratt (1969) and Gates & Bronskill (1974) indicated that this virus was not as common in Canada as in the UK. The virus was not detected by Barnett & Gibson (1975) in southern US pastures. The effect of RCVMV on yield of white clover has not been investigated, a serious omission considering its frequency of occurrence, particularly in the UK. Yield reductions in red clover have been noted by Denis (1969) who also found that plants infected with RCVMV were more susceptible to root rot fungi. There is hope that resistance to RCVMV can be developed in white clover cultivars, for Kham et al. (1978) discovered that resistance to this virus in red clover is controlled by a single dominant gene.

Red clover vein mosaic virus occurs naturally in plants belonging to the Amaranthaceae, Chenopodiaceae, Leguminosae and Solanaceae families (Varma 1970).

This virus is transmitted in a non-persistent manner by the following aphid species: *Acyrthosiphon pisum, Cavariella aegopodii* (Scop.), *C. theobaldi* (Gill and Bragg), *Therioaphis ononidis* (Kalt.) and *Myzus persicae*. Transmission by aphids is inefficient, but the virus is easily transmitted mechanically. Seed transmission of RCVMV has been reported for faba bean (Sander 1959).

The virus particles are flexuous rods 645 nm long and 12 nm wide. Several strains of the virus have been noted (Varma 1970, Hagedorn and Hanson 1951).

(F) Alfalfa mosaic virus (AMV) Group Alfalfa mosaic virus

White clover in most countries has been found to contain AMV, although Gibbs et al. (1966) were unable to detect it in UK pastures. Generally, the percentage of infected plants in pastures is not high (Pratt 1968, Fry 1952). However, Barnett & Gibson (1975) found that up to 24% of plants were infected with AMV in their survey of white clover pastures in the USA. The effects of virus infection on pasture yield have not been studied. Greenhouse-grown Ladino clover infected with AMV had yield reductions of 33% (Houston & Oswald 1953) and 0-30% (Miller & Halpin 1962) compared with healthy plants. Gibson et al. (1981) also studied greenhouse-grown Ladino clover and found that AMV-infected clover had fewer leaves, stolons and rooting nodes than virus-free plants. Infection with AMV also greatly reduced nodulation, which may explain why Roberts (1956) found that AMV-infected plants were less winter hardy than healthy clovers. Halisky et al. (1960) reported that infected white clover plants had 41% fewer flowerheads than healthy plants, indicating that seed production is affected by AMV. Symptoms develop 12-21 days after inoculation of white clover plants with the virus. Vein clearing and mottling of newly formed leaves are often the first signs of infection. These symptoms later develop into characteristic yellowish angular patches. Stolon internodes often become shortened and leaf petioles can be stunted. Symptoms are most striking at temperatures between 18 and 24°C, but during warmer cloudy weather, symptoms are less obvious (Kreitlow & Price 1949). Some of the many strains of the virus produce only mild mottling on white clover, others cause necrotic lesions (Houston & Oswald 1953).

The virus is readily transmitted mechanically to white clover. A number of aphid species transmit AMV in a non-persistent manner (Kennedy et al. 1962) and transmission by dodder has also been reported (Schmelzer 1956). Seed-transmission and pollen-transmission (Frosheiser 1974) of AMV has been demonstrated for some hosts but not for white clover.

The range of hosts susceptible to AMV after artificial inoculation is extremely wide (Hull 1969) and many species of plants have been found to be infected naturally with the virus. Most legumes are susceptible to AMV and Barnett & Gibson (1975) were able to infect 17 of 22 *Trifolium* species with this virus. They inoculated many white clover lines but only 1-5 plants of each of the other *Trifolium* species was tested. White clover lines varied considerably in the number of plants infected, suggesting that breeding for resistance may be possible. However, some of the five apparently resistant species of *Trifolium* might possibly prove susceptible if more plants were tested, or if more strains of the virus were used as inoculum. The presence of a large number of virus strains may itself create difficulties in breeding resistant white clover.

Alfalfa mosaic virus has up to six component particles which are bacilliform or spherical. The three largest components are essential for infection. Mixtures of the functional components of two different virus strains yield strains that are hybrids of both parents after infection (Bos & Jaspars 1971).

(G) Subterranean clover red leaf virus (SCRLV) [soybean dwarf virus (SDV)] Group Luteovirus

This virus was described first by Kellock (1971) in Australia where he found it to be widespread on subterranean clover. Virus-infected leaves turn bright red as they mature and are smaller than those on healthy plants. Severely affected plants may collapse and die, rendering virus-infected subterranean clover pastures totally unproductive (Anon. 1968).

Subterranean clover red leaf virus has been recorded on white clover and a number of other legumes in New Zealand (Wilson & Close 1973, Ashby 1976, Ashby et al. 1979, Ashby 1980), and on white clover in Tasmania (Johnstone 1978). The disease is completely symptomless in white and red clovers (Ashby et al. 1979). In New Zealand the virus is widespread in white clover and the percentage of infected plants in pastures can be high. Ashby (personal communication) sampled 100 white clover plants in each of 5 pastures and found that 28-40% of the plants were infected with SCRLV. Ashby (1976) has also reported that more than 50% of 5000 single-spaced alsike clover plants less than one year old showed symptoms of SCRLV. Affected plants showed leaf reddening, were smaller than healthy plants and regrowth after grazing was slower. Some plants showed tolerance or resistance to the virus, suggesting that it may be possible to develop resistant cultivars. Although SCRLV reduces the yield of subterranean and alsike clovers, Ashby was unable to detect any effect of the virus on yield of white clover in a glasshouse trial. Johnstone (1978) found a very high level of SCRLV-infected white clover in Tasmanian pastures, but he did not comment on the effects of the virus on plant production.

The virus causes economic damage in pea and faba bean crops in New Zealand and Australia and may be responsible for reduced sugar beet yields. Ashby et al. (1979), Johnstone (1978) and Johnstone & Munro (1980) regard white clover as the main reservoir of SCRLV for infecting these crops. The host range of this virus has not been investigated extensively, but in addition to the hosts already mentioned, the virus has been recovered from naturally infected *Phaseolus vulgaris* L.,*Glycine max* (L.) Merr., *Erodium* spp., *Rumex obtusifolius* L., *Lens esculenta* L., *Medicago sativa* L., 3 species of lupins and 7 species of clovers (Ashby et al. 1979, Wilson & Close 1973). The virus has also been transferred experimentally to several other species of plants (Ashby et al. 1979, Kellock 1971).

Attempts to transmit SCRLV by mechanical means have been unsuccessful. The aphid *Aulacorthum solani* is the main vector, but occasionally *Acyrthosiphon pisum* and *Macrosiphum euphorbiae* have been found to transmit the virus in a persistent manner. *Myzus persicae* appeared to transmit SCRLV from pea to subterranean clover when pea leaf roll virus was also present (Wilson & Close 1973). Kellock (1971) was unable to demonstrate seed transmission of the virus in subterranean clover.

Johnstone et al. (1982) found that isolates of SCRLV from New Zealand and Tasmania appeared identical on the basis of particle morphology and serological tests. Tasmanian SCRLV was distantly related serologically to Californian isolates of beet western yellows and legume yellows viruses. Ashby et al. (1979) discuss the marked relationship between SCRLV, the yellowing strains of soybean dwarf virus (SDV), and filaree red leaf virus. Ashby (personal communication) has now compared antisera of SCRLV and SDV and finds them similar. In view of this and the similarity in vector relationships and host range, he believes that SCRLV and SDV are the same.

The virus particle is isometric and 27 nm in diameter (Johnstone et al. 1982).

(H) Subterranean clover stunt virus (SCSV) Group Luteovirus

Subterranean clover stunt virus is found primarily in subterranean clover and was first described by Grylls & Butler (1956) in Australia. Symptoms on subterranean clover vary with the host cultivar but usually show first as a marginal necrosis of young leaflets which later become cupped and necrotic. Petioles of new leaves remain short and extreme stunting occurs in some varieties. The outer older leaves may become crimson. Early infestation in the field kills some plants, stunts others, and inhibits seed formation. Growth of plants infected at a late stage of growth is not seriously affected, but seed production is reduced.

The disease has not been reported outside Australia, but is widespread therein throughout the southern region. Over 75% of plants in swards have shown disease symptoms (Grylls & Butler 1959). The virus is of economic importance on subterranean clover and efforts have been made to breed resistant varieties (Grylls & Peak 1969).

Subterranean clover stunt virus has also been found naturally in white clover, in many other legumes and in wisteria and dahlia. Disease symptoms on white clover and the other legumes are similar to those described for subterranean clover (Grylls & Butler 1959). Very few of these legumes showed any resistance to the virus.

Subterranean clover stunt virus is transmitted in a persistent manner by four species of aphid: *Aphis craccivora* (Koch), *A. gossypii* Glov., *Macrosiphum euphorbiae* and *Myzus persicae*, the first named being the most important vector (Grylls & Peak 1969). The virus has not been transmitted mechanically nor has seed transmission been demonstrated.

There is no published information on the virus particle. Grylls & Peak (1969) note that there are several strains of SCSV which can give mild or severe stunting in subterranean clover.

(I) Peanut stunt virus (PSV) Group Cucumovirus

The symptoms of PSV on white clover are very distinct. Plants become stunted as the result of reduced elongation of petioles, stolons and roots. Fewer leaves, stems, flowers and root nodules are produced, decreasing yield (Gibson et al. 1981). Infected leaves show mosaic mottling and cup-like rolling of the leaflets which often proliferate to produce 4-6 leaflets. Choopanya (1968a) observed that symptoms were most severe at 30°C but that reductions in yield were greatest at 17-23°C, at which temperatures the virus-free control plants grew best.

Peanut stunt virus was described first on peanuts by Miller & Troutman (1966). Hebert (1967) discovered the virus in white clover in North Carolina, USA and he considered these infected plants to act as a reservoir host between successive crops of peanuts, as 50% of the clover he sampled contained PSV. Similarly, in Virginia in the USA, Tolin et al. (1970) found that 65% of white clover plants collected near peanut fields were infected with this virus. Barnett & Gibson (1975) sampled white clover pastures in 8 southern states of the USA and found the virus in 14 of 19 pastures. The percentage of plants infected in any one pasture varied from 0 to 74%, but the average was 21%. Choopanya (1968b) recovered the virus in all 42 counties of South Carolina where white clover is grown, and Lucas & Harper (1972) found PSV in 45% of white clover samples in their North Carolina survey. It thus appears that PSV is widespread and common on white clover in the southern USA but that it has not been reported elsewhere on this species.

The virus has a wide host range and can be transmitted to many dicotyledonous species (Mink 1972). Barnett & Gibson (1975) were able to infect 17 of 22 *Trifolium* species with PSV. They found a few clones of white clover which could not be infected, so that it may be possible to develop resistant varieties if need be.

Peanut stunt virus is non-persistent and is transmitted by the aphids *Aphis craccivora, A. spiraecola* (Patch) and *Myzus persicae* (Hebert 1967). Virus can be acquired by aphids in less than one minute and then inoculated in less than one minute (Mink 1972). It can also be transmitted mechanically in infected sap and by dodder (Miller & Troutman 1966). No transmission of the virus through white clover seed has been reported, although a low seed transmission rate occurs in peanuts (Troutman et al. 1967).

Peanut stunt virus has a multicomponent genome; the particles are isometric and about 30 nm in diameter. In physical and chemical properties it belongs to the same group as cucumber mosaic virus and tomato aspermy virus (Mink 1972).

(J) Red clover necrotic mosaic virus (RCNMV) Group Carnation ringspot

This virus causes a serious disease of red clover and infected plants show severe leaf distortion, mottling and necrosis. Red clover plants can be greatly stunted and may not survive the winter (Scott & Davies 1979). Hollings & Stone (1977a) report that they artificially infected a wide range of plants and note the existence of several different virus strains.

White clover was thought to be immune to RCNMV (Gerhardson & Lindsten 1973). However, Lynes et al. (1981) report that a virus isolated from white clover in Victoria, Australia in 1965 has recently been identified as RCNMV. This virus strain probably has a wide host range, for it infected 16 out of 17 species in 6 plant families. The virus was systemic in white clover but produced no symptoms. Not all attempts to infect white clover plants with this strain were successful and Lynes et al. (1981) believe this was because high temperatures and high light intensity inhibited infection. The virus has also been detected recently in white clover plants in New Zealand (R.A. Forster, personal communication).

Red clover necrotic mosaic virus is readily transmitted mechanically, but attempts to transmit it with insects and nematodes or through seed from infected red clover plants have been unsuccessful (Bowen & Plumb 1979). The virus has been transmitted experimentally from diseased to healthy red clover plants through soil. A vector relationship between the chytrid fungus *Olpidium* and the virus was suspected, and MacFarlane (1981) describes experiments which suggest that *O. radicale* may be the vector. This virus has a bipartite genome, particles being isometric and about 34 nm in diameter (Gould et al. 1981).

Several new virus diseases of white clover have been described recently. They are as follows:

(K) Clover primary leaf necrosis virus (CPLNV)

A red clover plant showing yellow leaf mottling and growing on a roadside in Vancouver, Canada, was the original source of this virus (Ragetli & Elder 1977). Sampling of red, white and alsike clover fields in British Columbia revealed that a few plants were infected with CPLNV. However, the diseased condition of the plants and the absence of flowers prevented absolute identification. Infected white clover leaves showed interveinal chlorosis and the leaf margins were yellow. When young white clover plants were mechanically inoculated with infected sap, the primary leaves developed necrotic lesions. Fifteen out of 43 plant species were mechanically infected with purified virus. No insect vector has been found. Virus particles are isometric and 36 nm in diameter.

(L) Clover yellows virus (CYV) Group Closterovirus

A rod-shaped virus has been found in white and crimson clovers in Japan (Ohki et al. 1976). Leaves of infected plants showed mild yellowing with marginal reddening on older leaves and plants were slightly stunted. The virus was transmitted to white clover and faba bean by *Aphis craccivora* Koch. Attempts to transmit the virus mechanically to several common virus-indexing plants were unsuccessful. Virus particles were very flexuous rods, 12 nm wide and 1700 nm long with helical symmetry 3.7 nm in pitch. These particles were found only in phloem cells, often in great numbers, where they caused a phloem necrosis.

(M) White clover streak mosaic virus (WCSMV)

Babovic (1975, 1978) in Yugoslavia isolated a virus from white clover plants which showed dark-green vein banding and interveinal chlorosis. The disease was mechanically transmitted to young white clover seedlings and to several common virus indicator plant species. Virus particles were isometric and 30 nm in diameter.

(N) Clover blotch virus (CBV)

This virus, first found in white clover in Czechoslovakia in 1970, produces irregular chlorotic blotches on leaves of diseased plants(Musil & Leskova 1973). Clover blotch virus is transmitted by the aphids *Acyrthosiphon pisum* and *Myzus persicae* in a non-persistent manner. It has also been mechanically transmitted to 6 species of clover and to members of several other plant families. The virus has isometric particles 26-28 nm in diameter.

Numerous viruses have been transmitted experimentally to white clover in the course of host range studies. However, few of these viruses have been isolated from white clover plants growing in pastures. Where present in field-grown clover, these viruses may be important, not so much for their effect on clover growth, but as a source of virus infecting valuable cash crops such as tobacco (Fulton 1948) or strawberries (Harrison & Winslow 1961).

All of the following viruses which have been reported as occurring occasionally in white clover are described in some detail in the C.M.I/A.A.B. Descriptions of Plant Viruses. The reference to this description is given at the end of the section on each virus disease.

(O) Strawberry latent ringspot virus (SLRSV) Group Nepovirus

This virus was found in two fields in the south-west region of the UK by Gibbs et al. (1966) in their clover virus survey. The few plants infected with SLRSV also contained arabis mosaic virus. Both viruses are transmitted by the nematode *Xiphinema diversicaudatum* (Micol.) Thorne. (Virus description, Murant 1974).

(P) Arabis mosaic virus (ArMV) Group Nepovirus

Harrison & Winslow (1961) reported that pastures in Devon, England, often contained white clover infected with ArMV. Infected plants had chlorotic blotches on their leaves and these diseased plants often occurred in patches in fields. Arabis mosaic virus is transmitted by dagger nematodes and these patches of virus-infected clover coincided with high populations of dagger nematodes in the soil (Harrison & Cadman 1959). Gibbs et al. (1966) found 4% of white clover plants in Britain were infected with ArMV. Only 8 of the 26 fields they sampled contained plants with ArMV, but all infected plants came from fields in the west and north of the country. (Virus description, Murant 1970b).

(Q) Tobacco ringspot virus (TRSV) Group Nepovirus

Tobacco ringspot virus has been isolated from white clover plants in the USA in Indiana (Tuite 1960) and in North Carolina (Lucas & Harper 1972). Infected plants were symptomless and rare. Lucas & Harper (1972) noted that only 2 of the 300 white clover samples taken in their survey were infected with TRSV. (Virus description, Stace-Smith 1970).

(R) Tomato black ring virus (TBRV) Group Nepovirus

The survey by Gibbs et al. (1966) revealed one symptomless white clover plant in Scotland which was infected with tomato black ring virus. (Virus description, Murant 1970a).

(S) Tomato spotted wilt virus (TSWV) Group Tomato spotted wilt

There appears to be only one record of TSWV occurring naturally in white clover. Paliwal (1974) isolated the virus from diseased white clover in a field in southern Ontario, Canada. These plants showed pronounced chlorotic and necrotic leaf spots and mosaic patterns. A thrip-transmitted virus with membrane-bound particles, it causes serious diseases in vegetables, tobacco and ornamentals. (Virus description, Ie 1970).

(T) Pea Streak Virus (PSV) Group Carlavirus

Pratt (1968) in a survey of clover fields in Ontario, Quebec and the Maritime provinces of Canada detected PSV in a few white clover fields in Ontario and New Brunswick. Pea streak virus does not appear to have been found in any other virus surveys of white clover. (Virus description, Bos 1973).

(U) Tobacco streak virus (TSV) Group Ilarvirus

This is another virus which appears to occur only rarely in white clover. Fulton (1948) found it in a few symptomless white clover plants growing near fields of tobacco in the USA in Wisconsin. He was able to transmit the virus by sap inoculation to 169 plant species, including 4 clovers. (Virus description, Fulton 1971).

(V) Cucumber mosaic virus (CMV) Group Cucumovirus

Cucumber mosaic virus was isolated from white clover plots at Rothamsted, England, over 40% of plants in one plot being infected (Govier & Cockbain 1977). After some initial difficulties, Govier (1979) was able to transmit the virus with *Myzus persicae* from infected clover to young seedling clovers. This virus was also found in white clover growing in Yugoslavia (Taraku et al. 1977). This isolate was readily sap-transmissible as well as being transmitted by *M. persicae*. (Virus description, Francki et al. 1979).

(W) Cymbidium ringspot virus (CyRSV) Group Tombusvirus

Cymbidium ringspot virus was obtained from white clover growing in a pasture in England. Infected plants showed pale green flecks and some younger leaves were distorted. Hollings et al. (1977) claimed that CyRSV was not common in white clover and their record appears to be the only one of this virus on clover. The other natural host of CyRSV is Cymbidium orchid (Hollings et al. 1977). Cymbidium ringspot virus has been transmitted by inoculation of sap to a wide range of hosts and the clover strain transferred to 64 plant species in 22 families. (Virus description, Hollings & Stone 1977b).

4. Diseases caused by mycoplasma-like, spiroplasma-like and rickettsia-like organisms

A number of diseases of white clover which were originally thought to have been caused by viruses are now considered to be caused by mycoplasma-like organisms (MLO), spiroplasma-like organisms (SLO) and rickettsia-like organisms (RLO). Examples of this type of disease which affect white clover are phyllody and witches' broom (Frazier & Posnette 1957), legume little leaf (Hutton & Grylls 1956), stolbur, parastolbur, and clover dwarf (Valenta & Musil 1963), clover proliferation (Chiykowski 1965), clover yellow edge (Chiykowski 1976), rugose leaf curl (Grylls 1954) and clover club leaf (Windsor & Black 1973a). As more information about symptoms, host range and transmission of these diseases became known, it was apparent that several of the disease agents were closely related if not identical (Bos & Grancini 1965).

Great interest was aroused when Doi et al. (1967) described mycoplasma-like bodies in sieve elements of diseased plants with four different yellows-type symptoms. In a companion paper, Ishiie et al. (1967) described the temporary remission of disease symptoms in these plants when tetracycline antibiotics, but not penicillin, were applied to them (virus diseases do not normally respond to antibiotics). These papers stimulated considerable research into the role of MLO and RLO in other plant diseases of unknown aetiology.

The mycoplasmas are the smallest and simplest self-replicating procaryotes. Unlike all other procaryotes they have no cell wall, being bounded by a single, triple-layered membrane. They are highly pleomorphic, coccoid to filamentous, with a tendency to form branched myeloid structures.

All MLO and half the RLO occur only in the phloem cells of their plant hosts; the remaining RLO are found only in the xylem. Translocation is disrupted in diseased plants and callose formation, together with degeneration of phloem tissue, may result in stunting, wilting or leaf yellowing. Witches' broom formations in shoots and phyllody of flowers can result from premature development of adventitious buds and from altered apical dominance caused by changes in the levels of growth regulators.

Because these organisms are restricted to the vascular system in plants, it is not surprising that the vectors are insects that feed on plant sap. Most vectors are leafhoppers (cicadellids), although cercopid, fulgorid and psyllid bugs can also be vectors. The organisms usually have an incubation period of several weeks while they multiply within the insect, which then becomes infective for the rest of its life. Experimental transmission may sometimes be accomplished by grafting or by dodder, but never mechanically with sap from diseased plants.

It is now suspected that MLO, SLO and RLO cause all the clover diseases described by Bos & Grancini (1965) as being of virus origin (Bowyer et al. 1969, Lombardo et al. 1970). Until it is possible to grow all the causal organisms of these diseases in pure culture, their differentiation must rely on host range and symptomatology.

Valenta (1964) studied over 200 isolates of yellows agents and distinguished four entities. These were named clover phyllody, clover dwarf, stolbur and parastolbur. Ploaie & Maramorosch (1969) found MLO present in plants suffering from these diseases. The most common mycoplasma disease in white clover is clover phyllody. This disease as well as several other mollicute diseases of white clover will be discussed.

(A) Clover phyllody

Phyllody is common in white clover and several other *Trifolium* species in many parts of the world, although it has not been reported in New Zealand. The host range is wide. Chiykowski (1974) noted that 79 species in 22 families were attacked and further testing will no doubt add many more susceptible species.

The most characteristic symptom of the disease is the phyllody affecting most or all of the flowers in the inflorescence. These are modified into leafy structures which do not produce seed. Leaves may show veinal necrosis and older leaves are mildly bronzed. Infected plants are slightly stunted, with some loss of vegetative vigour and are more prone to winter death than healthy plants.

Gaborjanyi & Sziraki (1978) noted lower cytokinin levels in leaves infected with phyllody and they suggested that mycoplasma infection causes a hormonal imbalance in the plant. Symptom development is affected by temperature. Nakamura et al. (1978) found that at high temperature (25°C day, 15°C night) 76% of flowers were abnormal, whereas at low temperature (18°C, 13°C) only 10% were. They observed that multiplication of the MLO was greatly depressed at temperatures below 18°C.

Clover phyllody was found in at least 40% of white clover seed crops in Britain (Carr and Large 1963), although it was uncommon for more than 10% of heads in a crop to be affected. Creelman (1961) noted that as many as 50% of plants were infected in some portions of a clover field in Canada.

MLO occur in mature sieve elements in all parts of diseased plants. Sinha (1976) found that these MLO showed a great diversity of form and size, the predominant type being spherical, 65-500 nm in diameter. Other forms present were filamentous, dumbell-shaped, or formed chains probably through replication by budding and/or binary fission.

Several workers have claimed to have cultured plant pathogenic mycoplasmas on synthetic media, but it appears that in most cases the mycoplasmas which grew were contaminants. Giannotti & Giannotti (1978) now claim to have isolated the MLO agent of phyllody. In culture it could take a spiral or non-spiral form depending on the technique of growing it. A second MLO was also isolated from clover phyllody plants. This MLO was not of the spiral type, but induced yellowing, proliferation, stunting and death of plants.

Many mycoplasmas are themselves infected by viruses and the phyllody MLO is no exception. Using electron microscopy, Gourret et al. (1973) saw rod-shaped virus-like particles associated with MLO in salivary glands of infected leafhoppers and in a few phloem cells in root nodules of white clover. Virus-like particles were also found as a rosette around apparently degenerated MLO.

At least 12 species of leafhoppers are able to transmit the clover phyllody MLO from diseased to healthy plants and they have been shown to multiply within the vector (Gouranton & Maillet 1973). The incubation period between leafhopper feeding and the onset of disease symptoms was recorded as 4-6 weeks by Frazier & Posnette (1957). However, more recent research has shown that the duration of the latent period is temperature-dependent, becoming longer when the temperature is lowered (Vago & Giannotti 1972). They reported a minimum threshold period of only 12-15 days. Transovarial passage of the phyllody organism has been reported by Posnette & Ellenberger (1963), but this has not yet been substantiated.

Phyllody can be transmitted experimentally by dodder and grafting (Frazier & Posnette 1957). Kleinhempel et al. (1975) were unable to demonstrate seed transmission when second-generation seedlings from late-infected plants were examined.

An important observation by Vanderveken (1964) was that roots of diseased plants were stunted and contained many whitish-yellow ineffective nodules. Joshi & Carr (1967) confirmed this and found that fresh seed, sown in soil that previously contained phyllody diseased plants, germinated poorly. These seedlings developed mainly small nodules and the rhizobia seemed less effective. Gross et al. (1974) attempted to repeat this work but were unable to detect a heritable alteration to the rhizobia in phyllody-infected plants. Although phyllody may not affect subsequent generations of rhizobia in clover roots, it does appear to reduce the efficiency of bacteroids in root nodules (Smets et al. 1977).

Little work has been done on the control of clover phyllody. As with most mycoplasma-induced diseases, tetracyclines delay the expression of disease symptoms but cannot cure the disease (Chiykowski 1973). Carr (1965) found that diseased plants lacked competitive ability and were replaced by healthy plants in the sward. Provided that extensive reinfection does not occur, seed yield can therefore be restored to that of uninfected swards. Immune clover plants have not been found, although some genotypes show only a slight loss of vigour after infection. Thus, the opportunity exists to breed for tolerance to the disease.

The three other types of yellows diseases distinguished by Valenta (1964) are found mainly in central Europe. The infectious agents can be transmitted to a wide range of plant species by leafhoppers, and experimentally by dodder or grafting. Disease symptoms in these plants may vary, depending on the strain of pathogen transmitted. Symptoms of these diseases in white clover are as follows (Valenta et al. 1961, Valenta & Musil 1963):

(i) Clover dwarf

Infected plants have shortened internodes and axillary buds proliferate into short shoots bearing leaves which are very much reduced in size. Older leaves become bronzed in summer. Flowers may not be formed and if present they are very rudimentary. Many infected plants die during winter.

(ii) Stolbur

All parts of infected white clover plants are reduced in size. The flowers appear almost normal but in fact are sterile. As with the other yellows diseases, the severity of symptoms depends on the strain of organism infecting the plant.

(iii) Parastolbur

There is a marked proliferation of axillary buds into short shoots. Leaf stalks are rather long and thin at first but later become much shorter than in healthy plants. Leaves are reduced in size, malformed and folded downwards. Inflorescences have fewer flowers and flower stalks are longer than in normal plants. Individual flowers may be virescent or phylloid.

(B) Clover yellow edge (CYE)

A disease agent, thought to be a mycoplasma, which causes yellowing of leaf margins on new clover leaves, has been described by Chiykowski (1969, 1976, 1981). The disease has been found in Canada throughout eastern Ontario on white, alsike and red clovers, although its incidence is low (Chiykowski & Bolton 1970).

The disease agent is transmitted by the leafhopper *Aphrodes bicinctus* (Schrank). Chiykowski (1976, 1981) was able to transmit the Clover Yellow Edge agent to 15 plant species in 6 families, most of the legumes tested being highly susceptible, symptoms appearing within 14 weeks of leafhopper feeding. Organisms resembling mycoplasma, and with considerable pleomorphism, were seen in petioles, leaves, roots and root nodules of Ladino clover infected with CYE. They also occurred in the intestine of infective leafhoppers. Many plants showed a mild chlorosis followed by moderate to severe plant stunting. New leaves and flowers were smaller than normal, but flowers were never virescent or phylloid.

Although CYE has some characteristics common to other leafhopper-transmitted diseases, Chiykowski (1976) gives reasons for believing it distinct from other yellows diseases.

(C) Clover club leaf

This disease has been described in North America (Windsor & Black 1973a), and white clover exhibiting similar symptoms has been noted in England (Markham et al. 1975), France (Benhamou et al. 1978), and in Canada (Benhamou & Sinha 1981). The disease agent was shown to be transmitted through the egg of its leafhopper vector (Black 1948), and to multiply within the insect (Black 1950). The pathogen has also been transmitted from diseased to healthy clover plants through dodder, in which multiplication of the pathogenic organism takes place (Benhamou et al. 1978, Benhamou & Sinha 1981).

White clover plants affected by this disease produce long petioles bearing dwarfed leaflets with chlorotic margins. Typically, one leaflet is smaller than the other two and severely affected leaves are twisted or clubbed. At high temperatures (23-30°C), Markham et al. (1975) found that plants reddened and died, while plants kept at 15°C or below survived.

The causal agent of clover club leaf is considered to be an RLO (Windsor & Black 1973b, Markham et al. 1975, Benhamou & Sinha 1981). The organism associated with the disease was localised in the phloem sieve tubes and companion cells of leaves, petioles, stems and roots of infected plants. These RLO were 1000-2000 nm × 200-300 nm and bounded by two trilaminar membranes, each about 8 nm thick.

Treatment of infected plants with penicillin produced a remission of symptoms (Windsor & Black 1973b). Penicillin is known to affect the synthesis of cell walls in bacteria and so this remission of symptoms in plants affected by clover club leaf provides further grounds for suspecting the agent to be an RLO. The ultimate proof will come when the organism can be cultured.

(D) Rugose leaf curl (RLC)

Rugose leaf curl was first recognised during laboratory experiments in which the leafhopper *Austrogallia torrida* (Evans), collected from a clover and a lucerne pasture in Australia, was found to transmit symptoms of a previously unknown disease to several test plants (Grylls 1954).

On white clover the symptoms of disease are stunting, marginal chlorosis, rugosity and curling of leaves. Anthocyanin pigmentation may be present in the older leaves. Often the whole plant is affected and some plants may die, but occasionally only small portions of the plant show symptoms. Plants may recover and appear healthy, although disease symptoms may reappear later.

Surveys on the incidence of RLC by Grylls et al. (1972) revealed that newly established pastures were affected most severely. Serious yield reductions occurred in white clover and red clover, and lucerne showed poor persistence. In fields where the disease had been noted previously as severe, many of the plants appeared to have regenerated and showed no symptoms. Grylls & Day (1966) described a field of young white clover in which the proportion of plants infected with RLC had declined from 90-100% in late autumn to only 10-15% by the following spring. The disease had almost disappeared by autumn and recovery continued for the next 3 years. This interesting phenomenon of recovery from infection was studied in the greenhouse (Grylls & Day 1966) and in the field by Grylls et al. (1972). They concluded that healthy plants arose from stolons which, although developing from severely affected parent plants, were separated from them by a necrotic barrier across which the pathogen was unable to move. The recovery of diseased plants contributed significantly to regeneration of clover species in pastures after outbreaks of RLC infection. Only a small proportion of regenerated plants became reinfected. The initial high rate of infection, subsequent recovery, and the development of apparently resistant plants from recovered stolons were also noted by Jones (1973) in field studies with *Trifolium semipilosum*.

Rugose leaf curl pathogen is transmitted by the leafhopper *A. torrida* and an interesting feature is that it is also transmitted through the egg of this vector. Grylls (1954) was able to infect plants by grafting them with infected material but not by inoculation with infected sap or by dodder transmission.

It was thought originally that RLC was caused by a virus, and Grylls et al. (1974) reported that they had seen virus particles in thin sections of salivary glands of *A. torrida* which had been reared on RLC-infected clover. They also found virus particles in partially purified extracts of diseased clover, but were unsuccessful in detecting virus in sections of infected tissue. More recently, however, Behncken & Gowanlock (1976) found that a bacterium-like organism was consistently present in thin sections of tissues taken from diseased clover plants. No similar bodies were seen in healthy clovers, nor were virus particles found in either healthy or diseased plants. The bacteria-like bodies were restricted to phloem sieve tubes and phloem parenchyma cells. The organism was bounded by a cell wall and plasma membrane and measured approximately 1000-2000 nm × 200 nm. Behncken & Gowanlock treated infected plants with penicillin and obtained remission of symptoms for several weeks, which, together with their observation of bacteria-like bodies in infected clovers, indicated strongly that the causal agent was not a virus.

Several RLO have now been cultured, but attempts to culture the organism from plants infected with RLC have as yet been unsuccessful. Until this has been achieved and Koch's postulates completed, an etiological role cannot be assigned to this pathogen.

5. General Discussion

Though we now have much information about white clover diseases, pasture pathologists are rarely able to give satisfactory answers to the questions most often asked by farmers, plant breeders and pasture ecologists who want to know, for instance:

1. Do particular diseases cause sufficient loss of productivity to justify a change of management, an extensive breeding programme or fungicide treatment?
2. What times of year do diseases cause most damage or accumulation of oestrogens and do these correspond to critical periods in the cycle of clover growth and animal fertility?
3. To what extent do natural selection and compensatory growth within and among component pasture species nullify or intensify the effects of disease damage?
4. What cumulative effects on clover growth do the interaction of diseases, pests and environment cause?

The many surveys of clover diseases in different countries provide qualitative information about the distribution of pathogens, but rarely give a guide to seasonal variations or those brought about by different management practices. Where quantitive data have been obtained it is often impossible to correlate these with other data to show the relationship between the amount of disease and yield loss. The approach of Garren (1954, 1955), who studied the role of a succession of 'pathogen societies' in the seasonal 'damage growth cycle' of white clover, deserves to be adopted more widely. Unfortunately, his data did not include root pathogens or obligate leaf parasites such as rust and sooty blotch. Recent studies on the seasonal incidence and causes of damage to white clover foliage in some New Zealand pastures (R.A. Skipp & M.G. Lambert unpublished) have suggested that the predominant fungal diseases in grazed pastures and breeders' plots differed. Breeders' plots, therefore, may not be the best places to assess the relative importance of diseases, or find resistant material. Carefully conducted surveys could do much to increase our knowledge of the real effect of disease in pastures, but this type of work can cost much time and labour.

The figures quoted by pathologists for probable losses in yield have often been obtained from greenhouse studies, so that while they are often impressive, they may not be very realistic representations of the pasture situation. The genetic heterogeneity of clover plants within swards extends to factors controlling disease resistance as well as agronomic characteristics. Furthermore, other pasture species, including closely related legumes, as well as grasses and weeds, may be completely immune to the specific biotype attacking white clover. Thus one might expect that resistant clover plants and other species would, by virtue of their increased competitiveness, compensate for the lost productivity of the diseased plants. However, the balance of species in the pasture could be affected adversely by such changes, which could have long-term repercussions on total production.

The amount of damage done by disease can be modified by management practices. For example, frequent defoliation would be expected to reduce epidemics of foliage pathogens but may increase root disease by imposing stress on the plant and reducing its capacity for vigorous root growth. It must also be remembered that clover roots are naturally invaded by *Rhizobium* and endomycorrhizal fungi, whereas much greenhouse work is carried out in sterile soil supplied with a complete nutrient mixture. Thus there are compelling reasons to study the effect of diseases in the field under typical conditions.

Valuable information about the effects of disease on pasture growth can be obtained by using simulated swards (Catherall 1966) in which, for instance, virus-free and virus-infected (or resistant and susceptible) clover plants are established among other pasture species and the plots defoliated at intervals typical of normal management practice. Screening clover germplasm for resistance to common diseases may be worthwhile as a means of finding resistant plants which can be used as tools in this type of study. Fungicides of known specificity can also be useful (Lancashire & Latch 1966).

The abundant information now available about the causes of root and stolon rotting diseases underlines the interaction between the stresses imposed by environment and management and the damage done by disease. The more that pasture agronomists and plant pathologists can work together to study problems such as poor persistence, the more likely it is that a meaningful appreciation of their cause will emerge (e.g. Halpin et al. 1963). The study of the interactions between members of the natural microflora of clover is an enormous and worthwhile study in its own right.

Pasture pathologists are far less numerous than those employed on arable crops yet they often have to deal with several plant species, large numbers of pathogens and a complex agronomic situation. There is now a need for imaginative new programmes to answer some of the remaining and fundamentally important questions of clover pathology.

References

Agnihothrudu, V. (1962) Notes on fungi from northeast India XVII. *Menisporella assamica* gen.et sp.nov. *Proceedings, Indiana Academy of Science*, Section B, **56**, 97-101.

Agrawal, H.; Bos, L.; Chessin, M. (1962) Distribution of clover yellow mosaic and white clover mosaic viruses on white clover in the United States. *Phytopathology* **52**, 517-519.

Åkerberg, E. (1958) Årsberättelse över Sveriges Utsädesforenings verksamhet år 1957. *Sveriges Utsädesförenings Tidskrift* **68**, 99-183.

Albrecht, H. (1942) Effect of diseases upon survival of white clover *Trifolium repens* L., in Alabama. *Journal of the American Society of Agronomy* **34**, 725-730.

Aldrich, D.T.A.; Doling, D.A. (1967) Varietal resistance to clover rot in white clover. *Nature, UK* **214**, 946-947.

Amelung, D. (1966) Krankheitserreger und Schädlinge auf Futterpflanzen bzw. Futtermitteln als Ursache für Gesundheitsstörungen bei Haustieren. *Wissenschaftliche Zeitschrift der Universität Rostock* **15**, 245-249.

Anon. (1968) Red-leaf - a new virus disease that can make subterranean clover totally unproductive. *Journal, Department of Agricultural, Victoria* **66**, 182-184.

Arthur, J.C. (1934) Manual of the rusts in United States and Canada. Lafayette, Indiana, USA; Purdue Research Foundation, 438 pp.

Ashby, J.W. (1976) Subterranean clover red leaf virus and bean yellow mosaic virus in alsike clover. *New Zealand Journal of Agricultural Research* **19**, 373-376.

Ashby, J.W. (1980) Virus diseases of annual legume crops. *Proceedings Agronomy Society of New Zealand* **10**, 77-80.

Ashby, J.W.; Teh, P.B.; Close, R.C. (1979) Symptomatology of subterranean clover red leaf virus and its incidence in some legume crops, weed hosts, and certain alate aphids in Canterbury, New Zealand. *New Zealand Journal of Agricultural Research* **22**, 361-365.

Aubé, C. (1966) Pathogenicity of *Fusarium* species from alsike clover. *Canadian Plant Disease Survey* **46**, 11-13.

Babović, M.V. (1975) Prugasti mozaik, novo virozno oboljenje bele deteline u Jugoslaviji *Zastita Bilja* **26**, 199-204.

Babović, M.V. (1978) Prečišćavanje i neka svojstva virusa prugastog mozaika bele deteline. *Acta Biologica Iugoslavica* **B 15**, 175-179.

Bancroft, J.B.; Tuite, J.; Hissong, G. (1960) Properties of white clover mosaic virus from Indiana. *Phytopathology* **50**, 711-717.

Barnett, O.W.; Gibson, P.B. (1975) Identification and prevalence of white clover viruses and the resistance of *Trifolium* species to these viruses. *Crop Science* **15**, 32-37.

Barnett, O.W.; Gibson, P.B. (1977) Effect of virus infection on flowering and seed production of the parental clones of Tillman white clover (*Trifolium repens*). *Plant Disease Reporter* **61**, 203-207.

Barrière, Y.; Massenot, M.; Raynal, G. (1974) Les maladies des légumineuses fourragères. II. Les maladies du feuillage taches foliaires à *Leptosphaerulina* (pepper spot). *Annales de Phytopathologie* **6**, 341-347.

Bayliss-Elliott, J.S.; Stansfield, O.P. (1924). The life history of *Polythrincium trifolii* Kunze. *Transactions of the British Mycological Society* **9**, 218-228.

Behncken, G.M.; Gowanlock, D.H. (1976). Association of a bacterium-like organism with rugose leaf curl disease of clovers. *Australian Journal of Biological Sciences* **29**, 137-146.

Benhamou, N.; Giannotti, J.; Louis, C. (1978) Transmission de germes de type rickettsöide par la plante parasite: *Cuscuta sabinctusa* Durr. et Hilg. *Acta Phytopathologica Academiae Scientiarum Hungaricae* **13**, 107-119.

Benhamou, N.; Sinha, R.C. (1981) Association of a rickettsia-like organism with a disease of white clover. *Canadian Journal of Plant Pathology* **3**, 191-196.

Bercks, R. (1971) White clover mosaic virus. *C.M.I./A.A.B. Descriptions of Plant Viruses*, No. 41.

Bercks, R.; Brandes, J. (1963) Elektronenmikroskopische und serologische Untersuchungen zur Klassifizierung des Clover yellow mosaic Virus. *Phytopathologische Zeitschrift* **47**, 380-390.

Berger, R.D.; Hanson, E.W. (1963a) Pathogenicity, host-parasite relationships, and morphology of some forage legume Cercosporae, and factors related to disease development. *Phytopathology* **53**, 500-508.

Berger, R.D.; Hanson, E.W. (1963b) Relation of environmental factors to growth and sporulation of *Cercospora zebrina*. *Phytopathology* **53**, 286-294.

Berkenkamp, B. (1977) Blackpatch of forage legumes. *Canadian Plant Disease Survey* **57**, 65-67.

Bjorling, K. (1942) Undersökningar rörande Klöverrötan. II. Studies av utvecklingshistoria och variation hos *Sclerotinia trifoliorum*. *Meddelanden Från Statens Växtskyddsanstalt, Stockholm* **37**, 154 pp.

Black, L.M. (1948) Transmission of clover club-leaf virus through the egg of its insect vector. (Abstr.) *Phytopathology* **38**, 2.

Black, L.M. (1950) A plant virus that multiplies in its insect vector. *Nature, UK* **166**, 852-853.

Blake, C.T.; Chamblee, D.S.; Woodhouse, W.W. (1966) Influence of some environmental and management factors on the persistence of Ladino clover in association with orchard grass. *Agronomy Journal* **58**, 487-489.

Booth, C. (1971) The genus *Fusarium*. Kew, U.K.; Commonwealth Mycological Institute, 237 pp.

Booth, C.; O'Rourke, C. (1973) *Cymadothea trifolii*. *C.M.I. Descriptions of Pathogenic Fungi and Bacteria*, No. 393.

Booth, C.; Pirozynski, K.A. (1967a) *Leptosphaerulina trifolii*. *C.M.I. Descriptions of Pathogenic Fungi and Bacteria*, No. 146.

Booth, C.; Pirozynski, K.A. (1967b) *Pleospora herbarum*. *C.M.I. Desciptions of Pathogenic Fungi and Bacteria*, No. 150.

Bos, L. (1970) Bean yellow mosaic virus. *C.M.I./A.A.B. Descriptions of Plant Viruses* No. 40.

Bos, L. (1973) Pea streak virus. C.M.I./A.A.B. Descriptions of Plant Viruses No. 112.

Bos, L.; Matt, D.Z.; Bancroft, J.B.; Gold, A.H.; Pratt, M.J.; Quantz, L.; Scott, H.A. (1960) Serological relationship of some European, American, and Canadian isolates of the white clover mosaic virus. *Tijdschrift Plantenziekten* **66**, 102-106.

Bos, L.; Grancini, P. (1965) Some experiments and considerations on the identification of witches' broom viruses, especially in clovers, in the Netherlands and Italy. *Netherlands Journal of Plant Pathology* **71**, Suppl. 1, 20 pp.

Bos, L.; Jaspars, E.M.J. (1971) Alfalfa mosaic virus. *C.M.I./A.A.B. Descriptions of Plant Viruses*, No. 46.

Bowen, R.; Plumb, R.T. (1979) The occurrence and effects of red clover necrotic mosaic virus in red clover (*Trifolium pratense*). *Annals of Applied Biology* **91**, 227-236.

Bowyer, J.W.; Atherton, J.G.; Teakle, D.S.; Ahern, G.A. (1969) Mycoplasma-like bodies in plants affected by legume little leaf, tomato big bud , and lucerne witches' bro diseases. *Australian Journal of Biological Sciences* **22**, 271-274.

Brooks, F.T. (1953) Plant Diseases; 2nd Edition. O ford, UK; Oxford University Press, 457 pp.

Butler, G.W.; Greenwood, R.M.; Soper, K. (1959) Effects of shading and defoliation on the turnover of root and nodule tissue of plants of *Trifolium repens*, *Trifolium pratense*, and *Lotus uliginosus*. *New Zealand Journal of Agricultural Research* **2**, 415-426.

Camp. R.R.; Whittingham, W.F. (1972) Host-parasite relationships in sooty blotch disease of white clover. *American Journal of Botany* **59**, 1057-1067.

Campbell, C.L. (1980) Root rot of ladino clover induced by *Codinaea fertilis*. *Plant Disease* **64**, 959-960.

Cappellini, R.A. (1960) Field inoculations of forage legumes and temperature studies with isolates of *Sclerotinia trifoliorum* and *Sclerotinia sclerotiorum*. *Plant Disease Reporter* **44**, 862-864.

Carr, A.J.H. (1957) *Pseudoplea trifolii* on lucerne. *Plant Pathology* **6**, 38.

Carr, A.J.H. (1965) Clover phyllody virus. *Report, Welsh Plant Breeding Station for 1964*, pp. 97-98.

Carr, A.J.H. (1971) Herbage legumes. In: Western, J.H. (Ed.) Diseases of crop plants. London, UK; MacMillan Press Ltd., pp. 254-271.

Carr, A.J.H.; Large, E.C. (1963) Surveys of phyllody in white clover seed crops, 1959-62. *Plant Pathology* **12**, 121-127.

Carroll, R.B.; Elliott, E.S. (1964) Similarity of species of fungi isolated from roots of alfalfa and red clover. (Abstr.) *Phytopathology* **54**, 746.

Catherall, P.L. (1966) Effects of barley yellow dwarf virus on the growth and yield of single plants and simulated swards of perennial ryegrass. *Annals of Applied Biology* **57**, 155-162.

Chi, C.C. (1965) Pathogenicity of *Fusarium* species from red clover. *Canadian Plant Disease Survey* **45**, 3-7.

Chi, C.C.; Hanson, E.W. (1962) Interrelated effects of environment and age of alfalfa and red clover seedlings on susceptibility to *Pythium debaryanum*. *Phytopathology* **52**, 985-989.

Chilton, S.J.P.; Henson, L.; Johnson, H.W. (1943) Fungi reported on species of *Medicago, Melilotus* and *Trifolium*. *Miscellaneous Publications, United States Department of Agriculture*, No. 499, 152 pp.

Chiykowski, L.N. (1965) A yellows-type virus of alsike clover in Alberta. *Canadian Journal of Botany* **43**, 527-536.

Chiykowski, L.N. (1969) A leafhopper transmitted clover disease in the Ottawa area. *Canadian Plant Disease Survey* **49**, 16-19.

Chiykowski, L.N. (1973) Effectiveness of antibiotics applied as postinoculation sprays against clover phyllody and aster yellows. *Canadian Journal of Plant Science* **53**, 87-91.

Chiykowski, L.N. (1974) Additional host plants of clover phyllody in Canada. *Canadian Journal of Plant Science* **54**, 755-763.

Chiykowski, L.N. (1976) Transmission characteristics and host range of the clover yellow edge agent. *Canadian Journal of Botany* **54**, 1171-1179.

Chiykowski, L.N. (1981) Association of mycoplasma-like organisms with clover yellow edge disease. *Canadian Journal of Plant Pathology* **3**, 139-144.

Chiykowski, L.N.; Bolton, A.T. (1970) Distribution of a newly reported leafhopper-transmitted clover disease in eastern Ontario. *Canadian Plant Disease Survey* **50**, 136-137.

Choopanya, D. (1968a) Studies on peanut stunt virus on *Trifolium* species, especially *Trifolium repens* L. *Dissertation Abstracts B* **29**, 835.

Choopanya, D. (1968b) Distribution of peanut stunt virus on white clover in South Carolina and its relationship to peanut culture. *Plant Disease Reporter* **52**, 926-928.

Clark, M.F.; Barclay, P.C. (1972) The use of immunoosmophoresis in screening a large population of *Trifolium repens* L. for resistance to white clover mosaic virus. *New Zealand Journal of Agricultural Research* **15**, 371-375.

Colville, W.L.; Torrie, J.H. (1962) Effects of management on food reserves, root rot incidence and forage yields of medium red clover, *Trifolium pratense* L. *Agronomy Journal* **54**, 332-335.

Creelman, D.W. (1961) A summary of the prevalence of plant diseases in Canada in 1960. Decline and phyllody (Virus). *Canadian Plant Disease Survey* **41**, 51.

Creelman, D.W. (1967). Summary of the prevalence of plant diseases in Canada in 1965. A compilation. *Canadian Plant Disease Survey* **47**, 31-71.

Cressman, R.M. (1967) Internal breakdown and persistence of red clover. *Crop Science* **7**, 357-361.

Crowder, L.V.; Craigmiles, J.P. (1960) The effect of soil temperature, soil moisture, and flowering on the persistency and forage production of white clover stands. *Agronomy Journal* **52**, 382-385.

Crush, J.R. (1976) Endomycorrhizas and legume growth in some soils of the MacKenzie Basin, Canterbury, New Zealand. *New Zealand Journal of Agricultural Research* **19**, 473-476.

Curl, E.A.; Weaver, H.A. (1958) Diseases of forage crops under sprinkler irrigation in the southeast. *Plant Disease Reporter* **42**, 637-644.

Davis, W.H. (1924) Summary of investigations on clover rusts. *Mycologia* **16**, 203-219.

Denis, S.J. (1969) The red clover vein mosaic virus-*Fusarium* root rot complex in red clover. *Dissertation Abstracts B* **29**, 3572.

Denis, S.J.; Elliott, E.S. (1967) Decline of red clover plants infected with red clover vein mosaic virus and *Fusarium* species. (Abstr.) *Phytopathology* **57**, 808-809.

Dijkstra, J. (1964) Inoculation with ascospores of *Sclerotinia trifoliorum* for detection of clover rot resistant red clover. *Euphytica* **13**, 314-319.

Dijkstra, J. (1966) Rotting by spread of mycelium from ascospore lesions of *Sclerotinia trifoliorum*. *Netherlands Journal of Plant Pathology* **72**, 279-283.

Dingley, J.M. (1969) Records of plant diseases in New Zealand. *Bulletin, New Zealand Department of Scientific and Industrial Research*, No. 192, 298 pp.

Dixon, G.R. (1975) Resistance of red and white clover cultivars to clover rot (*Sclerotinia trifoliorum*). *Annals of Applied Biology* **81**, 276-278.

Doi, Y.; Teranaka, M.; Yora, K.; Asuyama, H. (1967) [Mycoplasma- or PLT group-like microorganisms found in the phloem elements of plants infected with mulberry dwarf, potato witches' broom, aster yellows or Paulownia witches' broom]. *Annals of the Phytopathological Society of Japan* **33**, 259-266.

Dunn, G.M.; Kilpatrick, R.A.; Chow, H.W. (1964) Effects of dieldrin and methyl bromide on *Sitona* larvae and root rot of white clover. *Crop Science* **4**, 276-279.

Edmunds, J.E.; Mai, W.F. (1967) Effect of *Fusarium oxysporum* on movement of *Pratylenchus penetrans* toward alfalfa roots. *Phytopathology* **57**, 468-471.

Elliot, A.M.; Wilcoxson, R.D. (1964) Effect of temperature and moisture on formation and ejection of ascospores and on survival of *Leptosphaerulina briosiana*. *Phytopathology* **54**, 1443-1447.

Ellis, M.B. (1971) Dematiaceous Hyphomycetes. Kew, UK; Commonwealth Mycological Institute, 608 pp.

Ellis, M.B. (1976) More Dematiaceous Hyphomycetes. Kew, UK; Commonwealth Mycological Institute. 507 pp.

Evans, P.S. (1973) The effect of repeated defoliation to three different levels on root growth of five pasture species. *New Zealand Journal of Agricultural Research* **16, 31-34.**

Falloon, R.E. (1976) *Curvularia trifolii* as a high-temperature turfgrass pathogen. *New Zealand Journal of Agricultural Research* **19**, 243-248.

Falloon, R.E.; Skipp, R.A. (1982) Fungicide seed treatments improve lucerne establishment. *Proceedings of the 35th New Zealand Weed and Pest Control Conference*, 127-129.

Fezer, K.D. (1961) Common root rot of red clover: pathogenicity of associated fungi and environmental factors affecting susceptibility. *Memoirs, Cornell Agricultural Experiment Station*, No. 377, 38 pp.

Ford, R.E.; Baggett, J.R. (1965) Relative severity of legume viruses in peas measured by plant growth reduction. *Plant Disease Reporter* **49**, 627-629.

Francki, R.I.B.; Mossop, D.W.; Hatta, T. (1979) Cucumber mosaic virus. *C.M.I./A.A.B. Descriptions of Plant Viruses* No. 213.

Frandsen, K.J. (1942) Meddelelse om kløverens baegersvamp (*Sclerotinia trifoliorum*). *Nordisk Jordbruksforskning* **24**, 12-23.

Frazier, N.W.; Posnette, A.F. (1957) Transmission and host-range studies of strawberry green-petal virus. *Annals of Applied Biology* **45**, 580-588.

Frosheiser, F.I. (1960) A method for screening alfalfa plants for resistence to *Pseudopeziza medicaginis*. *Phytopathology* **50**, 568.

Frosheiser, F.I. (1974) Alfalfa mosaic virus transmission to seed through alfalfa gametes and longevity in alfalfa seed. *Phytopathology* **64**, 102-105.

Fry, P.R. (1952) Occurrence of lucerne-mosaic virus in New Zealand. *New Zealand Journal of Science and Technology* Sect. A,**34** , 320-326.

Fry, P.R. (1959) A clover mosaic virus in New Zealand pastures. *New Zealand Journal of Agricultural Research* **2**, 971-981.

Fulton, N.D.; Hanson, E.W.(1960) Studies on root rots of red clover in Wisconsin. *Phytopathology* **50**, 541-550.

Fulton, R.W. (1948) Hosts of the tobacco streak virus. *Phytopathology* **38**, 421-428.

Fulton, R.W. (1971) Tobacco streak virus. *C.M.I./A.A.B. Descriptions of Plant Viruses* No. 44.

Gáborjány, R.; Sziráki, I. (1978) Cytokinins in white clover plants infected by clover phyllody mycoplasma. *Acta Phytopathologica Academiae Scientiarum Hungaricae* **13**, 33-40.

Garren, K.H. (1954) Disease development and seasonal succession of pathogens of white clover. Part I. Leaf diseases. *Plant Disease Reporter* **38**, 579-582.

Garren, K.H. (1955) Disease development and seasonal succession of pathogens of white clover. Part II. Stolon diseases and the damage-growth cycle. *Plant Disease Reporter* **39**, 339-341.

Gates, L.F.; Bronskill, J.F. (1974) Viruses of clovers and alfalfa in Essex County, Ontario, 1970-73. *Canadian Plant Disease Survey* **54**, 95-100.

Gäumann, E. (1923) Beiträge zu einer Monographie der Gattung *Peronospora* Corda. *Beiträge zur Kryptogamenflora der Schweiz* **4**, 360 pp.

Gerdemann, J.W. (1953) An undescribed fungus causing a root rot of red clover and other Leguminosae. *Mycologia* **45**, 548-554.

Gerhardson, B.; Lindsten, K. (1973) Red clover mottle virus and red clover necrotic mosaic virus in Sweden. *Phytopathologische Zeitschrift* **76**, 67-79.

Giannotti, J.; Giannotti, D. (1978) Culture in vitro d'un mycoplasme de plante agent de prolifération et de nanisme chez *Trifolium repens* L. *Annales de Phytopathologie* **10**, 489-492.

Gibbs, A.J.; Varma, A.; Woods, R.D. (1966) Viruses occurring in white clover (*Trifolium repens* L.) from permanant pastures in Britain. *Annals of Applied Biology* **58**, 231-240.

Gibson, P.B.; Hollowell, E.A. (1966) White clover. *Handbook, US Department of Agriculture* No. 314, 33 pp.

Gibson, P.B.; Barnett, O.W.; Huddleston, L.H. (1979) Virus infections reduce yield of Yuchi arrowleaf clover. *Plant Disease Reporter* **63**, 297-300.

Gibson, P.B.; Barnett, O.W.; Skipper, H.D. (1980) Effects of alfalfa mosaic, clover yellow vein, and peanut stunt viruses on growth and nodulation of white clover. (Abstr.). *Phytopathology* **70**, 567.

Gibson, P.B.; Barnett, O.W.; Skipper, H.D.; McLaughlin, M.R. (1981) Effects of three viruses on growth of white clover. *Plant Disease* **65**, 50-51.

Goth, R.W. (1962) Aphid transmission of white clover mosaic virus. *Phytopathology* **52**, 1228.

Goto, M.; Okabe, N. (1965) Bacterial plant diseases in Japan. X. Bacterial spot disease of clover caused by a strain of *Pseudomonas andropogonis*. *Report, Faculty of Agriculture, Shizuoka University* **15**, 45-49.

Gough, F.J.; Elliott, E.S. (1956) Blackpatch of red clover and other legumes caused by *Rhizoctonia leguminicola* sp. nov. *Bulletin, West Virginia Agricultural Experiment Station*, No. 387, 23 pp.

Gould, A.R.; Francki, R.I.B.; Hatta, T.; Hollings, M. (1981) The bipartite genome of red clover necrotic mosaic virus. *Virology* **108**, 499-506.

Gouranton, J.; Maillet, P.L. (1973) High resolution autoradiography of mycoplasmalike organisms multiplying in some tissue of an insect vector for clover-phyllody. *Journal of Invertebrate Pathology* **21**, 158-163.

Gourret, J.P.; Maillet, P.L.; Gouranton, J. (1973) Virus-like particles associated with the mycoplasmas of clover phyllody in the plant and in the insect vector. *Journal of General Microbiology* **74**, 241-249.

Govier, D.A. (1979) Cucumber mosaic virus from white clover. *Report, Rothamsted Experimental Station for 1978*, Part 1, p. 218.

Govier, D.A.; Cockbain, A.J. (1977) Viruses in white clover. *Report, Rothamsted Experimental Station for 1976*, Part 1, p. 267.

Graham, J.H. (1957) A *Stemphylium* disease of ladino white clover. *Phytopathology* **47**, 213-215.

Graham, J.H.; Luttrell, E.S. (1961) Species of *Leptosphaerulina* on forage plants. *Phytopathology* **51**, 680-693.

Graham, J.H.; Newton, R.C. (1960) Relationship between injury by the clover root curculio and incidence of *Fusarium* root rot in ladino white clover. *Plant Disease Reporter* **44**, 534-535.

Graham, J.H.; Sprague, V.G.; Robinson, R.R. (1957) Damping-off of ladino clover and *Lespedeza* as affected by soil moisture and temperature. *Phytopathology* **47**, 182-185.

Grose, S.; Wells, B.; Beringer, J.E. (1974) *Annual Report, John Innes Institute* **75**, pp. 88.

Grylls, N.E. (1954) Rugose leaf curl - a new virus disease transovarially transmitted by the leafhopper *Austroagallia torrida*. *Australian Journal of Biological Sciences* **7**, 47-58.

Grylls, N.E.; Butler, F.C. (1956) An aphid-transmitted virus affecting subterranean clover. *Journal of the Australian Institute of Agricultural Science* **22**, 73-74.

Grylls, N.E.; Butler, F.C. (1959) Subterranean clover stunt, a virus disease of pasture legumes. *Australian Journal of Agricultural Research* **10**, 145-159.

Grylls, N. E. ; Day, A. W. (1966) Susceptibility, inhibition of infection in stolons, and induced resistance to rugose leaf curl virus infection in white clovers. *Australian Journal of Agricultural Research* **17**, 119-131.

Grylls, N.E.; Peak, J.W. (1969) A virus complex of subterranean clover. *Australian Journal of Agricultural Research* **20**, 37-45.

Grylls, N.E.; Galletly, J.C.; Campbell, R.C. (1972) A field study of rugose leaf curl virus infection in stoloniferous *Trifolium* species. *Australian Journal of Experimental Agriculture and Animal Husbandry* **12**, 293-298.

Grylls, N.E.; Waterford, C.J.; Filshie, B.K.; Beaton, C.D. (1974) Electron microscopy of rugose leaf curl virus in red clover, *Trifolium pratense* and in the leafhopper vector *Austroagallia torrida*. *Journal of General Virology* **23**, 179-183.

Guy, P.; Gibbs, A.; Harrower, K. (1980) The effect of white clover mosaic virus on nodulation of white clover, (*Trifolium repens* L. cv. Ladino). *Australian Journal of Agricultural Research* **31**, 307-311.

Hagedorn, D.J.; Bos, L.; Van der Want, J.P.H. (1959) The red clover vein-mosaic virus in the Netherlands. *Tijdschrift over Plantenziekten* **65**, 13-23.

Hagedorn, D.J.; Hanson, E.W. (1951) A comparative study of the viruses causing Wisconsin pea stunt and red clover vein mosaic. *Phytopathology* **41**, 813-819.

Halisky, P.M.; Houston, B.R.; Magie, A.R. (1960) Alfalfa mosaic virus in white clover and potatoes. *Plant Disease Reporter* **44**, 120-125.

Halpin, J.E.; Gibson, P.B.; Beinhart, G.; Hollowell, E.A. (1963) Selection and evaluation of white clover clones. II. The role of midsummer diseases. *Crop Science* **3**, 87-89.

Halpin, J.E.; Hanson, E.W. (1958) Effect of age of seedlings of alfalfa, red clover, ladino white clover, and sweetclover on susceptibility to *Pythium*. *Phytopathology* **48**, 481-485.

Halpin, J.E.; McCarter, S.M. (1961) Fungi associated with white clover stolons in selected areas of the southeast during mid-summer, 1959. *Plant Disease Reporter* **45**, 298-299.

Hampton, R.O. (1963) Seed transmission of white clover mosaic and clover yellow mosaic viruses in red clover. (Abstr.) *Phytopathology* **53**, 1139.

Hansen, J.D.; Curl, E.A. (1964) Interactions of *Sclerotium rolfsii* and other microorganisms isolated from stolon tissue of *Trifolium repens*. *Phytopathology* **54**, 1127-1132.

Hanson, A.A.; Graham, J.H. (1955) A comparison of greenhouse and field inoculation of ladino clover with *Sclerotinia trifoliorum*. *Agronomy Journal* **47**, 280-281.

Hanson, E.W. (1953) Relative prevalence and severity of the diseases of forage legumes in Wisconsin 1946-1952. *Plant Disease Reporter* **37**, 467-472.

Hanson, E.W.; Kreitlow, K.W. (1953) The many ailments of clover. *In*: Stefferud, A. (Ed.). Plant Diseases. US Department of Agriculture Yearbook of Agriculture 1953, 940 pp.

Hardison, J.R. (1948) Occurrence of anther mold in ladino clover in Oregon. *Plant Disease Reporter* **32**, 242.

Harrison, B.D.; Cadman, C.H. (1959) Role of a dagger nematode (*Xiphinema* sp.) in outbreaks of plant diseases caused by arabis mosaic virus. *Nature, UK* **184**, 1624-1626.

Harrison, B.D.; Winslow, R.D. (1961) Laboratory and field studies on the relation of arabis mosaic virus to its nematode vector *Xiphinema diversicaudatum* (Micoletzky). *Annals of Applied Biology* **49**, 621-633.

Harville, B.G.; Derrick, K.S. (1978) Identification and prevalence of white clover viruses in Louisiana. *Plant Disease Reporter* **62**, 290-292.

Hawksworth, D.L.; Chea, C.Y.; Sheridan, J.E. (1979) *Bimuria novae-zelandiae* gen. et sp. nov., a remarkable ascomycete isolated from a New Zealand barley field. *New Zealand Journal of Botany* **17**, 267-273.

Hebert, T.T. (1967) Epidemiology of the peanut stunt virus in North Carolina. (Abstr.) *Phytopathology* **57**, 461.

Hiratsuka, N. (1973) Contributions to the rust-flora of Eastern Asia. X. Revision of taxonomy of the genus *Uromyces* in the Japanese Archipelago. *Report, Tottori Mycological Institute, Japan*, No. 10, 98 pp.

Hollings, M.; Nariani, T.K. (1965) Some properties of clover yellow vein, a virus from *Trifolium repens* L. *Annals of Applied Biology* **56**, 99-109.

Hollings, M.; Stone, O.M. (1974) Clover yellow vein virus. *C.M.I./A.A.B. Descriptions of Plant Viruses* No. 131.

Hollings, M.; Stone, O.M. (1977a) Red clover necrotic mosaic virus. *C.M.I./A.A.B. Descriptions of Plant Viruses* No. 181.

Hollings, M.; Stone, O.M. (1977b) Cymbidium ringspot virus. *C.M.I./A.A.B. Descriptions of Plant Viruses* No. 178.

Hollings, M.; Stone, O.M.; Barton, R.J. (1977) Pathology, soil transmission and characterization of cymbidium ringspot, a virus from cymbidium orchids and white clover (*Trifolium repens*). *Annals of Applied Biology* **85**, 233-248.

Honda, Y.; Yunoki, T. (1973) [Studies on the sclerotial root rot of ladino clover. I. The damage and infection route of sclerotial root rot caused by *Sclerotinia trifoliorum* Erik. in ladino clover]. *Bulletin, Tohoku National Agricultural Experimental Station* **46**, 143-160.

Honda, Y.; Yunoki, T. (1975) [Studies on *Sclerotinia* root rot and crown rot of ladino clover. 2. Development of *Sclerotinia* root rot and crown rot under snow]. *Bulletin, Tohoku National Agricultural Experimental Station* **50**, 87-97.

Houston, B.R.; Oswald, J.W. (1953) The mosaic virus disease complex of ladino clover. *Phytopathology* **43**, 271-276.

Hughes, S.J.; Kendrick, W.B. (1968) New Zealand fungi. 12. *Menispora, Codinaea, Menisporopsis*. *New Zealand Journal of Botany* **6**, 323-375.

Hull, R. (1969) Alfalfa mosaic virus. *Advances in Virus Research* **15**, 365-433.

Hutton, E.M.; Grylls, N.E. (1956) Legume little leaf, a virus disease of subtropical pasture species. *Australian Journal of Agricultural*

Research **7**, 85-97.

Hutton, E.M.; Peak, J.W. (1954) Varietal reactions of *Trifolium subterraneum* L. to Phaseolus virus 2 Pierce. *Australian Journal of Agricultural Research* **5**, 598-607.

Ie, T.S. (1970) Tomato spotted wilt virus. *C.M.I./A.A.B. Description of Plant Viruses* No. 39.

Iida, W.; Takahashi, H.; Iizuka, N. (1969) [Studies on the diseases of forage crops. I. Diseases of forage grasses and legumes in the Tohoku district]. *Bulletin, Tohoku National Agricultural Experiment Station* **37**, 1-15.

Irwin, J.A.G.; Jones, R.M. (1977) The role of fungi and nematodes as factors associated with death of white clover (*Trifolium repens*) stolons over summer in southeastern Queensland. *Australian Journal of Experimental Agriculture and Animal Husbandry* **17**, 789-794.

Isawa, K.; Nishihara, N. (1975) [Influence of diseases on the chemical composition and nutritive value of forage crops. III. Chemical composition of ladino clover infected with *Curvularia* leaf blight]. *Bulletin, National Grasslands Research Institute (Japan)* No. 7, 99-106.

Ishiie, T.; Doi, Y.; Yora, K.; Asuyama, H. (1967) [Suppressive effects of antibiotics of tetracycline group on symptom development of mulberry dwarf disease]. *Annals of the Phytopathological Society of Japan* **33**, 267-275.

James, J.R.; Lucas, L.T.; Chamblee, D.S.; Campbell, W.V. (1980) Influence of fungicide and insecticide applications on persistence of Ladino clover. *Agronomy Journal* **72**, 781-784.

Jenkyn, J.F. (1980) Chemical control of *Sclerotinia* rot of clover. *Report, Rothamsted Experimental Research Station for 1979* Part 1, p. 174.

Johnson, F. (1942) The complex nature of white-clover mosaic. *Phytopathology* **32**, 103-116.

Johnson, H.W.; Keeling, B.L. (1969) Pathogenicity of *Phytophthora parasitica* isolated from Regal white clover roots. *Plant Disease Reporter* **53**, 446-450.

Johnstone, G.R. (1978) Diseases of broad bean (*Vicia faba* L. *major*) and green pea (*Pisum sativum* L.) in Tasmania caused by subterranean clover red leaf virus. *Australian Journal of Agricultural Research* **29**, 1003-1010.

Johnstone, G.R.; Duffus, J.E.; Munro, D., Ashby, J.W. (1982) Purification of a Tasmanian isolate of subterranean clover red leaf virus, and its serological interactions with a New Zealand isolate and other luteoviruses. *Australian Journal of Agricultural Research* **33**, 697-703.

Johnstone, G.R.; Munro, D. (1980) Yellows of sugar beet caused by subterranean clover red leaf virus. *Australasian Plant Pathology* **9**, 4-5.

Jones, F.R. (1919) The leaf-spot diseases of alfalfa and red clover caused by the fungi *Pseudopeziza medicaginis* and *Pseudopeziza trifolii* respectively. *Bulletin, U.S. Department of Agriculture* No. 759, 38 pp.

Jones, F.R.; Weimer, J.L. (1938) *Stagonospora* leaf spot and root rot of forage legumes. *Journal of Agricultural Research* **57**, 791-812.

Jones, L.R.; Williamson, M.M.; Wolf, F.A.; McCulloch, L. (1923) Bacterial leafspot of clovers. *Journal of Agricultural Research* **25**, 471-490.

Jones, R.J. (1973) The effect of cutting management on the yield, chemical composition and in vitro digestibility of *Trifolium semipilosum* grown with *Paspalum dilatatum* in a sub-tropical environment. *Tropical Grasslands* **7**, 277-284.

Jørstad, I. (1967) *Uromyces* on *Trifolium repens*. *Nytt Magasin for Botanikk* **14**, 19-30.

Joshi, H.U.; Carr, A.J.H. (1967) Effect of clover phyllody virus on nodulation of white clover (*Trifolium repens*) by *Rhizobium trifolii* in soil. *Journal of General Microbiology* **49**, 385-392.

Justham, M.; Ogilvie, L. (1950) Clover rot investigations. *Annals of Applied Biology* **37**, 328.

Kapoor, J.N. (1967) *Erysiphe trifolii*. *C.M.I. Descriptions of Pathogenic Fungi and Bacteria*, No. 156.

Kellock, A.W. (1971) Red-leaf virus - a newly recognized virus disease of subterranean clover (*Trifolium subterraneum* L.). *Australian Journal of Agricultural Research* **22**, 615-624.

Kennedy, J.S.; Day, M.F.; Eastof, V.F. (1962) A conspectus of aphids as vectors of plant viruses. London, UK; Commonwealth Institute of Entomology, 114 pp.

Khaleeva, Z.N.; Petrov, A.F. (1970) Rak Klevera v Leningradskoy oblasti. *Zashchita Rastenii*, **15**, 34.

Khan, M.A.; Maxwell, D.P.; Smith, R.R. (1978) Inheritance of resistance to red clover vein mosaic virus in red clover. *Phytopathology* **68**, 1084-1086.

Killian, C. (1923) Le *Polythrincium trifolii* Kunze parasite du trèfle. *Revue de Pathologie Végétale et d'Entomologie Agricole de France* **10**, 202-219.

Kilpatrick, R.A. (1958a) Fungi isolated from ladino white clover seeds (*Trifolium repens* L.) in New Hampshire. *Plant Disease Reporter* **42**, 142-144.

Kilpatrick, R.A. (1958b) Curvularia leaf blight of clovers and its causal agent, *Curvularia trifolii*. *Phytopathology* **48**, 513-515.

Kilpatrick, R.A. (1959a) Fungi associated with red and white clovers in New Hampshire. *Plant Disease Reporter* **43**, 1111-1113.

Kilpatrick, R.A. (1959b) A disease of ladino white clover caused by a yeast, *Rhodotorula glutinis* var. *rubescens*. *Phytopathology* **49**, 148-151.

Kilpatrick, R.A. (1960) Heat-induced stolon rot of ladino white clover in the greenhouse. *Plant Disease Reporter* **44**, 653-654.

Kilpatrick, R.A. (1961a) Fungi associated with larvae of *Sitona* spp. *Phytopathology* **51**, 640-641.

Kilpatrick, R.A. (1961b) Floral infection of ladino white clover, incited by *Curvularia trifolii*. *Plant Disease Reporter* **45**, 286-287.

Kilpatrick, R.A. (1962) Laboratory studies on *Leptosphaerulina trifolii* (Abstr.) *Phytopathology* **52**, 17.

Kilpatrick, R.A.; Dunn, G.M. (1956) Late season diseases of forage crops in New Hampshire. 1955. *Plant Disease Reporter* **40**, 384-386.

Kilpatrick, R.A.; Dunn, G.M. (1958) Observations on insects and fungi associated with tap root survival of white clover in New Hampshire. *Plant Disease Reporter* **42**, 819-820.

Kilpatrick, R.A.; Dunn, G.M. (1961) Fungi and insects associated with deterioration of white clover taproots. *Crop Science* **1**, 147-149.

Kilpatrick, R.A.; Judd, R.W.; Dunn, G.M.; Rich, A.E. (1966) Agents affecting survival of red and white clovers exposed to low temperatures in a freezing chamber.*Crop Science* **6**, 499-501.

Kitani, K.; Kuniyasu, K.; Natsume, T. (1968) [Studies on the southern blight, *Corticium rolfsii* (Saccardo) Curzi, of ladino clover, *Trifolium repens* L., cultivated in upland paddy fields]. *Bulletin, Shikoku Agricultural Experiment Station* **18**, 139-160.

Kleinhempel, H.; Müller, H.M.; Spaar, D. (1975) Zur Frage der Übertragung mykoplasmaähnlicher Organismen durch das Saatgut infizierter Pflanzen. *Archiv für Phytopathologie und Pflanzenschutz* **11**, 89-99.

Kobel, F. (1920) Zur Biologie der Trifolien bewohnenden *Uromyces*-Arten. *Zentralblatt für Bakteriologie, Parasitenkunde Infektionskrankheiten und Hygiene* (Abt. 2.) **52**, 215-235.

Kreitlow, K.W. (1949) *Sclerotinia trifoliorum*, a pathogen of ladino clover. *Phytopathology* **39**, 158-166.

Kreitlow, K.W. (1964) Effect of virus infection on persistence of ladino clover. (Abstr.) *Phytopathology* **54**, 898.

Kreitlow, K.W.; Graham, J.H.; Garber, R.L. (1953) Diseases of forage grasses and legumes in the northeastern states. *Bulletin, Pennsylvania Agricultural Experiment Station*, No. 573, 42 pp.

Kreitlow, K.W.; Hunt, O.J.; Wilkins, H.L. (1957) The effect of virus infection on yield and chemical composition of ladino clover. *Phytopathology* **47**, 390-394.

Kreitlow, K.W.; Price, W.C. (1949) A new virus disease of ladino clover. *Phytopathology* **39**, 517-528.

Kreitlow, K.W.; Yu, H.S. (1955) *Curvularia* leaf blight of red clover. *Plant Disease Reporter* **39**, 181-182.

Kristinsson, H. (1967) Untersuchungen zum sexuellen Entwicklungsgang von *Pseudopeziza medicaginis* (Lib.) Sacc. und *Drepanopeziza ribis* (Kleb.) v. Höhn. *Phytopathologische Zeitschrift* **60**, 1-40.

Lancashire, J.A.; Latch, G.C.M. (1966) Some effects of crown rust (*Puccinia coronata* Corda) on the growth of two ryegrass varieties in New Zealand.*New Zealand Journal of Agricultural Research* **9**, 628-640.

Latter, B.D.H. (1952) Australian clover rusts. *Proceedings of the Linnean Society of New South Wales*, **77**, 337-356.

Laundon, G.F. (1973) *Uromyces fallens* and *U. trifolii-repentis* in New Zealand. *Transactions of the British Mycological Society* **61**, 83-87.

Laundon, G.F. (1975) Taxonomy and nomenclature notes on Uredinales. *Mycotaxon* **3**, 133-161.

Leach, C.M. (1958) Additional evidence for seed-borne mycelium of *Sclerotinia sclerotiorum* associated with clover seed. *Phytopathology* **48**, 388-389.

Leach, C.M.; Dickason, E.A.; Gross, A.E. (1961) Effects of insecticides on insects and pathogenic fungi associated with alsike clover roots. *Journal of Economic Entomology* **54**, 543-546.

Leach, C.M.; Dickason, E.A.; Gross, A.E. (1963) The relationship of insects, fungi and nematodes to the deterioration of roots of *Trifolium hybridum* L. *Annals of Applied Biology* **52**, 371-385.

Leach, J.G.; Elliot, E.S. (1951) Black patch of red clover in West Virginia. *Plant Disease Reporter* **35**, 335-336.

Leath, K.T. (1971) Quality of light required for sporulation by *Leptosphaerulina*. *Phytopathology* **61**, 70-72.

Leath, K.T.; Byers, R.A. (1973) Attractiveness of diseased red clover roots to the clover root borer. *Phytopathology* **63**, 428-431.

Leath, K.T.; Byers, R.A. (1977) Interaction of *Fusarium* root rot with pea aphid and potato leafhopper feeding on forage legumes. *Phytopathology* **67**, 226-229.

Leath, K.T.; Kendall, W.A. (1978) Fusarium root rot of forage species: pathogenicity and host range. *Phytopathology* **68**, 826-831.

Leath, K.T.; Lukezic, F.L.; Crittenden, H.W.; Elliot, E.S.; Halisky, P.M.; Howard, F.L.; Ostazeski, S.A. (1971) The *Fusarium* root rot complex of selected forage legumes in the northeast. *Bulletin, Pennsylvania Agricultural Experiment Station*, No. 777, 64 pp.

Leath, K.T.; Zeiders, K.E.; Byers, R.A. (1973) Increased yield and persistence of red clover after a soil drench application of benomyl. *Agronomy Journal* **65**, 1008-1009.

Lehman, S.G. (1951) *Curvularia* leaf blight of ladino clover. *Plant Disease Reporter* **35**, 79-80.

Lindsten, K.; Brishhammer, S.; Tomenius, K. (1976) Investigations on relationship and variation of some legume viruses within the potyvirus group. *Meddelanden, Statens Växtskyddsanstalt* **16**, 289-322.

Lombardo, G.; Bassi, M.; Gerola, F.M. (1970) Mycoplasma development and cell alterations in white clover affected by clover dwarf. An electron microscopy study. *Protoplasma* **70**, 61-71.

Loveless, A.R. (1951) Observations on the biology of clover rot. *Annals of Applied Biology* **38**, 642-664.

Lucas, L.T.; Harper, C.R. (1972) Mechanically transmissible viruses from ladino clover in North Carolina. *Plant Disease Reporter* **56**, 774-776.

Lukezic, F.L.; Bloom, J.R.; Carroll, R.B. (1969) Development of *Fusarium* root rot and its effect on root and crown carbohydrate levels of clipped alfalfa plants grown in a gnotobiotic environment. *Phytopathology* **59**, 1575-1579.

Lynes, E.W.; Teakle, D.S.; Smith, P.R. (1981) Red clover necrotic mosaic virus isolated from *Trifolium repens* and *Medicago sativa* in Victoria. *Australasian Plant Pathology* **10**, 6-7.

McCarter, S.M.; Halpin, J.E. (1961a) The effect of availability of nutrients and frequency of cutting white clover on the pathogenicity of eight soil borne fungi. (Abstr.) *Bulletin, South Carolina Academy of Science* **23**, 55.

McCarter, S.M.; Halpin, J.E. (1961b) Studies on the pathogenicity of 4 species of soil fungi on white clover as affected by the presence of bean yellow mosaic virus under conditions of controlled temperature and light. (Abstr.) *Phytopathology* **51**, 644.

McCarter, S.M.; Halpin J.E. (1962) Effects of four temperatures on the pathogenicity of nine species of fungi on white clover. (Abstr.). *Phytopathology* **52**, 20.

Macfarlane, I. (1979) Clover rot. *Research Report. Rothamsted Experimental Station for 1978*, Part 1. pp. 217-218.

Macfarlane, I. (1981) Putative vectors of red clover necrotic mosaic virus. *Research Report, Rothamsted Experimental Station for 1980*, Part 1, p. 188.

McGee, D.C.; Kellock, A.W. (1974) *Fusarium avenaceum*, a seed-borne pathogen of subterranean clover roots. *Australian Journal of Agricultural Research* **25**, 549-557.

McLaughlin, M.R.; Barnett, O.W.; Gibson, P.B. (1978) Detection and differentiation of bean yellow mosaic virus and clover yellow vein virus by ELISA. (Abstr.) *Phytopathology News* **12**, 198.

McGlohon, N.E. (1959) Survey of fungi associated with white clover stolons. *Plant Disease Reporter* **43**, 22-24.

McVey,D.V.; Gerdemann, J.W. (1960) Host-parasite relations of *Leptodiscus terrestris* on alfalfa, red clover and birdsfoot trefoil. *Phytopathology* 50, 416-421.

Mains, E.B. (1935) Michigan fungi I. *Papers, Michigan Academy of Science, Arts and Letters* **20**, 81-93.

Mains, E.B. (1936) Host Specialization of *Uromyces trifolii*. *Papers, Michigan Academy of Science, Arts and Letters* **21**, 129-134.

Markham, P.G.; Townsend, R.; Plaskitt, K.A. (1975) A rickettsia-like organism associated with diseased white clover. *Annals of Applied Biology* **81**, 91-93.

Mence, M.J. (1964) The differential mortality of *Trifolium repens* and *Phleum pratense* seedlings in relation to temperature. *Plant and Soil* **21**, 113-124.

Menzies, S.A. (1973a) Factors increasing root rot and affecting persistence of clover. *Proceedings of the 26th New Zealand Weed and Pest Control Conference*, 122-125.

Menzies, S.A. (1973b) Root rot of clover caused by *Codinaea fertilis*. *New Zealand Journal of Agicultural Research* **16**, 239-245.

Miller, C.R.; Halpin, J.E. (1962) Yield responses of 12 genetic lines of white clover to infection by bean yellow mosaic virus and alfalfa mosaic virus. (Abstr.) *Phytopathology* **52**, 21.

Miller, L.I.; Troutman, J.L. (1966) Stunt disease of peanuts in Virginia. *Plant Disease Reporter* **50**, 139-143.

Mink, G.I. (1972) Peanut stunt virus. *C.M.I./A.A.B. Descriptions of Plant Viruses*, No. 92.

Moody, A.R.; Kilpatrick, R.A.; Rich, A.E. (1967) Effect of cutting height on development of root rot in clover. *Plant Disease Reporter* **51**, 826-828.

Moore, W.C. (1959) British parasitic fungi. Cambridge, UK; Cambridge University Press, 430 pp.

Mordue, J.E.M. (1974a) *Corticium rolfsii*. *C.M.I. Descriptions of Pathogenic Fungi and Bacteria*, No. 410.

Mordue, J.E.M. (1974b) *Thanatephorus cucumeris*. *C.M.I. Descriptions of Pathogenic Fungi and Bacteria*, No. 406.

Murant,A.F. (1970a)Tomato black ring virus. *C.M.I./A.A.B. Descriptions of Plant Viruses* No. 38.

Murant,A.F. (1970b)Arabis mosaic virus. *C.M.I./A.A.B. Descriptions of Plant Viruses* No. 16.

Murant, A.F. (1974) Strawberry latent ringspot virus. *C.M.I./A.A.B. Descriptions of Plant Viruses* No. 126.

Musil, M.; Lešková, O. (1973) Dva nove viry izolovane z jete lu v Ceskoslovensku. *Ochrona Rostlin* **9**, 259-264.

Nakamura, S.; Saito, T.; Yamamoto, M. (1978) [Studies on clover phyllody (part 2.). Effects of temperature upon the appearance of symptoms]. *Journal of Agricultural Science (Tokyo Nogyo Diagaku)* **23**, 97-101.

Narita, T. (1972) [Brown leaf spot of red clover and ladino clover, caused by *Stagonospora recedens* (Massal) Jones & Weimer.] *Annual Report of the Society of Plant Protection of North Japan* **23**, 19-24.

Nelson, C.J.; Smith, D. (1969) Growth of birdsfoot trefoil and alfalfa. IV. Carbohydrate reserve levels and growth analysis under two temperature regimes. *Crop Science* **9**, 589-591.

Newton, J.E.; Betts, J.E.; Drane, H.M.; Saba, N. (1970) The oestrogenic activity of white clover. In: White Clover Research. *Occasional Symposium of the British Grassland Society* No. 6, 309-314.

Niemann, E. (1962) Zur Prüfung von Rot- und Weissklee auf Krebsresistenz. *Nachrichtenblatt des Deutschen Pflanzenschutzdienstes* **14**, 5-9.

Nishihara, N. (1966) [A comparative study of *Curvularia trifolii* (Kauff.) Boedijn isolated from *Trifolium*]. *Bulletin, National Institute of Animal Industry* **12**, 15-23.

Nishihara, N. (1968) [A disease of ladino clover caused by *Choanephora cucurbitarum* in Japan]. *Transactions of the Mycological Society of Japan* **9**, 38-42.

Noble, M. (1948) Seed borne diseases of clover. *Transactions of the British Mycological Society* **30**, 84-91.

Ohki, S.T.; Doi, Y.; Yora, K. (1976) Clover yellows virus. *Annals of the Phytopathological Society of Japan* **42**, 313-316.

O'Rourke, C.J. (1970) White clover diseases in the Irish Republic. In: White clover research. *Occasional Symposium of the British Grassland Society* No. 6, 139-142.

O'Rourke, C.J. (1971) Host range studies of grass and clover pathogens. *Research Report, Plant Sciences and Crop Husbandry, An Foras Talúntais, Dublin for 1970,* pp. 44-47.

O'Rourke, C.J. (1972) Grassland disease survey. *Research Report, Plant Sciences and Crop Husbandry, An Foras Talúntais Dublin for 1971*, pp. 24-25.

O'Rourke, C.J. (1976) Diseases of grasses and forage legumes in Ireland. Dublin: An Foras Talúntais, 115 pp.

O'Rourke, C.J.; Millar, R.L. (1966) Root rot and root microflora of alfalfa as affected by potassium nutrition, frequency of cutting, and leaf infection. *Phytopathology* **56**, 1040-1046.

Osborn, H.T. (1937) Vein-mosaic virus of red clover. *Phytopathology* **27**, 1051-1058.

Paliwal, Y.C. (1974) Some properties and thrip transmission of tomato spotted wilt virus in Canada. *Canadian Journal of Botany* **52**, 1177-1182.

Pape, H. (1931) Das Auswintern des Klees. *Mitteilungen der Deutschen Landwirtschaftsgesellschaft* **46**, 233-235.

Peterson, P.J.; Latch, G.C.M. (1969) Polyacrylamide gel electrophoresis of cellular proteins of *Cercospora* isolates from some pasture legumes. *New Zealand Journal of Science* **12**, 3-12.

Phillips, J.M.; Hayman, D.S. (1970) Improved procedures for clearing roots and staining parasitic and vesicular-arbuscular mycorrhizal fungi for rapid assessment of infection. *Transactions of the British Mycological Society* **55**, 158-161.

Ploaie, P.; Maramorosch, K. (1969) Electron microscopic demonstration of particles resembling mycoplasma or psittacosis-lymphogranuloma-trachoma group in plants infected with European yellows-type diseases. *Phytopathology* **59**, 536-544.

Posnette, A.F.; Ellenberger, C.E. (1963) Further studies of green petal and other leafhopper-transmitted viruses infecting strawberry and clover. *Annals of Applied Biology* **51**, 69-83.

Pratt, M.J. (1961) Studies on clover yellow mosaic and white clover mosaic viruses. *Canadian Journal of Botany* **39**, 655-665.

Pratt, M.J. (1967) Reduced winter survival and yield of clover infected with clover yellow mosaic virus. *Canadian Journal of Plant Science* **47**, 289-294.

Pratt, M.J. (1968) Clover viruses in eastern Canada in 1967. *Canadian Plant Disease Survey* **48**, 87-92.

Pratt, M.J. (1969) Clover yellow vein virus in North America. *Plant Disease Reporter* **53**, 210-212.

Ragetli, H.W.J.; Elder, M. (1977) Characteristics of clover primary leaf necrosis virus, a new spherical isolate from *Trifolium pratense*. *Canadian Journal of Botany* **55**, 2122-2136.

Rainey, D.P.; Smalley, E.B.; Crump, M.H.; Strong, F.M. (1964) Isolation of salivation factor from *Rhizoctonia leguminicola* on red clover hay. *Nature, UK* **205**, 203-204.

Roane, M.K.; Miller, J.D. (1976) *Uromyces nerviphilus* on interspecific hybrids of *Trifolium repens* in Virginia. *Plant Disease Reporter* **60**, 432-433.

Roberts, D.A. (1956) Influence of alfalfa mosiac virus infection upon winter hardiness of ladino clover. (Abstr.) *Phytopathology* **46**, 24.

Roberts, D.A. (1957) Natural infection of ladino clover by the red clover vein mosaic virus. *Plant Disease Reporter* **41**, 928-929.

Robertson, G.I. (1980) The genus *Pythium* in New Zealand. *New Zealand Journal of Botany* **18**, 73-102.

Sampson, K.; Western, J.H. (1954) Diseases of British grasses and herbage legumes. Cambridge, UK; Cambridge University Press, 118 pp.

Sander, E. (1959) Biological properties of red clover vein mosaic virus. *Phytopathology* **49**, 748-754.

Schillinger, J.A.; Leffel, R.C. (1964) Persistence of ladino clover, *Trifolium repens* L. *Agronomy Journal* **56**, 11-14.

Schmiedeknecht, M. (1964a) Physiologische Beobachtung an *Psuedopeziza*-Arten. *Zentralblatt für Bakteriologie, Parasitenkunde, Infektionskrankheiten und Hygiene* (Abt. 2) **118**, 510-521.

Schmiedeknecht, M. (1964b) Resistenz einiger Futterleguminosen gegen *Pseudopeziza*-Arten. *Züchter* **34**, 67-76.

Schmiedeknecht, M. (1964c) Morphologie, Spezialisierung und Phylogenie einiger *Pseudopeziza*-Arten. *Biologisches Zentralblatt* **83**, 695-715.

Schmelzer, K. (1956) Kenntnis der Übertragbarkeit von Viren durch *Cuscuta*-Arten. *Phytopathologische Zeitschrift* **28**, 1-56.

Scott, S.W. (1981) Contamination of seed of white clover (*Trifolium repens*) with mycelium of a *Sclerotinia* species. *Transactions of the British Mycological Society* **77**, 671-674.

Scott, S.W.; Ellis Davies, W. (1979) Red clover necrotic mosaic virus. *Report, Welsh Plant Breeding Station for 1978*, p.200-201.

Scott, S.W.; Evans, D.R. (1980) *Sclerotinia trifoliorum* Erikss. on white clover. *Grass and Forage Science* **35**, 159-164.

Scott, S.W.; Hughes, S. (1980) Survey of viruses in upland white clovers. *Report, Welsh Plant Breeding Station for 1979*, p. 173-174.

Siddiqui, W.M.; Halisky, P.M.; Lund, S. (1968) Relationship of clipping frequency to root and crown deterioration in red clover. *Phytopathology* **58**, 486-488.

Singh, R.P.; López-Abella, D. (1971) Natural infection of coriander plants by a strain of clover yellow vein virus. *Phytopathology* **61**, 333-334.

Sinha, R.C. (1976) Ultrastructure of mycoplasma-like organisms purified from clover phyllody-affected plants. *Journal of Ultrastructure Research* **54**, 183-189.

Skipp, R.A.; Christensen, M.J. (1981) Invasion of white clover roots by fungi and other soil micro-organisms. I. Surface colonisation and invasion of roots growing in seived pasture soil in the glasshouse. *New Zealand Journal of Agricultural Research* **24**, 235-241.

Skipp, R.A.; Christensen, M.J.; Caradus, J.R. (1982) Invasion of white clover roots by fungi and other soil micro-organisms. II. Invasion of roots in grazed pastures. *New Zealand Journal of Agricultural Research* **25**, 87-95.

Skipp, R.A.; Christensen, M.J. (1982) Invasion of white clover roots by fungi and other soil micro-organisms. III. The capacity of fungi isolated from white clover roots to invade seedling root tissue. *New Zealand Journal of Agricultural Research* **25**, 97-101.

Smets, G.; Dekegel, D.; Vandérveken, J. (1977) Presence de structures anormales dans les bactéroïdes de trèfles blancs atteints de phyllodie. *Parasitica* **33**, 111-118.

Smith, W.H.; Peterson, J.L. (1966) The influence of the carbohydrate fraction of the root exudate of red clover, *Trifolium pratense* L. on *Fusarium* spp., isolated from the clover root and rhizosphere. *Plant and Soil* **25**, 413-424.

Snyder, W.C.; Hansen, H.N. (1945) The species concept in *Fusarium* with reference to Discolor and other sections. *American Journal of Botany* **32**, 657-666.

Stace-Smith, R. (1970) Tobacco ringspot virus. *C.M.I./A.A.B. Descriptions of Plant Viruses* No. 17.

Stovold, G.E. (1974) Root rot caused by *Pythium irregulare* Buisman, an important factor in the decline of established subterranean clover pastures. *Australian Journal of Agricultural Research* **25**, 537-548.

Stuckey, I.H. (1962) Factors affecting persistence of ladino white clover. *Crop Science* **2**, 173-174.

Sundheim, L.; Wilcoxon, R.D. (1965) *Leptosphaerulina briosiana* on alfalfa: infection and disease development, host-parasite relationships, ascospore germination and dissemination. *Phytopathology* **55**, 546-553.

Takahashi, H.; Iida, W. (1969) [Studies on diseases of forage crops. II. Curvularia leaf blight of ladino white clover]. *Bulletin, Tohoku National Agricultural Experiment Station* **37**, 15-41.

Tapio, E. (1970) Virus diseases of legumes in Finland and in the Scandinavian countries. *Annales Agriculturae Fenniae* **9**, 1-97.

Taraku, N.; Juretić, N.; Miličić, D. (1977) Properties of two virus isolates from clovers in Croatia. *Acta Botanica Croatica* **36**, 47-55.

Thornton, R.H. (1965) Studies of fungi in pasture soils. I. Fungi associated with live roots. *New Zealand Journal of Agricultural Research* **8**, 417-449.

Tiffany, L.H.; Gilman, J.C. (1954) Species of *Colletotrichum* from legumes. *Mycologia* **46**, 52-75.

Tolin, S.A.; Isakson, O.W.; Troutman, J.L. (1970) Association of white clover and aphids with peanut stunt virus in Virginia. *Plant Disease Reporter* **54**, 935-938.

Troutman, J.L.; Bailey, W.K.; Thomas, C.A. (1967) Seed transmission of peanut stunt virus. *Phytopathology* **57**, 1280-1281.

Tuite, J. (1960) The natural occurrence of tobacco ringspot virus. *Phytopathology* **50**, 296-298.

Vago, C.; Giannotti, J. (1972) Les mycoplasmes chez les végétaux et chez les vecteurs. *Physiologie Végétale* **10**, 87-101.

Valenta, V. (1964) A comparative study of European yellows-type viruses. *Proceedings of the 5th Conference of the Czechoslovak Plant Virologists, Prague 1962*, pp. 70-72.

Valenta, V.; Musil, M. (1963) Investigations on European yellows type virus. II. The clover dwarf and parastolbur viruses. *Phytopathologische Zeitschrift* **47**, 38-65.

Valenta, V.; Musil, M.; Mišiga, S. (1961) Investigations on European yellows-type viruses. I. The stolbur virus. *Phytopathologische Zeitschrift* **42**, 1-38.

Vanderveken, J. (1964) Influence d'un virus sur la nodulation chez *Trifolium repens*. *Annales de l'Institut Pasteur, Paris*. **107**, 143-148.

Varma, A. (1970) Red clover vein mosaic virus. *C.M.I./A.A.B. Descriptions of Plant Viruses* No. 22 .

Varma, A.; Gibbs, A.J.; Woods, R.D. (1970) A comparative study of red clover vein mosaic virus and some other plant viruses. *Journal of General Virology* **8**, 21-32.

Wadham, S.M. (1925) Observations on clover rot (*Sclerotinia trifoliorum* Eriks.). *New Phytologist* 29, 50-56.

Walker, J. (1978) The cause for *Uromyces trifolii*. *Mycotaxon* **7**, 423-435.

Watkins, G.M. (1961) Physiology of *Sclerotium rolfsii*, with emphasis on parasitism. *Phytopathology* **51**, 110-113.

Watson, R.D.; Guthrie, J.W. (1964) Virus-fungus relationships in a root rot complex in red clover. *Plant Disease Reporter* **48**, 723-727.

Weimer, J.L. (1950) Blackpatch of soybean and other forage legumes. *Phytopathology* **40**, 782-784.

Welsh, M.F.; Stace- Smith, R.; Brennan, E. (1973) Clover yellow mosaic virus from apple trees with leaf pucker disease. *Phytopathology* **63**, 50-57.

West, E. (1961) *Sclerotium rolfsii*, history, taxonomy, host range, and distribution. *Phytopathology* **51**, 108-109.

Westbrooks, F.E.; Tesar.M.B. (1955) Tap root survival of ladino clover. *Agronomy Journal* **47**, 403-410.

Willard, C.J. (1930) Root reserves of alfalfa with special reference to time of cutting and yield. *Journal of the American Society of Agronomy* **22**, 599-602.

Williams, G.H.; Western, J.H. (1965) The biology of *Sclerotinia trifoliorum* Erikss. and other species of sclerotium-forming fungi. I. Apothecium formation from sclerotia. *Annals of Applied Biology* **56**, 253-260.

Willis, C.B. (1965) Observations on the diseases of forage crops in Prince Edward Island. *Canadian Plant Disease Survey* **45**, 8-11

Willis, C.B.; Thompson, L.S. (1979) Effects of phenamiphos, methyl bromide and fallowing on *Pratylenchus penetrans*, yield of *Medicago sativa* and *Fusarium* infections. *Journal of Nematology* **11**, 265-269.

Wilson, J.; Close, R.C. (1973) Subterranean clover red leaf virus and other legume viruses in Canterbury. *New Zealand Journal of Agricultural Research* **16**, 305-310.

Wilson, M.; Henderson, D.M. (1966) British rust fungi. Cambridge, UK; Cambridge University Press, 344 pp.

Windsor, I.M.; Black, L.M. (1973a) Evidence that clover club leaf is caused by a Rickettsia-like organism. *Phytopathology* **63**, 1139-1148.

Windsor, I.M.; Black, L.M. (1973b) Remission of symptoms of clover club leaf following treatment with penicillin. *Phytopathology* **63**, 44-46.

Wolf, F.A. (1935) Morphology of *Polythrincium*, causing sooty blotch of clover. *Mycologia* **27**, 58-73.

Wong, E.; Flux, D.S.; Latch, G.C.M. (1971) The oestrogenic activity of white clover (*Trifolium repens* L.). *New Zealand Journal of Agricultural Research* **14**, 639-645.

Wong, E.; Latch, G.C.M. (1971a) Effect of fungal diseases on phenolic contents of white clover. *New Zealand Journal of Agricultural Research* **14**, 633-638.

Wong, E.; Latch, G.C.M. (1971b) Coumestans in diseased white clover. *Phytochemistry* **10**, 466-468.

Zwingli, W.; Grundbacher, F. (1958) Weisskleeversuch mit einigen bekannten Zuchtsorten. *Mitteilungen für die Schweizerische Landwirtschaft* **6**, 155-159.

Pests

D. L. Gaynor and R. A. Skipp
Entomology Division and Plant Diseases Division, DSIR,
Palmerston North, New Zealand

Contents

1. Introduction

The pests of white clover (*Trifolium repens*) have been studied less than those of other forage legumes such as lucerne and red clover, probably because white clover has a subsidiary role in the pastoral agriculture of Europe and North America, and only a few scientists work on recognising and assessing its pest problems. White clover may, in fact, be less troubled by pests than are other forage legumes because its stoloniferous growth habit reduces reliance on a single vulnerable tap root and so enables it to compensate for any damage. In addition, white clover is normally grown in a sward of mixed pasture species, rather than as a monoculture, which reduces the likelihood of damaging increases in pest populations and enables companion species to compensate for lost clover production.

There are some areas, however, in which white clover pests are extremely destructive. In New Zealand pasture, where white clover is the major legume component, pests probably cost hundreds of millions of dollars per annum in lost production. White clover is severely damaged by several native scarab beetles and hepialid moths which were originally limited to the native tussock grasslands but now thrive in the extensive and relatively recently established areas of permament pasture. Native scarabs and hepialids are also becoming serious pests in Australia as more improved pastures containing white clover are established (Hardy 1973b, 1974, Davidson et al. 1979). In addition to these endemic pests, others such as the lucerne flea and some aphids have become troublesome after their introduction into areas where they have no natural enemies. Some pests such as slugs, case-bearer moths and nematodes are damaging wherever white clover is grown.

Since the range of pests can differ from country to country, few generalizations can be made about the importance of each species. The need for control can be decided only by local assessment of the economic significance of a pest but in most cases threshold economic damage levels are not known. Therefore, we shall discuss only those invertebrates presently considered pests or potential pests of white clover, not those merely associated with white clover. White clover pests are better understood in New Zealand than elsewhere, and consequently there is a predominance of New Zealand examples.

The four main methods of pest control for pastures are: chemical, cultural, biological and varietal. Chemical control has traditionally been the main control method, particularly in high-value seed crops, but rising costs and environmental considerations have forced a move towards fewer, more accurately timed pesticide applications often with reduced success. Cultural control through changes in grazing and cropping management is now being used more widely as an alternative to chemical control. Many pests have increased in the absence of their natural enemies and renewed efforts are being made to control them biologically by introducing predators and parasites from other countries. The development of cultivars of white clover which can better tolerate or resist attack has been considered as a means of control, particularly for nematodes (Williams & Barclay 1972), and is being pursued with respect to insects. White clover can sometimes be replaced in pasture by other more resistant or tolerant species, such as *Lotus penduncculatus* or lucerne (*Medicago sativa*), although these plants also have pest problems and necessitate changing farm management practices. But perhaps the most important recent development has been the attempt to integrate a number of different control measures. This requires a thorough understanding of the biology of the pest, its relationship with white clover, the economic threshold of the pest and the cost-benefit analysis of control. Kain et al. (1982a) outline an integrated pest management system for the New Zealand grass grub *Costelytra zealandica*. They indicate that the integrated pest management system should provide the cheapest and most long-lasting way of removing pest constraints on increased pasture production.

2. Seed and flower feeding insects

(A) Clover seed weevils

The clover seed weevils *Apion* spp. (Coleoptera: Apionidae) are major pests of clover seed crops in Europe, particularly of red clover which may suffer up to 90% loss of yield. The major pest species on red clover are *A. apricans* Hbst., *A. trifolii* (L.) and to a less extent *A. dichroum* Bedel and *A. assimile* Kirby (Brudea 1973), whereas only *A. dichroum* is considered a pest of white clover seed crops. *A. dichroum* has been recorded on white clover in the British Isles (Morgan-Jones 1950; Freeman 1965, 1967), Finland (Markkula & Myllymäki 1957, 1958, 1964), Sweden (Ohlsson 1968), Germany (Buhl & Schütte 1964), and Romania (Perju & Palagesiu 1971), but probably exists wherever white clover is grown for seed in Europe. The other red clover seed weevils can also be found on white clover but are not pests, possibly because of reduced reproduction on this host (Shuster & Khrolinskiĭ 1973). *A. virens* Hbst., *A. difforme* Ahrens (Freeman 1967), *A. nigritarse* Kirby, (Perju & Palagesiu 1971), and *A. seniculus* Kirby (Tanasijević 1967) have been infrequently recorded on white clover.

Adult *A. dichroum* are blue, pear-shaped weevils 3-4 mm long with long rostrums, almost straight antennae and well developed legs. In England the adults hibernate in hedges and woodland litter and fly to neighbouring white clover fields in spring. Ovary development and mating coincide with initiation of flowering. Most eggs are laid in summer within unopened or recently opened flowers through holes cut by the mandibles. The legless larvae feed on the developing ovules and pupate in the old flowers after about 33 days. Emergent adults quickly return to the woodland and settle on the underside of leaves of trees and low herbage. They eventually fall to the litter layer where they hibernate (Morgan-Jones 1950; Freeman 1965, 1967). In Europe *A. dichroum* appears to undergo a further generation before hibernation (Freeman 1967; Brudea 1973).

Freeman & Wimble (1970) suggested that control of *A. dichroum* would increase seed yields of white clover by more than 20 kg/ha. Jones & Jones (1974) reported that over 25% of the seed in a white clover crop in East Anglia in the UK was destroyed by *A. trifolii* and *A. apricans*. Morgan-Jones (1950) indicated that control of *A. dichroum* in white clover seed crops is necessary in some years, and Buhl & Schütte (1964) in Germany recorded a heavy infestation of *A. dichroum* in white clover which required control. The paucity of information suggests that *A. dichroum* is a sporadic pest.

Chemical control appears to be more successful if insecticides are applied after the first early cut and just at the beginning of the second flowering, provided care is taken to avoid killing bees, although Ivanova & Malakhanova (1980) suggest that treatment at the budding stage gave effective control. Several insecticides have been used to control *Apion* spp. in clover seed crops. These include maldison (malathion), toxaphene, trichlorofon, dimethoate and fenitrothion (Buhl & Schütte 1964, Freeman & Wimble 1970, Nizhnikov 1975, Perju et al. 1975). Tereshchenko & Sobchuk (1980) and Perju (1982) mention numerous insecticides effective against red clover seed pests in the Ukraine and Romania.

Cultural control involves cutting crops grown for hay when 25% of the flower heads are in bloom, thus desiccating the vulnerable *Apion* larvae. Some natural control through parasitism does exist. The most important parasites of *Apion* spp. in Europe are *Stenomalina epistena* (Wlk.), *Triaspis caudatus* (Nees) and *T. floricola* (Wesm.) (Legowski 1955, Brudea 1973, Perju 1982). Tudor & Brudea (1979) have reported on five chalcidoid parasites of *Apion* spp. in Romania, including *Eurytoma curculimum* Mayr attacking *A. apricans* and *A. trifolii* on white clover. A braconid, possibly *Microctonus aethiops* (Nees), can parasitise and sterilise up to 70% of *A. dichroum* adults in England, and a chalcidoid wasp *Entedon* sp. can parasitise *A. dichroum* larvae (Freeman 1967). The effect on seed yields of *A. dichroum* control through parasitism is unknown but is generally believed not to be significant (Obrtel 1962).

Another clover seed weevil, *Miccotrogus picirostris* (F.), has caused up to 70% seed loss in crops of white clover in south-eastern Washington and Idaho (Johansen 1965, Yunus & Johansen 1967). It is also present in Canada (Moreland 1953), Finland (Markkula & Myllymäki 1964) and Romania (Perju & Palagesiu 1971), but is not generally considered a serious pest of white clover in these countries. According to Johansen (1965), insecticides for the control of this weevil should be applied when two or more weevils per sweep of a net are found thoughout the field and, if possible, applications should be timed to coincide with the first appearance of brown blooms. No effective natural enemies have been reported on *M. picirostris* in the Pacific Northwest region of the USA.

(B) Clover leaf and head weevils

The lesser clover leaf weevil, *Hypera nigrirostris* (F.) (Coleoptera: Curculionidae), and the clover head weevil, *H. meles* (F.), are persistent pests of red clover and occasionally serious pests of white clover in the USA, Great Britain, Europe, and Asia Minor (MacNay 1954, Markkula & Myllymäki 1962, Ioannisiani 1963, Sechriest & Treece 1963, App 1966, Munson & Enns 1968, Jones & Jones 1974, Smith et al. 1975, Schumann & Diederich 1983). Sakamoto (1980) reported 70% yield reduction in white clover in Japan because of *H. nigrirostris*. Both species have only one complete generation per year. The adult *H. nigrirostris* is 4 mm long, with iridescent, greenish, scale-like hairs and *H. meles* is 1 cm long, brown with three white longitudinal stripes. The adults of both species overwinter in plant debris on the ground, becoming active again in spring before mating. *H. nigrirostris* oviposits in leaves, petioles, stipules and stems, whereas *H. meles* oviposits in petioles and stems (Markkula & Tinnila 1957, Markkula & Myllymäki 1962, Sechriest & Treece 1963, App 1966).

The larvae are exposed feeders adhering to host plants by means of sticky secretions (Jones & Jones 1974). *H. nigrirostris* larvae are yellowish-green with a dorsal white line and feed in leaf axils, terminal buds and developing flower heads. In severe infestations many terminal and side buds may be destroyed. *H. meles* larvae are pearly-white or greyish with faint brown or blue longitudinal dorsal stripes. They feed on florets and developing seed.

Parasites of many *Hypera* species are known. In Ohio *H. nigrirostris* larvae on red clover were found to be parasitised by the braconids *Bathyplectes exiguus* (Grav.), *Bracon mellitor* (Say) and *B. tenuiceps* (Mues.) (Sechriest & Treece 1963), but exploitation of parasites for control appears to be ineffective. Chemical control, in areas where weevils are a persistent problem, involves spraying crops in spring when adults first become active (Jones & Jones 1974). Other *Hypera* species known to feed on white clover but which are not regarded as serious pests are the clover leaf weevil (*H. punctata* (F.)) and the alfalfa weevil (*H. postica* (Gyll.)) (Starks & Thurston 1958, App 1966, Byrne & Blickenstaff 1968, Keller et al. 1970).

(C) Clover case-bearer moths

Case-bearer moths, Coleophora spp. (Lepidoptera: Coleophoridae), are known to attack clover seed crops in New Zealand, Australia, the USA, Europe and Asia Minor, but are particularly damaging to commercial seed crops of white clover in New Zealand, Tasmania and the USA. Two species are found in New Zealand: *C. spissicornis* (Haw.) the banded clover case-bearer and *C. frischella* (L.), the white-tipped clover case-bearer, the former being slightly the more important. In California *C. spissicornis* is important and has spread to the eastern USA, while *C. frischella* is important in Tasmania (Stuart 1958, Dumbleton 1963, French 1972, Hardy 1973a, Pearson 1980).

The biology of *Coleophora* spp. in New Zealand has been reviewed by Stuart (1958), Dumbleton (1963) and Pearson (1982). *C. spissicornis* has one generation per year while *C. frischella* has two to three. Both species overwinter as larvae and pupation occurs in spring. *C. spissicornis* flies from November to January and *C. frischella* flies from late October to late March. Eggs are laid at the base of petals in open florets. The light-yellow first-instar larvae eat through seed pods into seeds and later larval instars cause tunnelling damage within florets. Fourth instar larvae form cigar-shaped cases of web and petals around themselves while continuing to search for seeds. In early generations, *C. frischella* larvae become attached to florets about 6 weeks after hatching and pupate. However, larvae of *C. spissicornis* and the last generation of *C frischella* move or fall to the soil surface and become dormant before pupating. In Tasmania the biology of *C. frischella* is similar to that in New Zealand but there are fewer generations per year because white clover flowering is later (Hardy 1983).

A single larva can consume all the seeds in 2-3 florets within 24 hours, so large populations of *Coleophora* will considerably decrease seed yields. Kelsey (1958) in South Canterbury, New Zealand, recorded 63% of white clover seed destroyed by case-bearer moths in one season, but the average loss appears closer to 10-15% (Dumbleton 1959). Hardy (1973a) in Tasmania found that *C. frischella* destroyed about 3% (9 kg/ha) of seed in the first year of seed set, 32% (90 kg/ha) in the second year and 46% (114kg/ha) in the third and subsequent years. French (1972) also noted less damage in the first year and recommended that farmers take seed that year rather than later. In Canterbury, Pearson (1980) reported 33.2 and 23.8% seed losses in crops of cv. Grasslands Huia and Grasslands Pitau white clover respectively. He also reported that less damage occurred when paddocks were closed for seed production in mid-November, possibly because this late closing date stimulated rapid inflorescence appearance, which produced more inflorescences in which the moths could oviposit during their period of peak abundance.

Biological and chemical control of *Coleophora* spp. have been investigated in New Zealand. All attempts at biological control before 1962 were unsuccessful but since then the parasitic braconid *Bracon variegator* Spinola, imported from France, has been successfully established on *Coleophora* spp. in seed producing regions of the South Island (Stuart 1958, Thomas 1969, French 1972). *B. variegator* and another parasitic wasp, *Neochrysocharis* sp., prefer to attack fourth instar *C. frischella* larvae which limits their effectiveness (Pearson 1982). Work on chemical control has been reviewed by French (1972). Bromophos and dichlorvos sprayed on to flowering crops on warm, calm evenings, when moth flight is at a maximum and honey bee activity at a minimum produced effective control. However, because of the risk of killing bees, insecticides should be used as a last resort, guided by proper assessment of moth populations in relation to white clover flower-head numbers. Pearson (1982) outlines an appropriate assessment technique for determining when to control *Coleophora*.

(D) Ladino clover seed midge

The Ladino clover seed midge, *Dasineura gentneri* Pritchard (Diptera: Cecidomyiidae), is a pest of Ladino white clover seed crops in Oregon and Washington State (Bishop 1954, App 1966) and of other white clovers in Great Britain (Milne 1960, 1962). It is present on white clover in Holland although of little economic importance (Nijveldt 1963). It is also a pest of red clover seed crops (Milne 1960).

The small flies emerge from the litter layer in spring to mate on flowering clover. One or two smooth shiny eggs are laid inside an open floret and give rise to light-yellow larvae. As these larvae feed on sap at the base of pods, developing seeds become shrunken and collapse. Once fully fed, the larvae fall to the ground and pupate inside white cocoons. There are three or four generations per year and larvae of the final generation become dormant in response to cool temperatures. Pupation is delayed then until the following spring.

Two hymenopterous parasites, *Tetrastichus* sp. near *marcovitchi* (Cwfd.) and *Ceraphron* sp.n. near *pacifica* (Whitt.), have been found in *D. gentneri* larvae in Oregon but do not provide effective control (Bishop 1954). Partial control can be achieved by cutting crops in early summer to desiccate larvae in flowering heads during haymaking.

Other midges have been recorded as attacking white clover. *Dasineura leguminicola* (Lintner) (clover seed midge) was recorded by Barnes (1946) as a pest of both red and white clover, but Milne (1960) showed clearly that it was only a pest of red clover seed heads. *D. trifolii* (F. Loew) (clover leaf midge) causes gall formation in leaves of red, white, zigzag and strawberry (*T. fragiferum*) clovers in the USA and Europe (Barnes 1946).

3. Foliage feeding insects

(A) Subterranean hepialid caterpillars

Numerous native subterranean caterpillars (Lepidoptera: Hepialidae) of Australia and New Zealand have become major pests of both grasses and white clover in improved pastures. In Australia the main species are the corbies, *Oncopera rufobrunnea* Tindale, *O. alboguttata* Tindale and *0. fasciculata* (Walk.), and the ghost moth, *Oxycanus fuscomaculatus* Walk., which can be found from Queensland to Tasmania (Madge 1954, Wallace 1970, Hardy 1971, 1973b, 1974, MacQuillan 1975, Milner 1977, Reid & Hardy 1978). In New Zealand most of the damage is done by species of the genus *Wiseana* (commonly called porina) which are found in southern North Island districts and throughout the South Island (Perrott 1974). Currently there are five species of *Wiseana* recognised: *W. cervinata* (Walk.), *W. umbraculata* (Gr.), *W. signata* (Walk.), *W. despecta* (Walk.), and *W. copularis* (Meyrick) (Dumbleton 1945, 1966), but the systematics of the *W. cervinata/despecta/copularis* species group is still unresolved (Dumbleton 1966). This *W. cervinata* complex is the most widely distributed species. Eleven other hepialid species exist in New Zealand but do not damage pasture (Dumbleton 1966). Because the hepialids generally have similar life histories, the *W. cervinata* complex in New Zealand, having been intensively studied for many years, will be discussed.

Details of porina biology can be found in Dick (1945), Dumbleton (1945), Cottier (1962), Pottinger (1968a, 1973, 1977), Esson (1970), French (1973), Perrott (1974), Carpenter & Wyeth (1980), and French & Pearson (1981). Porina have one generation per year. Eggs are laid on pasture or soil in spring (late October-November) or summer depending on weather conditions and species. Eggs are creamy-white, becoming black within a few hours of oviposition. Hatching occurs after 3-5 weeks if the relative humidity at the soil surface is suitable. The larvae spend the first 4-6 weeks under dense vegetation or in the top 1-2 cm of the soil. In January, they construct almost vertical tunnels 15-25 cm deep, the entrances of which are capped by loose accumulations of webbing and soil. During the day the larvae remain hidden but emerge at night to feed near the tunnel opening. The larvae take approximately 270 days to mature, passing through 6-12 instars depending on environmental conditions. Development is most rapid from February to April. By late April-early May they reach full size, and by July are fully fed. Larvae are limp, flabby, greenish-yellow caterpillars 6-7 cm long. In August they enter a prepupal stage and then pupate in September. Initially pupae are creamy in colour but soon turn dark red-brown. They move up and down their tunnels and at emergence the adults leave their pupal cases half protruding from the tunnel.

Early-flying adults of the *W. cervinata* complex emerge from September to December and late-flying adults emerge from November to February; *W. umbraculata* adults emerge from October to January. The timing depends on the preceding weather, with wetter weather delaying emergence. Adults are incapable of feeding, males and females living for an average of 7 and 9 days respectively. Mating occurs before females are capable of flight and egg deposition begins 2-30 minutes after mating. Females lay approximately 750 eggs, but fecundity appears to be correlated with larval nutrition and up to 2600 eggs can be laid. Most flight occurs immediately after sunset, particularly after days of high temperature. Although many eggs are laid close to where females emerge, flight can play an important part in spreading the pest. As females become lighter by depositing eggs, flights become longer. Emigration from bare soil to pasture with plenty of growth is particularly noticeable.

Pasture damage occurs during autumn and winter (February to August) and is first apparent when patches of bare ground appear around feeding tunnels. The bare soil often becomes invaded with composite weeds (Dumbleton & Dick 1942, McLaren & Crump 1969, Pottinger 1973, Perrott 1974). In general, an economic decrease in pasture production can occur when there are more than 20-40 larvae/m^2 (Savage & French 1981) and production losses greater than 50% can occur when populations reach 80-100 larvae/m^2 (Dumbleton & Dick 1942, McLaren & Crump 1969, Pottinger 1973). Porina populations and consequent pasture damage vary considerably with soil type, soil and sward moisture levels, sward composition, sward height and density, farm management and natural porina mortality (Pottinger 1973). With dry, slow growing, open pasture the threshold for damage is closer to 20 larvae/m^2 but with vigorously growing pasture the threshold is closer to 40 larvae/m^2. Porina densities can be easily assessed by counting the number of porina tunnels found under 32 x 32 cm boards placed over smooth, bare soil from which pasture has been scraped away (Savage & French 1981).

Porina populations can be considerably reduced by grazing pastures in October and November. The effect of removing rank pasture is to reduce moisture in the sward and thus decrease survival of eggs and first instar larvae (Taylor 1964, Pottinger 1968a, Fenemore & Allen 1969, Crawford & Kalmakoff 1978).

Rotational grazing at high stock density (500 sheep/ha for 2 days), timed to coincide with first instar larval activity in a particular region, is the most effective method of control by grazing management (French 1973). French & Savage (1981) give more details on porina control through mob stocking. Some of the losses from porina can be compensated for by applying nitrogenous fertilizer in winter to stimulate grass growth. However, the more nutritious herbage in fertilized pastures will increase growth and fecundity of porina, thus making it even more important to control eggs and first instar larvae in the summer/autumn (Ball 1969, Harris & Brock 1972, Harris 1973).

Harris (1969) suggested that white clover is preferentially grazed by porina, while Kain et al. (1979b) concluded that feeding is indiscriminate, the plants consumed being determined by their distance from the tunnels. Larvae feeding on white clover are known to have high survival rates and high weight gains (Farrell et al. 1974, Farrell 1976, 1977). Practices that increase ryegrass growth during winter may reduce pasture damage by decreasing the amount of white clover present and consequently removing the most favourable food source. In addition, older white clover plants that have begun stolon formation appear to be more tolerant and can recover more easily from defoliation (Harris 1969).

The possible use of resistant plants in pasture to inhibit porina development has been studied. Tall oat grass (*Arrhenatherum elatius*),sweet vernal (*Anthoxanthum odoratum*), crown vetch (*Coronilla varia*), *Phalaris tuberosa, Lotus pedunculatus, Lotus tenuis,* zig zag clover (*Trifolium medium*) and arrow leaf clover (*T. vesiculosum*) have shown resistance to porina attack in some trials (Farrell et al. 1974, Farrell 1976, 1977). Porina caused less damage to swards containing the legume *Lotus pedunculatus* plus perennial ryegrass, cocksfoot or a *Phalaris* spp. hybrid than to pure swards of the grasses (Farrell 1976), but successful grazing systems using these combinations have not yet been developed.

Natural control through disease, predators and parasites has also been investigated. The green muscardine fungus *Metarhizium anisopliae* (Metsch.) Sorokin is known to cause high larval mortality in some seasons and in some soils (Dumbleton 1945). The fungus may have already reached equilibrium in New Zealand soils (Latch 1965), but recent investigations suggest that greater proportions of porina populations may be killed if cereal baits covered with *Metarhizium* are spread near feeding tunnels (Latch & Kain 1983). Other diseases include the gregarine protozoan disease *Diplocystis oxycani* (Dumbleton) and an unknown microsporidian (Fleming et al. 1982). In addition granulosis virus, entomopox virus, iridescent virus and nuclear polyhedrosis virus (NPV) affect porina and these enzootic viruses could control larval populations below an economic threshold level if a stable host-virus equilibrium were established. This equilibrium could be achieved by management aimed at increasing and spreading viruses, such as oversowing rather than cultivating damaged pasture, intensive rotational grazing of young pasture and taking hay from older pastures (Fowler & Robertson 1972, Moore 1972, Moore et al. 1973, Kalmakoff & Moore 1975, Kalmakoff & Crawford 1976, Kalmakoff 1979, Fleming et al. 1982).

Little work has been done on predation and parasitism in porina. Eyles (1966) reported that the staphylinid *Thyreocephalus chloropterus* (Erichson) was a predator of larvae. Starlings, dotterels and seagulls feed on larvae, particularly after larvae have been flooded from their tunnels, but the influence of birds is small (Dumbleton 1945, Pottinger 1968a, Wood 1973). The tachinid flies *Pales usitata* (Hutton), *Protohystricia alcis* (Walk.), *Pseudoccisor versutus* (Hutton) and *Plagiomyia* sp., and the ichneumonid wasp *Pterocormus lotatorius* (F.), all parasitise some *Wiseana* species. Their effectiveness is reduced by hyperparasitism, the well protected nature of the larvae and the scarcity of suitable parasite habitats for the non-parasitic stages of their life-cycles (Eyles 1965). Three entomophagous nematodes, *Neoplectana glaseri* Steiner, *N. bibionis* Bovien and *Heterorhabditis heliothidis*, are being investigated as potential control agents (Jackson et al. 1981, Kain et al. 1981). Non-chemical control of porina is currently insufficient to significantly reduce damage, but work on increasing the effectiveness of all natural control measures is continuing.

Porina has been controlled by insecticides for many years (Waller 1968). Since the 1964 restriction on the use of DDT, the most successful chemicals for its control have been the organophosphates fenitrothion, diazinon, trichlorfon and trichloronat, applied as soon as damage becomes apparent (Pottinger 1973, French & Pottinger 1975, Barratt & Stewart 1982). Where there is spring flight, treatment should be carried out in March and where there is summer flight, in winter (July). Granular formulations should be used in wet and frosty areas. Synthetic pyrethroids appear to be detoxified by porina (Jordan & Chang 1981). Control will usually not exceed 80% and this, coupled with the cost of chemicals and the low profit margins on most farms, forces many farmers to live with, rather than control, porina populations. Recent work using kibble wheat baits coated with fenitrothion suggests that baits may provide an economically viable control (Carpenter 1982, Haack et al. 1982).

(B) Aphids

Heavy aphid (Homoptera: Aphididae) infestations can considerably reduce yields of seed crops of white clover. Trought (1977) found that blue-green lucerne aphid, *Acyrthosiphon kondoi* Shinji, caused seed yield losses of up to 20% in New Zealand. Blue-green lucerne aphid is widespread in North America, Australia and New Zealand, where it can devastate lucerne (Cox & Dale 1977, Gaynor et al. 1979, Hill 1979), and has been recorded as attacking white clover in Japan and Australia (Franzmann et al. 1979, Gonzalez et al. 1979). Pea aphid, *A. pisum* (Harris), also damages white clover seed crops in New Zealand (Trought 1980) and seedling white clover in Australia (Ridland & Berg 1981). As discussed in Chapter 11, Section 2.(D) stress owing to *A. pisum* infestation can increase the severity of fusarium root rot (Leath & Byers 1977). Pea aphid is also an important pest of lucerne and other legumes (Harper et al. 1978, Archibald et al. 1979, Gaynor et al. 1979). App (1966) reported that the clover aphid *Nearctaphis bakeri* (Cowen) may sometimes be abundant on white clover in the USA but gave no indication of the severity of damage.

Aphids are probably more significant on white clover as vectors of virus diseases (see Chapter 11, Section 3.) which can seriously affect other forage and arable crops. Groundnut crops in the USA can become infected with peanut stunt virus after the spring migration of viruliferous *Aphis craccivora* Koch from white clover (Tolin et al. 1970). Pea aphid can transmit alfalfa mosaic virus (Fry 1952, Manglitz & Kreitlow 1960), bean leaf roll virus (Cockbain & Gibbs 1973) and pea leaf roll virus (Thottappilly 1969) from infected clover to lucerne, bean and pea crops respectively. White clover is also a good host for *Aulacorthum solani* (Kalt.), which damages potato (Kazion 1971). This aphid has been shown to carry subterranean clover red leaf virus from white clover to pea and bean crops in Tasmania (Johnstone 1978).

Effective control of *A. kondoi* was achieved by Trought (1980) with bromophos and pirimicarb, the former being preferred when the spray programme was to control both aphids and clover case bearer, *Coleophora* spp. Wightmann & Whitford (1982) indicate that demeton-s-methyl and endosulfan are also effective. Frequent sampling was recommended to monitor aphid populations and spraying should be initiated when more than four aphids are found per flowering head. Numerous parasites, predators and diseases of *A. kondoi* and *A. pisum* are mentioned in the literature. Several *Aphidius* spp. have been introduced into New Zealand for biological control (Walker & Cameron 1981) and other parasites exist (Gonzales et al. 1979) but their effectiveness is often decreased by hyperparasitism. Many insects are predacious, particularly coccinellids, but they vary from country to country although their mode of action is similar and this is well discussed by Baumgaertner et al. (1981). Entomophorales fungi can significantly affect *A. kondoi* and *A. pisum* (Cameron & Milner 1981). Markkula & Roukka (1971) noted that white clover varied in its reaction to different biotypes of *A. pisum*, and Horber (1974) found some evidence of resistance to pea aphid in white clover.

(C) Leafhoppers

Potato leafhopper, *Empoasca fabae* (Harris) (Homoptera: Cicadellidae), damages clover in the USA, particularly when plant growth has been retarded by unfavourable conditions (App 1966). Leafhopper feeding causes reddening and browning of leaves. The insect overwinters in the extreme south of the country and migrates northwards in spring, producing several generations each year. Eggs are deposited in stems and large veins. Damage caused by *E. fabae* can also increase the severity of fungal root rots (Leath & Byers 1977).

Leafhoppers also transmit diseases caused by rickettsia-like organisms (rugose leaf curl) and mycoplasma-like organisms (clover phyllody), as discussed in Chapter 11, Section 4. The vector of rugose leaf curl in Australia is *Austrogallia torrida* Evans (Grylls 1954). Several leafhoppers have been reported to transmit clover phyllody and related disorders; these include *Euscelis plebejus* (Fallen) (Musil 1959, Cousin et al. 1968), *Macrosteles cristatus* Ribaut (Lehman 1973) and in North America *M. fascifrons* Stal, *Scaphytopius acutus* Say (Lee & Chiykowski 1963), *Aphrodes bicinctus* (Schr.) (Chiykowski 1970, 1975) and *Paraphlepsius irroratus* Say (Chiykowski 1965). The aster leafhopper, *M. fascifrons*, is one of the most common leafhoppers in pastures and forage crops in the USA and has a host preference for white clover (Kerr 1957, Hagel et al. 1973), while *E. plebejus* is common in European pastures. Hill (1950) reported transmission of yellow dwarf virus from *T. repens* to its tobacco host by *Orosius argentatus* (Evans).

(D) Lucerne flea

The lucerne flea, *Sminthurus viridis* (L.) (Collembola; Sminthuridae), is a serious pest of pasture legumes in Australia, New Zealand and South Africa (Dumbleton 1938, Wallace 1970, Roux 1981). It attacks white clover, subterranean clover and lucerne by removing irregular patches of the leaf surface between the leaf veins, leaving behind the cuticle. Plants develop a characteristic windowed mottling which appears as flecking under light infestations or as silvery leaves under heavy infestations. Repeated attacks can kill the clover and decrease pasture production and quality by producing nitrogen-deficient, grass-dominant weedy pasture (Pottinger 1983). It can be particularly damaging to autumn-sown emerging seedlings (Wallace 1970).

S. viridis has a characteristic globular body 3 mm long, with dark mottling and, like other springtails, an appendage which enables it to jump several centimeters. It is present from late autumn to early summer but produces a resistant egg stage during the dry summer. There are about 6 generations per year in northern New Zealand with the peak population occuring in the spring (Pottinger 1983). It is a pest in both the dry (Wallace & Mahon 1963) and cool-moist areas of Australia (Johnston 1960; Wallace 1970).

Chemical control has been investigated in Australia (Johnston 1960, Wallace 1963a, MacFarlane 1967, MacQuillan 1975) and in New Zealand (Townsend et al. 1979, Wrenn et al. 1983). Pottinger (1983) recommends the use of chlorpyrifos, dimethoate or fenitrothion whenever 50% or more of the clover leaves in a paddock show damage. MacQuillan et al. (1975) stress the importance of not depleting predator populations.

In Australia the predacious mite *Bdellodes lapidaria* (Kramer) is an effective natural check on lucerne flea populations but is limited in its distribution and is not effective in Tasmania (Ireson 1982). Another mite *Neomolgus capillatus* (Kramer) was introduced in 1969 into West Australia and gives good control but has a very slow rate of spread (Wallace 1981). Lucerne flea is not a pest in Europe (Kooistra 1964), possibly because of the influence of mites.

(E) Seed-harvesting ants

Local problems in establishing white clover and other pasture species after aerial oversowing of unploughable hill country in central and northern New South Wales, Australia, have been attributed largely to the removal of seed by ants of the species *Pheidole megacephala* (F.) and *P. bos* ssp. *baucis* Forel (Hymenoptera: Formicidae) (Campbell 1966). Both lime-pelleted and unpelleted seed are removed but losses can be reduced, and establishment increased, by insecticidal seed treatment. Campbell & Gilmour (1979) significantly reduced lucerne and *Phalaris aquatica* seed losses by coating seeds with permethrin and bendiocarb. Johns & Greenup (1976) found that seed removal is likely to be most important if oversowing is done in the summer. They have attempted to provide a model to predict seed losses in relation to plant species and environmental conditions.

(F) Crickets and grasshoppers

Crickets (Orthoptera: Gryllidae) are pests of white clover in the southern USA where *Nemobius fasciatus* De Green has caused stand depletion (Neilson & Bass 1967). Black field cricket, *Teleogryllus commodus* (Walk.), can cause considerable damage to pastures in the summer in South Australia (Browning 1954), Victoria (McDonald & Cummins 1976, Stahle & Williams 1982), New South Wales (Braithwaite 1959) and in the warm northern regions of New Zealand (Blank et al. 1982).

T. commodus adults are black to chestnut-brown, about 30 mm long, with wings folded to a point over their abdomen. Males make a distinctive loud chirping noise at night. Eggs are laid in late summer-autumn, enter a diapause during the winter and hatch in the spring. The nymphs go through about 12 instars before reaching maturity. The adults are present during summer-autumn and hide in cracks in the ground to avoid high temperatures. Dispersal flights occur in late summer (Blank 1979). *T. commodus* is particularly troublesome during oversowing because it eats the seed (Blank et al. 1980a). In established pasture, grasses are the preferred food but white clover can be completely removed in severe infestations (Blank & Olson 1979, 1981, Blank 1982a) and is preferred over many other pasture species (Blank et al. 1980c). Severe outbreaks occur during summer droughts caused by a dry spring and late autumn rains and can reduce production 23-28% (Blank 1982a).

Baits incorporating the insecticide maldison can significantly increase pasture, sheep and milk fat production during outbreaks (Blank et al. 1980b, Blank & Olson 1981, Blank 1982b, Blank et al. 1982). Control depends on the prediction of outbreaks and Blank (1982c) and Stahle & Williams (1982) discuss the problems of sampling crickets. Blank (1983) suggests that counting crickets flushed from cracks in the ground by using pyrethroid or detergent in water from early-January onwards, gives a good indication of when control is necessary. Hill (1983) recorded 43% parasitism of eggs by *Calliscelio teleogrylli* sp.n. and *Probaryconus dubius* (Nixon) in northern New Zealand. Scotti et al. (1981) summarise their work on cricket paralysis virus and Reinganum et al. (1981) discuss the possibility of using this virus and the fungus *Metarhizium anisopliae* for biological control (see also Williams et al. 1982).

Grasshoppers (Orthoptera: Acrididae) can be serious pests during outbreaks which occur when conditions are very dry. Feeding on white clover has been reported in the USA (App 1966). In Australia, the native wingless grasshopper *Phaulacridium vittatum* (Sjost) can damage improved pasture containing *T. subterraneum* (Clark 1967). More recently there have been reports of damage to pasture in the moister tablelands of northern New South Wales where white clover is included in pasture mixes (Roberts 1972). Baker (1982) suggests that *Trichospidea oestracea*, a nemestrid Dipteran parasite, may cause significant declines in *P. vittatum* populations in the Central Tablelands of New South Wales. Baker also reports that a mermithid nematode parasitizes *P. vittatum*. Chemical control has been considered by Casimir (1968) and Edge & Casimir (1975), but Davidson (1982) suggests that chemical control is not economic.

(G) Other foliage feeders

Many other insects have been recorded as feeding on, or as sporadic pests of, white clover. In Australia larvae of the pasture scarabs (Coleoptera: Scarabaeidae) *Aphodius tasmaniae* Hope in the south-east (Allen 1976) and *A. pseudotasmaniae* Given in Tasmania (Hardy 1970) are nocturnal foliage feeders. They live in tunnels in light soils and damage annual pastures and lucerne during winter. Subterranean clover is particularly susceptible to these grubs but perennial pastures containing white clover are also attacked, although they generally recover well (Carne 1956, Maelzer 1964). *A. tasmaniae* is also present in localised areas of New Zealand but is not regarded as a serious pest (Pottinger 1968b, East et al. 1980). Coles & Pinnock (1982) suggest that the fungal pathogens *Metarhizium anisopliae* and *Cordyceps aphodii* Mathieson have good potential for controlling *A. tasmaniae*.

Native New Zealand broad-nosed weevils (*Irenimus* spp., *Nicaeana cinerea* Broun) are suspected of causing significant damage to white clover oversown into high country pastures (Barratt 1982; Barratt & Johnstone 1984). Early damage to cotyledons and first true leaflets depresses plant growth and affects establishment (Barratt 1980).

Many lepidopterous larvae feed on white clover and occasionally reach pest proportions. East et al. (1978) recorded a severe outbreak of the common blue butterfly, *Zizina otis labradus* (Godt.) (Lycaenidae), on white clover in central North Island, New Zealand. *Z. o. labradus* also feeds on red clover and lucerne and is found in Australia (Myers 1921, Miller 1971). Many other lycaenid species are known to feed on clovers and other legumes throughout the world. White clover is an alternative host for the light-brown apple moth *Epiphyas postvittana* (Walk.) (Tortricidae), although it is not seriously damaged. This leaf roller is a serious pest of fruit trees in Australia and New Zealand (Thomas 1968, Danthanarayana 1975). Noctuid moths (Noctuidae), such as the green cloverworm, *Plathypena scabra* (F.), in the USA (Pedigo et al. 1973), greasy cutworm, *Agrotis ipsilon* (Hufn.), and the tropical armyworm, *Spodoptera litura* (F.), in Australia and New Zealand (Allan 1975, Cox 1976, Persson 1977) and armyworm, *Mythimna separata* (Walk.), in New Zealand (Kain et al. 1968) occasionally cause severe damage to pasture containing white clover.

Periodic outbreaks of thrips (Thysanoptera) can seriously affect seed yields of clover by feeding on the flowers and ovaries. The red clover thrips, *Haplothrips niger* (Osb.), a serious pest of red clover and an occasional pest of white clover, is present in Canada, Finland, Romania and New Zealand (Yates 1952, Loan & Holdaway 1955, Perju et al. 1975).

4. Root feeding insects

(A) Pasture scarabs

Many scarabs (Coleoptera: Scarabaeidae) have become major pests of improved and semi-improved pastures throughout the world. Most of the economically important pasture scarabs (often called cockchafers, curl grubs, white grubs, or grass grubs) are larval root feeders of the sub-families Melolonthinae, Rutelinae, Cetoniinae or Dynastinae (Ritcher 1958).

Throughout the USA *Phyllophaga* spp. are injurious to permanent pasture (Graber et al. 1931), while *Amphimallon majalis* (Raz.) (the European chafer) is a pest in eastern States (Burrage & Gyrisco 1954). In Europe larvae of *Melolontha* spp. and *Amphimallon* spp. are widespread pests of pasture (Ritcher 1958). In Great Britain, *Phyllopertha horticola* (L.) (the garden chafer), is a localised pest of pasture and the larvae are known to feed on white clover roots (Milne & Laughlin 1956, Laughlin 1957). However, the specific effects of pasture scarabs on *T. repens* in Europe are poorly documented.

Species of *Sericesthis, Rhopaea,* and *Anoplognathus* are known to damage perennial pastures in the northern tablelands of eastern Australia, where white clover is an important component (Davidson & Roberts 1969, Davidson et al. 1970). These species have life cycles of 1-3 years. Adults live for about a month, eggs take about 2 weeks to hatch and larvae remain active throughout the greater period of the life cycle, with a short pupation (Davidson et al. 1979). Damage occurs throughout the year but peaks in the autumn and winter. At present damage is not great but Davidson et al. (1979) predict that as pastures are improved in the northern tablelands, scarab populations, and hence damage to white clover-ryegrass pastures, will increase. Davidson (1982) suggests that insecticidal control would be uneconomic and that the only effective control is that exerted by weather, predators and parasites.

In New Zealand the major scarab pest is the native grass grub, *Costelytra zealandica* (White), while other native scarabs such as *Odontria* spp. and *Pyronota* spp. are important in some localities (Pottinger 1973, East 1979a, Barratt & Campbell 1982, Miln et al. 1983). Because of the severe damage to pastures in New Zealand, *C. zealandica* can probably be regarded as the most serious scarab pest of white clover in the world. Therefore a detailed account of its life history and control is given below.

Grass grub is an endemic species of the sub-family Melolonthinae present throughout New Zealand from sea level to 1200 m (Hoy 1963, Fenemore 1974). Its life cycle normally takes 1 year to complete although it can take 2 years, particularly in the south of the South Island (Perrott & Stockdill 1972, Stewart & Stockdill 1972). All phases of the life cycle remain in the soil with the exception of the adults which emerge for mating, feeding and dispersal.

The brown, 1 cm long adults emerge over 2-3 weeks from late October to early November in warmer northern soils, or from November to late December in colder southern soils (Kelsey 1968, Pottinger 1973, Fenemore 1974). They live for 3-4 weeks and sometimes damage native broadleaf and exotic deciduous trees and seedling brassica crops (Kelsey 1968). Mating occurs on the ground soon after emergence and the females then burrow back into the soil close to where they had emerged and lay clusters of small, shiny, white eggs 10-15 cm deep (Pottinger 1973, Fenemore 1974).

Larvae grow to 2 cm long, and are typical scarab larvae: white, soft bodied, with three pairs of legs, a hard, dark-brown head and are normally found curled into a C-shape in the soil (Hoy & Given 1952). Development involves three larval instars and begins with the eggs hatching in late December-early January. Pupation occurs in earthen cells 15 cm below the soil surface from September to November. Pasture damage is most noticeable from March to September when third instar larvae are present (Kelsey 1951, Pottinger 1973).

The root feeding larvae sever the roots from the aerial parts. This results in patches of plants which yellow and die in the winter and are easily pulled out by grazing animals. These patches are often soft and spongy underfoot, are easily invaded by weeds, and recover poorly in the spring (Pottinger 1973). Grass grub damage, through its effect on pasture production and hence animal production (Flay & Garrett 1942, Kain & Atkinson 1972), was estimated to cost $NZ60 million a year in 1967 (Harrison 1967).

Many natural factors can reduce damage locally. Prolonged drought, high temperatures and water saturation increase larval mortality (Wightman 1979a). Large autumn and winter starling (*Sturnus vulgaris* L.) populations can significantly reduce larval numbers (East & Pottinger 1975, Moeed 1980). Diseases also increase grub mortality. Milky disease (*Bacillus* spp.) can affect up to 58% of populations in localised areas but is generally regarded as ineffective and difficult to artificially rear (Fowler 1974). More recently a bacterium (tentatively named *Hafnia alvei*) has given good control in some situations (Jackson & Trought 1982, Trought et al. 1982).

Three protozoans, *Mattesia* sp., *Nosema takapauensis* Hall, Oliver and Given, and *Nosema costelytra* Hall, Oliver and Given, retard larval development and increase mortality, especially in late third instar larvae (Hall et al. 1976, 1977, Miln 1978, 1979). A high incidence of these diseases depends on a high grass grub density in the previous year. For this delayed density-dependent mortality to have a significant effect on reducing grass grub damage, management that increases the susceptibility of the grubs to the diseases, or pasture plants that can tolerate higher densities of grass grub, are needed (Miln 1982a,b). Other protozoan diseases (Miln 1978) and some viruses (Moore et al. 1974, Dearing et al. 1980, Wigley & Miln 1982) have also been isolated. The use of insecticides can seriously affect disease incidence and this varies with pathogen and time of application (Miln 1983).

Naturally occurring insect predators and parasites (Farrell 1972) and introduced parasites (Cameron et al. 1979) have had little effect. Merton (1982) suggests that in *Proscissio cana* Hutton, a tachinid parasite of grass grub, successful parasitism occurs only when the grass grub populations are hemivoltine and overlap. Normally, grass grub is univoltine with discrete yearly populations. Three entomophagous nematodes (*Neoplectana glaseri, Heterorhabditis heliothidis, H. bacteriophora* Poinar) are being investigated as potential control agents (Jackson et al. 1981; Jackson & Trought 1982). Up to 66% control was achieved by *N. glaseri* on small field plots (Kain et al. 1982b).

An important form of natural control is density-dependent larval mortality, which is caused primarily by larval combat at population densities above those which cause significant pasture damage (Kain 1975). East & Kain (1982), Kain et al. (1982a) and Miln (1982a) suggest that compensatory practices such as the application of fertiliser nitrogen, the planting of tolerant pasture species and the feeding of hay would enable pasture damage to be tolerated, thus allowing natural suppression to reduce grass grub numbers to a low level. East & Willoughby (1983) report that peak grass grub populations on old pastures in the central North Island declined naturally to low levels due to unexplained summer larval mortality. They found that where this natural suppression was disturbed by cultivation or using insecticides there were severe outbreaks in later years. They recommend that only small areas of a farm should be disturbed at any one time so as to avoid synchronising outbreaks on a large scale.

Management can also greatly influence grub numbers. Cultivation between late September and early October, while pupae are present, can reduce subsequent grub numbers by 90% but is limited to areas where early spring cultivation is possible (Kain & Atkinson 1970). Mob-stocking in winter or summer can significantly increase grub mortality in proportion to the stocking rate (Farrell 1972, Kain & Atkinson 1970, 1972, East 1979b). Summer hard grazing can raise soil temperatures to lethal levels for grass grub (East & Willoughby 1980). Kain et al. (1982a) suggest that treading and rolling pasture in late autumn-winter when the soil is at intermediate field water capacity can reduce grub numbers. Stewart & van Toor (1983) found rolling to be as effective as insecticide control and less expensive. But the most important management practices appear to be the use of insecticides and resistant or tolerant plants.

Trought (1979) has reviewed the extensive literature on the chemical control of grass grub. Since the banning of DDT for use in pastoral land in New Zealand in 1970, no chemical has been developed which gives reliable control in all New Zealand climates and soil types. Of the short-lived insecticides currently being used and which are broadcast as granules, diazinon and ethoprophos vary greatly in effectiveness from place to place, while fensulfothion can be more widely applied. Fensulfothion, however, is expensive, has a high mammalian toxicity which requires a long withholding period, and is therefore often difficult to fit into farm rotations. Lindane gives more consistent results, particularly when used against adult beetles (Henzell et al. 1981, Lauren et al. 1982), but is already restricted in use because of residue problems (Thomson & Lagan 1982) and some grub resistance to it is apparent (Trought 1979). Henzell et al. (1983) indicate that two synthetic pyrethroids and deltamethrin may be suitable alternatives to lindane in this context.

Withholding periods can be reduced and the effectiveness of the insecticide increased by drilling the granules into the pasture (Carpenter et al. 1981, Kain et al. 1982a), but Trought (1979) and East et al. (1981) suggest that diazinon liquid formulations applied in lines as a non-atomizing jet can give good, less expensive control. Kain & Atkinson (1970) suggest that as over 90% of a grass grub population is confined to the visibly affected area, or within a surrounding 1 metre margin, fensulfothion can give good control if applied over the damaged areas. A single application of deltamethrin made when pheromone traps indicate that adult flight has begun (Henzell & Lauren 1977), will reduce adult numbers before oviposition and consequently reduce grub numbers (Venning et al. 1982). Double and triple applications of synthetic pyrethroids give reliable control but are more expensive (Pottinger et al. 1977, 1979, Henzell & Lauren 1981, Lauren et al. 1978, 1981). The variability of results and increasing costs of chemicals has stimulated more work on resistant and tolerant plants.

The attractiveness of white clover to grass grub is well attested (Radcliffe 1970, 1971, Farrell & Sweney 1972, 1974, Kain & Atkinson 1970, 1977, Sutherland & Hillier 1974, Farrell 1975, Kain et al. 1979a,c, Wilson 1978a, b, c). The density of grass grub populations depends greatly on the density of white clover in the pasture. Larvae aggregate more under white clover than ryegrass; and during early winter, when larvae are most susceptible to environmental changes, white clover provides an important source of food (Kain & Atkinson 1977, Kain et al. 1979a, c). This has prompted research into species which can resist or tolerate attack better than white clover and ryegrass (Kain et al. 1979a). So far resistance has been found in *Lotus pendunculatus*, lucerne (*Medicago sativa*), *Phalaris aquatica*, Yorkshire fog (*Holcus lanatus*), *Poa pratensis* and blue lupin (*Lupinus angustifolius*) (Farrell & Sweney 1974, Farrell 1977, Kain & Atkinson 1977). Lucerne is now commonly used in rotation with pasture to reduce grass grub populations (Kain & Atkinson 1970, Farrell & Sweney 1972). Kain et al. (1982a) report that a white clover selection (Grasslands G18) appears to have a high degree of tolerance to grass grub. Tolerance has been found in cocksfoot (*Dactylis glomerata*) and tall fescue (*Festuca arundinacea*) (Farrell & Sweney 1974, East & King 1977, Kain et al. 1979c, East et al. 1982). Kain (1975) and East (1979a) suggest that tolerant pasture combinations may raise the threshold population density for density-dependent larval mortality to regulate populations.

Screening for resistance in white clover has so far been unsuccessful (Wilson 1978a,b,c, I.Verry personal communication) but further screening of white clover ecotypes is continuing (D.L. Gaynor, unpublished). Sutherland et al. (1982b) have reviewed 10 years investigation into the chemical relationship between plants and grass grub feeding and survival. Isoflavonoids have been isolated as grass grub feeding deterrents from the roots of the resistant legumes *Lotus pedunculatus* (Russell et al. 1978) and *Lupinus angustifolius* (Lane et al., in press). Aliphatic nitro compounds isolated from *Lotus pedunculatus* (Hutchins et al. 1984) and saponins from *Medicago sativa* (Sutherland et al. 1982a) are toxins as well as feeding deterrents to grass grub larvae. Isoflavonoids varying widely in feeding deterrent activity have been isolated from a range of legumes, including white clover (Russell et al. 1979, Sutherland et al. 1980). Work is currently directed towards the possibility of introducing effective feeding deterrent activity into white clover roots by modifying isoflavonoid biosynthesis in the plant, or introducing the ability to synthesise other defensive compounds. Increasing the concentration of biologically active compounds in the roots may be associated with an increase in their concentration in the foliage which could be detrimental to stock. Thus, adequate testing of animal performance will be necessary before the release of any new plant material.

Until suitable alternatives to susceptible ryegrass-white clover pastures are developed, grass grub will remain the major pest of white clover in New Zealand. Control will depend on the successful integration of all the control methods outlined above (Kain 1979). Kain et al. (1982a) outline an integrated pest management system for grass grub in an intensive dairy farm. They emphasise that accurate prediction of grub density and animal production losses caused by the grubs on individual farms is important in implementing the programme.

(B) *Sitona* weevils

Weevils of the genus *Sitona* (Coleoptera: Curculionidae) damage several economically important legumes, including peas, faba beans, lucerne and clovers. The adults eat characteristic semicircular notches in the edge of leaves (Morrison et al. 1974, Anon 1977). White clover supports populations of *S. lincella* (Bonsdorff), *S. lepida* Gyll., *S. hispidulus* (F.) and *S. lineatus* (L.) in Europe (Wagn 1954, Markkula 1959, Markkula & Köppä 1960), *S. discoideus* Gyll. in New Zealand (Wightman 1979b; Goldson & French 1983) and an unknown *Sitona* sp. in India (Hameed et al. 1978). *S. lineatus* feeds on white clover in the summer in the western USA but its preferred hosts are peas and beans (Fisher & O'Keefe 1979a, b). Only *S. hispidulus*, the clover root curculio, is considered a pest of white clover. Adults usually cause little damage but root feeding by the larvae may seriously affect productivity and persistence of white clover in North America (App 1966, Thompson & Willis 1967, James et al. 1979) and Japan (Sakamoto 1980).

The life cycle of *S. hispidulus* in central USA, where detailed studies have been carried out, is described by Bigger (1930), Marshall & Wilber (1934), Underhill et al. (1955), and Leibee et al. (1980). The black-brown adults (3-5 mm long) emerge in mid-summer after a pupation period of about 3 weeks but are present throughout the year. After emergence most adults move to field borders where they remain relatively inactive before returning in the autumn (Leibee et al. 1981). Oviposition begins in the autumn and eggs and possibly young larvae overwinter in the soil, but most eggs are laid in the spring when temperatures rise above 8°C (Brudea 1980). The white, fleshy larvae have prominent yellow-brown heads and grow through 5 instars from about 0.7 up to 5 mm in length during the spring and early summer. During this period they feed progressively on nodules, feeder roots and thicker tap roots in the top 15 cm of soil. Similar damage is done to other legumes such as lucerne, sweet clover, red clover and alsike clover. Problems of persistence in these legumes and white clover might be related to an interaction between the curculio and pathogenic micro-organisms (Dickason et al. 1958, 1968, Newton & Graham 1960, Thompson & Willis 1967, James et al. 1979) and this possibility is further discussed in Chapter 11, Section 2.(D). Thompson & Willis (1971) report that white clover has no resistance to *S. hispidulus* larvae or adults. Byers & Kendall (1982) confirm this finding and report resistance in *Lotus corniculatus, Coronilla varia* and *Vicia grandiflora* var. Kitaibeliana. Powell et al. (1983), however, report on 4 genotypes of Ladino white clover which appear to be resistant to *S. hispidulus.*

No systematic attempt appears to have been made to assess control methods or the economic importance of *S. hispidulus* on white clover. However, insecticidal treatments for use on lucerne (Underhill et al.

1955, Turner 1957, Neal & Ratcliffe 1975) and alsike clover (Dickason et al. 1958) have been studied. In Illinois one cricket and three carabid predators of *S. hispidulus* have been identified (Barney et al. 1979; Barney & Armbrust 1980). Yeargan & Shuck (1981), and Bloem & Yeargan (1982a, b) discuss possible reasons for the unsuccessful attempt to control *S, hispidulus* in the USA by importing the egg parasite *Patasson ameerei* (Debauche) from the Mediterranean.

(C) White-fringed weevils

The South American white-fringed weevils, *Graphognathus* spp. (Coleoptera: Curculionidae), have become pests of a wide range of legumes in South America, south-eastern USA, southern South Africa, south-eastern Australia and New Zealand (Perrott 1964, Wallace 1970, Gross et al. 1972, May 1975, Warner 1975, Ottens & Todd 1980). *G. leucoloma* (Boh.), the most widely distributed species, is the only species found in Australia and New Zealand. The weevils are very damaging to legume crops, such as groundnuts and lucerne, and can significantly reduce the white clover component of pastures, with corresponding increases in the weed or grass component, or both (Cumber 1960, Wallace 1970, King & East 1979, East 1982, King et al. 1982).

The large (10 x 4 mm) grey adult weevils have a whitish band along the margins of the fused elytra. They cannot fly, but are active on the ground from mid-summer to autumn. The adults feed on legume foliage but cause little damage. Reproduction is parthenogenetic, with the adult population composed entirely of females which can lay more than 1000 eggs each. The soft, white, oval eggs are laid in clusters of 10-80 in surface litter and are cemented together with mucus, enabling them to withstand drought. Hatching occurs 8 days after the eggs become moist and the creamy coloured, legless larvae then enter the soil. The larvae are root feeders and are the main damage-causing stage in crops and pasture. In warm areas the life cycle takes one year, larval damage occurring during autumn and early winter, but in colder areas larvae can take up to 17 months to develop. Pupation occurs in the spring (Todd 1964, May 1975, Steven 1979).

Adult fecundity and longevity is greatly increased by the presence of white clover in pasture (East 1976, 1977a, b, East & King 1977). The weevil population rapidly increases in clover-dominant pasture, and reductions of 18-58% in annual dry matter production of white clover have been recorded. Summer and autumn losses can be as high as 70% (King & East 1979, East 1982, King et al. 1982). New pastures, or legume crops sown immediately after heavily infested old pasture, are often difficult to establish because of the complete destruction of emerging seedlings by the larvae (Wallace 1970, East 1976, Morton & Roberts 1978). Consequently, control measures have concentrated on preventing re-establishment of the weevils, particularly in new legume crops, rather than the removal of existing populations. This can be achieved by intensive cultivation in spring or early summer, or by a summer-autumn fallow. A summer crop of sudax (*Sorghum bicolor* (L.) Moench x *S. sudanense* (Piper) Stapf) can also greatly decrease the carried-over weevil population (East & Parr 1977a).

Chemical control of larvae in lucerne is not considered economic (Todd 1964, 1969), but frequent applications of a range of chemicals during the main period of adult activity can give adequate control (East & Parr 1977b). Ottens & Todd (1979) suggest that diflubenzuron can cause reproductive inhibition when applied to eggs, larvae and adults.

Because white-fringed weevil outbreaks are sporadic, control in pasture is difficult. Lucerne and red clover can tolerate damage more than white clover and *Lotus* spp., and subterranean clover and *Desmodium* spp. are unfavourable for the weevils (East 1982, King et al. 1982, East 1983). In Argentina the larvae are attacked by nematodes, bacteria, fungi, mites and carabid beetles, and the adults are attacked by protozoa (Ahmad 1974). The possibility of biological control in New Zealand by insect predators and parasites has not been investigated, but three entomophagous nematodes, *Neoplectana glaseri, Heterorhabditis heliothidis* and *H. bacteriophora*, are being investigated as potential control agents (Jackson et al. 1981). In Australia a 79% reduction in a *G. leucoloma* population was believed to be associated with the presence of *Heterorhabditis* spp. (Sexton & Williams 1981).

(D) Clover root weevils

The clover root weevils *Amnemus quadrituberculatus* (Boh.) and *A. superciliaris* (Pasc.) (Coleoptera: Curculionidae) are indigenous Australian weevils which can be very damaging to many clovers, including white clover. They are present only in coastal New South Wales and southern Queensland on well drained soils (Braithwaite et al. 1958; Wallace 1970). The white, legless larvae damage a range of clovers during winter by severing the tap roots about 1 cm below the soil surface. Fibrous roots can develop above the severed tap root but the weakened plants are susceptible to moisture stress and often succumb to heavy grazing. The tap-rooted subterranean clover is more susceptible to damage than the stoloniferous white clover. Severely damaged white clover fails to respond to irrigation and fertilizers.

The adult weevils, 7-12 mm long and fawn to brick-red, do not fly but are active from spring to early winter, feeding on the leaves and petioles of a range of forage legumes but doing little damage. Eggs are laid in the central hollow of the petiole about 2 weeks after adult emergence. Emerging larvae fall to the soil and, from March to November, develop through 7 instars. Pupation occurs in the spring (Braithwaite et al. 1958).

Populations of the clover root weevils appear to exist in close association with white clover. Braithwaite et al. (1958) suggest that white clover, which occurs naturally in this area of Australia, is probably the natural food-plant of the weevil. Braithwaite & Jane (1963) and Braithwaite et al. (1965) showed that granular applications of dieldrin lowered larval numbers. Smith (1978) controlled *A. superciliaris* with dieldrin and heptachlor on *Desmodium* spp. but these insecticides are no longer recommended because of residue problems (Gilbert & Lewis 1982). Wilson (1980) reported that the subtropical pasture legume Bargoo jointvetch (*Aeschynomene falcata*) is highly resistant to *A. quadrituberculatus*.

5. Mites

Mites attack white clover and sometimes seriously affect seed yields by reducing plant vigour. The two-spotted mite, *Tetranychus urticae* Koch, and another spider mite, *T. turkestani* (Ugarov & Nikolski) (Acari: Tetranychidae), have caused serious damage to white clover in the USA and Japan, particularly in areas under irrigation (Klostermeyer 1954, Cagle 1956, App 1966, Mori & Moriyama 1970). They feed during summer on the underside of leaves and spin delicate webs which can eventually envelop the whole plant. Infested leaves have yellow blotches which range in size from small specks to large areas. Eventually, plant vigour and consequently seed production are affected. The predacious mite *Phytoseiulus persimilis* Athias-Henriot can control *T. urticae* on white clover, and *Amblyseius californicus* (McGregor) and *Typhlodromus occidentalis* Nesbitt can also control it on strawberry (a more seriously affected crop) (Oatman et al. 1977). Mori & Saito (1979) have controlled *T. urticae* on red clover in glasshouses using *P. persimilis, Neoseiulus longispinosus* (Evans) and *Amblyseius deleoni* Muma & Denamark.

The legume mite *Petrobia apicalis* (Banks) (Tetranychidae) has been recorded as a serious pest of white clover in the southern USA (Smith & Weber 1954, Zein-Eldin 1956). These mites feed during winter on the upper surface of the leaf and severely damaged plants are unattractive to grazing animals and produce little or no seed. Control can be effected by maldison and demeton (Smith & Weber 1954). The clover mite, *Bryobia praetiosa* Koch (Tetranychidae), feeds on many herbaceous plants, including grasses and clovers, but is not regarded as a serious pest of white clover (Jeppson et al. 1975). *B. repensi* Manson, has been reported as a pest of white clover in Australia (Hamilton 1972). The red-legged earth mite, *Halotydeus destructor* (Tucker) (Eupodidae), is a widespread pest of autumn-sown legume seedlings in areas with Mediterranean climates in south-eastern and south-western Australia (Wallace 1970). In these areas subterranean clover is the most widely grown legume, and establishment can be severely reduced by mite attack (Norris 1948). Control has concentrated on treating legume seed with systemic organophosphates (Wallace 1963b, Smith 1968). *H. destructor* has not been reported as a pest of white clover in Australia and is regarded as only a sporadic pest in New Zealand (Dumbleton 1947, Atkinson et al. 1956).

6. Slugs

Slugs (Gastropoda: Pulmonata) cause serious economic losses to potato, cereal and some horticultural crops in moist temperate climates (Stephenson & Bardner 1976). The major species of slugs found in pastures world-wide appear to be *Deroceras reticulatum* (Muller), *Arion hortensis* Ferussac and *Arion ater* L. (Runham & Hunter 1970, Barker 1979). Martin (1978) and Barker (1979) provided keys to the introduced slugs and snails of New Zealand and coloured illustrations of some common British slugs have been reproduced by Jones & Jones (1974). Extensive accounts of the biology, behaviour and economic importance of slugs have been given by Stephenson (1968), Runham & Hunter (1970), and Stephenson & Bardner (1976). Barker & Pottinger (1981) list eleven major slug species present in New Zealand and outline their biology and control. Banded snails, *Cepaea nemoralis* (L.) and *C. hortensis* (Muller), are troublesome in leguminous forage crops in Britain (Jones & Jones 1974), but snails are generally thought to be less damaging than slugs.

Although damaging effects on white clover in pastures are not well documented, slugs, or their characteristic slime trails and signs of feeding activity (irregular holes in the leaf lamina, often with a window of remaining cuticle), are frequently seen on white clover plants. Damage to Ladino white clover caused by the slugs *Deroceras agreste* L. and *Arion ater* was thought by Howitt (1961) to be responsible for a rapid decline in the clover component of irrigated cocksfoot dairy pastures in Washington State. In New Zealand Charlton (1978a, b, 1979) considered that slugs could be a major reason for the poor establishment of white clover in moist hill country after spring oversowing. Blank & Bell (1982) reported that the slugs *D. reticulatum* and *D. panormitanum* (Lesson & Pollonicra) cause low levels of legume seed loss during winter and early spring and show a preference for white clover seed. Barker et al. (1983) have shown that *D. reticulatum* prefer to eat white clover over most other pasture plants in New Zealand. Seedling loss in crops direct drilled into pastures can also be high, because slugs preferentially move into the slit furrow left by the drill and eat seedlings along the seeded rows (Charlton, personal communication). Ferguson & Barratt (1983) found that *D. reticulatum* seriously affected the establishment of the white clover component of ryegrass/white clover mixtures direct-drilled into pasture in the spring.

Slugs are adapted to live in a cool, moist environment. Most species lay 2-3 clusters of eggs in the soil, with up to 300 eggs per cluster. Eggs are laid during autumn, winter and early spring, which is the period of peak slug populations and feeding activity. Feeding occurs in conditions of 100% relative humidity and generally in temperatures of less than 20-30°C (Stephenson 1968, Runham & Hunter 1970). In summer, activity decreases as slugs move further down into the soil to maintain a moist environment, but may be resumed after heavy rain or irrigation (Howitt 1961). Some species feed more often above the ground (e.g. *D. reticulatum*) while others are more subterranean (e.g. *Arion* spp.) (Charlton, personal communication). *Deroceras* spp. may undergo two or more generations per year under favourable conditions.

A small proportion of plants within any natural population of white clover are cyanogenic. When damaged, the plants release hydrogen cyanide by enzymatic hydrolysis of cyanogenic glucosides. Mature cyanogenic white clover plants are reported to be less intensively grazed by slugs and other invertebrates than acyanogenic plants (Corkill 1952, Whiteman 1973, Angeesing 1974), but this deterrence by the plants seems ineffective at the seedling stage (Charlton 1979). Furthermore, cyanogenic plants show poor cold and drought tolerance (Angeesing & Angeesing 1973).

Two chemicals, metaldehyde and methiocarb, have been widely used for slug control, particularly in non-pastoral agriculture. Both act mainly as stomach poisons and are usually mixed with bran or bran extracts which act as feeding attractants. Van der Gulik & Springett (1980) have reported that, under laboratory conditions, several other chemicals have molluscicidal activity. However, thimet, which they reported to have 100% molluscicidal activity, has since proved ineffective in the field (Charlton, personal communication).

Stephenson & Bardner (1976) discussed improvements in formulation of slug baits to increase their kill, resistance to moulds and persistence on the soil surface during wet weather. Clearly, no formulation is ideal but the use of hardened gelatine pellets is showing promise. A new formulation being investigated in

England using metaldehyde mixed with protein extract appears particularly interesting because it can kill both adult slugs and developing embryos (Charlton, personal communication). It seems unlikely, however, that molluscicides will ever be used widely to reduce slug damage in established pastures, although Howitt (1961) recommended routine incorporation of metaldehyde in sprinkler irrigation water for slug control in the north west USA. Chemical control during white clover establishment may be economic. Charlton (1978a,b) showed that methiocarb applied as a seed coating in gum arabic will greatly increase the numbers of seedlings which survive in turf blocks bearing a mixed population of *D. reticulatum* and *A. hortensis*. Some improvement in formulation appears necessary to extend the active life of the protectant under hill country oversowing conditions (Charlton 1978b).

References

Ahmad, R. (1974) Studies on *Graphognathus leucoloma* (Boh.) (Col:Curculionidae) and its natural enemies in the central provinces of Argentina. *Technical Bulletin, Commonwealth Institute of Biological Control* **17**, 19-28.

Allan, D.J. (1975) Greasy cutworm, *Agrotis ipsilon* Hufnagel, life-cycle. *Information Series, Department of Scientific and Industrial Research, New Zealand* **No. 105/11**, 4 pp.

Allen, P.G. (1976) Alternative insecticides for the control of the pasture cockchafer (*Aphodius tasmaniae*) in pasture. *Australian Journal of Experimental Agriculture and Animal Husbandry* **16**, 755-757.

Angeesing, J.P.A. (1974) Selective eating of the acyanogenic form of *Trifolium repens*. *Heredity* **32**, 73-83.

Angeesing, J.P.A.; Angeesing, W.J. (1973) Field observations on the cyanogenesis polymorphism in *Trifolium repens*. *Heredity* **31**, 276-282.

Anonymous (1977) Pea, bean and clover weevils. *Advisory Leaflet, Ministry of Agriculture, Fisheries and Food* **No. 61**, 4 pp.

App, B. A. (1966) Insect pests. Gibson, P.B. and Hollowell, E.A. (Eds). In: White clover. *Agricultural Handbook, US Department of Agriculture*, **No. 314**, pp.19-21.

Archibald, R.D.; Cox, J.M.; Deitz, L. (1979) New records of plant pests in New Zealand. III. Six species of Homoptera. *New Zealand Journal of Agricultural Research* **22**, 201-207.

Atkinson, J.D.; Brien, R.M.; Chamberlain, E.E.; Cottier, W.; Dingley, J.M.; Jacks, H.; Reid, W.D.; Taylor, G.G. (1956) Plant Protection in New Zealand. Wellington, New Zealand; Government Printer, 288-290.

Baker, G.L. (1982) The seasonal abundance of Dipterous and Mermithid parasites of wingless grasshopper, *Phaulacridium vittatum* (Sojst) in the central tablelands of New South Wales. *Proceedings of the 3rd Australasian Conference of Grassland Invertebrate Ecology*, 169-176

Ball, R. (1969) Legume and fertilizer nitrogen in New Zealand pastoral farming. *Proceedings of the New Zealand Grassland Association* **31**, 117-126.

Barker, G.M. (1979) The introduced slugs of New Zealand (Gastropoda: Pulmonata). *New Zealand Journal of Zoology* **6**, 411-437.

Barker, G.M.; Pottinger, R.P. (1981) Slugs biology, damage and control in nurseries, vegetables, cereals and pasture. *Aglink FPP 439, MAF Information Services, Wellington, New Zealand*, 2 pp.

Barker, G.M.; Willoughby, B.E.; Pottinger, R.P. (1983) Feeding by slugs on some pasture plant species: laboratory experiments. *Proceedings of the 36th New Zealand Weed and Pest Control Conference*, 207-211.

Barnes, H.F. (1946) Gall Midges of Economic Importance. Vol. II. Fodder Crops. London, UK; Crosby Lockwood, 26-38.

Barney, R.J.; Armbrust, E.J. (1980) Field predation of alfalfa weevil and clover root curculio adults. *Journal of Economic Entomology* **73**, 599-560.

Barney, R.J.; Roberts, S.J.; Paush, R.D.; Armbrust, E.J. (1979) Insect predators of the alfalfa weevil and clover root curculio (Coleoptera: Curculionidae) during fall field reentry. *Great Lakes Entomologist* **12**, 153-155.

Barratt, B.I.P. (1980) Effect of simulated insect damage to white clover seedlings. *Proceedings of the 33rd New Zealand Weed and Pest Control Conference*, 49-51.

Barratt, B.I.P. (1982) Insect damage to white clover oversown into New Zealand native tussock grassland. *Proceedings of the 3rd Australasian Conference of Grassland Invertebrate Ecology*, 25-31.

Barratt, B.I.P.; Campbell, R.A. (1982) Biology of the striped chafer, *Odontria striata* (Coleoptera: Scarabaeidae). 1. The adult flight and ground surface activity, female reproductive maturation, and food plant selection. *New Zealand Journal of Zoology* **9**, 249-266.

Barratt, B.I.P.; Johnstone, P.D. (1984) Effects of insects, seeding rates, and insecticide seed dressing on white clover and Maku lotus in tussock grassland. *New Zealand Journal of Agricultural Research* **27**, 13-18.

Barratt, B.I.P.; Stewart, K.M. (1982) Chemical control of striped chafer and porina in improved tussock grassland. *Proceedings of the 35th New Zealand Weed and Pest Control Conference*, 107-109.

Baumgaertner, J.U.; Frazer, B.D.; Gilbert, N.; Gill, B.; Gutierrez, A.P.; Ives, P.M.; Nealis, V.; Raworth, D.A.; Summers, C.G. (1981) Coccinellids (Coleoptera) and aphids (Homoptera). The overall relationship. *Canadian Entomologist* **113**, 975-980.

Bigger, J.H. (1930) Notes on the life history of the clover root curculio *Sitona hispidula* Fab. in central Illinois. *Journal of Economic Entomology* **23**, 334-341.

Bishop, G.W. (1954) Life history and habits of a new seed midge *Dasyneura gentneri* Pritchard. *Journal of Economic Entomology* **47**, 141-147.

Blank, R.H. (1979) Black field cricket. Biology, damage and control. *Aglink FPP 223, MAF Information Services, Wellington, New Zealand*. 2 pp.

Blank, R.H. (1982a) The pest status of *Teleogryllus commodus* in New Zealand. *Proceedings of the 3rd Australasian Conference of Grassland Invertebrate Ecology*, 33-40.

Blank, R.H. (1982b) Control of *Teleogryllus commodus* using insecticidal baits and other agents. *Proceedings of the 3rd Australasian Conference of Grassland Invertebrate Ecology*, 355-363.

Blank, R.H. (1982c) A preliminary investigation of the prediction and monitoring of black field cricket, *Teleogryllus commodus*, outbreaks (Orthoptera: Gryllidae). *New Zealand Entomologist* **7**, 231-236.

Blank, R.H. (1983) Crickets. Assessment and control. *Aglink FPP 720, MAF Information Services, Wellington, New Zealand*, 3 pp.

Blank, R.H.; Bell, D.S. (1982) Seasonal patterns of pasture seed loss to black field cricket (*Teleogryllus commodus*) and other invertebrates. *New Zealand Journal of Agricultural Research* **25**, 123-129.

Blank, R.H.; Bell, D.S.; Page, C.R.; Olson, M.H. (1982) Cricket damage on the Kaipara clay flats, Ruawai in 1982. *Proceedings of the 35th New Zealand Weed and Pest Control Conference*, 113-118.

Blank, R.H.; Bell, D.S.; Morgan, H.M. (1980a) Black field crickets consume oversown pasture seed. *Proceedings of the 33rd New Zealand Weed and Pest Control Conference*, 37-40.

Blank, R.H.; Bell, D.S.; Olson, M.H. (1980b) Screening baits for use against black field crickets. *Proceedings of the 33rd New Zealand Weed and Pest Control Conference*, 41-45.

Blank, R.H.; Olson, M.H. (1979) An assessment of the damaging stages of *Teleogryllus commodus* (Orthoptera: Gryllidae). *Proceedings of the 2nd Australasian Conference of Grassland Invertebrate Ecology*, 121-123.

Blank, R.H.; Olson, M.H. (1981) The damage potential of the black field cricket *Teleogryllus commodus*. *New Zealand Journal of Agricultural Research* **24**, 251-258.

Blank, R.H.; Olson, M.H.; Cox, N.R.; Bell, D.S. (1980c) Black field cricket (*Teleogryllus commodus*) food preferences between sixteen pasture species. *New Zealand Journal of Agricultural Research* **23**, 409-415.

Bloem, K.A.; Yeargan, K.V. (1982a) Host-finding behaviour of *Patasson lameerei* (Hym: Chalcidoidea) a parasitoid of *Sitona* eggs (Col: Curculionidae). *Entomophaga* **27**, 93-97.

Bloem, K.A.; Yeargan, K.V. (1982b) Effects of extreme temperatures on *Patasson lameerei* survival and oviposition. *Journal of the Kansas Entomological Society* **55**, 159-169.

Braithwaite, B.M. (1959) Insect pests of pastures on the north coast of New South Wales. *Agricultural Gazette of New South Wales* **70**, 589-597.

Braithwaite, B.M.; Gilbert, W.S.; Jane, A. (1965) The effect of formulation and method of applying dieldrin on the control of clover root weevil (*Amnemus* spp.) and on residue levels on pastures. *Australian Journal of Experimental Agriculture and Animal Husbandry* **5**, 232-235.

Braithwaite, B.M.; Jane, A. (1963) Problems in assessing insecticidal control of *Amnemus* species (Coleoptera, Curculionidae) in clover pastures of north coastal N.S.W. *Bulletin of Entomological Research* **54**, 523-533.

Braithwaite, B.M.; Jane, A.; Swain, F.G. (1958) *Amnemus quadrituberculatus* (Boh.) a weevil pest of clover pastures on the north coast of N.S.W. *Journal of the Australian Institute of Agricultural Science* **24**, 146-154.

Browning, T.O. (1954) Observations on the ecology of the Australian field cricket, *Gryllulus commodus* Walker in the field. *Australian Journal of Zoology* **2**, 205-222.

Brudea, V. (1973) Date privind răspîndirea şi biologia principalelor specii de *Apion* dăunătoare roşu. *Probleme de Protectia Plantelor* **1**, 260-269.

Brudea, V. (1980) [Contributions to the biology of the weevil *Sitona hispidulus* F. (Order Coleoptera - Family Curculionidae), a pest of leguminous forage crops.] *Probleme de Protectia Plantelor* **6**, 105-116.

Buhl, C.; Schütte, F. (1964) Über Auftreten und Bekämpfung von *Apion flavipes* Payk. in Weissklee-Samenbeständen. *Zeitschrift für Pflanzenkrankheiten, Pflanzenpathologie und Pflanzenschutz* **71**, 189-197.

Burrage, R.H.; Gyrisco, G.G. (1954) Estimates of populations and sampling variance of European chafer larvae from samples taken during the first, second and third instar. *Journal of Economic Entomology* **47**, 811-817.

Byers, R.A.; Kendall, W.A. (1982) Effects of plant genotypes and root nodulation on growth and survival of *Sitona* spp. larvae. *Environmental Entomology* **11**, 440-443.

Byrne, H.D.; Blickenstaff, C.C. (1968) Host plant preference of the alfalfa weevil in the field. *Journal of Economic Entomology* **61**, 334-335.

Cagle, L.R. (1956) Life history of spider mite *Tetranychus atlanticus* McGregor. *Technical Bulletin, Virginia Agricultural Experiment Station* **No. 124**, 22 pp.

Cameron, P.J.; Milner, R.J. (1981) Incidence of *Entomophthora* spp. in sympatric populations of *Acyrthosiphon kondoi* and *A. pisum*. *New Zealand Journal of Zoology* **8**, 441-446.

Cameron, P.J.; Valentine, E.W.; Butcher, C.F. (1979) Prospects for biological control of pasture Scarabaeidae (Coleoptera) in New Zealand. *Proceedings of the 2nd Australasian Conference of Grassland Invertebrate Ecology*, 213-216.

Campbell, M.H. (1966) Theft by harvesting ants of pasture seed broadcast on unploughed land. *Australian Journal of Experimental Agriculture and Animal Husbandry* **6**, 334-338.

Campbell, M.H.; Gilmour, A.R. (1979) Reducing losses of surface sown seeds due to harvesting ants. *Australian Journal of Experimental Agriculture and Animal Husbandry* **19**, 706-711.

Carne, P.B. (1956) An ecological study of the pasture scarab *Aphodius howitti* Hope. *Australian Journal of Experimental Agriculture and Animal Husbandry* **4**, 259-314.

Carpenter, A. (1982) Field evaluation of baits for porina control. *Proceedings of the 35th New Zealand Weed and Pest Control Conference*, 110-112.

Carpenter, A.; Wyeth, T.K. (1980) Porina moth flight activity in the southern North Island. *Proceedings of the 33rd New Zealand Weed and Pest Control Conference*, 30-32.

Carpenter, A.; Wyeth, T.K.; Allan, P.J.T. (1981) Chemical control of grass grub in the southern North Island: results of 1977-79 trials. *Proceedings of the 34th New Zealand Weed and Pest Control Conference*, 225-228.

Casimir, M. (1968) Control of grasshoppers (*Phaulacridium vittatum*) by ultra low volume aerial application of technical maldison. *Australian Journal of Experimental Agriculture and Animal Husbandry* **8**, 631-636.

Charlton, J.F.L. (1978a) Control of slug damage in oversown pasture legumes. *Proceedings of the 31st New Zealand Weed and Pest Control Conference* **31**, 127-130.

Charlton, J.F.L. (1978b) Slugs as a possible cause of establishment failure in pasture legumes oversown in boxes. *New Zealand Journal of Experimental Agriculture* **6**, 313-317.

Charlton, J.F.L. (1979) Effects of slugs during establishment of oversown legumes in box experiments. *Proceedings of the 2nd Australasian Conference of Grassland Invertebrate Ecology*, 253-255.

Chiykowski, L.N. (1965) Transmission of clover phyllody virus by the leafhopper *Paraphlepsius irroratus* (Say). *Canadian Entomologist* **97**, 1171-1173.

Chiykowski, L.N. (1970) Notes on the biology of the leafhopper *Aphrodes bicincta* (Homoptera: Cicadellidae) in the Ottawa area. *Canadian Entomologist* **102**, 750-758.

Chiykowski, L.N. (1975) *Aphrodes bicincta* as a vector of the clover phyllody agent. *Annals of the Entomological Society of America* **68**, 645-648.

Clark, D.P. (1967) A population study of *Phaulacridium vittatum* Sjort (Acrididae). *Australian Journal of Zoology* **15**, 799-872.

Cockbain, A.J.; Gibbs, A.J. (1973) Host range and overwintering sources of bean leaf roll and pea enation mosaic viruses in England. *Annals of Applied Biology* **73**, 177-187.

Coles, R.B.; Pinnock, D.E. (1982) Control of the pasture cockchafer with the fungal pathogen *Metarhizium anisopliae*. *Proceedings of the 3rd Australasian Conference of Grassland Invertebrate Ecology*, 191-198.

Corkill, L. (1952) Cyanogenesis in white clover (*Trifolium repens* L.). VI. Experiments with high glucoside and glucoside-free strains. *New Zealand Journal of Science and Technology* **34A**, 1-16.

Cottier, W. (1962) Life-cycle of pasture insects. *Proceedings of the 15th New Zealand Weed and Pest Control Conference*, 216-233.

Cousin, M.T.; Moreau, J.P.; Grison, C. (1968) Mise en evidence de discontinuités dans la transmission de la phyllodie du tréfle par *Euscelis plebejus* Fall. *Annales des Epiphytes* **19,** 115-120.

Cox, J. (1976) Tropical army worm *Spodoptera litura* Fabr., life-cycle. *Information Series, Department of Scientific and Industrial Research, New Zealand* **No. 105/19**, 4 pp.

Cox, J.; Dale, P.S. (1977) New records of plant pests in New Zealand. II. *New Zealand Journal of Agricultural Research* **20**, 109-111.

Crawford, A.M.; Kalmakoff, J. (1978) Transmission of *Wiseana* spp. nuclear polyhedrosis virus in the pasture habitat. *New Zealand Journal of Agricultural Research* **21**, 521-526.

Cumber, R.A. (1960) The insect complex of sown pastures in the North Island. X. Insects revealed in studies of soil, turf and dung. *New Zealand Journal of Agricultural Research* **3**, 253-267.

Danthanarayana, W. (1975) The bionomics, distribution and host range of the light brown apple moth, *Epiphyas postvittana* (Walk.) (Tortricidae). *Australian Journal of Zoology* **23**, 419-437.

Davidson, R.L. (1982) An ecological view of pasture pests on the Northern Tablelands of New South Wales. *Proceedings of the 3rd Australasian Conference of Grassland Invertebrate Ecology*, 67-74.

Davidson, R.L.; Hilditch, J.A.; Wiseman, J.R.; Wolfe, V.J. (1979) Growth, fecundity and mortality responses of Scarabaeidae (Coleoptera) contributing to population increases in improved pastures. *Proceedings of the 2nd Australasian Conference of Grassland Invertebrate Ecology*, 66-69.

Davidson, R.L.; Roberts, R.J. (1969) Scarab damage to grass and clover as influenced by depth of feeding. *Bulletin of Entomological Research* **58**, 559-565.

Davidson, R.L.; Wiseman, J.R.; Wolfe, V.J. (1970) A systems approach to pasture scarab problems in Australia. *Proceedings of the 11th International Grassland Congress*, 681-684.

Dearing, S.C.; Scott, P.D.; Wigley, P.J.; Dhana, S.D. (1980) A small RNA virus isolated from the grass grub *Costelytra zealandica* (Coleoptera: Scarabaeidae). *New Zealand Journal of Zoology* **7**, 267-269.

Dick, R.D. (1945) Ecological observations on *Oxycanus cervinata*. *New Zealand Journal of Science and Technology* **27A**, 32-36.

Dickason, E.A.; Leach, C.M.; Gross, A.E. (1958) Control of the clover root curculio on alsike clover. *Journal of Economic Entomology* **51**, 554-555.

Dickason, E.A.; Leach, C.M.; Gross, A.E. (1968) Clover root curculio injury and vascular decay of alfalfa roots. *Journal of Economic Entomology* **61**, 1163-1165.

Dumbleton, L.J. (1938) The lucerne flea (*Sminthurus viridis*) in New Zealand. *New Zealand Journal of Science and Technology* **20A**, 197-211.

Dumbleton, L.J. (1945) Contribution to the ecology of *Oxycanus cervinata* Walk. *New Zealand Journal of Science and Technology* **27A**, 114-128.

Dumbleton, L.J. (1947) The red-legged earth mite. *New Zealand Journal of Agriculture* **74**, 9-13.

Dumbleton, L.J. (1959) Clover case-bearers. *New Zealand Journal of Agriculture* **99**, 523-528.

Dumbleton, L.J. (1963) The biology and control of *Coleophora* spp. (Lepidoptera: Coleophoridae) on white clover. *New Zealand Journal of Agricultural Research* **6**, 277-292.

Dumbleton, L.J. (1966) Genitalia, classification and zoogeography of the New Zealand Hepialidae (Lepidoptera). *New Zealand Journal of Science* **9**, 920-981.

Dumbleton, L.J.; Dick, R.D. (1942) Chemical control of *Oxycanus cervinata* Walker. II. Experiments in 1941 season. *New Zealand Journal of Science and Technology* **23A**, 284-293.

East, R. (1976) The effects of pasture species on white-fringed weevil populations. *Proceedings of the 29th New Zealand Weed and Pest Control Conference* 165-167.

East, R. (1977a) Effects of pasture and forage crop species on longevity, fecundity and oviposition rate of adult white-fringed weevils. *New Zealand Journal of Experimental Agriculture* **5**, 177-181.

East, R. (1977b) White-fringed weevil: biology and control. *New Zealand Ministry of Agriculture and Fisheries, FPP*. **No. 46**, 2 pp.

East, R. (1979a) Population studies of Australasian pasture Scarabaeidae (Coleoptera). *Proceedings of the 2nd Australasian Conference of Grassland Invertebrate Ecology*, 56-61.

East, R. (1979b) Effects of grazing management on *Costelytra zealandica* populations (Coleoptera: Scarabaeidae). *Proceedings of the 2nd Australasian Conference of Grassland Invertebrate Ecology*, 180-184.

East, R. (1982) Interactions between whitefringed weevil *Graphognathus leucoloma* and legume species in the northern North Island. *New Zealand Journal of Agricultural Research* **25**, 131-140.

East, R. (1983) Whitefringed weevil effects on pasture species in northern North Island. *Aglink FPP 478, MAF Information Services, Wellington, New Zealand.* 3 pp.

East, R.; Kain, W.M. (1982) Prediction of grass grub, *Costelytra zealandica*, (Coleoptera: Scarabaeidae) populations. *New Zealand Entomologist* **7**, 222-227.

East, R.; King, P.D. (1977) Effects of botanical composition of pastures on insect pest populations. *New Zealand Entomologist* **6**, 273-278.

East, R.; Koller, M.S.; Willoughby, B.E. (1982) Effects of grass grub density on production of tall fescue, cocksfoot and ryegrass sown with white clover. *Proceedings of the 35th New Zealand Weed and Pest Control Conference*, 82-85.

East, R.; Parr, J. (1977a) Cultural control of white-fringed weevil. *Proceedings of the 30th New Zealand Weed and Pest Control Conference*, 46-49.

East, R.; Parr, J. (1977b) Chemical control of white-fringed weevil in lucerne. *Proceedings of the 30th New Zealand Weed and Pest Control Conference*, 50-55.

East, R.; Pottinger, R.P. (1975) Starling (*Sturnus vulgaris* L.) predation on grass grub (*Costelytra zealandica* (White), Melolonthinae) populations in Canterbury. *New Zealand Journal of Agricultural Research* **18**, 417-452.

East, R.; Watson, R.N.; Welsh, R.D. (1978) Damage to pasture legumes by larvae of the common blue, *Zizina otis labradus* (Lepidoptera: Lycaenidae). *New Zealand Entomologist* **6**, 390-391.

East, R.; Willoughby, B.E. (1980) Effects of pasture defoliation in summer on grass grub (*Costelytra zealandica*) populations. *New Zealand Journal of Agricultural Research* **23**, 547-563.

East, R.; Willoughby, B.E. (1983) Grass grub (*Costelytra zealandica*) population collapse in the northern North Island. *New Zealand Journal of Agricultural Research* **26**, 381-390.

East, R.; Willoughby, B.E.; Haigh, N.A.; Wrenn, N.R. (1980) Chemical control of Tasmanian grass grub in the northern North Island. *Proceedings of the 33rd New Zealand Weed and Pest Control Conference*, 155-157.

East, R.; Willoughby, B.E.; Wrenn, N.R. (1981) Evaluation of the solid stream application technique for grass grub control in the northern North Island. *Proceedings of the 34th New Zealand Weed and Pest Control Conference*, 229-233.

Edge, V.E.; Casimir, M. (1975) Laboratory evaluation of five insecticides against wingless grasshopper (*Phaulacridium vittatum*). *Australian Journal of Experimental Agriculture and Animal Husbandry* **15**, 142-144.

Esson, M.J. (1970) Time lapse photography for the observation of porina caterpillar behaviour. *Proceedings of the 23rd New Zealand Weed and Pest Control Conference*, 200-204.

Eyles, A.C. (1965) Observations on some parasites of two *Wiseana* species (Lep.: Hepialidae). *New Zealand Journal of Agricultural Research* **8**, 951-958.

Eyles, A.C.; (1966) A predator of *Wiseana* (Lep.: Hepialidae). *New Zealand Journal of Agricultural Research* **9**, 699-703.

Farrell, J.A.K. (1972) Observations on soil-inhabiting insect populations of improved pasture in Nelson province, with particular reference to *Costelytra zealandica* (White) (Col., Scarabaeidae). *New Zealand Journal of Agricultural Research* **15**, 878-892.

Farrell, J.A.K. (1975) Effects of food plant species on the weight loss and survival and fecundity of adult *Costelytra zealandica* (Coleoptera:Scarabaeidae). *New Zealand Journal of Agricultural Research* **3**, 303-305.

Farrell, J.A.K. (1976) Field microplot trials on plant resistance to porina. *Proceedings of the 29th New Zealand Weed and Pest Control Conference*, 168-171.

Farrell, J.A.K. (1977) Plant resistance to insects and the selection of resistant lines. *New Zealand Entomologist* **6**, 244-261.

Farrell, J.A.K.; Sweney, W.J. (1972) Plant resistance to the grass grub, *Costelytra zealandica* (Col.:Scarabaeidae). I. Resistance in pasture legumes. *New Zealand Journal of Agricultural Research* **15**, 904-908.

Farrell, J.A.K.; Sweney, W.J. (1974) Plant resistance to the grass grub, *Costelytra zealandica* (Coleoptera: Scarabaeidae). III. Resistance in *Lotus* and *Lupinus*. *New Zealand Journal of Agricultural Research* **17**, 69-72.

Farrell, J.A.K.; Sweney, W.J.; Jones, A.E. (1974) Plant resistance to the porina caterpillar *Wiseana cervinata* (Lepidoptera: Hepialidae). I. Resistance in legumes and grasses. *New Zealand Journal of Agricultural Research* **17**, 373-378.

Fenemore, P.G. (1974) Grass grub, *Costelytra zealandica* (White), life-cycle. *Information Series, Department of Scientific and Industrial Research, New Zealand* **No. 105/4**, 2 pp.

Fenemore, P.G.; Allen, V.A.L. (1969) Oviposition preference and larval survival in *Wiseana cervinata* (Walker), (Hepialidae). *New Zealand Journal of Agricultural Research* **12**, 146-161.

Ferguson, C.M.; Barratt, B.I.P. (1983) Slug damage to pasture renovated by direct drilling. *Proceedings of the 36th New Zealand Weed and Pest Control Conference*, 212-215.

Fisher, J.R.; O'Keefe, L.E. (1979a) Host potential of some cultivated legumes for the pea leaf weevil, *Sitona lineatus* (Linnaeus) (Coleoptera: Curculionidae). *Pan-Pacific Entomologist* **55**, 199-201.

Fisher, J.R.; O'Keefe, L.E. (1979b) Food plants of the pea leaf weevil, *Sitona lineatus* (Linnaeus) (Coleoptera: Curculionidae) in northern Idaho and eastern Washington. *Pan-Pacific Entomologist* **55**, 202-207.

Flay, A.H.; Garrett, H.E. (1942) The grass grub problem. Farm management and economic significance in Canterbury farming. *New Zealand Journal of Science and Technology* **23A**, 322-329.

Fleming, S.; Archibald, R.; Stewart, K.; Kalmakoff, S. (1982) Natural regulation of *Wiseana* spp. by enzootic disease. *Proceedings of the 3rd Australasian Conference of Grassland Invertebrate Ecology*, 283-289.

Fowler, M. (1974) Milky disease (*Bacillus* spp.) occurrence and experimental infection in larvae of *Costelytra zealandica* and other Scarabaeidae. *New Zealand Journal of Zoology* **1**, 97-109.

Fowler, M.; Robertson, J.S. (1972) Iridescent virus infection in field populations of *Wiseana cervinata* (Lepidoptera: Hepialidae) and *Witlesia* sp. (Lepidoptera: Pyralidae) in New Zealand. *Journal of Invertebrate Pathology* **19**, 154-155.

Franzmann, B.A.; Scattini, W.J.; Rynne, K.P.; Johnson, B. (1979) Lucerne aphids effects on 18 pasture legumes in southern Queensland: a glasshouse study. *Australian Journal of Experimental Agriculture and Animal Husbandry* **19**, 59-63.

Freeman, B.E. (1965) An investigation of the distribution during winter of the white clover seed weevil *Apion dichroum* Bedel (Col., Curculionidae). *Journal of Applied Ecology* **2**, 105-113.

Freeman, B.E. (1967) The biology of the white clover weevil *Apion dichroum* Bedel (Col., Curculionidae). *Journal of Applied Ecology* **4**, 535-552.

Freeman, B.E.; Wimble, R.H. (1970) The use of malathion to control the white clover seed weevil, *Apion dichroum* Bedel (Col., Curculionidae). *Plant Pathology* **19**, 119-120.

French, R.A. (1972) Clover case-bearers in New Zealand. A review of the problem with special reference to control. *Proceedings of the 25th New Zealand Weed and Pest Control Conference*, 220-223.

French, R.A. (1973) Some aspects of the biology, population dynamics and economic status of *Wiseana cervinata* (Walker) (Hepialidae:Lepidoptera). Ph.D. Thesis, Lincoln College, New Zealand. 377 pp.

French, R.A.; Pearson, J.F. (1981) Food intake and feeding behaviour of porina (*Wiseana* spp.) (Lepidoptera: Hepialidae) larvae in the field. *New Zealand Journal of Experimental Agriculture* **9**, 383-386.

French, R.A.; Pottinger, R.P. (1975) Chemical control of pasture pests on the West Coast. *Proceedings of the 28th New Zealand Weed and Pest Control Conference*, 258-261.

French, R.A.; Savage, M.J. (1981) Porina. Control by mob stocking. *Aglink FPP 591, MAF Information Services, Wellington, New Zealand*, 2 pp.

Fry, P.R. (1952) Occurrence of lucerne mosaic virus in New Zealand. *New Zealand Journal of Science and Technology* **34A**, 320-326.

Gaynor, D.L.; Kain, W.M.; Atkinson, D.S.; Oliver, M.J. (1979) Observations on *Acyrthosiphon kondoi* and *A. pisum* on lucerne crops in the southern North Island, New Zealand (Hemiptera:Aphididae). *Proceedings of the 2nd Australasian Conference of Grassland Invertebrate Ecology*. 134-138.

Gilbert, W.S.; Lewis, C.E. (1982) Residues in soil, pasture and grazing dairy cattle after the incorporation of dieldrin and heptachlor into soil before sowing. *Australian Journal of Experimental Agriculture and Animal Husbandry* **22**, 106-115.

Goldson, S.L.; French, R.A. (1983) Age-related susceptibility of lucerne to sitona weevil, *Sitona discoideus* Gyllenhal (Coleoptera: Curculionidae), larvae and the associated patterns of adult infestation. *New Zealand Journal of Agricultural Research* **26**, 251-255.

Gonzalez, D.; Miyazaki, M.; White, W.; Takada, H.; Dickson, R.C.; Hall, J.C. (1979) Geographical distribution of *Acyrthosiphon kondoi* Shinji (Homoptera: Aphididae) and some of its parasites and hyperparasites in Japan. *Kontyu* **47**, 1-7.

Graber, L.F.; Fluke, C.L.; Dexter, S.R. (1931) Insect injury to blue grass in relation to the environment. *Ecology* **12**, 547-566.

Gross, H.R.; Shaw, A.Z.; Mitchell, J.A.; Padgett, G.R. (1972) Crop damage by larvae of the white-fringed beetle as related to time of larval introduction into soil. *Journal of Economic Entomology* **65**, 713-716.

Grylls, N.E. (1954) Rugose leaf curl - a new virus disease transovarially transmitted by the leafhopper *Austrogallia torrida*. *Australian Journal of Biological Science* **7**, 47-58.

Haack, N.A.; Kain, W.M.; Springett, B.P. (1982) The use of insecticidal baits for control of porina (*Wiseana* spp.). *Proceedings of the 3rd Australasian Conference of Grassland Invertebrate Ecology*, 349-354.

Hagel, G.T.; Landis, B.J.; Ahrens, M.C. (1973) Aster leafhopper: source of infestation host plant preference and dispersal. *Journal of Economic Entomology* **66**, 877-881.

Hall, I.M.; Oliver, E.H.A.; Given, B.B. (1976) *Nosema takapauensis* n. sp., a microsporidan parasite of larvae of *Costelytra zealandica* (Coleoptera: Scarabaeidae) in New Zealand. *New Zealand Journal of Zoology* **3**, 257-260.

Hall, I.M.; Oliver, E.H.A.; Given, B.B. (1977) *Nosema costelytrae* n. sp., a microsporidan parasite of larvae of *Costelytra zealandica* (Coleoptera: Scarabaeidae) in New Zealand. *New Zealand Journal of Zoology* **4**, 1-3.

Hameed, S.F.; Kaul, B.K.; Dogra, G.S. (1978) New record of a sitonid weevil on *Trifolium repens* Linn. in Kangra Valley, Himachal Pradesh. *Indian Journal of Entomology* **40**, 467-468.

Hamilton, J.T. (1972) *Bryobia repensi* Manson (Tetranychidae: Acar.), a new pasture pest and its control. *Journal of the Australian Institute of Agricultural Science* **38**, 62-63.

Hardy, R.J. (1970) Further investigations of the effectiveness of carbaryl and fenitrothion for the control of pasture cockchafer *Aphodius pseudotasmaniae* Given. *Tasmanian Journal of Agriculture* **41**, 76-78.

Hardy, R.J. (1971) Alternatives to DDT for the control of *Oxycanus fuscomaculatus* Walker (Lepidoptera: Hepialidae). *Australian Journal of Experimental Agriculture and Animal Husbandry* **11**, 362-364.

Hardy, R.J. (1973a) A survey of the level of injury to crops of white clover seed by clover seed moth *Coleophora frischella* (L.) in Tasmania. *Tasmanian Journal of Agriculture* **44**, 214-216.

Hardy, R.J. (1973b) Importance of the timing of chlorfenvinphos and fenitrothion sprays for control of corbie and winter corbie in Tasmania. *Australian Journal of Experimental Agriculture and Animal Husbandry* **13**, 208-212.

Hardy, R.J. (1974) The biology and pest status of *Oxycanus fuscomaculatus* Walker (Lepidoptera: Hepialidae) in Tasmania. *Journal of the Australian Entomological Society* **13**, 317-328.

Hardy, R.J. (1983) The biology and behaviour of the clover seed moth *Coleophora frischella* (Lepidoptera: Coleophoridae) in Tasmania. *Journal of the Australian Entomological Society* **22**, 7-14.

Harper, A.M.; Miska, J.P.; Manglitz, G.P.; Irwin, B.J.; Abbrust, E.J. (1978) The literature of arthropods associated with alfalfa. III. A bibliography of the pea aphid *Acyrthosiphon pisum* (Harris) (Homoptera: Aphididae). *Special Publication, Agricultural Experiment Station, University of Illinois, Urbana-Champaign* **No. 50**, 89 pp.

Harris, W. (1969) Some effects of porina caterpillar (*Wiseana* spp.) infestation on perennial ryegrass, cocksfoot and white clover. *New Zealand Journal of Agricultural Research* **12**, 543-552.

Harris, W. (1973) A comparison of the use of organophosphate insecticide or nitrogen fertiliser to compensate for pasture yield losses caused by porina caterpillar. *Proceedings of the 26th New Zealand Weed and Pest Control Conference* **26**, 189-193.

Harris, W.; Brock, J.L. (1972) Effect of porina caterpillar (*Wiseana* spp.) infestation on yield and competitive interactions of ryegrass and white clover varieties. *New Zealand Journal of Agricultural Research* **15**, 723-740.

Harrison, R.A. (1967) Insect pests and research in New Zealand. *SPAN* **10**, 114-116.

Henzell, R.F.; Lauren, D.R. (1977) Use of sex attractant traps to estimate the development stage of grass grub, *Costelytra zealandica* (White) (Coleoptera: Scarabaeidae), in the soil. *New Zealand Journal of Agricultural Research* **20**, 75-78.

Henzell, R.F.; Lauren, D.R. (1981) Development of methods for prediction of the starting date and duration of insecticide spraying programmes against adult grass grub, *Costelytra zealandica* (White) (Coleoptera: Scarabaeidae). *New Zealand Entomologist* **7**, 266-271.

Henzell, R.F.; Lauren, D.R.; Cox, N.R. (1979) Monitoring *Costelytra zealandica* flight with sex attractant traps (Coleoptera: Scarabaeidae). Proceedings of the 2nd Australasian Conference of Grassland Invertebrate Ecology, 125-128.

Henzell, R.F.; Lauren, D.R.; du Toit, G.D.G.; Wrenn, N.R. (1981) Do soil applied insecticides control grass grub beetles? *Proceedings of the 34th New Zealand Weed and Pest Control Conference*, 234-238.

Henzell, R.F.; Skinner, R.A.; Clements, R.O. (1983) Insecticides for control of adult grass grub, *Costelytra zealandica* (White). V. Screening and behaviour of insecticides in soil bioassays. *New Zealand Journal of Agricultural Research* **26**, 129-133.

Hill, A.V. (1950) Yellow dwarf of tobacco in Australia. IV. Some host plants of the virus. *Australian Journal of Agricultural Research* **1**, 141-143.

Hill, R.L. (1979) Bluegreen lucerne aphid, *Acyrthosiphon kondoi* Shinji and pea aphid, *A. pisum* (Harris), life-cycles. *Information Series, Department of Scientific and Industrial Research, New Zealand* **No. 105/28**, 2 pp.

Hill, R.L. (1983) Two egg parasites (Hymenoptera: Scelionidae) of the black field cricket, *Teleogryllus commodus* (Orthoptera: Gryllidae), in New Zealand. *New Zealand Journal of Zoology* **10**, 57-62.

Horber, E. (1974) Resistenz gegen Insekten an Futterleguminosen. *Schweizerische Landwirtschaftliche Forschung* **13**, 221-234.

Howitt, A.J. (1961) Chemical control of slugs in orchard grass-Ladino white clover pasture in the Pacific North West. *Journal of Economic Entomology* **54**, 778-781.

Hoy, J.M. (1963) Summation of grass grub research. *Proceedings of the 16th New Zealand Weed and Pest Control Conference*, 118-122.

Hoy,J.M.; Given,B.B.(1952) A revision of the Melolonthinae of New Zealand. Part II. Final instar larvae. *Bulletin, Department of Scientific and Industrial Research, New Zealand* **102**, 138-171.

Hutchins, R.F.N.; Sutherland, O.R.W.; Gnanasunderam, C.; Greenfield, W.J.; Williams, E.M.; Wright, H.J. (1984) Toxicity of nitro compounds from *Lotus pedunculatus* to grass grub *(Costelytra zealandica)*) (Coleoptera: Scarabaeidae). *Journal of Chemical Ecology* **10,** 81-93.

Ioannisiani, T.G. (1963) [The population dynamics of the beetles *Phytonomus meles* F. and *Phytonomus nigrirostris* F. on clover]. *Zoologicheskiĭ Zhurnal* **42**, 1810-1816.

Ireson, J.E. (1982) A re-examination of the distribution of the pasture snout mite, *Bdellodes lapidaria* (Kramer) (Ascari: Bdelidae) and the lucerne flea, *Sminthurus viridis* (L.) (Collembola: Sminthuridae) in Tasmania. *Journal of the Australian Entomological Society* **21**, 251-255.

Ivanova, S.G.; Malakhanova, E.L. (1980) [Protection of seed crops of clover from pests.] *Zashchita Rastenii* **5**, 31-32.

Jackson, T.A.; Bedding, R.A.; Trought, T.E.T.; Kain, W.M.; East, R. (1981) The potential use of nematodes for the control of pasture insects. *Proceedings of the 34th New Zealand Weed and Pest Control Conference*, 170-172.

Jackson, T.A.; Trought, T.E.T. (1982) Progress with the use of nematodes and bacteria for the control of grass grub. *Proceedings of the 35th New Zealand Weed and Pest Control Conference*, 103-106.

James, J.R.; Lucas, L.T.; Chamblee, D.S.; Campbell, W.V. (1979) The influence of fungicide and insecticide application on persistance of Ladino clover. (Abstract) *Phytopathology* **69**, 1-A5.

Jeppson, L.R.; Keifer, H.; Baker, E.W. (1975) Mites injurious to economic plants. Berkeley, USA; University of California Press, 614 pp.

Johansen, C. (1965) Bee poisoning versus clover seed weevil control in white Dutch clover grown for seed. *Journal of Economic Entomology* **58**, 857-863.

Johns, G.G.; Greenup, L.R. (1976) Pasture seed theft by ants in northern New South Wales. *Australian Journal of Experimental Agriculture and Animal Husbandry* **16**, 249-256.

Johnston, C.J.R. (1960) Biology and control of lucerne flea. *Journal of Agriculture, Victorian Department of Agriculture* **58**, 505, 507-509, 511-513.

Johnstone, G.R. (1978) Diseases of broad bean (*Vicia faba* L. major) and green pea (*Pisum sativum* L.) in Tasmania caused by subterranean clover red leaf virus. *Australian Journal of Agricultural Research* **29**, 1003-1010.

Jones, F.G.W.; Jones, M.G. (1974) Pests of field crops (2nd Ed.) London, UK; Edward Arnold, 448 pp.

Jordan, T.W.; Chang, C.K. (1981) Pyrethroid resistance mechanisms in porina caterpillars. *Proceedings of the 34th New Zealand Weed and Pest Control Conference*, 167-169.

Kain, W.M. (1975) Population dynamics and pest assessment studies of grass grub (*Costelytra zealandica* (White) Melolonthinae) in the North Island of New Zealand. Ph.D. Thesis, Lincoln College, University of Canterbury, New Zealand.

Kain, W.M. (1979) Pest management systems for control of pasture insects in New Zealand. *Proceedings of the 2nd Australasian Conference of Grassland Invertebrate Ecology*, 172-179.

Kain, W.M.; Atkinson, D.S. (1970) A rational approach to grass grub control. *Proceedings of the 23rd New Zealand Weed and Pest Control Conference*, 180-183.

Kain, W.M.; Atkinson, D.S. (1972) A farmlet study on the effects of grass grub damage in the Takapau District of central Hawkes Bay. *Proceedings of the 25th New Zealand Weed and Pest Control Conference*, 232-238.

Kain, W.M.; Atkinson, D.S. (1977) Development of resistant pasture and methods of pasture management of grass grub (*Costelytra zealandica* (White)) control. *New Zealand Journal of Agricultural Research* **20**, 507-517.

Kain, W.M.; Bedding, R.A.; Van der Mespel, C.S. (1982b) Preliminary evaluations of parasitic nematodes for grass grub (*Costelytra zealandica* (White)) control in central Hawkes Bay of New Zealand. *New Zealand Journal of Experimental Agriculture* **10**, 447-450.

Kain, W.M.; East, R.; Douglas, J.A. (1979a) *Costelytra zealandica* - pasture species relationships on the pumice soils of the central North Island of New Zealand. *Proceedings of the 2nd Australasian Conference of Grassland Invertebrate Ecology*, 88-91.

Kain, W.M.; Holland, T.V.; Henzell, R.F. (1979b) Feeding studies on *Wiseana* spp. *Proceedings of the 2nd Australasian Conference of Grassland Invertebrate Ecology*, 270-274.

Kain, W.M.; Slay, M.W.; Atkinson, D.S. (1979c) Evaluation of grass grub - plant interactions of grasses sown with and without white clover in central Hawkes Bay. *Proceedings of the 32nd New Zealand Weed and Pest Control Conference*, 86-91.

Kain, W.M.; Thomson, N.A.; Plunkett, I.G.; Atkinson, D.S.; Malcolm, J.R. (1982a) Development and implementation of integrated management of pasture pests in intensive dairy farms of the southern North Island region of New Zealand. *Proceedings of the 3rd Australasian Conference of Grassland Invertebrate Ecology*, 339-348.

Kain, W.M.; Wallis, M.J.; Payne, E.O. (1968) Control of army caterpillar *Pseudaletia separata*. *Proceedings of the 21st New Zealand Weed and Pest Control Conference*, 202-208.

Kain, W.M.; Wyeth, T.K.; Kale, A.J. (1981) Preliminary field studies on the use of an indigenous parasitic nematode for control of porina. *Proceedings of the 34th New Zealand Weed and Pest Control Conference*, 173-175.

Kalmakoff, J. (1979) Enzootic virus control of *Wiseana* spp. in the pasture environment. *Proceedings of the 2nd Australasian Conference of Grassland Invertebrate Ecology*, 202-204.

Kalmakoff, J.; Crawford, A.M. (1976) Virus control of porina. *New Zealand Journal of Agriculture* **133**, 41-42.

Kalmakoff, J.; Moore, S.G. (1975) The ecology of nucleopolyhedrosis virus in porina (*Wiseana* spp., Lepidoptera: Hepialidae). *New Zealand Entomologist* **6**, 73-76.

Kazion, V. (1971) [Studies on ecology of foxglove aphid *Aulacorthum solani* Kaltenbach. 1. On the development, reproduction and activity]. *Bulletin of Hokkaido Perfectural Agricultural Experiment Station* **23**, 97-104.

Keller, C.J.; Taylor, N.L.; Van Meter, C.L.; Pass, B.C. (1970) Feeding response of the adult alfalfa weevil to plant species phytogenetically related to alfalfa. *Journal of Economic Entomology* **63**, 302-303.

Kelsey, J.M. (1951) Grass grub and grass caterpillar control. *New Zealand Journal of Agriculture* **83**, 195-200.

Kelsey, J.M. (1958) Damage by clover case-bearer caterpillars. *New Zealand Journal of Agricultural Research* **1**, 525-526.

Kelsey, J.M. (1968) Flight and ground collection records of *Costelytra zealandica* (White) at a site in Canterbury, New Zealand. *New Zealand Journal of Science* **11**, 664-685.

Kerr, T.W. (1957) Leafhoppers associated with forage crops in Rhode Island. *Journal of Economic Entomology* **50**, 271-273.

King, P.D.; East, R. (1979) Effects of pasture composition on the dynamics of *Heteronychus arator* and *Graphognathus leucoloma* populations (Coleoptera: Scarabaeidae and Curculionidae). *Proceedings of the 2nd Australasian Conference of Grassland Invertebrate Ecology*, 79-82.

King, P.D.; Meekings, J.S.; Mercer, C.F. (1982) Effects of whitefringed weevil (*Graphognathus leucoloma*) and black beetle (*Heteronychus arator*) populations on pasture species. *New Zealand Journal of Agricultural Research* **25**, 405-414.

Klostermeyer, E.C. (1954) Mite control on legume seed crops. *Journal of Economic Entomology* **47**, 30-34.

Kooistra, G. (1964) Some data concerning the presence and behaviour of springtails (Collembola) on grass and white clover. *Netherlands Journal of Plant Pathology* **70**, 136-141.

Lane, G.A.; Sutherland, O.R.W.; Skipp, R.A. (in press) Isoflavonoids as insect feeding deterrents and antifungal components from the root of *Lupinus angustifolius*. *Journal of Chemical Ecology*.

Latch, G.C.M. (1965) *Metarhizium anisopliae* (Metschnikoff) Sorokin strains in New Zealand and their possible use for controlling pasture inhabiting insects. *New Zealand Journal of Agricultural Research* **8**, 384-396.

Latch, G.C.M.; Kain, W.M. (1983) Control of porina caterpillar (*Wiseana* spp.) in pasture by the fungus *Metarhizium anisopliae*. *New Zealand Journal of Experimental Agriculture* **11**, 351-354.

Laughlin, R. (1957) Biology and ecology of the garden chafer, *Phyllopertha horticola* (L.). III. The growth of the larvae. *Bulletin of Entomological Research* **48**, 127-154.

Lauren, D.R.; Henzell, R.F.; Wrenn, N.R. (1981) Comparison of two spray date prediction techniques for control of grass grub beetles. *Proceedings of the 34th New Zealand Weed and Pest Control Conference*, 243-246.

Lauren, D.R.; Henzell, R.F.; Wrenn, N.R. (1982) A comparison of lindane broadcast and drilled in spring for grass grub beetle control. *Proceedings of the 35th New Zealand Weed and Pest Control Conference*, 94-97.

Lauren, D.R.; Henzell, R.F.; Yortt, M.L. (1978) The importance of application time when synthetic pyrethroids are used to control adult grass grub. *Proceedings of the 31st New Zealand Weed and Pest Control Conference*, 154-159.

Leath, K.T.; Byers, R.A. (1977) Interaction of *Fusarium* root rot with pea aphid and potato leafhopper feeding on forage legumes. *Phytopathology* **67**, 226-229.

Lee, P.E.; Chiykowski, L.N. (1963) Mechanical transmission of clover phyllody virus to its leafhopper vectors. *Canadian Journal of Botany* **41**, 311-312.

Legowski, T.J. (1955) *Triaspis* (*Sigalphus*) *floricola* Wesm. (Hym. Braconidae) parasitising the larvae of clover seed weevils (*Apion* spp.) in East Anglia. *Entomological Monthly Magazine* **1955**, 273-274.

Lehman, W. (1973) *Euscelis plebejus* (Fallen) und *Macrosteles cristatus* Ribaut als Überträger von Pflanzenkrankheiten von der muteter Mykoplasma-Ätiologie. *Archiv für Phytopathologie und Pflanzenschutz* **9**, 363-370.

Leibee, G.L.; Pass, B.C.; Yeargan, K.V. (1980) Instar determination of clover root curculio, *Sitona hispidulus* (Coleoptera: Curculionidae). *Journal of the Kansas Entomological Society* **53**, 473-475.

Leibee, G.L.; Pass, B.C.; Yeargan, K.V. (1981) Seasonal abundance and activity of *Sitona hispidulus* adults in Kentucky. *Environmental Entomology* **10**, 27-30.

Loan, C.; Holdaway, F.G. (1955) Biology of the red clover thrips *Haplothrips niger* (Osborn) (Thysanoptera: Phloeothripidae). *Canadian Entomologist* **87**, 210-219.

MacFarlane, J.R. (1967) Imidan gives best control of lucerne flea. *Journal of Agriculture, Victorian Department of Agriculture* **65**, 334-336.

MacNay, C.G. (1954) New records of insects in Canada (1952) A review. *Canadian Entomologist* **86**, 55-60.

MacQuillan, M.J. (1975) Evaluation of chlorpyrifos and fenchlorphos for control of major insect pests of pastures in Australia. *Australian Journal of Experimental Agriculture and Animal Husbandry* **15**, 550-555.

McDonald, R.; Cummins, G. (1976) Renovating cricket-damaged pasture. *The Journal of Agriculture, Victoria* **74**, 235-237.

McLaren, G.F.; Crump, D.K. (1969) An economic study of the use of insecticides for the control of porina caterpillar. *Proceedings of the 22nd New Zealand Weed and Pest Control Conference*, 307-322.

Madge, P.E. (1954) A field study of the biology of the underground grass caterpillar, *Oncopera fasiculata* (Walker) (Lepidoptera:Hepialidae), in South Australia. *Australian Journal of Zoology* **2**, 193-204.

Maelzer, D.A. (1964) Rainfall and changes in abundance of *Aphodius tasmaniae* Hope (Coleoptera: Scarabaeidae) in the lower south-west of South Australia. *Australian Journal of Zoology* **12**, 263-278.

Manglitz, G.R.; Kreitlow, K.W. (1960) Vectors of alfalfa and bean yellow mosaic viruses in ladino white clover. *Journal of Economic Entomology* **53**, 113-115.

Markkula, M. (1959) The biology and especially the oviposition of the *Sitona* Germ. (Col., Curculionidae) species occurring as pests of grassland legumes in Finland. *Valtion Maatalouskoetoiminnan Julkaisuja* **No. 178**, 41-74.

Markkula, M.; Köppä, P. (1960) A composition of the *Sitona* (Col., Curculionidae) population on grassland legumes and other leguminous plants. *Annales Entomologici Fennici* **26**, 246-263.

Markkula, M.; Myllymäki, S. (1957) Investigation into the oviposition on red and alsike clover and alfalfa of *Apion apricans* Herbst. *A. assimile* Kirby, *A. flavipes* Payk., *A. seniculus* Kirby and *A. virens* Herbst. (Col., Curculionidae) *Annales Entomologici Fennici* **23**, 203-207.

Markkula, M.; Myllymäki, S. (1958) The composition of the *Apion* (Col., Curculionidae) population of grassland legumes and some wild leguminous plants. *Annales Entomologici Fenniae* **24**, 97-124.

Markkula, M.; Myllymäki, S. (1962) The distribution, abundance and biology of the clover head weevil *Phytonomus meles* Fabr. (Col., Curculionidae) in Finland. *Annales Entomologici Fenniae* **28**, 49-63.

Markkula, M.; Myllymäki, S. (1964) The abundance of seed pests of alsike and white clover in Finland. *Annales Agriculturae Fenniae* **3**, 235-243.

Markkula, M.; Roukka, K. (1971) Resistance of plants to the pea aphid *Acyrthosiphon pisum* Harris (Hom., Aphididae). IV. Fecundity on different alsike and white clover varieties. *Annales Agriculturae Fenniae* **10**, 111-113.

Markkula, M.; Tinnila, A. (1957) Studies of the biology of the lesser clover leaf weevil, *Phytonomus nigrirostris* Fabr. (Col., Curculionidae). *Valtion Maatalouskoetoiminnan Julkaisuja* **No. 152**, 62 pp.

Marshall, G.E.; Wilber, D.A. (1934) The clover root curculio (*Sitona hispidula*) in Kansas. *Journal of Economic Entomology* **28**, 807-814.

Martin, N.A. (1978) Introduced slugs and snails. *Proceedings of the 31st New Zealand Weed and Pest Control Conference*, 124-126.

May, B.M. (1975) White-fringed weevil *Graphognathus leucoloma* Boheman, life-cycle. *Information Series, Department of Scientific and Industrial Research, New Zealand* **No. 105/12**, 2 pp.

Merton, J.M. (1982) Interactions of the Tachinid, *Proscissio cana* with its host the New Zealand grass grub, *Costelytra zealandica*. *Proceedings of the 3rd Australasian Conference of Grassland Invertebrate Ecology*, 161-168.

Miller, D. (1971) Common Insects in New Zealand. Wellington, New Zealand; A.H. and A.W. Reed, 1-4.

Miln, A.J. (1978) Protozoan parasites of *Costelytra zealandica* (Coleoptera: Scarabaeidae) in New Zealand. *New Zealand Entomologist* **6**, 392-399.

Miln, A.J. (1979) The relationships between density, disease and mortality in *Costelytra zealandica* populations (Coleoptera: Scarabaeidae). *Proceedings of the 2nd Australasian Conference of Grassland Invertebrate Ecology*, 75-79.

Miln, A.J. (1982a) The role of disease in the population dynamics of grass grub (*Costelytra zealandica* (White)). *Proceedings of the 3rd Australasian Conference of Grassland Invertebrate Ecology*, 147-154.

Miln, A.J. (1982b) The effects of cultivation on disease incidence in grass grub populations. *Proceedings of the 35th New Zealand Weed and Pest Control Conference*, 86-89.

Miln, A.J. (1983) The effects of different times of insecticide application on disease incidence in grass grub populations. *Proceedings of the 36th New Zealand Weed and Pest Control Conference*, 199-202.

Miln, A.J.; Carpenter, A.; Bates, L.H. (1983) Insecticidal control of manuka beetle. *Proceedings of the 36th New Zealand Weed and Pest Control Conference*, 203-206.

Milne, A.; Laughlin, A. (1956) Biology and ecology of the garden chafer, *Phyllopertha horticola* (L.) I. The adult and egg production. *Bulletin of Entomological Research* **47**, 7-22.

Milne, D.L. (1960) The gall midges (Diptera:Cecidomyidae) of clover flower heads. *Transactions of the Royal Entomological Society of London* **112**, 73-108.

Milne, D.L. (1962) Gall midges (Cecidomyidae) causing seed losses in red and white clovers. *South African Journal of Agricultural Science* **5**, 89-94.

Milner, R.J. (1977) The role of disease during an outbreak of *Oncopera alboguttata* Tindale and *O. rufobrunnea* Tindale (Lepidoptera: Hepialidae) in the Ebor/Dorrigo region of N.S.W. *Journal of the Australian Entomological Society* **16**, 21-26.

Moeed, A. (1980) Diets of adult and nestling starlings (*Sturnus vulgaris*) in Hawkes Bay, New Zealand. *New Zealand Journal of Zoology* **7**, 247-256.

Moore, S.G. (1972) A virus outbreak kills porina. *New Zealand Journal of Agriculture* **133**, 41-42.

Moore, S.G.; Kalmakoff, J.; Miles, J.A.R. (1973) Virus diseases of porina (*Wiseana* spp., Lepidoptera: Hepialidae). I. Field isolation. *New Zealand Journal of Science* **16**, 139-153.

Moore, S.G.; Kalmakoff, J.; Miles, J.A.R. (1974) An iridescent virus and a rickettsia from the grass grub *Costelytra zealandica* (Coleoptera: Scarabaeidae). *New Zealand Journal of Zoology* **1**, 205-210.

Moreland, C.R. (1953) Some aspects of the ecology of the clover seed weevil *Miccotrogus picirostris* (F.) (Coleoptera: Curculionidae). *Report of the Entomological Society of Ontario* **84**, 91-101.

Morgan-Jones, J. (1950) Clover seed weevils. *Annals of Applied Biology* **37**, 313-320.

Mori, H.; Moriyama, S. (1970) Biological control of *Tetranychus urticae* Koch on white clover by *Phytoseiulus persimilis* Athias-Henriot (Acarina: Tetranychidae, Phytoseiidae). *Mushi* **43**, 29-38.

Mori, H.; Saito, Y. (1979) Biological control of *Tetranychus urticae* Koch (Acarina: Tetranychidae) populations by three species of phytoseiid mites (Acarina: Phytoseiidae). *Journal of the Faculty of Agriculture, Hokkaido University* **59**, 303-311.

Morrison, W.P.; Pass, B.C.; Nichols, M.P.; Armbrust, E.J. (1974) The literature of anthropods associated with alfalfa. II. A bibliography of the *Sitona* species (Coleoptera: Curculionidae). *Biological Notes, Natural History Survey Division, State of Illinois* **No. 88**, 24 pp.

Morton, J.D.; Roberts, E. (1978) Effects of soil moisture and white-fringed weevil larvae (*Graphognathus leucoloma* Boh.) on young lucerne (*Medicago sativa* L.) plants. *New Zealand Journal of Experimental Agriculture* **6**, 251-255.

Munson, R.E.; Enns, W.R. (1968) Missouri distribution of species of *Hypera* (Coleoptera: Curculionidae). *Journal of the Kansas Entomological Society* **41**, 266-270.

Musil, M. (1959) Ubertragung des Stolburvirus durch die Zikade *Euselis plebejus* (Fallen). *Biologia* **14**, 410-417.

Myers, J.G.E. (1921) Insect pests of lucerne and clover. *New Zealand Journal of Agriculture* **23**, 156-162.

Neal, J.W. Jr; Ratcliffe, R.H. (1975) Clover root curculio: control with granular carbofuran as measured by alfalfa regrowth, yield and root damage. *Journal of Economic Entomology* **68**, 829-831.

Newton, R.C.; Graham, J.H. (1960) Incidence of root-feeding weevils, root-rot, internal breakdown, and virus and their effect on longevity of red clover. *Journal of Economic Entomology* **53**, 865-867.

Nielson, R.J.; Bass, M.H. (1967) Seasonal occurrance and numbers of instars of *Nemobius fasciatus*, a pest on white clover. *Journal of Economic Entomology* **60**, 699-701.

Nijveldt, W. (1963) Observations on gall midges of white and red clover. *Netherlands Journal of Plant Pathology* **69**, 161-187.

Nizhnikov, M.I. (1975) [Metathion and Rogor on seed crops of clover]. *Zashchita Rasteniĭ* **No. 9**, 24 pp.

Norris, K.R. (1948) Seasonal activity of the attack of the red-legged earth mite (*Halotydeus destructor*) on subterranean clover. *Journal of the Council of Scientific and Industrial Research, Australia*, **21** 7-15.

Oatman, E.R.; McMurtry, J.A.; Gilstrap, F.E.; Voth, V. (1977) Effects of releases of *Amblyseius californicus, Phytoseiulus persimilis* and *Typhlodromus occidentalis* on twospotted spidermite on strawberry in Southern California. *Journal of Economic Entomology* **70**, 45-47.

Obrtel, R. (1962) Nosatčíci (*Apion* sp.), škudci semenných kultur červeného jetele. II. Část, možnosti ochranných opatření. *Roslinna Výroba* **8**, 1335-1352.

Ohlsson, B. (1968) Studies of *Apion* species on clover in Sweden (Col., Curculionidae). I. Short range dispersal and choice of hibernation sites of species injurious to cultivated clover. *Meddelanden från Statens Växtskyddsanstalt* **14**, 95-121.

Ottens, R.J.; Todd, J.W. (1979) Effect of diflubenzuron on reproduction and development of *Graphognathus peregrinus* and *G. leucoloma. Journal of Economic Entomology* **72**, 743-746.

Ottens, R.J.; Todd, J.W. (1980) Leaf area consumption of cotton, peanuts and soybeans by adult *Graphognathus peregrinus* and *G. leucoloma. Journal of Economic Entomology* **73**, 55-57.

Pearson, W.D. (1980) Management practices reduce case bearer (Lepidoptera: Coleophoridae) damage to white clover seed crops. *New Zealand Journal of Experimental Agriculture* **8**, 71-74.

Pearson, W.D. (1982) Clover casebearer. Biology, damage and control, a guide. *Aglink FPP 620, MAF Information Services, Wellington, New Zealand*, 4 pp.

Pedigo, L.P.; Stone, J.D.; Lentz, G.L. (1973) Biological synopsis of the green cloverworm (*Plathypena scabra*:Lep., Noctuidae) in Central Iowa [USA]. *Journal of Economic Entomology* **66**, 665-674.

Perju, T. (1982) Selective pesticides in controlling red clover seed pests. *Acta Phytopathologica Academiae Scientiarum Hungaricae* **17**, 171-178.

Perju, T.; Brudea, V.; Ghizdavu, I.; Mustea, D.; Musat, D.; Peteanu, S. (1975) Cercetări de bioecologie şi de combatere a dăunătorilor culturilor de trifoi soşu. *Probleme de Protecţia Plantelor* **3**, 417-457.

Perju, T.; Palageşiu, I. (1971) Curculionidae (Coleoptera) care se des voltă pe diferite specii spontone de trifoli (*Trifolium* sp.). *Studii şi Cercetări de Biologie, Seria Zoologie* **23**, 377-380.

Perrott, D.C.F. (1964) The white fringed weevil *Graphognathus leucoloma* (Boheman) in the South Island of New Zealand. *New Zealand Entomologist* **3**, 52.

Perrott, D.C.F. (1974) Porina moth, *Wiseana* species, life-cycle. *Information Series, Department of Scientific and Industrial Research, New Zealand* **No. 105/1**, 2 pp.

Perrott, D.C.F.; Stockdill, S.M.J. (1972) Variation in the life-cycle of the scarabaeid *Costelytra zealandica. New Zealand Entomologist* **5**, 216-221.

Persson, B. (1977) Distribution of catch in relation to emergence of adults in some Noctuid pest species in South Coastal Queensland. *Australian Journal of Zoology* **25**, 95-102.

Pottinger, R.P. (1968a) Comments on the ecology of grass grub and porina caterpillar. *Proceedings of the New Zealand Grassland Association* **30**, 102-113.

Pottinger, R.P. (1968b). Tasmanian grass grub *Aphodius tasmaniae* Hope, a pest of light land pasture and lucerne. *Agricultural Bulletin, Canterbury, New Zealand* **No. 465**, 7 pp.

Pottinger, R.P. (1973) Pests of grassland and small seed crops. In: *Pastures and Pasture Plants.* Langer, R.H.M. (Ed), Wellington, New Zealand; A.H. and A.W. Reed, 365-406.

Pottinger, R.P. (1977) Porina. Biology, damage and control. *Aglink FPP 43, MAF Information Services, Wellington, New Zealand*, 3 pp.

Pottinger, R.P. (1983) Recent developments with pasture pests. *Proceedings of the 35th Ruakura Farmers' Conference* 99-105.

Pottinger, R.P.; East, R.; Welsh, R.D.; Stratton, A.E.; Maxwell, B.C. (1977) Evaluation of cypermethrin for control of adult grass grub. *Proceedings of the 30th New Zealand Weed and Pest Control Conference*, 217-221.

Pottinger, R.P.; Welsh, R.D.; East, R.; Stratton, A.E. (1979) Control of *Costelytra zealandica* adults with cypermethrin (Coleoptera: Scarabaeidae). *Proceedings of the 2nd Australasian Conference of Grassland Invertebrate Ecology*, 165-167.

Powell, G.S.; Campbell, W.V.; Cope, W.A.; Chamblee, D.S. (1983) Ladino clover resistance to the clover root curculio (Coleoptera: Curculionidae). *Journal of Economic Entomology* **76**, 264-268.

Radcliffe, J.E. (1970) Some effects of grass grub (*Costelytra zealandica* (White)) larvae on pasture plants. *New Zealand Journal of Agricultural Research* **13**, 87-104.

Radcliffe, J.E. (1971) Effects of grass grub (*Costelytra zealandica* (White)) larvae on pasture plants. IV. Effect of grass grubs on perennial ryegrass and white clover. *New Zealand Journal of Agricultural Research* **14**, 625-632.

Reid, R.N.D.; Hardy, R.J. (1978) Corbie damage can be reduced by grazing management. *Tasmanian Journal of Agriculture* **49**, 23-24.

Reinganum, C.; Gagen, S.J.; Sexton, S.B.; Vellacott, H.P. (1981) A survey for pathogens of black field cricket, *Teleogryllus commodus*, in the western district of Victoria, Australia. *Journal of Invertebrate Pathology* **38**, 153-160.

Ridland, P.M.; Berg, G.N. (1981) Seedling resistance to pea aphid of lucerne, annual medic and clover species in Victoria. *Australian Journal of Experimental Agriculture and Animal Husbandry* **21**, 506-511.

Ritcher, P.D. (1958) Biology of Scarabaeidae. *Annual Review of Entomology* **3**, 311-334.

Roberts, R.J. (1972) A newly recognized form of pasture damage by *Phaulacridium vittatum* (Acrididae) on the Northern tablelands of N.S.W. *Journal of the Australian Entomological Society* **11**, 257-258.

Roux, D.P. le (Editor) (1981) Agricultural research 1980. *Report, Department of Agricultural Technical Services, South Africa* **7**, 239 pp.

Runham, N.W.; Hunter, P.J. (1970) Terrestrial Slugs. London, UK; Hutchinson, 184 pp.

Russell, G.B.; Sutherland, O.R.W.; Hutchins, R.F.N.; Christmas, P.E. (1978) Vestitol: a phytoalexin with insect feeding-deterrent activity. *Journal of Chemical Ecology* **4**, 571-579.

Russell, G.B.; Sutherland, O.R.W.; Lane, G.A.; Biggs, D.R. (1979) Is there a common factor for insect and disease resistance in pasture legumes? *Proceedings of the 2nd Australasian Conference of Grassland Invertebrate Ecology*, 95-97.

Sakamoto, Y. (1980) [Ecological studies of clover weevils.] *Journal of the College of Dairying* **8**, 113-148.

Savage, M.J.; French, R.A. (1981) Porina. Chemical control. *Aglink FPP 588, MAF Information Services, Wellington, New Zealand*, 3 pp.

Schumann, K.; Diederich, P. (1983) [Weevils causing damage to leaves in red clover stands.] *Nachrichtenblatt für den Pflanzenschutz in der DDR* **37**, 129-131.

Scotti, P.D.; Longworth, J.F.; Plus, N.; Croizier, G.; Reinganum, C. (1981) The biology and ecology of strains of an insect small RNA virus complex. *Advances in Virus Research* **26**, 117-143.

Sechriest, R.E.; Treece, R.E. (1963) The biology of the lesser clover leaf weevil *Hypera nigrirostris* (Fab.) (Coleoptera: Curculionidae) in Ohio. *Research Bulletin, Ohio Agricultural Experiment Station* **No. 956**, 28 pp.

Sexton, S.B.; Williams, P. (1981) A natural occurrence of parasitism of *Graphognathus leucoloma* (Boheman) by the nematode *Heterorhabditis* sp. *Journal of the Australian Entomological Society* **20**, 253-255.

Shuster, M.M.; Khrolinskii, L.G.(1973) The influence of food plants on the development of the clover weevils *Apion apricans* Hbst. and *A. aestivum* Germ.] *Trudy Vsesoyuznogo Nauchno-Issledovatel'skogo Instituta Zashchity Rastenii* **36**, 12-20.

Smith, C.F.; Weber, J.C. (1954) The legume mite *Petrobia* (Tetranychina) *apicalis* (Banks) a pest of several winter growing legumes. *Bulletin of the Louisiana Agricultural Experiment Station* **No. 493**, 25 pp.

Smith, C.M.; Knight, W.E.; Pitre, H.N. (1975) Feeding preference of the clover head weevil on clovers of the genus *Trifolium*. *Journal of Economic Entomology* **68**, 165-166.

Smith, D. (1978) Evaluating long term control of *Amnemus superciliaris* (Pascoe) with pre-sowing insecticidal treatments in desmodium pastures. *Queensland Journal of Agricultural and Animal Sciences* **35**, 83-88.

Smith, D.F. (1968) Pastures in Victoria. In *Pasture improvement in Australia*. Wilson, B. (Ed.) Sydney, Australia; K.C. Murray Publishing Co., 192-209.

Stahle, P.P.; Williams, P. (1982) Problems associated with determining the extent and importance of cricket infestations in the field. *Proceedings of the 3rd Australasian Conference of Grassland Invertebrate Ecology*, 47-50.

Starks, K.J.; Thurston, R. (1958) Chemical control of the clover leaf weevil larvae. *Journal of Economic Entomology* **51**, 195-198.

Stephenson, J.W. (1968) A review of the biology and ecology of slugs of agricultural importance. *Proceedings of the Malacological Society of London* **38**, 169-178.

Stephenson, J.W.; Bardner, R. (1976) Slugs in agriculture. *Report of the Rothamsted Experimental Station* **Part 2**, pp. 169-187.

Steven, D. (1979) Dispersal of *Graphognathus leucoloma* (Coleoptera: Curculionidae). *Proceedings of the 2nd Australasian Conference of Grassland Invertebrate Ecology*, 146-148.

Stewart, K.H.; Stockdill, S.M.J. (1972) Two year life-cycle in grass grub and its implications on control. *Proceedings of the 25th New Zealand Weed and Pest Control Conference*, 244-247.

Stewart, K.M.; Van Toor, R. (1983) Control of grassgrub (*Costelytra zealandica* (White)) by heavy rolling. *New Zealand Journal of Experimental Agriculture* **11**, 265-270.

Stuart, A.M. (1958) A review of the work in New Zealand on the clover case-bearers *Coleophora spissicornis* Haw. and *C. alcyonipennella*? Kollar (Coleophoridae, Lep.). *New Zealand Journal of Agricultural Research* **1**, 239-248.

Sutherland, O.R.W.; Hillier, J.R. (1974) Olfactory response of *Costelytra zealandica* (Coleoptera: Melolonthinae) to the roots of several pasture plants. *New Zealand Journal of Zoology* **1**, 365-369.

Sutherland, O.R.W.; Hutchins, R.F.N.; Greenfield, W.J. (1982a) Effect of lucerne saponins and lotus condensed tannins on survival of grass grub, *Costelytra zealandica*. *New Zealand Journal of Zoology* **9**, 511-513.

Sutherland, O.R.W.; Hutchins, R.F.N.; Russell, G.B.; Lane, G.A.; Biggs, D.R. (1982b) Biochemical plant resistance mechanisms: an evaluation of basic research. *Proceedings of the 3rd Australasian Conference of Grassland Invertebrate Ecology*, 245-254.

Sutherland, O.R.W.; Russell, G.B.; Biggs, D.R.; Lane, G.A. (1980) Insect feeding deterrent activity of phytoalexin isoflavonoids. *Biochemical Systematic Ecology* **8**, 73-75.

Tanasijević, N. (1967) Novi prilog proučavanju faune Apiona (Coleoptera: Curculionidae). *Zaštita Bilja* **18**, 301-310.

Taylor, R.L. (1964) Some observations on subterranean grass caterpillar. *Proceedings of the 17th New Zealand Weed and Pest Control Conference* 145-148.

Tereshchenko, V.E.; Sobchuk, N.P. (1980) [Protection of seed crops of clover from pests.] *Zashchita Rastenii* **5**, 31.

Thomas, W.P. (1968) An artificial diet for rearing the light brown apple moth *Epiphyas postvittana* (Walk.) (Lepidoptera: Tortricidae). *New Zealand Entomologist* **4**, 31-32.

Thomas, W.P. (1969) The establishment in New Zealand of *Bracon variegator* Nees (Hymenoptera: Braconidae) a parasite of *Coleophora* species (Lepidoptera: Coleophoridae). *New Zealand Journal of Science* **12**, 43-53.

Thompson, L.S.; Willis, C.B. (1967) Note on the incidence of *Sitona* spp. root injury and root rot in forage legumes in the Maritime Provinces. *Journal of Economic Entomology* **60**, 1181-1182.

Thompson, L.S.; Willis, C.B. (1971) Forage legumes preferred by the clover root curculio and preferences of the curculio and root lesion nematode for species of *Trifolium* and *Medicago*. *Journal of Economic Entomology* **64**, 1518-1520.

Thomson, N.A.; Lagan, J.F. (1982) An investigation into the possibility of using lindane for grass grub control on dairy farms. *Proceedings of the 35th New Zealand Weed and Pest Control Conference*, 98-102.

Thottappilly, G. (1969) Zur Kenntnis des Wirtspflanzenkreises beim Blattrollvirus der Erbse. *Zeitschrift für Pflanzenkrankheiten Pflanzenpathologie und Pflanzenschutz* **76**, 67-78.

Todd, D.H. (1964) Biology and control of white-fringed weevil. *Proceedings of the 17th New Zealand Weed and Pest Control Conference*, 125-129.

Todd, D.H. (1969) Results of laboratory screening tests with insecticides for control of white-fringed weevil larvae and adults. *Proceedings of the 22nd New Zealand Weed and Pest Control Conference*, 301-306.

Tolin, S.A.; Isakson, O.W.; Troutman, J.L. (1970) Association of white clover and aphids (Hem., Hom., Aphididae) with peanut stunt virus in Virginia (USA). *Plant Disease Reporter* **54**, 935-938.

Townsend, R.J.; Pottinger, R.P.; Rowe, S.J. (1979) Evaluation of alternative insecticides for lucerne flea on clover. *Proceedings of the 32nd New Zealand Weed and Pest Control Conference*, 68-71.

Trought, T.E.T. (1977) Control of blue-green lucerne aphids on white clover seed crops. *Proceedings of the 30th New Zealand Weed and Pest Control Conference*, 181-191.

Trought, T.E.T. (1979) Review of chemical control of grass grub, *Costelytra zealandica* (White) in New Zealand 1968-1978. *Proceedings of the 2nd Australasian Conference of Grassland Invertebrate Ecology*, 160-165.

Trought, T.E.T. (1980) Pests in herbage seed production. In: *Herbage Seed Production*. Lancashire, J.A. (Ed) New Zealand Grassland Association, Palmerston North, New Zealand, 41-45.

Trought, T.E.T.; Jackson, T.A.; French, R.A. (1982) Incidence and transmission of disease in grass grub (*Costelytra zealandica*) in Canterbury. *New Zealand Journal of Experimental Agriculture* **10**, 79-82.

Tudor, C.; Brudea, V. (1979) Parazitarae de catre chalcidoide a speciilor de *Apion* seminifage la culturile de trifoi din Moldova. *Probleme de Protecția Plantelor* **7**, 121-130.

Turner, E.C. Jr (1957) Control of the clover root curculio on alfalfa. *Journal of Economic Entomology* **50**, 645-648.

Underhill, G.W.; Turner, E.C. Jr; Henderson, R.G. (1955) Control of the clover root curculio on alfalfa with notes on life history and habits. *Journal of Economic Entomology* **48**, 184-187.

Van der Gulik, J.; Springett, J.A. (1980) The effect of commonly used biocides on slugs. *Proceedings of the 33rd New Zealand Weed and Pest Control Conference*, 225-229.

Venning, M.S.; Lauren, D.R.; Hanzell, R.F. (1982) The potential of one high rate application of synthetic pyrethroids for control of adult grass grub. *Proceedings of the 35th New Zealand Weed and Pest Control Conference*, 90-93.

Wagn, O. (1954) Iagttagelser over optraeden of bladrandbiller (*Sitona*-arter) i baelgplanteafgrøder. *Tidsskrift for Planteavl* **57**, 706-7

Walker, G.P.; Cameron, P.J. (1981) The biology of *Dendrocerus carpenteri* (Hymenoptera: Ceraphronidae), a parasite of *Aphidius* species, and field observations of *Dendrocerus* species as hyperparasites of *Acyrthosiphon* species. *New Zealand Journal of Zoology* **8**, 531-538.

Wallace, M.M.H. (1963a) Treatment of pastures with granular phorate to control the lucerne flea (*Sminthurus viridis*) and the red-legged earth mite (*Halotydeus destructor*). *Australian Journal of Experimental Agriculture and Animal Husbandry* **3**, 219-224.

Wallace, M.M.H. (1963b) Seed treatment with systematic insecticide for control of red-legged earth mite (*Halotydeus destructor*). *Australian Journal of Experimental Agriculture and Animal Husbandry* **3**, 357-364.

Wallace, M.M.H. (1970) Insects of grasslands. In: *Australian Grasslands*: Moore, R.M. (Ed.) Canberra, Australia; Australian National University Press, 361-380.

Wallace, M.M.H. (1981) Tackling the lucerne flea and red-legged earth mite. *Journal of Agriculture, Western Australia*. **22**, 72-74.

Wallace, M.M.H.; Mahon, J.A. (1963) Effect of insecticide treatment on the yield and botanical composition of sown pastures in Western Australia. *Australian Journal of Experimental Agriculture and Animal Husbandry* **3**, 39-50.

Waller, J.B. (1968) A review of chemical control measures for porina. *Proceedings of the 21st New Zealand Weed and Pest Control Conference*, 185-188.

Warner, R.E. (1975) New synonyms, key, and distribution of *Graphognathus*, whitefringed beetles (Coleoptera: Curculionidae) in North America. *Cooperative Economic Insect Report, US Department of Agriculture* **24**, 855-860.

Whiteman, R.J. (1973) Herbivore feeding and cyanogenesis in *Trifolium repens* L. *Heredity* **30**, 241-245.

Wightman, J.A. (1979a) *Costelytra zealandica* and its environment (Coleoptera: Scarabaeidae). *Proceedings of the 2nd Australasian Conference of Grassland Invertebrate Ecology*, 70-74.

Wightman, J.A. (1979b) *Sitona humeralis* in the South Island of New Zealand (Coleoptera: Curculionidae). *Proceedings of the 2nd Australasian Conference of Grassland Invertebrate Ecology*, 138-140.

Wightman, J.A.; Whitford, D.N.J. (1982) Integrated control of pests of legume seed crops. 1. Insecticides for mirid and aphid control. *New Zealand Journal of Experimental Agriculture* **10**, 209-215.

Wigley, P.J.; Miln, A.J. (1982) The distribution and ecology of Flock House virus in *Costelytra zealandica* (Coleoptera: Scarabaeidae). *Proceedings of the 3rd Australasian Conference of Grassland Invertebrate Ecology*, 155-160.

Williams, P.; Stahle, P.P.; Gagen, S.J.; Murphy, G.D. (1982) Strategies for control of the black field cricket, *Teleogryllus commodus* (Walker). *Proceedings of the 3rd Australasian Conference of Grassland Invertebrate Ecology*, 365-369.

Williams, W.M.; Barclay, P.C. (1972) The effect of clover stem eelworm on the establishment of pure swards of white clover. *New Zealand Journal of Agricultural Research* **15**, 356-362.

Wilson, E.R.L. (1978a) White clover and grass grub. I. Screening techniques for resistance and tolerance. *New Zealand Journal of Agricultural Research* **21**, 715-721.

Wilson, E.R.L. (1978b) White clover and grass grub. II. Screening 74 overseas lines for resistance and tolerance. *New Zealand Journal of Agricultural Research* **21**, 723-726.

Wilson, E.R.L. (1978c) Effects of grass grub on establishment of legume seedlings in pots. *New Zealand Journal of Agricultural Research* **21**, 727-731.

Wilson, G.P.M. (1980) Bargoo jointvetch: tough legume for tough country. *Agricultural Gazette of New South Wales* **91**, 51-53.

Wood, J. (1973) Arthropods which have been found in starling (*Sturnus vulgaris*) faeces. *New Zealand Entomologist* **5**, 159-162.

Wrenn, N.R.; Pottinger, R.P.; Dentener, P.R.; McGhie, R.A. (1983) Evaluation of insecticides for lucerne flea control in pasture. *Proceedings of the 36th New Zealand Weed and Pest Control Conference*, 185-189.

Yates, J. (1952) *Haplothrips niger* Osb., the red clover thrips. *New Zealand Journal of Science and Technology* **34B**, 166-172.

Yeargan, K.V.; Shuck, C.I. (1981) Longevity and reproductive rate of *Patasson lameerei* (Hym.: Mymaridae), a parasitoid of *Sitona* spp. (Col.: Curculionidae) eggs. *Entomophaga* **26**, 119-124.

Yunus, C.M.; Johansen, C.A. (1967) Bionomics of the clover seed weevil *Miccotrogus picirostris* (Fabricius) in southeastern Washington and adjacent Idaho. *Technical Bulletin, Washington Agricultural Experiment Station* **No. 53**, 16 pp.

Zein-Eldin, E.A. (1956) Studies on the legume mite, *Petrobia apicalis*. *Journal of Economic Entomology* **49**, 291-296.

Pests - Nematodes

R. A. Skipp and D. L. Gaynor
Plant Diseases Division and Entomology Division, DSIR,
Palmerston North, New Zealand

Contents

1. Introduction

The plant-parasitic nematodes are minute, soil-dwelling eelworms which cause damage while invading plant tissues or during feeding. They are often abundant in pasture soils, but the importance of their effects on herbage production can be insidious rather than dramatic. The nematode pests of white clover can adversely affect establishment, productivity, persistence, and the plant's tolerance of adverse conditions. The major pest species are distributed widely throughout the world. Some have a wide host range and attack many crop plants as well as white clover, whereas other species show host specialisation and exist as distinct biotypes, each with a narrow range of forage legume hosts.

This chapter summarises published information on the biology, ecology and pathogenicity of stem nematode, clover cyst nematode, root knot nematodes, root lesion nematodes and the free-living nematodes which attack white clover. Much of the literature available on clover nematodes concerns assessments of the density of nematode populations in soil, and of the effects of nematode damage on herbage yield. Evidence of the field importance of nematodes is clearly of great value to agronomists and plant breeders who have to decide if control measures should be developed. The necessary field experiments are difficult to perform and seldom give completely unambiguous results. The limitations of the methods used should therefore be considered when conclusions are being drawn.

2. Assessment of nematode populations and effects on clover growth

Standard methods of sampling, including nematode extraction or staining and counting, are used to estimate the density of eggs, larvae, adults or in some cases resistant egg-filled cysts, in soil or plant tissues (Thorne 1961, Whitehead & Hemming 1965, Seinhorst & Sen 1966, Barker & Nusbaum 1971). The data obtained can provide valuable information about population density and spatial distribution of nematodes within soil and tissues, and seasonal fluctuations in nematode numbers. All quantitative studies require adequate replication because of the patchy distribution of nematodes in soil.

Nematode populations in permanent pastures appear to remain fairly stable at an 'equilibrium density' (Nusbaum & Barker 1971). Soil fertility and clover content influence the population level reached at 'equilibrium density' (Yeates 1974, 1978a), and for given soil this value represents the amount of inoculum available for plant infection.

Seinhorst & Sen (1966) suggested that significant damage to the host plant occurs only when the infective inoculum exceeds the plant's 'tolerance limit'. This can be determined experimentally in glasshouse pot trials by growing plants in soil containing inoculum in a graded series of concentrations and ascertaining the lowest density of nematodes in the soil at which there is a measurable reduction in clover growth. Experimental values are then compared with nematode densities found in field soils to determine whether there are sufficient nematodes at particular sites to cause damage. Data of this type have been presented by several workers as evidence of the importance of cyst nematode in pastures.

Wallace (1973), Barker & Olthof (1976) and Jones & Kempton (1978) have considered theoretical aspects of population dynamics. Because of the difficulties presented by such studies, and in the mathematical modelling of the relationship between nematode numbers and effects on plant growth, concepts such as 'tolerance limit' should be regarded as of limited value for indicating the field significance of a particular nematode. The effect of particular nematode species on the growth of white clover has more usually been investigated by comparing the amounts of dry matter produced by clover growing in infested and uninfested soil. However, yield responses in soils sterilised, fumigated or treated with nematicides may be due as much to changes in soil fertility (Ross & McNeilly 1975, Yeates et al. 1977) and to the control of other pests and diseases (Parris 1958, Munnecke et al. 1978, Bergeson et al. 1978, Rodriguez-Kabana & Curl 1980) as to the destruction of the specific nematode under investigation. Some compounds such as methyl bromide as well as being nematicides, are in fact broad-spectrum biocides.

Glasshouse pot experiments have provided valuable information on nematode parasitism and reproduction, and on the physiology of infected clover. However, results from these experiments should be extrapolated to the field only with extreme caution, since the dynamics of field and glasshouse populations can differ markedly.

The accumulated evidence of a number of field nematicide trials supports the view that nematodes are damaging in pastures. However, because the nematicides used in these experiments lack specificity, it is difficult to be certain that the increases in clover growth observed were caused solely by control of a particular nematode. In future work, levels of soil organisms other than the target nematodes should be monitored.

More conclusive information will be gained from studies with experimental clovers resistant to particular nematode pests (Dijkstra 1971). Using near-isogenic lines of white clover, or at least clones which produce similar amounts of herbage in the absence of the pest, it should be possible to measure the effects of both nematode attack and the non-specific effects of chemical treatment. For instance, clover production can be measured in a four-way test in which 'nematicide' and 'no nematicide' treatments are applied to both the resistant and the susceptible lines. Attempts have been made at several research centres to select clover plants resistant to particular nematodes, and once resistant material is available it could be of great value in assessing the pest status of clover nematodes.

3. Stem Nematode

(A) Importance and distribution

The characteristic damage caused by *Ditylenchus dipsaci* (Kühn) Filipjev in red clover has been recognised since the early 19th century (Bingefors 1952) and this nematode is now regarded as one of the most devastating plant-parasitic nematodes in temperate regions (Hooper 1972). It is known to attack over 450 different plant species (Goodey et al. 1965), including daffodil, narcissus, onion, garlic, carrot, pea, faba bean, potato, strawberry, sugar beet, fodder beet, swede, rye, oat, lucerne, red clover, alsike clover and white clover (Thorne 1961, Eriksson 1972, Hooper 1972). Not all plants are attacked by the same biological race of *D. dipsaci*, and the 11 major races differ in their preference for particular crop plants (Seinhorst 1957). Separate races attack lucerne, red clover and white clover, the lucerne and red clover races being the more important economically (Eriksson 1972).

The white clover race of *D. dipsaci* is present in Europe (particularly Scandanavia and the UK), the USA, Australia and New Zealand (Bingefors 1960, 1969, Grandison 1965, Eriksson 1972). It does not survive in tropical regions (Hooper 1972). It is considered seriously damaging to white clover in Sweden and Denmark (Frederiksen 1950, Bingefors 1960), possibly the UK (Cook & York 1980) and during pasture establishment in New Zealand (Williams & Barclay 1972).

(B) Identification and life cycle

An illustrated description of *D. dipsaci* has been given by Hooper (1972), and further information provided by Hooper & Southey (1978a).

D. dipsaci is endoparasitic and feeds in the parenchymatous tissue of the host. The adult is 1-1.3 mm long, colourless and possesses a spear for feeding (Hooper 1972). Mating is essential for reproduction, and egg laying can begin about four days after the last moult. In onions grown at 15°C, females were found to produce 200-500 eggs over a 45- to 73-day period (Yuksel 1960). Under these conditions the egg-to-adult life cycle of four moults and four larval stages takes 19-23 days. The first moult occurs within the egg, and the second and third larval stages quickly develop, usually within the plant. Plants are infected usually by the migratory fourth-stage pre-adult larvae, in contrast to the cyst and root knot nematodes which have infective second stage larvae.

Forth-stage larvae of *D. dipsaci* leave the host when its surface is covered with a film of moisture and may pass into the soil or aggregate on the soil surface (Wallace 1962) before invading other host plants. If the soil dries out or the host plant dies, the fourth stage larvae can enter a dormant phase in which they can withstand desiccation for long periods until favouable cool, moist conditions return (Hastings 1942, Bingefors 1960, Wallace 1962). Fielding (1951) revived dormant larvae from teasel plants kept dry for 23 years. Stem nematode can also survive prolonged low temperature (Thorne 1961, Wallace 1973).

Infective larvae invade clover seedlings at the soil surface through stomata which open wide in response to nematode secretions (Thorne 1961). If the invaded plant is not a suitable host to allow nematode reproduction, the larvae cause localised necrotic lesions and some distortion of invaded tissues (Dijkstra 1957, Wallace 1973). If the clover plant is a suitable host, pectinase enzymes produced by the nematode dissolve the middle lamellae of parenchymatous cells, causing them to separate and become rounded. This creates an environment in which the nematodes move freely and reproduce rapidly (Jones & Jones 1974).

(C) Infection, dispersal and host damage

The first external sign of infection in young white clover seedlings is a swelling at the base of the shoot. Later, the plants develop dense groups of stunted shoots, abnormally small leaves, swollen petioles which are often transversely ridged, and swollen shoots just below the flower head. Infested plants produce fewer stolons than healthy plants. Heavily infested plants eventually die, leaving bare or weedy patches in the second or third year of infestation. Seedling losses may be high, particularly in autumn-sown pastures where low temperatures and accelerated grass growth in early winter can apply further stress to young clover plants (Frederiksen 1950, Bingefors 1960, Williams & Barclay 1972, MAFF 1972, Jones & Jones 1974).

Larvae can be dispersed by infested soil, hay or dead plant material which is wind-blown or carried by stock or farm machinery. In lightly infested plants of clover, lucerne, onion or teasel, the nematodes can be carried upwards by stem elongation. The infective larvae adhere to seed coats, become dormant, and are then dispersed with the seed (Frandsen 1951, Bingefors 1960, Hooper 1972). In the first year after sowing, stunted plants developing from infested seed may be seen in small patches. In subsequent years these patches increase in size and other foci develop as the nematode is dispersed within the field. Atkinson & Sykes (1981) found that in fields of lucerne sown with lightly infested seed there was a 4- to 5-fold increase in the number of foci, their mean size and the area of field damaged during the first few years of infestation. The rate of spread from infestation foci depends on soil type. Larvae remain near the surface in clay soils and are carried in runoff water, particularly on sloping ground, creating strips of damaged plants down a slope (Seinhorst 1957, Wallace 1962, MAFF 1972, Jones & Jones 1974).

There is little information about production losses caused by *D. dipsaci* attack. Seinhorst (1965) indicated that for onions, the proportion of exposed plants (rather than the degree of each attack) determines the extent of crop loss. Seinhorst (1956) showed that 10 nematodes/500 g soil can damage onions seriously, while 25/500 g soil would leave no healthy plants in the field. He also showed that 10-40 nematodes/500 g soil will damage rye greatly.

Grandison (1965) detected no significant stem nematode damage to white clover in established pastures in New Zealand. On the other hand, Williams & Barclay (1972) showed that *D. dipsaci* can affect severely the establishment of the plant. The yield of a susceptible cultivar over a 9-month period was increased from 1000 kg dry matter/ha to over 8000kg/ha by applying granular nematicide to the seedbed. Yield losses sustained by a less susceptible cultivar (Grasslands Huia) diminished as plants that survived the initial attack colonised the bare ground. It was suggested that rapid recovery might not be possible in mixed pasture because of competition from grasses. Yeates (1977) also considered that stem nematode was primarily an establishment problem in New Zealand pastures. Cook & York (1980) considered that the unpredictable performance of white clover in UK pastures could to some extent be due to stem nematode attack. They also noted that the least vigorous clover lines evaluated in breeders' plots were the most heavily infested with *D. dipsaci*.

(D) Control

Chemical treatment by soil fumigation is not generally economic, although thionazin has been used in trials with white clover, red clover and lucerne in New Zealand (Brewerton et al. 1969, Williams & Barclay 1972). Other nematicides have not been tested with clover crops, although aldicarb, oxamyl, fenamiphos and dazomet have all shown promise for use with onions (Brzeski 1970, Winfield et al. 1971, Whitehead et al. 1973). Generally, first grade, machine dressed lucerne and clover seed is free of *D. dipsaci* (Grandison, personal communication), but infested seed can be treated by methyl bromide fumigation. During fumigation the moisture content of seed must not rise above 12% or germination may be affected (Hague & Clark 1959, Hague 1968, MAFF 1972). Atkinson & Sykes (1981) pointed out that methyl bromide seed treatment may not be totally effective if a proportion of the seeds carry aggregates of nematodes.

Crop rotation using non-susceptible crops for several years may help avoid severe attacks of stem nematode. Cereals, beans and sugar beet have been recommended for use in rotation with lucerne (Griffin 1979). Successful control depends on the race of *D. dipsaci* present, the soil type, and the removal of weed hosts (Linnman 1945, Bingefors 1948, Frandsen 1951, Dijkstra 1956, Hooper 1972). Under a non-susceptible crop, nematode numbers decreased more in light and sandy soils than in heavy soils; therefore, crop rotation is recommended for nematode control only on light soils (Seinhorst 1956, 1957, Wallace 1962).

The development of nematode-resistant cultivars of white clover seems to be the most promising form of control, although the genetic variability of the nematode may enable it to overcome the resistance of any one cultivar (Sturhan 1964, Hesling 1966, Webster 1967, Viglierchio & Croll 1968, Viglierchio et al. 1973). Seinhorst (1957) recognised 11 distinct races in Europe, but Hesling (1966) names 21 races. The host range of each race overlaps considerably and they can interbreed (Webster 1967). Both the white and the red clover races infect and damage white clover, but only the white clover race is able to multiply on this host. White clover cultivars vary in their susceptibility to the clover races of *D. dipsaci*, but resistance to the red clover race does not necessarily accompany resistance to the white clover race (Williams 1972).

Cultivars of lucerne, red clover and white clover resistant to their respective races of *D. dipsaci* have been bred. Bingefors (1969) has summarised the development of resistant lucerne cultivars, including Nemastan and Lahontan. Highly resistant tetraploid red clover cultivars have been bred in Scandinavia, e.g. cv. Britta (Jönsson 1976), the Netherlands and the UK (Bingefors 1969). The Scandinavian white clover cultivars Pajbjerg Milka IIK, Lodi Øtofte IIK and Pajbjerg Zero IIK are resistant to stem nematode. A large-leaved English white clover and a Ladino white clover have been reported to be resistant in the UK (Frederiksen 1950, MAFF 1972). In plot trials in New Zealand, Ladino x Grasslands 4700 and Morocco x Grasslands 4700 selections were found to be resistant to stem nematode but were considered agronomically unsuitable for widespread use (Williams & Barclay 1972). In a more extensive laboratory study, Williams (1972) screened 27 introductions for nematode resistance and found that the six most resistant cultivars were all of an unsuitable Ladino type. A more extensive breeding programme appears necessary to incorporate resistance into better adapted types.

Although *D. dipsaci* is a serious pest of many crops and is potentially damaging to white clover, more information is required to determine the extent of damage in pastures.

4. Clover cyst nematode

(A) Importance and distribution

The clover cyst nematode, *Heterodera trifolii* Goffart, was described first from red clover roots but is regarded as a more serious pest of white clover (Eriksson 1972). In common with other root pests and pathogens, many of which may also invade roots attacked by *H. trifolii*, infestation causes a general root malfunction and an accompanying depression of foliage growth (Taha & Raski 1969, Widdowson et al. 1973, Yeates et al. 1975a). Declining clover productivity because of cyst nematode attack is difficult to detect in pastures but may occur through the death of weakened plants or the poor establishment of clover seedlings (Wilson 1978).

The nematode is distributed widely throughout the clover growing areas of Europe, North America, Australasia, the USSR, India and the Middle East (Grandison 1963, Mulvey & Anderson 1974, Cook & York 1980). Its capacity for widespread dissemination and tolerance of cold, heat and desiccation is attributed to the production of a resistant, egg-filled cyst by the female nematode (Ross 1960, Holtzmann & Aragaki 1963, Healy et al. 1973).

(B) Identification and life cycle

Hesling (1978) has provided a key for identifying species of *Heterodera* and a diagnostic description of *H. trifolii* is given by Mulvey & Anderson (1974). Additional information on identification and taxonomy is provided by Wouts & Sturhan (1978), Anderson & Mulvey (1979) and Hirschmann & Triantaphyllou (1979). The life cycle of *H. trifolii* has been described by Gerdeman & Linford (1953) and illustrated in detail by Mulvey (1959a).

The characteristic white, glistening, lemon shaped bodies of adult female nematodes (0.9 mm long) are filled with eggs and protrude from the surface of infected roots. There are no males and eggs are produced by parthenogenesis (Mulvey 1958, Wouts 1978). Mature females are released from root tissues into the soil, their cuticles become tanned and their bodies form impermeable cysts which protect the eggs within. Although most eggs remain within the female some are extruded in a gelatinous matrix before the cyst is formed (Gerdemann & Linford 1953).

Four juvenile stages precede the adult. First-stage larvae develop and moult while still remaining within the egg. The infective second-stage larvae can then remain dormant in eggs within the cyst for several years (Mulvey & Anderson 1974, Yeates & Visser 1979). Dormant second-stage larvae provide a continuous source of inoculum for infection, although host roots can be re-infected more rapidly by larvae from eggs in the gelatinous matrix (Gerdemann & Linford 1953). Root diffusates stimulate hatching of the eggs, and emergence of larvae from cysts and also attract free-swimming larvae to root tips (Mankau & Linford 1960). Leaf tissue (Ross 1960) and more mature root tissue (Gerdemann & Linford 1953) can also be invaded.

Once in roots of a susceptible host, second-stage larvae quickly become orientated parallel to the stele where their feeding activity promotes the dissolution of plant cell walls to form a multinucleate syncytium. The development of a syncytium is essential for the establishment of a parasitic relationship (Mankau 1958, Endo & Schaeffer 1967). Once this is formed, the nematode remains stationary while it grows in girth and develops through three further moults to the adult.

Eggs containing infective second-stage larvae which have been extruded from the female in the gelatinous matrix provide the first source of inoculum for reinfection of host roots, thus allowing completion of the nematode's life cycle. Mulvey (1959a) found that at 10-20°C, *H. trifolii* required 56 days to complete its life cycle. Generation times are much shorter at higher temperatures as reported by Mulvey (1959b, 31 days at 20°C) and Mankau & Linford (1960, 17 days at 21-27°C). Thus under conditions allowing continuous infection and reproduction, as in the glasshouse, eight of more generations per year are possible (Mulvey 1959b). In the field, however, fewer generations would be expected because of seasonal fluctuations in temperature and plant growth. Two generations per year have been recorded in Germany (Goffart 1944) and in New Zealand (Yeates 1973a, Yeates and Risk 1976) where spring and autumn were the periods of peak invasion. In the New Zealand studies, a dormant period in summer corresponded to a time of drought. In a moist site in the North Island, New Zealand, larval invasion did not occur in two distinct seasonal phases (Yeates 1973b) and in the USA in Illinois, Gerdemann & Linford (1953) found evidence of a series of overlapping generations. The factors controlling root invasion and reproductive activity in the field are unclear. Oostenbrink (1967) considered seasonal fluctuations in the emergence of larvae from cysts to reflect the influence of winter diapause and soil temperatures on hatching. This view is consistent with Yeates' (1973a) finding that larval invasion could be induced during winter by transferring plants from the field to a warm glasshouse. The two reproductive cycles of *H. trifolii* per year in New Zealand may be a specific characteristic of the nematode as suggested by Yeates & Risk (1976), but soil temperature and moisture and the root growth activity of the host plant also influence infection and reproduction (Yeates 1973a).

(C) Effects in the glasshouse

Pot experiments by Hidding et al. (1963), Holtzmann & Aragaki (1963), Wardojo et al. (1963), Chapman (1964), Stelter & Meinl (1972) and Yeates (1974, 1978a) showed that shoot dry matter production by white clover was substantially reduced in soil infested with *H. trifolii*. Seinhorst (1981) indicated that there were two mechanisms involved in the reduction of clover growth by *H. trifolii*; one was a direct effect of the nematode on the plant and the other related to reduced water uptake. Korteweg & s'Jacob (1980) found that the proportion of *T. repens* in a ryegrass-clover mixture declined progressively when *H. trifolii* was present in the potting mix. This decline occurred at both high and low levels of soil nitrogen, and was most rapid when a high level of inoculum was used (13 760 eggs/l soil).

In pot experiments, clover seedlings are affected much more severely than adult plants (Dijkstra 1971, Healy et al. 1973). Skipp & Christensen (1981) found that roots of white clover seedlings were stunted severely after *H. trifolii* infection. As the swollen female nematodes emerged from infected roots, epidermal and cortical tissues were ruptured and killed. This allowed invasion by soil fungi, some of which could be pathogenic to white clover roots (Skipp & Christensen 1982). The association between nematode damage and fungal invasion is well known (Armstrong et al. 1976) and is discussed later in connection with other clover nematodes.

The effects of soil moisture, liming, superphosphate application and mineral nutrition on plant damage and nematode reproduction have also been studied in pot experiments. Yeates (1974) found that *H. trifolii* greatly reduced clover growth in soils whose moisture content ranged from near field capacity to wilting point. Nematode activity was limited in dry soils but damage was still evident. Addition of lime to infested soil had little effect on white clover or on the reproduction of *H. trifolii* (Yeates 1974). This contrasts with results from previous pot experiments (Healy et al. 1973), indicating that liming could alleviate damage resulting from a combined infection of *H. trifolii* and *Meloidogyne hapla* Chitwood.

Pot-grown clover plants in phosphate-deficient soil responded to the addition of superphosphate fertiliser whether or not they were infected with *H. trifolii* (Yeates 1974, 1978a). However, the increases in yield which resulted from control of the nematode were far greater than could be obtained solely by adding phosphate. Under field conditions the poor utilisation of phosphate by infected plants could result in considerable wastage of applied fertiliser. Riispere & Krall (1970) found that clover tolerated *H. trifolii* best when grown under optimum conditions of water and nutrient supply.

The nitrogen fixing ability of clover roots also is affected by *H. trifolii* invasion (Wardojo et al. 1963, Taha & Raski 1969, Yeates et al. 1977), but it is not yet clear whether this is caused by a general disruption of root function or by more specific effects on the physiological interaction between root nodules and their symbiotic bacteria. Larvae of *H. trifolii* are attracted to and readily invade root nodules (Taha & Raski 1969).

(D) Effects in pastures

H. trifolii densities in pasture soils have been measured in several countries in an attempt to assess the field significance of this nematode. In the Netherlands, 75% of pastures sampled were infested with *H. trifolii* (Hidding et al. 1963). Ennik et al. (1965) found that in patches where clover was in the process of disappearing from swards infected with more than 13 nematode genera, only *H. trifolii* showed a population density in the patches consistent with its being the cause of the observed damage. Seinhorst & Sen (1966), however, disputed the importance of cyst nematode in Dutch pastures. They found that the 'tolerance limit' for white clover grown in pots was a nematode density of 50 eggs/g of soil; the nematode density of pasture soils was much lower and considered non-damaging. In New Zealand, Yeates (1975) found *H. trifolii* at all but one of 77 sites sampled. The mean cyst count for 54 pasture soils was 460 (0-2300)/kg dry soil (Yeates 1977). At 800 cysts/kg dry soil in pots, clover growth was depressed by 25% but when ryegrass was grown with the clover, only 150 cysts/kg dry soil were needed to depress clover growth by 50% (Yeates 1974). Therefore, in the mixed ryegrass-white clover pastures of New Zealand, *H. trifolii* might be expected to cause considerable damage.

Results from field trials in which plots of white clover were infested artificially with *H. trifolii* also provide some evidence that cyst nematode damage may occur in a pasture. Hidding et al. (1963) in the Netherlands found that clover disappeared rapidly from infested field plots, and plants that remained were of poor nutritional quality (crude protein levels were reduced from 27% to 21%). In artificially infested swards of *Dactylis glomerata* L. and *Trifolium repens* L. mixtures in Iowa, USA, the total annual dry matter production of Merit Ladino white clover fell by 47% and that of Dutch white clover by 14% (Norton 1967). Norton considered that since cyst nematode was present in most pastures in the State, it was probably important in the decline of white clover productivity.

Field trials using toxic chemicals have also been cited as evidence for the importance of *H. trifolii* damage in pastures (Yeates 1977). Ennik (1968) found that biocide applications increased the persistence of white clover in Dutch grasslands, and in a number of New Zealand trials fumigant treatment of field plots on soil containing *H. trifolii* were reported to increase the yield of clover herbage. Plots of white clover at a site in the Wairarapa (North Island, New Zealand) sown on seedbeds treated with methyl bromide, DD, DBCP and ethoprophos yielded 7-12% more dry matter over 18 months than untreated control plots (Yeates et al. 1975a). Inoculum of *H. trifolii* was monitored during the trial and found to decrease in treated plots. At a Taranaki site (North Island, New Zealand) on soil infested with *H. trifolii* and *Meloidogyne hapla* fumigant treatment of the seedbed also increased clover yield (by 6-15%, Yeates et al. 1975b). Greater responses to chemical treatment have been recorded from pasture with low productivity in Southland (South Island, New Zealand) where clover yields in oversown plots treated with ethoprophos and DBCP applied in spring and summer were up to 85% greater than in untreated plots (Risk & Yeates, personal communication). The magnitude of yield increases recorded in these trials varied with the season. Generally, these responses to fumigant treatment were greatest when *H. trifolii* larvae invaded at these sites (spring and autumn) (Yeates 1973a, Yeates & Risk 1976). Yeates (1977) suggested that severe attack by *H. trifolii* during critical periods of clover growth could have significant effects on stock production, particularly in dry areas.

(E) Host range and biotypes

The host range of *H. trifolii* must be known before adopting crop rotation control measures. Studies by Gerdemann & Linford (1953), Holtzmann & Aragaki (1963), Sen (1963), Norton & Isely (1967), Stehman & Norton (1969) and Townshend & Potter (1976) indicate that apart from *T. repens*, *T. pratense* is the only other important forage legume host for *H. trifolii*. Crop plant hosts for *H. trifolii* include bean, pea, pumpkin, squash, cucumber, spinach and Swiss chard. Carnation is damaged by this nematode in France (Cuany 1970) and weed hosts include *Rumex* spp. and *Chenopodium glaucum* L. (Gerdemann & Linford 1953, Mulvey 1959b).

Carrot, tomato, soybean, onion, peanut and red beet are non-hosts and can be used in rotations (Mulvey & Anderson 1974). Cyst nematode does not reproduce on grasses, brassicas and oats, so these can be used safely as alternative forage crops (Gerdemann & Linford 1953). Yeates et al. (1976) suggested that a 2-year break from pastures with a brassica crop might be as effective as chemicals in controlling *H. trifolii* and *M. hapla*.

Variations in host range exhibited by cyst nematodes collected from different parts of the world suggest the presence of a number of biotypes of *H. trifolii*. Singh & Norton (1970), who examined the relative abundance of cysts produced by different isolates of *H. trifolii* on species of *Trifolium*, found that variation between isolates was greatest on *T. pratense* and *T. fragiferum* L. On *T. pratense*, differences were as great among eight isolates from western Iowa as they were among isolates from widely separated localities in North America. The numbers of cysts produced by inoculum of individual isolates of *H. trifolii* also varied on the three cultivars of white clover tested.

To overcome subsequent attack by biotypes of different virulence, sceening programmes to detect sources of resistance to cyst nematode in white clover should include several isolates of nematode. However, Dijkstra (1971) found no evidence of different white clover biotypes in Holland and argued that more pathogenic biotypes would develop very infrequently because the parthenogenetic reproduction of *H. trifolii* does not favour genetic recombination.

Three other species of *Heterodera*, *H. humuli* Filipjev, *H. lespedezae* Golden & Cobb and *H. galeopsidis* Goffart, have been shown also to form cysts on *T. repens* (Sen & Jensen 1969, Edwards & Malek 1971, Stelter & Meinl 1972).

(F) Control

If white clover selections resistant to cyst nematode could be found they could be used both as a means of studying nematode damage in pasture as discussed earlier, and as a basis for control by host-plant resistance. In a screening test, Yeates et al. (1973) were unable to detect any resistance to *H. trifolii* in seven white clover seed lines growing in naturally infested soil as a source of inoculum. Kuiper (1960) and Dijkstra (1971), however, found a number of resistant clones among those selected for persistence in a pasture infested with *H. trifolii*. Resistance appeared to be controlled by more than one gene (Dijkstra 1971). Resistant clones came from seed originating in six countries (Kuiper 1960); thus it should be possible to make selections from within local cultivars. Selection procedures should take into account both the resistance of plants to nematode reproduction and tolerance to root damage (Yeates et al. 1973).

Control measures other than crop rotation and host resistance seem impracticable except for special purposes. Nematicides are expensive but may prove valuable as seed or seedbed treatments to overcome damage at establishment. Management practices can affect levels of nematode populations in the soil (Yeates 1976, 1978b), but it may be difficult to convince farmers to use radically different methods to overcome only one of the many limitations to livestock production. So long as we lack conclusive evidence that *H. trifolii* causes economically significant damage in pastures there will be little incentive to develop and use control measures other than breeding for resistance or tolerance.

5. Root knot nematodes

(A) Importance and distribution

Species of *Meloidogyne* attack and damage a wide range of economically important plants. White clover is host to *M. arenaria* (Neal) Chitwood, *M. hapla, M. incognita* (Kofoid & White) Chitwood and *M. javanica* Chitwood. The characteristic 'root knot' initiated after infection by *Meloidogyne* spp. is a gall produced by swelling (hypertrophy) and division (hyperplasia) of cells in the invaded root cortex (Bird 1979). Galls are found often on the tips of short roots or rootlets; infection near the root tip usually prevents further growth. Like the related clover cyst nematode (*H. trifolii*), root knot nematodes can affect establishment, productivity and persistance of white clover through a general disruption of root growth pattern and function (Colman 1964, Hijink & Kuiper 1964).

Meloidogyne spp. are more susceptible to desiccation and cold injury than *H. trifolii* because they cannot form cysts (Franklin 1965). This may explain why their distribution is restricted to moist subtropical and temperate regions (Orton Williams 1972, 1975). Greater cold tolerance is shown by *M. hapla*, the northern root knot nematode, than other species and this nematode is widespread on clover in northern Europe (Orton Williams 1974) and throughout New Zealand (Grandison, personal communication). All four species of *Meloidogyne* are present on white clover in the USA (Allison 1956) and Australia (Colman 1964).

(B) Identification and life cycle

A key for identifying 16 species of *Meloidogyne* based on the cuticular pattern in the tail region of the female has been given by Franklin (1965) and recent taxonomic reviews were provided by Esser et al. (1976), Taylor & Sasser (1978a), Franklin (1979), and Eisenback et al. (1980). Illustrated annotated descriptions of *M. javanica, M. incognita, M. hapla* and *M. arenaria* have been given by Orton Williams (1972, 1973, 1974, 1975). These authors also described the life cycle of *Meloidogyne* species.

Mature females remain embedded in the root gall tissue and lay eggs into a gelatinous matrix (egg sac) produced by rectal gland cells. When egg sacs are exposed on root surfaces the outer layers may dry and toughen, thus protecting the eggs from desiccation. *M. hapla* probably overwinters in the egg sac (Brzeski & Hendricks 1971). The first larval moult occurs within the egg, giving rise to infective second-stage larvae.

Root penetration, and to some extent hatching of the eggs, are thought to be stimulated by root diffusates. Larvae released into the soil penetrate the root, usually in the region of cell elongation, and move through the cortex to the pericycle, where cell walls are pierced by repeated thrusts of the mouth spear. The spear is hollow and serves as a channel for digestive secretions and ingestion of cell contents during feeding. The walls of several adjacent pierced cells break down and coalescence of the protoplasts, accompanied by repeated nuclear division, produces a so called giant cell (Bird 1979). The remaining cell walls which delimit the giant cell become thickened irregularly. Damage to the host results from a disruption of vascular conduction by the giant cells, which remain active throughout the life of the nematode provided there is continuous feeding activity. Galls are initiated in response to nematode secretions but do not appear to be essential for parasitism.

Third- and fourth-stage larvae develop within the spike-tailed cuticle of the second-stage larvae but do not feed because the mouth spear is lost during the second moult. The spear is formed again in the adult, which is an active feeder, growing rapidly in girth until mature. Adults of both sexes occur but the ratio of males to females is low except when conditions are unfavourable for nematode infection and reproduction. Asexual, parthenogenetic reproduction seems more common even when males do transfer sperm to the embedded females (Orton Williams 1972, 1973, 1974, 1975).

As with *Heterodera* spp., the time required for completion of the life cycle depends on temperature and moisture. Under glasshouse conditions or in warmer areas, many more generations (10 or more) may occur per year than in the cooler areas where clover is grown. Two generations of *M. hapla* are usual in central Europe (Orton Williams 1974). Yeates (1973b) reported that *M. hapla* invaded roots in the Taranaki area of New Zealand in spring and early summer, but the number of generations per year was unknown.

(C) Effects on clover growth

There is little information on the damage to clover caused by root knot nematodes in pastures. Pot experiments by Colman (1964), Hijink & Kuiper (1964) and Taha & Raski (1969) proved that *M. hapla* and *M. javanica* are pathogenic to *T. repens* and greatly reduced dry matter production. In the USA, experiments by Halpin (1963), using field bins of artificially infested soil, showed that *M. incognita* affected growth, vigour and persistence of white clover. Similarly, Baxter & Gibson (1959) established a number of clones of white clover within metal rims embedded in a field plot of fumigated soil and made separate inoculations with four *Meloidogyne* species. Four months after inoculation, all plots showed leaf yellowing and stand reduction, with *M. arenaria, M. incognita* and *M. javanica* having more effect than *M. hapla*. However, recovery from *M. hapla* damage during the winter was poor and after the second winter, most plants in the plots infected with *M. hapla* were dead. It was suggested that *M. hapla* is particularly damaging in the mild winters of the southern USA. In Canada, field trials with Ladino white clover, red clover, birdfoot trefoil and lucerne in microplots demonstrated the damaging effects of *M. hapla* on seedling establishment and growth of white clover during two seasons (Townshend & Potter 1978). Survival of clovers over two years was poor, even in nematicide treated plots. In New Zealand, glasshouse pot experiments and field trials showed that yields of white clover were less in non-fumigated soil containing both *M. hapla* and *H. trifolii* than in fumigated soil in which these nematodes had been killed (Healy et al. 1973, Widdowson et al. 1973, Yeates et al. 1975b). Yeates et al. (1977) found that *M. hapla* infestation reduced nitrogen fixation by clover grown in pots.

(D) Host range

Host range studies have shown that root knot nematodes can reproduce on some pasture grass species and have identified those grasses, fodder legumes and crop plants that can be used as alternatives to clover in rotation. Knowledge of the reactions of different plant species to root knot nematodes can also be used as an aid to identification (Sasser 1980).

McGlohon et al. (1961) and Townshend & Potter (1976) showed that *M. hapla* did not reproduce significantly on any of the grass species tested, but *M. javanica, M. incognita* and *M. arenaria* produced egg masses on *Bromus inermis* Leyss., *D. glomerata, Lolium multiflorum* Lam., *Phalaris aquatica* L., *Poa pratensis* L. and *Sorghum vulgare* L. Other important forage legumes, including red clover, lucerne, *Lotus* spp. and *Lespedeza* spp. are damaged seriously by root knot nematodes and thus cannot be considered suitable replacements for white clover in infested soils.

Root knot nematodes attack many crop plants but the host ranges of individual *Meloidogyne* spp. differ. Coffee, cotton, maize, watermelon and most cereals are resistant to *M. hapla* (Orton Williams 1974). Hijink & Kuiper (1964) considered that rotations that included cereals have controlled *M. hapla* in the Netherlands. Plants resistant to *M. javanica* included Brussels sprout, radish, squash, some varieties of maize, peanut and pepper; cereals generally show mild infection (Tanveer & Saad 1971). Cereals show some resistance to *M. incognita*, while safflower, Brussels sprout, radish, peanut and some pea and bean varieties are also non-hosts (Tanveer & Saad 1971). A list of crop cultivars resistant to root knot nematodes has been produced by Sasser & Kirby (1979). Where short pasture leys are used in an arable farming rotation, the possible build-up of root knot nematodes must be considered in relation to their possible effect on a subsequent crop.

(E) Control

The existence of several *Meloidogyne* species in the warmer areas of the world seems to preclude the practical use of clovers resistant to root knot nematode in these regions, although some selections of clones resistant to *M. incognita* has been attempted (McGlohon & Baxter 1958, Bain 1959, Shepherd 1965, Bingefors 1969). However, the predominance of *M. hapla* in cooler climates may make the search for material resistant or tolerant to this nematode more worthwhile. Yeates et al. (1973) found that all seven white clover lines tested were susceptible to *M. hapla*. A more extensive screening programme to detect individual white clover plants resistant to this nematode is being undertaken in New Zealand by Grandison (personal communication). Whenever practicable, screening programmes need to take account of the possible existence of biotypes (Sturhan 1971) and to distinguish between resistance and tolerance by studying both nematode reproduction and effects on the plant.

Crop rotation might provide short-term control of root knot nematode in pastoral farming, but chemicals will probably only have a limited application (e.g. during establishment). Detailed information about the chemical control of *Meloidogyne* species is given by Taylor & Sasser (1978b).

6. Root lesion nematodes

(A) Importance and distribution

Minute brown streaks caused by the tunnelling of root lesion nematodes (*Pratylenchus* spp.) can often be seen in the cortex of roots of white clover plants dug from the field. Like the cyst and root knot nematodes, they are obligate plant parasites, but rather than remaining in roots as sedentary feeders both larvae and adults of *Pratylenchus* can migrate to and from root tissues. Many species of *Pratylenchus*, including those found on white clover, have a wide host range and important crops such as potato, cotton, banana, forest trees, berry and tree fruits can be damaged severely. Of the species that attack white clover, *Pratylenchus penetrans* (Cobb) Filipjev & Schuurmans Stekhoven is the most common in temperate areas (Corbett 1973, Willis et al. 1976). *P. scribneri* Steiner and *P. brachyurus* Godfrey have been reported from the southern USA (Good & Blue 1954, Minton 1965), and *P. pratensis* (de Man) Filipjev has been recorded on Ladino clover in Oregon (Thorne 1961).

(B) Identification and life cycle

Pitcher (1965) listed the salient features of the genus *Pratylenchus* as: opaque; blunt head; stout spear;

lobed oesophagus; posterior vulva; conoid tail (female). Sher & Allen (1953) discussed the taxonomy of the genus. The diagnostic features of species have been summarised by Thorne (1961) and Lööf (1978). Corbett (1973) has given a comprehensive, illustrated account of *P. penetrans* and its life cycle.

Many *Pratylenchus* species, including *P. scribneri* and *P. brachyurus*, reproduce by parthenogenesis, but *P. penetrans* reproduces sexually (Triantaphyllou 1971). Eggs are laid singly in the soil or in root tissues. Larvae may be seen often curled within the egg and, on hatching, second-stage larvae are released. The second and two subsequent larval stages plus the adult can tunnel through roots with the aid of their mouth spears and feed on the contents of cortical cells. Freckman & Chapman (1972) found that individuals of *P. penetrans* moved no more than 2 mm from the site of penetration of red clover roots, which could be anywhere along the length of the root. However, Olthof (1982) found that the proportion of successful penetrations of lucerne roots decreased with the age of the root tissue being penetrated. Early larval stages often caused no visible damage, but the progress of older larvae and adults through roots is accompanied by necrosis and browning adjacent to the walls of the tunnel, thus causing the characteristic root lesion. Kimpinski (1975) showed that four generations of *P. penetrans* could occur per year on red clover and that reproduction could take place at low temperatures (3-4°C).

(C) Population density and effects on clover growth

Little is known about how much root lesion nematodes damage white clover in pastures. In Alabama, USA, Minton (1965) found that clover production had declined in areas where populations of *P. scribneri* were large. Yeates (1976) suggested that the poor response to phosphate fertiliser application in a New Zealand trial may have resulted from an accompanying increase in the population of *Pratylenchus* spp. Southey (1959) considered that poor white clover growth in Britain was caused by *Pratylenchus* spp.

Damage to forage legumes by *Pratylenchus* spp. is best documented in Canada. Townshend et al. (1973) found 80 to 15 700 *P. penetrans* nematodes per g dry root in samples taken from 13 white clover crops sampled in eastern Canada. Willis & Thompson (1967) similarly found large populations of *P. penetrans* in fields of white, red and alsike clovers and in *Lotus corniculatus* L. in the Maritime Provinces. In glasshouse experiments, Willis & Thompson (1969) demonstrated that white clover growth was depressed by *P. penetrans*. Application of the nematicides oxamyl and phenamiphos to stands of *L. corniculatus* reduced numbers of *P. penetrans* and increased forage yield (Thompson & Willis 1976).

Environmental factors can greatly influence populations of *Pratylenchus* spp. in soil and roots. Yeates (1978c) found *Pratylenchus* spp. to be far more numerous in irrigated than dryland pastures, suggesting that the nematode may be susceptible to desiccation. Nematode numbers may increase most rapidly when clover is growing actively (Wehunt 1957). Yeates (1978c, d) found that *Pratylenchus* populations were highest in spring and early autumn in irrigated sites and that increases in numbers of the nematode followed heavy applications of lime and phosphate (Yeates 1976). However, Yeates (1976) also found that *Pratylenchus* populations were increased significantly by raising the stocking rate, a factor that might tend to suppress root growth. Dolliver (1961) also found that mild stress such as defoliation would increase the numbers of *P. penetrans* in the roots of peas.

Several reviews (Thorne 1961, Pitcher 1965, Corbett 1973, Armstrong et al. 1976) have indicated that penetration wounds can act as sites for entry of pathogenic fungi and bacteria. Damage by nematodes may also lower the resistance of roots to disease. In some vascular and root rot diseases of important crop plants, *Pratylenchus* spp. are considered major constituents of a pathogen complex (Bookbinder et al. 1982). The association between fungal root rots of white clover and invasion by *Pratylenchus* spp. is unclear. Good & Blue (1954) found that fumigation of soil containing *P. brachyurus* and *Sclerotinia sclerotiorum* (Lib.) De Bary increased dry matter production of Ladino clover, but the relative importance of nematode and fungus was not known. Edmunds & Mai (1966a,b) found that *P. penetrans* populations in lucerne roots were greater in the presence of *Trichoderma viride* Pers. ex S.F. Gray and suggested that fungal enzymes may aid penetration. Willis & Thompson (1975) found that nematicides reduced the incidence of fusarium root rot in pure stands of *Lotus corniculatus* infested with *P. penetrans*, suggesting an association between nematodes and fungal attack. Similar results have been obtained with lucerne (Willis & Thompson 1979).

(D) Control

There has been no real incentive to find methods for controlling root lesion nematodes in pasture while evidence that they cause appreciable decreases in productivity is lacking. Thompson & Willis (1971) examined 50 *Trifolium* and 39 *Medicago* species for resistance to *P. penetrans*, but could find no resistance upon which to base a breeding programme. However, a search for resistant clovers may be warranted so that they can be used in grass-clover leys in rotation with crops susceptible to *Pratylenchus* spp.

7. Free-living nematodes

In addition to the root-invading nematodes discussed above, pasture soils also harbour many free-living, plant-parasitic nematodes of the orders Tylenchida and Dorylaimida (Yeates 1975, 1978c,d, Corbett 1978, Hooper & Southey 1978b). These nematodes also possess mouth spears but restrict their feeding to surface cells (e.g. *Trichodorus*) or deeper root tissues (e.g. *Xiphinema, Helicotylenchus*). Damage is aggravated often by the toxic effects of injected secretions. Experimental proof of pathogenicity of these nematodes is difficult to obtain (Pitcher 1957, 1965), but they probably cause significant damage in some pastures.

Investigations by McGlohon et al. (1961) provide most of the information available about the effect of these nematodes on white clover. They found that species of *Paratylenchus, Criconemoides, Helicotylenchus, Tylenchorhynchus, Trichodorus, Xiphinema* and *Hoplolaimus* were common on white clover in a number of forage crops in North Carolina, USA. Many of these nematodes also attacked other forage legumes and companion grasses (see also Jenkins et al. 1956, Townshend et al. 1973, Feldmesser et al. 1975, Willis et al. 1976). In glasshouse studies, lance nematode (*Hoplolaimus tylenchiformis* Daday), stubby root nematode (*Trichodorus christiei* Allen) and spiral root nematode (*Helicotylenchus nannus* Steiner) reduced foliage and root growth of Ladino clover, the damage being greatest when nematode populations were mixed. Although pin nematode (*Paratylenchus projectus* Jenkins) and stunt nematode (*Tylenchorhynchus claytoni* Steiner) had no effect on growth, these nematodes multiplied greatly on Ladino clover. In field studies by McGlohon et al. (1961), a one-year fallow reduced the numbers of stem (*Ditylenchus dipsaci*), spiral, lance, and pin nematodes substantially in a number of soils, the greatest reduction occuring in the spring. Spiral nematodes were found to be resistant to desiccation and remained viable in air-dry soil for more than 250 days; lance and stubby root nematode did not survive for long in dry soil.

Pot experiments by Minton & Cairns (1957), Rohde & Jenkins (1957) and Wood (1973) have confirmed that *T. repens* is a good host for *H. nannus, T. christiei* and *P. projectus*. Good & Blue (1954) found that sting nematode, *Belonolaimus gracilis* Steiner, reduced Ladino clover growth in glasshouse trials. Yeates (1975) reported that *Tylenchorhynchus* spp., *Helicotylenchus* spp. and *Paratylenchus* spp. were common in New Zealand pasture soils.

Patchiness in some UK pastures has been attributed to the frequent occurrence of dagger nematode, *Xiphinema diversicaudatum* Micoletsky (Harrison & Winslow 1961). This nematode is known to reproduce on white clover and is important agronomically as a vector of arabis mosaic virus and strawberry latent ringspot virus which cause important diseases of strawberry. As white clover is a host for the nematode and for both viruses (Thomas 1970), it may act as a virus reservoir in pasture soils which might later be used to grow strawberries. A similar reservoir situation exists with *Longidorus elongatus* de Man, which transmits raspberry ringspot virus and tomato black ring virus to and from *T. repens* (Thomas 1969).

8. General conclusions

The significance of any pest must be judged finally in economic terms, and its damage measured by techniques which reflect likely effects on farm production. Although the accumulated evidence from research on clover nematodes encourages the belief that these can be primary causes of lost production in pastures, much of this evidence must be regarded as circumstantial. We still have little quantitative information with which to judge the extent of the problem of nematode pests of white clover. As even less is known about the effects of diseases and all but a few of the invertebrate pests (see Chapters 11 and 12), we re-emphasise that imaginitive new programmes are required for thorough evaluation of the effects on pasture production of diseases and pests - including nematodes.

The constraints to maximum clover production in pastures are complex and nematode damage is only one of the factors involved, just as nematodes are only part of an enormous community of soil micro-organisms. Interactions among plant-parasitic nematodes, other organisms in the soil, and plant roots could have even greater effects on clover growth than the damage caused by the nematodes themselves. As we have seen, some nematodes can disrupt translocation patterns in roots and may affect the development and function of root nodule and mycorrhizal symbioses. Nematode infection can increase the severity of damage by root-rotting fungi, and exudates from rotting roots may influence their attractiveness to nematode larvae. The influence of mycorrhizal fungi on nematode infection and damage (Cooper & Grandison 1979), and the development of large populations of nematode-trapping fungi in soil, are examples of microbial interrelationships which might be manipulated to benefit the clover plant.

The effective study of complex interactions demands co-operation from specialists of different disciplines. Such a multidisciplinary approach might also lead to a better understanding of the underlying causes of the perennial problems of pasture agronomy such as poor establishment, productivity and persistence or, indeed, of the superior performance of selected breeding material at particular trial sites, or of the effectiveness of certain management practices.

Acknowledgements

We are grateful to Dr G. Grandison, Dr M. Foot and Mr C. Mercer for their helpful comments and constructive criticism.

References

Allison, J.L. (1956) Root knot of perennial forage legumes. (Abstr.) *Phytopathology* **46**, 6.

Anderson, R.V.; Mulvey, R.M. (1979) Plant-parasitic nematodes in Canada. Part 1. An illustrated key to the genera. *Monograph, Research Branch, Agriculture Canada,* **No. 20**, 152 pp.

Armstrong, J.M.; Jatala, P.; Jensen, H.J. (1976) Bibliography of nematode interactions with other organisms in plant disease complexes. *Bulletin, Oregon Agricultural Experiment Station,* **No. 623**, 19 pp.

Atkinson, H.J.; Sykes, G.B. (1981) An analysis from aerial photography of the spread of seed-borne *Ditylenchus dipsaci* in lucerne. *Nematologica* **27**, 235-241.

Bain, D.C. (1959) Selection for resistance to root knot of white and red clover. *Plant Disease Reporter* **43**, 318-322.

Barker, K.R.; Nusbaum, C.J. (1971)Diagnostic and advisory programs.In: Zuckerman, B.M.; Mai, W.F.; Rohde, R.A. (Eds). Plant parasitic nematodes. New York, USA; Academic Press, **Vol. 1**, 281-301.

Barker, K.R.; Olthof, T.H.A. (1976) Relationship between nematode population densities and crop responses. *Annual Review of Phytopathology* **14**, 327-353.

Baxter, L.W.; Gibson, P.B. (1959) Effect of root-knot nematodes on persistence of white clover. *Agronomy Journal* **51**, 603-604.

Bergeson, G.B.; Griffin, G.D.; Lembright, H.W.; Lownsbery, B.F.; Norton, D.C. (1978) Nematicide evaluation as affected by nematode and nematicide interactions with other organisms. In: Zehr, E.I.; Bird, G.W.; Fisher, K.D.; Hickey, K.D.; Lewis, F.M.; Line, R.F.; Rickard, S.F. (Eds) Methods for evaluating plant fungicides, nematicides, and bactericides. St. Paul, Minnesota, USA; American Phytopathological Society, 128-134.

Bingefors, S. (1948)Växtföljden och 'klövertrötthet' orsakad av nematodangrepp. *Försök och Forskning* **5**, 94-95.

Bingefors, S.(1952) Nematode resistance in clover and lucerne. *Proceedings of the 6th International Grassland Congress,* **Vol. 2**, 1591-1596.

Bingefors, S. (1960) Stem nematodes in clovers and lucerne and their control by breeding for resistance. *Proceedings of the 8th International Grassland Congress,* 78-81.

Bingefors, S. (1969) The use of nematode-resistant varieties of grasses and legumes. *Herbage Abstracts* **39**, 107-111.

Bird, A.F. (1979) Histopathology and physiology of syncytia. In: Lamberti, F.; Taylor, C.E. (Eds) Root knot nematodes (*Meloidogyne* species) systematics, biology and control. London, UK; Academic Press, 155-170.

Bookbinder, M.G.; Bloom, J.R.; Lukezic, F.L. (1982) Interactions among selected endoparasitic nematodes and three pseudomonads on alfalfa. *Journal of Nematology* **14**, 105-109.

Brewerton, H.V.; McGrath, H.J.W.; Grandison, G.S. (1969)Thionazin residues on lucerne and red clover (Experiments for control of *Ditylenchus dipsaci*). *New Zealand Journal of Agricultural Research* **12**, 171-176.

Brzeski, M.W. (1970) Chemiczne zwalczanie wegorka niszczyka na cebuli. *Mater. IV. Ogólnopolsk. Zjazdu Warzyw. Skierniew.* 34-35.

Brzeski, M.W.; Hendricks, E.K. (1971) The overwintering and hatching of *Meloidogyne hapla* Chitw. *Zeszyty Problemowe Postepow Nauk Rolniczych,* **No. 121**, 235-244.

Chapman, R.A. (1964) Effect of clover cyst nematode on growth of red and white clovers. *Phytopathology* **54**, 417-418.

Colman, R.L. (1964) Root knot nematodes as a cause of poor growth of clover on the far north coast of N.S.W. *Agricultural Gazette of New South Wales* **75**, 1367-1368.

Cook, R.; York, P.A. (1980) Nematodes and herbage improvement. *Annual Report, Welsh Plant Breeding Station for 1979*, 177-207.

Cooper, K.M.; Grandison, G.S. (1979) Interaction of mycorrhizas and root-knot nematode (Abstr.) *49th Congress of Australian and New Zealand Society for the Advancement of Science* **1**, 405.

Corbett, D.C.M. (1973) *Pratylenchus penetrans. C.I.H. Descriptions of Plant-parasitic nematodes,* **Set 2**, No. 25.

Corbett, D.C.M. (1978) Migratory soil Tylenchida. In: Southey, J.F. (Ed.) Plant nematology; 3rd Edition (G.D.1). London. UK; Ministry of Agriculture, Fisheries and Food, 188-206.

Cuany, A. (1970) Étude de l'action de l'aldicarbe vis-à-vis de l'*Heterodera* de l'oeillet dans le Midi méditerranéen. *Phyiatrie-Phytopharmacie* **19**, 177-184.

Dijkstra, J. (1956) Experiences with the breeding of red clover resistance to the stem eelworm. *Euphytica* **5**, 298-307.

Dijkstra, J. (1957) Symptoms of susceptibility and resistance in seedlings of red clover attacked by the stem nematode *Ditylenchus dispsaci* (Kühn) Filipjev. *Nematologica* **2**, 228-236.

Dijkstra, J. (1971) Breeding for resistance to *Heterodera* in white clover. *Euphytica* **20**, 36-46.

Dolliver, J.S. (1961) Population levels of *Pratylenchus penetrans* as influenced by treatments affecting dry weight of Wando pea plants. *Phytopathology* **51**, 364-367.

Edmunds, J.E.; Mai, W.F. (1966a) Effect of *Trichoderma viride, Fusarium oxysporum*, and fungal enzymes upon the penetration of alfalfa roots by *Pratylenchus penetrans. Phytopathology,* **56**, 1132-1135.

Edmunds, J.E.; Mai, W.F. (1966b) Population increase of *Pratylenchus penetrans* in alfalfa and celery roots infected with *Trichoderma viride. Phytopathology* **56**, 1320-1321.

Edwards, D.I.; Malek, R.B. (1971) Cyst production by *Heterodera lespedezae* on selected legumes. *Plant Disease Reporter* **55**, 927-929.

Eisenback, J.D.; Hirschmann, H.; Triantaphyllou, A.C. (1980) Morphological comparison of *Meloidogyne* female head structures, perineal patterns, and stylets. *Journal of Nematology* **12**, 300-313.

Endo, B.Y.; Schaeffer, G.W. (1967) Response of *Heterodera trifolii* in red clover roots to azauracil and other inhibitors. *Phytopathology* **57**, 576-579.

Ennik, G.C. (1968) Growth-promoting effect of some biocides on grass. *Netherlands Journal of Agricultural Science* **16**, 132-135.

Ennik, G.C.; Kort, J.; Bund, C.F. (1965) The clover cyst nematode (*Heterodera trifolii* Goffart) as the probable cause of death of white clover in a sward. *Journal of the British Grassland Society* **20**, 258-262.

Eriksson, K.B. (1972) Nematode diseases of pasture legumes and turf grasses. In: Webster, J.M. (Ed.) Economic nematology. London, UK; Academic Press, pp. 66-96.

Esser, R.P.; Perry, V.G.; Taylor, A.L. (1976) A diagnostic compendium of the genus *Meloidogyne* (Nematoda:Heteroderidae). *Proceedings of the Helminthological Society of Washington* **43**, 138-150.

Feldmesser, J.; Ostazeski, S.A.; Golden, A.M. (1975) Nematodes in forage grasses. (Abstr.) *Journal of Nematology* **7**, 321.

Fielding, M.J. (1951) Observations on the length of dormancy in certain plant infecting nematodes. *Proceedings of the Helminthological Society of Washington* **18**, 110-112.

Frandsen, K.J. (1951) Studies on the clover stem nematode (*Tylenchus dipsaci* Kühn). *Acta Agriculturae Scandinavica* **1**, 203-270.

Franklin, M.T. (1965) *Meloidogyne* - root-knot eelworms. In: Southey, J.F. (Ed.) Plant nematology. *Technical Bulletin, Ministry of Agriculture, Fisheries and Food,* **No. 7**, pp. 59-88.

Franklin, M.T. (1979) Taxonomy of the genus *Meloidogyne.* In: Lamberti, F.; Taylor, C.E. (Eds) Root-knot nematodes (*Meloidogyne* species); systematics, biology and control. London, UK; Academic Press, pp. 37-52.

Freckman, D.W.; Chapman, R.A. (1972) Infection of red clover seedlings by *Heterodera trifolii* Goffart and *Pratylenchus penetrans* (Cobb) *Journal of Nematology* **4**, 23-28.

Frederiksen, H. (1950) Undersøgelser og Forsøg over Kloveraalens Udbredelse, Skadevirkning, Smitteforhold og Bekaempelse. *Beretning om Landboforeningernes Virksomhed for Planteavlen paa Sjealland,* Year 1949. pp. 245-279.

Gerdemann, J.W.; Linford, M.B. (1953) A cyst-forming nematode attacking clovers in Illinois. *Phytopathology* **43**, 603-608.

Goffart, H. (1944) Beobachtung über das Auftreten von *Heterodera schachtii* an Klee. *Zeitschrift für Pflanzenkrankheiten und Pflanzenschutz* **54**, 12-18.

Good, J.M.; Blue, W.G. (1954) Relationships between plant parasitic nematodes, pathogenic fungi, and Ladino clover yields in experimental pot studies. *Proceedings of the Soil Science Society of Florida* **14**, 159-166.

Goodey, J.B.; Franklin, M.T.; Hooper, D.J. (1965) T. Goodey's The nematode parasites of plants catalogued under their hosts; 3rd Edition. Farnham Royal, U.K.; Commonwealth Agricultural Bureaux, 214 pp.

Grandison, G.S. (1963) The clover cyst nematode (*Heterodera trifolii* Goffart) in New Zealand. (A note). *New Zealand Journal of Agricultural Research* **6**, 460-462.

Grandison, G.S. (1965) The stem nematode (*Ditylenchus dipsaci*) in clovers in New Zealand. (A note) *New Zealand Journal of Agricultural Research* **8**, 1090-1091.

Griffin, G.D (1979) Alfalfa stem nematode. In: Graham, J.H.; Stuteville, D.L.; Frosheiser, F.L.; Erwin, D.L. (Eds) A compendium of alfalfa diseases. St. Paul, Minnesota, USA; American Phytopathological Society, pp. 50-51.

Hague, N.G.M. (1968) Fumigation of shallot seed with methyl bromide to control stem eelworm, *Ditylenchus dipsaci. Plant Pathology* **17**, 127-128.

Hague, N.G.M.; Clark, W.C. (1959) Fumigation with methyl bromide and chloropicin to control seed-borne infestations of the stem-eelworm (*Ditylenchus dipsaci*) on lucerne (*Medicago sativa*). *Mededelingen van de Landbouwhogeschool en de Opzoekringsstations van de Staat te Gent* **24**, 628-636.

Halpin, J.E. (1963) The effect of soil fungi and root-knot nematodes on the growth of white clover in field bins (Abstr.)*Phytopathology* **53**, 877.

Harrison, B.D.; Winslow, R.D. (1961) Laboratory and field studies on the relation of arabis mosaic virus to its nematode vector *Xiphinema diversicaudatum* (Micoletsky). *Annals of Applied Biology* **49**, 621-633.

Hastings, R.J. (1942) Longevity of congelations of bulb nematode, *Ditylenchus dipsaci* (Kühn) Filipjev, from Narcissus. *Science in Agriculture* **23**, 1-3.

Healy, W.B.; Widdowson, J.P.; Yeates, G.W. (1973) Effect of root nematodes on the growth of seedling and established white clover on a yellow-brown loam. *New Zealand Journal of Agricultural Research* **16**, 70-76.

Hesling, J.J. (1966) Biological races of stem eelworm. *Annual Report, Glasshouse Crops Research Institute for 1965,* 132-141.

Hesling, J.J. (1978) Cyst nematodes: morphology and identification of *Heterodera, Globodera* and *Punctodera.* In: Southey, J.F. (Ed.) Plant nematology; 3rd. Edition (GDI). London, UK. Ministry of Agriculture, Fisheries & Food, 125-155.

Hidding, J.; Hijink, M.J.; Oostenbrink, M. (1963) Openbrengst-en-kwaliteitsverlies van witte klaver in ein grasklavermengsel door her klavercystenaaltje, *Heterodera trifolii. Mededelingen van de Landbowhogeschool en de Opzoekrings-stations van de Staat to Gent* **28**, 679-683.

Hijink, M.J.; Kuiper, K. (1964) Crop rotation effects in Leguminosae due to *Meloidogyne hapla.* (Abstr.) *Nematologica* **10**,64.

Hirschmann, H.; Triantaphyllou, A.C. (1979) Morphological comparison of members of the *Heterodera trifolii* species complex. *Nematologica* **25**, 458-481.

Holtzmann, O.V.; Aragaki, M. (1963) Clover cyst nematodes in Hawaii. *Plant Disease Reporter* **47**, 886-889.

Hooper, D. J. (1972) *Ditylenchus dipsaci. CIH Descriptions of Plant-parasitic Nematodes* Set 1, No. 14.

Hooper, D.J.; Southey, J.F. (1978a) *Ditylenchus, Anguina* and related genera. In: Southey, J.F. (Ed.) Plant nematology; 3rd Edition (GD1) London, UK; Ministry of Agriculture, Fisheries and Food, 78-97.

Hooper, D.J.; Southey, J.F. (1978b) Plant parasitic Dorylaimida, morphology and identification. In; Southey, J.F. (Ed.) Plant nematology; 3rd. Edition (GD1) London, UK; Ministry of Agriculture, Fisheries & Food, 207-231.

Jenkins, W.R.; Taylor, D.P.; Rohde, R.A. (1956) Nematodes associated with clover, pasture, and forage crops in Maryland. *Plant Disease Reporter* **40**, 184-186.

Jones, F.G.W.; Jones, M.G. (1974) Pests of field crops; 2nd Edition. London, UK; Edward Arnold, 448 pp.

Jones, F.G.W.; Kempton, R.A. (1978) Population dynamics, population models and integrated control. In: Southey, J.F. (Ed.) Plant nematology; 3rd Edition (GD1). London, UK; Ministry of Agriculture, Fisheries and Food, 333-361.

Jönsson, H.A. (1976) Britta - en rödklöversort med betydande resistensanlag. In: Weibulls Årsbok, Weibullsholm, Sweden, 10-12.

Kimpinski, J. (1975) Population dynamics of *Pratylenchus penetrans* in red clover (Abstr.) *Journal of Nematology* **7**, 325.

Korteweg, C.R.; s'Jacob, J.J. (1980) Influence of *Heterodera trifolii* and nitrogen on white clover in a clover-grass mixture. *Meledelingen van de Faculteit Landbouwwetenschappen Rijksuniversiteit Gent* **45**, 761-767.

Kuiper, K. (1960) Resistance of white clover varieties to the clover cyst eelworm *Heterodera trifolii* Goffart. *Nematologica,* **Suppl. 2**, 95-96.

Linnman, N. (1945) Klövertrötthet och klöverål. *Lantmannen* **29**, 567-569.

Lööf, P.P.A. (1978) The genus *Pratylenchus* Filipjev, 1936 (Nematoda: Pratylenchidae): a review of its anatomy, morphology, distribution, systematics, and identification. Uppsala, Sweden; Swedish University of Agricultural Sciences, Research Information Centre, 50 pp.

McGlohon, N.E.; Baxter, L.W. (1958) The reaction of *Trifolium* species to the southern root-knot nematode *Meloidogyne incognita* var. *acrita. Plant Disease Reporter* **42**, 1167-1168.

McGlohon, N.E.; Sasser, J.N.; Sherwood, R.T. (1961) Investigations of plant-parasitic nematodes associated with forage crops in North Carolina. *Technical Bulletin, North Carolina Agricultural Experiment Station,* **No. 148**, 39 pp.

Mankau, R. (1958) Pathological disturbances caused by *Heterodera trifolii* in susceptible and resistant plants. (Abstr.). *Phytopathology* **48**, 395.

Mankau, R.; Linford, M.B. (1960) Host-parasitic relationships of the clover cyst nematode, *Heterodera trifolii* Goffart. *Bulletin, Illinois Agricultural Experiment Station,* **No. 667**, 50 pp.

Ministry of Agriculture, Fisheries and Food, UK (1972) Stem eelworm on clover. *Advisory Leaflet, Ministry of Agriculture, Fisheries and Food,* **No. 409**, 5 pp.

Minton, N.A. (1965) Reaction of white clover and five other crops to *Pratylenchus scribneri. Plant Disease Reporter,* **49**, 856-859.

Minton, N.A.; Cairns, E.J. (1957) Suitability of soybeans var. Ogden and twelve other plants as hosts of the spiral nematode (Abstr.) *Phytopathology* **47**, 313.

Mulvey, R.H. (1958) Parthenogenesis in a cyst-forming nematode, *Heterodera trifolii* (Nematoda: Heteroderidae). *Canadian Journal of Zoology* **36**, 91-93.

Mulvey, R.H. (1959a) Investigations on the clover cyst nematode, *Heterodera trifolii* (Nematoda: Heteroderidae). *Nematologica* **4**, 147-156.

Mulvey, R.H. (1959b). Susceptibilities of plants to the clover cyst nematode *Heterodera trifolii*, and the period required to complete a life cycle. *Nematologica* **4**, 132-135.

Mulvey, R.H.; Anderson, R.V. (1974) *Heterodera trifolii. C.I.H. Descriptions of Plant-Parasitic Nematodes,* **Set 4**, No. 46.

Munnecke, D.E.; Bricker, J.L.; Kolbezen, M.J. (1978) Comparative toxicity of gaseous methyl bromide to ten soilborne phytopathogenic fungi. *Phytopathology* **68**, 1210-1216.

Norton, D.C. (1967) Relationships of *Heterodera trifolii* to some forage legumes. *Phytopathology* **57**, 1305-1308.

Norton, D.C.; Isely, D. (1967) Cyst production of *Heterodera trifolii* on some leguminosae. *Plant Disease Reporter* **51**, 1017-1020.

Nusbaum, C.J.; Barker, K.R. (1971) Population dynamics. In: Zuckerman, B.M.; Mai, W.F.; Rohde, R.A. (Eds.) Plant parasitic nematodes. New York, USA; Academic Press, **Vol. 1**, 303-323.

Oostenbrink, M. (1967) Studies on the emergence of encysted *Heterodera* larvae. *Mededelingen van de Faculteit Landbouwwetenschappen, Rijksuniversiteit Gent* **32**, 503-539.

Olthof, T.H.A. (1982) Effect of age of alfalfa root on penetration by *Pratylenchus penetrans. Journal of Nematology* **14**, 100-105.

Orton Williams, K.J. (1972) *Meloidogyne javanica. C.I.H. Descriptions of Plant-parasitic Nematodes* **Set. 1**, No. 3.

Orton Williams, K.J. (1973) *Meloidogyne incognita. C.I.H. Descriptions of Plant-parasitic Nematodes,* **Set 2**, No. 18.

Orton Williams, K.J. (1974) *Meloidogyne hapla. C.I.H. Descriptions of Plant-parasitic Nematodes,* **Set 3**, No. 31.

Orton Williams, K.J. (1975) *Meloidogyne arenaria. C.I.H. Descriptions of Plant-parasitic Nematodes,* **Set 5**, No. 62.

Parris, G.K. (1958) Soil fumigants and their use: a summary. *Plant Disease Reporter* **42**, 273-278.

Pitcher, R.S. (1957) A critical review of current techniques for the study of migratory root nematodes as etiological agents. *Nematologica,* **2 Suppl. 1,** 413-423.

Pitcher, R.S. (1965) Migratory soil nematodes. In: Southey, J.F. (Ed.) Plant nematology. *Technical Bulletin, Ministry of Agriculture, Fisheries and Food,* **No. 7**, 142-157.

Riispere, U.; Krall, E. (1970) (Results of some soil and sand culture experiments for examination of relationships in the parasite-host system on heteroderosis of leguminous plants). *Zeszyty Problemowe Postepow Nauk Rolniczych,* **No. 92**, 267-272.

Rodriguez-Kabana, R.;Curl, E.A. (1980) Nontarget effects of pesticides on soilborne pathogens and disease. *Annual Review of Phytopathology* **18**, 311-332.

Rohde, R.A.; Jenkins, W.R. (1957) Host range of a species of *Trichodorus* and its host-parasite relationships on tomato. *Phytopathology* **47**, 295-298.

Ross, J.P. (1960) *Heterodera trifolii*, a foliage pathogen of white clover. *Phytopathology* **50**, 866-867.

Ross, D.J.; McNeilly, B.A. (1975) Influence of four nematicides on soil nitrogen mineralisation and nitrogen uptake by white clover in a yellow-grey earth. *New Zealand Journal of Agricultural Research* **18**, 155-162.

Sasser, J.N. (1980) Root knot nematodes: a global menace to crop production. *Plant Disease* **64**, 36-41.

Sasser, J.N.; Kirby, M.F. (1979) Crop cultivars resistant to root knot nematodes, *Meloidogyne* species, with information on seed sources. Raleigh, North Carolina, USA; North Carolina State University Press, 24 pp.

Seinhorst, J.W. (1956) Populations studies of stem eelworms (*Ditylenchus dipsaci*). *Nematologica* **1**, 159-164.

Seinhorst, J.W. (1957) Some aspects of the biology and ecology of stem eelworms. *Nematologica* **2 Suppl.1**, 355-361.

Seinhorst, J. W. (1965) The relations between nematode density and damage to plants. *Nematologica* **11**, 137-154.

Seinhorst, J.W. (1981) Water consumption of plants attacked by nematodes and mechanisms of growth reduction. *Nematologica* **27**, 34-51.

Seinhorst, J.W.; Sen, A.K. (1966) The population density of *Heterodera trifolii* in pastures in the Netherlands and its importance for the growth of white clover. *Netherlands Journal of Plant Pathology* **72**, 169-183.

Sen, A.K. (1963) Three unlisted cucurbit host plants of *Heterodera trifolii*. *Plant Disease Reporter* **47**, 942.

Sen, A.K.; Jensen, H. J. (1969) Host-parasite relationships of various plants and the hop cyst nematode, *Heterodera humuli*, *Plant Disease Reporter* **53**, 37-40.

Shepherd, R.L. (1965) Reaction of *Trifolium repens* L. to root-knot nematodes, *Meloidogyne incognita acrita*. *Dissertation Abstracts B* **26**, 1271.

Sher, S.A.; Allen, M.W. (1953) Revision of the genus *Pratylenchus* (Nematoda: Tylenchidae). *University of California Publications in Zoology* **57**, 441-470.

Singh, N.D.; Norton, D.C. (1970) Variability in host-parasite relationships of *Heterodera trifolii*. *Phytopathology* **60**, 1834-1837.

Skipp, R.A.; Christensen, M.J. (1981) Invasion of white clover roots by fungi and other soil microorganisms. I. Surface colonisation and invasion of roots growing in sieved pasture soil in the glasshouse. *New Zealand Journal of Agricultural Research* **24**, 235-241.

Skipp, R.A; Christensen, M.J. (1982) Invasion of white clover roots by fungi and other soil microorganisms. III. The capacity of fungi isolated from white clover roots to invade seedling root tissue. *New Zealand Journal of Agricultural Research* **25**, 97-101

Southey, J.F. (1959) Some records of root lesion eelworms, *Pratylenchus* spp., in England. *Plant Pathology* **8**, 130-132.

Stehman, E.V.; Norton, D.C. (1969) Invasion of the Hedysareae by *Heterodera trifolii*. *Plant Disease Reporter* **53**, 121-123.

Stelter, H.; Meinl, G. (1972) Die Auswirkung des Befalles von Rot- und Weisklee durch *Heretodera trifolii* und *Heterodera galeopsidis*. *Archiv für Pflanzenschutz* **8**, 463-470.

Sturhan, D. (1964) Kreuzungsversuche mit biologischen Rassen des Stengelälchens (*Ditylenchus dipsaci*). *Nematologica* **10**, 328-334.

Sturhan, D. (1971) Biological races. In: Zuckerman, B.M.; Mai, W.F.; Rohde, R.A. (Eds) Plant parasitic nematodes. New York, USA; Academic Press, **Vol. 2**, 51-71.

Taha, A.H.Y.; Raski, D.J. (1969) Interrelationships between root-nodule bacteria, plant-parasitic nematodes and their leguminous host. *Journal of Nematology* **1**, 201-211.

Tanveer, M.; Saad, A.T. (1971) Reaction of some cultivated crops to two species of root-knot nematodes. *Plant Disease Reporter* **55**, 1082-1084.

Taylor, A.L.; Sasser, J.N. (1978a) Biology, identification, and control of root-knot nematodes (*Meloidogyne* species). A cooperative publication of the Department of Plant Pathology, North Carolina State University, and the United States Agency for International Development. Raleigh, North Carolina, USA; North Carolina State University Graphics, 111 pp.

Taylor, A.L.; Sasser, J.N. (1978b) Experimental and agronomic use of nematicides. Raleigh, North Carolina, USA; North Carolina State University Press, 20 pp.

Thomas, P.R. (1969) Crop and weed plants compared as hosts of viruliferous *Longidorus elongatus* (de Man). *Plant Pathology* **18**, 23-28.

Thomas, P.R. (1970) Host status of some plants for *Xiphinema diversicaudatum* (Micol.) and their susceptibility to viruses transmitted by this species. *Annals of Applied Biology* **65**, 169-178.

Thompson, L.S.; Willis, C.B. (1971) Forage legumes preferred by the clover root curculio and preferences of the curculio and root lesion nematodes for species of *Trifolium* and *Medicago*. *Journal of Economic Entomology* **64**, 1518-1520.

Thompson, L.S.; Willis, C.B. (1976) Efficacy of multiple applications of oxamyl and phenamiphos for control of *Pratylenchus penetrans* in birdsfoot trefoil. *Journal of Nematology* **8**, 342-346.

Thorne, G. (1961) Principles of nematology. New York, USA; McGraw-Hill Book Co. Inc., 553 pp.

Townshend, J.L.; Potter, J.W. (1976) Evaluation of forage legumes, grasses, and cereals as hosts of forage nematodes. *Nematologica* **22**, 196-201.

Townshend, J.L.; Potter, J.W. (1978) Yield losses among forage legumes infected with *Meloidogyne hapla*. *Canadian Journal of Plant Science* **58**, 939-943.

Townshend, J.L.; Willis, C.B.; Potter, J.W.; Santerre, J. (1973) Occurrence and population densities of nematodes associated with forage crops in Eastern Canada. *Canadian Plant Disease Survey* **53**, 131-136.

Triantaphyllou, A.C. (1971) Genetics and Cytology. In: Zuckerman, B.M.; Mai, W.F.; Rohde, R.A. (Eds) Plant parasitic nematodes, New York, USA, Academic Press, **Vol. 2,** pp. 1-34.

Viglierchio, D.R.; Croll, N.A. (1968) Host resistance reflected in differential population structures.*Science*, New York **161**, 271-272.

Viglierchio, D.R.; Siddiqui, I.A.; Croll, N.A. (1973) Culturing and population studies of *Ditylenchus dipsaci* under monaxenic conditions. *Hilgardia* **42**, 177-213.

Wallace, H.R. (1962) Observations on the behaviour of *Ditylenchus dipsaci* in soil. *Nematologica* **7**, 91-101.

Wallace, H.R. (1973) Nematode ecology and plant disease. London, UK; Edward Arnold, 228 pp.

Wardojo, S.; Hijink, M.J.; Oostenbrink, M. (1963) Schade bij witte klaver door inokulatie met *Heterodera trifolii, Meloidogyne hapla* en *Pratylenchus penetrans. Mededelingen van de Landbouwhogeschool en de Opzoekringsstations van de Staat te Gent* **28,** 672-678.

Webster, J.M. (1967) The significance of biological races of *Ditylenchus dipsaci* and their hybrids. *Annals of Applied Biology* **59**, 77-83.

Wehunt, E.J. (1957) Population trends of nematodes associated with white clover in Louisiana. (Abstr.) *Phytopathology* **47**, 36.

Whitehead, A.G.; Hemming, J.R. (1965) A comparison of some quantitative methods of extracting small vermiform nematodes from soil. *Annals of Applied Biology* **55**, 25-38.

Whitehead, A.G.; Tite, D.J.; Fraser, J.E. (1973) Chemical control of stem nematode. *Report, Rothamsted Experimental Station for 1972* **Part 1** , pp. 166-167.

Widdowson, J.P.; Yeates, G.W.; Healy, W.B. (1973) The effect of root nematodes on the utilisation of phosphorus by white clover on a yellow-brown loam. *New Zealand Journal of Agricultural Research* **16**, 77-80.

Williams, W.M. (1972) Laboratory screening of white clover for resistance to stem nematode. *New Zealand Journal of Agricultural Research* **15**, 363-370.

Williams, W.M.; Barclay, P.C. (1972) The effect of clover stem eelworm on the establishment of pure swards of white clover. *New Zealand Journal of Agricultural Research* **15**, 356-362.

Willis, C.B.; Thompson, L.S. (1967) Root-lesion nematodes associated with forage legumes in the Maritime Provinces. *Canadian Plant Disease Survey* **47**, 87-88.

Willis, C.B.; Thompson, L.S. (1969) Effect of root-lesion nematode on yield of four forage legumes under greenhouse conditions. *Canadian Journal of Plant Science* **49**, 505-509.

Willis, C.B.; Thompson, L.S. (1975) Influences of carbofuran and benomyl on yield and persistence of birdsfoot trefoil. *Canadian Journal of Plant Science,* **55**, 95-99.

Willis, C.B.; Thompson, L.S. (1979) Effects of phenamiphos, methyl bromide, and fallowing on *Pratylenchus penetrans*, yield of *Medicago sativa*, and *Fusarium* infection. *Journal of Nematology* **11**, 265-269.

Willis, C.B.; Townshend, J.L.; Anderson, R.V.; Kimpinski, J.; Mulvey, R.H.; Potter, J.W.; Santerre, J.; Wu, L.Y. (1976) Species of plant-parasitic nematodes associated with forage crops in Eastern Canada. *Plant Disease Reporter* **60**, 207-210.

Wilson, H.G. (1978) Clover cyst nematodes; management problem. *New Zealand Fertilizer Journal,* **Feb. 1978**, 18.

Winfield, A.L.; Murdoch, G.; John, M.E. (1971) Chemical control of stem nematodes (*Ditylenchus dipsaci*) of bulb onions. *Proceedings of the 6th British Insecticides and Fungicides Conference* **1**, 141-149.

Wood, F. H. (1973) Biology and host range of *Paratylenchus projectus* Jenkins, 1956 (Nematoda, Criconematidae) from a sub-alpine tussock grassland. *New Zealand Journal of Agricultural Research* **16,** 381-384.

Wouts, W.M. (1978) On the males of *Heterodera trifolii* Goffart, 1932 (Nematoda: Heteroderidae). *Nematologica* **24**, 115-120.

Wouts, W.M.; Sturhan, D. (1978) The identity of *Heterodera trifolii* Goffart, 1932 and the description of *H. daverti* n. sp. (Nematoda: Tylenchida). *Nematologica* **24**, 121-128.

Yeates, G.W. (1973a) Annual cycle of root nematodes on white clover in pasture. I. *Heterodera trifolii* in a yellow-grey earth. *New Zealand Journal of Agricultural Research* **16**, 569-574.

Yeates, G.W. (1973b) Annual cycle of root nematodes on white clover in pasture. II. *Meloidogyne hapla* and *Heterodera trifolii* in a yellow-brown loam. *New Zealand Journal of Agricultural Research* **16**, 575-578.

Yeates, G.W. (1974) Effects of *Heterodera trifolii* on the growth of clover in a yellow-grey earth under greenhouse conditions. *New Zealand Journal of Agricultural Research* **17**, 379-385.

Yeates, G.W. (1975) Nematode genera from some New Zealand pastures. *Scientific Report, New Zealand Soil Bureau* **No. 21**, 22 pp.

Yeates, G.W. (1976) Effect of fertilizer treatment and stocking rate on pasture nematode populations in a yellow-grey earth. *New Zealand Journal of Agricultural Research* **19**, 405-408.

Yeates, G.W. (1977) Soil nematodes in New Zealand pastures. *Soil Science* **123**, 415-422.

Yeates, G.W. (1978a) Influence of *Heterodera trifolii* cyst inoculum on white clover growth and final cyst count in pots of a yellow-brown pumice soil. *New Zealand Journal of Agricultural Research* **21**, 149-151.

Yeates, G.W. (1978b) Reinfestation of small plots by clover cyst nematode - a note. *New Zealand Journal of Agricultural Research* **21**, 147.

Yeates, G.W. (1978c) Populations of nematode genera in soils under pasture. I. Seasonal dynamics in dryland and irrigated pastures on a southern yellow-grey earth. *New Zealand Journal of Agricultural Research* **21**, 321-330.

Yeates, G.W. (1978d) Populations of nematode genera in soils under pasture. II. Seasonal dynamics in dryland and effluent-irrigated pastures on a central yellow-grey earth. *New Zealand Journal of Agricultural Research* **21**, 331-340.

Yeates, G.W.; Crouchley, G.C.; Witchalls, J.T. (1975a) Effect of soil fumigation on white clover growth in a yellow-grey earth infested with clover cyst nematode. *New Zealand Journal of Agricultural Research* **18**, 149-153.

Yeates, G.W.; Healy, W.B.; Widdowson, J.P.; Thomson, N.A.; MacDiarmid, B.N. (1975b) Influence of nematicides on growth of plots of white clover on a yellow-brown loam. *New Zealand Journal of Agricultural Research* **18**, 411-416.

Yeates, G.W.; Healy, W.B.; Widdowson, J.P. (1973) Screening of legume verieties for resistance to the root nematodes *Heterodera trifolii* and *Meloidogyne hapla*. *New Zealand Journal of Agricultural Research* **16**, 81-86.

Yeates, G.W.; Healy, W.B.; Widdowson, J.P.; Thomson, N.A.; MacDiarmid, B.N. (1976) Effect of a soil fumigant on the establishment and growth of a grazed pasture on a yellow-brown loam. *New Zealand Journal of Agricultural Research* **19**, 397-403.

Yeates, G.W.; Risk, W.H. (1976) Annual cycle of root nematodes on white clover in pasture. III. *Heterodera trifolii* in a yellow-brown earth. *New Zealand Journal of Agricultural Research,* **19**, 393-396.

Yeates, G.W.; Ross, D.J.; Bridger, B.A.; Visser, T.A. (1977) Influence of the nematodes *Heterodera trifolii* and *Meloidogyne hapla* on nitrogen fixation by white clover under glasshouse conditions. *New Zealand Journal of Agricultural Research* **20**, 401-413.

Yeates, G.W.; Visser, T.A. (1979) Persistence of *Heterodera trifolii* (Nematoda) cysts in the absence of host plants. *New Zealand Journal of Agricultural Research* **22**, 649-651.

Yuksel, H.S. (1960) Observations on the life cycle of *Ditylenchus dipsaci* on onion seedlings. *Nematologica* **5**, 289-296.

Herbicide effects

M. P. Rolston
Grasslands Division, DSIR, Palmerston North, New Zealand

White clover (*Trifolium repens* L.) is a valuable pasture species, as well as being a weed in some horticultural and lawn situations. Herbicides are used to control weeds in white clover seed crops and weeds in pasture with little or no damage to clover. In other situations herbicides are used to kill weeds, including white clover.

Some herbicides used in pasture (2,4-D, MCPA, 2,4,5-T) can injure white clover reducing clover growth (Honore et al. 1980) and nitrogen fixation (Rolston et al. 1982), and causing a decline in animal production (Lewis 1957, Hartley & Thomson 1982). White clover injury in pasture can be minimised by spot treatment of weeds (e.g. motorised knapsack treatment) rather than broadcast application (Rolston et al. 1982). Changing the time of application of 2,4-D from spring to autumn-winter reduced clover injury (Brewster et al. 1978; Honore et al. 1980). The butyric analogues (2,4-DB and MCPB) of 2,4-D and MCPA cause little or no damage to white clover (Matthews 1965, Honore et al. 1980) although seed crops can be injured (Lee 1972), and should be used in preference if the weed species are susceptible.

This review lists a wide range of herbicides (Table 1) and the response of white clover at rates of the product commonly used in commercial practice. Four response classes are used: R = resistant; MR = moderately resistant, temporary suppression may occur; MS = moderately susceptible, growth is checked, sometimes severely; S = susceptible. There is considerable unpublished information on white clover responses to herbicides. Where there is no published data, personal communications are presented.

The list of herbicides (Table 1) should be considered as a guide, and not a recommendation, because the susceptibility of white clover to herbicides is influenced by various factors. Formulation; 2,4-D ester is more active than 2,4-D amine (Freyer & Makepeace 1978). Growth stage; white clover tolerance to herbicides tends to increase as the seedling develops from unifoliate, through to the 4th trifoliate leaf stage (Freyer & Makepeace 1978). Table 1 does not attempt to separate the seedling growth stages. Environmental factors such as soil moisture status, organic matter, clay content, pH, rainfall, temperature, frost and relative humidity can influence and modify the activity of herbicides (Muzik, 1976) and therefore the response of plants. In other species, cultivars can differ in their tolerance and susceptibility to herbicides, but no reports on white clover cultivar comparisons were found.

TABLE 1: RESPONSE OF SEEDLING AND ESTABLISHED WHITE CLOVER TO HERBICIDES

Herbicide	*Seedling* Rate (kg/ha)	*Seedling* Response	*Established* Rate (kg/ha)	*Established* Response	Reference
alachlor	1.5 – 3.4	MS/S			13, 15
allidochlor	4.0	MS/S			28
alloxydim	2.0	R	2.0	R	29
amitrole	1.0	S	1.0	MS	42
AMS	3.0	MR	3.0	R	16
asulam			1.0 – 3.0	MR	20, 22, 59, 66
atrazine	0.5 – 1.0	S	1.0 – 2.2	MR/MS	24, 34, 57
aziprotyrne	1.5 – 2.8	MR/MS			15, 64
benazolin	0.2 – 0.4	R	0.4 – 1.12	R	9, 19
benfluralin	1.1 – 1.25	R/MR			13
bensulide	3.4 – 5.6	R/MR			13
bentazone	1.4	R	1.4 – 3.4	R	9, 18
benzoylprop-ethyl	1.5 – 2.0	R	1.5 – 2.0	R	1
bifenox	1.1 – 2.2	S			28
bromacil	0.5	S	1.12 – 1.68	MS/S	34, 57
bromofenoxim	0.5 – 1.0	MS			28
	1.0 – 4.0	S			28
bromoxynil	0.6 – 0.8	MR/MS	0.8 – 1.6	R	13, 41, 42
buthidazole	0.5 – 1.0	S	0.5 – 1.0	MR/MS	21, 24
			3.0 – 5.0	S	39
butylate	4.5 – 6.7	MR			13
cacodylic acid			2.2 – 6.7	MS/S	34
carbetamide	2.5 – 2.8	R	2.5 – 4.5	R	9, 38, 60, 69
chloramben	2.0 – 3.0	S			28
chlorbufam	0.9	MS	0.9	R	3
chloridazon	3.6	MR	3.6	R	3
chloroxuron	3.0	MS/S			15
chlorpropham	2.7	R	5.0	R	64, 71
chlorsulfuron	0.02	MS	0.02	MR	16
chlorthiamid			5.0 – 9.0	MR	62
chlorthal-dimethyl	10.0	R			43
clopyralid	0.015 – 0.03	S	0.015 – 0.03	R/MS	29, 48, 58
			0.03 – 0.10	MS	29
			0.15 – 0.20	S	29
4-CPA	4.0	MR/MS	2.0 – 4.0	MR/MS	31, 42
cyanazine	1.0	S	2.0	MR/MS	21, 24
cycloate	2.2 – 4.4	MS/S			28
cyprazine	0.3	S			55
2,4-D amine	0.6	MS	1.0 – 2.0	MR	19
2,4-D ester	0.6	S	1.0 – 2.0	MR	26, 58
dalapon			1.0 – 2.8	MR	2, 72
2,4-DB	1.5 – 2.2	R	2.2	R	9, 19
desmedipham	2.0	S			28
desmetryn	0.3	MS			15
dicamba	0.2	S	0.3	MS/S	21, 67

Herbicide	Seedling		Established		
	Rate (kg/ha)	Response	Rate (kg/ha)	Response	Reference
dichlobenil	1.1 – 2.2	R	4.0 – 9.0	R	34, 63
	4.0	MS			57
dichlorprop	1.0	S	1.0	MR/MS	42, 65
diclofop-methyl	1.0	R	2.1	R	9,25
difenzoquat	1.0	R	1.0	R	9
dimefuron			0.8	R	46
dinitramine	0.4 – 0.5	R			9
dinoseb ammonium	0.8	R/MR	0.8	R/MR	9, 19
dinoseb amine	1.2 – 2.2	R/MR	1.2 – 2.2	R/MR	9, 19
dinoseb acetate	2.5 – 2.8	R/MR	2.5 – 2.8	R/MR	9, 19
diphenamid	1.7 – 2.2	MS/S			13, 27
diquat			0.25 – 1.0	MR	9, 42, 67
diuron	1.0 – 2.0	S	2.0	MR/S	21, 52, 57
EPTC	3.4 – 5.6	R			13
ethalfluralin	8.0	S	8.0 – 10.0	S	4
ethofumesate	0.2 – 0.5	S	1.2 – 2.0	MS/S	7, 8
fenoprop	1.0	MS/S	4.5	S	5, 68
fenoxaprop			0.24	R	72
fenthiaprop			0.24	R	72
flamprop-methyl	0.7 – 1.0	R			9
fluazifop	0.4 – 0.5	R	1.5	R	21, 61
fosamine	4.8 – 9.6	MR	4.8 – 9.6	MR	16
glufosinate-ammonium			1.5	S	73
glyphosate	0.4	S	0.4 – 1.5	MR	10, 22, 46
hexazinone			0.9	S	32
ioxynil	0.8 – 1.1	R/MR	0.8	R/MR	19, 42
isoproturon			2.5	R	19, 46
karbutilate	5.0	S	5.0	S	12
lenacil	0.3 – 1.2	R			55
linuron	0.3 – 0.5	R/MR	1.0	MS/S	62, 64, 70
	1.0 – 2.25	MR/MS			
MCPA	0.6	MS/S	1.0 – 2.0	MR	19, 26, 58
MCPB	1.5 – 2.2	R	1.5 – 2.2	R	9, 19
mecoprop	1.5	S	1.0 – 2.2	MR/MS	42, 65
			2.2 – 4.5	S	5, 17
mefluidide			0.5 – 1.0	MR	30
methabenzthiazuron	1.4	MR			9
methazole	0.5 – 1.5	MS/S			15, 64
metolachlor			3.0	R	12
metoxuron	3.5	S			15
metribuzin	0.4 – 0.8	S	0.5 – 0.8	MR	14, 15, 21
MH	4.0	R	4.0	R	42, 70
			5.6 – 8.0	MS	
napropamide	2.0 – 4.0	MS	1.5	R	28, 35
nitralin	0.6	R			13
nitrofen	1.0 – 4.0	R/MS	1.0	MR	42, 62

Herbicide	Seedling		Established		Reference
	Rate (kg/ha)	Response	Rate (kg/ha)	Response	
oxadiazon	1.5	S	1.5 – 3.0	MR	44
oxyfluorfen			0.25 – 0.8	R	21, 36
paraquat	0.2 – 0.4	MS/S	0.4 – 0.6	R/MR	21, 22, 38
pebulate	3.0 – 4.0	MR			15, 28
pendimethalin	1.5 – 2.0	R			21, 49
phenmedipham	2.0	MS/S			28
picloram	0.2	S	0.2	S	17, 67
PP 333			1.5 – 2.0	MR/MS	70
prometryne	0.35 – 0.75	R/MR	0.5	R	12, 53, 64
	0.5 – 1.0	MS/S	3.3 – 5.6	S	34
propachlor	4.5				13
propazine	1.5	S			15
propham	4.4	R	4.8	R	34, 71
propyzamide	1.2	R	1.2 – 3.4	R	9, 34, 54
sethoxydim			0.75	R	51
simazine	0.5 – 1.0	MS/S	1.0 – 2.0	MR	13, 24, 52, 57
2,4,5-T	0.55	MS/S	2.0 – 4.0	MR/S	66, 68
2,3,6-TBA	4.5	S	4.5	S	5
TCA			6.0 – 12.5	MR	45, 47
terbacil	1.0	S	1.0 – 2.24	MS/S	15, 34
terbumeton } terbuthylazine }	5.0	S	5.0 – 8.0	MS/S	13
terbutryne	0.5 – 1.0	S	0.5	MR	13
tetrapion			1.0 – 1.33	MR/MS	11
thiobencarb	6.0	R			55
tri-allate	1.4 – 1.7	R			9
triclopyr	0.5	S	0.25 – 1.0	MS	29, 31, 40
trifluralin	0.8 – 1.6	R			9, 21
vernolate	3.4 – 4.5	R			13

References

1. Allen, F.C.; Smallridge, T.J.; Bourdot, G.W. (1974) Wild oat control in ryegrass for seed production. *Proceedings of the 27th New Zealand Weed and Pest Control Conference*, 67-71.
2. Atkinson, G.C. (1970) Chemical control of barley grass. *Proceedings of the 23rd New Zealand Weed and Pest Control Conference*, 93-96.
3. BASF New Zealand Ltd. Personal communication from B. Smith, 12 April 1982.
4. Bayer New Zealand Ltd. Ustilan herbicide, specimen label.
5. Black, W.N. (1976) Effects of herbicide rates and time of application on the control of tansy ragwort in pastures. *Canadian Journal of Plant Science* **56**, 605-610.
6. Brewster, B.D.; Rolston, M.P.; Appleby, A.P. (1978) Control of tansy ragwort in western Oregon pastures. Circular of Information 665. Agriculture Experiment Station, Oregon State University, Corvallis, 10pp.
7. Brock, J.L.; Henderson, J.D. (1976) Control of legume weeds in lotus seed crops by ethofumesate. *Proceedings of the 29th New Zealand Weed and Pest Control Conference*, 93-96.
8. Brock, J.L.; Henderson, J.D. (1980) The tolerance of established 'Grasslands Maku' lotus to four herbicides. *Proceedings of the 33rd New Zealand Weed and Pest Control Conference*, 79-81.
9. Butler, J.H.B. (1980) Weed control in clovers, lotus and sainfoin. *Aglink FPP 398, Media Services, Ministry Agriculture and Fisheries, New Zealand.* 4p.

10. Campbell, M.H.; Gilmour, A.R. (1979) Effect of time and rate of application of herbicides on serrated tussock (*Nassella trichotoma*) and improved pasture species. 1. Glyphosate and 2,2-DPA. *Australian Journal of Experimental Agriculture and Animal Husbandry*, **19**, 472-475.

11. Campbell, M.H.; Gilmour, A.R.; Vere, D.T. (1979) Effects of time and rate of application of herbicides on serrated tussock (*Nassella trichotoma*) and improved pasture species. 2. Tetrapion. *Australian Journal of Experimental Agriculture and Animal Husbandry* **19**, 476-480.

12. Ciba-Geigy New Zealand Ltd, Personal communication from T.M. Patterson. 29 June 1982.

13. Cope, W.A.; Waldrep, T.W.; Chamblee, D.S.; Lewis, W.M. (1973) Evaluation of herbicides in the establishment of alfalfa, ladino white clover and crown vetch. *Agronomy Journal* **65**, 820-825.

14. Cox, T.I. (1974) The use of metribuzin in potatoes and tomatoes. *Proceedings of the 27th New Zealand Weed and Pest Control Conference*, 100-104.

15. Cox, T.I.; Ingles, A.; Kerr, R.M. (1979) Herbicides in commercial vegetable crops - a guide. *Aglink HPP 78, Media Services, Ministry of Agriculture and Fisheries.* 6pp.

16. Du Pont (New Zealand) Ltd. Personal communication from M.J. Cornwell, 13 April 1982.

17. Erskine, D.S.C. (1970) Chemical control of broadleaved dock (*Rumex obtusifolius*). *Experimental Work, Edinburgh School of Agriculture*, 1969. pp. 35.

18. Fellowes, R.W. (1973) Preliminary trials with bentazon on pasture. *Proceedings of the 26th New Zealand Weed and Pest Control Conference*, 45-47.

19. Fryer, J.D.; Makepeace, R.J. (1978) Recommendations for weed control in grassland and herbage legumes. In: Fryer, J.D.; Makepeace, R.J. (Ed). Weed Control Handbook Vol II. Recommendations, British Crop Protection Council, pp. 216-247.

20. Grant, D.A. (1968) Some effects of two benzenesulphonylcarbamate herbicides on a mixed pasture. *Proceedings of the 21st New Zealand Weed and Pest Control Conference*, 143-148.

21. Grasslands Division, DSIR, Palmerston North. Unpublished data from J.D. Henderson and M.P. Rolston.

22. Haggar, R.J.; Bastian, C.J. (1980) Regulating the content of white clover in mixed swards using grass suppressing herbicides. *Grass and Forage Science* **35**, 129-137.

23. Hartley, M.J.; Thomson, N.A. (1982) Effect and control of Californian thistle in dairy pasture. *Proceedings of the New Zealand Grassland Association* **43**, 104-108.

24. Henderson, J.D. (1977) Weed control in legumes. *Proceedings of the 30th New Zealand Weed and Pest Control Conference*, 31-33.

25. Hoechst (1981) Hoegrass 36EC. New Zealand Farmers Fertiliser Co., Production Bulletin. 4pp.

26. Honore, E.N.; Rahman, A.; Dyson, C.B. (1980) Effects of winter application of phenoxy herbicides on white clover and pasture production in lower Northland. *Proceedings of the 33rd New Zealand Weed and Pest Control Conference*. 55-58.

27. Hoyert, J.N. (1975) Effects of diphenamid on weed control and tobacco seedling production in Maryland. *Tobacco International* **177**, 51-52 (cited in Weed Abstracts 26: 2898).

28. International Plant Protection Centre, Corvallis, Oregon. Unpublished data and personal communication from L.C. Burrill, 1 March 1983.

29. Ivon Watkins-Dow Ltd. Personal communication from G.W. Woon, 11 June 1982.

30. Jackson, I.F.; O'Connor, B.P.; Jacobson, D.J.I. (1980) Mefluidide: a plant growth regulator. *Proceedings of the 33rd New Zealand Weed and Pest Control Conference, 67-73.*

31. James, T.K.; Hartley, M.J. (1977) Control of musky storksbill in pasture. *Proceedings of the 30th New Zealand Weed and Pest Control Conference*, 42-45.

32. Lane, P.M.S.; Cornwell, M.J. (1981) Control of certain hard to kill weed species in lucerne with hexazinone. *Proceedings of the 34th New Zealand Weed and Pest Control Conference*, 17-20

33. Lee, W.O. (1972) Effects of 2,4-DB on white clover seed production. *Weed Science* **20**, 330-331.

34. Lee, W.O. (1975) Control of winter annuals weeds in white clover raised for seed. *Weed Science* **23**, 441-442.

35. Lee, W.O. (1979) Control of winter annual broadleaf weeds in an established white clover seed field. Unpublished research report, Agricultural Research Service, USDA, Crop Science Department, Oregon, State University, Corvallis, Oregon, pp. 71-76.

36. Lee, W.O. (1980) Control of weed in established white clover raised for seed. Unpublished research report, Agricultural Research Service, USDA, Crop Science Department, Corvallis, Oregon, pp. 90-95.

37. Lewis, K.H.C. (1957) The effects of spraying on fat lamb production. *Proceedings of the 10th New Zealand Weed Control Conference*, 44-47.

38. Logan, I.C.; Arnst, R.N. (1974) White clover seed crop: paraquat and carbetamide mixtures for weed control. *Proceedings of the 27th New Zealand Weed and Pest Control Conference*, 63-66.

39. MacDiarmid, B.N. (1975) Vel 5026 - a new herbicide for non-selective weed control. *Proceedings of the 28th New Zealand Weed and Pest Control Conference*, 154-159.

40. MacDiarmid, B.N. (1977) Properties and activity of triclopyr on some New Zealand bush weeds. *Proceedings of the 30th New Zealand Weed and Pest Control Conference*, 66-70.

41. Marco, O.N. (1974) Effects of 2,4-DB and bromoxynil, singly or mixtures, on legumes and weeds during establishment of pastures. *Malezas y su Control* **3**, 10-18. (cited in *Weed Abstracts* **28:**, 1522, 1979).

42. Matthews, L.J. (1965) Tolerance of clovers to herbicides. *Proceedings of the 18th New Zealand Weed and Pest Control Conference*, 37-41.

43. Matthews, L.J. (1975) Weed control by chemical means. *New Zealand Ministry of Agriculture and Fisheries Bulletin* **No. 329**. 710 p.

44. May and Baker New Zealand Ltd. Personal communciation from C. Surman, 14 April 1982.

45. McLean, J.R.F.; Dixon, A.T. (1972) Selective aerial treatment of barley grass. *Proceedings of the 25th New Zealand Weed and Pest Control Conference*, 51-55.

46. Mead, H.; Livingston, D.; Ross, B.L. (1976) Herbicide tolerance studies in white clover. *Proceedings 13th British Crop Protection Conference - Weeds*, 677-679.

47. Meeklah, F.A.; McRobb, H. (1974) The reliability of herbicides for barley grass control. *Proceedings of the 27th New Zealand Weed and Pest Control Conference*, 79-81.

48. Meeklah, F.A.; Mitchell, R.B.; McMillan, H. (1981) Chemical control of daisies in Otago/Southland pastures. *Proceedings of the 34th New Zealand Weed and Pest Control Conference*, 137-140.

49. Moore, R.W. (1975) Penoxalin, a new selective herbicide for weed control in peas, beans and other crops. *Proceedings of the 28th New Zealand Weed and Pest Control Conference*, 185-188.

50. Muzik, T.J. (1976) Influence of environmental factors on toxicity to plants. In: L.J. Andus (Ed). Herbicides; Physiology, Biochemistry, Ecology Vol 2. London; Academic Press, pp. 204-243.

51. Naish, R.W.; Iwataki, I.; Ishikawa, M. (1982) Sethoxydim - a new post-emergence selective grass herbicide. *Proceedings of the 35th New Zealand Weed and Pest Control Conference*, 274-278.

52. Palmer, P.C.; Williams P.P. (1970) An evaluation of additives to paraquat for the control of barley grass. *Proceedings of the 23rd New Zealand Weed and Pest Control Conference*, 103-106.

53. Patterson, T.M. (1973) Control of chickweed in pasture. *Proceedings of the 26th New Zealand Weed and Pest Control Conference*, 48-49.

54. Paxman, R.N.; Forgie, C.D. (1972) Effects of pronamide on various pasture grass species. *Proceedings of the 25th New Zealand Weed and Pest Control Conference*, 56-61.

55. Richardson, W.G.; Dean, M.L. (1973) The pre-emergence selectivity of some recently developed herbicides: lenacil, RU12068, metribuzin, cyprazine, EMT-IT5914 and benthiocarb. *Technical Report, Weed Research Organisation*, Oxford, UK., **No. 25**, 57p.

56. Rolston, M.P.; Lambert, M.G.; Clark, D.A. (1982) Weed, control options in hill country. *Proceedings of the New Zealand Grassland Association* **43**, 196-204.

57. Rolston, M.P.; Lee, W.O.; Appleby, A.P. (1979) Volunteer legume control in legume seed crops with carbon bands and herbicides. I. White Clover. *Agronomy Journal* **71**, 665-670.

58. Rutherford, G.R.; Bell, P.D.; Forgie, C.D. (1981) 3,6-dicloropicolinic acid/MCPB for control of nodding thistle in pasture. *Proceedings of the 34th New Zealand Weed and Pest Control Conference*, 122-125.

59. Soper, D. (1970) The tolerance of pasture grasses to asulam. *Proceedings of the 10th British Weed Control Conference*, 465-475.

60. Soper, D.; Hutchison, A.S. (1974) The regulation of grasses in clover seed crops and pasture using carbetamide. *Proceedings of the 12th British Weed Control Conference*, 779-786.

61. Stewart, T.M.; Jacobson, D.J.I.; Kilsby, K.B. (1982) Fluazifop-butyl for the selective control of many grass weeds in broadleaf crops. *Proceedings of the 35th New Zealand Weed and Pest Control Conference*, 279-283.

62. Tasmanian Department of Agriculture (1978) Recommendations for the chemical control of weeds in peas and beans. Chart 1p.

63. Tasmanian Department of Agriculture (1979) Recommendations for the chemical control of weeds in berry fruits and vines. Chart 1p.

64. Tasmanian Department of Agriculture (1980) Recommendations for the chemical control of weeds in onion crops. Chart compiled by the Weeds Section of the Plant Production Division.

65. Taylor, R.L. (1973) Control of ragwort. *Proceedings of the 26th New Zealand Weed and Pest Control Conference*, 50-52.

66. Thompson, A. (1973) The effect of herbicides on gorse and pasture species. *Proceedings of the 26th New Zealand Weed and Pest Control Conference*, 13-16.

67. Thompson, A. (1974) Herbicide effects on ragwort and pasture. *Proceedings of the 27th New Zealand Weed and Pest Control Conference*, 90-93.

68. Thompson, F.B. (1956) Clover tolerance. *Proceedings of the 9th New Zealand Weed Control Conference*. 89-93.

69. Wasmuth, A.G.; Miles, K.B. (1972) The suppression of grass weeds in white clover seed crops with carbetamide. *Proceedings of the 25th New Zealand Weed and Pest Control Conference*, 210-213.

70. Weed Research Organisation, Oxford, UK. Personal communication from R.J. Haggar, 2 June 1982.

71. Williams, R.D. (1982) Seed Crops. In: R.D. Williams (Ed). Oregon Weed Control Handbook, Oregon State University, Corvallis, Oregon, pp. 51-52.

72. Kerse, G.W.; Meeklah, F.A.; Sunde, R.I. (1983) Hoe 35609 and Hoe 33171: two new selective grass herbicides. *Proceedings of the 36th New Zealand Weed and Pest Control Conference*, 265-268.

73. Rahman, A.; James, T.K.; Mortimer, J. (1983) A comparison of some post-emergence herbicides for control of grass weeds. *Proceedings of the 36th New Zealand Weed and Pest Control Conference*, 269-272.

Index

Printed and bound by CPI Group (UK) Ltd, Croydon, CR0 4YY
16/09/2026
14963434-0001